Problems in
Applied
Mathematics
Selections from *SIAM Review*

Problems in Applied Mathematics

Selections from *SIAM Review*

Edited by Murray S. Klamkin

®   *Philadelphia*

Society for Industrial and Applied Mathematics

Dedicated to the memory of
Henry E. Fettis and Yudell L. Luke,
formerly associate problem editors

Library of Congress Cataloging-in-Publication Data

Problems in applied mathematics : selections from SIAM review / edited
  by Murray S. Klamkin.
    p. cm.
  Includes bibliographical references.
  ISBN 0-89871-259-9
  1. Mathematics—Problems, exercises, etc. I. Klamkin, Murray S.
II. Society for Industrial and Applied Mathematics. III. SIAM
review.
QA43.P76 1990
510'.76—dc20   90-49670
                                                                                       CIP

All rights reserved. Printed in the United States of America. No part of this book may
be reproduced, stored, or transmitted in any manner without the written permission
of the publisher. For information, write the Society for Industrial and Applied Mathematics, 3600 University City Science Center, Philadelphia, Pennsylvania 19104-2688.

Copyright ©1990 by the Society For Industrial and Applied Mathematics.

Preface

The Problem Section of *SIAM Review*, which began with the first issue back in 1959, was designed to offer classroom instructors, students, and other interested problemists a set of problems, solved or unsolved, illustrating various applications of mathematics. In many cases, the unsolved problems were eventually solved.

In all other of the many journal problem sections of which I am aware, only brief descriptions of the problems are given, probably to save space. In *Siam Review*, however, contributors are encouraged to include motivational material describing how their problem arose. (See, for example, the problem "steady-state diffusion-convection,"on page 241.) Despite this encouragement, many contributors only include a one-line description e.g., "this problem arose in a study of heat conduction." This is partly understandable for what is being preferred is almost equivalent to a short note and no doubt if the author did this he would rather submit it as a note and get credit for it as a paper rather than as a problem since it does not carry the same weight, if any, with faculty salary and promotions committees. Incidentally, there are quite a number of problems and their solutions which are more substantial than quite a number of published papers and many of these problems and solutions have been cited in quite a number of papers, e.g., "ohmic heating" on page 245. Our preference is for applied problems with motivational material, but we do use other problems from time to time. Also, five reprints of the corresponding problem section are mailed gratis to all proposers and solvers whose contributions are published. Other contributors can receive one reprint by enclosing a stamped self-addressed envelope (no stamp necessary outside the U.S.A. and Canada).

A problem with a number such as 80-6* indicates it was published in 1980. Since, with few exceptions, their are 20 new problems published each year in four issues, this problem occured in the second issue. The * indicates a problem submitted without sollution. In some cases the author does have a solution but since it is not particularly nice or awkward to write up he does not submit it and hopes to see a more elegant solution (and this does occur).

Since 1959 some 600+ problems, with and without solutions, in diverse fields have been published in *SIAM Review*. Only a few of these problems have been reviewed in the literature. Since it is our belief that an organized selection of these problems would be of interest and useful to various groups of mathematicians and non-mathematicians, SIAM is publishing this book.

What is this big interest in problems? Like Mallory, who replied to the question of why climb mountains, mathematicians and non-mathematicians solve problems because "they are there." The mountains present a physical challenge whereas the mathematical problems present an intellectual challenge. As evidence for this desire for challenge consider the large number of problem and puzzle sections not only in mathematical journals but also in various magazines and newspapers.

Throughout the centuries starting with the Sumerians and Babylonians (roughly 2500 BC) one finds no end of problems and questions since problems and questions beget more problems and questions in an unending cycle. For a history of the earliest problems see references [1]–[7] at the end of the book. A specific example of an early challenging problem is the famous "cattle problem" attributed to Archimedes [3]:

> Compute the number of cattle of the Sun, O Stranger, and if you are wise apply your wisdom and tell me how many once grazed on the plains of the island of Sicilian Thrinacia, divided into four herds by differences in the colour of their skin—one milk-white, the second sleek and dark-skinned, the third tawny-coloured, and fourth dappled.

In each herd there was a great multitude of bulls, and there were these ratios. The number of white bulls, mark well, O Stranger, was equal to one-half plus one-third the number of number of dark-skinned, in addition to all the tawny-coloured. The number of dappled bulls, observe, was equal to one-sixth plus one-seventh the white, in addition, to all the tawny-coloured.

Now for the cows there were these conditions: the number of white cows was exactly equal to one-third plus one-fourth of the whole dark-skinned herd; the number of dark-skinned cows, again, was equal to one-fourth plus one-fifth of the whole dappled herd, bulls included; the number of dappled cows was exactly equal to one-fifth plus one-sixth of the whole tawny-coloured herd as it went to pasture; and the number of tawny-coloured cows was equal to one-sixth plus one-seventh of the whole white herd.

Now if you could tell me, O Stranger, exactly how many cattle of the Sun, not only the number of well-fed bulls, but the number of cows as well, of each colour, you would be known as one neither ignorant nor unskilled in numbers, but still you would not be reckoned among the wise. But come now, consider these facts, too, about the cattle of the Sun.

When the white bulls were mingled with the dark-skinned, their measure in length and depth was equal as they stood unmoved, and the broad plains of Thrinacia were all covered with their number. And, again, when the tawny-coloured bulls joined with the dappled ones they stood in perfect triangular form beginning with one and widening out, without the addition or need of any of the bulls of other colours.

Now if you really comprehend this problem and solve it giving the number in all the herds, go forth a proud victor, O Stranger, adjudged, mark you, all-powerful in this field of wisdom.

For a subsequent history and solutions to this problem which corresponds to the following simultaneous Diophantine set of 9 equations in 10 unknowns, see [8]:

$$(1)\ W = (\tfrac{1}{2} + \tfrac{1}{3})X + Z, \qquad\qquad (2)\ X = (\tfrac{1}{4} + \tfrac{1}{5})Y + Z,$$

$$(3)\ Y = (\tfrac{1}{6} + \tfrac{1}{7})W + Z, \qquad\qquad (4)\ w = (\tfrac{1}{3} + \tfrac{1}{4})(X + x),$$

$$(5)\ x = (\tfrac{1}{4} + \tfrac{1}{5})(Y + y), \qquad\qquad (6)\ y = (\tfrac{1}{5} + \tfrac{1}{6})(Z + z),$$

$$(7)\ z = (\tfrac{1}{6} + \tfrac{1}{7})(W + w), \quad (8)\ W + X = \blacksquare, \quad (9)\ Y + Z = \blacktriangle,$$

where $\blacksquare$ and $\blacktriangle$ denote a square and a triangular number, respectively.

Jumping to 1900, David Hilbert gave an address to the International Congress of Mathematicians in Paris on "Mathematical Problems" [9]–[11]. His address included his famous list of 23 problems whose solutions would in his view make significant progress in mathematics. In my view, the essay itself should be required reading for all mathematicians. It is also the forerunner of ideas later appearing in the highly recommended works of George Polya [12]–[14] on problem proposing and solving. Here are some significant paragraphs from the essay:

The deep significance of certain problems for the advance of mathematical science in general and the important role which they play in the work of the individual investigator are not to be denied. As long as a branch of science offers an abundance of problems, so long is it alive; a lack of problems foreshadows extinction or the cessation of independent development. Just as every human undertaking pursues certain objects, so also mathematical research requires its problems. It is by the solution of problems that the investigator tests the temper of his steel, he finds new methods and new outlooks, and gains a wider and freer horizon.

"Having now recalled to mind the general importance of problems in mathematics, let us turn to the question from what sources this science derives its problems. Surely the first and oldest problems in every branch of mathematics spring from experience and are suggested by the world of external phenomenon. Even the rules of calculation with integers must have been discovered in this fashion in a lower stage of human civilization, just as the child of today learns the application of these laws by empirical methods. The same is true of the first problems of geometry, the problems bequeathed us by antiquity, such as the doubling of the cube, the squaring of the circle; also the oldest problems in the theory of the solution of numerical equations, in the theory of curves and the differential and integral calculus, in the calculus of variations, the theory of Fourier series and the theory of potential—to say nothing of the further abundance of problems properly belonging to mechanics, astronomy and physics.

But in the further development of a branch of mathematics, the human mind, encouraged by the success of its solutions, becomes conscious of its independence. It evolves from itself alone, often without appreciable influence from without, by means of logical combination, generalization, specialization, by separating and collecting ideas in fortunate ways, new and fruitful problems, and appears then itself as the real questioner. Thus arose the problem of prime numbers and the other problems of number theory, Galois's theory of equations, the theory of algebraic invariants, the theory of abelian and automorphic functions; indeed almost all the nicer questions of modern arithmetic and function theory arise this way.

In the meantime, while the creative power of pure reason is at work, the outer world again comes into play, forces upon us new questions from actual experience, opens up new branches of mathematics, and while we seek to conquer these new fields of knowledge for the realm of pure thought, we often find the answers to old unsolved problems and thus at the same time advance most successfully the old theories. And it seems to me that the numerous and surprising analogies and that apparently prearranged harmony which the mathematician so often perceives in the questions, methods and ideas of the various branches of his science, have their origin in this ever-recurring interplay between thought and experience.

"If we do not succeed in solving a mathematical problem, the reason frequently consists in our failure to recognize the more general standpoint from which the problem before us appears only as a single link in a chain of related problems. After finding this standpoint, not only is this problem frequently more accessible to our investigation, but at the same time we come into possession of a method which is applicable also to related problems. The introduction of complex paths of integration by Cauchy and the notion of ideals in number theory by Kummer may serve as examples. This way for finding general methods is certainly the most practicable and the most certain; for he who seeks for methods without having a definite problem in mind seeks for the most part in vain.

In dealing with mathematical problems, specialization plays, as I believe, a still more important part than generalization. Perhaps in most cases where we seek in vain the answer to a question, the cause of the failure lies in the fact that problems simpler and easier than the one in hand have been either not at all or incompletely solved. All depends, then, on finding out these easier problems, and solving them by means of devices as perfect as possible and of concepts capable of generalization. This rule is one of the most important levers for overcoming mathematical difficulties and it seems to me that it is used always, though perhaps unconsciously.

This book can be used to augment the small number of applied problems from the *SIAM Review* Problem Section included in the SIAM publication, *Mathematical Modelling: Classroom Notes in Applied Mathematics*. To be independent of that book, we have duplicated these applied problems in this book. Since there are quite a number of other type problems as well, contestants of various mathematical competitions, in particular, the Mathematical Competition in Modeling [15] and the William Lowell Putnam Competition [16] can use this collection for practice.

In this book the problems have been classified into 22 broad sections, Mechanics, Electrical Resistance, Probability, Combinatorics, Series, Special Functions, Ordinary Differential Equations, Partial Differential Equations, Definite Integrals, Integral Equations, Matrices and Determinants, Numerical Approximation and Asymptotics, Inequalities, Optimization, Graph Theory, Geometry, Polynomials, Simultaneous Equations, Identitites, Zeros, Functional Equations and Miscellaneous. A number of the problems are multi-faceted (e.g., a problem on a series of special functions) so their placements are somewhat arbitrary in that a problem could appear in several sections. Since there are quite a few problems of the latter type and they have been placed in the Series section, the next section is Special Functions. Also, since there are quite a few geometric inequality problems in the Geometry section, this is noted as a footnote in the Inequalities section. Each problem has a title (occasionly more descriptive than the original title) with key words. The titles listed in the table of contents in the appropriate section together with the name(s) of the proposer(s) (first) and then the author(s) of the accompanying solution(s) and/or comments. Also included are asterisks (*) indicating that the problem is still unsolved. As an encouragement to reduce our backlog of unsolved problems, every person who submits the first received acceptable solution to any * problem in the book will receive one SIAM book of his/her choice.

In many cases, all the solutions and comments have not been used. For these cases, there will be a date and page number at the end of the problem title indicating where these other solutions and/or comments can be found in *SIAM Review*. We also have not included the names of the other solvers of the problems.

For the convenience of the reader, I have included a list of supplementary references at the end of almost all the sections. Some of these lists, where applicable, were taken from the *Mathematical Modelling* book. The others list more recent books and, even if not exhaustive, should provide most readers with a good start into the pertinent literature. Some of these references also have extensive reference lists. For example, the Chester reference in partial differential equations has 303 references and the Carlson reference in special functions has 194. Also at the end of the book, I have included a list of problem collection books as well as a list of some journal problem sections.

I thank all the many contributors and referees over the past 32 years for their concern and involvement with *Problem Proposing and Solving*. I am very grateful to the associate editors I have had and still have for the Problem Section over the years: William F. Trench, 1969–1977; Yudell L. Luke, 1969–1983; Henry E. Fettis, 1969–1984; Cecil C. Rousseau 1973–; and Otto G. Ruehr, 1978 –. Without these editors to share the burden of editing the section, I would have given up my association with the section a long time ago. I am particularly grateful to the SIAM management, in particular Ed Block, for having a unique Problem Section, and to him and Vickie Kearn for encouraging the publication of this book. I would also like to thank SIAM's editorial and production staff, in particular, Nancy Hagan Abbott and Dennis Michael Thomas for their help in producing and polishing this volume . Last but not least, I am very grateful to my wife Irene for her assistance with this book in many ways.

Murray S. Klamkin
University of Alberta

Contents

1. Mechanics

The * indicates a problem submitted without solution.

2. Electrical Resistance[†]

3. Probability[**]

[†]   Also see pp. 356, 360, 435, 460.
[**] Also see pp. 255, 258, 264, 302, 373, 435, 446, 569.

4. Combinatorics

5. Series[†]

[†]   Also see pp. 129, 144, 145, 146, 149.

6. Special Functions[†]

[†] Also see pp. 154, 172, 175, 176, 178–79, 260–63, 344, 396–97, 400, 567.

7. Ordinary Differential Equations[†]

[†] Also see pp. 8, 13–15, 500

8. Partial Differential Equations

9. Definite Integrals[†]

[†] Also see pp. 17, 406–7, 411–12, 414–15, 422–23, 428, 435–36, 491–92, 496, 567.

10. Integral Equations

11. Matrices and Determinants

12. Numerical Approximations and Asymptotic Expansions

13. Inequalities†

---

† Also see pp. 21, 23–24, 92–93, 202, 309, 470, 506, 509–10, 512–14, 521–22, 524.

14. Optimization[†]

[†]   Also see pp. 2, 26, 63, 103, 119, 324, 474, 482, 487, 511.

15. Graph Theory

16. Geometry[†]

---

† Also see p. 67.

17. Polynomials[†]

18. Simultaneous Equations

[†]   Also see p. 418, 451.

19. Identities[†]

20. Zeros[**]

21. Functional Equations[§]

22. Miscellaneous

[†]   Also see pp. 123, 125–27, 132–34, 141, 148, 197–98, 208, 534.
[**]   Also see pp. 322, 345, 380, 526–27, 529, 531.
[§]   Also see p. 18

1. MECHANICS

On the Motion of a Trailer-Truck

Problem 84-19, *by* J. C. ALEXANDER (University of Maryland).

Consider a semi rig, i.e., a truck cab pulling a trailer. Assume for simplicity that the cab and the trailer have single rear axles. Assume the support for the trailer is directly over the cab's rear axle. Show that at any instant of motion, the center of curvature of the track of the trailer's rear wheels is at the point where the extension of the rear axles of the cab and trailer intersect. (See diagram.)

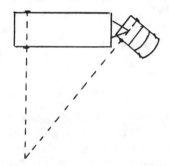

Editorial note: The proposer notes that his proof is analytic and that he would like to see a more elegant (geometric) one.

Solution by B. HALPHEN (Laboratoire Central des Ponts et Chaussees, Paris, France).

At any time the velocity of any point I of the trailer rear axle AB is perpendicular to AB: the length AI is constant and AI is the common normal to the trajectories of A and I. It is then easy to prove that:

•At points corresponding to the same time, trajectories of A and any point I of AB have the same center of curvature O, on the extension of AB.

•The ratio of the increments of curvilinear abcissas on trajectories of A and I is equal to OA/OI, which implies that the ratio of the velocities of A and I is

(1)
$$\frac{V_A}{V_I} = \frac{OA}{OI}.$$

On the other hand, the velocity of point C considered as belonging to the truck is perpendicular to the truck rear axle. As C is also a point of the trailer, the instantaneous rotation center of the trailer is at the point ω where the extension of the rear axles of the cab and trailer intersect. Then:

(2)
$$\frac{V_A}{V_I} = \frac{\omega A}{\omega I}.$$

From (1) and (2), one concludes that O and ω are at the same point.

1

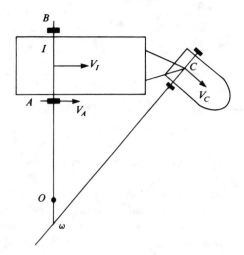

Editorial note: Also noted was a paper by J. C. Alexander and J. Maddocks, *On the maneuvering of vehicles*, which was submitted to the SIAM J. of Applied Mathematics and which treats the kinetic problem of a vehicle with several steerable or fixed axles and also solves a simple optimal steering problem.

Maximal Stability

Problem 86-16*, by DAVID SINGMASTER (Polytechnic of the South Bank, London.)
    Consider a can of beer or soda. Assuming the can is a right circular cylinder, it is known that the center of gravity (CG) is at its lowest when it coincides with the top of the liquid in the can. However, riding on British Rail made me wonder what level of fluid in the can would make the can maximally stable. We can measure the stability as the energy required to raise the CG, from its position when the can is vertical, to its position when the CG is just above the point of support of a tilted can. When is this position maximized?
    For standardization, let the can have radius R, height H, mass M, and let it contain mass m of liquid when full.

Solution by M. ORLOWSKI *and* M. PACHTER (NRIMS/CSIR, Pretoria, South Africa).
    A fluid level of $1.094R$, where R is the radius of the can, would make a standard beer can maximally stable.
    Here we assume the thickness of the walls of the can to be negligible. Let x be the level of the fluid in the upright can (see Fig. 2). Then, the altitude of the CG of the (upright) can is

(1)
$$y_1(x) = \frac{MH^2 + mx^2}{2(MH + mx)}.$$

Next we calculate the location of the CG of a tilted can whose fluid level was x when the can was in the upright position. The angle of tilt of the can is θ (see Fig. 3). The first step is to calculate the CG of the fluid mass in the tilted can. This is done by considering rectangular slices of fluid that are at a distance r from the axis of symmetry of the can (see Fig. 2). Hence, the height of the r-slice of fluid is $x + r \tan \theta$ and its width is $2\sqrt{R^2 - r^2}$. Thus, the area of the r-slice is $2(x + r \tan \theta)\sqrt{R^2 - r^2}$ and

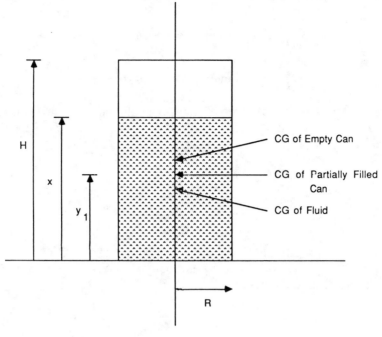

FIG. 2

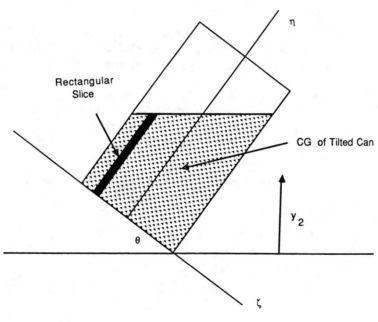

FIG. 3

its volume is $2(x + r \tan \theta)\sqrt{R^2 - r^2}dr$. The (ξ, η) coordinates of the CG of the r-slice are $\xi = r$, $\eta = \frac{1}{2}(x + r \tan \theta)$, and therefore the (ξ, η) coordinates of the CG of the fluid are

(2)
$$\bar{\xi}=\frac{2}{\pi R^2 x}\int_{-R}^{R}(x+r\tan\theta)r\sqrt{R^2-r^2}\,dr=\frac{R^2\tan\theta}{4x},$$

(3)
$$\bar{\eta}=\frac{1}{\pi R^2 x}\int_{-R}^{R}(x+r\tan\theta)^2\sqrt{R^2-r^2}\,dr=\frac{4x^2+R^2\tan^2\theta}{8x},$$

respectively. The (ξ, η) coordinates of the CG of the empty can (of mass M) are $(0, H/2)$, and hence, in view of (2) and (3), the (ξ, η) coordinates of the CG of the partially filled can are

(4)
$$\tilde{\xi}=\frac{mR^2\tan\theta}{4(MH+mx)},$$

(5)
$$\tilde{\eta}=\frac{4(MH^2+mx^2)+mR^2\tan^2\theta}{8(MH+mx)}.$$

Let $y_2(x, \theta)$ denote the altitude CG of the partially filled can (see Fig. 2). If the CG of the partially filled can is directly above the point of support, then $y_2\cos\theta = \tilde{\eta}$ and $y_2\sin\theta = R - \tilde{\xi}$. Thus, the altitude is given by

(6)
$$y_2=\frac{R-\tilde{\xi}}{\sin\theta}=\frac{R}{\sin\theta}\left[1-\frac{mR\tan\theta}{4(MH+mx)}\right],$$

where x and θ are related by the constraint

(7)
$$\tan\theta=\frac{R-\tilde{\xi}}{\tilde{\eta}}=2R\frac{4(MH+mx)-mR\tan\theta}{4(MH^2+mx^2)+mR^2\tan^2\theta}.$$

Next we introduce the dimensionless problem parameters $\mu \doteq M/m\ (\ll 1)$ and $h \doteq H/R$, and we turn x, y_1, and y_2 into dimensionless variables by scaling each by R. Also, we let $z = \tan\theta$. In dimensionless form, (1), (6), and (7) become

(8)
$$y_1=\frac{\mu h^2+x^2}{2(\mu h+x)},$$

(9)
$$y_2=\frac{\sqrt{1+z^2}}{z}\left[1-\frac{z}{4(\mu h+x)}\right],$$

(10)
$$z=2\frac{4(\mu h+x)-z}{4(\mu h^2+x^2)+z^2},$$

respectively.

We are interested in the solution of the optimization problem

(11)
$$\max_{x,z}\{y_2-y_1\},$$

subject to (10). Note that the above analysis applies provided the optimal x, z satisfy the condition

(12)
$$z\leq x\leq h-z.$$

If it so happens that the problem parameters μ, h eventually yield an optimal solution x^*, y^* of the constrained optimization problem such that (12) does not hold, i.e., if the range of validity of our calculation is then violated, we must separately consider in our rederivation of (2), (3) the extreme cases of low fluid level $x < z$ and high fluid level $x > h - z$. Equation (10) is a quadratic equation in x, namely

$$zx^2 - 2x + \tfrac{1}{4}[z^3 + 2(1 + 2\mu h^2)z - 8\mu h] = 0.$$

Hence, x is related to z by

(13)
$$x = \frac{2 \pm \sqrt{4 - z^4 - 2(1 + 2\mu h^2)z^2 + 8\mu hz}}{2z},$$

provided $z > 0$ and

(14)
$$4 \geqq z^4 + 2(1 + 2\mu h^2)z^2 - 8\mu hz.$$

Substituting

$$zx^2 = 2x - \tfrac{1}{4}[z^3 + 2(1 + 2\mu h^2)z - 8\mu h]$$

in (8), we obtain

$$y_1 = \frac{1}{z}\left[1 - \frac{z^3 + 2z}{8(\mu h + x)}\right].$$

Thus

$$y_2 - y_1 = \frac{\sqrt{1 + z^2} - 1}{z}\left[1 + \frac{z(\sqrt{1 + z^2} - 1)}{8(\mu h + x)}\right].$$

Finally, substituting for x from (11), we end up considering the simple unconstrained scalar optimization problem

(15)
$$\max_z \frac{\sqrt{1 + z^2} - 1}{z}\left[1 + \frac{z^2(\sqrt{1 + z^2} - 1)}{4\{2(1 + \mu hz) \pm \sqrt{4 - z^4 - 2(1 + 2\mu h^2)z^2 + 8\mu hz}\}}\right].$$

Also recall that z lies in the range (14) and the optimal z must satisfy (12), namely

(16)
$$z^2 \leqq 1 \pm \tfrac{1}{2}\sqrt{4 - z^4 - 2(1 + 2\mu h^2)z^2 + 8\mu hz} \leqq hz - z^2.$$

Finally, the optimal fluid level x is found by substituting the optimal z in (13).

The problem parameters are $\mu = 0.1132$ and $h = 3.4375$. For these parameters the quartic equation associated with (14) has a unique real positive solution $\bar{z} = 0.914$ and therefore (14) yields the range $0 < z \leqq \bar{z} = 0.914$. Furthermore, the maximum in (15) is then attained at the boundary $\bar{z}$ of the z-range, i.e., $z^* = \bar{z} = 0.914$ and thus $x^* = 1/\bar{z} = 1.094$. Finally, note that (12) is then satisfied since $0.914 < 1.094 < 2.523$.

The insight gained above affords us the possibility of ascertaining the range of validity (in parameter space) of our solution. In particular, if μ is sufficiently small, then $\bar{z} > 1$, $x^* = 1/\bar{z} < 1$ and the left-hand part of (12) is violated. Hence, the smallest possible value of the parameter μ for our analysis to be valid is that which makes $\bar{z} = 1$ the unique real, positive solution of the quartic solution

$$z^4 + 2(1 + 2\mu h^2)z^2 - 8\mu hz - 4 = 0.$$

In other words $\mu \geqq \bar{\mu} = (1/4h(h - 2))$. Furthermore, it is required that $h > \bar{h} = 2$. Hence, the parameter space (μ, h) is partitioned by the curve $\bar{\mu}(h) = (1/4h(h - 2))$, $h > \bar{h} = 2$.

Problem 74-17, *A Stability Problem*, by O. BOTTEMA (University of Delft, the Netherlands).

Unit masses are fixed at each of the 2^n vertices of a hypercube in $E^n (n \geq 1)$. A test particle which can move freely in space is attracted by the unit masses according to Newton's inverse square law of gravitation. The center of the cube is obviously a position of equilibrium for the test particle. Is this a position of stable or unstable equilibrium?

Solution by O. P. LOSSERS (Technological University, Eindhoven, the Netherlands).

We consider the more general case where the test particle is attracted with a force proportional to the kth power of the inverse of the distance. Here k is an integer satisfying $k \geq 2$. Then the total force acting on the test particle is proportional to

$$\mathbf{F}(\mathbf{x}) = - \sum_{i=1}^{N} (\mathbf{x} - \mathbf{a}_i) |\mathbf{x} - \mathbf{a}_i|^{-k-1},$$

where $N = 2^n$, and $\mathbf{a}_1, \cdots, \mathbf{a}_N$ are the vertices of the unit cube. We have $\mathbf{F}(\mathbf{x}) = \nabla \varphi(\mathbf{x})$, where the potential φ is given by

$$\varphi(\mathbf{x}) = (k - 1)^{-1} \sum_{i=1}^{N} |\mathbf{x} - \mathbf{a}_i|^{-(k-1)}.$$

The origin is stable if and only if φ has a local maximum at $\mathbf{x} = \mathbf{0}$. We calculate the Hessian $(\varphi_{x_i x_j})$, that is, the functional matrix $\mathbf{F}_\mathbf{x}(\mathbf{x}) = (\partial F_i / \partial x_j)$ at $\mathbf{x} = \mathbf{0}$. If $\mathbf{f}(\mathbf{x}) = (\mathbf{x} - \mathbf{a})|\mathbf{x} - \mathbf{a}|^{-k-1}$, then

$$\mathbf{f}_\mathbf{x}(\mathbf{x}) = |\mathbf{x} - \mathbf{a}|^{-k-1} I - (k+1)|\mathbf{x} - \mathbf{a}|^{-k-3}(\mathbf{x} - \mathbf{a})(\mathbf{x} - \mathbf{a})^T.$$

Consequently, using $|a_i| = \sqrt{n}, \sum_{i=1}^{N} a_i a_i^T = NI$ (as is easily verified), we obtain

$$\mathbf{F}_\mathbf{x}(\mathbf{0}) = - \sum_{i=1}^{N} |\mathbf{a}_i|^{-k-3} \{ |\mathbf{a}_i|^2 I - (k+1) \mathbf{a}_i \mathbf{a}_i^T \}$$

$$= -n^{-(k+3)/2} N(n - k - 1) I.$$

If $n > k + 1$, then $F_\mathbf{x}(\mathbf{0})$ is negative definite, hence the origin is stable. If $n < k + 1$, then $F_\mathbf{x}(\mathbf{0})$ is positive definite, which implies the instability of the origin. If $n = k + 1$, the potential φ is a harmonic function, $\Delta \varphi = 0$. It is well known that a harmonic function does not have maxima or minima. Hence $\mathbf{0}$ is unstable.

Safe Car Following

Problem 87-14*, *by* VINCENT SALMON (Menlo Park, California).

It has been recommended by some highway safety officials that for reasonable safety in a lane of traffic with auto A_1 following A_2, A_1 should remain at least t_0 seconds behind A_2. Thus starting at any given time, A_1 should take a time interval of at least t_0 seconds to reach the location that A_2 occupied at that time. Now suppose that the speeds and spacing are initially constant and optimal, and that A_2 accelerates. What should be the acceleration of A_1 so as to maintain the minimum safe separation?

Solution by STEVEN C. PINAULT (AT&T Engineering Research Center, Princeton, New Jersey).

We let $x_i(t)$ denote the position of car A_i at time t, for $i = 1, 2$. Then the rule that "starting at any given time, A_1 should take a time interval of at least t_0 to reach the location that A_2 occupied at that time" can be expressed as

$$x_1(t + t_0) \leqq x_2(t)$$

for all t. Thus the minimum separation is maintained at all times if $x_1(t) = x_2(t - t_0)$.

This, however, does not seem like the proper way to extend the rule from the constant-speed case to the variable-speed case. In particular, it presents the problem that when A_2 suddenly decelerates, A_1 will continue as usual for t_0 seconds and then decelerate just as abruptly, which defeats the purpose of the rule.

Note that the time and work required to stop a car are primarily a function of its speed—the acceleration stops as soon as the foot leaves the gas pedal, and the speed determines the amount of momentum which needs to be dissipated by the braking action, as well as how far the car will travel in the time it takes the foot to reach the brake pedal. Thus, a more appropriate way to extend the rule to the variable speed case is to interpret it in terms of instantaneous speed; at any given time, the instantaneous speed of A_1 should be such that it would take t_0 seconds to reach the current location of A_2. This rule has the advantage that it leads to timely and appropriate action when A_2 is decelerating. Now the condition becomes: given the function $x_2(t)$, require that $x_1(t)$ satisfy the first-order differential equation

$$x_1(t) + t_0 \dot{x}_1(t) = x_2(t).$$

Solving this equation, we obtain

$$x_1(t) = e^{-t/t_0} \left[x_1(0) + \frac{1}{t_0} \int_0^t e^{s/t_0} x_2(s) \, ds \right].$$

For example, if $x_2(t) = \beta t$ for $t \leqq 0$, and $x_2(t) = \beta t + \gamma t^2$ for $t \geqq 0$, then

$$x_1(t) = \begin{cases} \beta(t - t_0) & \text{for } t \leqq 0, \\ (\beta - 2\gamma t_0)(t - t_0) + \gamma(t^2 - 2t_0^2 e^{-t/t_0}) & \text{for } t \geqq 0. \end{cases}$$

●

Malevolent Traffic Lights

Problem 82-16, *by* J. C. LAGARIAS (Bell Laboratories, Murray Hill, NJ).

Can the red-green pattern of traffic lights separate two cars, originally bumper-to-bumper, by an arbitrary distance? We suppose that:

(1) Two cars travel up a semi-infinite street with traffic lights set at one block intervals. Car 1 starts up the street at time 0 and car 2 at time $t_0 > 0$.

(2) Both cars travel at a constant speed 1 when in motion. Cars halt instantly at any intersection with a red light, and accelerate instantly to full speed when the light turns green. If the car has entered an intersection as the light turns red, it does not stop.

(3) Each light cycles periodically, alternately red and green with red time μ_j, green time λ_j and initial phase θ_j (i.e., phase at time 0) at intersection j.

Can one define triplets $(\lambda_j, \mu_j, \theta_j)$ ($j = 1, 2, \cdots$) so that $\{(\lambda_j, \mu_j): j = 1, 2, \cdots\}$ is a finite set and so that car 1 gets arbitrarily far ahead of car 2?

Solution by O. P. LOSSERS (Eindhoven University of Technology, Eindhoven, The
   Netherlands).

The answer is yes as we shall now prove. We consider two types of traffic lights
(I) $\lambda = \mu = a/2$, (II) $\lambda = \mu = b/2$, where a and b are chosen in such a way that a/b is
irrational. The phases of the lights are chosen in such a way that the first car never has to
stop. This is easily accomplished. However, we can manipulate more with the phases.
Suppose that the second car arrives at a light of type I with a delay of $ra + \theta$ seconds
($r \in N$, $0 < \theta < a$). The phase is chosen in such a way that the light changes $\frac{1}{2}\theta$ seconds
after car 1 passes. So the delay of car 2 increases to $(r + \frac{1}{2})a + \frac{1}{2}\theta$. If we had only used
lights of type I the delay would monotonically increase to $(r + 1)a$. A similar statement is
true for lights of type II. At each intersection we still have the choice of the type of light.
We choose this type in such a way that the increase is maximal. Since a/b is irrational, no
multiple of a is a multiple of b and the time delay will tend to ∞.

Editorial note. The proposer shows that if all the cycle times are commensurable
then any two cars remain within a bounded distance of each other, no matter how the
lights are specified. [C.C.R.]

Shape of a Roller-Coaster Track

Problem 88-6, by JOHN S. LEW (IBM Research Division, T. J. Watson Research
   Center, Yorktown Heights, New York).

A roller-coaster car under the influence of gravity moves without friction along
a track which lies in a vertical plane. Determine the equation of the track if the
magnitude of the force exerted by the car on the track is constant. In particular, take
this force to be a given positive multiple c of the car's weight.

Solution by LAWRENCE E. FLYNN (Harris–Stowe State College, St. Louis).

Describe the path of the track by $x = x(z)$. Choose z so that the car will have zero
velocity when $z = 0$ [$v(0) = 0$]. (This is a conservative system and can be set up so
that the kinetic energy is 0 when $z = 0$.) By letting the positive z-direction be down,
we have $|v(z)|^2 = 2mgz$. We first consider paths with $x'(z) \geq 0$ (' means d/dz) with
the car on the inside of the track. The forces on the track due to gravity and curvature
are given by

$$(1) \qquad f_g(z) = \frac{mgx'(z)}{\sqrt{1 + x'(z)^2}} \quad \text{and} \quad f_c(z) = \frac{|v(z)|^2}{r_c} = \frac{2mgzx''(z)}{[1 + x'(z)^2]\sqrt{1 + x'(z)^2}}$$

respectively, with r_c = the radius of curvature. By setting the total force [$= f_g(z) + f_c(z)$]
equal to a constant multiple of the car's weight, mgc, we find that

$$(2) \qquad \frac{x'(z)}{\sqrt{1 + x'(z)^2}} + \frac{2zx''(z)}{[1 + x'(z)^2]\sqrt{1 + x'(z)^2}} = c.$$

Given $x'(z) = \tan \phi(z)$ [$\phi(z)$ = the angle the track makes with the vertical], (2) becomes

$$(3) \qquad 2z\phi'(z) \cos \phi(z) = c - \sin \phi(z).$$

Equation (3) is separable. Its solution is

$$(4) \qquad c - \sin \phi(z) = \frac{A}{\sqrt{z}} \quad \text{or} \quad x'(z) = \frac{c - (A/\sqrt{z})}{\sqrt{1 - (c - (A/\sqrt{z}))^2}}.$$

Given an initial slope $x'(z_0) = \tan \phi_0$, then $A = \sqrt{z_0}(c - \sin \phi_0)$. Since $\sqrt{z_0}$ appears as a linear factor in A, and A and $1/\sqrt{z}$ always appear together in (4), by a change of the z-scale, relative to x, we can recover the solutions for other z_0 values. From (4) we find that for any $c > -1$ there is a minimum value of z, which we denote by z_{min}, given by

$$(5) \qquad \sqrt{z_{min}} = \frac{A}{c+1} = \sqrt{z_0}\,\frac{c - \sin \phi_0}{c+1},$$

and that for $c > 1$ there is a maximum value given by

$$(6) \qquad \sqrt{z_{max}} = \frac{A}{c-1} = \sqrt{z_0}\,\frac{c - \sin \phi_0}{c-1}.$$

Equation (2) is to be solved for four cases.

Case 1. $c = 0$. The only solutions for this case are the free-fall trajectories

$$(7) \qquad x(z) = -2 z_{min} \sqrt{\frac{z}{z_{min}} - 1} + D,$$

with $z_{min} = A^2$. If we want $x(z_{min}) = 0$, then $D = 0$. If $z_{min} = 0$, then the solutions are just vertical lines. The graph of (7) for $z_{min} = 1$ and $D = 0$ and its reflection about $x = 0$ (the solution for $x''(z) < 0$) appear in Figs. 1–3 with the label #1.

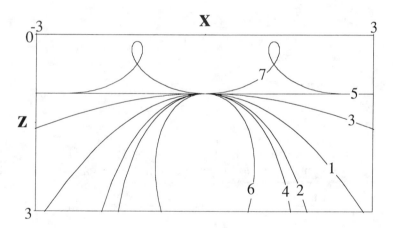

FIG. 1

Case 2. $0 < |c| < 1$. Define θ by $c = \sin \theta$, and work with $\phi_0 < \theta$. Then from (4), we see that $\lim_{z \to \infty} x'(z) = \tan \theta$ and

$$x'(z) \sim \tan \theta - \frac{1}{(1 - \sin \theta) \cos \theta} \sqrt{\frac{z_{min}}{z}} \qquad (z \to \infty).$$

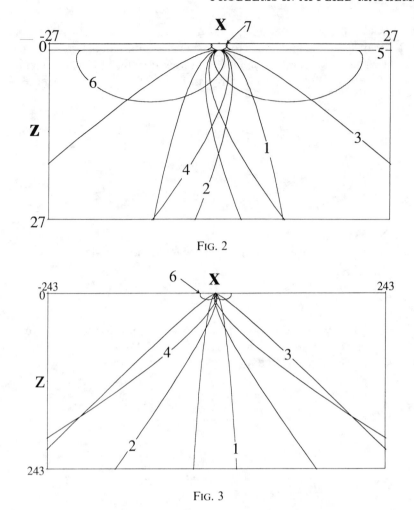

FIG. 2

FIG. 3

From (2), the straight-track solution is $x(z) = (\tan \theta)z + D$ as expected ($z_{min} = 0$ and $\phi_0 = \theta$). The other solutions, which follow by integrating equation (4), are

(8) $$x(z) = C\left[\frac{\sec \omega \tan \omega}{2} - (\csc \theta + \sin \theta) \tan \omega + \frac{3}{2}\ln (\sec \omega + \tan \omega)\right] + D,$$

with

$$C = \frac{2z_{min} \tan \theta}{(1 - \sin \theta)^2} \quad \text{and} \quad \sec \omega = (1 - \sin \theta)\sqrt{\frac{z}{z_{min}} + \sin \theta}.$$

If we want $x = 0$ when the track reaches its highest point ($= z_{min}$), then $D = 0$. A careful check of signs will show that (by using symmetry and $-\pi/2 < \phi_0 < 0$) this solution can produce the curves for all sign cases of $x'(0)$ and $x''(0)$.

For large z, the highest-order terms in the difference between the solutions given in (8) and the linear (straight-track) solution are like $\sqrt{z}$. The two solutions for $c = \pm\sqrt{2}/2$ with $z_{min} = 1$ and $D = 0$ (and their reflections) appear in Figs. 1–3 with labels #2 and #3. Note that one curve has a loop and the other does not. For any

c $(0 < |c| < 1$ and $z_{min} > 0)$, there will always be a nonlooping solution outside of the free-fall curve $(c = 0)$ and a looping one with the loop inside the free-fall curve. The looping solutions eventually cross the free-fall curve. Of course, as $c \to 0$ these curves will remain closer to the free-fall curve for larger z. For $0 < c < 1$, the car rides on the inside of the looping track and the outside of the nonlooping track. For $-1 < c < 0$, the cars ride on the "wrong side of the tracks."

Case 3. $|c| = 1$. There are two sets of solutions for this case. The first is the flat straight-track solution, $z = D$. The second is obtained from (4). The value of z_{min} is given in (5) with $c = 1$. We find $\lim_{z \to \infty} x'(z) = \infty$ and

$$x'(z) \sim \frac{1}{2} \sqrt[4]{\frac{z}{z_{min}}} \qquad (z \to \infty).$$

Notice that these solutions will be superlinear in z. The solution of (4) is

(9) $$x(z) = z_{min} \left\{ \frac{2}{5} \left[\sqrt{\frac{z}{z_{min}}} - 1 \right]^{5/2} - 2 \left[\sqrt{\frac{z}{z_{min}}} - 1 \right]^{1/2} \right\} + D.$$

This solution was obtained and analyzed by l'Hôpital in [1]. The curve given by (9) for $z_{min} = 1$ and $D = 0$ and its reflection appear in Figs. 1–3 with the label #4. The flat-track solution, $z = 1$, appears in Figs. 1 and 2 with label #5.

Case 4. $c > 1$. We no longer have straight-track solutions. Define θ by $c = \csc \theta$. The values of z_{min} and z_{max} are given in (5) and (6), respectively. Integrating (4), we obtain

(10) $$x(z) = C \left[\frac{\omega}{2} - (\csc \theta + \sin \theta) \cos \omega - \frac{\cos \omega \sin \omega}{2} \right] + D,$$

with

$$C = \frac{2 z_{min} \sec \theta}{(\csc \theta - 1)^2} \quad \text{and} \quad \sin \omega = (\csc \theta - 1) \sqrt{\frac{z}{z_{min}}} - \csc \theta.$$

If we want $x = 0$ at $z = z_{min}$, then $D = C\pi/4$. These solutions loop repeatedly. Two periods of two solutions for $c = \pm 2$ that are level at $z = 1$ and $x = 0$ appear in Figs. 1–3 with the labels #6 and #7. The upper curve has $z_{min} = \frac{1}{9}$ and the lower curve has $z_{max} = 9$. The two solutions are identical up to a linear transformation.

The two curves for $c = 1$ (#4 and #5) and the free-fall trajectory (#1) form boundaries separating the multiple-loop curves, the single-loop curves, and the non-looping curves from each other as pictured in the figures. In Figs. 1–3, if the roller-coaster is on top of the track (the positive z-direction is down) at $z = 1$ and $x = 0$, then we have $c = 2, 1, \sqrt{2}/2, 0, -\sqrt{2}/2, -1$, and -2 for curves #7, #5, #3, #1, #2, #4, and #6, respectively.

Straight lines are obtained as solutions for cases with $\theta = \phi_0$ [$c = \sin \phi_0$]. For these cases, the roller-coaster goes up to $z = 0$ (i.e., $z_{min} = 0$) stops, and comes back down the same track. As $z_{min} \to 0$ the parabolic solution collapses to a line; looping solutions and solutions for $|c| > 1$ are lost.

REFERENCE

[1] G. F. A. DE L'HÔPITAL, *Solution d'un problème physico-mathématique*, Mém. Acad. Royale des Sciences (1700), pp. 11–27.

Falling Dominoes

Problem 71-19*, *by* D. E. DAYKIN (University of Reading, England).

"How fast do dominoes fall?"

The "domino theory for Southeast Asia" says that if Vietnam falls, then Laos falls, then Cambodia falls, and so on. Hearing a discussion of the theory led me to wonder about the proposed physical problem. The reader is invited to set up his own "reasonable" simplifying assumptions, such as perfectly elastic dominoes, constant coefficient of friction between dominoes and the table, an initial configuration with the dominoes equally spaced in a straight line, and so on.

Solution by B. G. MCLACHLAN, G. BEAUPRE, A. B. COX *and* L. GORE (Stanford University).

A similarity rule can be derived for the speed of the wave of falling of dominoes under the following simplifying assumptions: the dominoes are equally spaced in a row; the domino width does not appreciably affect the domino motion; and, except at close spacings, the domino thickness has negligible affect on the domino motion. From the physics of the problem it is further assumed that the wave speed V is only dependent on the following parameters: domino spacing d, domino height h, and gravity g—refer to

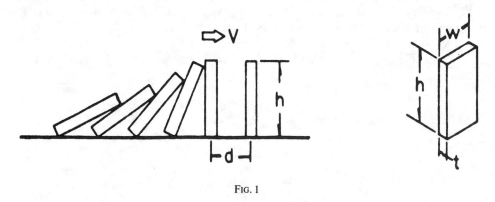

FIG. 1

Fig. 1 for nomenclature. Therefore the wave speed is given by

$$V = V(d, h, g).$$

By inspection, the dimensionless form is

$$\frac{V}{\sqrt{gh}} = f(d/h).$$

The normalized wave speed is only a function of the relative spacing.

To validate this relationship, experiments were performed using typical dominoes (dimensions: $h = 4.445$ cm, $t = .794$ cm, $w = 2.223$ cm). Two heights were tested (1 and $2 \times h$). The double height ($2 \times h$) was obtained by taping two dominoes together. Testing was conducted on a flat surface, the dominoes (≈ 100) being equally spaced in a straight row. Timing was done by hand utilizing a stopwatch: the timing start and stop was based on the initial impulse (provided by hand) and the fall of the last domino in the row.

The experimental results are shown in Fig. 2, where the normalized wave speed is plotted as a function of relative spacing. The data confirms the derived similarity rule: a good correlation being shown between the single and double height domino normalized wave speeds for the same relative spacing. Evident is a variation of the normalized wave speed over the two domino heights tested for large relative spacings (close to one): at such large relative spacings the dominoes tend not to fall straight making scatter in the data inevitable. Also, of note is that the normalized wave speed displays the appropriate asymptotic behavior for relative spacings approaching the domino thickness to height ratio (As $d/h \rightarrow t/h$, $V \rightarrow \infty$ due to the solid body character of the domino row).

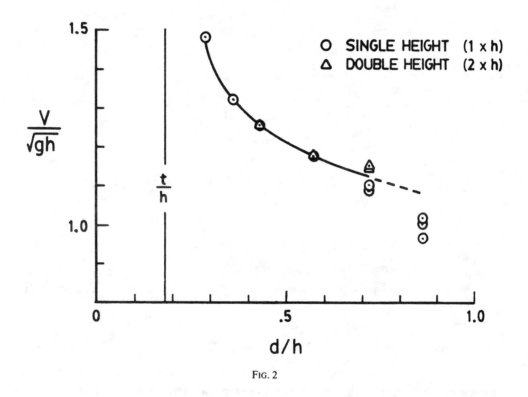

FIG. 2

Birthday Candles Blowout Problem

Problem 88-16[†], *by* M. PACHTER (National Research Institute for Mathematical Sciences, Pretoria, South Africa).

A straight-line high-pressure front approaches at constant speed and sweeps through a given array of n lit candles (supposedly arranged on top of a birthday cake). The candles are extinguished one at a time at the instant of passage through the moving front (see Fig. 1). The blowout sequence of the candles is observed (this is a permutation of n elements), and on the basis of this "measurement," it is required that we estimate the direction from which the front came. Obviously, the estimate consists of a confidence "interval" C, which is an angular sector.

For a given array of n candles, we have the following questions.

(a) From the numerical sequence information, how should we calculate the uncertainty C in the direction of arrival of the front?

---

†The author's solution is subject to revision.

(b) If there is no a priori information on the possible direction of arrival of the front, how should we arrange n available candles to minimize the maximal possible measure of uncertainty C in the estimated direction of movement of the front?

(c) If it is known that the possible direction of arrival of the front is confined to an a priori known angular sector of extent $2\alpha\pi$ radius, where $0 < \alpha < 1$, then how should n available candles be arranged to minimize the maximal possible measure of uncertainty C in the direction of movement of the front?

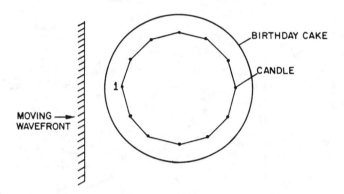

FIG. 1. n birthday candles are arranged at the n vertices of the regular n-gon.

Pendulum with Variable Length

Problem 83-14, *by* M. A. ABDELKADER (Alexandria, Egypt).

A simple pendulum with variable string length and subject to viscous damping is described by

$$\theta'' + \left\{2\frac{r'}{r} + \frac{c}{m}\right\}\theta' + \frac{g}{r}\sin\theta = 0,$$

where $\theta(t)$ is the angle the string makes with the vertical, m the mass of the bob, c the damping coefficient and g the acceleration due to gravity. Determine the exact expression for $r(t)$ such that the bob moves horizontally for all time, the initial conditions being $\theta = \theta_0$ and $d\theta/dt = 0$ at $t = 0$, where $0 < \theta_0 < \pi/2$. (When r is constant, no exact solution of the differential equation is known.)

Solution by DEBORAH FRANK LOCKHART (Michigan Technological University).

For purely horizontal motion, $r\cos\theta \equiv k =$ constant. Letting $u = k\tan\theta$ leads to

$$u'' + \frac{c}{m}u' + \frac{g}{k}u = 0,$$

$$u(0) = k\tan\theta_0, \qquad u'(0) = 0.$$

If $\beta = \sqrt{|\alpha|}$, where $\alpha = 1 - 4gm^2/kc^2$ and $\gamma = c/2m$, then

$$u = \begin{cases} \dfrac{k}{\beta} e^{-\gamma t} (\sinh \gamma \beta t + \beta \cosh \gamma \beta t) \tan \theta_0, & \alpha > 0, \\[2ex] \dfrac{k}{\beta} e^{-\gamma t} (\sin \gamma \beta t + \beta \cos \gamma \beta t) \tan \theta_0, & \alpha < 0, \\[2ex] k e^{-\gamma t} (1 + \gamma t) \tan \theta_0, & \alpha = 0, \end{cases}$$

and $r^2 = k^2 + u^2$.

Paul Bunyan's Washline

Problem 78-17, *by* J. S. Lew (IBM T.J. Watson Research Center).

It is well known that if a uniform thin flexible cord is suspended freely from its endpoints in a uniform gravitational field, then the shape of the cord will be an arc of a catenary. Determine the shape of the cord if we use a very long one which requires the replacement of the uniform gravitational field approximation by the inverse square field.

Solution by the proposer.

We may determine a unique plane through the earth's center via the two fixed supports, and may suppose the required curve in this plane by symmetry arguments. Introduce polar coordinates (r, θ) in this plane, and locate the origin at the earth's center. Let s denote the cord arclength, l the constant linear density, $U(r)$ the gravitational potential, P and Q the fixed supports. Then $\int_P^Q ds$ is the same for any admissible curve, and $\int_P^Q lU(r) \, ds$ is a minimum for the equilibrium curve. However if the arclength is constant while the potential energy is either minimal or maximal, then clearly $\delta \int_P^Q L \, d\theta = 0$, where

(1) $$L(r, r', \theta) = [lU(r) - F](r^2 + r'^2)^{1/2}, \qquad r' = \frac{dr}{d\theta}.$$

Here F is an unknown constant having the dimensions of a force, a Lagrange multiplier representing the tension in the cord. We might now derive the Euler–Lagrange equation for this problem,

(2) $$(d/d\theta)(\partial L/\partial r') = \partial L/\partial r,$$

but we observe its translation-invariance in the variable θ [1, p. 262], and we obtain a first integral with no intermediate steps:

(3) $$r'(\partial L/\partial r') - L = E = \text{constant}.$$

Here E has the dimensions of energy, whence E/F has the dimensions of length. Direct substitution into (3) gives a first-order equation for the unknown curve:

(4) $$E(r^2 + r'^2)^{1/2} = r^2[F - lU(r)].$$

Moreover we expect a "turning point" on this curve where we achieve minimal or maximal distance from the origin. Thus we let (r_0, θ_0) be the coordinates of this point, and we note that $dr/d\theta = 0$ at (r_0, θ_0). Evaluating (4) at this point yields a relation among these unknown constants:

(5) $$lU(r_0) = F - (E/r_0).$$

If we introduce the dimensionless quantities

(6) $$\rho = r/r_0, \qquad \gamma = 1 - (Fr_0/E),$$

then we obtain the differential equation

(7) $$\rho^{-2}(\rho^2 + \rho'^2)^{1/2} = 1 - \gamma + \gamma U(r_0\rho)/U(r_0).$$

Recalling the standard formula $U(r) = -k/r$ for a suitable constant k, we can rewrite (7) as

(8) $$(\rho^2 + \rho'^2)^{1/2} = (1 - \gamma)\rho^2 + \gamma\rho.$$

Setting $u = 1/\rho$, we can simplify (8) to

(9) $$(u^2 + u'^2)^{1/2} = 1 - \gamma + \gamma u.$$

We square equation (9) and differentiate the result, cancel the factor $2u'$ and obtain

(10) $$u'' + (1 - \gamma^2)u = \gamma(1 - \gamma).$$

Since the solution curve in the original variables has a turning point at (r_0, θ_0), the corresponding function in these new variables satisfies the conditions

(11) $$u = 1, \qquad u' = 0, \quad \text{at } \theta = \theta_0.$$

To find the unknown curve we need only solve (10)–(11); to recover the original variables we need only use (6). The form of the solution depends on the value of γ. The results are:

(12) $\gamma = 1$: $\qquad \rho \equiv 1,$

(13) $\gamma = -1$: $\qquad \rho = [1 - (\theta - \theta_0)^2]^{-1},$

(14) $|\gamma| < 1$: $\qquad \rho = (1 + \gamma)[\gamma + \cos((1 - \gamma^2)^{1/2}(\theta - \theta_0))]^{-1},$

(15) $|\gamma| > 1$: $\qquad \rho = (1 + \gamma)[\gamma + \cosh((\gamma^2 - 1)^{1/2}(\theta - \theta_0))]^{-1}.$

Thus the solution is a circular arc for $\gamma = 1$, a straight line for $\gamma = 0$, and an elementary function for all values of γ.

Moreover the standard expression for curvature in polar coordinates is

(16) $$\kappa(\theta) = (\rho^2 + \rho'^2)^{-3/2}[\rho^2 + 2\rho'^2 - \rho\rho''];$$

whence the value of this curvature at the turning point is

(17) $$\kappa(\theta_0) = 1 - \rho''(\theta_0) = 1 + u''(\theta_0) = \gamma.$$

If γ is positive then the solution curve is concave at θ_0 and the potential energy is maximal for this curve. If γ is negative then the solution curve is convex at θ_0 and the potential is minimal for this curve. Thus the desired solutions have negative γ.

REFERENCE

[1] R. COURANT AND D. HILBERT, *Methods of Mathematical Physics I*, Interscience, New York, 1953.

Problem 64–6*, *Gravitational Attraction*, by MORTON L. SLATER (Sandia Corporation).

Determine the gravitational attraction between a uniform solid torus and a unit mass particle located on its axis.

Solution by C. J. BOUWKAMP (Technological University, Eindhoven, Netherlands).

Let a be the radius of the cross-section, b the radius of the central line, and z the distance from the unit mass particle to the central plane of the torus. Assume the torus to be of unit mass density. Then the total mass of the torus is

$$(1) \qquad M = 2\pi^2 a^2 b.$$

Further, let ρ and ϕ denote polar coordinates in the plane of the cross-section with center located in the central line. Then the gravitational potential of the torus at the location of the unit mass particle is

$$V = 2\pi \int_0^a \rho \, d\rho \int_{-\pi}^{\pi} d\phi \, \frac{b + \rho \cos \mu}{\sqrt{(z - \rho \sin \phi)^2 + (b + \rho \cos \phi)^2}} \, .$$

We have

$$(2) \qquad V = \frac{\partial}{\partial b} W,$$

where

$$(3) \qquad W = 2\pi \int_0^a \rho \, d\rho \int_{-\pi}^{\pi} d\phi [(z - \rho \sin \phi)^2 + (b + \rho \cos \phi)^2]^{1/2}.$$

The expression between square brackets can be transformed into

$$z^2 + b^2 + \rho^2 - 2\rho \sqrt{z^2 + b^2} \sin \left(\phi - \tan^{-1} \frac{b}{z} \right).$$

Since in (3) the integration with respect to ϕ is over a full period of the sine function, we can drop the phase factor $\tan^{-1}(b/z)$ and we can also replace the sine by the cosine. Hence,

$$(4) \qquad W = 2\pi \int_0^a \rho \, d\rho \int_{-\pi}^{\pi} d\phi [z^2 + b^2 + \rho^2 - 2\rho \sqrt{z^2 + b^2} \cos \phi]^{1/2}.$$

If we now set

$$u = \sqrt{z^2 + b^2}, \qquad v = \frac{\rho}{u} = \frac{\rho}{\sqrt{z^2 + b^2}},$$

(4) is transformed into

$$(5) \qquad W = 2\pi u \int_0^a \rho \, d\rho \int_{-\pi}^{\pi} d\phi [1 - 2v \cos \phi + v^2]^{1/2}.$$

The second integral in (5) can be expressed in terms of complete elliptic integrals, or better still, in terms of a hypergeometric function, viz.,

$$\int_{-\pi}^{\pi} d\phi \, [1 - 2v \cos \phi + v^2]^{1/2} = 4(1 + v)E\left(\frac{2\sqrt{v}}{1 + v}\right)$$

$$= 4[2E(v) - (1 - v^2)K(v)] = 2\pi F(-\tfrac{1}{2}, -\tfrac{1}{2}; 1; v^2).$$

The remaining integration over ρ can now easily be performed:

$$W = 2\pi^2 a^2 u F(-\tfrac{1}{2}, -\tfrac{1}{2}; 2; a^2/u^2).$$

In view of (1) and (2), we then have

$$\frac{V}{M} = \frac{1}{u} F\left(-\frac{1}{2}, \frac{1}{2}; 2; \frac{a^2}{u^2}\right).$$

Since the attractive force equals $-\partial V/\partial z$, the required force is then found to be given by

$$\frac{M z}{(z^2 + b^2)^{3/2}} F\left(-\frac{1}{2}, \frac{3}{2}; 2; \frac{a^2}{z^2 + b^2}\right).$$

Problem 74-20*, *Gravitational Attraction*, by H. R. AGGARWAL (NASA, Ames Research Center).

Determine explicitly the mutual force of gravitational attraction between two congruent spherical segments forming a "dumb-bell" shaped body whose central cross section is given by the figure.

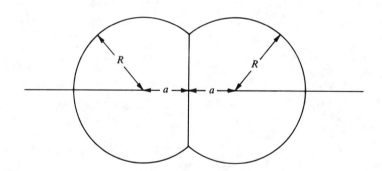

Characterizations of Parabolic Motion

Problem 88-5, by M. S. KLAMKIN (University of Alberta).

By a parabolic motion, we mean the motion which ensues when a particle is projected in a uniform gravitational field and is not subject to any frictional forces. Equivalently, the equations of motion are of the form $x = at$, $y = bt + ct^2$ and the trajectory is a parabola.

Let RP and RQ be two tangents to the trajectory and let RT be a vertical segment as in Fig. 1. The following properties of the motion are known:

(1) $$PR = V_P t \quad \text{and} \quad RT = gt^2/2$$

where V_P is the speed of the particle at P and t is the time for the particle to go from P to T.

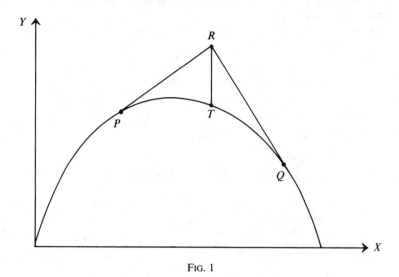

FIG. 1

(2) $PR/RQ = V_P/V_Q$ and it takes the particle the same time to go from P to T as from T to Q.

Show that either properties (1) or (2) characterize parabolic motion, i.e., if either (1) or (2) holds for all points on a smooth trajectory, the motion must be parabolic.

Solution by FRANK MATHIS (Baylor University).

Let $x = x(t)$ and $y = y(t)$ be a smooth trajectory, i.e., x and y are continuously differentiable functions of t. We may assume without loss of generality that there is an interval I containing zero with $x'(t) > 0$ for t in I. Select P, R, and T as in Fig. 1 with coordinates $(x(t_P), y(t_P))$, $(x(t_T), y_R)$ and $(x(t_T), y(t_T))$, respectively, so that t_P and t_T are in I. Then

(1) $$V_P = \sqrt{x'(t_P)^2 + y'(t_P)^2}$$

and the tangent line PR may be parameterized by

$$x = x(t_P) + x'(t_P)s, \qquad y = y(t_P) + y'(t_P)s$$

for $0 \leqq s \leqq (x(t_T) - x(t_P))/x'(t_P)$. It follows that

(2) $$y_R = y(t_P) + \frac{y'(t_P)}{x(t_P)}[x(t_T) - x(t_P)],$$

(3) $$PR = \frac{x(t_T) - x(t_P)}{x'(t_P)} \sqrt{x'(t_P)^2 + y'(t_P)^2}.$$

First we will assume that property (1) holds. By simple translations we may take $t_P = x(t_P) = y(t_P) = 0$. Set $t = t_T$ so that t is the time for the particle to go from P to T. Then since $PR = tV_P$, we have from (1) and (3) that

(4) $$x(t) = x'(0)t.$$

Thus x is linear for any t in I. But then x' is constant on I and so by the continuity of x', we may take I to be all of $\mathbb{R}$.

Now since $QR = \frac{1}{2}gt^2$, $y_R - y(t) = \frac{1}{2}gt^2$, and using (2) and (4), we find that

$$y(t) = y'(0)t - \frac{1}{2}gt^2.$$

Therefore, property (1) implies parabolic motion.

Next we assume property (2). Let Q be as in Fig. 1 with coordinates $(x(t_Q), y(t_Q))$ and t_Q in I. Then as above, we have

$$V_Q = \sqrt{x'(t_Q)^2 + y'(t_Q)^2},$$

(5) $$RQ = \frac{x(t_Q) - x(t_T)}{x'(t_Q)} \sqrt{x'(t_Q)^2 + y'(t_Q)^2},$$

$$y_R = y(t_Q) + \frac{y'(t_Q)}{x'(t_Q)}[x(t_Q) - x(t_T)].$$

Because the particle takes the same time to go from P to T as from T to Q, we may set $t_T = t$, $t_P = t - s$, and $t_Q = t + s$. Since $PR/RQ = V_P/V_Q$, we then have

(6) $$x'(t+s)[x(t) - x(t-s)] - x'(t-s)[x(t+s) - x(t)] = 0.$$

This condition holds for all t and s with $t \pm s$ in I. In particular, if we let $t = s$,

$$x'(2t) = x'(0)\left[\frac{x(2t) - x(t)}{x(t) - x(0)}\right]$$

and it follows that x is at least twice continuously differentiable for t in I, $t \neq 0$.

Returning to (6), we may now differentiate both sides with respect to s and then divide by s to obtain

$$x''(t+s)\left[\frac{x(t) - x(t-s)}{s}\right] + x''(t-s)\left[\frac{x(t+s) - x(t)}{s}\right] = 0.$$

Then letting $s \to 0$ and using the continuity of x'', we see that

$$2x''(t)x'(t) = 0.$$

But $x'(t) > 0$ for t in I. So x'' must be identically zero for t in I, $t \neq 0$. Hence, using the continuity of x' as before, we conclude that x is linear for all t.

Finally, to show that y is quadratic in t, let us again set $t_P = x(t_P) = y(t_P) = 0$ and let $t = t_Q$, so that $t_T = t/2$, $x(t_T) = x'(0)t/2$, and $x(t_Q) = x'(0)t$. Then from (2) and (5) we have

$$y'(0)\frac{t}{2} = y_R = y(t) - y'(t)\frac{t}{2}.$$

That is, y satisfies the differential equation

$$y' = \frac{2}{t}y - y'(0),$$

which has the solution

$$y = y'(0)t + ct^2$$

where c is an arbitrary constant. Thus, we conclude that property (2) also implies a motion that is parabolic.

> *Problem 60-11', A Minimum Time Path*, by W. L. BADE (Avco Research and Advanced Development Division).
>
> A projectile starting from rest, subject to a monotonically decreasing acceleration, traverses a distance s_0, attaining a final velocity v_0. Determine the maximum time of traverse.
>
> Solution by F. ZETTO (Chicago, Illinois).
>
> Since $v = \int_0^t a\, dt$ and a is monotonically decreasing, it follows that v is a convex function of t. Since the area under the v-t curve is s_0, the time duration will be maximum when v is a minimum subject to being convex. Whence, $v = kt$ where $v_0 = kt_{max}$, $kt_{max}^2 = 2s_0$. Thus, $t_{max} = 2s_0/v_0$, (it is assumed that s_0, $v_0 > 0$)

> *Problem 59-6*, The Smallest Escape Asteroid*, by M. S. KLAMKIN (AVCO Research and Advanced Development Division).
>
> A problem which was solved in the American Mathematical Monthly (May 1953, p. 332) was to determine the largest asteroid that one could jump "clear" off (escape). A more interesting and more difficult problem would be to determine the smallest asteroid that one could jump "clear" off. The difficulty arises in the reaction of the asteroid. For a large one the reaction is negligible. But this is not true for a small one.

> *Problem 64-3, Escape Velocity with Drag*, by D. J. NEWMAN (Yeshiva University).
>
> If we assume that the frictional force (or drag) retarding a missile is proportional to the density of the air, $\rho(x)$, at altitude x (above the earth) and to the square of the velocity, then the differential equation of the motion of the missile can be written in the form

$$\ddot{x} + \rho(x)\dot{x}^2 + (x+1)^{-2} = 0; \qquad x(0) = 0, \qquad \dot{x}(0) = V_0$$

(after a proper normalization of the constants).

a) Show that escape is not always possible (e.g., if $\rho(x) \geq (2x+2)^{-1}$).

b) Find the necessary and sufficient condition on $\rho(x)$ in order to allow escape.

c) Give an explicit formula for the escape velocity when it exists.

Solution by J. ERNEST WILKINS, JR., (General Dynamics/General Atomic Div.)

Let v be the velocity $\dot{x}$, and let E be the kinetic energy $v^2/2$. Then

$$\frac{dE}{dx} = \frac{dv}{dt} = -2\rho(x)E - (x+1)^{-2},$$

so that

(1)
$$E = \frac{1}{\phi(x)}\left\{ V_0^2 - \int_0^x \frac{\phi(\lambda)\, d\lambda}{(\lambda+1)^2} \right\},$$

where

$$\phi(x) = 2 \exp \int_0^x 2\rho(y)\, dy.$$

Suppose that

$$I \equiv \int_0^\infty \frac{\phi(\lambda)\,d\lambda}{(\lambda + 1)^2} = \infty.$$

Then for any positive finite V_0, there exists a unique positive finite value ξ such that

$$V_0^2 = \int_0^\xi \frac{\phi(\lambda)\,d\lambda}{(\lambda + 1)^2}.$$

The kinetic energy E, and hence the velocity v, vanishes when $x = \xi$. Since the acceleration $\ddot{x}$ is always negative (the air density $\rho(x)$ is assumed to be nonnegative) it follows that ξ is the maximum value attained by x and that the missile falls back to earth after reaching the height ξ.

If, on the other hand, the integral I is finite, and the initial velocity V_0 is not less than $I^{1/2}$, then it follows from (1) that $E > 0$, and hence that $v > 0$, throughout the missile trajectory. It then follows that $x \to \infty$ as $t \to \infty$ and the missile will escape the earth.

We conclude that (a) escape will not occur if $\rho(x) \geq (2x + 2)^{-1}$, for in that case $I = \infty$, (b) the finiteness of the integral I is a necessary and sufficient condition for escape to occur, and (c) the minimum initial velocity V_0 which the missile must possess in order to escape is $I^{1/2}$.

Editorial Note: The term $P(x)\dot{x}^2$ in the differential equation should have been $k\rho(x)\dot{x}|\dot{x}|$ since the drag force changes sign with $\dot{x}$ and also one cannot normalize all the proportionality constants away.

Morduchow notes that the above analysis has been based on a strictly "vertical" (radial) motion of the missile and raises the question of finding the escape condition when the initial velocity V_0 is not in a radial direction. To answer that question one would have to consider the following system of equations:

$$\ddot{r} - r\dot{\theta}^2 = -gr^{-2}k^2 - k\rho(r)r\dot{\theta}\sqrt{\dot{r}^2 + r^2\dot{\theta}^2},$$

$$r\ddot{\theta} + 2\dot{r}\dot{\theta} = -k\rho(r)\dot{r}\sqrt{\dot{r}^2 + r^2\dot{\theta}^2}.$$

In connection with this, there are some interesting remarks with regard to the general effect of a resisting medium on the motion of a comet (e.g., Encke's Comet) in *A Treatise on Dynamics of a Particle*, by E. J. Routh, Dover, New York, 1960, pp. 246–249.

Also, it would be of interest to determine the escape condition in the original problem if the drag force is replaced by $\rho(x)\dot{x}^n$, $n > 0$. For the special case of $\rho(x) = k(x + 1)^{-2}$, we have escape if and only if

$$\int_0^\infty \frac{v\,dv}{kv^n + 1} > 1.$$

This holds for all $n \leq 2$. For $n > 2$, the condition reduces to

$$\pi > nk^{2/n} \sin \frac{2\pi}{n}.$$

Problem 61-4, *Flight in An Irrotational Wind Field*, by M. S. KLAMKIN (AVCO) AND D. J. NEWMAN (Yeshiva University).

If an aircraft travels at a constant air speed, and traverses a closed curve in a horizontal plane (with respect to the ground), the time taken is always less when there is no wind, then when there is any constant wind. Show that this result is also valid for any irrotational wind field and any closed curve (the constant wind case is due to T. H. Matthews, Amer. Math. Monthly, Dec. 1945, Problem 4132).

Solution by G. W. Veltkamp (Technical University, Eindhoven, Netherlands).

Let the speed of the aircraft (with respect to the air) be 1 and let at any point of the (oriented) path the tangential and normal components of the velocity of the wind be u and v, respectively. Then the velocity of the aircraft along its path is easily seen to be

(1) $$ w = u + \sqrt{1 - v^2}. $$

Of course the desired flight can be performed only if

(2) $$ w > 0 $$

everywhere along the path. We have to prove that for any closed curve C along which (2) is satisfied

(3) $$ \int_C \frac{ds}{w(s)} > \int_C ds, $$

unless $u = v = 0$ everywhere on C. It follows from (1) by some algebra that

$$ \frac{1}{w} = \frac{1}{2}\left(w + \frac{1}{w}\right) - u + \frac{u^2 + v^2}{2w}. $$

Hence (since (2) is assumed to hold) we have everywhere on C

(4) $$ \frac{1}{w} \geqq 1 - u $$

with equality only if $u = v = 0$. Since the irrotationality of the wind field implies that $\int_C u\, ds = 0$, the assertion (3) directly follows from (4).

Solution by the proposers.

If we let

W = wind velocity,

V = actual plane velocity (which is tangential to the path of flight), then $|\mathbf{V} - \mathbf{W}|$ is the constant air speed of the airplane (without wind) and will be taken as unity for convenience.

We now have to show that

(1) $$ \oint \frac{ds}{|\mathbf{V}|} \geqq \oint \frac{ds}{1}. $$

By the Schwarz inequality,

(2) $$ \oint |\mathbf{V}|\, ds \cdot \oint \frac{ds}{|\mathbf{V}|} \geqq \left\{\oint ds\right\}^2. $$

Since

$$\oint |\mathbf{V}|\, ds = \oint \mathbf{V} \cdot d\mathbf{R} = \oint (\mathbf{V} - \mathbf{W}) \cdot d\mathbf{R} + \oint \mathbf{W} \cdot d\mathbf{R},$$

and

$$\oint \mathbf{W} \cdot d\mathbf{R} = 0 \qquad\qquad (\mathbf{W} \text{ is irrotational}),$$

(3) $$\qquad \oint |\mathbf{V}|\, ds \le \oint |\mathbf{V} - \mathbf{W}|\, |dR| = \oint ds.$$

(1) now follows from (2) and (3).

Flight in an Irrotational Wind Field. II

Problem 82-15, *by* M. S. KLAMKIN (University of Alberta).

It is a known result (see *Problem* 61-4, SIAM Rev., 4 (1962), p. 155) that if an aircraft traverses a closed curve at a constant air speed with respect to the wind, the time taken is always less when there is no wind than when there is any bounded irrotational wind field.

(i) Show more generally that if the wind field is $k\mathbf{W}$ ($\mathbf{W}$ bounded and irrotational and k is a constant), then the time of traverse is a monotonic increasing function of k ($k \ge 0$).

(ii) Let the aircraft be subject to the bounded irrotational wind field $\mathbf{W}_i$, $i = 1, 2$, and let T_i denote the time of flight over the same closed path. If $|\mathbf{W}_1| \le |\mathbf{W}_2|$ at every point of the traverse, does it follow that $T_1 \le T_2$?

Solution by the proposer.

(i) Let the arc length s denote the position of the plane on its path and let $w(s)$, $\theta(s)$ denote, respectively, the speed and the direction of the wind with respect to the tangent line to the path at position s. It is assumed that the wind field is continuous and that $1 > kw$ where the plane's speed is taken as 1. By resolving $k\mathbf{W}$ into components along and normal to the tangent line of the plane's path, the aircraft's ground speed is

$$\sqrt{1 - k^2 w^2 \sin^2 \theta} + kw \cos \theta,$$

and then the time of flight is given by

$$T(k) = \oint \frac{ds}{\sqrt{1 - k^2 w^2 \sin^2 \theta} + kw \cos \theta}.$$

From Problem 61-4, it is known that $T(k) \ge T(0)$ with equality iff $k\mathbf{W} = 0$. We now show that $T(k)$ is a strictly convex function of k which implies the desired result. Differentiating $T(k)$, we get

$$\frac{dT}{dk} = -\oint \left\{ w \cos \theta - \frac{kw^2 \sin^2 \theta}{\sqrt{1 - k^2 w^2 \sin^2 \theta}} \right\} \left\{ \sqrt{1 - k^2 w^2 \sin^2 \theta} + kw \cos \theta \right\}^{-2} ds.$$

Then $T'(0) = -\oint w \cos \theta\, ds = 0$ since $\mathbf{W}$ is irrotational. On differentiating again, $T''(k) > 0$ since the integrand consists of positive terms. Thus $T(k)$ is strictly convex (for $\mathbf{W} \ne 0$).

(ii) The answer here is negative. Just consider two constant wind fields, both having the same wind speeds. Since the times of the traverses will in general be different, we cannot have both $T_1 \leq T_2$ and $T_2 \leq T_1$.

A problem related to part (i) is that the aircraft flies the same closed path twice with the second time around in the reverse direction. All the other conditions of the problem are the same as before except that the wind field need not be irrotational. Then the total time of flight is an increasing function of k ($k\mathbf{W} \neq 0$). In this case if the aircraft only flew one loop, the time of flight could be less then the time of flight without wind (just consider a whirlwind). Here the total time of flight is

$$T(k) = \oint \frac{ds}{\sqrt{1 - k^2 w^2 \sin^2 \theta} + kw \cos \theta} + \oint \frac{ds}{\sqrt{1 - k^2 w^2 \sin^2 \theta} - kw \cos \theta}.$$

By the A.M.–G.M. inequality, the sum of the integrands is $\geq 2(1 - k^2 w^2)^{-1/2} \geq 2$ which shows that $T(k) \geq T(0)$ with equality iff $k\mathbf{W} = 0$. Then as before $T''(k) > 0$.

Problem 75–21, n-dimensional Simple Harmonic Motion, by I. J. SCHOENBERG (University of Wisconsin).

In R_n, we consider the curve

(1) $\Gamma : x_i = \cos (\lambda_i t + a_i),$ $i = 1, \cdots, n,$ $-\infty < t < \infty.$

which represents an n-dimensional simple harmonic motion entirely contained within the cube $U : -1 \leq x_i \leq 1, i = 1, \cdots, n$. We want Γ to be truly n-dimensional and will therefore assume without loss of generality that

(2) $\lambda_i > 0$ for all i.

We consider the open sphere

$$S : \sum_{i=1}^{n} x_i^2 < r^2$$

and want the motion (1) to take place entirely outside of S, hence contained in the closed set $U - S$. What is the largest sphere S such that there exist motions Γ entirely contained in $U - S$? Show that the largest such sphere S_0 has the radius $r_0 = \sqrt{n}/2$, and that the only motions Γ within $U - S_0$ lie entirely on the boundary $\sum x_i^2 = r_0^2$ of S_0.

Solution by A. M. FINK (University of Minnesota).

Let $f(t) = \sum_{k=1}^{n} \cos^2(\lambda_k t + a_k) - n/2, n \geq 2$. Then $f(t) = \frac{1}{2}\sum_{k=1}^{n} \cos 2(\lambda_k t + a_k)$ is an almost periodic function with mean value $M(f) = 0$. Now $f \geq 0$ and $M(f) = 0$ implies [1, Thm. 5.8] that $f \equiv 0$. Thus $\min f(t) \leq 0$ and $r_0^2 \leq n/2$. To see that $r_0^2 \geq n/2$, consider $f(t) = \frac{1}{2}\sum_{k=1}^{n} \cos 2(\lambda t + (\pi/n)k)$ for any $\lambda > 0, n \geq 2$. Then $f(t) = \frac{1}{2}b_1 \cos \lambda t - \frac{1}{2}b_2 \sin \lambda t$, where $b_1 = \sum_{k=1}^{n} \cos(2\pi/n)k$ and $b_2 = \sum_{k=1}^{n} \sin (2\pi/n)k$. But $b_1 + ib_2 = \sum_{k=1}^{n} e^{2\pi ki/n} = 0$. Thus $f \equiv 0$ and $r_0^2 \geq n/2$. Now the statement that $f \geq 0$ with $M(f) = 0$ implies $f \equiv 0$ is the last conclusion. For $n = 1$, the statement fails.

Comment. If, as is usual, we consider the $2n$-dimensional motion where the velocities are the other n-coordinates, then the above arguments show that the minimum of $g(t) = \sum_{i=1}^{n}[\cos^2(\lambda_i t + a_i) + \lambda_i^2 \sin^2(\lambda_i t + a_i)] \leq \frac{1}{2}(n - \sum_{i=1}^{n}\lambda_i^2) < r_0^2$. Thus the $2n$-dimensional problem has the same solution in the sense that all spheres with radii $< r_0$ have a Γ outside them but the sphere with radius r_0 does not. The motions also need not be truly n-dimensional unless the λ_i are independent over the rationals.

REFERENCE

[1] A. M. FINK, *Almost periodic differential equations*, Springer-Verlag Lecture Notes, vol. 377, Springer-Verlag, New York.

Editorial note. For other related results, see the forthcoming paper·of the proposer and C. PISOT, *Extremum problems for the motions of a billiard ball.*

Problem 59–7 (Corrected), A Minimum Time Path, by D. J. NEWMAN (Yeshiva University).

Determine the minimal time path of a jet plane from take-off to a given point in space. We assume the highly idealized situation in which the total energy of the jet (kinetic + potential + fuel) is constant and, what is reasonable, that the jet burns fuel at its maximum rate which is also constant. We also assume that $v = y = x = t = 0$ at take-off which gives

$$v^2 + 2gy = at$$

as the energy equation.

Solution by the proposer.

The problem here is to minimize $\int dt$ subject to the constraints

$$\dot{x}^2 + \dot{y}^2 + 2gy = at,$$

$$\int \dot{x}\, dt = X, \qquad \int \dot{y}\, dt = Y.$$

Defining $\tan \gamma = \dot{y}/\dot{x} = \dfrac{dy}{dx}$, the problem becomes: minimize $\int dt$ subject to the constraints

(1) $$\dot{y}^2/\sin^2\gamma + 2gy = at,$$

$$\int \dot{y} \cot \gamma\, dt = X, \qquad \int \dot{y}\, dt = Y,$$

(note: by using γ instead of x, we obtain a simplification).

The usual Lagrange multiplier technique yields the variational problem:

$$\text{Minimize} \int \{\lambda(t)(\dot{y}^2/\sin^2\gamma + 2gy - at) - C\dot{y} \cot \gamma\}\, dt.$$

The resulting Euler equations are

(2) $$\frac{d}{dt}(2\lambda\dot{y}/\sin^2\gamma - C \cot \gamma) = 2gy,$$

(3) $$2\lambda\dot{y} = C\tan\gamma,$$

(note: (3) has already been integrated and simplified). Substituting (3) into (2) yields

(4) $$2\lambda g = C\dot{\gamma}\sec^2\gamma$$

and then (3) becomes

(5) $$\dot{y} = gt'\sin\gamma\cos\gamma, \qquad \left(t' = \frac{dt}{d\gamma}\right).$$

From (1) and (5), we obtain

(6) $$g^2t'^2\cos^2\gamma + 2gy = at,$$

which on differentiating gives

(7) $$2g^2t't''\cos^2\gamma - g^2t'^2\sin 2\gamma + 2g\dot{y}t' = at'.$$

Replacing $\dot{y}$ from (5) into (7) leads to

$$2g^2t'' = a\sec^2\gamma,$$

which on integration is

(8) $$2g^2t' = a(\tan\gamma - \tan\gamma_0).$$

Since $\dfrac{dy}{d\gamma} = y' = gt'^2\sin\alpha\cos\alpha, \; x' = y'\cot\gamma$, we obtain y' and x' parametrically;

(9) $$4g^3y' = a^2\tan\gamma\,(\sin^2\gamma + k\sin 2\alpha + k^2\cos^2\gamma),$$

(10) $$4g^3x' = a^2\,(\sin^2\gamma + k\sin 2\alpha + k^2\cos^2\gamma),$$

where $k = -\tan\gamma_0$. Noting that initially $x = y = t = 0, \gamma = -\tan^{-1}k$, we can integrate (8), (9), and (10) to obtain

(16) $$\frac{2g^2t}{a} = \ln\,(\sec\gamma)/(\sec\gamma_0) - \gamma\tan\gamma_0 + \gamma_0\tan\gamma_0,$$

(17) $$\frac{16g^3x}{a^2} = 2(1 + k^2)(\gamma - \gamma_0) + (k^2 - 1)(\sin 2\gamma - \sin 2\gamma_0)$$
$$- 2k(\cos 2\gamma - \cos 2\gamma_0),$$

(18) $$\frac{16g^3y}{a^2} = 4\ln\frac{\sec\gamma}{\sec\gamma_0} + 4k(\gamma - \gamma_0) - (k^2 - 1)(\cos 2\gamma - \cos 2\gamma_0)$$
$$- 2k(\sin 2\gamma - \sin 2\gamma_0).$$

If γ (final) and γ_0 are chosen properly, we can insure that $x = X$ and $y = Y$ (finally). Then from (16), (17), (18) we have the parametric equations for the optimal path plus the minimum time of flight.

Editorial note: For more realistic and more complicated minimum time-to-climb problems, it is very improbable that the variational equations can be explicitly integrated (as in the above equations). In general, it would be more practical to use the techniques of dynamic programming. In particular, see R. E. Bellman and S. E. Dreyfus, "Applied Dynamic Programming", Princeton University Press, New Jersey, 1962, pp. 209–219.

Problem 72-7*, *A Parametric Linear Complementarity Problem*, by G. Maier
(Polytechnic of Milan, Italy).

The following question arose in analyzing elastoplastic structures. Consider
the set of relations

(1)
$$y = Mx + a\alpha + b;$$
$$y \geq 0, \qquad x \geq 0, \qquad x^T y = 0,$$

where a and b are given real n-vectors, all components of b are positive, M is a
given real $n \times n$ matrix, x and y are variable vectors, α is a parameter. For any
α, (1) represents a "linear complementarity problem," which is fundamental in
linear and quadratic programming and in related topics.

Let $\bar{x}(\alpha), \bar{y}(\alpha)$ be its solution. Under which conditions are all the components
of the $\bar{x}$ part of the solution nondecreasing functions of the parameter α for $\alpha \geq 0$?

Necessary and/or sufficient conditions for this fact would provide a better
insight in an important field of structural mechanics. In most cases of interest in
this field M is positive definite or at least positive semidefinite; in many cases it
is also asymmetric.

The mechanical relevance of this problem can be briefly explained as follows.
An elastoplastic structure subjected to proportionally increasing loads may or
may not undergo "local unloading." In the former case, the analysis should be
carried out step-by-step, applying an "incremental theory"; in the latter, for many
structures, a simpler "holonomic, deformation theory" can provide the solution
in a single step [1]. Criteria for assessing the applicability of the simpler theory are
clearly desirable. A restrictive sufficient condition was given for continua [2].
On the basis of suitable discretizations [3], the absence of local unloading corresponds to the monotonicity in this problem proposal.

REFERENCES

[1] B. Budianski, *A reassessment of deformation theories of plasticity*, J. Appl. Mech., 26 (1959),
     pp. 259–264.
[2] A. A. Ilyushin, *Plasticité*, Editions Eyrolles, Paris, 1956.
[3] G. Maier, *A matrix structural theory of piecewise-linear plasticity with interacting yield planes*,
     Meccanica—J. Italian Assoc. Theoret. Appl. Mech., 5 (1970), pp. 54–66.

Solution by R. W. Cottle (Stanford University).

As the problem is stated, it is implied that the relations

$$y = Mx + c;$$
$$y \geq 0, \qquad x \geq 0, \qquad x^T y = 0,$$

have a unique solution for every $c \in R^n$. This is the case if and only if M has positive
principal minors [3]. This assumption on M will play an important role throughout
the present discussion. A more general treatment of the parametric linear complementarity problem is developed elsewhere [1].

Let M be a fixed $n \times n$ matrix with positive principal minors. For $\alpha \geq 0$, $a \in R^n$, and $b \in R^n_+$ (i.e., $b \geq 0$) there is a unique solution $\bar{x} = \bar{x}(\alpha; a, b)$ of the system

$$Mx + \alpha a + b \geq 0,$$

$$x \geq 0,$$

$$x^T[Mx + \alpha a + b] = 0.$$

Thus $\bar{x}$ is a well-defined mapping of $R_+ \times R^n \times R^n_+$ into R^n_+. We shall say that $\bar{x}$ is *uniformly monotone* if for every $(a, b) \in R^n \times R^n_+$, all the components of $\bar{x}(\alpha; a, b)$ are nondecreasing functions of α. For a fixed pair (a, b), we let $\bar{x}(\alpha) = \bar{x}(\alpha; a, b)$.

We are asked for necessary and sufficient conditions on M such that the mapping $\bar{x}$ is uniformly monotone. The solution to the problem is the following result which simultaneously furnishes an algorithm for computing $\bar{x}(\alpha)$. This computational procedure has a direct geometric interpretation. For each α, the vector $\alpha a + b$ is expressed as an element of a complementary cone [2] using the largest possible number of columns of $-M$. Precisely such an expression is required, and this makes the method "natural."

THEOREM. *The mapping $\bar{x}$ is uniformly monotone if and only if M has positive principal minors and nonpositive off-diagonal entries.* (Such an M is known as a *Minkowski matrix*.)

Proof. Suppose M has positive principal minors and nonpositive off-diagonal entries. Let $(a, b) \in R^n \times R^n_+$ be arbitrary. The triple $(\alpha, x, y) = (0, 0, b)$ satisfies the system of linear relations

(1) $$y = Mx + \alpha a + b,$$

(2) $$x \geq 0, \quad y \geq 0,$$

(3) $$\alpha \geq 0,$$

and the orthogonality condition

(4) $$x^T y = 0.$$

Thus, we have $\bar{x}(0) = 0$. Now with $x = 0$, increase the parameter α in (1). There are two possibilities. Either $\min_i a_i \geq 0$, in which case $\bar{x}(\alpha) = 0$ for all $\alpha \geq 0$, or else $\min_i a_i < 0$. The former case is trivial whereas the latter means there is a critical value

$$\alpha_1 = \min_i \left\{ \frac{-b_i}{a_i} : a_i < 0 \right\}$$

such that for $\alpha \in [0, \alpha_1]$, $\bar{x}(\alpha) = 0$. (With $x = 0$ and $\alpha > \alpha_1$, y is no longer nonnegative.) Under the more general assumption $b \geq 0$ (rather than $b > 0$) it could happen that $\alpha_1 = 0$, but this presents no difficulty. For purposes of exposition, it is fruitful, but not restrictive, to assume

$$\alpha_1 = \frac{-b_i}{a_i}, \qquad\qquad i = 1, \cdots, k_1,$$

and

$$\alpha_1 < \frac{-b_i}{a_i} \quad \text{if } k_1 < i \leq n \quad \text{and} \quad a_i < 0.$$

With α fixed at α_1, carry out a block pivot on the leading $k_1 \times k_1$ principal sub-matrix of M. For convenience, denote it by M_{11} and partition the system (1) as

$$y^1 = M_{11}x^1 + M_{12}x^2 + \alpha a^1 + b^1,$$
$$y^2 = M_{21}x^1 + M_{22}x^2 + \alpha a^2 + b^2.$$

Like M, M_{11} is a Minkowski matrix and hence has a nonnegative inverse. The block pivot on M_{11} leads to the system

$$x^1 = M_{11}^{-1}y^1 - M_{11}^{-1}M_{12}x^2 - \alpha M_{11}^{-1}a^1 - M_{11}^{-1}b^1,$$
$$y^2 = M_{21}M_{11}^{-1}y^1 + (M_{22} - M_{21}M_{11}^{-1}M_{12})x^2 + \alpha(a^2 - M_{21}M_{11}^{-1}a^1)$$
$$+ b^2 - M_{21}M_{11}^{-1}b^1,$$

which in tabular form has the sign pattern shown in Fig. 1.

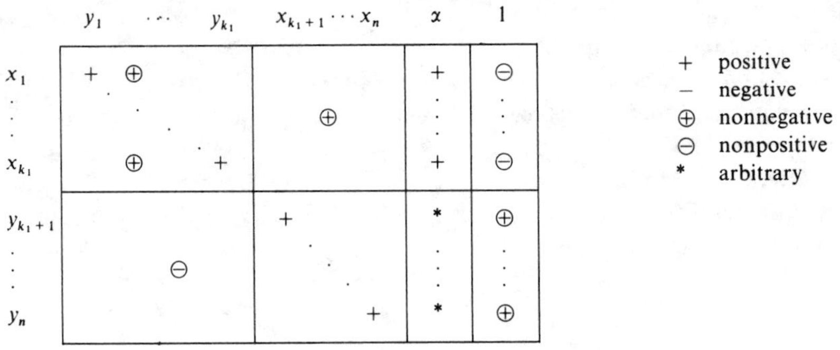

FIG. 1

Indeed, it is well known that $M_{22} - M_{21}M_{11}^{-1}M_{12}$ (the so-called Schur complement of M_{11} in M) is a Minkowski matrix [2].

If the α column contains no more negative entries, we are done, so suppose otherwise. The value of y_i corresponding to any such entry a_i' must be positive when $\alpha = \alpha_1$. As α increases, a critical value α_2 will be reached for which at least one y-variable equals zero. (Meanwhile the variables $x_1, \cdots, x_{k_1}$ have increased!) At this point, another principal (block) pivot is to be made. The sign pattern of the corresponding tableau is like that of its predecessor except that (again without loss of generality) k_1 is replaced by a larger integer k_2. This type of argument can be repeated until the transformed α column becomes nonnegative.

Now for the converse. As remarked earlier, uniform monotonicity makes sense only when the matrix M has positive principal minors. It therefore remains to show that an arbitrary off-diagonal entry of M is nonpositive. We shall employ the natural method of the first part of the proof to show that it can lead to uniform monotonicity only if M has nonpositive off-diagonal entries. The validity of this approach is based on the fact that the unique solution to the current problem (i.e., for the present value of α) is always at hand.

It will suffice to show that $m_{12} \leq 0$, for any entry of M can be shifted to the $(1, 2)$ position. Let us assume for contradiction that $m_{12} > 0$. Our freedom to choose any $a \in R^n$ and $b \in B^n_+$ means we may assume $n = 2$. Indeed, the last $n - 2$ components of these vectors can be chosen so as to render the behavior of $y_3, \cdots, y_n$ irrelevant to the proof. Moreover, we may chooose $a < 0$ and $b > 0$ so that

(i) $-b_1/a_1 < -b_2/a_2,$

(ii) $a_2 m_{11} - a_1 m_{21} < 0,$

(iii) $a_2 m_{12} - a_1 m_{22} < 0.$

Assumption (i) implies the first pivot will be made on m_{11}. It yields the tableau

	y_1	x_2	α	1
x_1	$1/m_{11}$ $\quad -m_{12}/m_{11}$		$-a_1/m_{11}$	$-b_1/m_{11}$
y_2	m_{21}/m_{11} $\quad (m_{11}m_{22} - m_{12}m_{21})/m_{11}$		$(a_2 m_{11} - a_1 m_{21})/m_{11}$	$(b_2 m_{11} - b_1 m_{21})/m_{11}$

From assumption (ii), we conclude that a second pivot will be required when α reaches the second critical value

$$\alpha_2 = -\frac{b_2 m_{11} - b_1 m_{21}}{a_2 m_{11} - a_1 m_{21}}.$$

After the exchange of y_2 with x_2 it follows from (iii) that the increase of α beyond the value α_2 will make x_1 *decrease*, in violation of the monotonicity property of $\bar{x}$. This completes the proof.

As a bonus, we get the following corollary.

COROLLARY. *If in the parametric linear complementarity problem, the matrix M is a Minkowski matrix, all the $\bar{x}_i$ are nondecreasing convex functions of the parameter α.*

REFERENCES

[1] R. W. COTTLE, *Monotone solutions of the parametric linear complementarity problem*, Mathematical Programming, 3 (1972), pp. 210–224.
[2] D. E. CRABTREE, *Applications of M-matrices to non-negative matrices*, Duke Math. J., 33 (1966), pp. 197–208.
[3] H. SAMELSON, R. M. THRALL AND O. WESLER, *A partition theorem for Euclidean n-space*, Proc. Amer. Math. Soc., 9 (1958), pp. 805–807.

Supplementary References
Mechanics

[1] R. N. ARNOLD AND L. MAUNDER, *Gyrodynamics and its Engineering Applications*, Academic, London, 1961.
[2] K. L. ARORA AND N. X. VINH, "Maximum range of ballistic missiles," SIAM Rev. (1965) pp. 544–550.
[3] C. D. BAKER AND J. J. HART, "Maximum range of a projectile in a vacuum," Amer. J. Phys. (1955) pp. 253–255.
[4] R. M. L. BAKER, JR. AND M. W. MAKEMSON, *An Introduction to Astrodynamics*, Academic, N.Y., 1967.
[5] S. BANACH, *Mechanics*, Hafner, N.Y., 1951.
[6] R. R. BATE, D. D. MUELLER AND J. E. WHITE, *Fundamentals of Astrodynamics*, Dover, N.Y., 1971.
[7] D. C. BENSON, "An elementary solution of the brachistochrone problem," AMM (1969) pp. 890–894.
[8] A. BERNHART, "Curves of general pursuit," Scripta Math. (1959) pp. 189–206.
[9] ———, "Curves of pursuit," Scripta Math. (1954) pp. 125–141.
[10] ———, "Curves of pursuit," Scripta Math. (1957) pp. 49–65.
[11] ———, "Polygons of pursuit," Scripta Math. (1959) pp. 23–50.

[12] L. BLITZER AND A. D. WHEELON, *"Maximum range of a projectile in vacuum on a spherical earth,"* Amer. J. Phys. (1957) pp. 21–24.

[13] R. L. BORRELLI, C. S. COLEMAN AND D. D. HOBSON, *"Poe's pendulum,"* MM (1985) pp. 78–83.

[14] B. BRADEN, *"Design of an oscillating sprinkler,"* MM (1985) 29–38.

[15] F. BRAUER, *"The nonlinear simple pendulum,"* AMM (1972) pp. 348–354.

[16] M. N. BREARLY, *"Motorcycle long jump,"* Math. Gaz. (1981) pp. 167–171.

[17] W. BURGER, *"The yo-yo: A toy flywheel,"* Amer. Sci. (1984) pp. 137–142.

[18] D. N. BURGHES, *"Optimum staging of multistage rockets,"* Int. J. Math. Educ. Sci. Tech. (1974) pp. 3–10.

[19] D. N. BURGHES AND A. M. DOWNS, *Modern Introduction to Classical Mechanics and Control,* Horwood, Chichester, 1975.

[20] W. E. BYERLY, *An Introduction to the Use of Generalized Coordinates in Mechanics and Physics,* Dover, N.Y., 1944.

[21] A. C. CLARKE, *Interplanetary Flight, An Introduction to Astronautics,* Harper, N.Y.

[22] J. M. A. DANBY, *Fundamentals of Celestial Mechanics,* Macmillan, N.Y., 1962.

[23] D. E. DAYKIN, *"The bicycle problem,"* MM (1972) p. 1 (also see (1973) pp. 161–162).

[24] R. W. FLYNN, *"Spacecraft navigation and relativity,"* Amer. J. Phys. (1985) pp. 113–119.

[25] G. GENTA, *Kinetic Energy Storage. Theory and Practice of Advanced Flywheel Systems,* Butterworths, Boston, 1985.

[26] H. GOLDSTEIN, *Classical Mechanics,* Addison-Wesley, Reading, 1953.

[27] A. GREY, *A Treatise on Gyrostatics and Rotational Motion,* Dover, N.Y., 1959.

[28] B. HALPERN, *"The robot and the rabbit — a pursuit problem,"* AMM (1969) pp. 140–144.

[29] V. G. HART, *"The law of the Greek catapult,"* BIMA (1982) pp. 58–63.

[30] M. S. KLAMKIN, *"Dynamics: Putting the shot, throwing the javelin,"* UMAP J. (1985) pp. 3–18.

[31] ———, *"Moving axes and the principle of the complementary function,"* SIAM Review (1974) pp. 295–302.

[32] ———, *"On a chainomatic analytical balance,"* AMM (1955) pp. 117–118.

[33] ———, *"On some problems in gravitational attraction,"* MM (1968) pp. 130–132.

[34] M. S. KLAMKIN AND D. J. NEWMAN, *"Cylic pursuit or the three bugs problem,"* AMM (1971) pp. 631–638.

[35] ———, *"Flying in a wind field I, II,"* AMM (1969) pp. 16–22, 1013–1018.

[36] ———, *"On some inverse problems in potential theory,"* Quart. Appl. Math. (1968) pp. 277–280.

[37] ———, *"On some inverse problems in dynamics,"* Quart. Appl. Math. (1968) pp. 281–283.

[38] J. M. J. KOOY AND J. W. H. UYENBOGAART, *Ballistics of the Future,* Stam, Haarlem.

[39] C. LANCZOS, *The Variational Principles of Mechanics,* University of Toronto Press, Toronto, 1949.

[40] D. F. LAWDEN, *"Minimal rocket trajectories,"* Amer. Rocket Soc. (1953) pp. 360–367.

[41] ———, *Optimal Trajectories for Space Navigation,* Butterworths, London, 1963.

[42] ———, *"Orbital transfer via tangential ellipses,"* J. Brit. Interplanetary Soc. (1952) pp. 278–289.

[43] J. E. LITTLEWOOD, *"Adventures in ballistics,"* BIMA (1974) pp. 323–328.

[44] S. L. LONEY, *Dynamics of a Particle and of Rigid Bodies,* Cambridge University Press, Cambridge, 1939.

[45] W. F. OSGOOD, *Mechanics,* Dover, N.Y., 1965.

[46] W. M. PICKERING AND D. M. BURLEY, *"The oscillation of a simple castor,"* BIMA (1977) pp. 47–50.

[47] H. PRESTON-THOMAS, *"Interorbital transport techniques,"* J. Brit. Interplanetary Soc. (1952) pp. 173–193.

[48] D. G. MEDLEY, *An Introduction to Mechanics and Modelling,* Heinemann, London, 1982.

[49] A. MIELE, *Flight Mechanics I, Theory of Flight Paths,* Addison-Wesley, Reading, 1962.

[50] N. MINORSKY, *Introduction to Nonlinear Mechanics,* Edwards, Ann Arbor, 1946.

[51] F. R. MOULTON, *An Introduction to Celestial Mechanics,* Macmillan, N.Y., 1962.

[52] ———, *Methods in Exterior Ballistics,* Dover, N.Y., 1962.

[53] A. S. RAMSEY, *Dynamics I, II,* Cambridge University Press, Cambridge, 1946.

[54] R. M. ROSENBERG, *"On Newton's law of gravitation,"* Amer. J. Phys. (1972) pp. 975–978.

[55] L. I. SEDOV, *Similarity and Dimensional Methods in Mechanics,* Academic, N.Y., 1959.

[56] M. J. SEWELL, *"Mechanical demonstration of buckling and branching,"* BIMA (1983) pp. 61–66.

[57] D. B. SHAFFER, *"Maximum range of a projectile in a vacuum,"* Amer. J. Phys. (1956) pp. 585–586.

[58] S. K. STEIN, *"Kepler's second law and the speed of a planet,"* AMM (1967) pp. 1246–1248.

[59] J. J. STOKER, *Nonlinear Vibrations,* Interscience, New York, 1950.

[60] J. L. SYNGE AND B. A. GRIFFITH, *Principles of Mechanics,* McGraw-Hill, N.Y., 1959.

[61] J. L. SYNGE, *"Problems in mechanics,"* AMM (1948) pp. 22–24.

[62] H. S. TSIEN, *"Take-off from satellite orbit,"* Amer. Rocket Soc. (1953) pp. 233–236.

[63] W. G. UNRUH, *"Instability in automobile braking,"* Amer. J. Phys. (1984) pp. 903–908.

[64] P. VAN DE KAMP, *Elements of Astromechanics,* Freeman, San Francisco, 1964.

[65] J. WALKER, *"The amateur scientist: In which simple questions show whether a knot will hold or slip,"* Sci. Amer. (1983) pp. 120–128.

[66] L. B. WILLIAM, *"Fly around a circle in the wind,"* AMM (1971) pp. 1122–1125.

[67] E. T. WHITTAKER, *A Treatise on the Analytical Dynamics of Particles and Rigid Bodies,* Dover, N.Y., 1944.

[68] C. WRATTEN, *"Solution of a conjecture concerning air resistance,"* MM (1984) pp. 225–228.

[69] W. WRIGLEY, W. M. HOLLISTER AND W. G. DENHARD, *Gyroscopic Theory, Design and Instrumentation,* M.I.T. Press, Cambridge, 1969.

[70] J. ZEITLIN, *"Rope strength under dynamic loads: The mountain climber's surprise,"* MM (1978) pp. 109–111.

2. ELECTRICAL RESISTANCE*

Problem 63-14, *A Resistance Problem*, by RON L. GRAHAM (Bell Telephone Laboratories).

A regular n-gon is given such that each vertex is connected to the center and to its two neighboring (nearest) vertices by means of unit resistors. Determine the equivalent resistance R_n between two adjacent vertices.

Solution. By C. J. BOUWKAMP (Philips Research Laboratories, and Technological University, Eindhoven, Netherlands).

Let A and B be two neighboring vertices; let an electric current I enter the "wheel" at A, and let it leave at B, as due to an applied voltage V across AB. Assuming $n > 2$ and $0 \leq k < n$, let currents i_{2k} flow from the center C to the successive vertices along the circumference, and let currents i_{2k+1} flow in the circumferential resistors. The current from C to A is i_0, that from C to B is $i_{2n-2} = -i_0$, and that from B to A is $i_{2n-1} = -2i_0 = V$. The input current at A is $I = i_1 - 3i_0$, hence

$$(1) \qquad\qquad R_n = \frac{-2i_0}{i_1 - 3i_0}.$$

Now, by applying alternately Kirchhoff's mesh-voltage and node-current laws, we easily obtain the recurrence relation

$$(2) \qquad\qquad i_k = i_{k-1} + i_{k-2},$$

holding for $1 < k < 2n - 1$.

Let $x = \frac{1}{2}(1 + \sqrt{5})$ and $y = -1/x$ denote the two zeros of the characteristic polynomial $z^2 - z - 1$ of (2). Then, with $u_k = (x^k - y^k)/(x - y)$, i_k $(k > 1)$. can be linearly expressed in terms of i_0 and i_1, as follows:

$$i_k = u_{k-1}i_0 + u_k i_1.$$

Here $\{u_k\}$, $k = 1, 2, \cdots$, is the well-known sequence of Fibonacci numbers:

$$1, 1, 2, 3, 5, 8, 13, 21, 34, 55, 89, \cdots.$$

Since

$$-i_0 = i_{2n-2} = u_{2n-3}i_0 + u_{2n-2}i_1,$$

we find

$$\frac{-i_1}{i_0} = \frac{1 + u_{2n-3}}{u_{2n-2}},$$

---

* Also see chapter 13 – Inequalities.

and substitution of this in (1) gives, after some transformation,

$$(3) \qquad R_n = \frac{2u_{2n-2}}{1 + u_{2n-2} + u_{2n}},$$

which solves the problem in question.

However, (3) can be simplified considerably if we distinguish between even and odd values of n. Let $v_k = x^k + y^k$; then $\{v_k\}$, $k = 1, 2, \cdots$, is the sequence of Lucas numbers:

$$1, 3, 4, 7, 11, 18, 29, 47, 76, 123, 199, \cdots .$$

As is well-known, the Fibonacci and Lucas sequences constitute two linearly independent solutions of the difference equation (2). There exist many relations between these numbers; for example,

$$(4) \qquad u_{n-1} + u_{n+1} = v_n, \qquad v_{n-1} + v_{n+1} = 5u_n,$$

which can easily be proved by induction.

Now, if n is odd we have $i_{n-1} = 0$ in virtue of symmetry; this leads to

$$\frac{-i_1}{i_0} = \frac{u_{n-2}}{u_{n-1}} = \frac{v_{n-3} + v_{n-1}}{v_{n-2} + v_n}.$$

If this is inserted in (1) we get after a few manipulations the set of formulas:

$$(5) \qquad R_n = \frac{2u_{n-1}}{u_{n-1} + u_{n+1}} = \frac{2u_{n-1}}{v_n} = \frac{2}{5}\left(1 + \frac{v_{n-2}}{v_n}\right),$$

$$\frac{1}{R_n} = \frac{1}{2}\left(1 + \frac{u_{n+1}}{u_{n-1}}\right), \qquad \qquad n \text{ odd.}$$

On the other hand, if n is even we have $i_n = -i_{n-2}$, again by symmetry, and thus

$$\frac{-i_1}{i_0} = \frac{u_{n-3} + u_{n-1}}{u_{n-2} + u_n} = \frac{v_{n-2}}{v_{n-1}}.$$

Therefore, the analog of (5) is obtained from (5) by interchange of u and v, except for an extra factor of 5 in the middle:

$$(6) \qquad R_n = \frac{2v_{n-1}}{v_{n-1} + v_{n+1}} = \frac{2v_{n-1}}{5u_n} = \frac{2}{5}\left(1 + \frac{u_{n-2}}{u_n}\right),$$

$$\frac{1}{R_n} = \frac{1}{2}\left(1 + \frac{v_{n+1}}{v_{n-1}}\right), \qquad \qquad n \text{ even.}$$

There is yet another way to evaluate R_n which deserves mention. Let n be an odd number; take (6) for $n + 1$ and (5) for n; then

$$\frac{5}{2}R_{n+1} - 1 = \frac{u_{n-1}}{u_{n+1}} = \left(\frac{2}{R_n} - 1\right)^{-1}.$$

If n is even, these relations hold if u is replaced by v. Therefore, if n is arbitrary we have the recurrence relation

$$(7) \qquad R_{n+1} = \frac{4}{5(2 - R_n)},$$

which with $R_3 = \frac{1}{2}$ allows the numerical evaluation of R_n independently of the Fibonacci and Lucas numbers.

The behavior of R_n for large values of n is determined by

$$R_\infty = \lim_{n\to\infty} R_n = 1 - 5^{-1/2} = 0.5527\ 8640\ 4\cdots.$$

The limit is attained monotonically from below. For $n \geq 20$, R_n equals R_∞ up to 8 decimals; see Table 1.

TABLE 1

Numerical values

n	u_n	v_n	R_n exact	R_n approximate
1	1	1		
2	1	3		
3	2	4	1/2	.5000 0000
4	3	7	8/15	.5333 3333
5	5	11	6/11	.5454 5455
6	8	18	11/20	.5500 0000
7	13	29	16/29	.5517 2414
8	21	47	58/105	.5523 8095
9	34	76	21/38	.5526 3158
10	55	123	152/275	.5527 2727
11	89	199	110/199	.5527 6382
12	144	322	199/360	.5527 7778
13	233	521	288/521	.5527 8311
14	377	843	1042/1885	.5527 8515
15	610	1364	377/682	.5527 8592
16	987	2207	2728/4935	.5527 8622
17	1597	3571	1974/3571	.5527 8633
18	2584	5778	3571/6460	.5527 8638
19	4181	9349	5168/9349	.5527 8639
20	6765	15127	18698/33825	.5527 8640

$$R_\infty = .5527\ 8640\ 4\cdots$$

Generalization of the problem. In what follows we determine the equivalent resistance between any two vertices of the wheel.

First, let r_n denote the equivalent resistance between the vertex A and the center C. Then

(8) $r_n = 1 - R_n.$

To prove (8), let S_n denote the equivalent resistance between A and B if the unit resistor between A and B is deleted. Then, by the parallel-connection theorem, $R_n^{-1} = 1 + S_n^{-1}$. Similarly, we have $r_n^{-1} = 1 + s_n^{-1}$, where s_n is the equivalent resistance between A and C if the unit resistor between A and C is deleted. Now, the wheel is not only highly symmetric but it is also a self-dual network; this implies $s_n S_n = 1$ which, with the two identities above, is (8).

Secondly, let $R_{n,m}$ denote the equivalent resistance between A and some vertex D along the circumference of the wheel such that there are m unit resistors between A and D, $0 < m < n$. Obviously, $R_{n,1} = R_n$, and in this case the input current is $i_1 - 3i_0$, while the applied voltage is $-2i_0$. Now, let this very current enter the wheel at A, let it leave at B, let it again enter at B, let it leave at the next vertex, and so on, until it leaves the wheel at D. The new set of currents is obtained by simply adding the partial currents of the $m - 1$ steps (at each step the subscripts of the currents increase by 2). For example, the current from C to A becomes

$$i_0' = i_0 + i_2 + \cdots + i_{2m-2}.$$

If we know i_0', we can calculate $R_{n,m}$ because

(9) $$R_{n,m} = \frac{-2i_0'}{i_1 - 3i_0}.$$

Evidently, i_0' can be linearly expressed in i_0 and i_1, as follows:

$$i_0' = i_0 + \sum_{k=1}^{m-1} i_{2k} = i_0 \sum_{k=0}^{m-1} u_{2k-1} + i_1 \sum_{k=1}^{m-1} u_{2k}$$

$$= (1 + u_{2m-2})i_0 + (-1 + u_{2m-1})i_1.$$

If this is substituted in (9) we get, after some transformation,

$$R_{n,m} = 2\,\frac{1 + u_{2n-1} - u_{2m-1} + u_{2n-2}\,u_{2m-2} - u_{2m-1}\,u_{2n-3}}{1 + u_{2n-2} + u_{2n}}.$$

To eliminate the product terms in the numerator, we replace them by their explicit expressions in terms of x and y, so as to obtain

$$u_{2n-2}u_{2m-2} - u_{2m-1}u_{2n-3} = -u_{2n-2m-1},$$

and therefore

(10) $$R_{n,m} = R_{n,n-m} = 2\,\frac{1 + u_{2n-1} - u_{2m-1} - u_{2n-2m-1}}{1 + u_{2n-2} + u_{2n}}.$$

This formula holds for $n \geq 3$ and $0 \leq m \leq n$. In fact, the numerator vanishes if m is either n or 0, as it should. From (10) may be derived

(11) $$R_{n,m} = (u_{2m-1} + u_{2m+1} - 2)R_n + 2(1 - u_{2m-1}),$$

which is useful for numerical calculations if a list of R_n (see Table 1) is available. In particular, (11) gives

$$R_{n,2} = 5R_n - 2, \qquad R_{n,3} = 8(2R_n - 1),$$

$$R_{n,4} = 3(15R_n - 8), \qquad R_{n,5} = 11(11R_n - 6).$$

Again, the general expression (10) can be much simplified if we distinguish between the four possibilities as to the parity of n and of m. Without going into details of proof, we state the final result:

(12) $$R_{n,m} = \begin{cases} 2u_m u_{n-m}/u_n, & n \text{ even}, m \text{ even},\\ 2v_m v_{n-m}/5u_n, & n \text{ even}, m \text{ odd},\\ 2u_m v_{n-m}/v_n, & n \text{ odd}, m \text{ even},\\ 2v_m u_{n-m}/v_n, & n \text{ odd}, m \text{ odd}. \end{cases}$$

Equivalently, we have

(13) $$R_{n,m} = \frac{2}{\sqrt{5}}\,\frac{(p^m - 1)(p^{n-m} - 1)}{p^n - 1}, \qquad p = x^2 = \frac{1}{2}(3 + \sqrt{5}).$$

Equation (13) was independently obtained by N. G. de Bruijn. In fact, it was his formula (13) that led us to the establishment of (12) given (10).

Problem 64-14, Resistance of a Ladder Network, by WILLIAM D. FRYER (Cornell Aeronautical Laboratory).

Find the input resistance R_n to the n-section ladder network as shown in the figure. Also, determine $\lim_{n\to\infty} R_n$.

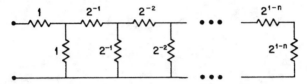

Solution by C. J. BOUWKAMP (Technological University, Eindhoven, Netherlands).

Divide all the resistances of the n-section ladder network by 2. Then the resulting network has input resistance $R_n/2$ and the $(n+1)$-section ladder network is a series connection of a unit resistor and a network consisting of two resistors in parallel, one of value 1 and the other of value $R_n/2$. Consequently, we must have the recurrence relation

$$R_1 = 2, \qquad R_{n+1} = \frac{2 + 2R_n}{2 + R_n}, \qquad n = 1, 2, 3, \cdots .$$

One method of finding an explicit expression for R_n is to let $R_n = p_n/q_n$, where

$$p_{n+1} = 2p_n + 2q_n , \quad q_{n+1} = p_n + 2q_n , \quad p_1 = 2q_1 = 2.$$

It then follows that p_n and q_n are two linearly independent solutions of the recurrence relation $f_{n+2} - 4f_{n+1} + 2f_n = 0$ whose solutions must be of the form $f_{n+2} = A(2 + \sqrt{2})^n + B(2 - \sqrt{2})^n$. Hence,

$$p_n = \frac{1}{2}[(2 + \sqrt{2})^n + (2 - \sqrt{2})^n], \quad q_n = \frac{\sqrt{2}}{4}[(2 + \sqrt{2})^n - (2 - \sqrt{2})^n].$$

Finally,

$$R_n = \sqrt{2}\,\frac{(2 + \sqrt{2})^n + (2 - \sqrt{2})^n}{(2 + \sqrt{2})^n - (2 - \sqrt{2})^n}$$

and $\lim_{n\to\infty} R_n = \sqrt{2}$.

Editorial Note. The recurrence relation for R_n can also be solved by finding the nth power of the matrix $\begin{bmatrix} 2 & 2 \\ 1 & 2 \end{bmatrix}$.

Waligorski also determines $\lim R_n$ for the network in which the elements of the rth section, 2^{1-r} and 2^{1-r}, are replaced by $\beta\alpha^{1-r}$ and α^{1-r}, $r = 1, 2, \cdots , n$. The limit is

$$R_\infty = \tfrac{1}{2}\{\beta + 1 - \alpha + \sqrt{(\beta + 1 - \alpha)^2 + 4\alpha\beta}\}$$

and is evaluated by means of the periodic continued fraction which is obtained from the recurrence relation.

Navot refers to A. C. BARTLETT, *The Theory of Electrical Artifical Lines and Filters*, Chapman and Hall, London, 1930, Chap. 3, where ladder networks are analyzed in terms of continuants and to A. M. MORGAN-VOYCE, *Ladder network analysis using Fibonacci numbers*, IRE Trans. Circuit Theory, CT-6 (1959), pp. 321–322, for a similar problem.

A Resistance Problem

Problem 84-14, by E. N. GILBERT *and* L. A. SHEPP (Bell Telephone Laboratories, Murray Hill, New Jersey).

The sides of polygon $A_1, \cdots, A_n$ are resistances $r_1, \cdots, r_n$. One measures the equivalent resistances $0 < \rho_i = r_i(R - r_i)/R$, where $R = r_1 + \cdots + r_n$, between consecutive pairs of vertices A_i, A_{i+1} ($i = 1, \cdots, n$ and $A_{n+1} = A_1$). Show that the r_i's are uniquely determined from the ρ_i's.

This problem arose in electrical tomography where an object is regarded as an unknown resistive network, and the unknown internal resistors in the network are deduced from external electrical measurements, analogously to x-ray tomography. The ring of resistors was proposed to learn whether or not electrical tomography is possible for this simple network topology. Although the problem remains unsolved for general networks, more general topologies are considered in a forthcoming paper.

Solution by J. A. WILSON (Iowa State University).

We change variables, letting $x_i = r_i/R$, so that $\sum_1^n x_i = 1$ and $0 < x_i < 1$. To show uniqueness, we prove that if $\sum_1^n y_i = 1, 0 < y_i < 1$, and $y_i(1 - y_i) = Kx_i(1 - x_i)$ for $1 \le i \le n$, K a positive constant, then $x_i = y_i$ for $1 \le i \le n$.

By symmetry, we may assume $0 < K \le 1$. Solving for y_i gives $y_i = \frac{1}{2}(1 \pm \sqrt{1 - 4Kx_i(1 - x_i)})$. Let $\phi(x) = \frac{1}{2}(1 - \sqrt{1 - 4Kx(1 - x)})$. It is easily checked that $\phi(x) = \phi(1 - x)$, that

$$\min(y_i, 1 - y_i) = \phi(x_i) \le x_i \le 1 - \phi(x_i) = \max(y_i, 1 - y_i),$$

with strict inequalities if $K \ne 1$, that ϕ is strictly increasing on $[0, \frac{1}{2}]$, and that $\phi(x)/x$ is decreasing on $(0, \frac{1}{2}]$, strictly if $K \ne 1$.

Note that $x_i + x_j < 1$, since $\sum x_i = 1$ and all x_i are positive. One consequence of this is that at most one x_i is $\ge \frac{1}{2}$. The same is true for the y_i's. Without loss of generality, assume $x_n = \max x_i$ and $x_1, \cdots, x_{n-1} < \min(\frac{1}{2}, 1 - x_n)$. We claim $y_1, \cdots, y_{n-1} < \frac{1}{2}$. Indeed, if one of these, say y_k, is $\ge \frac{1}{2}$, then $y_n < \frac{1}{2}$, and $\sum y_i > y_k + y_n = (1 - \phi(x_k)) + \phi(x_n) \ge 1$, a contradiction. We conclude that for $1 \le i \le n - 1$,

$$y_i = \phi(x_i) \le x_i < \min(\tfrac{1}{2}, 1 - x_n) \quad \text{and} \quad x_i \le x_n.$$

Next, we prove that $K = 1$. Suppose $K < 1$. Then $x_1 > y_1, \cdots, x_{n-1} > y_{n-1}$, and since $\sum x_i = \sum y_i$, we must have $x_n < y_n$ and $\phi(x_n) = 1 - y_n$. The fact that $\phi(x)/x$ is decreasing for $0 < x \le \frac{1}{2}$, shows that

$$\frac{y_i}{x_i} = \frac{\phi(x_i)}{x_i} \ge \frac{\phi(x_n)}{\min(x_n, 1 - x_n)} \ge \frac{1 - y_n}{1 - x_n}, \qquad 1 \le i \le n - 1.$$

The first inequality is strict if $x_n \ge \frac{1}{2}$, while the second is strict if $x_n < \frac{1}{2}$. Therefore,

$$\sum_1^n y_i > \frac{1-y_n}{1-x_n} \sum_1^{n-1} x_i + y_n = 1,$$

a contradiction. We conclude that $K = 1$.

With $K = 1$, we have $y_i = \phi(x_i) = x_i$ for $1 \le i \le n-1$. But then $y_n = x_n$, since $\sum y_i = \sum x_i$.

Editorial note: The proposers note that for $n = 3$, a simple explicit inversion formula exists, namely

$$r_i = \rho_i + \frac{\sigma_1 \sigma_2 \sigma_3}{2\sigma_i^2}$$

where

$$\sigma_i = \rho_1 + \rho_2 + \rho_3 - 2\rho_i, \qquad i = 1, 2, 3.$$

Coincidentally, this particular case has arisen in a characterization of the orthocenter of an acute triangle in, *A problem of equivalent resistances and a generalization of the Pythagorean theorem*, by G. D. Chakerian (University of California at Davis) *and* M. S. K., *Math. Mag., 59 (1986) 149-153*

Resistances in an n-Dimensional Cube

Problem 79-16*, *by* D. SINGMASTER (Polytechnic of the South Bank, London, England).

Determine the resistances $R(n, i)$ between two nodes a distance i apart in an n-cubical network if all the edges are of unit resistance.

Partial solution by B. C. RENNIE (James Cook University, Queensland, Australia). The following can be proved:

(A) $R(n, 1) = R(n + 1, 2)$.

(B) $R(n, n) = \frac{1}{2}R(n - 1, n - 1) + \frac{1}{n}$.

(C) $R(n, n) = \sum_{r=0}^{n-1} \frac{2^{-r}}{n-r}$.

(D) $R(n, 1) + R(n, n) = R(n, n - 1) + \frac{2}{n}$.

(E) $R(n, 2) + R(n, n) = R(n, n - 2) + \frac{2}{n-1}$.

(F) $2nR(n, 1) = (n - 1)R(n, 2) + 2$.

(G) $R(n, 1) = R(n + 1, 2) = \frac{2 - 2^{1-n}}{n}$.

(H) $R(n, n - 1) + \frac{2^{1-n}}{n} = R(n, n - 2) + \frac{2^{2-n}}{n-1} = \sum_{r=0}^{n-1} \frac{2^{-r}}{n-r} = R(n, n)$.

The first few numerical values of $R(n, k)$ are as follows:

$(n = 1)$ 1
$(n = 2)$ $\frac{3}{4}$ 1
$(n = 3)$ $\frac{7}{12}$ $\frac{3}{4}$ $\frac{5}{6}$
$(n = 4)$ $\frac{15}{32}$ $\frac{7}{12}$ $\frac{61}{96}$ $\frac{2}{3}$
$(n = 5)$ $\frac{31}{80}$ $\frac{15}{32}$ $\frac{241}{480}$ $\frac{25}{48}$ $\frac{8}{15}$

Proof of (A). Using Cartesian coordinates with 0 or 1 for each coordinate of each vertex, consider

$$P = (0, 0, 0, \cdots),$$

$$Q = (1, 0, 0, \cdots),$$

$$R = (0, 1, 0, \cdots),$$

$$S = (1, 1, 0, \cdots),$$

which are the corners of a two dimensional face of an $n+1$-dimensional cube (see Fig. 1). Consider the system as a four-terminal network, with unit currents flowing in at P and Q and flowing out at R and S. By symmetry P is at the same potential as Q, and R is at the same potential as S. We ask what is the voltage of P and Q above R and S; there are two ways to find the answer.

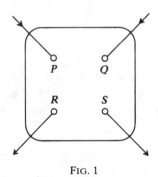

FIG. 1

Firstly, consider the $n+1$-dimensional cube as formed from two n-dimensional cubes with a typical vertex $(0, x)$ of the first joined by a one-ohm resistor to the corresponding vertex $(1, x)$ of the second. Temporarily we cut these resistors, then put unit current from P to R in the first cube and from Q to S in the second, the potentials being $R(n, 1)$ at P and Q, and zero at R and S. Any point $(0, x)$ will be at the same potential as the corresponding point $(1, x)$ and so the joining resistors between the two cubes may be restored without changing the currents. Therefore, our first answer to the question asked above is $R(n, 1)$.

Secondly, consider unit current passing in at P and out at S with S at earth potential. P will be at $R(n + 1, 2)$, and by symmetry Q and R will both be at potential $\frac{1}{2}R(n + 1, 2)$. By linearity, we may superimpose a similar pattern with unit current in at Q and out at R with potentials $R(n + 1, 2)$ at Q, $\frac{1}{2}R(n + 1, 2)$ at P and S, and zero at R. This gives the answer $R(n + 1, 2)$ to the question above, and so proves the required result.

Proof of (B). Consider the n-dimensional cube as a network with four terminals (Fig. 2),

$$A = (0, 0, 0, \cdots 0) \quad \text{(all coordinates zero)},$$

$$B = (1, 0, 0, \cdots 0) \quad \text{(all zero but the first)},$$

$$C = (0, 1, 1, \cdots 1) \quad \text{(all 1 but the first)},$$

$$D = (1, 1, 1, \cdots 1) \quad \text{(all 1)}.$$

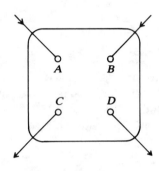

FIG. 2

With unit current flowing in at A and B and flowing out at C and D, as in the proof of (A), we observe that A and B will be at one potential and C and D at another. We ask what will be the potential difference of A and B above C and D, and we find the answer in two ways.

Firstly, consider the effect of unit current in at A and out at D with D earthed. By symmetry the n resistors joined to A and the n joined to D will all have currents $1/(n-1)$, and the potentials at the four terminals will be: $R(n, n)$ at A, $R(n, n) - 1/n$ at B, $1/n$ at C, and zero at D. Another possible current distribution is one unit in at B and out at C, with potentials $R(n, n)$ at B, $R(n, n) - 1/n$ at A, $1/n$ at D, and zero at C. By linearity, we may superimpose the potentials and subtract $1/n$ from all of them, giving potentials $2R(n, n) - 2/n$ at A and B, and zero at C and D.

Secondly, we may answer the question by considering the cube as made up of two $n-1$-dimensional cubes. As in the proof of (A), it makes no difference whether the cubes are connected, and so our second answer is that A and B are at a potential of $R(n-1, n-1)$ above C and D. Equating the two answers gives our result.

Proof of (C). By induction from (B).

Proof of (D). Consider the system as a three-terminal network with the vertices A, C and D (as in the proof of (B)) as terminals (Fig. 3). Let μ be the mutual resistance of AC with DC, that is the voltage produced in AC by unit current from D to C. By Kelvin's reciprocal theorem it is also the voltage in DC produced by unit current in AC. Note that the mutual resistance of DA with DC is $1/n$ (because unit current going from D to A divides equally among the n resistors connected to D). Now with C earthed we

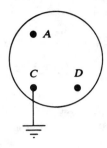

FIG. 3

can set up the following patterns of current and voltage in the system.

	Inward current		Potential	
	A	D	A	D
	1	0	$R(n, n-1)$	μ
	0	-1	$-\mu$	$-R(n, 1)$
Now by linearity:	1	-1	$R(n, n-1)-\mu$	$\mu - R(n, 1)$
But also:	1	-1	$R(n, n)-1/n$	$-1/n$

As the currents at A and D determine the potentials uniquely, this gives two equations; eliminating μ gives the result required.

Proof of (E). This is almost the same as that of (D), with the vertex C one edge away from D replaced by one two edges away.

Proof of (F). Consider the three-terminal network (Fig. 4) with terminals P, Q and S in the notation of the proof of (A). Let Q be earthed; then with currents x and y in at P and S respectively, the potentials are $ax + by$ at P and $bx + ay$ at S, where $a = R(n, 1)$ is the resistance between P and Q or between Q and S, and where b is the mutual resistance of PQ with SQ. To find a relation between a and b, consider $x = 1$ and $y = 0$. Of the unit current flowing out at Q, an amount a flows directly from P to Q along the resistor joining these two points. The remainder, $1 - a$, is equally divided among the $n-1$ other resistors connected to Q, and so S and all the other neighbors of Q (excluding P) are at potential $(1-a)/(n-1)$; and this must equal b. To find another equation, consider $x = 1$ and $y = -1$. The potentials at P and S are $a - b$ and $b - a$ respectively, but their difference has to be $R(n, 2)$. From the equations $2(a-b) = R(n, 2)$ and $1 - a = (n-1)b$ and $a = R(n, 1)$ the result follows.

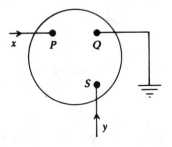

FIG. 4

Proof of (G) *and* (H). From (A) and (F) by induction we get (G). (H) comes from (C), (D), (E), (F) and (G).

Similar calculations will lead to the result:

$$n(n-1)(n-2)R(n, 3) = 2n(n-2) + 4 - 2^{1-n}(3n(n-1)+2).$$

Editorial note. There were two other solutions by G. EDMUNDSON (Schlumberger–Doll Research Center, Ridgefield, Connecticut), and J. MOULIN-OLLAGNIER AND D. PINCHON (Université Pierre et Marie Curie, Paris, France). However, these solutions were for the infinite cubic lattice in E^n. They obtained the resistance for this case as

$$R(n, i) = \frac{1}{\pi^n} \int_0^\pi \int_0^\pi \cdots \int_0^\pi \frac{(1-\cos ix_1)\, dx_1\, dx_2 \cdots dx_n}{n - (\cos x_1 + \cos x_2 + \cdots + \cos x_n)}$$

by Fourier transforms. Note that for $i = 1$, we obtain $R(n, 1) = 1/n$, which also could be gotten by symmetry considerations. [M.S.K.]

Problem 62–4, Resistance of a Cut-out Right Circular Cylinder, by ALAN L. TRIT-TER (Data Processing, Inc.).

Determine the resistance of the cut-out right circular cylinder (Fig. 1) between unsymmetrically placed perfectly conducting cylindrical electrodes E_1 and E_2 which intersect it orthogonally. Show that the resistance depends on the radii r_1 and r_2 of the electrodes and the distance d between their centers, but not on the radius of the cut-out cylinder.

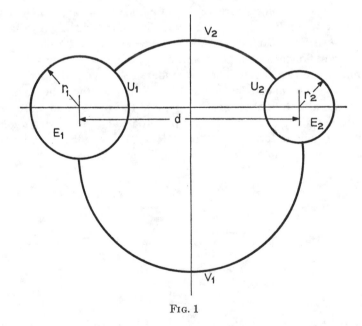

FIG. 1

Editorial Note: A solution of this problem was published in the October, 1964, issue. The following is a generalization and solution by SAMUEL P. MORGAN (Bell Telephone Laboratories).

This problem is essentially solved by W. R. Smythe, *Static and Dynamic Electricity*, 2nd ed., McGraw-Hill, New York, 1950, pp. 76–78 and 233–234.

Introduce the complex potential function

$$W = U + iV = \log \frac{a + x + iy}{a - x - iy} = 2 \tanh^{-1} \frac{x + iy}{a},$$

where $U(x, y)$ is the potential function and $V(x, y)$ is the stream function. Then

$$x + iy = a \tanh \tfrac{1}{2}(U + iV),$$

so

$$x = \frac{a \sinh U}{\cosh U + \cos V}, \qquad y = \frac{a \sin V}{\cosh U + \cos V}.$$

A typical equipotential is a circle given by

$$(x - a \coth U)^2 + y^2 = a^2 \operatorname{csch}^2 U,$$

with center at the point $(a \coth U, 0)$ and radius $a|\operatorname{csch} U|$. Similarly, a typical streamline is a circle given by

$$x^2 + (y + a \cot V)^2 = a^2 \csc^2 V.$$

The upper and lower arcs of the same circle correspond, say, to V_2 and V_1, where $V_2 = V_1 + \pi$.

If r_1, r_2, and d are given, we have $r_1 = -a \operatorname{csch} U_1$, $r_2 = a \operatorname{csch} U_2$, $d = a(\coth U_2 - \coth U_1)$, and $\cosh(U_2 - U_1) = (d^2 - r_1^2 - r_2^2)/(2r_1 r_2)$.

Finally, if the cut-out cylinder is of height H and conductivity k, the total resistance $\Re$ between the electrodes is

$$\Re = \frac{U_2 - U_1}{Hk(V_2 - V_1)} = \frac{1}{Hk\pi} \cosh^{-1} \left[\frac{d^2 - r_1^2 - r_2^2}{2r_1 r_2} \right].$$

Problem 70-14*, *Conductors of Unit Resistance*, by M. S. KLAMKIN (Ford Scientific Laboratory).

A known result due to Rayleigh [1], [2] is that conjugate conductors have reciprocal resistances.

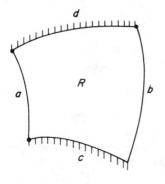 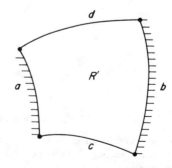

FIG. 1

Here the conductor is a two-dimensional simply connected region R with boundary arcs a and b as terminals. The complementary part of the boundary consists of two arcs c and d which are insulated. (See Fig. 1.) The conjugate conductor consists of a region congruent to R but now the arcs c and d are the terminals and the arcs a and b are insulated.

A self-conjugate conductor is one in which the region R is a reflection of itself in the straight line connecting the initial point A of terminal a with the initial point C of terminal b. (See Fig. 2.) It follows immediately by Rayleigh's result and symmetry that a self-conjugate conductor has unit resistance.

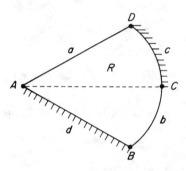

FIG. 2

If a given region has unit resistance for arbitrary chords BD which are perpendicular to a given chord AC, it is then conjectured that the conducter is self-conjugate (i.e., $\overline{AC}$ is an axis of symmetry). (See Fig. 3.)

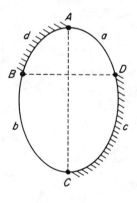

FIG. 3

REFERENCES

[1] J. W. S. RAYLEIGH, *On the approximate solution of certain problems relating to the potential*, Proc. London Math. Soc., 7 (1876), pp. 70–75; Scientific Papers, Vol. 1, No. 39.

[2] R. J. DUFFIN, *Distributed and lumped networks*, J. Math. Mech., 8 (1959), pp. 816–819.

Supplementary References
Electrical Networks

[1] W. N. ANDERSON, JR. AND R. J. DUFFIN, *"Series and parallel addition of matrices,"* J. Math. Anal. Appl. (1969) pp. 576–594.

[2] S. L. BASIN, *"The appearance of Fibonacci numbers and the Q matrix in electrical network theory,"* MM (1963) pp. 84–97.

[3] H. W. BODE, *Network Analysis and Feedback Amplifier Design,* Van Nostrand, N.Y., 1949.

[4] F. T. BOESCH, ed., *Large Scale Networks: Theory and Design,* IEEE Press, N.Y., 1976.

[5] R. BOTT AND R. J. DUFFIN, *"On the algebra of networks,"* Trans. AMS (1953) pp. 99–109.

[6] R. K. BRAYTON AND J. K. MOSER, *"Nonlinear networks I, II,"* Quart. Appl. Math. (1964) pp. 1–33, 81–104.

[7] R. L. BROOKS, C. A. B. SMITH, A. H. STONE AND W. T. TUTTE, *"The dissection of rectangles into squares,"* Duke Math. J. (1940) pp. 312–340.

[8] O. BRUNE, *"Synthesis of a finite two-terminal network whose driving-point impedance is a prescribed function of frequency,"* J. Math. Phys. (1931) pp. 191–236.

[9] W. CHEN, *"Boolean matrices and switching nets,"* MM (1966) pp. 1–8.

[10] C. CLOS, *"A study of nonblocking switching networks,"* Bell System Tech. J. (1953) pp. 406–424.

[11] G. C. CORNFIELD, *"The use of mathematics at the electricity council research centre in problems associated with the distribution of electricity,"* BIMA (1977) pp. 13–17.

[12] H. CRAVIS, *Communications Network Analysis,* Heath, Lexington, 1981.

[13] R. J. DUFFIN, *"An analysis of the Wang algebra of networks,"* Trans. AMS (1959) pp 114–131.

[14] ———, *"Distributed and lumped networks,"* J. Math. Mech. (1959) pp. 793–826.

[15] ———, *"The extremal length of a network,"* J. Math. Anal. Appl. (1962) pp. 200–215.

[16] ———, *"Nonlinear networks,"* Bull. AMS (1947) pp. 963–971.

[17] ———, *"Potential theory on a rhombic lattice,"* J. Comb. Th. (1968) pp. 258–272.

[18] ———, *"Topology of series-parallel networks,"* J. Math. Anal. Appl. (1965) pp. 303–318.

[19] R. J. DUFFIN, D. HAZONY AND N. MORRISON, *"Network synthesis through hybrid matrices,"* SIAM J. Appl. Math. (1966) pp. 390–413.

[20] A. M. ERISMAN, K. W. NEVES AND M. H. DWARAKANATH, eds., *Electric Power Problems: The Mathematical Challenge,* SIAM, 1980.

[21] D. GIVONE, *Introduction to Switching Theory,* McGraw-Hill, N.Y., 1970.

[22] E. A. GUILLEMIN, *Communication Networks II: The Classical Theory of Long Lines, Filters, and Related Networks,* Wiley, N.Y., 1935.

[23] F. HOHN, *"Some mathematical aspects of switching,"* AMM (1955) pp. 75–90.

[24] F. E. HOHN AND L. R. SCHISSLER, *"Boolean matrices in the design of combinatorial switching circuits,"* Bell System Tech. J. (1955) pp. 177–202.

[25] Z. KOHVAI, *Switching and Finite Automata Theory,* McGraw-Hill, N.Y., 1970.

[26] R. W. LIU, *"Stabilities of nonlinear networks,"* J. Elec. Eng. (1970) pp. 1–21.

[27] D. E. MULLER, *"Boolean algebras in electric circuit design,"* AMM (1954) pp. 27–29.

[28] P. PENFIELD, JR., R. SPENCE AND S. DUINKER, *Tellegen's Theorem and Electrical Networks,* M.I.T. Press, Cambridge, 1970.

[29] N. PIPPENGER, *"On crossbar switching networks,"* IEEE Trans. Comm. (1975) pp. 646–659.

[30] F. M. REZA, *"A mathematical inequality and the closing of a switch,"* Arch. Elec. Comm. (1981) pp. 349–354.

[31] ———, *Modern Network Analysis,* McGraw-Hill, N.Y., 1959.

[32] ———, *"Schwarz's lemma for n-ports,"* J. Franklin Inst. (1984) pp. 57–71.

[33] J. RIORDAN AND C. E. SHANNON, *"The number of two-terminal series-parallel networks,"* J. Math. Phys. (1942) pp. 83–93.

[34] S. SESHU AND M. B. REED, *Linear Graphs and Electrical Networks,* Addison-Wesley, Reading, 1961.

[35] SIAM-AMS *Proceedings, Mathematical Aspects of Electrical Network Analysis,* AMS, 1971.

[36] M. E. VAN VALKENBERG, *Introduction to Modern Network Synthesis,* Wiley, N.Y., 1960.

[37] L. WEINBERG, *Network Analysis and Synthesis,* McGraw-Hill, N.Y., 1962.

[38] H. S. WILF AND F. HARARY, eds., *Mathematical Aspects of Electrical Network Analysis,* AMS, 1971.

[39] A. N. WILLSON, JR., *Nonlinear Newtorks: Theory and Analysis,* Wiley, Chichester, 1974.

3. PROBABILITY

Problem 60-8*, On a Switching Circuit, by HERBERT A. COHEN (Space Instrumentation Division, Acton Laboratories).

Each of n switches connected in series are activated by clocks set to go off at a fixed time. The clocks, however, are imperfect in that they are in error given by a normal distribution with mean 0 and standard deviation 1. What is the standard deviation for the distribution of all switches being activated?

In particular, how does it behave asymptotically for large n?

Solution by A. J. BOSCH (Technological University, Eindhoven, the Netherlands).

The time that clock i goes off (later than the fixed time) is t_i $(i = 1, \cdots, n)$ with $f(t) \sim N(0, 1)$. All switches have been activated if and only if the "latest" has been activated. Let $t = \max(t_1, \cdots, t_n)$, $G(t_0) = P(t < t_0) = F^n(t_0)$ (all clocks are independent), hence $g(t) = nF^{n-1}(t)f(t)$.

The variance $\sigma_n^2 = n\int_{-\infty}^{\infty} (t - \mu_n)^2 F^{(n-1)}(t)f(t)\,dt$. There is no explicit formula for it, but there are many tables: D. Teichroew, *Tables of expected values of order statistics and products of order statistics for samples of size twenty and less from the normal distribution*, Ann. Math. Statist., 27 (1956), pp. 410–426.

For the asymptotic distribution, see references in H. A. David, *Order Statistics*, John Wiley, New York, 1970, and in E. J. Gumbel, *Statistics of Extremes*, Columbia Univ. Press, New York, 1960.

Problem 60-12, *A Sorting Problem*, by WALTER WEISSBLUM (AVCO Corporation).

The first step in a method commonly used in computing machines for sorting a sequence $x_1, x_2, \cdots, x_n$ of random numbers is to break the sequence into strings

$$(x_1, \cdots, x_{n_1}), (x_{n_1+1}, \cdots, x_{n_2}), \cdots$$

such that $(x_1, \cdots, x_{n_1})$ is the longest initial monotone string (either increasing or decreasing), $(x_{n_1+1}, \cdots, x_{n_2})$ is the longest subsequent monotone string, etc. In order to estimate the time required for the sorting process it is necessary to know the expected number of strings produced. For large n, this amounts to knowing the limit of the expected length of the kth string when an infinite sequence $x_1, x_2, \cdots$ is decomposed as above. If we assume that the x_i are identically distributed and independent and that the distribution is continuous, then it is clear that the actual distribution does not matter, and we might as well take it to be uniform in $(0, 1)$.

If we denote the expected length of the kth string by L_k, show that

$$L = \lim_{k \to \infty} L_k = \frac{1 + \cot \frac{1}{2}}{3 - \cot \frac{1}{2}} = 2.4203.$$

Solution by the proposer.

Let $f(x) = \lim f_k(x)$, where f_k is the density function for the first element of the kth string; $A(u, x)$ be the conditional density, given that a string starts with u and that the next string starts with x; and $E(x)$ be the expected length of a string, given that it starts with x. Then,

(1) $$f(x) = \int_0^1 f(u)A(u, x)\, du,$$

(2) $$L = \int_0^1 f(x)E(x)\, dx.$$

The formulas

$$A(u, x) = e^u + e^{1-u} - 1 - e^{|u-x|},$$

$$E(x) = e^x + e^{1-x} - 1$$

follow. Differentiating (1) twice now yields

$$f''(x) + f(x) = -L.$$

Since f is symmetric about $\frac{1}{2}$, we can write

$$f(x) = A \cos (x - \tfrac{1}{2}) - L.$$

Also, since f is a density function,

$$2A \sin \tfrac{1}{2} - L = 1.$$

Another equation in A and L is obtainable from (2), i.e.,

$$2L = (\sin \tfrac{1}{2} + \cos \tfrac{1}{2})A.$$

Solving this equation gives

$$L = \frac{1 + \cot \frac{1}{2}}{3 - \cot \frac{1}{2}}.$$

Problem 62–8, Spatial Navigation, by WALTER WEISSBLUM (AVCO Research and Advanced Development Division).

The following problem has arisen in classifying stars in connection with spatial navigation:

N points are chosen independently and uniformly in a square, forming a set S. For p and q in S, let the pair (p, q) be called "good" if there exists a circle containing p and q and no other members of S. If E_N is the expected number of good pairs, show that $E_N \sim 3N$.

Solution by JOHN L. MARTIN (National Physical Laboratory, Teddington, England).

The result $E_N \sim 3N$ does not depend strongly on the independence and uniformity of the distributions of the N points; in fact, it has a *combinatorial* character. Disregard for the moment the possibly exceptional cases where three or more points are collinear or four or more concyclic (these cases occur with vanishing probability anyway). Disregard also the case $N = 2$. Then:

If the convex hull of the N points is an n-gon, the number of "good" pairs is $3N - n - 3$.

A summarized proof follows:

The circumcircle of three of the N points A, B, C is said to be "good" if the circle has none of the N points in its interior. For this case the triangle ABC is also said to be "good" and it follows that AB, BC, and CA are all "good" pairs.

Focus attention on any good pair AB. This good pair will fall into one or the other of two categories.

(i) There may be *two* good circles through A and B (these will be the extreme members of the single infinity of circles through A, B, with none of the remaining $N - 2$ points in their interiors). In this case, AB is the common edge of two good triangles.

(ii) There may be *one* good circle through A and B, and AB will be the edge of one good triangle. For this to happen, *all* remaining $N - 2$ points must lie on the same side of the line AB; in fact, the good pairs in this category form a convex polygon, the convex hull of the N points.

It may be concluded that the joins of good pairs are the edges of a network of good triangles, whose boundary is the convex hull of the N points. The desired result follows from Euler's formula ($V - E + F = 2$) applied to this network. We know

$$\text{Number of vertices in network} = N,$$

$$\text{Number of edges in network} = E_N \, ;$$

on observing that each of $E_N - n$ edges is shared by two triangles, while each of n edges belongs to just one, we obtain

$$\text{Number of faces in network} = 1 + \tfrac{1}{3}(2E_N - n),$$

including the n-gon *external* to the convex hull. Euler's formula now gives

$$E_N = 3N - n - 3.$$

It is possible to preserve this result for the exceptional cases mentioned in the first paragraph by applying an appropriate convention.

Another result associated with Problem 62–8. When the N points are chosen independently and uniformly with mean density of 1 point per area l^2, the mean separation of the points of a good pair is asymptotically $32l/9\pi$.

We assume N very large, and neglect edge effects. Denote by $p(R)$ the probability that a randomly placed circle of radius R has none of the points in its interior; a standard argument shows that

$$\frac{dp}{dR} = -\frac{2\pi Rp}{l^2}.$$

This, with the condition $p(0) = 1$, gives

(1) $$p(R) = \exp(-\pi R^2/l^2).$$

Now choose from the N points a particular one, O say, and consider the distribution of triangles OAB, with A, B two other points of the set. The points A for which OA lies in $(a, a + da)$ must lie in an annulus of area $2\pi a\, da$; their expected number is thus $2\pi a\, da/l^2$. The expected number of points B for which OB lies in $(b, b + db)$ is $2\pi b\, db/l^2$. For any such pair A, B, the angle $\theta = AOB$ is uniformly distributed in $(0, \pi)$, by virtue of the independence of the distributions. It follows from these facts that the expected number of triangles OAB for which OA, OB, and AOB lie in $(a, a + da)$, $(b, b + db)$, $(\theta, \theta + d\theta)$ is

$$\tfrac{1}{2}(4\pi ab/l^4)\, da\, db\, d\theta$$

(the extra $\tfrac{1}{2}$ is present since this reckoning counts each triangle twice, once as OAB and once as OBA).

The probability that OAB is a *good* triangle is just the probability that its circumcircle is a good circle, namely expression (1) with

$$R^2 = \frac{a^2 + b^2 - 2ab\cos\theta}{4\sin^2\theta}.$$

Therefore the expected number of *good* triangles with vertex O and OA, OB, and AOB in the ranges specified earlier is

$$f(a, b, \theta)\, da\, db\, d\theta,$$

where

$$f(a, b, \theta) = (2\pi ab/l^4)\exp\left\{-\frac{\pi(a^2 + b^2 - 2ab\cos\theta)}{4l^2\sin^2\theta}\right\}.$$

The remainder of the work is a matter of evaluating certain elementary integrals. We give only the results. First, the expected number of good triangles with vertex at O is

$$\int_0^\infty da \int_0^\infty db \int_0^\pi d\theta\, f(a, b, \theta) = 6,$$

as might have been guessed; this leads immediately to the now familiar fact that $E_N \sim 3N$. Also, the average length of the edge OA of a good triangle OAB is

$$\frac{1}{6}\iiint a\, da\, db\, d\theta\, f(a, b, \theta) = 32l/9\pi,$$

and, as O is in no way a distinguished point, we conclude that this is in fact the mean separation of the points of a good pair, as originally asserted.

Problem 63-6, Asymptotic Distribution of Lattice Points in a Random Rectangle, by WALTER WEISSBLUM (AVCO Corporation).

An $n \times n^{-1}$ rectangle is thrown at random on the plane with angle uniformly distributed in $0 \leq \theta \leq 2\pi$ and center of rectangle also uniformly distributed in $0 \leq x \leq 1$, $0 \leq y \leq 1$. Find the limit as $n \to \infty$ of the distribution of the number of lattice points contained in the rectangle.

Solution by N. G. DE BRUIJN (Technological University, Eindhoven, Netherlands).

Let $p_j(n)$ be the probability of catching j lattice points; p_j denotes its limit as $n \to \infty$. We have

(1) $$\sum_{j=0}^\infty p_j(n) = 1, \qquad \sum_{j=0}^\infty jp_j(n) = 1,$$

since the expectation of j equals the area of the rectangle. We shall show that

(2) $$p_j = 3\pi^{-2}((j-1)^{-2} - 2j^{-2} + (j+1)^{-2}), \qquad j = 2, 3, \cdots,$$

and then (1) produces the values of p_0 and p_1 :

$$p_0 = 3\pi^{-2}, \qquad p_1 = 1 - 21/(4\pi^2).$$

If three lattice points form a proper triangle, then its area is $\geq \frac{1}{2}$. So if our rectangle catches all three, two of them lie in vertices of the triangle. As this happens with probability zero, we may assume that if more than two lattice points are caught, then they are all on a line.

Let $\mathbf{v}$ be a nonzero vector with integral components. To each lattice point P we let correspond a "needle", viz., the line segment $(P, P + \mathbf{v})$. The probability that our rectangle catches a needle is easily shown to be $K(|\mathbf{v}|)$, where $|\mathbf{v}|$ is the length of $\mathbf{v}$, and

$$K(d) = 0 \quad \text{if} \quad d > n,$$

$$K(d) = 2\pi^{-1} \int_0^{\text{arc sin}(1/nd)} (n^{-1} - d \sin \phi)(n - d \cos \phi)\, d\phi \quad \text{if} \quad 0 < d \leq n.$$

Let $t_j(n; \mathbf{v})$ be the probability that the rectangle catches exactly j lattice points $(j \geq 2)$ with difference vector $\mathbf{v}$ (that is, the probability that there exists a lattice point P such that $P + \mathbf{v}, \cdots, P + j\mathbf{v}$ are inside, but P and $P + (j+1)\mathbf{v}$ outside). The probability for catching a needle corresponding to a vector $k\mathbf{v}$ can now be expressed as follows:

$$K(k|\mathbf{v}|) = t_{k+1}(n, \mathbf{v}) + 2t_{k+2}(n, \mathbf{v}) + 3t_{k+3}(n, \mathbf{v}) + \cdots,$$

whence, for $j = 2, 3, \cdots$,

$$t_j(n, \mathbf{v}) = K((j+1)|\mathbf{v}|) - 2K(j|\mathbf{v}|) + K((j-1)|\mathbf{v}|).$$

So for $j = 2, 3, \cdots$, we have $p_j(n) = s_{j+1}(n) - 2s_j(n) + s_{j-1}(n)$, where for $k = 1, 2, \cdots$,

(3) $$s_k(n) = \frac{1}{2}\sum_{\mathbf{v}}{}^* K(k|\mathbf{v}|),$$

where * denotes that the summation is taken over all primitive vectors with integral components ("primitive" means that the components have g.c.d. 1; in other words, that the vector is not a multiple of a smaller integral vector). The factor $\frac{1}{2}$ arises from the fact that if a sequence of lattice points can be described by a vector $\mathbf{v}$, then it can also be described by $-\mathbf{v}$.

It remains to show that $s_k(n) \to 3\pi^{-2}k^{-2}$ as $n \to \infty$. We easily evaluate

$$K(\rho) = \pi^{-1}n^{-2}\rho^{-1}\{n - \rho + O((\rho+1)^{-1})\}.$$

Taking into account that the probability of a lattice vector being primitive equals $6\pi^{-2}$, we obtain

$$\frac{1}{2}\sum_{\mathbf{v}}{}^* K(k|\mathbf{v}|) \sim \frac{1}{2}\cdot 6\pi^{-2}\int_0^{2\pi} d\phi \int_0^{n/k} \rho K(k\rho)\, d\rho \sim \frac{1}{2}\cdot 6\pi^{-2}\cdot 2\pi\cdot \pi^{-1}n^{-2}\cdot \frac{1}{2}n^2 k^{-2} = 3\pi^{-2}k^{-2}.$$

We remark that it follows from (2) that $\sum_0^\infty j(j-1)p_j = 1$, whence the expectation of the square of the number of lattice points in the rectangle equals 2.

Also, if b is a constant, $0 < b \leq 1$, and if we throw a rectangle $n \times bn^{-1}$, then we obtain $p_0 = 1 - b + 3b^2/\pi^2$, $p_1 = b - 21b^2/4\pi^2$, $p_j = 3\pi^{-2}b^{-2}((j-1)^{-2} - 2j^{-2} + (j+1)^{-2})$. If $b > 1$, however, the problem gets more difficult, for then there is a positive probability to catch nontrivial triangles.

Problem 60-11, A Parking Problem, by M. S. KLAMKIN (AVCO), D. J. NEWMAN (Yeshiva University) and L. SHEPP (University of California, Berkeley).

Let $E(x)$ denote the expected number of cars of length 1 which can be parked on a block of length x if cars park randomly (with a uniform distribution in the available space).

Show that $E(x) \sim cx$ and determine the constant c.

Solution by ALAN G. KONHEIM (I.B.M.) and LEOPOLD FLATTO (Reeves Instrumentation Corp.).

We first show that $E(x)$ satisfies the integral equation

$$\text{(1)} \qquad E(x) = 1 + \frac{2}{x-1} \int_0^{x-1} E(t)\, dt \qquad (x > 1).$$

The probability that the rear end of the first car lies in the interval $(t, t + dt)$ is $1/(x-1)\, dt$ (for $0 \leq t \leq x - 1$). The interval $[0, x]$ is then decomposed into the two intervals $[0, t]$ and $[t + 1, x]$ in which the expected numbers of cars are $E(t)$ and $E(x - 1 - t)$, respectively. Integrating t over $[0, x - 1]$, we obtain (1)

Multiplying (1) by $(x - 1)$ and differentiating, we obtain

$$\text{(2)} \qquad (x-1)E'(x) + E(x) = 1 + 2E(x-1), \qquad (x > 1),$$

If

$$\text{(3)} \qquad F(s) = \int_0^\infty e^{-sx} E(x)\, dx = \int_1^\infty e^{-sx} E(x)\, dx, \qquad R(s) > 0,$$

then the Laplace transform satisfies the differential equation

$$\text{(4)} \qquad F'(s) + (1 + 2e^{-s}/s)F(s) = -e^{-s}/s^2, \qquad R(s) > 0,$$

which has as general solution

$$\text{(5)} \qquad \begin{aligned} F(s) = \; & K \exp\left\{ -\int_1^s (1 + 2e^{-t}/t)\, dt \right\} \\ & + \int_s^\infty \frac{e^{-t}}{t^2} \exp\left\{ \int_s^t (1 + 2e^{-u}/u)\, du \right\} dt. \end{aligned}$$

Since $E(t) = 0, 0 \leq t < 1$ and $E(t) \leq t$ $(1 \leq t < \infty)$ it follows that $F(s) < (2/s)e^{-s}$ (s real and greater than 1), so that $K = 0$. Therefore,

$$\text{(6)} \qquad F(s) = \int_s^\infty \frac{e^{-t}}{t^2} \exp\left\{ \int_s^t (1 + 2e^{-u}/u)\, du \right\} dt,$$

which gives upon several changes of variables

$$\text{(7)} \qquad F(s) = \frac{e^{-s}}{s^2} \int_s^\infty \exp\left\{ -2\int_s^t \frac{1 - e^{-u}}{u}\, du \right\} dt.$$

From (7), we see that

$$F(s) \sim \frac{c}{s^2} \quad \text{as} \quad s \to 0^+$$

where

$$c = \int_0^\infty \exp\left\{-2\int_0^t \frac{1-e^{-u}}{u}\,du\right\} dt.$$

Using a well known Tauberian theorem (Widder, *The Laplace Transform*, p. 192), we have

$$\int_1^x E(t)\,dt \sim cx^2/2 \quad \text{as} \quad x \to \infty,$$

and a standard argument shows

$$E(x) \sim cx \quad \text{as} \quad x \to \infty.$$

Also solved by the proposers.

Editorial Note:

This problem was obtained third-hand by the proposers and attempts were made to track down the origin of the problem. These efforts were unsuccessful until after the problem was published. Subsequently, H. Robbins, Stanford University, has informed me that he had gotten the problem from C. Derman and M. Klein of Columbia University in 1957 and that in 1958 he had proven jointly with A. Dvoretzky that

$$(8) \qquad\qquad E(x) = cx - (1-c) + O(x^{-n}), \qquad\qquad n \geq 1,$$

plus other results like asymptotic normality of x, etc. They had intended to publish their results but did not when they found that A. Renyi had published a paper proving (8) in 1958, i.e., *"On a One-Dimensional Problem Concerning Random Place Filling,"* Mag. Tud. Akad. Kut. Mat. Intézet. Közleményei, pp. 109–127. Also, (8) is proven by P. Ney in his Ph.D. thesis at Columbia.

A reference to the Renyi paper was also sent in by T. Dalenius (University of California, Berkeley).

An abstract of the Renyi paper was sent in anonymously from the National Bureau of Standards. The abstract appears in the International Journal of Abstracts: Statistical Theory and Method, Vol. I, No. 1, July 1959, Abstract No. 18. According to the abstract, there is a remark due to N. G. DeBruijn in the Renyi paper stating that a practical application of the Renyi result is in the parking problem that was proposed here. In addition, the constant c has been evaluated to be 0.748.

> *Problem 62–3, An Unfriendly Seating Arrangement,* by DAVE FREEDMAN (University of California, Berkeley) and LARRY SHEPP (Bell Telephone Laboratories).

There are n seats in a row at a luncheonette and people sit down one at a time at random. They are unfriendly and so *never* sit next to one another (no moving over). What is the expected number of persons to sit down?

Solution by HENRY D. FRIEDMAN (Sylvania Applied Research Laboratory).

Let E_n be the desired expected number and number the seats consecutively

from the left. The first person sits down in the ith seat, $i = 1, 2, \cdots, n$, leaving
$i - 2$ and $n - i - 1$ seats (interpret as zero if negative) to his left and right,
respectively, available to the unfriendly second person. This permits us to write
recursively, for fixed i, $E_n = 1 + E_{i-2} + E_{n-i-1}$. Taking an average over all
i, we obtain

$$(1) \qquad E_n = 1 + \frac{1}{n}\sum_{i=1}^{n} (E_{i-2} + E_{n-i-1}) = 1 + \frac{2}{n}\sum_{i=1}^{n} E_{i-2}$$

with $E_{-1} = E_0 = 0$.

It now follows from (1) that the generating function for E_n, $F(x) = \sum_{n=1}^{\infty} E_n x^n$, satisfies the differential equation

$$F'(x) - 2x(1 - x)^{-1}F(x) = (1 - x)^{-2}$$

with the initial condition $F(0) = 0$. Whence,

$$(2) \qquad F(x) = (1 - e^{-2x})(1 - x)^{-2}/2.$$

By expanding (2) into a power series in x, we find that

$$(3) \qquad E_n = \sum_{i=0}^{n-1} (n - i)(-2)^i/(i + 1)!$$

For numerical calculations, we rewrite (3) into

$$E_n = \sum_{i=0}^{\infty} S_i - \sum_{i=n}^{\infty} S_i$$

where S_i is the summand of (3). It now follows that $-\sum_{i=n}^{\infty} S_i$ lies between
0 and $(-2)^{n+1}/(n + 2)!$ and may be taken as an error term which is negligible for
large n. Then

$$E_n \sim \sum_{i=0}^{\infty} S_i = n(1 - e^{-2})/2 + (1 - 3e^{-2})/2.$$

DAVID ROTHMAN (Harvard University) considered the more general problem
where m seats are vacated on each side of a seated person and establishes a
recurrence relation for the expected number $E_{2m}(n)$ of persons seated. He obtains the same results as before for $m = 1$ and gives the asymptotic expression

$$E_{2m}(n) \sim (n + 2m + 1)A_m - 1$$

where

$$A_m = \int_0^1 \exp\left\{2\left[\sum_{i=1}^{m} \frac{(t^i - 1)}{i}\right]\right\} dt.$$

For $m = 0$ and $m = 1$,

$$A_m = \frac{(1 - e^{-2})^{2m/(m+1)}}{(m + 1)}$$

and this is also a good approximation for $m \geq 2$. By letting $m \to \infty$ and normalizing one obtains the solution to the Parking Problem (Problem 60–11, July 1962,
pp. 257–258). The latter problem which is the continuous version of the problem
here is to find the expected number, $E(x)$, of cars of length 1 which can be parked
on a block of length x if cars park randomly (with a uniform distribution in the

available space). Asymptotically, $E(x) \sim cx$ where

$$c = \int_0^\infty \exp\left\{-2 \int_0^t \frac{1 - e^{-u}}{u}\, du\right\} dt \cong 0.748.$$

The approximation for c here is

$$\lim_{m \to \infty} (m + 1)A_m \sim (1 - e^{-2})^2 = 0.747646.$$

Editorial note. J.K. MacKenzie (RIAS) has solved this problem and the generalization given by Rothman above in his paper *Sequential Filling of a Line by Intervals Placed at Random and its Application to Linear Adsorption*, Jour. of Chem. Physics, Vol. 37, Aug. 1962, pp. 723-728. In this paper he notes that the case $m = 1$ has been considered by E.S. Page, J. Roy. Statist. Soc., B21, 364 (1959) and F. Downton, J. Roy. Stat. Soc., B23, 207 (1961). The related problem where unit intervals are placed not discretely but continuously on a line has been considered by A. Rényi, Magyar Tudományos Akad. Mat. Kutató Int. Közleményei 3, 109 (1958) (see Math. Rev. 21, 577 (1960). The related problem where all possible nonoverlapping configurations of intervals are assumed equally likely has been discussed by J.L. Jackson and E.W. Montroll, J. Chem. Phys. 28, 1101 (1958) for the case $m = 1$ and by H.S. Green and R. Leipnik, Revs. Modern Phys. 32, 129 (1960) for the cases $m = 1$ and $m = \frac{3}{2}$. Jackson and Montroll also give some approximations to the two-dimensional problem.

Problem 72-20, Expected Number of Stops for an Elevator, by D. J. NEWMAN (Yeshiva University).

P persons enter an elevator at the ground floor. If there are N floors above the ground floor and if the probability of each person getting out on any floor is the same, determine the expected number of stops until the elevator is emptied.

Solution by PETER BRYANT and PATRICK E. O'NEIL (IBM Cambridge Scientific Center).

Define the random variable x_i to have the value 1 if the ith floor is a stop and 0 otherwise. Note that $E(x_i) = P(x_i = 1)$. The probability that no one stops at floor i, $P(x_i = 0)$, is the probability that P persons acting independently make a choice other than floor i, each with probability $(1 - 1/N)$. Thus

$$P(x_i = 0) = (1 - 1/N)^P.$$

Obviously,

$$P(x_i = 1) = 1 - P(x_i = 0) = 1 - (1 - 1/N)^P.$$

Now the expected number of stops is given by

$$E\left(\sum_{i=1}^N x_i\right) = \sum_{i=1}^N E(x_i) = N[1 - (1 - 1/N)^P].$$

A few remarks are in order. Note that the solution $N[1 - (1 - 1/N)^P]$ is asymptotic to $N[1 - \exp(-A)]$ if P/N tends to a limit A as N tends to infinity. Second, since the argument depends only on the independence of each individual in the elevator, the result may be generalized. Assume that the probability of person k stopping at floor i is p_{ik}, $i = 1, \cdots, N$, $k = 1, \cdots, P$. Then the expected number of stops is given by

$$\sum_{i=1}^{N} \left[1 - \prod_{k=1}^{P} (1 - p_{ik}) \right].$$

If $p_{ik} = p_i$, $i = 1, \cdots, N$, $k = 1, \cdots, P$, then the expected number of elevator stops is

$$\sum_{i=1}^{N} [1 - (1 - p_i)^P].$$

Using Jensen's inequality, it is easy to show that this expression is maximized when $p_i = 1/N, i = 1, \cdots, N$. For the case where there is no restriction on p_{ik}, a simple argument shows that the maximum expected number of stops is equal to $\min(N, P)$.

   Solution by F. W. STEUTEL (Technische Hogeschool Twente, Enschede, the Netherlands).

   The description allows for several models. We assume the following: the P supposedly indistinguishable persons entering the elevator at the ground floor are labeled $x_1, x_2, \cdots, x_P$, indicating the number of the floor at which they will get out. All possible combinations are equally likely. This model is equivalent to the distribution of P indistinguishable balls over N cells (cf. W. Feller, *An Introduction to Probability Theory and its Applications*, vol. 1, John Wiley, New York, 1968, p. 38). The number of distributions equals $\binom{N + P - 1}{P}$.

   Denoting by K the number of floors where people leave the elevator, we have $P(K = k) =$ prob (all balls are distributed over k cells, none of which are empty). Hence

$$P(K = k) = \binom{N}{k}\binom{P - 1}{P - k} \div \binom{N + P - 1}{P} \qquad k = 1, \cdots, M,$$

where $M = \min(N, P)$. We therefore have

$$\binom{N + P - 1}{P} E(K) = \sum_{k=1}^{M} k \binom{N}{k}\binom{P - 1}{P - k} = N \sum_{j=0}^{M-1} \binom{N - 1}{j}\binom{P - 1}{P - j - 1}$$

$$= N \binom{N + P - 2}{P - 1}.$$

It follows that

$$E(K) = \frac{NP}{N + P - 1}.$$

Higher moments can be obtained in the same way. Other occupancy models can be treated similarly.

   Editorial note. As several of the solvers pointed out, the combinatorial part of this problem is simply a version of the classical occupancy problem. The elements of the sample space are distributions of P elevator passengers among N

exit floors. If the P passengers are considered to be distinguishable, the sample space contains N^P points, and in the equally likely model, the expected number of elevator stops is $N[1 - (1 - 1/N)^P]$. If the passengers are considered to be indistinguishable, the sample space contains $\binom{N + P - 1}{P}$ points, and in the equally likely model, the expected number of elevator stops is $NP/(N + P - 1)$. One solver, S. H. Saperstone, considered both cases. Several solvers discussed the probability distribution for the number of stops and/or higher moments of the distribution. Other solvers pointed out that these results are available in standard references [1, p. 101], [2, Chap. 5]. Several solvers obtained the expected value for the floor on which the elevator is emptied. For the distinguishable case, the expected value of the terminal floor is $N - \sum_{k=1}^{N-1} (k/N)^P$, and for the indistinguishable case, the result is $(NP + 1)/(P + 1)$.

B. A. Powell has considered more realistic models of the elevator problem in [3]. In a forthcoming book, *An Elementary Description of the Combinatorial Basis of Thermodynamics*, T. A. Ledwell uses the elevator problem as an example to develop the techniques needed in statistical thermodynamics. In particular, he discusses in detail the "thermodynamic limit" of large P and N.

F. C. Roesler (Imperial Chemical Industries Limited) points out that Schrödinger considered the classical occupancy problem (distinguishable case) in the analysis of experimental data from cosmic ray counters [4]. Schrödinger asks the following question. If P cosmic rays bombard an assembly of N cosmic ray counters, what is the probability that exactly k of the counters go off? The problem considered by Schrödinger is especially interesting, since in this context special importance is given to the inverse problem of estimating P given the observed value of k. Roesler phrases the inverse problem for the elevator in the following way. "Having observed from the indicator lights that the elevator has stopped k times, what is the best guess for the number of passengers who entered on the ground floor?" [C.C.R.]

REFERENCES

[1] W. FELLER, *Introduction to Probability Theory and Its Applications*, vol. 1, John Wiley, New York, 1968.
[2] J. RIORDAN, *An Introduction to Combinatorial Analysis*, John Wiley, New York, 1958.
[3] D. P. GAVER AND B. A. POWELL, *Veriability in Round-Trip Times for an Elevator Car During Up-Peak*, Transportation Sci., 5 (1971), p. 169.
[4] E. SCHRÖDINGER, *A Combinatorial Problem in Counting Cosmic Rays*, Proc. Phys. Soc., A, LXVII (1951), p. 1040.

Problem 64-11, On a Permutation Sequence, by PAUL BROCK (General Electric Co.).

For a permutation sequence of the numbers 1, 2, 3, $\cdots$, N, what is the probability that the maximum element of the first J elements is also the maximum for the first K elements, $J < K < N$?

The solutions by THOMAS S. ENGLAR, JR. (Research Institute for Advanced Studies), E. S. LANGFORD (North American Aviation), JUDY RICHMAN (Drexel Institute of Technology) and SIDNEY SPITAL (California State Polytechnic College) were the same and are given as follows:

Consider the maximum of the first K elements in the permutation sequence. With equal probability, it may occupy any one of the first K positions. Also, for it to be the maximum of the first J elements, however, it is limited to the first J positions. Therefore the desired probability is J/K. Also, the restriction $J < K < N$ can be weakened to $J \leq K \leq N$, trivially.

Editorial Note. The other solutions first expressed the desired probability as a sum equivalent to

$$\frac{J}{N!} \sum_{r=1}^{N-K+1} \frac{(N-r)!\,(N-K)!}{(N-K-r+1)!} \, ,$$

which reduces to J/K using the identity

$$\sum_{s=k}^{N} \binom{s-1}{k-1} = \binom{N}{k} .$$

Problem 61–10, The Expected Value of a Product, by L. A. SHEPP (Bell Telephone Laboratories).

Let E_n be the expected value of the product $x_1 x_2 x_3 \cdots x_n$, where x_1 is chosen at random (with a uniform distribution) in $(0, 1)$ and x_k is chosen at random (with a uniform distribution) in $(x_{k-1}, 1)$, $k = 2, 3, \cdots, n$. Show that

$$\lim_{n \to \infty} E_n = \frac{1}{e} .$$

Solution by J. VAN YZEREN (Technische Hogeschool Eindhoven, Netherlands).

The random variates $x_1, x_2, x_3, \cdots$ can be produced by

$$x_i = 1 - u_1 u_2 u_3 \cdots u_i$$

where the u are independent choices from a uniform distribution.

Consider the formal product

(1)
$$\prod_{i=1}^{\infty} (1 + t u_1 u_2 u_3 \cdots u_i)$$

The coefficient of t^n is the sum of all products $u_1{}^n u_2{}^{n_2} u_3{}^{n_3} \cdots$ with $n \geq n_2 \geq n_3 \geq \cdots$. Evidently $E(u_1{}^n u_2{}^{n_2} u_3{}^{n_3} \cdots) = \dfrac{1}{n+1} \dfrac{1}{n_2+1} \dfrac{1}{n_3+1} \cdots$. The sum of all these expected values is

$$\frac{1/(n+1)}{1 - 1/(n+1)} \frac{1/n}{1 - 1/n} \cdots \frac{\frac{1}{2}}{1 - \frac{1}{2}} = \frac{1}{n!} .$$

Hence,

$$E\left\{ \prod_{i=1}^{m} (1 + t u_1 u_2 u_3 \cdots u_i) \right\}$$

converges for $m \to \infty$ to e^t, for any value of t! Putting $t = -1$ gives the required result.

Solution by J. H. VAN LINT (Technological University, Eindhoven, Netherlands).

Let $E_n(x)$ be the expected value of the product $x_1 x_2 \cdots x_n$, where x_1 is chosen at random (with a uniform distribution) in $(x, 1)$ and x_k is chosen at random (with a uniform distribution) in $(x_{k-1}, 1)$, $k = 2, 3, \cdots, n$.

The function $E_n(x)$ can be expressed as a multiple integral:

$$
(1) \quad E_n(x) = \frac{1}{1 - x} \int_x^1 \frac{x_1}{1 - x_1} \, dx_1
$$

$$
\int_{x_2}^1 \frac{x_2}{1 - x_2} \, dx_2 \cdots \int_{x_{n-2}}^1 \frac{x_{n-1}}{1 - x_{n-1}} \, dx_{n-1} \int_{x_{n-1}}^1 x_n \, dx_n .
$$

We define $E_0(x) = 1$, $E_1(x) = \dfrac{1 + x}{2}$. We then have for $n = 1, 2, \cdots$ the differential equation

$$
(2) \qquad\qquad \{ (1 - x) E_n(x) \}' = -\, x E_{n-1}(x).
$$

We now prove the inequality

$$
(3) \qquad\qquad e^{x-1} \leq E_n(x) \leq e^{x-1} + \frac{1 - x}{2^n} .
$$

For $n = 0$, the inequality is true. If the inequality is true for some value of n we find an inequality for E_{n+1} by applying (2). We find,

$$
e^{x-1} \leq E_{n+1}(x) \leq e^{x-1} + \frac{1}{2^n}\left\{ \frac{1}{6} + \frac{1}{6}x - \frac{1}{3}x^2 \right\} \leq \frac{1}{2^{n+1}}(1 - x) + e^{x-1}.
$$

By induction (3) is true for all n. A consequence of (3) is

$$
(4) \qquad\qquad \lim_{n \to \infty} E_n(x) = e^{x-1}
$$

(If we take $x = 0$, we find the theorem that was to be proved).

Problem 71-14, An Expected Value, by M. S. KLAMKIN (Ford Motor Company).

n numbers are chosen independently at random, one from each of the n intervals $[0, L_i]$, $i = 1, 2, \cdots, n$. If the distribution of each random number is uniform with respect to length in the interval it is chosen from, determine the expected value of the smallest of the n numbers chosen.

Solution by O. G. RUEHR (Michigan Technological University).

If the random variable x_i be associated with the interval $[0, L_i]$ and if $X = \min(x_i)$, $L = \min(L_i)$, then

$$
F(x) = \Pr\{X \leq x\} = 1 - \Pr\{X > x\} = 1 - \prod(1 - x/L_i).
$$

The expected value of X is then

$$
E(X) = \int_0^L x \, dF(x) = xF(x) \Big]_0^L - \int_0^L F(x) \, dx = \int_0^L \{1 - F(x)\} \, dx
$$

$$
= \frac{1}{S_n}\left\{ LS_n - \frac{L^2}{2}S_{n-1} + \frac{L^3}{3}S_{n-2} - \cdots (-1)^n \frac{L^{n+1}}{n + 1} \right\},
$$

where the S_i are the elementary symmetric functions of the L_i, e.g., $S_1 = \sum L_i$, $S_2 = \sum_{i \neq j} L_i L_j$.

Comment by the proposer. The special case $n = 3$ occurs in a problem in the 31st William Lowell Putnam Mathematical Competition (Amer. Math. Monthly, Sept. 1971). A more involved set of problems is to determine $E\{M_i(x_1, x_2, \cdots, x_n)\}$ where M_i denotes the ith smallest of the x_i's. For $i > 1$, the above method apparently is not applicable. Here, we indicate how to solve this general class of problems by determining the result for the two cases $i = 2$ and n. Letting

$$\Phi(r, n) = \int_0^{L_n} \cdots \int_0^{L_n} \int_0^{L_{n-1}} \cdots \int_0^{L_1} \max_{i=1}^{n+r-1} (x_i) \prod_{i=1}^{n+r-1} dx_i,$$

we obtain symbolically that

$$\Phi(r, n) = \left\{ \int_{L_{n-1}}^{L_n} + \int_0^{L_{n-1}} \right\}^r \int_0^{L_{n-1}} \cdots \int_0^{L_1} \max (x_i) \prod dx_i$$

$$= \Phi(r + 1, n - 1) + F(r, n),$$

where

$$F(r, n) = \sum_{j=0}^{r-1} \binom{r}{j} \left\{ \int_{L_{n-1}}^{L_n} \right\}^{r-j} \left\{ \int_0^{L_{n-1}} \right\}^{j+1} \int_0^{L_{n-2}} \cdots \int_0^{L_1} \max (x_i) \prod dx_i.$$

It is to be noted that $\Phi(r, n)$ and $F(r, n)$ are also functions of $L_1, L_2, \cdots, L_n$ but we have left them out for convenience. The summand for F equals

$$\binom{r}{j} L_1 L_2 \cdots L_{n-2} L_{n-1}^{j+1} \int_{L_{n-1}}^{L_n} \cdots \int_{L_{n-1}}^{L_n} \max_{k=1}^{r-j} (x_k) \prod_{k=1}^{r-j} dx_k.$$

In a manner similar to the above solution or otherwise, we obtain

$$\int_0^a \cdots \int_0^a M_r(x_i) \, dx_1 \, dx_2 \cdots dx_m = \frac{ra^{m+1}}{m+1}.$$

Thus,

$$F(r, n) = \sum_{j=0}^{r-1} \binom{r}{j} L_1 L_2 \cdots L_{n-2} L_{n-1}^{j+1} (L_n - L_{n-1})^{r-j} \frac{(r-j)L_n + L_{n-1}}{r-j+1}.$$

It follows from the recurrence equation that

$$\Phi(r, n) = \Phi(r + n - 1, 1) + F(r + n - 2, 2) + F(r + n - 3, 3) + \cdots + F(r, n),$$

where

$$\Phi(r, 1) = rL_1^{r+1}/(r + 1).$$

Finally,

$$E\{\max (x_1, x_2, \cdots, x_n)\} = \Phi(1, n)/L_1 L_2 \cdots L_n.$$

To obtain $E\{M_2(x_1, x_2, \cdots, x_n)\}$, we first consider

$$\psi(r, n) = \int_0^{L_n} \int_0^{L_{n-1}} \cdots \int_0^{L_2} \cdots \int_0^{L_2} \int_0^{L_1} M_2(x_i) \prod_{i=1}^{n+r-1} dx_i.$$

Then using, $\int_0^{L_n} = \int_0^{L_2} + \int_{L_2}^{L_n}$, we obtain

$$\psi(r, n) = \psi(r + 1, n - 1) + (L_n - L_2)\psi(r, n - 1).$$

Whence,

$$\psi(r, n) = \psi(r + n - 2, 2) + S_1\psi(r + n - 3, 2) + S_2\psi(r + n - 4, 2) + \cdots$$
$$+ S_{n-2}\psi(r, 2),$$

where the S_i are the elementary symmetric functions of

$$L_3 - L_2, L_4 - L_2, \cdots, L_n - L_2.$$

It now remains to determine

$$\psi(r, 2) = \int_0^{L_2} \cdots \int_0^{L_2} \int_0^{L_1} M_2(x_i) \, dx_1 \, dx_2 \cdots dx_{r+1}.$$

Symbolically,

$$\psi(r, 2) = \sum_{j=0}^{r} \binom{r}{j} \left\{ \int_{L_1}^{L_2} \right\}^{r-j} \left\{ \int_0^{L_1} \right\}^{j+1} M_2(x_i) \, dx_1 \, dx_2 \cdots dx_{r+1}$$

or

$$\psi(r, 2) = L_1 \int_{L_1}^{L_2} \cdots \int_{L_1}^{L_2} \min(x_i) \, dx_1 \, dx_2 \cdots dx_r$$

$$+ \sum_{j=1}^{r} \binom{r}{j} (L_2 - L_1)^{r-j} \int_0^{L_1} \cdots \int_0^{L_1} M_2(x_i) \, dx_1 \, dx_2 \cdots dx_{j+1}$$

$$= \frac{L_1(L_2 - L_1)^r(rL_1 + L_2)}{r + 1} + \sum_{j=1}^{r} \binom{r}{j} \frac{2L_1^{j+2}(L_2 - L_1)^{r-j}}{j + 2}.$$

Then,

$$E\{M_2(x_1, x_2, \cdots, x_n)\} = \psi(1, n)/L_1 L_2 \cdots L_n.$$

The previous problems become even more involved if we replace the intervals $[0, L_i]$ by $[K_i, L_i]$.

Problem 60-9′, A Pie Problem, by LARRY SHEPP (University of California, Berkeley).

A man gets a random percentage (uniformly distributed) of a pie on the first cut. In each succeeding cut he gets a random percentage of what he got the previous cut. Show that the probability that he remains frustrated, i.e., never gets the entire pie is

$$P = e^{-\gamma} > \tfrac{1}{2}.$$

Solution by Henry McKean (Massachusetts Institute of Technology).

Consider the successive fractions of pie $p_n(n \geq 1)$. Then $P[a \leq p_n < b] = b - a, 0 \leq a < b \leq 1$. If $P = p_1 + p_1 p_2 + p_1 p_2 p_3 + \cdots$ then $q = p_2 + p_2 p_3 + \cdots$ is identical in law to P. Consequently,

$$(1) \quad E[e^{-\alpha p}] = E[e^{-\alpha p_1(1+q)}] = \int_0^1 e^{-\alpha t} E[e^{-\alpha t p}] \, dt = \alpha^{-1} \int_0^{\alpha} e^{-\beta} E[e^{-\beta p}] \, d\beta.$$

Converting (1) into a differential equation for $\alpha E[e^{-\alpha p}]$ and solving, one finds that

(2) $$E[e^{-\alpha p}] = c_1 \alpha^{-1} \exp\left\{ -\int_\alpha^\infty \beta^{-1} e^{-\beta}\, d\beta \right\}$$

where the constant c_1 can be evaluated as $e^{-\gamma}$ by noting that $P[p < \infty] = 1$ implies

(3)
$$1 = \lim_{\alpha \downarrow 0} E[e^{-\alpha p}] = c_1 \lim_{\alpha \downarrow 0} \exp\left\{ \int_\alpha^1 \beta^{-1}\, d\beta - \int_\alpha^\infty \beta^{-1} e^{-\beta}\, d\beta \right\}$$

$$= c_1 \exp\left\{ \int_1^\infty t^{-1}(1 - e^{-t} - e^{-1/t})\, dt \right\} = c_1 e^\gamma.$$

On the other hand, if $0 < t < 1$, then

$$Q(t) \equiv P[p \leq t] = P[p_1(1 + q) \leq t],$$

$$Q(t) \leq P[q \leq t/p_1 - 1] = \int_0^t P[q \leq t/\theta - 1]\, d\theta = c_2 t.$$

Thus the desired probability is

$$P[p \leq t] = c_2 = Q'(0) = \lim_{\alpha \uparrow \infty} \alpha \int_0^\infty e^{-\alpha t}\, dQ = \lim_{\alpha \uparrow \infty} \alpha E[e^{-\alpha p}] = c_1 = e^{-\gamma}.$$

Problem 63–9*, An Optimal Search, by RICHARD BELLMAN (The Rand Corporation).

Suppose that we know that a particle is located in the interval $(x, x + dx)$, somewhere along the real line $-\infty < x < \infty$ with a probability density function $g(x)$. We start at some initial point x_0 and can move in either direction. What policy minimizes the expected time required to find the particle, assuming a uniform velocity and

(a) assuming that the particle will be recognized when we pass x,

or

(b) assuming that there is a probability $p > 0$ of missing the particle as we go past it?

Also, what would be the optimum starting point x_0 ?

Editorial Note: A related class of two dimensional search problems are the following "swimming in a fog" problems. A person has been shipwrecked in a fog and wishes to determine the optimal path of swimming to get to shore (in the least expected time—assuming a uniform rate of swimming). The boundary conditions can be any of the following:

1. The ocean is a half-plane,
2. Condition (1) plus the knowledge that the initial distance to shore is $\leq D$ (with a uniform distribution),
3. The ocean boundary is a given closed curve, i.e., a circle, rectangle, or possibly not closed (a parabola),
4. Condition (2) and (3), etc.

Solution. See the research paper, *An optimal search problem*, by Wallace Franck, this issue, pp. 503-512, + errata 8 (1966) 254.

Problem 64-15, On a Probability of Overlap, by MURRAY S. KLAMKIN (Ford Scientific Laboratory).

Prove directly or by an immediate application of a theorem in statistics that the conjecture in the following abstract from Mathematical Reviews, March, 1964, p. 589, is valid:

"OLESKIEWICZ, M. *The probability that three independent phenomenon of equal duration will overlap.* (Polish, Russian and English summaries) Prace Mat. 64 (1960), 1–7.

The value P_3 of the probability that 3 stochastically independent phenomenon of equal duration t_0 which all occur during the time $t + t_0$ will overlap is shown by geometrical methods to be equal to $(3tt_0^2 - 2t_0^3)/t^3$. The author makes the conjecture that a similar formula holds for n independent events, namely $P_n = (ntt_0^{n-1} - (n - 1)t_0^n)/t^n$."

Solution by P. C. HEMMER (Norges Tekniske Høgskole, Trondheim, Norway).

We note that an overlap has to start simultaneously with the onset of one of the n events. The probability of overlap when this one event is assumed to be number 1 gives $1/n$ of the desired probability P_n. Denote the whole time by $(0, t + t_0)$ and let event number 1 occur in $(\lambda, \lambda + t_0)$, where λ is uniformly distributed in $(0, t)$. An overlap exists if and only if the other $n - 1$ events occur at time λ. Independence and uniform distribution guarantee the following probability for this to happen:

$$P(\lambda) = \begin{cases} (\lambda/t)^{n-1} & \text{if } 0 \leq \lambda \leq t_0, \\ (t_0/t)^{n-1} & \text{if } t_0 \leq \lambda \leq t. \end{cases}$$

Thus,

$$\frac{P_n}{n} = \int_0^t \frac{P(\lambda)\, d\lambda}{t} = \left(\frac{t_0}{t}\right)^n \left[\frac{t - t_0}{t_0} + \frac{1}{n}\right]$$

in agreement with the conjecture.

Solution by A. J. BOSCH (Technological University, Eindhoven, Netherlands).

Set $t_0/t = p$ and $1 - p = q$. Consider the arrival times of the n events. The probability that an event occurs before time $t - t_0$ is equal to q. The probability that the first event occurs before time $t - t_0$ and the $(n - 1)$ remaining events within time t_0 after the first is clearly equal to $\binom{n}{1} qp^{n-1}$. The probability that all events occur after time $t - t_0$ is p^n. Hence,

$$P^n = nqp^{n-1} + p^n = \{ntt_0^{n-1} - (n - 1)t_0^n\}/t^n.$$

Solution by W. OETTLI (IBM Research Laboratory, Zürich).

Let $x_1, x_2, \cdots, x_n$ be an ordered sample of a random variable with probability distribution $F(x)$ and probability density $f(x)$. As is known from the theory of order statistics, the range $r = x_n - x_1$ has the probability density

$$g(y) = n(n - 1) \int_{-\infty}^{\infty} \{F(z + y) - F(z)\}^{n-2} f(z) f(z + y)\, dz,$$

(cf. L. SCHMETTERER, *Einführung in die mathematische Statistik,* Springer, Vienna, 1956, p. 348). For the special case

$$f(x) = \begin{cases} 1/t & \text{if } 0 \leq x \leq t, \\ 0 & \text{elsewhere,} \end{cases}$$

we have

$$g(y) = n(n-1)t^{-n}y^{n-2}(t-y).$$

Overlapping takes place if $y \leq t_0$. Thus the probability of overlapping is given by

$$\int_0^{t_0} g(y)\, dy = \{ntt_0^{n-1} - (n-1)t_0^n\}/t^n,$$

as conjectured.

Editorial Note. Hemmer also gives two related problems:

1. Determine the probability of overlap of n independent events of durations $t_1, t_2, \cdots, t_n$, which all occur during the time t.

2. Determine the distribution of the duration of the overlap.

The solution of both these problems when boundary effects are neglected ($t \gg \max t_i$) is given in his paper, *A problem in geometrical probabilities*, Norske Vid. Selsk. Forh. (Trondheim), 32 (1959), pp. 117–120.

The desired probability can also be expressed as the n-fold integral

$$P_n = \frac{n!}{t^n} \int_0^t dx_1 \int_{x_1}^a dx_2 \cdots \int_{x_{n-1}}^a dx_n,$$

where $a = \min(t, x_1 + t_0)$. The integral reduces easily to

$$P_n = \frac{n}{t^n} \left\{ \int_0^{t-t_0} t_0^{n-1}\, dx_1 + \int_{t-t_0}^t (t - x_1)^{n-1}\, dx_1 \right\},$$

which gives the desired answer.

Using this integral solution, it is easy to give a solution to Hemmer's problem (1) assuming that the starting points of each duration are uniformly distributed in the constant interval $[0, t]$.

Let the smallest starting point be denoted by x_1 and let its corresponding duration be t_i. Then the probability of simultaneous overlap is

$$\frac{(n-1)!}{t^n} \int_0^t dx_1 \int_{x_1}^b dx_2 \cdots \int_{x_{n-1}}^b dx_n,$$

where $b = \min\{t, x_1 + t_i\}$. This reduces to

$$p_i = \left(\frac{t_i}{t}\right)^n \left[\frac{t - t_i}{t_i} + \frac{1}{n}\right].$$

Thus the total probability for overlap is

$$P_n = \sum_{i=1}^n p_i.$$

If, instead, we assume each duration is uniformly distributed in $[0, t]$, the calculations become much more involved.

Assume the starting points of each duration, in order, are $x_1, x_2, \cdots, x_n$ and that the corresponding durations are $\lambda_1, \lambda_2, \cdots, \lambda_n$ where the λ_i's are some ordering of the t_i's. Previously, each different ordering had the same probability of occurring but now they could be all different. The probability of overlap for this particular ordering is given by

$$\prod_{i=1}^{n} (t - \lambda_i)^{-1} \int_0^{t-\lambda_1} dx_1 \int_{x_1}^{b_2} dx_2 \cdots \int_{x_{n-1}}^{b_n} dx_n ,$$

where $b_r = \min\{x_1 + \lambda_1 , t - \lambda_r\}$. The probability that the particular order $x_1 \leq x_2 \leq \cdots \leq x_n$ occurs is given by

$$\prod_{i=1}^{n} (t - \lambda_i)^{-1} \int \int_{x_1 \leq x_2 \leq \cdots \leq x_n} \int dx_1 \, dx_2 \cdots dx_n$$

$$= \prod_{i=1}^{n} (t - \lambda_1)^{-1} \int_0^{M_1} dx_1 \int_{x_1}^{M_2} dx_2 \cdots \int_{x_{n-1}}^{M_n} dx_n ,$$

where $M_r = \min_{r \leq i \leq n}\{t - \lambda_i\}$. The desired probability is now obtained by multiplying the latter two probabilities and summing over all orders. While this can be done easily enough for small n, the calculations are apparently too troublesome to be pursued for large n.

Subsequent editorial note. A published solution to this problem appeared prior to the appearance of the problem and solution in this Review as indicated in the following abstract from Mathematical Reviews, November, 1966, p. 1115: "Zubrzycki, S. A problem concerning simultaneous duration of several phenomena. (Polish, Russian and English summaries) Prace Mat. 7 (1962), 7-9

The author proves a conjecture formulated by M. Oleszkiewicz [same Prace 4 (1960), 1-7; MR 27 ⋕ 3006]. The theorem proved by the author reads as follows: Let n independent events start at moments randomly chosen in $0 < t < T$ under the assumption of a uniform probability distribution. Let each event last t_0 time units $(t_0 < T)$. Then the probability P_n that the n events will have a common interval of duration is given by

$$P_n = \frac{nTt_0^{n-1} - (n-1)t_0^n}{T^n} . "$$

Problem 75-8, *Accident Probability*, by L. J. DICKSON (Randwick, New South Wales, Australia).

The city has only one hospital and all ambulances are based there. Streets are so crowded that any ambulance on duty, whether going or returning, strikes and injures pedestrians at an average of one per mile. More precisely, an ambulance traveling over a stretch of Δx miles always has the probability $\exp(-\Delta x)$ of injuring nobody, and the probability zero of hitting more than one person at once. Each downed pedestrian is picked up by a different ambulance.

A man at a distance d miles from City Hospital has a heart attack and calls for an ambulance.

(a) Find the probability $P_k(d)$ that exactly k pedestrians will be injured by all the ambulances.
(b) What is the expected value of k?

Solution by I. D. S. TAYLOR and J. G. C. TEMPLETON (University of Toronto, Canada).

Since we are concerned only with the number of the injured, and not with their distribution in space and time, we may replace the original model by the following simpler one. All pedestrians are now located at the hospital entrance. The jth ambulance A_j on leaving the hospital draws a value of a random variable X_j with density $f_{X_j}(x) = e^{-x}$, where $X_1, X_2, \cdots$ are mutually independent. If $X_j < 2d$, the jth ambulance drives through the now-empty streets, returns to the hospital after X_j miles, hits a pedestrian at the hospital entrance, and leaves the system. The process terminates when for the first time an ambulance draws a value X_{k+1} which exceeds $2d$. This ambulance A_{k+1} is able to pick up the original heart patient and return to the hospital without injuring a pedestrian. Since there are k ambulances for the k injured pedestrians, and since each ambulance has a free path between injuries drawn from the correct exponential distribution, the number of injured pedestrians k will have the same distribution as in the original model.

The problem is now easy. $P[X_j < 2d] = 1 - e^{-2d}$ for $j = 1, 2, \cdots, k$. It follows that k has a geometric distribution, with

(a) $$P[k = r] = e^{-2d}(1 - e^{-2d})^r,$$

(b) $$E[k] = \sum_{r=1}^{\infty} r e^{-2d}(1 - e^{-2d})^r$$
$$= e^{-2d}(1 - e^{-2d})/[1 - (1 - e^{-2d})]^2 = e^{2d} - 1.$$

Problem 76-4, *Geometric Probability*, by IWAO SUGAI (Applied Physics Laboratory, Johns Hopkins University).

Two points are chosen at random, uniformly with respect to area, one each from the plane regions $0 \leq x^2 + y^2 \leq a^2$ and $(a-b)^2 \leq x^2 + y^2 \leq a^2$, respectively. Find the probability P that the distance between the two points is at most b $(0 < b < a)$.

This problem may have application in ASW sonar buoy deployment, triangulation tactics of disabled vehicles for automatic monitors [1], mutual visibility for satellite communications [2], and rocket re-entry impact distribution.

REFERENCES

[1] S. RITER, W. B. JONES, JR. AND H. DOZIER, *Speeding the deployment of emergency vehicles*, IEEE Spectrum, 10(1973), pp. 56–62.
[2] I. SUGAI, *Probability of isotropic link connectivity using comsats in elliptic orbits*, Proc. IEEE (Correspondence), 53(1965), pp. 541–542.

Solution by the proposer.

Let us choose the first point A along the x-axis in the interval $[a - b, a]$. Then the differential probability is given by

(1) $$dP = \frac{2\pi x \, dx}{\pi a^2} \cdot \frac{S(x)}{\pi a^2}$$

where $S(x)$ is the shaded area in the Fig. 1.

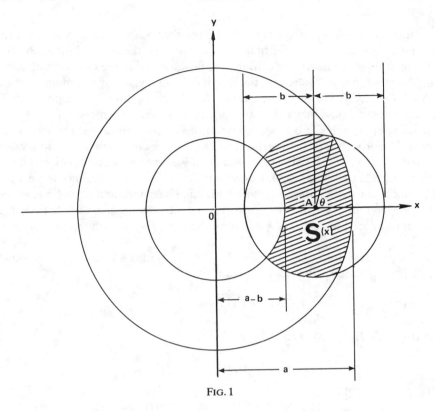

FIG. 1

For $x = a - b$, $\theta = 0$ and for $x = a$, $\theta = \theta_0 \equiv \sin^{-1} \sqrt{1 - k^2/4}$ where $k = b/a$. Whence,

$$(2) \qquad \pi^2 a^4 P = \int_{a-b}^{b} x S(x) \, dx.$$

By using the relations $x = a\Delta - b \cos \theta$ where $\Delta = \sqrt{1 - k^2 \sin^2 \theta}$, we obtain

$$(3) \qquad \begin{aligned} S(x) &= \pi(a^2 + 2b^2)/2 + (b^2 \sin \theta) \cos \theta - b^2\theta - ab\Delta \sin \theta - a^2 \sin^{-1} \Delta, \\ x \, dx &= \{ab(\Delta^2 + k^2 \cos^2 \theta) \sin \theta - b^2\Delta \sin 2\theta\} \, d\theta/\Delta. \end{aligned}$$

If $I(\theta)$ denotes the indefinite integral with respect to θ obtained by substituting (3) into (2), we get

$$\pi^2 a^4 P = I(\theta_0) - I(0).$$

The integration of $I(\theta)$ can be carried out by means of formulas contained in P. F. Byrd, M. D. Friedman, *Handbook of Elliptic Integrals for Engineers and Physicists*, Springer-Verlag, Berlin, 1951. Thus we find

$$I(\theta) = \frac{a^4}{2} [\sin^{-1}(k \sin \theta) - k\Delta \sin \theta] - \left[a\Delta \cos^2 \theta - \frac{b}{8} \sin 4\theta \right] b^3 \sin \theta$$

$$- \left[\left(\frac{\pi b}{2} \right)(a^2 + 2b^2) - \theta b^3 \right] \left[a\Delta \cos \theta - \frac{b}{2} \cos 2\theta \right]$$

$$+[a\Delta \cos \theta + b \sin^2 \theta]a^2 b \sin^{-1} \Delta.$$

In particular,

$$I(0) = \frac{\pi b^2}{4}[a^2 + 2b^2 - 4ab].$$

Whence,

$$P = \frac{N}{8\pi^2}$$

where

$$N = 4 \sin^{-1}(k\lambda) - 4k\beta\lambda + 2k^2[(2\beta + \lambda^2)\sin^{-1}\beta - \pi(\beta + 2)] + 8\pi k^3$$
$$- [3\pi + 4(\pi - \sin^{-1}\lambda)(\beta + 1)]k^4 - 2\beta k^5 + 2(\pi - \sin^{-1}\lambda - \lambda)k^6 + \lambda^2 k^7,$$
$$\beta = \sqrt{1 - k^2\lambda^2}, \qquad \lambda = \sqrt{1 - (k^2/4)}.$$

One can also show that

$$\bar{S} = \int_{a-b}^{a} xS(x)\, dx \bigg/ \int_{a-b}^{a} x\, dx = 2\pi^2[I(\theta_0) - I(\theta)][a^2 - (a-b)^2]^{-1}$$

and that

$$\bar{x} = \int_{a-b}^{a} xS(x)\, dx \bigg/ \int_{a-b}^{a} S(x)\, dx$$

can be expressed in terms of elliptic integrals.

Expected Values for Random Regions of a Circle

Problem 78-13*, *by* T. D. ROGERS (University of Alberta).

Given n points distributed uniformly in the unit circle, associate with each such point the region in the circle whose points are closer to it than the remaining $n-1$ *a priori* given points. If $A_1 \leq A_2 \leq \cdots \leq A_n$ is the ordered enumeration of the areas of these regions, what are the expected values of the A_i's? The result even in the case $n = 2$ is not known. This problem and related ones have biological applications to problems involving intraspecific competition and territoriality (see [1]).

REFERENCE

[1] J. E. LEWIS AND THOMAS D. ROGERS, *Some remarks on S-mosaics*, Lecture Notes in Biomathematics, vol. 2, Springer-Verlag, New York, 1974, pp. 146–151.

Partial solution by D. M. BOROSON (Massachusetts Institute of Technology, Lincoln Laboratory).

Suppose the two randomly chosen points are $\mathbf{w}_i = (x_i, y_i)$, $i = 1, 2$. Let $\mathbf{p}$ denote the vector $\overline{AB}$ where A and B denote, respectively, the point $\frac{1}{2}(\mathbf{w}_1 + \mathbf{w}_2)$ and the closest point to the origin on the perpendicular bisector of the line segment joining $\mathbf{w}_1$ and $\mathbf{w}_2$. The distance a of this bisector from the origin clearly satisfies the relation

$$\frac{\mathbf{w}_1 + \mathbf{w}_2}{2} = a \frac{\mathbf{w}_1 - \mathbf{w}_2}{|\mathbf{w}_1 - \mathbf{w}_2|} + \mathbf{p}.$$

Thus,

$$a = \frac{|(\mathbf{w}_1 - \mathbf{w}_2) \cdot (\mathbf{w}_1 + \mathbf{w}_2)|}{2|\mathbf{w}_1 - \mathbf{w}_2|}.$$

Since $\mathbf{w}_1$ and $\mathbf{w}_2$ were chosen randomly, the probability density function of the pair is

$$f(\mathbf{w}_1, \mathbf{w}_2) = \frac{1}{\pi^2} I(|\mathbf{w}_1| \le 1) I(|\mathbf{w}_2| \le 1)$$

where

$$I(A) = \begin{cases} 1, & A \text{ is true,} \\ 0, & A \text{ is false.} \end{cases}$$

Now performing the random variable transformation

$$\mathbf{z}_1 = \frac{\mathbf{w}_1 - \mathbf{w}_2}{2},$$

$$\mathbf{z}_2 = \frac{\mathbf{w}_1 + \mathbf{w}_2}{2},$$

and then setting

$$\mathbf{z}_i = (r_i \cos \theta_i, \; r_i \sin \theta_i), \qquad i = 1, 2,$$

we find that the new probability density function is

$$f(r_1, r_2, \theta_1, \theta_2) = k_1 r_1 r_2 I(r_1^2 + r_2^2 + 2 r_1 r_2 |\cos (\theta_1 - \theta_2)| \le 1)$$
$$\cdot I(r_1, r_2 \ge 0) I(\theta_1, \theta_2 \varepsilon [0, 2\pi))$$

for some normalizing constant k_1 and

$$a = r_2 |\cos (\theta_1 - \theta_2)|.$$

Transforming to the variables (r_1, r_2, θ_1, a) requires the Jacobian $|J| = \sqrt{r_2^2 - a^2}$ and we have, after integrating out θ_1,

$$f(a, r_1, r_2) = k_2 \frac{r_1 r_2}{\sqrt{r_2^2 - a^2}} I((r_1 + a)^2 + (r_2^2 - a^2) \le 1) I(r_2^2 \ge a^2) I(r_1 \ge 0).$$

Therefore, the marginal density of a is

$$f(a) = k_2 \int_0^{1-a} r_1 \int_a^{\sqrt{1+a^2-(r_1+a)^2}} \frac{r_2}{\sqrt{r_2^2-a^2}} \, dr_2 \, dr_1$$

$$= k_3 \int_0^{1-a} r_1 \int_0^{1-(r_1+a)^2} \frac{dt}{\sqrt{t}}$$

$$= k_4 \int_0^{1-a} r_1 \sqrt{1-(r+a)^2} \, dr_1$$

$$= k_4 \int_a^1 (r_1-a)\sqrt{1-r_1^2} \, dr_1.$$

Now, the area of the smaller of the two regions on either side of the bisector is seen to be

$$A_1(a) = 2 \int_a^1 \int_0^{\sqrt{1-y^2}} dx \, dy$$

$$= 2 \int_a^1 \sqrt{1-y^2} \, dy.$$

Also,

$$\frac{d}{da} f(a) = -\frac{k_4}{2} A_1(a).$$

Therefore, the expected value of the area is

$$EA_1(a) = -\frac{2}{k_4} \int_0^1 f(a) f'(a) \, da$$

$$= \frac{1}{k_4} f^2(0) = k_4 \left[\int_0^1 r\sqrt{1-r^2} \, dr \right]^2 = \frac{k_4}{9}.$$

Finally, letting $g(a) = f(a)/k_4$ and integrating by parts:

$$\frac{1}{k_4} = \int_0^1 g(a) \, da$$

$$= -\int_0^1 a g'(a) \, da$$

$$= \int_0^1 a \int_a^1 \sqrt{1-r^2} \, dr \, da$$

$$= \frac{1}{2} \int_0^1 r^2 \sqrt{1-r^2} \, dr = \frac{1}{4} B(\tfrac{3}{2}, \tfrac{3}{2}) = 2^{-5}\pi.$$

Hence,

$$EA_1 = \frac{32}{9\pi}$$

which is approximately 36.025% of the unit circle.

Problem 67–15*, Probability Distribution of a Network of Triangles, by MARY
BETH STEARNS (Ford Scientific Laboratory).

A domain wall in ferromagnetic materials can be considered as a two-dimensional membrane which, when subjected to an r.f. field, will oscillate in a manner
determined by the boundary conditions. One possible set of boundary conditions
would correspond to pinning the wall at impurities whose positions are random
in the wall. In describing the wall motion, we must know the area distributions
of triangles formed from three impurity sites. These triangles will contain no
other impurity pinning points in their interior and will be called "good" triangles.
What is the probability distribution of the areas of the resulting network of
"good" triangles formed by choosing N points distributed uniformly in a given
area?

Solution by ROGER E. MILES (The Australian National University, Canberra).

N random points ("particles") are distributed independently and uniformly
in the subset X, of two-dimensional Lebesgue measure $|X|$, of the plane E_2
($N \geq 3$). What is the probability distribution of the "good," or *empty*, triangles so
formed? Empty triangles are triangles with particle vertices containing none of
the remaining $N - 3$ particles within their interiors.

For a triangle t, let A denote its area and τ, some additional partial "description"; as extreme cases, τ may be void or may, together with A, completely
specify t. Then two equally natural interpretations of the distribution are (i) as
the distribution of (A, τ) for a random one of the *random* a.s. (almost surely) positive
number M of empty triangles so formed; and (ii) as the limiting empiric distribution, as $n \to \infty$, of (A, τ) for *all* empty triangles formed in n independent repetitions
of the experiment. Let the corresponding p.d.f.'s be $f_j^{(0)}(A, \tau)$, $j = 1, 2$. If $p_m = P(M$
$= m)$ and $g(A, \tau|M)$ is the p.d.f. of (A, τ) for a random empty triangle conditional on
$M = m$, then

(1)
$$f_1^{(0)}(A, \tau) = \sum_{m=1}^{\binom{N}{3}} p_m g(A, \tau|m),$$

and it may be shown that

(2)
$$f_2^{(0)}(A, \tau) = \sum_{m=1}^{\binom{N}{3}} m p_m g(A, \tau|m) \bigg/ \sum_{m=1}^{\binom{N}{3}} m p_m.$$

Thus (i) and (ii) are not the same. Both $f_1^{(0)}$ and $f_2^{(0)}$ depend on both N and the
shape of X, and, apart from special choices of X with small N values, their evaluation seems a hopeless undertaking. To illustrate this, let $h(A, \tau)$ be the p.d.f. for the
single triangle formed when $N = 3$. Then, taking X to be convex, use of the "Bayes'
formula" $f(A, \tau|C)P(C) = P(C|A, \tau)f(A, \tau)$ yields

(3)
$$f_2^{(0)}(A, \tau) \propto \{1 - (A/|X|)\}^{N-3} h(A, \tau).$$

However, knowledge simply of $h(A)$ would solve Sylvester's difficult classical

problem of determining the probability $1 - \left(\int Ah(A)\,dA/|X| \right)$ that, when $N = 4$,

the particles form a convex quadrilateral (see [2, pp. 42–46]).

The following approach overcomes both the ambiguity of definition and the shape dependence by supposing X replaced by E_2 itself, i.e., a homogeneous Poisson point (= particle here) process $\mathscr{P}$ in E_2 with intensity $\rho = N/|X|$. Then the number of particles in $Y \subset E_2$ has a Poisson distribution with mean $\rho|Y|$, and the numbers in disjoint subsets are independent. Let $\mathscr{T}$ denote the aggregate of triangles determined by ordered particle-triples of $\mathscr{P}$, and $\mathscr{T}^{(0)}$ the subaggregate of empty members of $\mathscr{T}$. A triangle in E_2 may be parametrized by

(4) $\qquad t = (x, y, \theta, l, \lambda, \alpha), \qquad (x, y) \in E_2, \quad 0 < l, \lambda < \infty, \quad 0 \le \theta, \alpha < 2\pi,$

so that its vertices are $(x, y), (x + l \cos \theta, y + l \sin \theta)$ and $(x + \lambda l \cos \overline{\theta + \alpha},$ $y + \lambda l \sin \overline{\theta + \alpha})$. Write δt for a small six-dimensional interval at the point t in t-space with sides $\delta x, \cdots, \delta \alpha$. Routine calculations, utilizing the extreme independence properties of $\mathscr{P}$, show that

(5) $\qquad P(\text{there exists a member of } \mathscr{T} \text{ in } \delta t) = (\rho l)^3 \lambda |\delta t| + o(|\delta t|).$

(x, y), θ, l and (λ, α) represent the position, orientation, size and shape of t, respectively. Set $\xi = (\theta, l, \lambda, \alpha)$. If $\#_X(\delta \xi)$ is the number of members of $\mathscr{T}$ with $(x, y) \in X \subset E_2$ and within orientation/size/shape limits $\delta \xi$ (a small four-dimensional interval), then clearly, by (5),

(6) $\qquad E\#_X(\delta \xi) = |X|(\rho l)^3 \lambda |\delta \xi| + o(|\delta \xi|).$

In fact, by ergodic theory it may be proved [6] that

(7) $\qquad \{\#_X(\delta \xi) - E\#_X(\delta \xi)\}/|X| \overset{\text{a.s.}}{\to} 0$

as $|X| \to \infty$. Equations (6) and (7) imply that

(8) $\qquad f(\xi) \propto l^3 \lambda$

expresses the *ergodic relative frequency* (e.r.f.) of the members of $\mathscr{T}$. It is unique up to a constant factor. Equation (8) indicates that θ, l, λ and α are *ergodically independent*, with $f(\theta) \propto 1$, $f(l) \propto l^3$, $f(\lambda) \propto \lambda$ and $f(\alpha) \propto 1$. If an e.r.f. is normalizable, then we may speak of the corresponding *ergodic probability distribution* (e.p.d.). Thus θ and α, unlike l and λ, both possess e.p.d.'s; these distributions are both uniform.

Let now A and α, β, γ denote the area and angles of t (with β opposite the side of length l), and set $\xi' = (\theta, A, \beta, \gamma)$. Note that each "$\xi$ triangle" corresponds to two distinct "ξ' triangles," symmetrical about the side of length l. Routine manipulations show that on each branch the Jacobian

(9) $\qquad |\partial \xi/\partial \xi'| = (2A \sin \alpha \sin \beta \sin \gamma)^{-1/2}(\sin \alpha/\sin \beta).$

Equations (6), (7) and (9) imply that

(10) $\qquad \begin{aligned} \#_X(\delta \xi')/|X| &\overset{\text{a.s.}}{\to} 2(\rho l)^3 \lambda |\partial \xi/\partial \xi'| \, |\delta \xi'| + o(|\delta \xi'|) \\ &= (4\rho^3 A/(\sin \alpha \sin \beta \sin \gamma))|\delta \xi'| + o(|\delta \xi'|) \end{aligned}$

as $|X| \to \infty$. This yields an alternative form (cf. (8))

(11) $\qquad f(\xi') \propto A/(\sin (\beta + \gamma) \sin \beta \sin \gamma)$

of the e.r.f. of $\mathscr{T}$. θ, A and (β, γ) are ergodically independent. Only θ has an e.p.d., which is uniform. Intuitively, the symmetry of $f(\xi')$ in α, β and γ is to be expected.

Turning now to $\mathcal{T}^{(0)}$, and again utilizing the powerful Poisson independence, we may similarly show (cf. (10)) that

(12) $\#_X^{(0)}(\delta\xi')/|X| \overset{\text{a.s.}}{\to} e^{-\rho A} \cdot (4\rho^3 A/(\sin\alpha\sin\beta\sin\gamma))|\delta\xi'| + o(|\delta\xi'|)$

as $|X| \to \infty$. In this case, as one might expect, the e.r.f. of A is normalizable. The ergodic independence of θ, A and (β, γ) is maintained, with

(13) $f^{(0)}(\theta) = (2\pi)^{-1}$ (uniform),

(14) $f^{(0)}(A) = \rho^2 A e^{-\rho A}$ (gamma; $2\rho A$ is χ^2 with 4 d.f.),

(15) $f^{(0)}(\beta, \gamma) \propto \{\sin(\beta + \gamma)\sin\beta\sin\gamma\}^{-1}$.

Equation (15) indicates that "almost all" triangles of $\mathcal{T}^{(0)}$ have angles 0, 0 and π. However, this does not vitiate the role of $f^{(0)}(A)$ as the e.p.d.f. of A for members of $\mathcal{T}^{(0)}$ within given arbitrary orientation/shape limits.

Returning to the original problem, $f^{(0)}(A)$ is clearly the common limit of both $f_1^{(0)}(A)$ and $f_2^{(0)}(A)$ as $N, |X| \to \infty$ such that $N/|X| = \rho$. Thus $f^{(0)}(A)$ may be expected to furnish a good approximation to both these p.d.f.'s provided N is sufficiently large and X is sufficiently "rotund," so that, among other things, edge effects on the boundary of X are negligible. Of course, its tail at ∞ is to be disregarded, certainly for $A > |X|$. However, the probability in the tail beyond $|X|$ will anyway usually be small. Again, $f^{(0)}(\beta, \gamma)$ for values of α, β and γ well away from zero will also, under similar conditions, furnish a good approximation to the relative frequency of empty triangle shapes in X.

From an analytic viewpoint, it turns out to be preferable to define an empty, or *Delaunay, triangle* as one whose circum-circle is empty. The Delaunay triangles (with respect to $\mathcal{P}$) fit together to cover E_2 without overlapping. The set of points closer to a particle than any other particle is a *Voronoi* (convex) *polygon*. The Voronoi polygons, clearly, similarly partition E_2. The two aggregates are dual in the sense that, for each particle, the number of sides of its associated Voronoi polygon equal the number of Delaunay triangles with the particle as vertex (see [1], [6] and [7, Chaps. 7, 8]). In fact, in his elegant solution [3] of the closely related Problem 62–8, Martin utilized the Delaunay triangles determined by a *finite* random point set.

The author is at present engaged in writing a series of papers on Poisson s-flats in E_d, $\mathcal{P}$ being the special case $s = 0$, $d = 2$ (this case is treated in much greater detail in [6]). See [5] for the statement of a general theorem on Poisson ergodic distributions and [4] for a summary of the main results when $s = 1$, $d = 2$. $s = 0$, $d = 1$ is of course the standard linear Poisson process.

REFERENCES

[1] E. N. Gilbert, *Random subdivision of space into crystals*, Ann. Math. Statist., 33 (1962), pp. 958–972.

[2] M. G. Kendall and P. A. P. Moran, *Geometrical Probability*, Hafner, New York, 1963.

[3] J. L. Martin, *Solution to Problem 62–8*, this Review, 7 (1965), pp. 132–134.

[4] R. E. Miles, *Random polygons determined by random lines in a plane*, Proc. Nat. Acad. Sci. U.S.A., 52 (1964), pp. 901–907 and 1157–1160.

[5] ———, *A wide class of distributions in geometrical probability*, Ann. Math. Statist., 35 (1964), pp. 1407–1415.

[6] ———, *On the homogeneous planar Poisson point process*, Math. Biosci., in press.

[7] ———, C. A. Rogers, *Packing and covering*, Math. Tract No. 54, Cambridge Univ. Press, Cambridge, 1964.

Comment by JOHN L. MARTIN (University of London King's College).

From the described physical situation, I believe that the proposer really wants to use "good" in the sense of Problem 62–8. And in this case the expected number of good triangles (assuming the region is convex and $N \to \infty$) follows from an elementary integration of

$$\frac{2\pi ab}{l^4} \exp\{-\tfrac{1}{2}ab \sin \theta\} \, da \, db$$

(in the notation of Problem 62–8).

An Average Maximum Distance

Problem 89-8*, by PETER SENN (ETH-Zentrum, Zurich, Switzerland).

Consider a cube with edges of length $2a$ whose center is located on a given plane. We now randomize the orientation of the cube relative to the plane keeping its center fixed. Let the resulting distances of the eight vertices of the cube from the plane be $d_1, d_2, \cdots, d_8$ and let

$$D = \max(d_1, d_2, \cdots, d_8).$$

It follows easily that $a \leq D \leq a\sqrt{3}$. Prove or disprove that the expected value for D is $3a/2$. This value agrees with random walk simulations.

The problem has arisen in the design for a Monte Carlo algorithm for computing areas of irregularly shaped surfaces.

Solution by JOHN G. WATSON (Applied Mathematics, Menlo Park, California).

The following calculation gives a formula for the average maximum distance of an n-dimensional cube. The formula confirms the three-dimensional conjecture. For this calculation, we take the cube to be fixed, and average over all possible orientations of a plane passing through its center.

Consider the n-dimensional cube centered at the origin and with edge length $2a$. We position the cube so that its vertices are generated by the n-dimensional vector $a(\pm 1, \pm 1, \cdots, \pm 1)$ using each of the 2^n possible choices of signs. Generally, the distance between any point and a plane through the origin with unit normal $\mathbf{u}$ is equal to the magnitude of the dot product of $\mathbf{u}$ and the position vector from the origin to the point. In particular, the distance between a vertex and such a plane is given by

$$d = a \, | \pm u_1 \pm u_2 \pm \cdots \pm u_n |,$$

where the us are the components of the unit vector $\mathbf{u}$, i.e., they are direction cosines. The maximum distance is equal to

$$D(\mathbf{u}) = a(| u_1 | + | u_2 | + \cdots + | u_n |),$$

since a vertex can be chosen to reinforce the signs of the components of $\mathbf{u}$. We note that $a \leq D \leq a\sqrt{n}$.

The expected value of D is obtained by averaging over all planar orientations $\mathbf{u}$. The realizations of $\mathbf{u}$ cover the n-dimensional unit sphere uniformly, and the density distribution of $\mathbf{u}$ is given by dA/S_n, where dA is an element of area and S_n is the total surface area of the n-dimensional unit sphere. Hence the expected value of D is

$$\langle D \rangle = \frac{1}{S_n} \int D(\mathbf{u}) \, dA$$

$$= \frac{an}{S_n} \int |\mathbf{k} \cdot \mathbf{u}| \, dA,$$

where the second equality follows from the first by symmetry and $\mathbf{k}$ is any conveniently chosen unit vector.

The plane through the center of the unit sphere with normal $\mathbf{k}$ splits the plane into two equal hemispheres. Now consider the projection of either hemisphere onto the plane. The integral appearing in the second expression for $\langle D \rangle$ is equal to twice the area of this projection A_p. The projection is the $(n-1)$-dimensional unit sphere and it follows that $A_p = V_{n-1}$, where

$$V_m = \frac{\pi^{m/2}}{\Gamma((m/2)+1)}$$

is the volume of the m-dimensional unit sphere. Since $S_n = nV_n$, the expected value of D simplifies to

$$\langle D \rangle = \frac{2anV_{n-1}}{S_n}$$

$$= \frac{2aV_{n-1}}{V_n}$$

$$= \frac{2a}{\sqrt{\pi}} \frac{\Gamma((n+2)/2)}{\Gamma((n+1)/2)}.$$

In particular, $\langle D \rangle = 4a/\pi$ for $n = 2$ and $\langle D \rangle = 3a/2$ for $n = 3$.

Problem 75-12, An Average Distance, by H. J. OSER (National Bureau of Standards).

Evaluate the 4-fold integral

$$F = \int_0^1 \int_{-1}^0 \int_{-1/2}^{1/2} \int_{-1/2}^{1/2} \{(x_1 - x_2)^2 + (y_1 - y_2)^2\}^{1/2} \, dx_1 \, dx_2 \, dy_1 \, dy_2$$

which gives the average distance between points in two adjacent unit squares.

Editorial note. The proposer notes that the problem was suggested by C. R. Johnson and that the result should be of interest to workers in transportation modeling and similar fields.

Comment by J. D. MURCHLAND (University College, London, England).

Ghosh [1] gives the average distance between points in two unit squares, adjacent like a black square and a white square on the chessboard, as 1.088, between squares adjacent like two black squares, as 1.473, within a unit square as 0.521 and within an oblong formed of two unit squares as 0.805. In fact, he gives the actual probability density function for the first two problems and indicates its form in some detail for the general case of two rectangles. He also provides the explicit density function and first four moments for two points within a general rectangle.

Ghosh's paper is included in the bibliography of Kendall and Moran's book [2]. They give the average distance and density function between two points within the circular disc of unit area, 0.511, and refer to Borel [3] for the triangle, square and general convex polygons.

M. J. de Smith of this College has available a bibliography of some thirty more recent papers on this topic, though the problem of greatest interest to transportation modelers apparently remains unsolved. This is the average distance between points in two nonoverlapping circular discs, as a function of their radii and center to center distance.

REFERENCES

[1] B. GHOSH, *Random distances within a rectangle and between two rectangles*, Bull. of the Calcutta Math. Soc., 43 (1951), pp. 17–24.
[2] M. G. KENDALL AND P. A. P. MORAN, *Geometrical Probability*, Griffin, London, 1963, pp. 41–42.
[3] E. BOREL, Principes et formules classique du Calcul des Probabilités. Traité du Calcul des Probabilités et de ses Applications, tom. 1, fasc. 1, Gauthier-Villars, Paris, 1925.

Solution by D. J. DALEY (Australian National University, Canberra, Australia).

Let $D(1, a)$ denote the mean distance between two points located independently and with uniform probability distributions over a rectangle with sides of length 1 and a. Then since

$$(1) \qquad D(1, 2) = \tfrac{1}{2}[D(1, 1) + F],$$

we can find F from the more general and more simply computed

$$(2) \qquad D(1, a) = 4a^{-2} \int_0^1 (1-v)\, dv \int_0^a (a-u)(u^2+v^2)^{1/2}\, du.$$

This double integral equals $E((U^2+V^2)^{1/2})$, where U and V are independent random variables with symmetric triangular probability distributions on $(-a, a)$ and $(-1, 1)$, respectively (the factor 4 at (2) comes from the symmetry). Either by direct evaluation or by reference to tables of integrals (e.g., § 230 of Dwight [2]), the inner integral at (2) equals

$$\tfrac{1}{2}a\{a(a^2+v^2)^{1/2} + v^2 \log[(a+(a^2+v^2)^{1/2})/v]\} - \tfrac{1}{3}\{(a^2+v^2)^{3/2} - v^3\}.$$

The second integration (using, e.g., § 635 of [2], writing $r = (a^2+1)^{1/2}$, and including the factor $4a^{-2}$) yields on simplification

$$(3) \qquad \begin{aligned} D(1, a) &= \{1 + a^5 + (a^2 - (a^2-1)^2)r\}/15a^2 \\ &\quad + \{a^3 \log((1+r)/a) + \log(a+r)\}/6a. \end{aligned}$$

Substituting $a = 1$ and $a = 2$ and using (1) then leads to

$$(4) \qquad \begin{aligned} F &= (29 - 2\sqrt{2} - 5\sqrt{5})/30 + (8 \log((1+\sqrt{5})/2) + \log(2+\sqrt{5}) - 2\log(1+\sqrt{2}))/6 \\ &\approx 1.08814. \end{aligned}$$

Remarks. 1. By using the identity

(5) $\qquad (a+b)^2 D(1, a+b) = a^2 D(1, a) + b^2 D(1, b) + 2ab D_{ab},$

we can find the mean distance D_{ab} between two points distributed uniformly over two adjacent rectangles with sides of length 1 and a, and 1 and b, respectively. A further identity then gives the mean distance between points in two such rectangles separated by a third rectangle of length c. The ultimate extension of mean distances obtainable by such a decomposition method is for points in rectangles with sides parallel to the axes such that the projections of the sides parallel to (say) the x-axis, $(0, a)$ and (b, c) say, must have both b/a and c/a rational (i.e., 0, a, b and c lie on a lattice); the location and side lengths in the other direction can be arbitrary.

2. $D(1, 1)$ is given in [1] (the expression for M_2 there).

3. Let the random variable Y with mean μ and variance σ^2 define another random variable X by the relation $X = f(Y)$ for some function f. Then (see, e.g., [1] for comment and references)

(6) $\qquad E(X) \approx f(\mu) + \tfrac{1}{2}\sigma^2 f''(\mu).$

With $f(y) = y^{1/2}$, application of this Taylor series approximation method gives

$$D(1, a) = E((U^2 + V^2)^{1/2}) \approx \{(1 + a^2)/6\}^{1/2}\{1 - 7(1 + a^4)/40(1 + a^2)^2\},$$

and for points distributed independently and uniformly in unit squares with centers at the origin and (c, d), the mean distance between them equals

$$E(\{(c + V_1)^2 + (d + V_2)^2\}^{1/2}) \approx (R^2 + \tfrac{1}{3})^{1/2}\{1 - (7 + 60R^2)/80(1 + 3R^2)^2\},$$
$$R^2 \equiv c^2 + d^2,$$

where V_1 and V_2 are distributed independently and identically as V. In the case of F, $R^2 = 1$, and the approximation gives

$$F \approx (4/3)^{1/2}(1 - 67/1280) = 1.09426.$$

4. In view of the results in [1], it seems unlikely that explicit closed form expressions exist for mean distances between points distributed randomly within higher-dimensional (≥ 3) hypercuboids.

REFERENCES

[1] R. S. ANDERSSEN, R. P. BRENT, D. J. DALEY AND P. A. P. MORAN, *Concerning* $\int_0^1 \cdots \int_0^1 (x_1^2 + \cdots + x_k^2)^{1/2} dx_1 \cdots dx_k$ *and a Taylor series method*, SIAM J. Appl. Math., 30 (1976), pp. 22–30.

[2] H. B. DWIGHT, *Tables of Integrals and Other Mathematical Data*, 4th ed., Macmillan, New York, 1961.

R. J. VAUGHAN (University of Newcastle, New South Wales, Australia) solves the problem by first establishing the probability density function. He also gives the following traffic engineer's method for approximating the solution plus a number of references to average distances with respect to circles.

The "route factor" is defined as

$$R_f = \frac{\text{average distance by a given route}}{\text{average direct distance}}.$$

We know that the average rectangular distance between two random points in adjacent squares is $\tfrac{4}{3}$ (average horizontal distance = 1, average vertical distance = $\tfrac{1}{3}$). Hence, if we can estimate the route factor for rectangular routing between adjacent squares, R_a, then we can estimate the average direct distance. Consider R_s, the route factor for rectangular routing in the same square. Now

$$R_s = \frac{2/3}{.5214} = 1.279.$$

It is interesting to note that provided the two points are selected at random from the same region, then the route factor for rectangular routing is always approximately 1.27.

Now to find R_a, we must modify R_s to adjust for the fact that the horizontal distance is on average greater than the vertical distance. Consider the worst case, in each case:

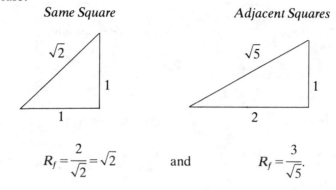

| Same Square | Adjacent Squares |

Here, $R_f = \dfrac{2}{\sqrt{2}} = \sqrt{2}$ and $R_f = \dfrac{3}{\sqrt{5}}.$

Hence

$$R_a \approx R_s \times \frac{1}{\sqrt{2}} \times \frac{3}{\sqrt{5}} = 1.213.$$

Therefore the average direct distance between two random points in adjacent squares $\approx \frac{4}{3} \times (1/1.213) \approx 1.10$, the exact value being 1.088.

REFERENCES

[1] H. G. APSIMON, *Mathematical note 2754—A repeated integral*, Math. Gaz., 42 (1958), p. 52.
[2] D. FAIRTHORNE, *The distance between random points in two circles*, Biometrika, 51 (1964), pp. 275–277.
[3] F. GARWOOD AND J. C. TANNER, *Note 2800 on Note 2754—A repeated integral*, Math. Gaz., 42 (1958), pp. 292–293.

Problem 78-8, Average Distance in a Unit Cube, by TIMO LEIPÄLÄ (University of Turku, Turku, Finland).

Determine (a) the probability density, (b) the mean, and (c) the variance for the Euclidean distance between two points which are independently and uniformly distributed in a unit cube.

The numerical value of the mean is given in [1] and it is conjectured in [3] that an explicit closed form expression for it does not exist. The probability distribution for the distance in the interval [0, 1] is given in [2].

REFERENCES

[1] R. S. ANDERSSEN, R. P. BRENT, D. J. DALEY AND P. A. P. MORAN, *Concerning* $\int_0^1 \cdots \int_0^1 (x_1^2 + \cdots + x_k^2)^{1/2} dx_1 \cdots dx_k$ *and a Taylor series method*, SIAM J. Appl. Math., 30 (1976), pp. 22–30.
[2] É. BOREL, *Principes et Formules Classiques du Calcul des Probabilités*, Gauthier-Villars, Paris, 1925, pp. 88–90.
[3] D. J. DALEY, *Solution to problem* 75-12, this Review, 18 (1976), pp. 498–499.

Solution by the proposer.

(a) *The probability density.* The distribution of distance of two points indepen-dently and uniformly distributed in the interval $[0, 1]$ is $1 - (1-x)^2$. Because its probability density is $2(1-x)$, we obtain the desired probability density in the unit cube as

$$f(z) = 8 \iiint_{x \in C} (1-x_1)(1-x_2)(1-x_3)\, dx_1\, dx_2\, dx_3$$

where

$$C = \{x \in \mathbb{R}^3 \mid x_1^2 + x_2^2 + x_3^2 = z^2, 0 \le x_i \le 1, i = 1, 2, 3\}.$$

Transferring to three-dimensional polar coordinates and writing

$$h(\phi, \psi) = (1 - z \cos \phi \cos \psi)(1 - z \cos \phi \sin \psi)(1 - z \sin \psi),$$

we can write the density

$$f(z) = \begin{cases} 8z^2 \displaystyle\int_0^{\pi/2} \cos \phi \int_0^{\pi/2} h(\phi, \psi)\, d\phi\, d\psi, & 0 \le z \le 1, \\[2ex] 8z^2 \displaystyle\int_0^{\arccos 1/z} \cos \phi \int_{\arccos(1/(z \cos \phi))}^{\arcsin(1/(z \cos \phi))} h(\phi, \psi)\, d\phi\, d\psi \\[2ex] \qquad + 8z^2 \displaystyle\int_{\arccos 1/z}^{\arcsin 1/z} \cos \phi \int_0^{\pi/2} h(\phi, \psi)\, d\phi\, d\psi, & 1 \le z \le \sqrt{2}, \\[2ex] 8z^2 \displaystyle\int_{\arccos \sqrt{2}/z}^{\arcsin 1/z} \cos \phi \int_{\arccos(1/(z \cos \phi))}^{\arcsin(1/(z \cos \phi))} h(\phi, \psi)\, d\phi\, d\psi, & \sqrt{2} \le z \le \sqrt{3}. \end{cases}$$

Performing the integration, we obtain finally

$$f(z) = \begin{cases} 4\pi z^2 - 6\pi z^3 + 8z^4 - z^5, & 0 \le z \le 1, \\[2ex] (6\pi - 1)z - 8\pi z^2 + 6z^3 + 2z^5 - 8z(1 + 2z^2)\sqrt{z^2 - 1} + 24z^3 \arccos \dfrac{1}{z}, \\[1ex] \hspace{8cm} 1 \le z \le \sqrt{2}, \\[2ex] (6\pi - 5)z + 6(\pi - 1)z^3 - z^5 + 8z(1 + z^2)\sqrt{z^2 - 2} - 16z^2 \arcsin \dfrac{1}{z^2 - 1} \\[2ex] \qquad - 24(1 + z^2)\arccos \dfrac{1}{\sqrt{z^2 - 1}} + 16z^2 \arccos \dfrac{z}{\sqrt{2z^2 - 2}}, & \sqrt{2} \le z \le \sqrt{3}. \end{cases}$$

(b) *The mean.* Using the probability density just obtained, we get the mean

$$\mu = \tfrac{1}{105}[4+17\sqrt{2}-6\sqrt{3}-7\pi+21 \ln[(1+\sqrt{2})(7+4\sqrt{3})]] \approx 0.66171.$$

(c) *The variance.* The value $\tfrac{1}{2}$ of the second moment is directly obtained without using the density $f(z)$. Thus, the variance is

$$\sigma^2 = \frac{1}{2} - \frac{1}{105^2}[4+17\sqrt{2}-6\sqrt{3}-7\pi+21 \ln[(1+\sqrt{2})(7+4\sqrt{3})]]^2 \approx 0.06214.$$

(d) *The probability distribution.* Using $f(z)$ it is also possible to obtain the probability distribution of the distance:

$$F(z) = \begin{cases} \dfrac{4\pi}{3}z^3 - \dfrac{3\pi}{2}z^4 + \dfrac{8}{5}z^5 - \dfrac{1}{6}z^6, & 0 \le z \le 1, \\[2mm] \dfrac{1}{10} - \dfrac{\pi}{2} + \left(3\pi - \dfrac{1}{2}\right)z^2 - \dfrac{8\pi}{3}z^3 + \dfrac{3}{2}z^4 + \dfrac{1}{3}z^6 + \dfrac{2}{5}(2-9z^2-8z^4)\sqrt{z^2-1} \\[2mm] \qquad\qquad\qquad\qquad\qquad +6z^4 \operatorname{arc cos}\dfrac{1}{z}, & 1 \le z \le \sqrt{2}, \\[2mm] \dfrac{9}{10} - \dfrac{\pi}{2} + \left(3\pi - \dfrac{5}{2}\right)z^2 + \dfrac{3}{2}(\pi-1)z^4 - \dfrac{1}{6}z^6 + \dfrac{2}{5}(1+9z^2+4z^4)\sqrt{z^2-2} \\[2mm] \qquad +2(1-6z^2-3z^4)\operatorname{arc cos}\dfrac{1}{\sqrt{z^2-1}} - \dfrac{16}{3}z^3 \operatorname{arc sin}\dfrac{1}{z^2-1} \\[2mm] \qquad +\dfrac{16}{3}z^3 \operatorname{arc cos}\dfrac{z}{\sqrt{2z^2-2}}, & \sqrt{2} \le z \le \sqrt{3}. \end{cases}$$

The value of $F(z)$ in the interval $[0, 1]$ is given in [2].

(e) *The distribution of the distance from the corner.* If we perform the same calculations with $h(\phi, \psi) = \tfrac{1}{8}$, we obtain results for the Euclidean distances between the corner of the square and a point uniformly distributed in it. Thus, we obtain the probability density

$$f(z) = \begin{cases} \dfrac{\pi}{2}z^2, & 0 \le z \le 1, \\[2mm] \dfrac{3\pi}{2}z - \pi z^2, & 1 \le z \le \sqrt{2}, \\[2mm] \dfrac{3\pi}{2}z - 6z \operatorname{arc cos}\dfrac{1}{\sqrt{z^2-1}} - 2z^2 \operatorname{arc sin}\dfrac{1}{z^2-1} + 2z^2 \operatorname{arc cos}\dfrac{z}{\sqrt{2z^2-2}}, \\ & \sqrt{2} \le z \le \sqrt{3}; \end{cases}$$

the probability distribution

$$F(z) = \begin{cases} \dfrac{\pi}{6}z^3, & 0 \le z \le 1, \\[2mm] -\dfrac{\pi}{4} + \dfrac{3\pi}{4}z^2 - \dfrac{\pi}{3}z^3, & 1 \le z \le \sqrt{2}, \\[2mm] -\dfrac{\pi}{4} + \dfrac{3\pi}{4}z^2 + \sqrt{z^2-2} + (1-3z^2)\operatorname{arc cos}\dfrac{1}{\sqrt{z^2-1}} - \dfrac{2z^3}{3}\operatorname{arc sin}\dfrac{1}{z^2-1} \\[2mm] \qquad +\dfrac{2z^3}{3}\operatorname{arc cos}\dfrac{z}{\sqrt{2z^2-2}}, & \sqrt{2} \le z \le \sqrt{3}; \end{cases}$$

the mean

$$\mu = \tfrac{1}{24}[6\sqrt{3} - \pi + 12 \ln (2 + \sqrt{3})] \approx 0.96059$$

given also in [1] and the variance

$$\sigma^2 = 1 - \tfrac{1}{576}[6\sqrt{3} - \pi + 12 \ln (2 + \sqrt{3})]^2 \approx 0.07726.$$

Editorial note. The mean distance is also found for a rectangular box in E2629 by T. S. Bolis and D. P. Robbins (Amer. Math. Monthly, 85 (1978), pp. 277–278). [M.S.K.]

A Random Knockout Tournament

Problem 86-2, by D. E. KNUTH (Stanford University).

A random tournament of order $n > 0$ consists of 2^n players who are paired at random; they play 2^{n-1} matches, and the 2^{n-1} winners play a random knockout tournament of order $n - 1$. In $n = 0$ the sole player is declared the champion.

Suppose the players $x_1, \cdots, x_{2^n}$ have the property that x_i always beats x_j whenever $j > i + 1$; but x_i beats x_{i+1} with probability p, independent of the outcome of all other matches.

Show that x_i will be the eventual champion with probability

$$1 + \sum_{m=1}^{n} (p-1)^m \prod_{j=1}^{m} \frac{(2^n - 2^{j-1})}{(2^n - j)(2^j - 1)},$$

where $q = 1 - p$; and show that the limiting value as $n \to \infty$ is

$$(1 - \tfrac{1}{2}q)(1 - \tfrac{1}{4}q)(1 - \tfrac{1}{8}q)(1 - \tfrac{1}{16}q) \cdots .$$

Solution by the proposer.

Let us generalize the problem by assuming that x_i beats x_{i+1} with probability p_i.

The *normal play* of a random tournament occurs when there are no upsets, i.e., when the better-ranked player wins each match. A tournament is said to have a *ladder* of length m if its normal play has x_1 beating x_2 after x_2 beats x_3 after $\cdots$ after x_m beats x_{m+1}, and if x_{m+1} does not meet x_{m+2} in the normal play. (The latter proviso implies that the ladder length is unique.)

Let P_m be the probability that x_1 will be champion, given a random knockout tournament whose ladder length is m. Then $P_0 = 1$ since a ladder of length 0 occurs only when $n = 0$ and x_1 is unopposed. In a ladder of length m, players x_m and x_{m+1} will be undefeated until they play each other; then the probability is p_m that x_{m+1} is eliminated (whence the subsequent play is essentially a ladder of length $m - 1$), and the probability is $1 - p_m$ that x_m is eliminated (whence the subsequent play is essentially a ladder of length $m - 2$, if $m \geq 2$). It follows that $P_1 = p_1$, and

$$P_m = p_m P_{m-1} + (1 - p_m) P_{m-2}, \qquad m \geq 2.$$

This recurrence implies that $P_m - P_{m-1} = (p_m - 1)(P_{m-1} - P_{m-2})$; hence we have

(∗) $$P_m - P_{m-1} = (p_1 - 1) \cdots (p_m - 1), \qquad m \geq 1.$$

Notice that P_m depends only on m, not n.

Let L_{mn} be the probability that the ladder length is m or more, in a random knockout tournament of n rounds. Then the probability that x_i is champ can be written

(**)
$$C_n = \sum_{k=0}^{n} P_m(L_{mn} - L_{m+1,n})$$

$$= 1 + \sum_{k=1}^{n} L_{mn}(P_m - P_{m-1}),$$

since $P_0 L_{0n} = 1$ and $L_{n-1,n} = 0$; the ladder length cannot exceed n since the tournament has only n rounds.

A random knockout tournament can be specified by a random permutation $a_1 \cdots a_{2^n}$ of $\{1, \cdots, 2_n\}$; the first round pits a_1 against a_2, a_3 against a_4, etc., and subsequent rounds similarly pair up the surviving players from left to right. The number of such permutations that define a tournament in which normal play has x_i beating x_{i+1} in round r_i, given $n \geq r_1 > r_2 > \cdots > r_m > 0$, is

$$2^n \cdot 2^{r_1-1} \cdot 2^{r_2-1} \cdot \cdots \cdot 2^{r_m-1} \cdot (2^n - m - 1)!,$$

since there are 2^n ways to place x_1, after which there are 2^{r_i-1} ways to place x_{i+1} for $i = 1, 2, \cdots, m$, after which the remaining $2^n - m - 1$ players can be placed arbitrarily. Hence

$$L_{mn} = \frac{(2^n - m - 1)!}{2^m(2^n - 1)!} S_{mn}, \qquad S_{mn} = \sum_{0 < r_m < \cdots < r_1 \leq n} 2^{r_1 + \cdots + r_m}.$$

Since $S_{m,n+1} - S_{mn} = 2^{n+1} S_{m-1,n}$ and $S_{mn} = 0$ when $m > n \geq 0$, it is not difficult to prove by induction on n that

$$S_{mn} = 2^m \frac{(2^n - 2^0)(2^n - 2^1) \cdots (2^n - 2^{m-1})}{(2^1 - 1)(2^2 - 1) \cdots (2^m - 1)};$$

therefore (*) and (**) yield the desired formula

$$C_n = \sum_{m=0}^{n} \prod_{j=1}^{m} \frac{(2^n - 2^{j-1})(p_j - 1)}{(2^n - j)(2^j - 1)}.$$

Let R_{mn} be the product

$$\prod_{1 \leq j \leq m} \frac{2^n - 2^{j-1}}{2^n - j} = \prod_{1 \leq j \leq m} \frac{(1 - 2^{j-1}/2^n)}{(1 - j/2^n)}.$$

Using the fact that $(1 - t_1) \cdots (1 - t_m)$ always lies between 1 and $1 - (t_1 + \cdots + t_m)$ when $0 \leq t_1, \cdots, t_m \leq 1$, we know that for $1 \leq m \leq n$ the numerator of R_{mn} is between 1 and $1 - 2^m/2^n$ and the denominator is between 1 and $1 - \binom{m+1}{2}/2^n$. Hence if we define $R_{mn} = 0$ for $m > n$ we have the estimate $R_{mn} = 1 + O(2^{m-n})$, uniformly in m and n.

Now let $Q_m = (2^1 - 1)(2^2 - 1) \cdots (2^m - 1)$. The infinite product $(1 - \frac{1}{2})(1 - \frac{1}{4})(1 - \frac{1}{8}) \cdots$ converges to 0.2888 (cf. [1, exercise 6.3-26]); hence $Q_m = 2^{1 + \cdots + m}(1 - 2^{-1}) \cdots (1 - 2^{-m}) > 0.2888 \cdot 2^{(m^2+m)/2}$.

The mth summand for C_n is $(-q)^m R_{mn}/Q_m = (-q)^m/Q_m + O(2^{m-n}/Q_m)$; we have $C_n = \sum_{m \geq 0} (-q)^m/Q_m + O(2^{-n})$ because the remainder term converges.

The remaining sum can be expressed as the stated infinite product because of a well-known identity due to Euler. (See, for example, [1, exercise 5.1.1-16], where the substitution $(z, q) := (-q/2, \frac{1}{2})$ yields the desired result.) Notice that even when $p = 0$, player x_1 will be champion with probability approaching 0.2888.

REFERENCE

[1] DONALD E. KNUTH, *The Art of Computer Programming, Vol. 3: Sorting and Searching*, Addison-Wesley, Reading, MA, 1973.

Problem 77-11, *A Coin Tossing Problem*, by DANNY NEWMAN (Stanford University).

If one tosses a fair coin until a head first appears, then the probability that this event occurs on an even numbered toss is exactly $\frac{1}{3}$. For this procedure, the expected number of tosses equals 2. Can one design a procedure, using a fair coin, to give a success probability of $\frac{1}{3}$ but have the expected number of tosses less than 2?

Solution by R. D. FOLEY (Virginia Polytechnic Institute).

Let N be a random variable which represents the number of coin tosses. If there exists a finite number n such that

$$\Pr(N < n) = 1,$$

(i.e., we know for certain we make only a finite number of tosses), then the probability of success is $i2^{-n}$, where i is the number of successful sequences of n coin tosses. However $i2^{-n}$ does not equal $\frac{1}{3}$ for any integers i and n. Hence, we know that for any n there exists at least one sequence of n heads and tails in which we would not have stopped. Thus for all n,

$$\Pr(N > n) \geq 2^{-n}.$$

Now,

$$E(N) = \sum_{n=0}^{\infty} \Pr(N > n) \geq \sum_{n=0}^{\infty} 2^{-n} = 2.$$

Thus it is impossible to develop a procedure using a fair coin to give a success probability of $\frac{1}{3}$ but have the expected number of tosses less than 2.

Solution by J. C. BUTCHER (University of Auckland, Auckland, New Zealand).

Let p be given in $[0, 1]$; then amongst all procedures based on tossing of fair coins which result in a probability of success equal to p, we define $N(p)$ as the infimum of the expected number of tosses. We will show that $N(p) = 2$ unless $p = 0$ or 1, in which case $N(p) = 0$, or $p = n/2^m$ for m and n positive integers with n odd. In this last case we will show that $N(p) = 2(1 - 2^{-m})$. Furthermore, we show that an expected number of tosses equal to $N(p)$ can be achieved. From our result, it follows that a success probability of $\frac{1}{3}$ cannot be achieved with a lower expected number of tosses than 2.

We first outline a procedure that gives the expected number we have quoted and then prove it is optimal. The procedure is to construct a binary fraction whose rth digit is 0 for a tail and 1 for a head on the rth toss. As soon as a digit differs from the binary representation of p we terminate the procedure and deem it a success if the binary fraction formed by the toss is less than the binary representation of p. In the case when $p = n/2^m$, we also terminate if the fraction found after m tosses equals p. Clearly, the expected number of tosses required by this procedure is 2 in the general case or $2(1 - 2^{-m})$ in the case $p = n/2^m$.

To prove that a smaller expected number is not possible, we use the fact that $N(p) \leq 2$ in all cases, and consider the subprocedures after a first toss when $p \neq 0, \frac{1}{2}$ or 1. If we were to require further tosses if either a head or a tail was recorded in the first toss, then the overall procedure would require at least two tosses. Hence, in the case of one of the outcomes we must terminate the experiment with a success (if $p > \frac{1}{2}$) or a failure (if $p < \frac{1}{2}$) and for the other outcome the subprocedure must have a probability of success $2p - 1$ (if $p > \frac{1}{2}$) or $2p$ (if $p < \frac{1}{2}$). If this is carried out in the optimal way, we see that

$$N(p) = \begin{cases} 1 + \frac{1}{2}N(2p) & (0 < p < \frac{1}{2}), \\ 1 + \frac{1}{2}N(2p-1) & (\frac{1}{2} < p < 1), \end{cases}$$

and iterated use of this gives the required result.

Comment by M. R. Brown (Yale University).

There is no better procedure, in a suitable "decision tree" model. This is an immediate consequence of Theorem 2.1 of D. E. Knuth and A. C. Yao, *The complexity of nonuniform random number generation*, Algorithms and Complexity, J. F. Traub, ed., Academic Press, New York, 1976, pp. 357–428.

Problem 67-18, *A Generating Function*, by P. I. Richards (Technical Operations, Inc.).

Let n objects be randomly distributed with equal probabilities into m boxes. If $P_k(n, m)$ is the probability that some box receives more than k objects, show that

$$\sum_{n=0}^{\infty} \frac{P_k(n, m)z^n}{n!} = e^z - \left\{ \sum_{i=0}^{k} \frac{1}{i!} \left(\frac{z}{m} \right)^i \right\}^m.$$

The problem arises in a simple model of a photographic emulsion, where each grain may have several photoelectron traps.

Solution by D. K. Guha (The Port of New York Authority).

When n objects are randomly distributed into m boxes with equal probabilities, the probability that boxes 1, 2, $\cdots$, m get x_1, x_2, $\cdots$, x_m number of objects, respectively, is given by the multinomial distribution

$$\frac{n!}{x_1! x_2! \cdots x_m!} \left(\frac{1}{m} \right)^{x_1} \left(\frac{1}{m} \right)^{x_2} \cdots \left(\frac{1}{m} \right)^{x_m},$$

where $x_1 + x_2 + \cdots + x_m = n$ and $1/m$ is the probability that a box gets the ball when it is distributed among m boxes with equal probabilities.

It is clear that for $n > mk$, there is at least one box which receives more than k objects in the above distribution.

Thus, $P_k(n, m) = 1$ whenever $n > mk$.

For $n \leq mk$, the probability $P_k(n, m)$ is given by the probability of the complement of the situation where each of the m boxes receives at most k objects.

Hence,

$$P_k(n, m) = 1 - \sum_{\substack{x_i \leq k \\ \Sigma x_i = n}} \frac{n!}{x_1! x_2! \cdots x_m!} \left(\frac{1}{m} \right)^{x_1 + x_2 + \cdots + x_m}$$

for $n \leq mk$. The generating function is then given by

$$\sum_{n=0}^{\infty} \frac{P_k(n, m)z^n}{n!} = \sum_{n=0}^{mk} \frac{z^n}{n!}\left\{1 - \sum_{\substack{x_i \leq k \\ \Sigma x_i = n}} \frac{n!}{x_1! x_2! \cdots x_m!}\left(\frac{1}{m}\right)^{x_1+\cdots+x_m}\right\} + \sum_{n=mk+1}^{\infty} \frac{z^n}{n!}$$

$$= \sum_{n=0}^{\infty} \frac{z^n}{n!} - \sum_{n=0}^{mk} \sum_{\substack{x_i \leq k \\ \Sigma x_i = n}} \frac{1}{x_1! x_2! \cdots x_m!}\left(\frac{z}{m}\right)^{x_1+\cdots+x_m}$$

$$= e^z - \left\{\sum_{i=0}^{k} \frac{1}{i!}\left(\frac{z}{m}\right)^i\right\}^m.$$

Comment by S. KULLBACK (George Washington University).

F. N. David and D. E. Barton (*Combinational Chance*, Griffin, London, 1962, pp. 221–223) consider the complementary problem that no box receives more than k objects and give the generating function from which the desired function can be immediately derived.

Additionally the proposer notes that if the objects are distributed unequally, entering the jth box with probability p_j, the corresponding result is

$$\sum_{n=0}^{\infty} P_k(n, m)z^n/n! = e^z - \prod_{j=1}^{m}\sum_{i=0}^{k}(1/i!)(p_j z)^i.$$

Problem 66-8, *A Variation of a Theorem of Kakutani*, by L. SHEPP (Bell Telephone Laboratories).

Suppose X_1, X_2, $\cdots$ are independent and positive random variables with $EX_i = 1$, $i = 1, 2, \cdots$. Show that $\prod_{n=1}^{\infty} X_n$ converges a.s. (to a nonzero value) if and only if $\prod_{n=1}^{\infty} E\sqrt{X_n}$ converges.

Solution by WALTER WEISSBLUM (AVCO Corporation).

First note that $E\sqrt{X_n} = \sqrt{EX_n} - \text{Var}\sqrt{X_n} = \sqrt{1} - \text{Var}\sqrt{X_n} = 1 - \epsilon_n$ with $\epsilon_n \geq 0$. If $\prod X_n$ converges to a positive value with probability one, the same holds for $\prod \sqrt{X_n}$. Letting $Z_n = \prod_{i=1}^{n}\sqrt{X_i}$, we have $Z_n \to Z > 0$. By Fatou's lemma, we have $\underline{\lim} EZ_n \lim = \prod_{i=1}^{n} E\sqrt{X_i} \geq E(\underline{\lim} Z_n) = EZ > 0$. Since $E\sqrt{X_n} \leq 1$ this proves that $\prod E\sqrt{X_n}$ is convergent. To prove the other implication, suppose $\prod E\sqrt{X_n}$ is convergent. Then $\sum(1 - E\sqrt{X_n}) = \sum \epsilon_n$ is convergent. To show that $\prod X_n$ converges with probability one is the same as to show that $\prod \sqrt{X_n}$, or that $\sum(1 - \sqrt{X_n}) = \sum Y_n$, converges with probability one. By Theorem 2.3 in Doob's *Stochastic Processes*, this would follow from $\sum EY_n$ and $\sum \text{Var } Y_n$ both being convergent. But $EY_n = 1 - E\sqrt{X_n} = \epsilon_n$ so the first series converges, and $\text{Var } Y_n = EX_n - (E\sqrt{X_n})^2 = 1 - (1 - \epsilon_n)^2 = 2\epsilon_n - \epsilon_n^2$, so the second series is also convergent.

Credibility Functions

Problem 79-10*, *by* Y. P. SABHARWAL *and* J. KUMAR (Delhi University, India).

Determine the general form of a function $F(x)$ satisfying the following conditions for $x \geq 0$:

(a) $0 \leq F(x) \leq 1$;

(b)
$$\frac{d}{dx}F(x)>0;$$

(c)
$$\frac{d}{dx}\{F(x)/x\}<0.$$

These conditions arise in the construction of credibility formulae in casualty insurance work [1], [2, p. 83].

REFERENCES

[1] F. S. PERRYMAN, *Experience rating plan credibilities*, Proc. Casualty Actuarial Soc., 24 (1937), p. 60; also reproduced in 58 (1971), p. 143.
[2] H. L. SEAL, *Stochastic Theory of a Risk Business*, John Wiley, New York, 1969.

Editorial note. The proposers note that $F(x)=x/(x+k)$ is a particular solution. However, it isn't difficult to extend this to $F(x)=G(x)/(G(x)+k)$ where $G(x)$ is a nonnegative increasing concave function of x, e.g., $G(x)=x^\alpha(0\le\alpha<1)$, $\log(1+x)$, etc. [M.S.K.].

Solution by JAN BRINK-SPALINK (University of Münster, Federal Republic of Germany).

We restrict ourselves to C^1-functions $F(x)$, though the pointwise conditions are defined on a larger classs of functions. We first note that $\lim_{x\to\infty} F(x)=:\mu$ exists, and that $0<F(x)<\mu\le1$ if $x>0$. A characterization of $F(x)$ is obtained by considering the simple transformation

$$\phi(x)=\frac{F'(x)}{F(x)}\qquad(x>0),$$

$$F(x)=\mu\,\exp\left(-\int_x^\infty \phi(t)\,dt\right).$$

LEMMA 1. *$F(x)$ is C^1 on $(0,\infty)$ and satisfies (a)–(c) for $x>0$ iff*

$$F(x)=\mu\,\exp\left(-\int_x^\infty \phi(t)\,dt\right)$$

with $\phi(x)$ continuous on $(0,\infty)$ and

(A)
$$0<\mu\le1,$$

(B)
$$0<\phi(x)<1/x\quad for\ x>0,$$

(C)
$$\int_1^\infty \phi(t)\,dt<\infty.$$

Proof. The proof is trivial. Condition (C) follows from the fact that $F(1)>0$. Note that $F(0)=0$ iff $\int_0^1 \phi(t)\,dt=\infty$.

On the other hand, if the problem is really to determine C^1-functions $F(x)$ on the closed half-line $[0,\infty)$, then we need some additional properties concerning the behavior of $\phi(x)$ near $x=0$. Because condition (c) at $x=0$ makes sense only if $F(0)=0$, we have

(D)
$$\int_0^1 \phi(t)\, dt = \infty.$$

Then existence and positivity of $F'(0)$ is equivalent to $0 < \lim_{x\downarrow 0} F(x)/x = F(1)\lim_{x\downarrow 0}\exp\left(\int_x^1 [1/t - \phi(t)]\, dt\right)$, i.e.,

(E)
$$\int_0^1 [1/t - \phi(t)]\, dt < \infty.$$

If we want continuity of $F'(x)$ at $x = 0$, we must also have $\lim_{x\downarrow 0}\phi(x)F(x) = \lim_{x\downarrow 0} F(x)/x (>0)$, i.e.,

(F)
$$\lim_{x\downarrow 0} x\phi(x) = 1.$$

Note that (F) implies (D). Finally, condition (c) at $x = 0$ means

$$0 > \lim_{x\downarrow 0}\frac{1}{x}\left\{\frac{F(x)}{x} - F'(0)\right\} = \lim_{x\downarrow 0}\frac{F(x)}{x^2}\left\{1 - \exp\left(\int_0^x [1/t - \phi(t)]\, dt\right)\right\}$$

$$= -F'(0)\lim_{x\downarrow 0}\frac{1}{x}\left\{\int_0^x [1/t - \phi(t)]\, dt\right\},$$

that is,

(G)
$$\lim_{x\downarrow 0}\frac{1}{x}\int_0^x [1/t - \phi(t)]\, dt \text{ exists and is in } (0, \infty).$$

Obviously (G) implies (E). Thus we have

LEMMA 2. $F(x)$ is C^1 on $[0, \infty)$ and satisfies (a)–(c) for $x > 0$ iff

$$F(x) = \mu\,\exp\left(-\int_x^\infty \phi(t)\, dt\right)$$

with $\phi(x)$ continuous on $(0, \infty)$ and satisfying (A), (B), (C), (F) and (G).
It should be noted that $F(x) = ax - bx^2 + o(x^2)$ as $x \to 0$ with $a, b > 0$.
Remark. Lemma 2 can be reformulated by the transformation

$$\phi(x) = \frac{1}{x}v(\ln x), \qquad x > 0.$$

Thus we have

$$F(x) = \mu\,\exp\left(-\int_{\ln x}^\infty v(s)\, ds\right),$$

and one easily verifies that

(B)
$$\Leftrightarrow 0 < v(s) < 1, \qquad s \in R,$$

(C)
$$\Leftrightarrow \int_0^\infty v(s)\, ds < \infty,$$

(F)
$$\Leftrightarrow \lim_{s\to\infty} v(-s) = 1,$$

(G)
$$\Leftrightarrow \lim_{s\to\infty} e^s \int_{-\infty}^{-s} [1 - v(t)]\, dt \in (0, \infty).$$

Problem 62–9, Normal Functions of Normal Random Variables, by LARRY SHEPP (Bell Telephone Laboratories).

a) If X and Y are normal independent random variables with zero means, show that

$$Z = \frac{XY}{\sqrt{X^2 + Y^2}}$$

is normal.

b) If in addition $\sigma^2(X) = \sigma^2(Y)$, show that

$$W = \frac{X^2 - Y^2}{X^2 + Y^2}$$

is normal and that Z and W are independent.

Solution by the proposer.

b) $X = R \cos \theta$, $Y = R \sin \theta$ on an appropriate sample space where θ is uniform on $(0, 2\pi)$ and R and θ are independent. Then R and 2θ are independent and have the same distribution (2θ taken modulo 2π). Thus, $Z = R \sin 2\theta$ and $W = R \cos 2\theta$ are independent normal variables.

a) If the variances of X and Y are different, the above fails since θ is no longer uniform on $(0, 2\pi)$. However, it is known that X^{-2} and Y^{-2} are random variables with a stable distribution ($\alpha = \frac{1}{2}$). Consequently, $X^{-2} + Y^{-2} = Z^{-2}$ on an appropriate sample space by the stable law property where Z must be normal.

Problem 72-25, Mean and Variance of a Function of a Random Permutation, by E. HAUER and J. G. C. TEMPLETON (University of Toronto, Ontario, Canada).

Let $(p_1, p_2, \cdots, p_n)$ denote a random permutation of $(1, 2, \cdots, n)$. Also let

$$N_j = \max(p_1, p_2, \cdots, p_j) - p_j, \qquad M_n = \sum_{j=1}^{n} N_j.$$

Determine the mean and variance of M_n given that the probability of each random permutation is the same.

The problem arose in a study of queuing in lanes, with service in random order and departure in order of arrival.

Solution by FARHAD MEHRAN (Rutgers University).

Let $X_j = \max(p_1, \cdots, p_j)$; then $N_j = X_j - p_j$ and

$$M_n = \sum_{j=1}^{n} X_j - \sum_{j=1}^{n} p_j = \sum_{j=1}^{n} X_j - (n/2)(n + 1).$$

Thus, to obtain the mean and the variance of M_n, it is sufficient to consider the X_j's.

The probability that $X_j = k$ is given by

(1) $$P(X_j = k) = \binom{k - 1}{j - 1} \div \binom{n}{j}, \qquad 1 \le k \le n.$$

Thus, the mean of X_j is given by

$$\mu_j = E(X_j) = \sum_{k=1}^{n} P(X_j = k) = j\frac{n+1}{j+1},$$

from which we derive the mean of M_n:

$$E(M_n) = \sum_{j=1}^{n} E(X_j) - (n/2)(n+1)$$

$$= (n/2)(n+1)(1 - 2/\alpha_n),$$

where $\alpha_n = n \left(\sum_{r=1}^{n} 1/(r+1)\right)^{-1}$. (For large n, $\alpha_n \sim n(\log(n+1) + \gamma - 1)^{-1}$, where γ is Euler's constant, $\gamma = .57722\ldots$).

Now assume, without loss of generality, that $i < j$. The conditional probability that $X_j = k$, given that $X_i = m$, is zero if $k < m$, while if $k \geq m$, it is given by

(2) $$P(X_j = k | X_i = m) = \binom{k - i - 1 + \delta_{km}}{j - i - 1 + \delta_{km}} \bigg/ \binom{n - i}{j - i},$$

where δ_{km} denotes the Kronecker delta. The joint distribution of X_i and X_j can be obtained from (1) and (2), and then it can be shown that

$$c_{ij} = \text{cov}(X_i, X_j) = \mu_i\left(\frac{n+1}{j+1} - \frac{n+2}{j+2}\right), \quad i \leq j.$$

Thus, the variance of M_n is given by

$$V(M_n) = \sum_{i=1}^{n} c_{ii} + 2\sum_{i=1}^{n-1} \sum_{j=i+1}^{n} c_{ij}.$$

For large values of n, the mean and variance of M_n can be approximated by

$$E(M_n) = \frac{n}{2}(n+1)\left[1 - O\left(\frac{\log n}{n}\right)\right]$$

and

$$V(M_n) = 2(n+1)^2 \log(n)[1 - O(1/\log n)].$$

Also solved by A. A. JAGERS (Technische Hogeschool Twente, Enschede, the Netherlands) and the proposers.

Editorial note. A. A. Jagers gives more precise asymptotic estimates:

$$E(M_n) = \frac{n}{2}(n+1) - (n+1)(\gamma - 1 + \log(n+1)) + O(1),$$

and

$$V(M_n) = 2(n+1)^2 \log(n+1) - (6 - 2\gamma + \pi^2/6)(n+1)^2$$

$$+ (n+1)\log^2(n+1) + (2\gamma + 5)(n+1)\log(n+1) + O(n).$$

[W.F.T.]

Estimation of the Median Effective Dose in Bioassay

Problem 81-12, *by* D. E. RAESIDE (University of Oklahoma, Health Sciences Center).

Consider a system subjected to a single stimulus intensity (dose level) x, and assume that the system responds in just one of two ways, "positively" or "negatively". Let the probability of a positive response be denoted by π, and assume that π is related to x through the quantal logistic dose response function

$$\pi = [1 + e^{-\beta(x-\gamma)}]^{-1}.$$

Use Bayesian estimation theory with a catenary loss function and a conjugate prior to determine an optimal estimate of the parameter γ (the parameter β is assumed known). Show that for certain parameter values and for large numbers of positive and negative responses this Bayes estimator is approximately the maximum likelihood estimator. See the paper *Optimal Bayesian sequential estimation of the median effective dose* by P. R. Freeman (Biometrika, 57 (1970), pp. 79–89) for a treatment of this problem using a quadratic loss function. See the paper *A class of loss function of catenary form* by D. E. Raeside and R. J. Owen (J. Statist. Phys., 7 (1973), pp. 189–195) for a summary of properties of the catenary loss function.

Solution by P. W. JONES *and* P. SMITH (Keele University, Staffordshire).

For a single dose level x let the number of responses in n' trials be r'. The likelihood is then

$$\frac{e^{r'\beta(x-\gamma)}}{[1 + e^{\beta(x-\gamma)}]^{n'}}, \qquad r' \leq n'.$$

Suppose that γ is assigned a natural conjugate prior distribution with density

(1) $$f(\gamma) \, \alpha \, \frac{e^{r_0\beta(x-\gamma)}}{[1 + e^{\beta(x-\gamma)}]^{n_0}}, \qquad 1 \leq r_0 \leq n_0, \quad n_0 \geq 2,$$

involving parameters r_0, n_0. Then by Bayes' theorem the posterior distribution of γ is (1) with (r_0, n_0) replaced by (r, n) where $r = r_0 + r'$ and $n = n_0 + n'$.

We require the minimum with respect to d of the integral

$$I = K \int_{-\infty}^{\infty} \frac{e^{r\beta(x-\gamma)} \cosh\,(a(\gamma - d)) \, d\gamma}{(1 + e^{\beta(x-\gamma)})^n} - 1,$$

where

$$\frac{1}{K} = \int_{-\infty}^{\infty} \frac{e^{r\beta(x-\gamma)} \, d\gamma}{(1 + e^{\beta(x-\gamma)})^n}.$$

This gives the Bayes estimator $d*$ of γ with respect to the catenary loss function $\cosh\,[a(\gamma - d)] - 1$, where a is a positive constant.

The integral can be written

(2) $$I = F_n(a, \beta, r, x)e^{-ad} + F_n(-a, \beta, r, x)e^{ad} - 1,$$

with

(3) $$F_n(a, \beta, r, x) = \frac{1}{2} K e^{r\beta x} \int_{-\infty}^{\infty} \frac{e^{\gamma(a-r\beta)} \, d\gamma}{(1 + e^{\beta(x-\gamma)})^n}.$$

The minimizer of (2) is

(4) $$d = d* = \frac{1}{2a} \ln\,[F_n(a, \beta, r, x)/F_n(-a, \beta, r, x)].$$

The substitution $e^{-\beta\gamma} = u$ transforms (3) into

$$F_n(a, \beta, r, x) = \frac{K}{2\beta} e^{r\beta x} \int_0^\infty u^{r-1-a/\beta}(1 + e^{\beta x}u)^{-n}\, du$$

$$= \frac{K}{2\beta} e^{ax} B\left(r - \frac{a}{\beta}, n - r + \frac{a}{\beta}\right), \qquad r > \frac{a}{\beta},$$

from [1, p. 285]. Express the beta function $B(\cdot, \cdot)$ in terms of gamma functions ([1], p. 950) and substitute for F_n in (4):

(5) $$d^* = \frac{1}{2a} \ln \left[\frac{e^{2ax}\Gamma(r - a/\beta)\Gamma(n - r + a/\beta)}{\Gamma(r + a/\beta)\Gamma(n - r - a/\beta)}\right],$$

with (Bayes) risk given by substituting $d = d^*$ in (2). In order to obtain the asymptotic behavior of (5) for large n we require

$$\ln \Gamma\left(n - r + \frac{a}{\beta}\right) = n \ln n - n - \left(r - \frac{a}{\beta} + \frac{1}{2}\right) \ln n + \ln \sqrt{2\pi} + O\left(\frac{1}{n}\right),$$

which has been adapted from the asymptotic expansion of the logarithm of the gamma function given in [1, p. 940]. With a similar expression for $\ln \Gamma(n - r - a/\beta)$, the expansion for d^* becomes

$$d^* = \frac{1}{\beta} \ln n + x + \frac{1}{2a} \ln \left[\frac{\Gamma(r - a/\beta)}{\Gamma(r + a/\beta)}\right] + O\left(\frac{1}{n}\right)$$

as $n \to \infty$ (n_0 finite). The parameter values must satisfy $r > a/\beta$.

The maximum likelihood estimator of γ is given by $x + \ln (n'/r' - 1)/\beta$, which has the same leading term as d^* for large n'.

REFERENCE

[1] I. S. GRADSHTEYN AND I. M. RYZHIK (1965), *Tables of Integrals, Series and Products*, Academic Press, New York.

Problem 76-18, A Monotonicity Property for Moments, by A. A. JAGERS (Technische Hogeschool Twente, Enschede, the Netherlands).

Let ϕ be a nonnegative function defined on $[0, \infty)$ with $\phi(0) = 0$. Let $\mathcal{F}$ be the class of all nonnegative random variables X such that $0 < E(X^s) < \infty$ for all $s > 0$. For $X \in \mathcal{F}$, $s > 0$, put

$$q_x(s) = [E(\phi(X)^s)/E(X^s)]^{1/s}.$$

(i) Prove that if ϕ is convex, then q_x is nondecreasing for each $X \in \mathcal{F}$.
(ii) Determine a necessary and sufficient condition on ϕ in order that q_x is nondecreasing for each $X \in \mathcal{F}$.

Part (i) of the problem, suggested by W. ALBERS, arose in finding the corresponding moment inequalities in some problems of robust estimation.

Solution by I. OLKIN (Imperial College of Science and Technology, England).

A more general result appears in *Monotonicity of ratios of means and other applications of majorization*, by A. W. Marshall, I. Olkin and F. Proschan, pp. 177–190 in *Inequalities*, ed. by O. Shisha, Academic Press, 1967.

The relevant results from that paper are as follows.

THEOREM. *Let F and G be distribution functions with* $F(0) = G(0) = 0$ *and define* $\bar{F} = 1 - F$, $\bar{G} = 1 - G$. *If*

(1) $$\bar{F}^{-1}(p)/\bar{G}^{-1}(p) \quad \text{is increasing in } p,$$

then

$$\left(\frac{\int x^r \, dG(x)}{\int x^r \, dF(x)} \right)^{1/r}$$

is increasing in r.

The condition (1) is equivalent to $F(x) = G[\psi(x)]$ for some nonnegative starshaped function ψ.

Remark. A real function ψ defined on $[0, \infty)$ is said to be *starshaped* if $\psi(x)/x$ is increasing in x. Every convex function ϕ on $[0, \infty)$ with $\phi(0) \leq 0$ is starshaped.

Condition (1) is also equivalent to the condition: if X is a random variable with distribution F, then $\psi(X)$ has distribution G for some nonnegative starshaped ψ.

In particular, if X has distribution F and $Y = \psi(X)$ has distribution G, where $\psi \geq 0$ is starshaped, then

$$\left[\frac{\int (\psi(x))^r \, dF(x)}{\int x^r \, dF(x)} \right]^{1/r}$$

is increasing in r.

Problem 63-8, A Monotonic Distribution Function, by CARL EVANS and JOHN FLECK (Cornell Aeronautical Laboratory).

Let $\xi(x_1, y_1)$, $\mathbf{n}(x_2, y_2)$ be independent samples of the elliptic normal distribution in the plane

$$\Phi(x, y) = \frac{1}{2\pi ab} \exp\left[-\frac{1}{2}\left(\frac{x-m}{a}\right)^2 - \frac{1}{2}\left(\frac{y-n}{b}\right)^2 \right], \qquad a, b > 0.$$

Show that the function

$$F(a, b, m, n) \equiv \text{Prob.}\{\xi \cdot \mathbf{n} < 0\}$$

is a decreasing function of $|m|$, $|n|$, and an increasing function of a, b, provided m and n are not both zero.

Solution by the proposers. First we shall establish the following representation:

$$F(a, b, m, n) = \frac{1}{\pi} \int_0^{\pi/2} \exp\left\{ -\frac{m^2}{a^2 + P(\theta)} - \frac{n^2}{b^2 + P(\theta)} \right\} d\theta,$$

where

$$P(\theta) = \sqrt{a^4 \cos^2\theta + b^4 \sin^2\theta}.$$

The orthogonal substitution

$$x_1 = (x_1' - x_2')/\sqrt{2}, \qquad x_2 = (x_1' + x_2')/\sqrt{2},$$
$$y_1 = (x_3' - x_4')/\sqrt{2}, \qquad y_2 = (x_3' + x_4')/\sqrt{2},$$

transforms the integral

$$F = \frac{1}{(2\pi ab)^2} \int_{x_1 x_2 + y_1 y_2 < 0} \exp\left[-\frac{1}{2}\left\{ \left(\frac{x_1 - m}{a}\right)^2 + \left(\frac{y_1 - n}{b}\right)^2 \right. \right.$$
$$\left. \left. + \left(\frac{x_2 - m}{a}\right)^2 + \left(\frac{y_2 - n}{b}\right)^2 \right\} \right] dx_1\, dy_1\, dx_2\, dy_2$$

into

$$F = \frac{1}{(2\pi ab)^2} \int_{x_2{}^2 + x_4{}^2 > x_1{}^2 + x_3{}^2} \exp\left[-\frac{1}{2}\left\{ \left(\frac{x_1 - m\sqrt{2}}{a}\right)^2 \right. \right.$$
$$\left. \left. + \left(\frac{x_3 - n\sqrt{2}}{b}\right)^2 + \left(\frac{x_2}{a}\right)^2 + \left(\frac{x_4}{b}\right)^2 \right\} \right] dx_1\, dx_2\, dx_3\, dx_4 .$$

Using polar coordinates for the second two variables and integrating on r gives

$$F = \frac{1}{(\pi ab)^2} \int_{B_2} \exp\left[-\frac{1}{2}\left\{ \left(\frac{x_1 - m\sqrt{2}}{a}\right)^2 + \left(\frac{x_3 - n\sqrt{2}}{b}\right)^2 \right\} \right] dx_1\, dx_3$$
$$\cdot \int_0^{\pi/2} \exp\left\{ - G(\theta)\, \frac{x_1{}^2 + x_3{}^2}{2} \right\} \frac{d\theta}{G(\theta)}$$

where

$$G(\theta) = a^{-2} \cos^2\theta + b^{-2} \sin^2\theta.$$

By interchanging the order of integrations and completing squares in the exponent, we get

$$F = \frac{1}{(\pi ab)^2} \int_0^{\pi/2} \exp\left\{ -\frac{m^2 G(\theta)}{1 + a^2 G(\theta)} - \frac{n^2 G(\theta)}{1 + b^2 G(\theta)} \right\} \frac{d\theta}{G(\theta)}$$
$$\cdot \int_{B_2} \exp\left[-\frac{1}{2}\left\{ \left(\frac{x_1 - \mu(\theta)}{\sigma_1(\theta)}\right)^2 + \left(\frac{x_3 - \nu(\theta)}{\sigma_2(\theta)}\right)^2 \right\} \right] dx_1\, dx_3 ,$$

where

$$\mu(\theta) = \frac{m\sqrt{2}}{1 + a^2 G(\theta)}, \qquad \sigma_1(\theta) = \frac{a}{\sqrt{1 + a^2 G(\theta)}},$$
$$\nu(\theta) = \frac{n\sqrt{2}}{1 + b^2 G(\theta)}, \qquad \sigma_2(\theta) = \frac{b}{\sqrt{1 + b^2 G(\theta)}}.$$

Hence,

$$F = \frac{2}{\pi ab} \int_0^{\pi/2} \frac{H^2(\theta) \exp\left\{ -\dfrac{m^2}{a^2 + H(\theta)} - \dfrac{n^2}{b^2 + H(\theta)} \right\} d\theta}{\sqrt{(a^2 + H(\theta))(b^2 + H(\theta))}} ,$$

where

$$H(\theta) = \frac{1}{G(\theta)}.$$

The substitution $\tan \theta = (b/a) \tan \phi$ gives

$$F = \frac{2}{\pi} \int_0^{\pi/2} \frac{I(\phi) \exp\left\{-\dfrac{m^2}{a^2 + I(\phi)} - \dfrac{n^2}{b^2 + I(\phi)}\right\} d\phi}{\sqrt{(a^2 + I(\phi))(b^2 + I(\phi))}},$$

where

$$I(\phi) = a^2 \cos^2 \phi + b^2 \sin^2 \phi.$$

Finally, the substitution

$$I(\phi) = P(\theta) = \sqrt{a^4 \cos^2 \theta + b^4 \sin^2 \theta}$$

gives the desired form from which the monotonic character of F is then evident.

Problem 74-21*, Two-Dimensional Discrete Probability Distributions, by P. BECKMANN (University of Colorado).

Let $q_n(k)$ be the nth-degree orthonormalized Chebyshev, Krawtchouk or Charlier polynomial of a discrete variable and let $P(k)$ be the corresponding jump (weighting) function. Let $0 \leq R \leq 1$.
It is conjectured that

$$F(k, m) = P(k)P(m) \sum_{n=0}^{\infty} R^n q_n(k)q_n(m) \geq 0$$

for all integers k, m in the pertinent orthogonality interval.
If the conjecture is valid, then $F(k, m)$ is a two-dimensional probability distribution with R the correlation coefficient of the random variables, and with identical marginal distributions $P(k)$, $P(m)$, which are uniform, binomial or Poisson for Chebyshev, Krawtchouk or Charlier polynomials, respectively. Other two-dimensional discrete distributions with given marginals and correlation coefficients can then be constructed by simple transformations.
An analogous theorem for probability distributions of continuous random variables has recently been demonstrated by the proposer.

Solution by GEORGE GASPER (Rheinisch-Westfälische Technische Hochschule Aachen, West Germany, and Northwestern University).

Using the notation in the Bateman Manuscript Project [3, Chap. 10] for the Charlier polynomials $c_n(x; a)$, $a > 0$, and the Krawtchouk polynomials $k_n(x) = k_n(x; p, N)$, $0 < p < 1$, $N = 0, 1, \cdots$, we find that proving the conjecture for these polynomials is equivalent to proving the nonnegativity of the Poisson kernels

(1) $$\sum_{n=0}^{\infty} R^n a^n c_n(x; a) c_n(y; a)/n!, \qquad x, y = 0, 1, \cdots,$$

(2) $$\sum_{n=0}^{N} R^n \left\{\binom{N}{n} p^n (1-p)^n\right\}^{-1} k_n(x) k_n(y), \qquad x, y = 0, 1, \cdots,$$

for $0 \leq R \leq 1$. In [4, p. 450] the author pointed out that a limit case of formulas (3.3) and (4.7) derived in that paper is the following formula of Meixner [8]

$$\sum_{n=0}^{\infty} \frac{a^n b^n}{n!} t^n c_n(x; a) c_n(y; b)$$

(3)
$$= e^{abt}(1 - at)^y (1 - bt)^x {}_2F_0[-x, -y; \cdot ; t/(1 - at)(1 - bt)].$$

Since the right side of (3) is clearly nonnegative when $0 \leq t < \min(a, b)$, the nonnegativity of (1) follows from the case $t = R/a, b = a$ of (3). It was also pointed out in [4] that analogues of (3) for the Krawtchouk polynomials and the Meixner polynomials $m_n(x; \beta, c)$, $\beta > 0$, $0 < c < 1$ [3, 10.24] follow from another formula of Meixner in [3, 2.5 (12)]. In fact, this formula and the hypergeometric representations

$$k_n(x) = k_n(x; p, N) = (-p)^n \binom{N}{n} {}_2F_1[-n, -x; -N; 1/p],$$

$$m_n(x; \beta, c) = (\beta)_n {}_2F_1[-n, -x; \beta; 1 - c^{-1}],$$

where $(\beta)_n = \beta(\beta + 1) \cdots (\beta + n - 1) = \Gamma(n + \beta)/\Gamma(\beta)$, give

$$\sum_{n=0}^{N} R^n \left\{ \binom{N}{n} r^n (1 - p)^n \right\}^{-1} k_n(x; p, N) k_n(y; r, N)$$

(4)
$$= (1 + Rp/(1 - p))^{N - x - y} (1 - R)^x (1 - Rp(1 - r)/r(1 - p))^y$$

$$\cdot {}_2F_1[-x, -y; -N; -R/(1 - R)(r(1 - p) - Rp(1 - r))],$$

$$\sum_{n=0}^{\infty} R^n \frac{c^n}{n!(\beta)_n} m_n(x; \beta, c) m_n(y; \beta, d)$$

(5)
$$= (1 - Rc)^{-\beta - x - y} (1 - R)^x (1 - Rc/d)^y$$

$$\cdot {}_2F_1[-x, -y; \beta; R(1 - c)(1 - d)/(1 - R)(d - Rc)].$$

From these two formulas it is obvious that the left sides of (4) and (5) are nonnegative for $0 \leq R \leq 1$ when $0 < p \leq r < 1$, $x, y = 0, 1, \cdots, N$ and when $0 < c \leq d < 1$, $\beta > 0$, $x, y = 0, 1, \cdots$, respectively.

Thus it only remains to consider the Chebyshev polynomials of a discrete variable $t_n(x)$, [3, 10.23]. However, since these polynomials are (except for a different normalization and a shift in N) the special case $\alpha = \beta = 0$ of the Hahn polynomials [5], [4]

$$Q_n(x) = Q_n(x; \alpha, \beta, N) = {}_3F_2\left[\begin{array}{ccc} -n, & n + \alpha + \beta + 1, & -x; \\ & \alpha + 1, & -N \end{array}\right], \quad \alpha, \beta > -1,$$

which satisfy the orthogonality relation

$$\sum_{x=0}^{N} \rho(x) Q_n(x) Q_m(x) = \begin{cases} 0 & \text{if } n \neq m, \\ 1/\pi_n & \text{if } n = m, \end{cases}$$

for $n, m = 0, 1, \cdots, N$, with

$$p(x) = p(x; \alpha, \beta, N) = \frac{\binom{x + \alpha}{x}\binom{N - x + \beta}{N - x}}{\binom{N + \alpha + \beta + 1}{N}},$$

$$\pi_n = \pi_n(\alpha, \beta, N) = \frac{(-1)^n(-N)_n(\alpha + 1)_n(\alpha + \beta + 1)_n}{n!(N + \alpha + \beta + 2)_n(\beta + 1)_n} \cdot \frac{2n + \alpha + \beta + 1}{\alpha + \beta + 1},$$

we shall consider the more general problem of proving that

(6)
$$\sum_{n=0}^{N} R^n \pi_n(\alpha, \beta, N) Q_n(x; \alpha, \beta, N) Q_n(y; \alpha, \beta, M) \geq 0$$

when $0 \leq R \leq 1$, $x = 0, 1, \cdots, N$, $y = 0, 1, \cdots, M$ and $M \geq N$.

In [4] it was shown that a formula of Watson for the product of two terminating hypergeometric functions has a generalization of the form

(7)
$$_3F_2\left[\begin{matrix} -n, n + a, b; \\ c, d \end{matrix}\right]{_3F_2}\left[\begin{matrix} -n, n + a, e; \\ c, f \end{matrix}\right]$$

$$= \frac{(-1)^n(a - c + 1)_n}{(c)_n} \sum_{r=0}^{n} \sum_{s=0}^{n-r} \frac{(-n)_{r+s}(n + a)_{r+s}(b)_r(e)_r(d - b)_s(f - e)_s}{r!s!(d)_{r+s}(f)_{r+s}(c)_r(a - c + 1)_s},$$

and then (7) was used to derive a formula which gives the nonnegativity of a sum which is obtained from the sum (6) by replacing the factor R^n by $(-z)_n/(-N)_n$. By using (7) in the sum (6) and proceeding as on page 444 of [4] with $\alpha + \beta + 1 = 2a$, we find that the sum (6) equals

$$\sum_{r=0}^{N} \sum_{s=0}^{N-r} \frac{R^{r+s}(2a)_{2r+2s}(a + 1)_{r+s}(M - r - s)!(-x)_r(-y)_r}{r!s!M!(N + 2a + 1)_{r+s}(a)_{r+s}(\alpha + 1)_r}$$

$$\cdot \frac{(x - N)_s(y - M)_s}{(\beta + 1)_s}{_3F_2}\left[\begin{matrix} 2r + 2s + 2a, r + s + a + 1, r + s - N; R \\ r + s + a, N + r + s + 2a + 1 \end{matrix}\right]$$

$$= \sum_{r=0}^{N} \sum_{s=0}^{N-r} \frac{R^{r+s}(1 - R)^{2N - 2r - 2s - 1}(2a)_{2r+2s}(a + 1)_{r+s}(M - r - s)!}{r!s!M!(N + 2a + 1)_{r+s}(a)_{r+s}}$$

$$\cdot \frac{(-x)_r(-y)_r(x - N)_s(y - M)_s}{(\alpha + 1)_r(\beta + 1)_s}{_3F_2}\left[\begin{matrix} 2N + 2a - 1, N + a + 1/2, N - r - s - 1; R \\ N + a - 1/2, N + r + s + 2a + 1 \end{matrix}\right]$$

by means of the transformation formula [3, 4.5 (3)], from which it follows that inequality (6) holds under the stated conditions.

Similarly, by means of the transformation formula [3, 2.9, (2)], we find that

$$\sum_{n=0}^{N} R^n \frac{(\alpha+1)_n(\alpha+\beta+1)_n(-1)^n(-N)_n}{n!(N+\alpha+\beta+2)_n(\beta+1)_n} Q_n(x;\alpha,\beta,N)Q_n(y;\alpha,\beta,M)$$

$$= \sum_{r=0}^{N}\sum_{s=0}^{N-r} \frac{R^{r+s}(1-R)^{2N+1-2r-2s}(2a)_{2r+2s}(M-r-s)!}{r!s!M!(N+2a+1)_{r+s}}$$

$$\cdot \frac{(-x)_r(-y)_r(x-N)_s(y-M)_s}{(\alpha+1)_r(\beta+1)_s} {}_2F_1\left[\begin{array}{c} 2N+2a+1, N+1-r-s; R \\ N+r+s+2a+1 \end{array}\right] \geq 0,$$

when $0 \leq R \leq 1$, $2a = \alpha + \beta + 1 \geq 0$, $x = 0, 1, \cdots, N$, $y = 0, 1, \cdots, M$ and $M \geq N$.

Remarks. Bailey's formula [1, (2.3)] for the Poisson kernel for Jacobi series and his formula [1, (2.1)] are limit cases of the above two formulas. The nonnegativity of the sums (1), (2) and the case $c = d$ of the left side of (5) can also be obtained by using the observations in [6, Chap. III], [9, Prob. 82] and the symmetry relations $c_n(x;a) = c_x(n;a)$, $k_n(x)/k_n(0) = k_x(n)/k_x(0)$ and $(\beta)_x m_n(x;\beta,c) = (\beta)_n m_x(n;\beta,c)$. For the positivity (and total positivity) of some other sums, see [7].

In [2, Appendix I] Beckmann used a maximum principle for elliptic equations to prove the nonnegativity of the Poisson kernel for Jacobi series. It would be of interest if a similar proof (or some other proof) could be found for (6).

REFERENCES

[1] W. N. BAILEY, *The generating function of Jacobi polynomials*, J. London Math. Soc., 13 (1938), pp. 8–12.

[2] P. BECKMANN, *Orthogonal Polynomials for Engineers and Physicists*, The Golem Press, Boulder, Colorado, 1973.

[3] A. ERDÉLYI, W. MAGNUS, F. OBERHETTINGER AND F. G. TRICOMI, *Higher Transcendental Functions*, vols. I and II, McGraw-Hill, New York, 1953.

[4] G. GASPER, *Nonnegativity of a discrete Poisson kernel for the Hahn polynomials*, J. Math. Anal. Appl., 42 (1973), pp. 438–451.

[5] S. KARLIN AND J. L. MCGREGOR, *The Hahn polynomials, formulas and applications*, Scripta Math. 26 (1961), pp. 33–46.

[6] ———, *The differential equations of birth-and-death processes and the Stieltjes moment problem*, Trans. Amer. Math. Soc., 85 (1957), pp. 489–546.

[7] ———, *Classical diffusion processes and total probability*, J. Math. Anal. Appl., 1 (1960), pp. 163–183.

[8] J. MEIXNER, *Erzeugende Funktionen der Charlierschen Polynome*, Math. Z., 44 (1939), pp. 531–535.

[9] G. SZEGÖ, *Orthogonal Polynomials*, Colloquium Publications, vol. 23, American Mathematical Society, New York, 1967.

The Entropy of a Poisson Distribution

Problem 87-6*, by C. ROBERT APPLEDORN (Indiana University).

The following problem arose during a study of data compression schemes for digitally encoded radiographic images. Specifically, Huffman optimum (minimum) codes are employed to reduce the storage requirements for high resolution gray-level images, for example, 2048×2048 pixels by 10 bits per pixel for a single image. Entropy estimates provide a lower bound for the average storage in terms of bits per pixel that can be achieved using these coding methods.

Due to the nature of the problem, the Poisson probability density function with mean value parameter m is of particular interest:

$$p(k) = \frac{m^k e^{-m}}{k!}.$$

The entropy H associated with this probability density function is given by

$$H = -\sum_{k=0}^{\infty} p(k) \log p(k).$$

The problem that develops during the entropy calculation is to determine a closed-form solution to the following summation:

$$\sum_{k=0}^{\infty} \log (k!) \frac{m^k e^{-m}}{k!}.$$

It is *conjectured* by the author that for large m (say, $m > 10$), the entropy H can be approximated by

$$H = \tfrac{1}{2} \log (2\pi em).$$

Prove or disprove.

Solution by RONALD J. EVANS (University of California, San Diego).

Let $\varphi(x) = \log \Gamma(x + 1)$. By definition of the entropy H,

$$H = \sum_{k=0}^{\infty} p(k)(\varphi(k) + m - k \log m) = m(1 - \log m) + \varphi_{\infty},$$

where

$$\varphi_{\infty} = \sum_{k=0}^{\infty} p(k)\varphi(k).$$

By [2, Chap. 3, Entry 10], [3, Entry 10], as (real) $m \to \infty$,

$$\varphi_{\infty} = \varphi(m) + \frac{m\varphi^{(2)}(m)}{2} + \frac{m\varphi^{(3)}(m)}{6} + \frac{m^2\varphi^{(4)}(m)}{8} + O(m^{-2}).$$

(Several minor errors in [2, Chap. 3, §10] and [3, §10] are corrected in [4].) Substituting in the well-known asymptotic expansions for the derivatives $\varphi^{(n)}(m)$ [1, (6.1.41), (6.3.18), (6.4.11)], we deduce that, as $m \to \infty$,

$$H = \frac{1}{2} \log (2\pi em) - \frac{1}{12m} + O(m^{-2}).$$

REFERENCES

[1] M. ABRAMOWITZ AND I. STEGUN, eds., *Handbook of Mathematical Functions*, Dover, New York, 1965.

[2] B. C. BERNDT, *Ramanujan's Notebooks, Part I*, Springer-Verlag, New York, 1985.

[3] B. C. BERNDT, R. J. EVANS, AND B. M. WILSON, *Chapter 3 of Ramanujan's Second Notebook*, Adv. in Math., 49 (1983), pp. 123–169.

[4] R. J. EVANS, *Ramanujan's Second Notebook: Asymptotic expansions for hypergeometric series and related functions*, in Proc. Ramanujan Centenary Conference, Academic Press, New York, 1988.

Solution by J. BOERSMA (Eindhoven University of Technology, Eindhoven, the Netherlands).

It is easily seen that the entropy H associated with the Poisson probability density function is given by

$$H=-\sum_{k=0}^{\infty} p(k) \log p(k)=m-m \log m+\sum_{k=0}^{\infty} \log (k!)\frac{m^k e^{-m}}{k!}.$$

In the latter series replace $\log (k!)$ by Malmstén's representation [1, Formula 1.9(1)]

$$\log (k!)=\int_0^{\infty}\left[k-\frac{1-e^{-kt}}{1-e^{-t}}\right]\frac{e^{-t}}{t} dt;$$

then it is found that

(*)
$$H=m-m \log m+\int_0^{\infty}\left\{m-\frac{1-\exp [m(e^{-t}-1)]}{1-e^{-t}}\right\}\frac{e^{-t}}{t} dt$$

$$=m-m \log m-\int_0^1\left[m-\frac{1-e^{-ms}}{s}\right]\frac{ds}{\log (1-s)}$$

by the substitution $1-e^{-t}=s$.

The final integral in (*) is decomposed into

$$-m\int_0^1\left[\frac{1}{\log (1-s)}+\frac{1}{s}\right]ds+\int_0^1\left(\frac{m}{s}-\frac{1-e^{-ms}}{s^2}\right)ds+\frac{1}{2}\int_0^1\frac{1-e^{-ms}}{s} ds$$

$$+\int_0^1\frac{1}{s}\left[\frac{1}{\log (1-s)}+\frac{1}{s}-\frac{1}{2}\right]ds-\int_0^1\frac{e^{-ms}}{s}\left[\frac{1}{\log (1-s)}+\frac{1}{s}-\frac{1}{2}\right]ds,$$

whereby the successive integrals are shortly denoted by $I_1, I_2, \cdots, I_5$. By back substitution $s=1-e^{-t}$ it is found that

$$I_1=m\int_0^{\infty}\left(\frac{1}{t}-\frac{1}{1-e^{-t}}\right)e^{-t} dt=-m\gamma,$$

where γ denotes Euler's constant (cf. [1, Formula 1.7(19)]). Through an integration by parts the second integral reduces to

$$I_2=\frac{1-e^{-ms}}{s}\bigg|_0^1+m\int_0^1\frac{1-e^{-ms}}{s} ds=1-e^{-m}-m+m\int_0^m\frac{1-e^{-t}}{t} dt$$

$$=1-e^{-m}-m+m[E_1(m)+\log m+\gamma],$$

in which $E_1(m)=\int_m^{\infty} e^{-t}t^{-1} dt$ stands for the exponential integral (cf. [1, Formulae 9.7(1), (5)]). In the same manner we have

$$I_3=\frac{1}{2}\int_0^m\frac{1-e^{-t}}{t} dt=\frac{1}{2}[E_1(m)+\log m+\gamma].$$

In the fourth integral we substitute $s=1-e^{-t}$, which yields

$$I_4=\int_0^{\infty}\left[\frac{e^{-t}}{(1-e^{-t})^2}-\frac{1+\frac{1}{2}t}{1-e^{-t}}\frac{e^{-t}}{t}\right]dt;$$

next, I_4 is decomposed into

$$I_4 = \int_0^\infty \left[\frac{e^{-t}}{(1-e^{-t})^2} - \frac{1}{t^2} \right] dt + \int_0^\infty \left[\frac{1}{t^2} - \frac{1+\frac{1}{2}t}{1-e^{-t}} \frac{e^{-t}}{t} \right] dt$$

$$= \left(\frac{-1}{1-e^{-t}} + \frac{1}{t} \right) \Big|_0^\infty + \int_0^\infty \left[\frac{1}{t^2} - \left(1 + \frac{1}{2}t \right) \frac{e^{-t}}{t} \frac{1+\frac{1}{2}t}{e^t-1} \frac{e^{-t}}{t} \right] dt$$

$$= -\frac{1}{2} + \int_0^\infty \left[\frac{1-e^{-t}}{t^2} - \frac{e^{-t}}{t} - \frac{1}{2} e^{-t} \right] dt - \int_0^\infty \left[\frac{1}{e^t-1} - \frac{1}{t} + \frac{1}{2} \right] \frac{e^{-t}}{t} dt$$

$$- \frac{1}{2} \int_0^\infty \left(\frac{1}{e^t-1} - \frac{1}{t} \right) e^{-t} dt.$$

The latter integrals can be evaluated directly and by use of [1, Formulae 1.9(3), 1.7(19)], viz.

$$\int_0^\infty \left[\frac{1-e^{-t}}{t^2} - \frac{e^{-t}}{t} - \frac{1}{2} e^{-t} \right] dt = -\frac{1-e^{-t}}{t} \Big|_0^\infty - \frac{1}{2} = \frac{1}{2},$$

$$\int_0^\infty \left[\frac{1}{e^t-1} - \frac{1}{t} + \frac{1}{2} \right] \frac{e^{-t}}{t} dt = 1 - \frac{1}{2} \log (2\pi),$$

$$\int_0^\infty \left(\frac{1}{e^t-1} - \frac{1}{t} \right) e^{-t} dt = \int_0^\infty \left[-1 + \frac{1}{1-e^{-t}} - \frac{1}{t} \right] e^{-t} dt = \gamma - 1.$$

Thus we find

$$I_4 = -\frac{1}{2} + \frac{1}{2} \log (2\pi) - \frac{1}{2} \gamma.$$

By inserting the previous results into (∗), we obtain the representation

$$H = \frac{1}{2} \log (2\pi em) + \left(m + \frac{1}{2} \right) E_1(m) - e^{-m} - \int_0^1 \frac{e^{-ms}}{s} \left[\frac{1}{\log(1-s)} + \frac{1}{s} - \frac{1}{2} \right] ds.$$

It does not seem possible to evaluate the latter integral in closed form; however, an evaluation by numerical integration is certainly feasible. The asymptotics of the integral, as $m \to \infty$, is readily established by use of Watson's lemma. To that end we insert the Taylor series

$$\frac{1}{\log(1-s)} + \frac{1}{s} - \frac{1}{2} = \sum_{k=1}^\infty a_k s^k, \qquad |s| < 1$$

with coefficients $a_1 = \frac{1}{12}$, $a_2 = \frac{1}{24}$, $a_3 = \frac{19}{720}$, etc., and integrate term by term. Then the complete asymptotic expansion of the entropy H is found as

$$H \sim \frac{1}{2} \log (2\pi em) - \sum_{k=1}^\infty a_k (k-1)! m^{-k} \qquad (m \to \infty)$$

apart from exponentially small terms of order $O(e^{-m})$.

REFERENCE

[1] A. ERDÉLYI, W. MAGNUS, F. OBERHETTINGER, AND F. G. TRICOMI, *Higher Transcendental Functions*, Vols. I, II, McGraw–Hill, New York, 1953.

N. M. BLACHMAN (GTE Government Systems Corp., Mountain View, California) notes that since the Poisson distribution has variance m and becomes approximately normal for large m, its entropy becomes that part of the $N(m, n)$ distribution, viz., $\log \sqrt{2\pi e m}$.

A. A. JAGERS (Universiteit Twente, Enschede, the Netherlands) shows that

$$\tfrac{1}{2}\log(2\pi m) < H < \tfrac{1}{2}\log(2\pi m) + 1.$$

Problem 74-13, *A Probabilistic Inequality*, by F. W. STEUTEL (Technische Hogeschool Eindhoven, the Netherlands).

It is known that if X is a random variable with $0 \leq X \leq 1$ a.s., then var $X \leq \tfrac{1}{4}$, with equality only when $P(X=0) = P(X=1) = \tfrac{1}{2}$. Prove more generally that

$$\sum_{n=0}^{\infty} \{\mu_{n+2} - \mu_{n+1}^2/\mu_n\} \leq \tfrac{1}{4},$$

where $\mu_k = E(X^k)$, $0 \leq X \leq 1$ a.s., and $P(X>0) > 0$.

The inequality arose in a problem on Bayesian estimation.

Solution by DANIEL KLEITMAN (Massachusetts Institute of Technology).

The desired inequality and a stronger result follow from performing an Abel partial summation on the left-hand expression, after subtracting one from each term in the bracket; thus

$$\sum_{n=0}^{\infty} \mu_{n+1}\left(\frac{\mu_{n+2}}{\mu_{n+1}} - \frac{\mu_{n+1}}{\mu_n}\right)$$

$$= \sum_{n=0}^{\infty} \mu_{n+1}\left(\frac{\mu_{n+2} - \mu_{n+1}}{\mu_{n+1}} - \frac{\mu_{n+1} - \mu_n}{\mu_n}\right)$$

$$= \mu_1(1 - \mu_1) - \sum_{n=0}^{\infty} \frac{(\mu_{n+1} - \mu_{n+2})^2}{\mu_{n+1}}$$

$$\leq \mu_1(1 - \mu_1) - \frac{(\mu_1 - \mu_2)^2}{\mu_1} \leq \mu_1(1 - \mu_1) \leq \tfrac{1}{4}.$$

It is clear that $(\mu_1 - \mu_2)^2/\mu_1$ vanishes only if $P(X=0) + P(X=1) = 1$, and $\mu_1(1 - \mu_1) = \tfrac{1}{4}$ only if $\mu_1 = \tfrac{1}{2}$, so that equality holds only if $P(X=0) = P(X=1) = \tfrac{1}{2}$.

A Probabilistic Inequality

Problem 78-16, *by* L. A. SHEPP *and* A. M. ODLYZKO (Bell Laboratories).

Let $X_1, X_2, \cdots, X_n$ be independent random variables and let $Y_i = f_i(X_i)$ where $f_i(x)\uparrow$, $i = 1, 2, \cdots, n$. Prove or disprove that if $A = \{X_1 + X_2 + \cdots + X_i \geq 0$, $i = 1, 2, \cdots, n\}$ and $B = \{Y_1 + Y_2 + \cdots Y_i \geq 0, i = 1, 2, \cdots, n\}$, then $P(A|B) \geq P(A)$. This problem arose in an economic model of bankruptcies among dependent firms, in an earlier version due to R. W. Rosenthal.

Solution by D. MCLEISH (University of Alberta).

The problem is a special case of the following: Let $X_1, X_2, \cdots, X_n$ be independent random variables and $S(X_1, X_2, \cdots, X_n)$ and $T(X_1, X_2, \cdots, X_n)$ be any two statistics which are nondecreasing functions of each component. Then,

$$E(ST) \geq (ES)(ET).$$

The proof is based on the well-known inequality: If $f(x)$ and $g(x)$ are nondecreasing functions and X an arbitrary random variable, then

(†) $E\{f(X)g(X)\} \geq \{Ef(X)\}\{Eg(X)\}.$

This inequality follows from the fact that if Y is independent of X having the same distribution,

$$0 \leq E[f(X) - f(Y)][\mathring{g}(X) - g(Y)] = 2E\{f(X)g(X)\} - 2\{Ef(X)\}\{Eg(X)\}.$$

The proof is by induction: the statement of the theorem is verified by (†) in the case $n = 1$. Suppose the theorem is true for $n - 1$. Then, observe that

$$E[S | X_1, X_2, \cdots, X_{n-1}] = \int_{-\infty}^{\infty} S(X_1, X_2, \cdots, X_{n-1}, y) \, dP_n(y)$$

where P_n is the distribution of X_n, and this is clearly a nondecreasing function of the components $X_1, X_2, \cdots, X_{n-1}$. Therefore applying (†) with expectation replaced by conditional expectation, we obtain

$$EST = E[E(ST | X_1, X_2, \cdots, X_{n-1})] \geq E\{E(S | X_1, X_2, \cdots, X_{n-1})E(T | X_1, X_2, \cdots, X_{n-1})$$

$$\geq \{ES\}\{ET\}$$

where the last step follows from the induction hypothesis.

To obtain the result as stated in the problem, let S be the indicator function of the set A and T the indicator of B. This argument applies to obtaining the "Kimball" inequality for bounding the level of significance for a family of tests in multifactor analysis of variance.

A Dice Problem

Problem 80-5* *by* M. S. KLAMKIN *and* A. LIU (University of Alberta).
    Given n identical polyhedral dice whose corresponding faces are numbered identically with arbitrary integers:
    (a) **Prove or disprove** that if the dice are tossed at random, the probability that the sum of the bottom n face numbers is divisible by n is at least $1/2^{n-1}$.
    (b) **Determine** the maximum probability for the previous sum being equal to $k \pmod{n}$ for $k = 1, 2, \cdots, n-1$.
    In (a), the special case for $n = 3$ was set by the first author as a problem in the 1979 U.S.A. Mathematical Olympiad.

An Optimal Error Distribution Problem

Problem 86-6* *by* RICHARD D. SPINETTO (University of Colorado).
    In many audit populations items may have partial errors. Suppose each item in the population has an error size known only to be in the interval $[0, 1]$. Suppose the mean population error is m where $0 < m < 1$. A simple random sample of size n is drawn with replacement from that population. Let S be the random variable representing the sum

of the error sizes of the n sampled items. Given a constant $c < nm$ how should the error sizes be distributed in the population to maximize $P(S \leq c)$?

It is conjectured that for each m and c, there is a population with just two error sizes, one of which is 0 or 1, such that $P(S \leq c)$ is maximized. Prove or disprove.

If this conjecture is true then it will be possible to determine simple bounds on upper confidence limits for some audit sampling problems.

On a Unique Minimum

Problem 86-11*, *by* TZE-SAN LEE (Western Illinois University).

Let $f(x) = \sum_{i=1}^{n} (c^2 b_i + x^2 a_i^2)(x + b_i)^{-2}$, $n \geq 2$, and $b_i > 0$ for all i. It can be shown that $f(x)$ always has a local minimum in $(0, \infty)$. Determine necessary and sufficient conditions for $f(x)$ to have a unique minimum.

This problem arises from computing the optimal ridge parameter in statistics.[1] Here x denotes the ridge parameter, $f(x)$ the mean squared error function, a_i and b_i the regression coefficients and eigenvalues, respectively, of $X^t X$ where X is the design matrix of the regression model. The constant c^2 is the common variance of the error term in the model.

Supplementary References
Probability

[1] J. R. BAILEY, *"Estimation from first principles,"* Math. Gaz. (1973) pp. 169–174.

[2] R. E. BARLOW AND F. PROSCHAN, *Statistical Theory of Reliability and Life Testing,* Holt, Rinehart & Winston, N.Y., 1975.

[3] G. A. BARNARD, *"Two aspects of statistical estimation,"* Math. Gaz. (1974) pp. 116–124.

[4] M. N. BARBER AND B. W. NINHAM, *Random and Restricted Walks: Theory and Applications,* Gordon and Breach, N.Y., 1970.

[5] D. R. BARR, *"When will the next record rainfall occur?"* MM (1972) pp. 15–19.

[6] M. S. BARTLETT, *An Introduction to Stochastic Processes with Special References to Methods and Applications,* Cambridge University Press, Cambridge, 1980.

[7] V. E. BENES, *Mathematical Theory of Connecting Networks and Telephone Traffic,* Academic, N.Y., 1965.

[8] A. T. BHARUCHA-REID, *Elements of the Theory of Markov Processes and their Applications,* McGraw-Hill, N.Y., 1960.

[9] D. BLACKWELL AND M. A. GIRSCHICK, *The Theory of Games and Statistical Decisions,* Wiley, N.Y., 1954.

[10] L. BREIMAN, *Probability and Stochastic Processes,* Houghton Mifflin, Boston, 1969.

[11] A. CHARNES AND W. W. COOPER, *"Deterministic equivalents for optimizing and satisfying under chance constraints,"* Oper. Res. (1963) pp. 18–39.

[12] L. E. CLARKE, *"How long is a piece of string?"* Math. Gaz. (1971) pp. 404–407.

[13] A. C. COLE, *"The development of some aspects of teletraffic theory,"* BIMA (1975) pp. 85–93.

[14] E. P. COLEMAN, *"Statistical decision procedures in industry I: Control charts by variables,"* MM (1962) pp. 129–143.

[15] R. B. COOPER, *Introduction to Queueing Theory,* Collier-Macmillan, N.Y., 1972.

[16] D. COX, *Renewal Theory,* Methuen, London, 1962.

[17] D. R. COX AND H. D. MILLER, *The Theory of Stochastic Processes,* Chapman and Hall, London, 1977.

[18] H. CRAMER, *Mathematical Methods of Statistics,* Princeton University Press, Princeton, 1946.

[19] H. A. DAVID, *Order Statistics,* Wiley, N.Y., 1970.

[20] M. DEGROOT, *Optimal Statistical Decisions,* McGraw-Hill, N.Y., 1970.

[21] J. DERDERIAN, *Maximin hedges,* MM (1978) 188–192.

[22] P. DIACONIS AND B. EFRON, *"Computer-intensive methods in statistics,"* Sci. Amer. (1983) pp. 116–130, 170.

[23] J. M. DOBBIE, *"Search theory: A sequential approach,"* Naval Res. Logist. Quart. (1963) pp. 323–334.

[24] S. F. EBEY AND J. J. BEAUCHAMP, *"Larval fish, power plants, and Buffon's needle problem,"* AMM (1977) pp. 534–541.

[25] B. EFRON AND C. MORRIS, *"Steins' paradox in statistics,"* Sci. Amer. (1977) pp. 119–127.

[26] R. A. EPSTEIN, *The Theory of Gambling and Statistical Logic,* Academic, N.Y., 1967.

[27] W. B. FAIRLEY AND F. MOSTELLER, eds., *Statistics and Public Policy,* Addison-Wesley, Reading, 1977.

[28] W. F. FELLER, *An Introduction to Probability Theory and its Applications I, II,* Wiley, N.Y., 1961, 1971.

[29] J. FORD, "How random is a coin toss?" Phys. Today (1983) pp. 40–47.

[30] D. P. GAVER AND G. L. THOMPSON, Programming and Probability Models in Operations Research, Brooks/Cole, Monterey, 1973.

[31] R. GEIST AND K. TRIVEDI, "Queueing network models in computer system design," MM (1982) pp. 67–80.

[32] N. GLICK, "Breaking records and breaking boards," AMM (1978) pp. 2–26.

[33] B. K. GOLD, "Statistical decision procedures in industry II: Control charts by attributes," MM (1962) pp. 195–210.

[34] V. L. GRAHAM AND C. I. TULCEA, Casino Gambling, Van Nostrand Reinhold, 1978.

[35] J. S. GROWNEY, "Planning for interruptions," MM (1982) pp. 213–219.

[36] E. J. GUMBEL, Statistics of Extremes, Columbia University Press, New York, 1958.

[37] D. HAGHIGHI-TALAB AND C. WRIGHT, "On the distributions of records in a finite sequence of observations with an application to a road traffic problem," J. Appl. Probability (1973) pp. 556–571.

[38] A. HALD, Statistical Theory with Engineering Applications, Wiley, N.Y., 1962.

[39] N. A. J. HASTINGS AND J. B. PEACOCK, Statistical Distributions: A Handbook for Students and Practitioners, Halsted Press, N.Y. 1975.

[40] J. M. HOWELL, "Statistical decision procedures in industry III: Acceptance sampling by attributes," MM (1962) pp. 259–268.

[41] R. ISAACS, "Optimal horse race bets," AMM (1953) pp. 310–315.

[42] R. JAGANNATHAN, "Chance-constrained programming with joint constraints," Oper. Res. (1974) pp. 358–372.

[43] N. L. JOHNSON AND S. KOTZ, Urn Models and their Applications, Wiley, N.Y., 1977.

[44] K. JORDAN, Chapters on the Classical Calculus of Probability, Akademiai Kiado, Budapest, 1972.

[45] F. P. KELLY, Reversibility and Stochastic Networks, Wiley, N.Y., 1979.

[46] M. G. KENDALL AND P. A. P. MORAN, Geometrical Probability, Griffin, London, 1963.

[47] M. S. KLAMKIN, "On the uniqueness of the distribution function for the Buffon needle problem," AMM (1953) pp. 677–680.

[48] M. S. KLAMKIN AND D. J. NEWMAN, "Extensions of the birthday surprise," J. Comb. Th. (1967) pp. 279–282.

[49] M. S. KLAMKIN AND J. H. VAN LINT, "An asymptotic problem in renewal theory," Stat. Neerlandica (1972) pp. 191–196.

[50] L. KLEINROCK, Communication Nets: Stochastic Message Flow and Delay, Dover, New York, 1972.

[51] D. E. KNUTH, "The toilet paper problem," AMM (1984) pp. 465–470.

[52] B. O. KOOPMAN, "Search and its optimization," AMM (1979) pp. 527–540.

[53] L. H. LIYANGE, C. M. GULATI AND J. M. HILL, A Bibliography on Random Walks, Math. Dept., University of Wollongong, Australia.

[54] C. L. MALLOWS AND N. J. A. SLOANE, "Designing an auditing procedure, or how to keep bank managers on their toes," MM (1984) pp. 142–151.

[55] N. R. MANN, R. E. SCAFER AND N. D. SINGPURWALLA, Methods for Statistical Analysis of Reliability and Life Data, John Wiley, Chichester, 1974.

[56] B. L. MILLER AND H. M. WAGNER, "Chance constrained programming with joint constraints," Oper. Res. (1965) pp. 930–945.

[57] R. G. MILLER, et al, Biostatistics Casebook, Wiley, N.Y., 1980.

[58] M. H. MILLMAN, "A statistical analysis of casino black jack," AMM (1983) pp. 431–436.

[59] O. B. MOAN, "Statistical decision, procedures in industry IV: Acceptance sampling by variables," MM (1963) pp. 1–10.

[60] S. G. MOHANTY, Lattice Path Counting and Applications, Academic Press, N.Y., 1979.

[61] F. MOSTELLER, et al, eds., Statistics by Example. Vol. I, Exploring Data; Vol. II, Weighing Changes; Vol. III, Detecting Patterns; Vol. IV, Finding Models; Addison-Wesley, Reading, 1973.

[62] G. F. NEWELL, Applications of Queueing Theory, Chapman and Hall, London, 1971.

[63] D. J. NEWMAN AND L. SHEPP, "The double dixie cup problem," AMM (1960) pp. 58–61.

[64] D. J. NEWMAN AND W. E. WEISSBLUM, "Expectation in certain reliability problems," SIAM Rev. (1967) pp. 744–747.

[65] G. C. PAPANICOLAOU, "Stochastic equations and their applications," AMM (1973) pp. 526–544.

[66] A. RENYI, Probability Theory, North-Holland, Amsterdam, 1970.

[67] H. ROBBINS, "Optimal stopping," AMM (1970) pp. 333–343.

[68] M. ROSENBLATT, ed., Studies in Probability Theory, MAA, 1978.

[69] H. SOLOMON, Geometric Probability, SIAM, 1978.

[70] F. SPITZER, Principles of Random Walks, Springer-Verlag, N.Y., 1976.

[71] L. D. STONE, "Search theory: A mathematical theory for finding lost objects," MM (1977) pp. 248–256.

[72] ———, Theory of Optimal Search, Academic, N.Y., 1975.

[73] J. L. SYNGE, "The problem of the thrown string," Math. Gaz. (1970) pp. 250–260.

[74] R. SYSKI, Introduction to Congestion Theory in Telephone Systems, Oliver and Boyd, London, 1960.

[75] L. TAKARS, Combinatorial Methods in the Theory of Stochastic Processes, Krieger, N.Y., 1977.

[76] K. TRIVEDI, Probability and Statistics with Reliability, Queueing, and Computer Science Applications, Prentice-Hall, N.J., 1982.

[77] S. VAJDA, *Probabilistic Programming,* Academic, N.Y., 1972.
[78] G. H. WEISS AND R. J. RUBIN, *"Random walks: Theory and selective applications,"* Adv. in Chem. Phys. (1982) pp. 363–505.
[79] S. S. WILKS, *Mathematical Statistics,* Wiley, N.Y., 1962.
[80] A. WUFFLE, *"The pure theory of elevators,"* MM (1982) pp. 30–37.
[81] S. ZACKS, *The Theory of Statistical Inference,* Wiley, N.Y., 1971.
[82] N. ZADEH, *"Computation of optimal poker strategies,"* Oper. Res. (1977) pp. 541–562.

4. COMBINATORICS

Problem 68-16*, A Combinatorial Problem, by MELDA HAYES (Ocean Technology, Inc.).

In how many ways can n identical balls be distributed in r boxes in a row such that each pair of adjacent boxes contains at least 4 balls?

This problem arose in some work on submarine detection.

Editorial note. It would also be of interest if any results could be given for the more general problem: In two dimensions, consider an $m \times n$ array of boxes and some particular polydominoe:

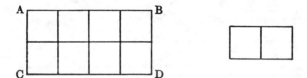

How many different ways can one distribute N identical balls in the mn boxes such that each polydominoe of the given form contains at least r balls and at most s balls. If we wish to symmetrize the problem by getting rid of some or all end effects, we can assume AB is connected to DC as in a cylinder, or AB is connected to DC and AD is connected to BC as in a torus. In the original problem, this corresponds to having the r boxes in a circle instead of a row. [M.S.K.]

Solution[2] by D. R. BREACH (University of Toronto).

This problem belongs to a class of combinatorial problems in which ordered selections, repetitions allowed, are to be made from a supply of objects of various types under the conditions that an object of one type may not immediately follow objects of certain other types. In § 1 a general method for constructing generating functions and recurrence relations for these problems is described. It depends on simple matrix algebra. In § 2 an analysis of the problem is given leading to a table of solutions for up to 10 boxes and 20 balls.

1. Generating functions for ordered selections. Given an unlimited supply of objects of p various types $x_1, x_2, \cdots, x_p$, these are to be placed in an ordered row of boxes with each box containing one object. The content of the nth box depends on the content of the $(n - 1)$th box; that is to say, there are adjacency restrictions. These restrictions are specified by the $p \times p$ adjacency matrix A with entries a_{ij} where

$$(1.1) \qquad\qquad a_{ij} = x_i,$$

if an x_i can follow an x_j, and

$$(1.2) \qquad\qquad\qquad\qquad a_{ij} = 0,$$

if an x_i cannot follow an x_j.

If the first n boxes are filled according to these restrictions, then we have an ordered selection of n objects with repetitions allowed from a supply of p types with adjacency restrictions. Call this an n-selection for brevity. The set of all n-selections for a particular n can be divided into subsets according to the content of the nth box. Let $f(n, i)$ be the enumerator of n-selections with type i in the last box. Then $f(n, i)$ will be a polynomial of degree n in the x_p in which the coefficient of any term gives the number of ways of ordering, under the given conditions, an n-combination of the x_p.

The $f(n, i)$ can be taken as the p components of a column vector $\mathbf{f}_n$, i.e.,

$$(1.3) \qquad\qquad \mathbf{f}_n = [f(n, 1), f(n, 2), \cdots, f(n, p)]^{\mathsf{T}}.$$

The total number $F(n)$ of n-selections is the sum of the components of $\mathbf{f}_n$. The summation is conveniently done by forming the scalar product $\mathbf{c} \cdot \mathbf{f}_n$ where

$$(1.4) \qquad\qquad\qquad \mathbf{c} = (1, 1, \cdots, 1)$$

and perhaps deserves the name "counting vector." Thus

$$(1.5) \qquad\qquad\qquad F(n) = \mathbf{c} \cdot \mathbf{f}_n.$$

Now in forming $(n + 1)$-selections, objects are added to the $(n + 1)$th box according to the restrictions specified by the matrix A. This relates $\mathbf{f}_{n+1}$ to $\mathbf{f}_n$ by

$$(1.6) \qquad\qquad\qquad \mathbf{f}_{n+1} = A\mathbf{f}_n.$$

Let

$$(1.7) \qquad\qquad G(t) = \mathbf{f}_1 t + \mathbf{f}_2 t^2 + \cdots + \mathbf{f}_n t^n \cdots$$

be a vector generating function for the $\mathbf{f}_i$. Then the recurrence relation (1.6) gives

$$(1.8) \qquad\qquad\qquad G(t) = \mathbf{f}_1 t + tAG(t),$$

whose solution for $G(t)$ is

$$(1.9) \qquad\qquad\qquad G(t) = [I - tA]^{-1}\mathbf{f}_1 t,$$

where I is the $p \times p$ identity matrix. On introducing the scalar generating function $G(t) = \sum F(n)t^n$, one has from (1.9)

$$(1.10) \qquad\qquad G(t) = \mathbf{c} \cdot \frac{\operatorname{adj}[I - tA]}{|I - tA|} \mathbf{f}_1 t = \sum_{n=1}^{\infty} F(n)t^n,$$

where $\operatorname{adj}[I - tA]$ is the adjoint matrix of $[I - tA]$. This is a generating function for the number of ordered n-selections with adjacency restrictions.

Assuming that an object of any type may occupy the first box, the vector $\mathbf{f}_1$ is

$$(1.11) \qquad\qquad\qquad \mathbf{f}_1 = (x_1, x_2, \cdots, x_p)^{\mathsf{T}}$$

and since it starts the generating function it may be called a starting vector. The term $c \cdot \text{adj}\,[I - At]f_1$ in (1.10) is then a linear combination of the cofactors of all the elements of $|I - At|$.

The counting vector and the starting vector may have other forms. For example, if $c = (0, 0, \cdots, 0, 1, 0, \cdots, 0)$, the nonzero entry being in the jth place, and $f_1 = (0, 0, \cdots 0, x_i, 0, \cdots, 0)$, then only the cofactor of the ijth element of $|I - At|$ appears in the product. This gives an enumeration of n-selections beginning with x_i and ending with x_j.

As a by-product of these methods a recurrence formula for $F(n)$ follows from the Cayley-Hamilton theorem; if

(1.12a) $$\phi(\lambda) \equiv |\lambda I - A| = 0,$$

then

(1.12b) $$\phi(A) = 0.$$

For n large enough this gives

(1.13) $$c \cdot \phi(A)A^{n-p-1}f_1 = 0,$$

which by (1.6) gives

(1.14) $$F(n) = \sum a(p - i)F(n - i), \qquad 1 \leq i \leq p,$$

where $a(p - i)$ is the coefficient of λ^{p-i} in $\phi(\lambda)$. Thus, once $|\lambda I - A|$ is evaluated a recurrence formula can be written down immediately.

2. Problem 68-16. This problem asks for the number of ways in which n balls can be placed in r boxes if each neighboring pair of boxes is to contain at least 4 balls. A generating function for the solution will be constructed by the above methods.

Suppose that an extra condition is added: that no box is to contain more than $p(> 4)$ balls. Then the objects in the supply may be taken as collections of $0, 1, 2, \cdots, p$ balls labelled according to their number so that x^i stands for a collection of i balls. The superscript notation is used because the total number of balls used must be counted. A 0-box can follow an i-box only if $i \geq 4$; a 1-box can follow an i-box only if $i \geq 3$, and so on. The adjacency matrix is then

(2.1)
$$A = \begin{bmatrix}
0 & 0 & 0 & 0 & 1 & 1 & \cdots & 1 \\
0 & 0 & 0 & x & x & x & \cdots & x \\
0 & 0 & x^2 & x^2 & x^2 & x^2 & \cdots & x^2 \\
0 & x^3 & x^3 & x^3 & x^3 & x^3 & \cdots & x^3 \\
x^4 & x^4 & x^4 & x^4 & x^4 & x^4 & \cdots & x^4 \\
x^5 & x^5 & x^5 & x^5 & x^5 & x^5 & \cdots & x^5 \\
\vdots & \vdots & \vdots & \vdots & \vdots & \vdots & \cdots & \vdots \\
x^p & x^p & x^p & x^p & x^p & x^p & \cdots & x^p
\end{bmatrix}.$$

Two extra 4-boxes may be introduced, one to appear first and one to appear last in the row. These may subsequently be removed without violating the conditions of the problem. This simplifies the algebra considerably. For the counting and starting vectors one then has

(2.2a) $$\mathbf{c} = (0,0,0,0,1,0,\cdots,0)$$

and

(2.2b) $$\mathbf{f}_1 = (0,0,0,0,x^4,0,\cdots,0)^{\mathsf{T}}.$$

The generating function is then

(2.3) $$G(t,x,p) = x^4 t\, D_{55}(|I - tA|)^{-1},$$

where D_{55} is the cofactor of the fifth term on the leading diagonal of $|I - At|$. The two extra 4-boxes introduced at the beginning and the end are represented by a term $x^8 t^2$, so the coefficient of $x^{n+8} t^{r+2}$ in (2.3) is the number of ways of distributing n balls into r boxes with no more than p balls per box and with each pair of neighbors containing at least 4.

By elementary operations on the last $p - 5$ rows and columns,

(2.4) $$|I - At| = \begin{bmatrix} 1 & 0 & 0 & 0 & -t \\ 0 & 1 & 0 & -tx & -tx \\ 0 & 0 & 1 - tx^2 & -tx^2 & -tx^2 \\ 0 & -tx^3 & -tx^3 & 1 - tx^3 & -tx^3 \\ -tzx^4 & -tzx^4 & -tzx^4 & -tzx^4 & 1 - tzx^4 \end{bmatrix}$$

(2.5) $$= 1 - (x^2 + x^3 + zx^4)t - (x^4 + zx^4 + zx^5)t^2$$
$$+ (x^6 + xz^6 + 2zx^7)t^3 + zx^8 t^4 - zx^{10} t^5,$$

where $z = \sum x^i$, $0 \le i \le p - 4$. The cofactor D_{55} has a similar expansion. If $|x| < 1$, then as $p \to \infty$, $z \to (1 - x)^{-1}$ and there is no restriction on the maximum number of balls in each box. After some elementary algebra it is found that

$$G(t,x,\infty) = x^4 t + \frac{x^8 t^2(1 - x)[1 + (1 + x)t - (2x + 1)x^2 t^2 - x^4 t^3 + x^6 t^4]}{(1 - x^2 t)[1 - x - x^3 t - (2 + x)x^4 t + x^8 t^4]}$$

(2.6) $$= x^4 t + x^8 t^2 + \frac{x^8 t^3}{1 - x} + x^{12}\frac{(5 - 4x)}{(1 - x)^2}t^4 + x^{12}\frac{(1 + x - x^5)}{(1 - x)^3}t^5$$

$$+ \frac{x^{16}}{(1 - x)^4}(15 - 20x + 6x^2)t^6$$

$$+ \frac{x^{20}}{(1 - x)^5}(1 + 3x + x^2 - 11x^5 + 7x^6)t^7 + \cdots$$

in which the coefficient of $x^{n+8}\, t^{r+2}$ is the solution to Melda Hayes' problem.
Also from (2.5) after letting $p \to \infty$ one has for the coefficient, $F(r)$, of $x^8 t^{r+2}$, for r sufficiently large:

(2.7) $$(1 - x)F(r) = x^2 F(r - 1) + 2x^4 F(r - 2) - (2x^6 + x^7)F(r - 3)$$
$$- x^8 F(r - 4) + x^{10} F(r - 5).$$

The expressions for $F(1), \cdots, F(5)$ are displayed in (2.6) and the generating function for r boxes may be determined recursively.

Now $F(r)$ will be a rational function of x with an expansion of the form $a(r, n)x^n$ where (r, n) is the number of ways of putting n balls into r boxes under the given conditions. Consideration of the coefficients of x^n in (2.7) then gives

$$a(r, n) - a(r, n-1) = a(r-1, n-2) + 2a(r-2, n-4) - 2a(r-3, n-6)$$

(2.8) $$- a(r-3, n-7) - a(r-4, n-8) + a(r-5, n-10).$$

TABLE 1

Boxes \ Balls	1	2	3	4	5	6	7	8	9	10
1	1									
2	1									
3	1									
4	1	5	1							
5	1	6	4							
6	1	7	9							
7	1	8	16							
8	1	9	25	15	1					
9	1	10	35	40	8					
10	1	11	46	76	31					
11	1	12	58	124	85					
12	1	13	71	185	190	35	1			
13	1	14	85	260	360	154	13			
14	1	15	100	350	610	424	76			
15	1	16	116	456	956	930	295			
16	1	17	133	579	1415	1775	889	70	1	
17	1	18	151	720	2005	3080	2188	448	19	
18	1	19	170	880	2745	4985	4652	1669	155	
19	1	20	190	1060	3658	7650	8891	4718	805	
20	1	21	211	1211	4762	11256	15686	11201	3136	126

This relation together with the expansions of $F(1), \cdots, F(5)$ were used to construct Table 1 where the results for $r \leq 10$, $n \leq 20$ are given.

If the restriction that each neighboring pair of boxes contain 4 balls is relaxed, then the number of ways of disposing of n balls is $\binom{n+r-1}{r-1}$ which is an upper bound for $a(r, n)$. If r is fixed and n allowed to become very large, then the dominant term on the right of (2.8) is $a(r-1, n-2)$. If the other terms on the right are neglected, the resulting equation is indeed satisfied by $a(r, n) = \binom{n+r-1}{r-1}$ which suggests that this is an asymptotic value for $a(r, n)$. This is reasonable, for if r is fixed then as $n \to \infty$ the proportion of configurations not satisfying the adjacency restrictions must decrease rapidly.

Acknowledgment. I am grateful to J. Riordan for his helpful advice.

Problem 71-6, Triangle on a Checkerboard, by D. J. NEWMAN (Yeshiva University).

Given a triangle and an infinite black and white checkerboard, show that the triangle can be placed on the board with all its vertices strictly in the black.

Solution by the proposer.

First tilt the triangle so that none of the sides project, in either board direction, onto a whole number of boxes. Now place one vertex on a horizontal line and slide till a second vertex is on a vertical one. By our initial precaution, it follows that neither of these two vertices is at a corner of a box and that the third one is interior to a box. Therefore, a slight motion vertically and a slight motion horizontally will not change the color for the third vertex but they can reverse either or both the colors of the other two vertices.

Problem 75-2, The Regiment Problem Revisited, by G. J. SIMMONS (Sandia Laboratories).

Is it possible to form a marching column of two's with $n - 1$ members from each of n regiments in such a way that every regiment is paired with every other regiment and no two members of the same regiment have fewer than the obvious maximum-minimum of $[(n - 3)/2]$ ranks separating them?

Solution by Class 18.325 (Massachusetts Institute of Technology).

The odd n case is most interesting since one can obtain a solution for the even, $2k$, case from one for the $2k + 1$ case by omitting all pairs involving any one regiment.

We seek an ordering of all pairs of positive integers up to $2k + 1$ such that the interval between successive occurrences of any integer is either $k - 2$ or $k - 1$.

A general construction for the odd k case (and in fact, one to which all possible constructions are isomorphic, as can be shown by detailed analysis) can be described as follows: Arrange the first $2k$ digits with uniform separations around a circle in the plane—and place the digit $2k + 1$ on the circle at both its easternmost and westernmost points. This circle is to be rotated (say clockwise) along with all the digits except $2k + 1$. When any digit lies due north of another and/or a digit comes directly in contact with a $2k + 1$, the circle is to be stopped and the pairs of integers having identical east-west coordinates are to be read off from west to east, and the rotation then resumed. The resulting sequences is as desired. Here is the $k = 3$ example:

$$7 \rightarrow \begin{array}{ccc} 1 & 2 & 3 \\ & & \\ 6 & 5 & 4 \end{array} \rightarrow 7.$$

The pairs are (16), (25), (34), R(76), (15), (24), (37), R(65), (14), (23), R(57), (64), (13), (27), R(54), (63), (12), R(47), (53), (62), (17). The symbol R indicates where rotations take place. It is clear that when one integer is involved in a pair so are all but perhaps the integer $2k + 1$, that between two occurrences of an integer in a pair the desired interval exists and that all pairs will appear before there are duplications.

Solution by the proposer.

Let $\mathscr{P}$ be the permutation

$$\mathscr{P} = (1)(3, 5, 7, \cdots, 6, 4, 2);$$

then the sequence of pairs

$$S_n, \mathscr{P}S_n, \mathscr{P}^2 S_n, \cdots, \mathscr{P}^L S_n$$

is a solution where $L = (n - 3)/2$ and S_n is the set of n pairs

$$(1, 2)(3, 4) \cdots (n - 2, n - 1)(1, n)(2, 3) \cdots (n - 1, n)$$

if n is odd, or else $L = n - 2$ and S_n is the set of $n/2$ pairs

$$(1, 2)(3, 4) \cdots (n - 1, n)$$

if n is even.

It is to be noted that for n odd, except for permutations of the regiment's · colors or reversing end for end the column of two's, the marching order is uniquely determined. On the other hand, for $n = 6$ there are 7,360 inequivalent columns of two's possible, and for $n = 8$, even a CDC 7600 has thus far been unable to compute the number of possible orderings of pairs.

A Covering Problem

Problem 82-2*, by D. J. NEWMAN (Temple University).

Given any collection of squares with total area 3, prove that they can cover the unit square. If the sides of the covering squares are all to be parallel to the sides of the unit square, then 3 is best possible. For the analogous problem with hypercubes in E_n, 3 is replaced by $2^n - 1$.

Solution by A. MEIR (University of Alberta).

The result follows, as a special case, from [1, Thm. 3], which states: A collection of n-dimensional cubes of sides $x_1 \geq x_2 \geq \cdots \geq x_N$ can cover a rectangular parallelpiped of size $a_1 \times a_2 \times \cdots \times a_n$ if

$$(1) \qquad \sum_{i=1}^{N} x_i^n \geq \prod_{j=1}^{n} (a_j + x_1) - x_1^n$$

The covering is possible with the sides of the cubes parallel to the sides of the parallelpiped.

If we set $a_1 = a_2 = \cdots = a_n = 1$, then (1) becomes

$$(2) \qquad \sum_{i=1}^{N} x_i^n \geq (1 + x_1)^n - x_1^n,$$

and since we may, obviously, assume that $x_1 \leq 1$, (2) will certainly be satisfied if

$$\sum_{i=1}^{N} x_i^n \geq 2^n - 1.$$

Since no proof of the result was given in [1] we give the proof here for the case $n = 2$. (A proof can be given similarly for E^n).

Thus we assume that

(3)
$$\sum_{i=1}^{N} x_i^2 \geq ab + (a + b)s_1,$$

and will prove that the squares with sides $x_1 \geq x_2 \geq \cdots \geq x_N$ can cover a rectangle of size $a \times b$, $a \leq b$. If $N = 1$, then $x_1 \geq b$, so the statement is trivial. Suppose we proved the statement for $N = 1, \cdots, m - 1$. Now let $N = m$ and also that j be the smallest integer such that $x_1 + x_2 + \cdots + x_j \geq a$. We cover the rectangle (see figure) $a \times x_j$ by squares of sides $x_1, \cdots, x_j$. The remaining rectangle of size $a \times (b - x_j)$ can be covered by the squares

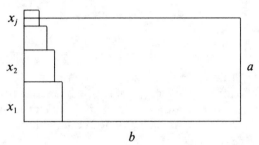

$x_{j+1}, \cdots, x_m$ according to our inductive hypothesis, if

(4)
$$\sum_{i=j+1}^{m} x_i^2 \geq a(b - x_j) + (a + b - x_j)x_{j+1}.$$

Now from (3) and by the choice of j

$$\sum_{i=j+1}^{m} x_i^2 \geq ab + (a + b)x_1 - \sum_{i=1}^{j} x_i^2$$

$$\geq ab + (a + b)x_1 - ax_1 - x_j^2$$

$$\geq ab + bx_j - x_j^2$$

$$\geq ab + (b - x_j)x_{j+1},$$

which proves (4) since $x_j \geq x_{j+1}$.

We remark that condition (3) is best possible in the following sense: For given $\varepsilon > 0$ let $M^2 + 1 > 2/\varepsilon$, $N = M^2 + 2M$, $x_1 = x_2 = \cdots = x_N = (M^2 + 1)^{-1/2}$. Then $\sum_{i=1}^{N} x_i^2 > 1 + 2x_1 - \varepsilon$ but squares with sides $x_1, \cdots, x_N$ cannot (by parallel covering) cover the unit square.

REFERENCE

[1] A. MEIR AND L. MOSER, *On packing of squares and cubes.* J. Combin. Theory, 5 (1968), pp. 126–134.

Comments by ANDRAS BEZDEK (Mathematical Institute of the Hungarian Academy of Sciences) *and* KAROLY BEZDEK (L. Eötvös University) Budapest, Hungary.

L. Fejes Tóth, independently of Newman, phrased a related problem in Oct, 1981. Namely:

Let K be a convex body of volume 1 in E_n. Denote by $f_n(K)$ the smallest positive number which has the following property. Given any collection of bodies, which are homothetic to K, with total volume $f_n(K)$, they can cover K. The problem is to determine

$f_n(K)$. L. Fejes Tóth conjectured that $2 \leq f_2(K) \leq 3$ and $f_n(C) = 2^n - 1$, where C denotes the unit hypercube in E_n.

Our paper *Eine hinreichende Bedingung für die Überdeckung des Einheitswürfels durch homothetische Exemplare in dem n-dimensionalen euklidischen Raum*, Beiträge zur Algebra und Geometrie, to appear, contains the following statements:

THEOREM 1. $f_n(C) = 2^n - 1$.

THEOREM 2. $2 \leq f_2(K) \leq 12$.

Also solved by the proposer.

Problem 76-17*, *A Reverse Card Shuffle*, by DAVID BERMAN and M. S. KLAMKIN (University of Waterloo).

The following problem, originating somewhere in England, was brought to our attention by G. Cross.

A deck of n cards is numbered 1 to n in random order. Perform the following operations on the deck. Whatever the number of the top card is, count down that many in the deck and turn the whole block over on top of the remaining cards. Then, whatever the number of the (new) top card, count down that many cards in the deck and turn this whole block over on top of the remaining cards. Repeat the process. Show that the number 1 will eventually reach the top.

Consider the following set of related and more difficult problems:

1. Determine the number of $N(k)$ of initial card permutations, so that the 1 first appears on top after k steps of the process. In particular, show that $N(0) = N(1) = N(2) = (n-1)!$ and that

$$N(3) = \begin{cases} (n-1)! - \frac{1}{2}(n-1)(n-3)(n-4)!, & n \text{ odd}, \\ (n-1)! - \frac{1}{2}(n-2)^2(n-4)!, & n \text{ even}. \end{cases}$$

(The method of the authors is apparently too unwieldy to determine $N(k)$ for $k > 3$).

2. Estimate the maximum number of steps it takes to get the 1 to the top.
3. For what n is there a unique permutation giving the maximum number of steps?
4. It is conjectured that the last step of a maximum step permutation leaves the cards in order (i.e., $1, 2, \ldots, n$).

Computer calculations give the following partial results:

n	Maximum number of steps	Number of maximum permutations
1	0	1
2	1	1
3	2	2
4	4	2
5	7	1
6	10	5
7	16	2
8	22	1
9	30	1

The first four steps of the maximum step permutation for $n = 9$ are:

Step	Permutation
0	615 97 28 34
1	279 51 68 34
2	729 51 68 34
3	861 59 27 34

Editorial note. D. E. KNUTH (Stanford University) notes that the card shuffle game here was shown to him in 1973 by J. H. CONWAY (Cambridge University) who proposed it and named it "topswaps." In the next year Knuth included part 2 on a take-home examination in the following form (also included is his solution):

Problem 3. Let $\pi = \pi[1]\pi[2] \ldots \pi[n]$ be a permutation of $\{1, 2, \ldots, n\}$, and consider the following algorithm:

```
begin integer array A[1:n]; integer k;
     (A[1], ... , A[n]) ← (π[1], ... , π[n]);
loop: print (A[1], ... , A[n]);
      k ← A[1];
      if k = 1 then go to finish;
      (A[1], ... , A[k]) ← (A[k], ... , A[1]);
      go to loop;
finish: end
```

For example, when $n = 9$ and $\pi = 314592687$, the algorithm will print

$$
\begin{array}{ccccccccc}
3 & 1 & 4 & 5 & 9 & 2 & 6 & 8 & 7 \\
4 & 1 & 3 & 5 & 9 & 2 & 6 & 8 & 7 \\
5 & 3 & 1 & 4 & 9 & 2 & 6 & 8 & 7 \\
9 & 4 & 1 & 3 & 5 & 2 & 6 & 8 & 7 \\
7 & 8 & 6 & 2 & 5 & 3 & 1 & 4 & 9 \\
1 & 3 & 5 & 2 & 6 & 8 & 7 & 4 & 9 \\
\end{array}
$$

and then it will stop.

Let $m = m(\pi)$ be the total number of permutations printed by the above algorithm. Prove that n never exceeds the Fibonacci number F_{n+1}. (In particular, the algorithm always halts.)

Extra credit problem. Let $M_n = \max \{m(\pi) | \pi$ a permutation of $\{1, \ldots, n\}\}$. Find the best upper and lower bounds on M_n that you can.

Problem 3 solution. If array element $A[1]$ takes on k distinct values during the (possibly infinite) execution of the algorithm, we will show that $m \leq F_{k+1}$ (hence m is finite). This is obvious for $k = 1$, since $k = 1$ can occur only when $\pi[1] = 1$.

If $k \geq 2$, let the distinct values assumed by $A[1]$ be $d_1 < d_2 < \cdots < d_k$. Suppose that $A[1] = d_k$ occurs first on the rth permutation, and let $t = \pi[d_k]$. Then the $(r+1)$st permutation will have $A[1] = t$ and $A[d_k] = d_k$. All subsequent permutations will also have $A[d_k] = d_k$ (they leave $A[j]$ untouched for $j \geq d_k$), hence at most $k - 1$ values are assumed by $A[1]$ after the rth permutation has been passed. By induction, $m - r \leq F_k$, so m is finite and $d_1 = 1$.

Interchanging d_k with 1 in π produces a permutation π' such that $m(\pi') = r$, and for which the values d_k and t never appear in position $A[1]$ unless $t = 1$. If $t = 1$ we have $r \leq F_k$, since $A[1]$ assumes at most $k - 1$ values when processing π', hence $m = r + 1 \leq F_{k+1}$. If $t > 1$ we have $r \leq F_{k-1}$ since $A[1]$ assumes at most $k - 2$ values when processing π' (note that $t = d_j$ for $j < k$) hence $m \leq F_k + r \leq F_{k+1}$.

Three hours of further concentration on this problem lead to the hypothesis that it is difficult either to prove or to disprove the conjecture $M_n = O(n)$; the upper bound F_{n+1} is exact only for $n \leq 5$.

[The upper bound applies more generally to any algorithm that sets $(A[1], \ldots, A[k]) \leftarrow (A[k], A[p_2], \ldots, A[p_{k-1}], A[1])$ when $p_2 \ldots p_{k-1}$ is an arbitrary permutation of $\{2, \ldots, k-1\}$.

Computer calculations show that $M_6 = 11$, $M_7 = 17$, $M_8 = 23$, $M_9 = 31$, so $M_{n+1} - M_n$ may possibly increase without limit. This search is speeded up slightly by restricting consideration to permutations without fixed points.

The long-winded permutations on 7, 8, 9 elements are 3146752, 4762153; 61578324; 615972834.

When $n \geq 3$ and $1 \leq k \leq 3$, exactly $(n-1)!$ permutations π satisfy $m(\pi) = k$. It is conjectured that exactly $(n-1)!$ permutations π will satisfy "$A[1] = n$ at some stage."]

Subsequent editorial note. It was shown by a counterexample that the conjecture in 4. is false.

Problem 75-1*, Idealized Optical Fibre Mode Conversion, by R. W. ALLEN (Allen Clark Research Centre, Northants, England).

An optical fibre carries power in two modes represented by 0 and 1. The path of one photon is represented by an N-bit binary number. The sequence 0 1 or 1 0 is counted as one transition. Thus the path 1 0 0 0 1 1 1 contains two transitions and three zeros. One problem here is to determine the number of paths which contain T transitions and M zeros. Prove whether or not the following formula, verified on a computer for $N \leq 20$, is valid for all N:

$$\text{no. of paths} \quad S(N, T, M) = 2H(N, T)\binom{M-1}{U}\binom{N-M-1}{U}.$$

where

$$H(2N, T) = \binom{N-1}{V} \Big/ \binom{N-1}{U},$$

$$H(2N+1, T) = \binom{2N}{2V} \Big/ \binom{2N}{T-1}$$

and

$$U = \left[\frac{T-1}{2}\right], \qquad V = \left[\frac{T}{2}\right]$$

and $[x]$ denotes the greatest integer $\leq x$.

Solution by A. S. LESLIE (Hughes Aircraft Company).

The total number of paths is $S(N, T, M) = S_0(N, T, M) + S_1(N, T, M)$, where the first term represents the paths with first bit 0, and the second represents those with first bit 1. It will be shown that

$$S_0(N, T, M) = \binom{M - 1}{V}\binom{N - M - 1}{U}$$

and

$$S_1(N, T, M) = \binom{M - 1}{U}\binom{N - M - 1}{V}.$$

A path of length N and first bit 0 may be viewed as formed by adjoining 0 to a path of length $N - 1$. The shorter path will have one less zero than the original path. The shorter path will have one less transition if it has first bit 1. Therefore

$$S_0(N, T, M) = S_0(N - 1, T, M - 1) + S_1(N - 1, T - 1, M - 1).$$

Similarly,

$$S_1(N, T, M) = S_0(N - 1, T - 1, M) + S_1(N - 1, T, M).$$

The expressions for S_0 and S_1 will be proved by induction on N. They can be easily verified directly for small N. Assume both expressions are true for $N - 1$. Then

$$S_0(N, T, M) = S_0(N - 1, T, M - 1) + S_1(N - 1, T - 1, M - 1)$$

$$= \binom{M - 2}{V}\binom{N - M - 1}{U} + \binom{M - 2}{V - 1}\binom{N - M - 1}{U}$$

$$= \binom{M - 1}{V}\binom{N - M - 1}{U},$$

and

$$S_1(N, T, M) = S_0(N - 1, T - 1, M) + S_1(N - 1, T, M)$$

$$= \binom{M - 1}{U}\binom{N - M - 2}{V - 1} + \binom{M - 1}{U}\binom{N - M - 2}{V}$$

$$= \binom{M - 1}{U}\binom{N - M - 1}{V}.$$

Note that replacing T with $T - 1$ means V is replaced by U, and U by $V - 1$.

For a combinatorial solution, consider a path with first bit 0 and T transitions as consisting of $V + 1$ sets of 0's alternating with $U + 1$ sets of 1's. Paths may differ in the allocation of $M - U - 1$ 0's in the $V + 1$ sets, and (independently) in the allocation of $N - M - V - 1$ 1's in the $U + 1$ sets. The number of combinations of these choices are $\binom{M - 1}{V}$ and $\binom{N - M - 1}{U}$, respectively.

It is easily verified that this solution,

$$S(N, T, M) = \binom{M - 1}{V}\binom{N - M - 1}{U} + \binom{M - 1}{U}\binom{N - M - 1}{V},$$

is equivalent to the formula stated in the problem.

Editorial note. D.J. Kleitman notes that the solution to the analogous problem for any number of symbols where the number of transitions from each symbol to each other one is given can be obtained by similar reasoning to the combinatorial solution along with the so-called B.E.S.T. theorem that relates the number or Eulerian circuits in a graph to a minor of the determinant of a version of its adjacency matrix. A number of solvers noted that the problem is a known one in the theory of runs and more general results exist. In particular, J.C. Bailar, L. Carlitz, C.L. Mallows and Marimont, respectively, refer to:

M. Fisz, *Probability Theory and Mathematical Statistics*, John Wiley, New York, 1963; Eq. 11.3.12;
   Annals of Math. Statistics, 11 (1940), pp. 147;162, pp. 367–392;
   Biometrika, 36 (1949), pp. 305–316;
W. Feller, *Introduction to Probability Theory and its Applications I*, John Wiley, New York, 1950,
   pp. 56–58.

Problem 59-3, *Optimum Sorting Procedure*, by Paul Brock.

A fundamental procedure in all business operations is that of filing information. Whenever the information in the file is to be updated, the updating items are first sorted in accordance with the key of the file. This is the standard alternative to a direct use of the file which is a random access procedure. In many cases, particularly in multiple file problems, the sorting procedure must be done by a computer. This is an expensive and time consuming operation. Many different procedures have been suggested, and each takes a certain amount of time. It would be useful to determine a minimum computer time procedure.

A general investigation of this problem is under way. In this investigation the following general problem arises: Given a sequence of positive integers, what is the expected length of its maximal monotonic nondecreasing subsequence.

The solution depends upon the length of the sequence and the number of allowable integers. The special case for two integers has been solved. The solution depends upon the following lemma which is proposed as a problem: Among sequences of fixed length M consisting the p 1's and q 2's, the number of sequences whose maximal monotonic nondecreasing subsequences are of length n is the same for all p, q such that $0 < p, q < n \leq M$.

Solution by the proposer.

Let $F(p, q, n)$ be the totality of F sequences of length n for a given p, q choice where an F sequence of length n is defined as a sequence containing a maximum monotone nondecreasing subsequence of length n. The proof is by induction on M. The smallest value of M for which

$$(1) \qquad\qquad\qquad 0 < p, q < n \leq M$$

holds is $M = 2$. For this case the theorem is obvious. Now consider the theorem for sequences of length $\bar{M} < M$: choose a p, q satisfying $p + q = M$ and (1), and fix $n \leq M$.

Case 1: $n = M$: For any values of $p, q > 0$, there is only one F sequence, namely,
$$11 \cdots 122 \cdots 2.$$
Case 2: $n < M$: Assume $q < n - 1$. Since $n < M$, it is clear that $p, q > 2$ for

(1) to hold. Consider those sequences starting with a 2. Since $q < n$, the 2 cannot be counted towards any F sequence. Thus, the total number of F sequences starting with a 2 is $F(p, q - 1, n)$. Now consider those sequences starting with a 1. The 1 must be contained in every F sequence. Hence, the totality with this condition satisfied is $F(p - 1, q, n - 1)$. Consequently, $F(p, q, n) = F(p, q - 1, n) + F(p - 1, q, n - 1)$.

Since

$$p + q - 1 < M,$$

$$p, q - 1 < n < M,$$

$$p - 1, q < n - 1 < M,$$

the conditions on the right satisfy the inductive hypothesis. To remove the condition $q < n - 1$, we consider the case of sequence ending in 1 or 2. This yields:

$$F(p, q, n) = F(p - 1, q, n) + F(p, q - 1, n - 1).$$

Here, the condition $p < n - 1$ must be imposed to satisfy the inductive hypothesis. Finally if $p = q = n - 1$, this would be the only combination of p 1's and q 2's satisfying (1) and hence the theorem is true trivially.

J. Gilmore refers to the paper *Sorting, trees and measures of order*, by W. H. Burge, Information and Control, vol. 1 (1958), pp. 181–197, which is concerned with finding the sorting method which takes the least time to keep a large quantity of data in an orderly array for easy reference. Burge finds that the optimal strategy for sorting a set of data depends upon the amount of order (in the information-theoretic sense) already existing in the data; and he shows how to find an optimal sorting strategy given a measure of this order.

Problem 60-8, *Another Sorting Problem*, by J. H. VAN LINT (Technische Hogeschool, Te Eindhoven, Nederland).

Consider all sequences of length M consisting of p_1 1's, p_2 2's, $\cdots$ p_k k's ($k \geq 2$) whose maximal monotonic non-decreasing subsequences of contiguous numbers are of length n where n satisfies the inequalities

(1) $$n > M - p_i \qquad i = 1, 2, \cdots, k.$$

Determine the number $N(p_1, p_2, \cdots, p_k; n)$ of sequences with this property.

Editorial note: This problem is similar to that of Brock's Optimum Sorting Procedure, Problem 59-3. However, in the latter problem, contiguity is not required.

Solution by the proposer. As a consequence of (1) the subsequence must contain every number 1, 2, $\cdots$, k at least once and hence it starts with the number 1 and ends with the number k.

First consider $n \leq M - 2$.

If the monotonic subsequence of length n is at the beginning (at the end) of the sequence there are $M - n$ elements left, of which $M - n - 1$ can have arbitrary values, because $M - n < p_i$ for all i, while the first (last) of these elements can have all values except k (except 1). If we choose these elements the monotonic subsequence is fully determined. Hence there are $2 \cdot k^{M-n-1} \cdot (k-1)$ sequences of this type.

Now consider the case where the monotonic subsequence is preceded by elements and followed by $M - n - l$ elements. Here l can have the values 1, 2 $\cdots$, $M - n - 1$. The last of the preceding elements can have all values except 1. The first of the elements following the subsequence can have all values except k. The other $M - n - 2$ preceding and following elements can have arbitrary values. (We again use (1)). After choosing these elements the subsequence is fully determined. We can still choose l.

So there are $(M - n - 1) \cdot (k - 1)^2 \cdot k^{M-n-2}$ sequences of this type. Adding the two results we find:

$$
\begin{aligned}
(2) \quad N &= 2 \cdot k^{M-n-1}(k - 1) + (M - n - 1) \cdot (k - 1)^2 \cdot k^{M-n-2} \\
&= (k - 1) \cdot k^{M-n-2} \cdot \{2k + (k - 1)(M - n - 1)\}
\end{aligned}
$$

We see that N is the same for all p_1, p_2, $\cdots$, p_k as long as (1) is satisfied and of course $p_1 + p_2 + \cdots + p_k = M$.

For $k = 2$ we find for (2): $N = 2^{M-n-2}(M - n + 3)$.

We still have 2 cases to discuss: $n = M$ and $n = M - 1$. Both are trivial. For $n = M$ we have $N = 1$ and for $n = M - 1$ the above holds but we only have the first term i.e., $N = 2(k - 1)$.

Problem 66–20, A Variation of the Game of "Twenty Questions," by D. J. NEWMAN (Yeshiva University).

In the usual game of "Twenty Questions," essentially one is to determine a chosen integer in a given range 1 to N by asking not more than twenty questions to which only "yes" or "no" answers are given. For this case, it is easy to determine the maximum N such that this can be done. If we introduce the variation that the person answering the questions is allowed to lie at most once, determine the maximum N.

Editorial note. Other variations arise by allowing the person answering to make at most n lies, exactly n lies, two consecutive lies, etc.

Solution by the proposer.

First let us literally treat the given problem.

UPPER BOUND. There are 2^{20} possible outcomes to the 20 questions. Because of the possibility of a lie in any place (or no lie at all), each number (from 1 to N) gives rise to 21 possible outcomes. If these N numbers are to be distinguishable, then we must have $21N \leq 2^{20}$ or $N \leq 49{,}932$.

LOWER BOUND. We show how to distinguish the numbers from 1 to 2^{15} ($= 32{,}768$). This gives $N \geq 32{,}768$.

First 15 questions: Is the kth binary digit a 1 ($k = 1, 2, \cdots, 15$)?

16th question: Have you lied yet (in your first 15 answers)?

Note. A "no" answer to this question is automatically true! a "yes" answer, on the other hand, insures that the lie has already occurred and so the next 4 questions must be truthfully answered.

Final 4 questions (it is known that a lie has occurred): Is the kth digit of the number of the question to which you lied a 1 ($k = 1, 2, 3, 4$)?

Now we turn to the general problem of n questions instead of 20. We treat, particularly, the case of n of the form $2^k - 1$ (i.e., $n = 1, 3, 7, 15, 31, \cdots$), and, in this case, we achieve total success by determining N exactly as 2^{2^k-k-1}, i.e., $N = 1, 2, 16, 2048, \cdots$.

Of course, as before, we have the upper bound $N \leq 2^n/(n+1) = 2^{2^k-k-1}$, so we need only display $2^k - 1$ questions which do distinguish between 2^{2^k-k-1} integers.

We must, therefore, determine $2^k - k - 1$ binary digits. Break up

$$2^k - k - 1 = (2^{k-1} - 1) + (2^{k-2} - 1) + \cdots$$

and ask questions as follows:

First $2^{k-1} - 1$ questions: What are the first $2^{k-1} - 1$ binary digits?
Next: Did you lie?
Next $2^{k-2} - 1$ questions: Next $2^{k-2} - 1$ binary digits?
Next question: Did you lie?
etc.

It is understood that if ever you get a "yes" answer to a "Did you lie?" you *then* locate the place where the lie occurred and you never again ask "Did you lie?"

The total number of questions asked is then always $2^k - 1$ and we have accomplished just what we set out to do.

A less picturesque, but still quite interesting, formulation of the problem can be given. By associating the "yes" answer with a 1 and the "no" with a 0, we obtain a correspondence between the possible n answers and the vertices of an n-cube. In these terms our problem can be stated as follows: For each vertex of the n-cube, consider the set formed by it and all its nearest neighbor vertices. What is the largest number of disjoint sets of this type which we can have?

Our previous remarks may now be interpreted as saying that, for $n = 2^k - 1$, the n-cube can be perfectly decomposed into such sets.

Postscript. For a related note on this problem, see Joel Spencer, *Guess a number-with lying*, Math. Mag. 57 (1984) 105-108 and the references therein. Also, Andy Liu notes that this problem is related to coding theory and he refers to his joint paper with N. Alon, *An application of set theory to coding theory*, Math. Mag. 62 (1989) 233-237, and the references therein.

Problem 78-9*, *A Variant of Silverman's Board of Directors Problem*, by W. AIELLO and T. V. NARAYANA (University of Alberta, Edmonton, Alberta, Canada).

Suppose we assign positive integer weights to the vote of each member of a board of directors that consists of n members so that the following conditions apply:

(1) Different subsets of the board always have different total weights so that there are no ties in voting (tie-avoiding).

(2) Any subset of size k will always have more weight than any subset of size $k-1$ ($k = 1, \cdots, n$) so that any majority carries the vote, abstentions allowed (nondistorting).

Kreweras, Wynne and Narayana [1] have given a solution (shown in table 1) for $n = 1, \cdots, 7$ that can be extended very easily from any n to $n+1$. It is conjectured that this is a *minimal dominance* solution. Here, an increasing sequence $(y_1, \cdots, y_n)$ is said to *dominate* another increasing sequence $(x_1, \cdots, x_n)$ if $y_i \geq x_i$ ($i = 1, \cdots, n$). So a solution $(x_1, \cdots, x_n)$ is minimal dominant if no other solution $(y_1, \cdots, y_n)$ exists such that $x_i \geq y_i$ ($i = 1, \cdots, n$).

TABLE 1
Nondistorting, tie-avoiding integer vote weights W_n.

Members, n	1	2	3	4	5	6	7
Totals, S_n	1	3	9	21	51	117	271

Column vectors of vote weights, W_n							
	1	2	4	7	13	24	46
		1	3	6	12	23	45
			2	5	11	22	44
				3	9	20	42
					6	17	39
						11	33
							22

The underlined values along the diagonal of vector elements are the I_n values, where:

$$I_1 = I_2 = 1 \quad \text{and} \quad I_{2n+1} = 2I_n,$$

$$I_{2n+2} = 2I_{2n+1} - I_n.$$

REFERENCE

[1] B. WYNNE AND T. V. NARAYANA, *Tournament configuration and weighted voting*, Cahiers du BURO, Paris, to appear.

Two Algebraic Problems

Problem 86-7 by JOHN H. HALTON (University of North Carolina).

The following problems arise out of a study of the effect of the Indian government's encouragement of voluntary sterilization in an attempt to stem the excessive population growth, on sex-ratio, based on practices adopted by certain Indian communities.

If each couple is assumed to have N opportunities to conceive and give birth to a child, with probabilities ξ and η of producing a son or a daughter, respectively, on each such opportunity; and if all couples producing n sons immediately undergo sterilization, without regard to the number of daughters produced; then the statistics of couples having less than n sons cannot be retroactively affected by the possible subsequent sterilization; so that the probability of their having less than n sons is

(A)
$$1 - U_n(\xi) = \sum_{k=0}^{n-1} \binom{N}{k} \xi^k (1 - \xi)^{N-k},$$

where the summation variable k denotes the number of sons produced. Therefore the probability of a couple having exactly n sons (more than n being impossible) must be $U_n(\xi)$ as defined above. Since

(B)
$$\sum_{k=0}^{N} \binom{N}{k} \xi^k (1 - \xi)^{N-k} = [\xi + (1 - \xi)]^N = 1,$$

it follows that

(C)
$$U_n(\xi) = \sum_{k=n}^{N} \binom{N}{k} \xi^k (1-\xi)^{N-k}.$$

On the other hand, by direct calculation, this is also the sum over all possible k (from n through N) of the probability of having the nth (and last) son on the kth opportunity, so that

(D)
$$U_n(\xi) = \sum_{k=n}^{N} \binom{k-1}{n-1} \xi^n (1-\xi)^{k-n}.$$

The *first problem* is to prove the identity of (C) and (D), for general ξ, without appeal to this probabilistic approach.

Among sterilized couples, there are, by our postulate, always exactly n sons. In this sub-population, the expected number of daughters (a conditional expectation) is then

$$E[d] = \sum_{k=n}^{N} \sum_{d=0}^{k-n} d \binom{k-1}{n-1} \binom{k-n}{d} \xi^n \eta^d (1-\xi-\eta)^{k-n-d} \Big/ U_n(\xi)$$

$$= \sum_{k=n}^{N} \binom{k-1}{n-1} \xi^n \sum_{d=0}^{k-n} d \binom{k-n}{d} \eta^d (1-\xi-\eta)^{k-n-d} \Big/ U_n(\xi)$$

(E)
$$= \sum_{k=n+1}^{N} \binom{k-1}{n-1} (k-n) \xi^n \eta \sum_{d=1}^{k-n} \binom{k-n-1}{d-1} \eta^{d-1} (1-\xi-\eta)^{k-n-d} \Big/ U_n(\xi)$$

$$= n \frac{\eta}{\xi} \sum_{k=n+1}^{N} \binom{k-1}{n} \xi^{n+1} [\eta + (1-\xi-\eta)]^{k-n-1} \Big/ U_n(\xi)$$

$$= n \frac{\eta}{\xi} \sum_{k=n+1}^{N} \binom{k-1}{n} \xi^{n+1} (1-\xi)^{k-n-1} \Big/ U_n(\xi) = n \frac{\eta}{\xi} U_{n+1}(\xi) \Big/ U_n(\xi),$$

by (D). Thus, the ratio of the number of boys (n) to the number of girls (which, by the strong law of large numbers, converges almost surely, i.e., with probability one, to $E[d]$) converges as N tends to infinity to

(F)
$$\frac{\xi}{\eta} V_n(\xi) = \frac{\xi}{\eta} U_n(\xi) \Big/ U_{n+1}(\xi).$$

The *second problem* is to prove that this quantity increases with n, for any given $0 < \xi < 1$ and $\eta \neq 0$.

Solution of the first problem by SEPPO KARRILA (University of Wisconsin).
Consider ξ fixed for the moment and form

$$F(t) := \sum_{n=1}^{N} U_n t^n.$$

After inserting (C) or (D) for U_n, one can reverse the order of summation and evaluate the resulting sum by making use of the binomial theorem and the summation rule for a finite geometric series. In each case

$$F(t) = \frac{t}{1-t}\{1 - [1 + \xi(t-1)]^N\}.$$

Now the uniqueness of the coefficients of a polynomial implies that (C) and (D) are equivalent.

Solution of the second problem by A. A. JAGERS (Technische Hogeschool Twente, Enschede, the Netherlands).

Let

$$p_k = \binom{N}{k} \xi^k (1-\xi)^{N-k}.$$

Then $U_n = U_n(\xi) = p_n + p_{n+1} + \cdots + p_N$ and

$$\frac{U_n}{U_{n+1}} = \frac{p_n + U_{n+1}}{p_{n+1} + U_{n+2}},$$

a *convex* combination of p_n/p_{n+1} and U_{n+1}/U_{n+2}. Hence it suffices to show that p_n/p_{n+1} increases with n. This is easy:

$$\frac{p_n p_{n+2}}{(p_{n+1})^2} = \binom{N}{n}\binom{N}{n+2} \Big/ \binom{N}{n+1}^2 = \frac{(n+1)(N-n-1)}{(n+2)(N-n)} < 1.$$

Problem 72-22, A Binomial Expansion Identity, by BARRY WOLK (University of Manitoba, Winnipeg, Canada).

If a and b are real numbers such that $n = a + b + 1$ is a positive integer, show that the sums of the first n terms in the binomial expansions of $(2-1)^a$ and $(\frac{1}{2} + \frac{1}{2})^b$ are equal.

  Editorial note. The problem here is a generalization of problem B-5 of the 1967 William Lowell Putnam Mathematical Competition.

  Solution by A. A. JAGERS (Technische Hogeschool Twente, Enschede, the Netherlands).

  We use the integral form for the remainder in a Taylor series. If $f_1(t) = (2-t)^a$, $f_2(t) = (\frac{1}{2} + t)^b$ and

$$f(h) = f(0) + \sum_{m=1}^{n-1} \frac{h^m}{m!} f^{(m)}(0) + R_n(f; h),$$

we have

(1) $$R_n(f_1; 1) = n\binom{a}{n}(-1)^n \int_0^1 (1-t)^{n-1}(2-t)^{a-n} \, dt$$

and

$$R_n(f_2; \tfrac{1}{2}) = n\binom{b}{n} 2^{-b} \int_0^1 (1-t)^{n-1}(1+t)^{b-n} \, dt.$$

Now since $(2-1)^a = 1 = (\frac{1}{2} + \frac{1}{2})^b$, and the binomial expansions in question coincide with the Taylor expansions of f_1 and f_2 around 0, we only have to prove that $R_n(f_1; 1)$ and $R_n(f_2; \frac{1}{2})$ are equal. To do this, it suffices to substitute $t = 2\tau/(1+\tau)$ and $a = n - b - 1$ in (1).

Editorial note. Under the same hypotheses, it was shown more generally by R. C. Cavanagh, H. W. Gould, F. D. Parker and the proposer, that the sum of the first n terms in the binomial expansions of

$$\{p - (p - 1)\}^a \quad \text{and} \quad \{1/p + (p - 1)/p\}^b$$

are equal. Gould also shows the equivalence of the latter to the known identity

$$\sum_{k=0}^{n} \binom{n-x}{k}(1 + y)^{n-k}(-y)^k = \sum_{k=0}^{n} \binom{x}{k}y^k.$$

(See H. W. Gould, *Combinatorial Identities*, Morgantown, West Virginia, 1972, p. 9.)

Another proof of the generalization can be given using the method of Jagers. To show that the remainders after n terms of the Taylor series expansions of the above two binomial expressions in p are equal, one has to prove that

$$p^b \int_0^1 (1 - x)^{n-1}\{p - (p - 1)x\}^{a-n} dx = \int_0^1 (1 - x)^{n-1}\{1 + (p - 1)x\}^{b-n} dx.$$

This follows by letting $x = pt/(1 + (p - 1)t)$ in the first integral. [M.S.K.]

An Identity

Problem 82-10, *by* JO ANN CLARK *and* O. G. RUEHR (Michigan Technological University).

Prove that

$$\sum_{j=0}^{n} \frac{1}{\binom{2n}{2j}} = \frac{2n + 1}{2^{2n+1}} \sum_{j=0}^{2n+1} \frac{2^j}{j + 1}.$$

Solution by ROBERT E. SHAFER (Berkeley, CA).

For Re $(\alpha) > 0$, Re $(\beta) > 0$ we establish the two identities by induction:

(1) $I = 2^{2n+2} \displaystyle\sum_{j=0}^{n} \frac{\Gamma(2j + \alpha)\Gamma(2n - 2j + \beta)}{\Gamma(2n + \alpha + \beta)} = \sum_{j=0}^{2n+1} 2^j \left[\frac{\Gamma(j + \alpha)\Gamma(\beta)}{\Gamma(j + \alpha + \beta)} + \frac{\Gamma(\alpha)\Gamma(j + \beta)}{\Gamma(j + \alpha + \beta)} \right],$

$$J = 2^{2n+2} \sum_{j=0}^{n-1} \frac{\Gamma(2j + \alpha + 1)\Gamma(2n - 2j + \beta - 1)}{\Gamma(2n + \alpha + \beta)}$$

(2)

$$= \sum_{j=0}^{2n} 2^j \left[\frac{\Gamma(j + \alpha)\Gamma(\beta)}{\Gamma(j + \alpha + \beta)} + \frac{\Gamma(\alpha)\Gamma(j + \beta)}{\Gamma(j + \alpha + \beta)} \right]$$

$$- 2^{2n+1} \left[\frac{\Gamma(2n + \alpha + 1)\Gamma(\beta)}{\Gamma(2n + \alpha + \beta + 1)} + \frac{\Gamma(\alpha)\Gamma(2n + \beta + 1)}{\Gamma(2n + \alpha + \beta + 1)} \right].$$

The integral representation of (1) is

(3)

$$2^{2n+2} \int_0^1 \sum_{j=0}^{n} (1 - x)^{2j+\alpha-1} x^{2n-2j+\beta-1} dx$$

$$= 2^{2n+2} \int_0^1 (1 - x)^{\alpha-1} x^{\beta-1} \frac{(1 - x)^{2n+2} - x^{2n+2}}{(1 - x)^2 - x^2} dx,$$

and of (2)

(4) $\quad 2^{2n+2} \int_0^1 \sum_{j=0}^{n-1} (1-x)^{2j+\alpha} x^{2n-2j+\beta-2} \, dx = 2^{2n+2} \int_0^1 (1-x)^\alpha x^\beta \dfrac{(1-x)^{2n} - x^{2n}}{(1-x)^2 - x^2} \, dx.$

Subtracting series (2) from series (1) one derives the integral

(5)
$$I - J = 2^{2n+2} \int_0^1 (1-x)^{\alpha-1} x^{\beta-1} \dfrac{(1-x)^{2n+1} + x^{2n+1}}{(1-x) + x} \, dx$$

$$= 2^{2n+2} \left[\dfrac{\Gamma(2n+\alpha+1)\Gamma(\beta)}{\Gamma(2n+\alpha+\beta+1)} + \dfrac{\Gamma(\alpha)\Gamma(2n+\beta+1)}{\Gamma(2n+\alpha+\beta+1)} \right].$$

From (4), we get

(6)
$$J = 2^{2n} \int_0^1 (1-x)^{\alpha-1} x^{\beta-1} \dfrac{(1-x)^{2n} - x^{2n}}{(1-x)^2 - x^2} [1 - (1-2x)^2] \, dx,$$

from which, using (3), we obtain (2). Combining (5) and (2) we complete the induction of (1). The author's problem is the special case $\alpha = \beta = 1$.

Comment by the proposers. Several similar identities are given in [1] (p. 18, l. 2.1 and p. 21, l. 2.25). A more general result,

$$\sum_{j=0}^n \dfrac{1}{\binom{pn}{pj}} = \dfrac{(pn+1)}{p} \sum_{j=0}^{(n+1)p-1} \dfrac{1}{[(n+1)p - j]2^{j-p}} \left| \dfrac{1}{2} + \sum_{s=1}^{[(p-1)/2]} \dfrac{(-1)^s \cos(\pi sj/p)}{(\cos(\pi s/p))^{j+2-p}} \right|,$$

is discussed in [2].

REFERENCES

[1] H. E. GOULD, *Combinatorial Identities*, Morgantown, West Virginia.
[2] O. G. RUEHR, *Two generalizations of Staver's sum identity*, to appear.

Problem 70–21, *A Combinatorial Identity*, by V. K. ROHATGI (Catholic University of America).

For any nonnegative integers n and k, show that

(1) $\quad \dfrac{1}{nk+1} \binom{nk+n}{n} = \sum_{i=1}^n \dfrac{k}{(ik+i-1)(nk-ki+1)} \binom{ik+i-1}{i-1} \binom{(n-i)(k+1)}{n-i}.$

Solution by the proposer.

It is not difficult to see (see, for example, L. Takács, J. Amer. Statist. Assoc., 57 (1962), pp. 327–337) that the number of lattice paths from $(0,0)$ to (m,n) that lie entirely below the line $x = ky$ is

$$\dfrac{m - kn}{m+n} \binom{m+n}{n}$$

and the number of those paths which touch or lie below $x = ky$ is

$$\frac{m + 1 - kn}{m + n}\binom{m + n}{n}.$$

The left side of the identity can be recognized as the number of lattice paths from $(0, 0)$ to (nk, n) which touch or lie below $x = ky$. Now we only have to observe that every such path must touch $x = ky$ for the *first* time (excluding $(0, 0)$) at (ik, i), $i = 1, 2, \cdots, n$, and from there on it must touch or lie below $x = ky$.

Comments by H. W. GOULD (West Virginia University).

The stated formula is a disguised form of a very special case of the general addition theorem which I have written so much about under the title of the "generalized Vandermonde convolution." It is interesting to me that the formula occurs as the answer to the present enumeration, however. I shall show below that the formula is equivalent to an interesting orthogonality relation which is immediate from the generalized Vandermonde convolution. For details of the literature see my papers in Duke Math. J. (1960 to present).

The addition theorem (Vandermonde generalization) is given by

$$(12) \quad \sum_{k=0}^{n} A_k(a, b)A_{n-k}(c, b) = A_n(a + c, b) \quad \text{with } A_k(a, b) = \frac{a}{a + bk}\binom{a + bk}{k},$$

valid for all real (or complex) a, b, c (being a polynomial identity in a, b, c). I shall show that Rohatgi's formula is equivalent to the case $a = 1$, $c = -1$, when we have clearly

$$(13) \quad \sum_{k=0}^{n} A_k(1, b)A_{n-k}(-1, b) = A_n(0, b) = \begin{cases} 0, & n \geq 1, \\ 1, & n = 0. \end{cases}$$

The steps I give below are not necessarily the most elegant.

Rohatgi's formula may be written in the form (changing a few letters)

$$(14) \quad \begin{aligned} &\sum_{k=1}^{n} \binom{bk + k - 1}{k - 1}\binom{bn - bk + n - k}{n - k}\frac{b(bn + 1)}{(bk + k - 1)(bn - bk + 1)} \\ &= \binom{bn + n}{n} \quad \text{for } n \geq 1. \end{aligned}$$

In view of the transformation $\binom{-x}{k} = (-1)^k\binom{x + k - 1}{k}$, this in turn may be written as

$$\sum_{k=1}^{n} \binom{-1 - bk}{k - 1}\binom{-1 - bn + bk}{n - k}\frac{b(bn + 1)}{(bk + k - 1)(-1 - bn + bk)} = \binom{-1 - bn}{n}.$$

Then after further elementary algebraic manipulation, using the relations

$$\frac{x}{k}\binom{x - 1}{k - 1} = \binom{x}{k}, \qquad \frac{1}{k}\binom{bk}{k - 1} = \frac{1}{bk + 1}\binom{bk + 1}{k},$$

and the previous transformation, we obtain, successively,

$$\sum_{k=1}^{n} \binom{bk}{k}\binom{-1+bn-bk}{n-k}\frac{-1+bn}{(-1+bn-bk)(-1+k-bk)} = \binom{-1+bn}{n}$$

(b has been replaced by $-b$ since it will turn out that (1) is valid for all real k),

$$\sum_{k=1}^{n} (-1)^k \frac{1}{k}\binom{-bk+k-2}{k-1}\binom{-1+bn-bk}{n-k}\frac{-1}{-1+bn-bk}$$

$$= \frac{-1}{-1+bn}\binom{-1+bn}{n},$$

$$\sum_{k=0}^{n} \frac{1}{1+bk}\binom{1+bk}{k}\cdot\frac{-1}{-1+b(n-k)}\binom{-1+b(n-k)}{n-k} = 0 \quad \text{for } n \geq 1.$$

The latter follows from (12) which also gives the value for $n = 0$. Also, (12) is a polynomial identity in a, b, c and thus is valid for all real or complex a, b, c.

Editorial note. Gould also notes that he has never seen anything "essentially" new obtained by lattice-point enumerations which could not be obtained from (12). This is analogous to the view that any integral which is obtained by contour integration can also be obtained without it. However, it is always advantageous to have more than one method for solving a class of problems. [M.S.K.]

A Binomial Summation

Problem 83-3, by GENGZHE CHANG *and* ZUN SHAN (University of Science and Technology of China, Hefei, Anhui, China).
Determine the sum

$$S_n = \sum_{k=0}^{n-1}\left\{\binom{n}{0}+\binom{n}{1}+\cdots+\binom{n}{k}\right\}\left\{\binom{n}{k+1}+\binom{n}{k+2}+\cdots+\binom{n}{n}\right\}.$$

Solution by D. R. BREACH (University of Canterbury, New Zealand).
The coefficient of x^k in $(1+x)^n(1+x+x^2+\cdots)$ is $\sum_{i=0}^{k}\binom{n}{i}$. Since $\binom{n}{m}=\binom{n}{n-m}$, the coefficient of x^{n-1-k} in $(1+x)^n(1+x+x^2+\cdots)$ is $\sum_{i=k+1}^{n}\binom{n}{i}$. The given sum is therefore the coefficient of x^{n-1} in

$$(1+x)^{2n}(1+x+x^2+\cdots)^2 = \frac{(1+x)^{2n}}{(1-x)^2} = \frac{d}{dx}\frac{(1+x)^{2n}}{(1-x)} - 2n\frac{(1+x)^{2n-1}}{1-x}.$$

Hence the sum is the coefficient of x^n in

$$n(1+x)^{2n}(1-x)^{-1} - 2n\,x(1+x)^{2n-1}(1-x)^{-1}.$$

Now the coefficient of x^n in $(1+x)^{2n}(1-x)^{-1}$ is

$$\sum_{i=0}^{n}\binom{2n}{i} = \frac{1}{2}\sum_{i=0}^{2n}\binom{2n}{i}+\frac{1}{2}\binom{2n}{n} = 2^{2n-1}+\frac{1}{2}\binom{2n}{n}.$$

Likewise the coefficient of x^n in $x(1+x)^{2n-1}(1-x)^{-1}$ is 2^{2n-2}. Therefore the given sum is equal to

$$\frac{n}{2}\binom{2n}{n}.$$

Solution by H. PRODINGER (Vienna University, Vienna, Austria).

The desired sum may be written as

$$\sum_{0\leq k<n}\ \sum_{0\leq i\leq k<j\leq n}\binom{n}{i}\binom{n}{j}$$

$$=\sum_{0\leq i<j\leq n}\binom{n}{i}\binom{n}{j}\cdot(j-i)$$

$$=n\sum_{0\leq i<j\leq n}\binom{n}{i}\binom{n-1}{j-1}-n\sum_{0<i\leq j\leq n}\binom{n-1}{i-1}\binom{n}{j}$$

$$=n\sum_{0\leq i\leq j<n}\binom{n}{i}\binom{n-1}{j}-n\sum_{0\leq n-j<n-i<n}\binom{n}{n-j}\binom{n-1}{n-i}$$

$$=n\sum_{0\leq i<n}\binom{n}{i}\binom{n-1}{i}=n\sum_{0\leq i<n}\binom{n}{i}\binom{n-1}{n-1-i}=n\binom{2n-1}{n-1}.$$

Editorial note. D.E. Knuth notes in his solution that this problem is closely related to the evaluation of

$$\overline{S}_n=\sum_{j,k}\binom{n}{j}\binom{n}{k}\max(j,k),\qquad \underline{S}_n=\sum_{j,k}\binom{n}{j}\binom{n}{k}\min(j,k).$$

Indeed, since $\overline{S}_n-\underline{S}_n=2S_n$ and $\overline{S}_n+\underline{S}_n=\sum\binom{n}{j}\binom{n}{k}(j+k)=n\cdot 2^{2n}$, we have $\overline{S}_n=n\cdot 2^{2n-1}+S_n$ and $\underline{S}_n=n\cdot 2^{2n-1}-S_n$.

Can one explicitly sum the analogous extensions of $\overline{S}_n$ and $\underline{S}_n$ to mth order summations? [M.S.K.]

An Identity

Problem 87-8, by JOHN W. MOON (University of Alberta).

Show that

$$\sum_{n=1}^{\infty}\frac{56n^2+33n-8}{(n+2)(n+1)}f_n^2=1$$

where

$$f_n=\frac{4^{-n}}{n}\binom{2n-2}{n-1}\qquad\text{for }n\geq 1.$$

Background. A *branch* of a rooted tree T_n is a maximal subtree that does not contain the root. A branch B with i nodes is a *primary* branch of T_n if $n/2\leq i\leq n-1$; if T_n has a primary branch B with i nodes, then a branch C with j nodes is a *secondary* branch if $(n-i)/2\leq j\leq n-1-i$. For many families F of rooted trees, the fraction of trees T_n in F that have a primary branch tends to 1 as $n\to\infty$. (See A. Meir and J.W. Moon, *On major and minor branches of rooted trees*, Canad. J. Math., 39 (1987) 673-693). It can be shown that the fraction of plane trees T_n that have a secondary branch tends to a limit p as $n\to\infty$, where

$$p = 3 - 12 \sum_{n=1}^{\infty} \frac{13n^2 + 5n - 2}{(n+1)(n+2)} f_n^2.$$

If we appeal to the proposed identity then we obtain the more rapidly converging expression

$$p = \frac{3}{14} + \frac{3}{14} \sum_{n=1}^{\infty} \frac{149n + 8}{(n+1)(n+2)} f_n^2$$

from which we find that $p = .59 \cdots$.

Solution by S. LJ. DAMJANOVIĆ (TANJUG Telecommunications Center, Belgrade, Yugoslavia).

We will show that

$$S(A, B, C) \equiv \sum_{n=1}^{+\infty} \frac{An^2 + Bn + C}{(n+1)(n+2)} f_n^2$$

$$= \frac{1}{15^2 \pi} \{9A + 8B + 96C\} - \frac{C}{8},$$

i.e., $S(56, 33, -8) = 1$ and $p = 3 - 12 S(13, 5, -2) = 28/15\pi$.

Proof.

$$S(A, B, C) = \frac{A}{2^4 3!} (x_1 + 1)(x_2 + 1)_5 F_4 \left[\begin{matrix} x_1 + 2, & x_2 + 2, & \frac{1}{2}, & \frac{1}{2}, & 1, \\ x_1 + 1, & x_2 + 1, & 4, & 2 & \end{matrix} ; 1 \right],$$

where $x_1 + x_2 = B/A$, $x_1 x_2 = C/A$.

More generally, starting from relations

$$_{p+1}F_{q+1} \left[\begin{matrix} x+1, & a_1, \cdots, a_p, \\ x, & b_1, \cdots, b_q \end{matrix} ; z \right] = {}_pF_q \left[\begin{matrix} a_1, \cdots, a_p, \\ b_1, \cdots, b_q \end{matrix} ; z \right]$$

$$+ \frac{a_1 \cdots a_p z}{b_1 \cdots b_q x} {}_pF_q \left[\begin{matrix} a_1 + 1, \cdots, a_p + 1, \\ b_1 + 1, \cdots, b_q + 1 \end{matrix} ; z \right]$$

and

$$_{z+1}F_{q+1} \left[\begin{matrix} a_1, \cdots, a_p, \\ b_1, \cdots, b_q, 2 \end{matrix} ; z \right] = \frac{\prod_{i=1}^{q}(b_i - 1)}{\prod_{i=1}^{p}(a_i - 1)} \left\{ {}_pF_q \left[\begin{matrix} a_1 - 1, \cdots, a_p - 1, \\ b_1 - 1, \cdots, b_q - 1 \end{matrix} ; z \right] - 1 \right\},$$

we obtain

$$(x_1 + 1)(x_2 + 1)_5 F_4 \left[\begin{matrix} x_1 + 2, & x_2 + 2, & a, & b, & 1, \\ x_1 + 1, & x_2 + 1, & c, & 2 & \end{matrix} ; 1 \right]$$

$$= \frac{x_1 x_2 (c-1)}{(a-1)(b-1)} {}_2F_1[-1] + (x_1 + x_2 + 1)_2 F_1 + \frac{ab}{c} {}_2F_1[1] - \frac{x_1 x_2(c-1)}{(a-1)(b-1)}$$

$$= \frac{\Gamma(c)\Gamma(c-a-b)}{\Gamma(c-a)\Gamma(c-b)} \left\{ \frac{(a-1)(b-1) + c - 2}{c-a-b-1} + \frac{c-a-b}{(a-1)(b-1)} x_1 x_2 + x_1 + x_2 \right\}$$

$$- \frac{(c-1)x_1 x_2}{(a-1)(b-1)}, \quad a \neq 1, \quad b \neq 1, \quad R(c-a-b) > 1,$$

where

$$_2F_1 = {}_2F_1\left[\begin{matrix}a,b;\\c\end{matrix}\ 1\right] = \frac{\Gamma(c)\Gamma(c-a-b)}{\Gamma(c-a)\Gamma(c-b)},$$

$R(c-a-b) > 0$ and

$$_2F_1[d] = {}_2F_1\left[\begin{matrix}a+d,\ b+d;\\c+d\end{matrix}\ 1\right].$$

For $a = b = \frac{1}{2}$, $c = 4$, $x_1 x_2 = C/A$, and $x_1 + x_2 = B/A$.

All the solvers except the proposer used hypergeometric functions. The proposer used the recurrences $f_{n+1} = f_n(2n-1)/(2n+2)$ and $r_n = f_n + r_{n+1}$, where $r_n = 2nf_n$, and elementary manipulations.

Problem 64-9, A Pair of Combinatorial Identities, by JOEL L. BRENNER (Stanford Research Institute).

Establish the identities:

$$\sum_{i=0}^{n}\binom{2n+1}{2i+1}\binom{2n-2i}{n-i}(x^2+y^2)^{2i+1}(xy)^{2n-2i}$$
$$= \sum_{i=0}^{2n+1}\binom{2n+1}{i}^2 (x^2)^i(y^2)^{2n+1-i},$$

$$\sum_{i=0}^{n}\binom{2n}{2i}\binom{2n-2i}{n-i}(x^2+y^2)^{2i}(xy)^{2n-2i} = \sum_{i=0}^{2n}\binom{2n}{i}^2 (x^2)^i(y^2)^{2n-i}.$$

The solutions by M. G. MURDESHWAR (University of Alberta) and V. K. ROHATGI (University of Alberta) were essentially the same and are as follows:

Let T denote the sum of all terms of degree $4n$ in the expansion of $(1 + x^2)^{2n}(1 + y^2)^{2n}$. Multiplying the binomial expansions of $(1 + x^2)^{2n}$ and $(1 + y^2)^{2n}$, we obtain

$$T = \sum_{i=0}^{2n}\binom{2n}{i}^2 (x^2)^i(y^2)^2 n^{-i}.$$

Since also

$$(1 + x^2 + y^2 + x^2y^2)^{2n} = \sum_{i=0}^{2n}\binom{2n}{i}(1 + x^2 + y^2)^{2n-i}(x^2y^2)^i,$$

we get by collecting terms of degree $4n$,

$$T = \sum_{i=0}^{n}\binom{2n}{i}(x^2y^2)^i\binom{2n-i}{2n-2i}(x^2+y^2)^{2n-2i}$$
$$= \sum_{i=0}^{n}\binom{2n}{n-i}\binom{n+i}{2i}(x^2+y^2)^{2i}(x^2y^2)^{n-i}.$$

Since

$$\binom{2n}{n-i}\binom{n+i}{2i} = \binom{2n}{2i}\binom{2n-2i}{n-i},$$

we now have the second identity. The first identity can be proved similarly by considering the terms of degree $(4n+2)$ in $(1 + x^2)^{2n+1}(1 + y^2)^{2n+1}$.

Problem 78-6, *A Combinatorial Identity,* by PETER SHOR (U.S.A. Mathematical Olympiad Team, 1977).

A function $S(m, n)$ is defined over the nonegative integers by
(A) $S(0, 0) = 1$,
(B) $S(0, n) = S(m, 0) = 0$ for $m, n \geq 1$,
(C) $S(m+1, n) = mS(m, n) + (m + n)S(m, n-1)$.
Show that

$$\sum_{n=1}^{m} S(m, n) = m^m.$$

Solution by OTTO G. RUEHR (Michigan Technological University).
Introduce the generating function

$$F_m(x) = \sum_{n=1}^{m} S(m, n)x^n$$

which is easily seen to satisfy the differential recurrence equation

$$F_{m+1} = [m + x(m + 1)]F_m + x^2 F_m'$$

whose solution can be written in operator form

$$F_m = \frac{e^{-m(1-1/x)}}{x^{m+1}} \left[x^3 e^{1-1/x} \frac{d}{dx} \right]^m x.$$

Now make the substitutions $x = 1/(1-\lambda)$ and $y = \lambda e^{-\lambda}$, to obtain the representation

$$F_m = e^{-m\lambda}(1-\lambda)^{m+1} \frac{d^m}{dy^m} \frac{1}{1-\lambda}.$$

By a well-known application of Lagrange's generalization of Taylor's theorem, we have that when $y = \lambda e^{-\lambda}$,

$$\lambda = \sum_{n=1}^{\infty} \frac{n^{n-1}}{n!} y^n$$

and it follows that

$$\frac{1}{1-\lambda} = 1 + y\frac{d\lambda}{dy} = 1 + \sum_{n=1}^{\infty} \frac{n^n}{n!} y^n.$$

Finally, we calculate

$$\sum_{n=1}^{m} S(m, n) = F_m(1) = \lim_{\lambda \to 0} \frac{d^m}{dy^m} \frac{1}{1-\lambda} = m^m$$

as required.

(11) $\qquad \sum_{j=0} S(i, j) = i^i \quad$ for $i = 1, 2, 3, \cdots$.

Comment by A. MEIR (University of Alberta).
By essentially the same inductive argument one can show that the more general

recursion relation

$$S(m+1, n) = (m+z)S(m, n) + (m+n)S(m, n-1)$$

has a solution satisfying

$$\sum_{n=1}^{m} S(m, n) = (m+z)^m.$$

Editorial note. Ruehr notes that his method can be applied to Meir's generalization. The differential recurrence equation becomes

$$F_{m+1} = [(m+z) + x(m+1)]F_m + x^2 F'_m$$

whose solution is

$$F_m = \frac{1}{x^{m+1}} e^{(m+z)/x} \left[x^3 e^{-1/x} \frac{d}{dx} \right]^m x e^{-z/x}.$$

The final step then requires the identity

$$\sum_{n=1}^{\infty} \frac{(n+z)^n \lambda^n e^{-n\lambda}}{n!} = \frac{e^{\lambda z}}{1-\lambda}$$

which he also establishes.

It would be of interest to determine all pairs of polynomials $P(m, n)$, $Q(m, n)$ (in particular, linear ones) such that if

$$S(m+1, n) = P(m, n)S(m, n) + Q(m, n)S(m, n-1)$$

and subject to conditions (A) and (B), then also

$$\sum_{n=1}^{m} S(m, n) = m^m. \qquad\qquad \text{[M.S.K.]}$$

A Combinatorial Identity

Problem 79-13, by T. V. NARAYANA *and* M. ÖZSOYOGLU (University of Alberta).
Prove that

$$\sum_{i=1}^{n} i \frac{m-n+2i+1}{m+n+1} \binom{m+n+1}{n-i} = \sum_{i=1}^{n} i2^{i-1} \frac{m-n+i}{m+n-i} \binom{m+n-i}{m},$$

where m, n are integral with $m > n > 0$.

Solution by A. A. JAGERS (Technische Hogeschool Twente, Enschede, The Netherlands).

Denote the left-hand side of the identity by A and the right-hand side by B. Then

$$A = \sum_{i=1}^{n} i \left\{ \binom{m+n}{m+i} - \binom{m+n}{m+i+1} \right\}$$

and

$$B = \sum_{i=1}^{n} i2^{i-1} \left\{ \binom{m+n-i-1}{m-1} - \binom{m+n-i-1}{m} \right\}.$$

Hence, it suffices to prove the generalization

$$\sum_{i=1}^{n} i\binom{s+1}{r+i+1} = \sum_{i=1}^{n} i2^{i-1}\binom{s-i}{r}, \qquad 0 \le s \le n+r, \quad r, n \ge 0.$$

This is accomplished by substituting $i2^{i-1} = \sum_{k=1}^{i} k \, (i \ k)$, changing the order of summation, and applying the convolution identity

$$\sum_{i=0}^{s}\binom{i}{k}\binom{s-i}{r} = \binom{s+1}{k+r+1}, \qquad k, r \ge 0,$$

which is equivalent to $f_k(x)f_r(x) = x^{-1}f_{k+r+1}(x)$, with $f_p(x) = x^p(1-x)^{-p-1}$ $(p \ge 0)$.

The Rogers–Ramanujan Identities

Problem 74-12, *by* G. E. ANDREWS (Pennsylvania State University).
    It is well known that the identity

$$\theta_n(x) = \sum_{\lambda=-\infty}^{\infty} (-1)^{\lambda} x^{\lambda(3\lambda-1)/2}\begin{bmatrix} 2n \\ n+\lambda \end{bmatrix} = (1-x^{n+1}) \cdots (1-x^{2n}),$$

where

$$\begin{bmatrix} A \\ B \end{bmatrix} = \prod_{j=1}^{B} (1-x^{A-j+1})(1-x^{j})^{-1} \text{ for } 0 \le B \le A,$$

$$\begin{bmatrix} A \\ B \end{bmatrix} = 0 \quad \text{otherwise},$$

may be used to prove the Rogers–Ramanujan identities [G. H. Hardy, *Ramanujan*, pp. 95–98]. This suggests the importance of the polynomials

$$g_n(x) = \sum_{\lambda=-\infty}^{\infty} (-1)^{\lambda} x^{\lambda(5\lambda-1)/2}\begin{bmatrix} 2n \\ n+\lambda \end{bmatrix}$$

and

$$h_n(x) = \sum_{\lambda=-\infty}^{\infty} (-1)^{\lambda} x^{\lambda(5\lambda+3)/2}\begin{bmatrix} 2n+1 \\ n+1+\lambda \end{bmatrix}.$$

    (a) Prove that $\theta_n(x)$ divides both $g_n(x)$ and $h_n(x)$.
    (b) Find formulas for the polynomials $r_n(x)$ and $s_n(x)$ defined by $r_n(x) = g_n(x)/\theta_n(x)$, and $s_n(x) = h_n(x)/\theta_n(x)$ so that one may directly deduce

$$\lim_{n\to\infty} r_n(x) = \sum_{n=0}^{\infty} \frac{x^{n^2}}{(1-x) \cdots (1-x^{n})}$$

and

$$\lim_{n\to\infty} s_n(x) = \sum_{n=0}^{\infty} \frac{x^{n^2+n}}{(1-x) \cdots (1-x^{n})}.$$

    (c) Prove the Rogers–Ramanujan identities

$$1 + \sum_{n=1}^{\infty} \frac{x^{n^2+\alpha n}}{(1-x)(1-x^{2}) \cdots (1-x^{n})} = \prod_{n=0}^{\infty} \frac{1}{(1-x^{5n+1+\alpha})(1-x^{5n+4-\alpha})},$$

where $\alpha = 0$ or 1.

Editorial note. The proposer notes that his solution relies on "heavy artillery" from the theory of hypergeometric functions, and it is his hope that someone will be able to solve the problem in a more elementary manner.

Solution by D. M. BRESSOUD (Institute for Advanced Study, Princeton NJ).

Let $(x)_n = \prod_{i=1}^{n}(1-x^i)$, $(x)_0 = 1$, and for $n < 0$, $(x)_n^{-1} = 0$; then

$$r_n(x) = \frac{g_n(x)}{\theta_n(x)} = \sum_{\lambda=-\infty}^{\infty}(-1)^{\lambda}x^{\lambda(5\lambda-1)/2}\frac{(x)_n}{(x)_{n+\lambda}(x)_{n-\lambda}}$$

and

$$s_n(x) = \frac{h_n(x)}{\theta_n(x)} = (1-x^{2n+1})\sum_{\lambda=-\infty}^{\infty}(-1)^{\lambda}x^{\lambda(5\lambda+3)/2}\frac{(x)_n}{(x)_{n+1+\lambda}(x)_{n-\lambda}}.$$

We shall use the identity for $\theta_n(x)$ with both sides divided by $(x)_{2n}$:

(1)
$$\frac{1}{(x)_n} = \sum_{\lambda=-\infty}^{\infty}(-1)^{\lambda}\frac{x^{\lambda(3\lambda-1)/2}}{(x)_{n+\lambda}(x)_{n-\lambda}},$$

and its companion:

(2)
$$\frac{1}{(x)_n} = \sum_{\lambda=-\infty}^{\infty}(-1)^{\lambda}\frac{x^{\lambda(3\lambda+1)/2}}{(x)_{n+\lambda+1}(x)_{n-\lambda}}.$$

[G. H. Hardy, *Ramanujan*, (6.16.5) and (6.16.6)]. We shall also need:

(3)
$$\frac{1}{(x)_{n+k}} = \sum_{m=0}^{\infty}\frac{x^{m^2+2km}}{(x)_{m+2k}}\begin{bmatrix}n-k\\m\end{bmatrix}, \qquad 0 \leq k \leq n,$$

(4)
$$\frac{1}{(x)_{n+k+1}} = \sum_{m=0}^{\infty}\frac{x^{m^2+2km+m}}{(x)_{m+2k+1}}\begin{bmatrix}n-k\\m\end{bmatrix}.$$

Proof of (3) *and* (4). We shall demonstrate that both sides generate the same partition function. It is well known that the left side of (3) generates all partitions with parts $\leq n+k$ [G. Andrews, *Theory of Partitions* (TOP), Thm. 1.1] and that $\begin{bmatrix}n-k\\m\end{bmatrix}$ generates all partitions into at most m parts, each of which is $\leq n-k-m$ (TOP, Thm. 3.1). Now, x^{m^2+2km} generates m copies of $m+2k$, so that

$$x^{m^2+2km}\begin{bmatrix}n-k\\m\end{bmatrix}$$

generates all partitions with exactly m parts, each of which is $\geq m+2k$ and $\leq n+k$. Thus,

$$\frac{x^{m^2+2km}}{(x)_{m+2k}}\begin{bmatrix}n-k\\m\end{bmatrix}$$

generates all partitions with parts $\leq n+k$, such that the mth largest part is $\geq m+2k$ and the $(m+1)$th largest part is $\leq m+2k$. Given any partition into parts $\leq n+k$, there exists a unique integer $m \geq 0$ such that the mth largest part is $\geq m+2k$ and the $(m+1)$th largest part is $\leq m+2k$, namely:

$$m = \max\left(\{0\} \cup \{i \geq 1 \mid i\text{th largest part is} \geq i + 2k\}\right).$$

Therefore,

$$\sum_{m=0}^{\infty} \frac{x^{m^2+2km}}{(x)_{m+2k}} \begin{bmatrix} n-k \\ m \end{bmatrix}$$

counts only partitions with parts $\leq n + k$, and counts each such partition precisely once. This proves (3). (4) follows from a similar argument.

Now,

$$r_n(x) = \frac{1}{(x)_n} + \sum_{\lambda=1}^{\infty} (-1)^\lambda x^{\lambda(5\lambda-1)/2} \frac{(1+x^\lambda)(x)_n}{(x)_{n-\lambda}(x)_{n+\lambda}}$$

$$= \frac{1}{(x)_n} + \sum_{\lambda=1}^{\infty} (-1)^\lambda x^{\lambda(5\lambda-1)/2} \frac{(1+x^\lambda)(x)_n}{(x)_{n-\lambda}} \sum_{m=0}^{\infty} \frac{x^{m^2+2m\lambda}}{(x)_{m+2\lambda}} \begin{bmatrix} n-\lambda \\ m \end{bmatrix}$$

$$= \frac{1}{(x)_n} + \sum_{\lambda=1}^{\infty} \sum_{m=0}^{\infty} (-1)^\lambda x^{(m+\lambda)^2+\lambda(3\lambda-1)/2} \frac{(1+x^\lambda)(x)_n}{(x)_{m+2\lambda}(x)_m(x)_{n-\lambda-m}}$$

$$= \sum_{s=0}^{\infty} x^{s^2} \frac{(x)_n}{(x)_{n-s}} \sum_{\lambda=-s}^{s} (-1)^\lambda \frac{x^{\lambda(3\lambda-1)/2}}{(x)_{s+\lambda}(x)_{s-\lambda}}$$

$$= \sum_{s=0}^{\infty} x^{s^2} \frac{(x)_n}{(x)_{n-s}(x)_s} \qquad \text{(by (1))}$$

$$= \sum_{s=0}^{\infty} x^{s^2} \begin{bmatrix} n \\ s \end{bmatrix}.$$

Thus, $r_n(x)$ is a polynomial in x and for $|x| < 1$:

$$\sum_{s=0}^{\infty} \frac{x^{s^2}}{(x)_s} = \lim_{n\to\infty} r_n(x)$$

$$= \frac{1}{(x)_\infty} \sum_{\lambda=-\infty}^{\infty} (-1)^\lambda x^{\lambda(5\lambda-1)/2}$$

$$= \prod_{i=1}^{\infty} \frac{(1-x^{5i-3})(1-x^{5i-2})(1-x^{5i})}{(1-x^i)} \qquad \text{(TOP, Thm. 2.8)}$$

$$= \prod_{i=1}^{\infty} \frac{1}{(1-x^{5i-4})(1-x^{5i-1})}.$$

Similarly, if we rewrite $s_n(x)$ as

$$s_n(x) = (1-x^{2n+1}) \sum_{\lambda=0}^{\infty} (-1)^\lambda x^{\lambda(5\lambda+3)/2} \frac{(1-x^{2\lambda+1})(x)_n}{(x)_{n-\lambda}(x)_{n+\lambda+1}},$$

then (4) and (2) imply that

$$s_n(x) = (1-x^{2n+1}) \sum_{s=0}^{\infty} x^{s^2+s} \begin{bmatrix} n \\ s \end{bmatrix},$$

which is a polynomial in x and implies that

$$\sum_{s=0}^{\infty} \frac{x^{s^2+s}}{(x)_s} = \prod_{i=1}^{\infty} \frac{1}{(1-x^{5i-3})(1-x^{5i-2})}.$$

Note. Equations (1) and (2) are nontrivial. However, it is unreasonable to look for a proof of these finite forms of the Rogers–Ramanujan identities which is simpler than the proof for the finite forms of Euler's pentagonal number theorem. What this proof establishes is that we don't need very much more. Incidentally, this proof is a boiled down version of Bailey's proof of Watson's q-analogue of Whipple's theorem.

Truncation of the Rogers–Ramanujan Theta Series[1]

Problem 83-13*, *by* G. E. ANDREWS (Pennsylvania State University).

In 1926, G. Szegö [4] investigated the family of orthogonal polynomials $K_n(x)$ defined by

$$K_n(x) = \sum_{\nu=0}^{n} \begin{bmatrix} n \\ \nu \end{bmatrix} q^{\nu(\nu+1)} x^\nu,$$

where

$$\begin{bmatrix} n \\ \nu \end{bmatrix} = \begin{cases} \dfrac{(1-q^n)(1-q^{n-1})\cdots(1-q^{n-\nu+1})}{(1-q^\nu)(1-q^{\nu-1})\cdots(1-q)}, & 0 \le \nu \le n, \\[2mm] 0, & \nu < 0, \quad \nu > n. \end{cases}$$

Recently (cf. [1], [2]) formulae have been found for $K_n(q^{-1})$ and $K_n(1)$ which imply the Rogers–Ramanujan identities in the limit. Prove the following formulae expressing the truncated Rogers–Ramanujan theta series as a sum of Szegö polynomials.

(1)
$$1 + \sum_{j=1}^{N} (-1)^j q^{j(5j-1)/2}(1+q^j)$$
$$= \sum_{k=0}^{N} (-1)^k q^{Nk+k(3k+1)/2} K_{N-k}(q^{2k-1}) \prod_{s=k+1}^{N} (1-q^s),$$

and

(2)
$$\sum_{j=0}^{N} (-1)^j q^{j(5j+3)/2}(1-q^{2j+1})$$
$$= \sum_{k=0}^{N} (-1)^k q^{Nk+k(3k+5)/2} K_{N-k}(q^{2k}) \prod_{s=k+1}^{N+1} (1-q^s).$$

Deduce the Rogers–Ramanujan identities [1] from (1) and (2). The truncation of Euler's pentagonal number theorem has been treated in a similar manner by D. Shanks [3].

REFERENCES

[1] G. E. ANDREWS, *Problem* 74-12, this Review, 16 (1974), p. 390.
[2] D. M. BRESSOUD, *Some identities for terminating q-series,* Math. Proc. Comb. Phil. Soc., 89 (1981), pp. 211–223.
[3] D. SHANKS, *A short proof of an identity of Euler,* Proc. Amer. Math. Soc., 2 (1951), pp. 747–749.

---

[1]Problem 83-13 as originally stated contained two typographical errors. In (1), the second q term appeared as $q^{Nk+K(3k+1)/2}$. In (2) the product term appeared as a summation.

[4] G. SZEGÖ, *Ein Beitrag zur Theorie der Thetafunktionen*, Sitz. Ber. Preuss. Akad. Wiss. Phys.-Math. Kl., pp. 242–252 (also in G. Szegö, *Collected Papers*, Vol. 1, R. Askey, ed., Birkhäuser, Boston, 1982, pp. 795–805).

Solution by W. B. JORDAN (Scotia, NY).

For brevity, write (u, v) for $\Pi_{j-u}^{v} (1 - q^j)$. We derive a recurrence for K. Assume it is of the form

$$K_{n+1}(x) = AK_n(ax) + BxK_n(bx).$$

Equate coefficients of corresponding powers of x:

For x^0: $1 = A$.

For x^{n+1}: $q^{(n+1)(n+2)} = Bb^n q^{n(n+1)}$, $B = b^{-n}q^{2n+2}$.

For x^r, $0 < r < n + 1$:

$$\begin{bmatrix} n + 1 \\ r \end{bmatrix} q^{r(r+1)} = \begin{bmatrix} n \\ r \end{bmatrix} a^r q^{r(r+1)} + B \begin{bmatrix} n \\ r - 1 \end{bmatrix} b^{r-1} q^{r(r-1)},$$

which reduces to

$$b^{-n+r-1} q^{2n-2r+2}(1 - q^r) = 1 - a^r + q^{n-r+1}(a^r - q^r)$$

which is to be an identity in r. There are two solutions:

$$a = 1, \quad b = q, \quad B = q^{n+2}, \qquad a = q, \quad b = q^2, \quad B = q^2,$$

so $K_{n+1}(x) = K_n(x) + q^{n+2}xK_n(qx) = K_n(qx) + q^2 xK_n(q^2x)$. From these we can get

$$(1 - q^{n+2}x) K_{n+1}(x) - K_n(x) = -q^{n+4}x^2 K_n(q^2x),$$

which is the form used in the reductions below.

Use R, S to denote the left and right sides of (1), and T, U of (2). We prove $R = S$ and $T = U$ by computing difference equations and solving them.

$$S_N = \sum_{k=0}^{N} S_{Nk}, \qquad S_N - S_{N-1} = S_{NN} + \sum_{k=0}^{N-1} (S_{Nk} - S_{N-1,k}),$$

where

$$S_{Nk} = (-1)^k (k + 1, N) q^{Nk+k(3k+1)/2} K_{N-k}(q^{2k-1}).$$

We prove

$$\sum_{r=0}^{k-1} (S_{Nr} - S_{N-1,r}) = (-1)^k q^{Nk+k(3k-1)/2} (k, N - 1) K_{N-k}(q^{2k-1})$$

which we do by induction on k. When $k = 1$:

$$\begin{aligned} S_{N0} - S_{N-1,0} &= (1, N) K_N(q^{-1}) - (1, N - 1) K_{N-1}(q^{-1}) \\ &= (1, N - 1)[(1 - q^N) K_N(q^{-1}) - K_{N-1}(q^{-1})] \\ &= -(1, N - 1) q^{N+1} K_{N-1}(q), \end{aligned}$$

so the formula is true for $k = 1$. Assuming it true for k, then for $k + 1$ we have

$$\sum_{r=0}^{k} (S_{Nr} - S_{N-1,r}) = (-1)^k q^{Nk+k(3k-1)/2} (k, N-1) K_{N-k}(q^{2k-1}) + S_{Nk} - S_{N-1,k}$$

$$= (-1)^k (k+1, N-1) q^{Nk+k(3k-1)/2}$$

$$\cdot [(1-q^k) K_{N-k}(q^{2k-1}) + (1-q^N) q^k K_{N-k}(q^{2k-1}) - K_{N-k-1}(q^{2k-1})]$$

On using the recurrence formula the square bracket becomes $-q^{N+3k+1} K_{N-k-1}(q^{2k+1})$, giving

$$\sum_{r=0}^{k} (S_{Nr} - S_{N-1,r}) = (-1)^{k+1} (k+1, N-1) q^{N(k+1)+(k+1)(3k+2)/2} K_{N-k-1}(q^{2k+1})$$

which completes the induction. Then

$$S_N - S_{N-1} = S_{NN} + \sum_{r=0}^{N-1} (S_{Nr} - S_{N-1,r})$$

$$= (-1)^N q^{N(5N+1)/2} + (-1)^N q^{N(5N-1)/2}$$

which, by inspection, is $R_N - R_{N-1}$. Since $S_0 = 1 = R_0$ it follows by induction that $S_N = R_N$, and (1) is proved.

In the limit as N increases S_N and R_N both converge and

$$R_\infty = S_\infty = S_{\infty,0} = (1, \infty) K_\infty(q^{-1}) = (1, \infty) \sum_{r=0}^{\infty} q^{r^2}/(1, r).$$

The Rogers–Ramanujan identities are

$$1 + \sum_{1}^{\infty} q^{n^2+pn}/(1, n) = \prod_{0}^{\infty} (1 - q^{5n+1+p})^{-1} (1 - q^{5n+4-p})^{-1}$$

with $p = 0$ or 1. We will have proved the first one ($p = 0$) if we can show that

$$R_\infty = \prod_{n=0}^{\infty} (1 - q^{n+1})(1 - q^{5n+1})^{-1}(1 - q^{5n+4})^{-1}$$

i.e., if $R_\infty = \prod_{1}^{\infty} (1 - q^{5n})(1 - q^{5n-2})(1 - q^{5n-3})$.

But the series and product expansions of $\theta_4(z, Q)$ are

$$1 + \sum_{1}^{\infty} (-1)^n 2Q^{n^2} \cos 2nz = \prod_{1}^{\infty} (1 - Q^{2n})(1 - Q^{2n-1} e^{2iz})(1 - Q^{2n-1} e^{-2iz}).$$

Take $Q = q^{5/2}$ and $e^{2iz} = q^{1/2}$, so $2 \cos 2nz = q^{-n/2} + q^{n/2}$. The right side becomes our infinite product and the left becomes R_∞, proving the first identity.

The proofs of (2) and of the second identity are identical to the above. We find

$$\sum_{r=0}^{k-1} (U_{Nr} - U_{N,r-1}) = (-1)^k q^{Nk+3k(k+1)/2} (k, N) K_{N-k}(q^{2k})$$

which leads to $U_N = T_N$, proving (2). Then

$$T_\infty = U_\infty = (1, \infty) K_\infty(1) = (1, \infty) \sum_{0}^{\infty} q^{n(n+1)}/(1, n).$$

The second identity will be proved if

$$\sum_0^\infty (-1)^n q^{n(5n+3)/2}(1 - q^{2n+1}) = \prod_0^\infty (1 - q^{n+1})(1 - q^{5n+2})^{-1}(1 - q^{5n+3})^{-1}.$$

These are the series and product expansions of $\theta_4(z, Q)$ with $Q = q^{5/2}$ and $e^{2iz} = q^{3/2}$, and the proof is complete.

Also solved by the proposer, as follows. In [1, Thm. 4], let $k = 3$, $b_1 = aq^{N+1}$, and c_1, c_2, b_2, c_3, b_3 all $\to \infty$. (1) follows if $a \to 1$; (2) follows if $a = q$.

REFERENCE

[1] G. E. ANDREWS, *Problems and prospects for basic hypergeometric functions*, from Theory and Applications of Special Functions, R. Askey, ed., Academic Press, New York, 1975, p. 199.

Problem 60-2, On a Binomial Identity Arising from a Sorting Problem, by PAUL BROCK (Hughes Aircraft Company).

If

$$H(A, B) = \sum_{i=0}^A \sum_{j=0}^B \binom{i + j}{j}\binom{A - i + j}{j}\binom{B + i - j}{B - j}\binom{A - i + B - j}{B - j},$$

show that

$$H(A, B) - H(A - 1, B) - H(A, B - 1) = \binom{A + B}{B}^2.$$

This problem has arisen by considering the $N!$ permutations of the integers 1 through N; each permutation will have a maximum length monotonic increasing and decreasing subsequence (see Problem 59-3). If one tries to count the number of those permutations whose maximum monotonic increasing subsequence is of length r and whose maximum monotonic decreasing subsequence is of length $N - r + 1$, the total count is $\binom{n - 1}{r - 1}^2$ sequences. The proof of this can be shown to be equivalent to demonstrating the above identity.

Solution by DAVID SLEPIAN (Bell Telephone Laboratories).

Let

$$G(u, v) = \sum_{A=0}^\infty \sum_{B=0}^\infty H(A, B)u^A v^B$$

$$= \sum_{i,j,k,l=0}^\infty \binom{i + j}{j}\binom{j + k}{k}\binom{k + l}{l}\binom{l + i}{i} u^{i+k} v^{j+l},$$

where $k = A - i$, $l = B - j$. The identity

(1)
$$\frac{1}{(1 - x)^j} = (1 - x) \sum_{i=0}^\infty \binom{i + j}{j} x^i,$$

will be used in the proof repeatedly.

In (1), set $x = ut$, multiply by $\binom{j + k}{k} v^j$ and sum over j, yielding

$$\sum_{j=0}^{\infty} \binom{j+k}{k}\left(\frac{v}{1-ut}\right)^{j} = (1-ut)\sum_{i,j=0}^{\infty}\binom{i+j}{j}\binom{j+k}{k}u^{i}v^{j}t^{i}.$$

Multiplying the latter equation by $1 - v/(1 - ut)$ and applying (1), yields

$$(2) \qquad \left(\frac{1-ut}{1-v-ut}\right)^{k} = (1-v-ut)\sum_{i,j=0}^{\infty}\binom{i+j}{j}\binom{j+k}{k}u^{i}v^{j}t^{i}.$$

Now multiply (2) by $\binom{k+l}{l}u^{k}$ and sum over k by reapplying (1);

$$\left(\frac{1-v-ut}{1-u-v-ut+u^{2}t}\right)^{l}$$

$$= (1-u-v-ut+u^{2}t)\sum_{i,j,k=0}^{\infty}\binom{i+j}{j}\binom{j+k}{k}\binom{k+l}{l}u^{i+k}v^{j}t^{i}.$$

Multiplying the latter equation by $\binom{l+m}{m}v^{l}$ and summing over l yields

$$(3) \qquad \frac{1}{R}\left(\frac{S}{R}\right)^{m} = \sum_{i,j,k,l=0}^{\infty}\binom{i+j}{j}\binom{j+k}{k}\binom{k+l}{l}\binom{l+m}{m}u^{i+k}v^{j+l}t^{i}$$

where

$$R = 1 - u - 2v + v^{2} + ut(u+v-1),$$

$$S = 1 - u - v + ut(u-1).$$

Multiplying (3) by $(2\pi i t^{m+1})^{-1}$, summing over m, and integrating about a proper contour, yields

$$(4) \quad G(u,v) = \frac{1}{2\pi i}\oint\frac{dt}{tR-S} = \frac{1}{2\pi i(u+v-1)}\oint\frac{dt}{ut^{2}+t(v-u-1)+1}.$$

Multiply (2) by t^{-k-1} and sum over k to obtain

$$\frac{1}{1-ut}\sum_{k=0}^{\infty}\left\{\frac{1-ut}{t(1-v-ut)}\right\}^{k+1} = \sum_{i,j,k=0}^{\infty}\binom{i+j}{j}\binom{j+k}{k}u^{i}v^{j}t^{i-k-1}.$$

Multiplying the latter equation by $(2\pi i)^{-1}$ and integrating about a proper contour;

$$(5) \quad F(u,v) = \sum_{i,j=0}^{\infty}\binom{i+j}{j}^{2}u^{i}v^{j} = -\frac{1}{2\pi i}\oint\frac{dt}{ut^{2}+t(v-u-1)+1}.$$

From (4) and (5), it follows that

$$F(u,v) = -(u+v-1)G(u,v).$$

Whence,

$$\sum_{A=0}^{\infty}\sum_{B=0}^{\infty}\binom{A+B}{B}^{2}u^{A}v^{B} \equiv \sum_{A=0}^{\infty}\sum_{B=0}^{\infty}\{H(A,B)[u^{A}v^{B} - u^{A+1}v^{B} - u^{A}v^{B+1}]\}.$$

Since, $H(A,-1) = H(-1,B) = 0$, the desired identity follows.

Also solved by the proposer.

Editorial note: Starting with the generating function of $H(A,B)$ (i.e., $G(u,v)$) and the given recurrence relation on $H(A,B)$, it can be shown after many elementary manipulations and transformations, that also

$$H(A, B) = \sum_{r=0}^{A} \sum_{s=0}^{B} \binom{r+s}{s}^2 \binom{A-r+B-s}{B-s}.$$

It would be of interest to give a direct proof of the equality for the two different expressions for $H(A, B)$, i.e., that

$$\sum_{i=0}^{A} \sum_{j=0}^{B} \binom{i+j}{j} \binom{A-i+j}{j} \binom{B+i-j}{B-j} \binom{A-i+B-j}{B-j}$$

$$= \sum_{i=0}^{A} \sum_{j=0}^{B} \binom{i+j}{j}^2 \binom{A-i+B-j}{B-j}.$$

Problem 75-10, *A Combinatorial Problem*, by G. E. ANDREWS (Pennsylvania State University).

Some years ago, P. Brock (Problem 60-2, this Journal, 4 (1962), pp. 396–398) posed the problem of showing that if

$$H(m, n) = \sum_{i=0}^{m} \sum_{j=0}^{n} \binom{i+j}{i} \binom{m-i+j}{j} \binom{i+n-j}{i} \binom{m+n-i-j}{n-j},$$

then

$$H(m, n) - H(m-1, n) - H(m, n-1) = \binom{m+n}{n}^2.$$

Prove also that

$$(2m+1)H(m, n) = (n+m+1)\left\{2H(m-1, n) + \binom{m+n}{n}^2\right\}.$$

Solution by the proposer.

L. Carlitz [2, p. 23] has shown that

$$H(m, n) = \frac{(m+n+1)!}{m!n!} \sum_{j\geq 0} \binom{m}{j}\binom{n}{j} \frac{1}{2j+1}$$

$$= \frac{(m+n+1)!}{m!n!} {}_3F_2\left[\begin{matrix} -m, -n, \frac{1}{2}; 1 \\ 1, \frac{3}{2} \end{matrix}\right].$$

Hence

$$H(m, n) = \frac{(m+2n+1)!(\frac{3}{2})_n}{m!n!^2(2n+1)(n+m+2)_n} {}_3F_2\left[\begin{matrix} -n, m+\frac{3}{2}, \frac{1}{2}; 1 \\ \frac{1}{2}-n, \frac{3}{2} \end{matrix}\right]$$

$$\text{(by}[1; p. 33, eq. (1), b = m+\tfrac{3}{2}, c = \tfrac{1}{2}, d = n-m-\tfrac{1}{2}, w = \tfrac{3}{2}])$$

$$= \frac{(m+n+1)!}{m!n!^2 2^n} \sum_{j\geq 0} \frac{1}{2j+1} \binom{n}{j}(2m+3)(2m+5)\cdots(2m+2j+1)$$

$$\cdot (2n-2j-1)(2n-2j-3)\cdots 3 \cdot 1.$$

We now split the above sum into two sums by splitting $(2m+2j+1)$ into $(2m)+(2j+1)$. Hence

$$H(m, n) = \frac{2(m+n+1)}{2m+1} H(m-1, n)$$

$$+ \frac{(m+n+1)!}{(2m+1)m!n!^2 2^n}(2n-1)(2n-3)\cdots 3\cdot 1 \sum_{j=0}^{n} \frac{(m+1/2)_j(-n)_j}{j!(-n+1/2)_j}$$

$$= \frac{2(m+n+1)}{2m+1} H(m-1, n) + \frac{(m+n+1)!(2n-1)(2n-3)\cdots 3\cdot 1(-n-m)_n}{(2m+1)m!n!^2 2^n(-n+1/2)_n}$$

$$\text{(by the Vandermonde sum } [1, \text{p. 3}])$$

$$= \frac{2(m+n+1)}{2m+1} H(m-1, n) + \frac{(m+n+1)}{2m+1}\binom{m+n}{m}^2.$$

REFERENCES

[1] W. N. BAILEY, *Generalized Hypergeometric Series*, Cambridge University Press, Cambridge, England, 1935.
[2] L. CARLITZ, *A binomial identity arising from a sorting problem*, this Review, 6 (1964), pp. 20–30.

Problem 71-16*, *A Double-Sum Identity*, by T. E. PHIPPS, JR. (U.S. Naval Ordnance Laboratory).

Prove that

$$S = \sum_{s=0}^{m} \sum_{r=0}^{m-s} \frac{(-1)^{r+s}(n+p+q-r-s-1)!}{r!s!(m-r-s)!(p-r)!(q-s)!} \equiv 0,$$

where $m \geq n > 0$ and p, q are each $\geq m$.

Solution by J. W. RIESE (Kimberly-Clark Corporation).

The sum under consideration may be grouped as follows:

$$S = \sum_{s=0}^{m} \sum_{r=0}^{m-s} \frac{(-1)^{r+s}}{r!(p-r)!s!(q-s)!} \cdot \left[\frac{(n-1+p+q-r-s)!}{(m-r-s)!} \right].$$

Note first that the factor in brackets may be expressed as:

$$\lim_{x\to 1} \left(\frac{\partial}{\partial x}\right)^{p+q+n-1-m} x^{p+q+n-1-r-s}.$$

Over the stated summation range $(r+s \leq m)$, this factor will always be positive. Beyond this range $(r+s > m)$, this factor will be zero; therefore, the summation range can be extended without changing its value:

$$S = \lim_{x \to 1} \left(\frac{\partial}{\partial x}\right)^{p+q+n-1-m} x^{n-1} \sum_{s=0}^{q} \frac{(-1)^s x^{q-s}}{s!(q-s)!} \sum_{r=0}^{p} \frac{(-1)^r x^{p-r}}{r!(p-r)!}.$$

The two sums are now independent binomial expansions, and can be summed immediately to $(x - 1)^q$, $(x - 1)^p$, respectively, yielding:

$$S = \lim_{x \to 1} \left(\frac{\partial}{\partial x}\right)^{p+q+n-1-m} x^{n-1}(x - 1)^{p+q}.$$

By hypothesis, $n \leq m$, or $n - m - 1 < 0$; therefore the exponent of $\partial/\partial x$ is lower than the exponent of $(x - 1)$. Therefore the differentiation will always leave some positive power of $(x - 1)$, which will go to zero under the subsequent limiting process (i.e., as $x \to 1$).

Therefore, $S = 0$.

Editorial note. Many of the solvers used the Vandermonde convolution and also noted that the restrictions on the parameters could be relaxed to obtain the sum in general, i.e.,

$$S = (-1)^{p+q} \frac{(n-1)!}{m!} \binom{p+q}{q}\binom{m-n}{p+q}. \qquad \text{[M.S.K.]}$$

Problem 67–4, A Double Sum, by L. CARLITZ (Duke University).

Show that

$$\sum_{r=0}^{m} \sum_{s=0}^{n} \binom{r+s}{r}^2 \binom{m+n-r-s}{m-r}^2 = \frac{1}{2}\binom{2m+2n+2}{2m+1}.$$

Solution by the proposer.

We have

$$\{(1 - x - y)^2 - 4xy\}^{-1/2} = \sum_{r=0}^{\infty} \binom{2r}{r} x^r y^r (1 - x - y)^{-2r-1}$$

$$= \sum_{r=0}^{\infty} \binom{2r}{r} x^r y^r \sum_{k=0}^{\infty} \binom{2r+k}{k}(x + y)^k$$

$$= \sum_{r=0}^{\infty} \binom{2r}{r} x^r y^r \sum_{m,n=0}^{\infty} \frac{(2r+m+n)!}{(2r)!\,m!\,n!} x^m y^n$$

$$= \sum_{m,n=0}^{\infty} x^m y^n \sum_{r=0}^{\min(m,n)} \frac{(m+n)!}{(2r)!(m-r)!(n-r)!}$$

$$= \sum_{m,n=0}^{\infty} \binom{m+n}{n} x^m y^n \sum_{r=0}^{\min(m,n)} \binom{m}{n}\binom{n}{r}$$

$$= \sum_{m,n=0}^{\infty} \binom{m+n}{n}^2 x^m y^n.$$

Similarly,

$$\{(1 - x - y)^2 - 4xy\}^{-1} = \sum_{r=0}^{\infty} 2^{2r} x^r y^r (1 - x - y)^{-2r-2}$$

$$(1) \qquad = \sum_{r=0}^{\infty} 2^{2r} x^r y^r \sum_{m,n=0}^{\infty} \frac{(2r + m + n + 1)!}{(2r + 1)!\, m!\, n!} x^m y^n$$

$$= \sum_{m,n=0}^{\infty} \frac{(m + n + 1)!}{m!\, n!} x^m y^n \sum_{r=0}^{\min(m,n)} \frac{(-m)_r(-n)_r}{(2r + 1)!}.$$

Now

$$\sum_{r=0}^{\min(m,n)} \frac{(-m)_r(-n)_r}{(2r + 1)!} 2^{2r} = \sum_{r=0}^{\min(m,n)} \frac{(-m)_r(-n)_r}{r!\,(3/2)_r}$$

$$= \frac{(3/2)_{m+n}}{(3/2)_m (3/2)_m} = \frac{(2m + 2n + 1)!}{(2m + 1)!(2n + 1)!} \frac{m!\, n!}{(m + n)!},$$

so that

$$(2) \qquad \frac{1}{2} \sum_{m,n=0}^{\infty} \binom{2m + 2n + 2}{2m + 1} x^m y^n = \{(1 - x - y)^2 - 4xy\}^{-1}.$$

Comparing (1) and (2) we obtain the stated result.

Remark. In exactly the same way we can prove that

$$\sum_{m,n=0}^{\infty} C(m, n; \lambda) x^m y^n = \{(1 - x - y)^2 - 4xy\}^{-\lambda},$$

where

$$C(m, n; \lambda) = \frac{(2\lambda)_{m+n}}{m!\, n!} \frac{(\lambda + \frac{1}{2})_{m+n}}{(\lambda + \frac{1}{2})_m (\lambda + \frac{1}{2})_n}.$$

This implies

$$\sum_{r=0}^{m} \sum_{s=0}^{n} C(r, s; \alpha) C(m - r, n - s; \beta) = C(m, n; \alpha + \beta).$$

Problem 76-14*, *Three Multiple Summations*, by L. CARLITZ (Duke University).

The following formulas appear incidentally in the paper: *Some binomial coefficient identities* (Fibonacci Quarterly, 4 (1966), pp. 323–331):

$$\sum_{i=0}^{m} \sum_{j=0}^{n} (-1)^{i+j} \binom{i+j}{i} \binom{m-i+j}{m-i} \binom{i+n-j}{i} \binom{m-i+n-j}{m-i}$$

$$(1)$$

$$= \begin{cases} \left(\dfrac{\frac{1}{2}(m+n)}{\frac{1}{2}m} \right)^2 & (m, n \text{ both even}), \\ 0 & \text{otherwise}, \end{cases}$$

$$(2) \qquad \sum_{i=0}^{m} \sum_{j=0}^{n} (-1)^{i+j} \frac{\binom{m}{i}^2 \binom{n}{j}^2}{\binom{m+n}{i+j}} = \delta_{mn},$$

(3)†
$$\sum_{r=0}^{\min(i,j,k)} \frac{\binom{i}{r}\binom{j}{r}\binom{k}{r}}{\binom{i+j+k}{r}} = \frac{(j+k)!(k+i)!(i+j)!}{i!j!k!(i+j+k)!}.$$

Simpler proofs of these formulas would be desirable.

We remark that the LHS of (3) is equal to

$$_4F_3\left[\begin{matrix} -i, -j, -k, \frac{1}{2}; 1 \\ 1, -\frac{1}{2}(i+j+k), -\frac{1}{2}(i+j+k)+\frac{1}{2} \end{matrix}\right].$$

Solution (of (2) and (3)) by D. R. Breach (University of Canterbury, Christchurch, New Zealand).

For (2): The orthogonality integral $I(m, n)$ for the Legendre polynomials, when expressed through Rodrigue's formula is

$$I(m, n) = \frac{1}{2}(m+n+1) \int_{-1}^{+1} \frac{2^{-n-m}}{n!m!} \cdot \left(\frac{d}{dx}\right)^n (x^2-1)^n \cdot \left(\frac{d}{dx}\right)^m (x^2-1)^m \, dx = \delta_{mn}.$$

As is known, the result can be established by integration of parts. Writing $x^2 - 1$ as $(x-1)(x+1)$ and applying Leibniz's formula to each derivative term in the integrand, we get

$$I(m, n) = \sum_{r=0}^{m} \sum_{s=0}^{n} \int_{-1}^{+1} 2^{-n-m-1}(m+n+1)\binom{m}{r}^2\binom{n}{s}^2 (x-1)^{r+s}(x+1)^{m+n-r-s} \, dx.$$

Integrating by parts $(r+s)$ times yields

$$\int_{-1}^{+1} (x-1)^{r+s}(x+1)^{m+n-r-s} \, dx = (-1)^{r+s} \frac{2^{m+n+1}}{(m+n+1)}\binom{m+n}{r+s}^{-1}.$$

Therefore,

$$I(m, n) = \sum_{r=0}^{m} \sum_{s=0}^{n} (-1)^{r+s}\binom{m}{r}^2\binom{n}{s}^2\binom{m+n}{r+s}^{-1} = \delta_{mn}.$$

For (3): Suppose $i \leq j \leq k$. Then

$$\sum_{i=0}^{\min(i,j,k)} \binom{i}{r}\binom{j}{r}\binom{k}{r}\binom{i+j+k}{r}^{-1}$$

$$= \sum_{i=0}^{\min(i,j,k)} \binom{i}{r}\binom{j}{r}\frac{(i+j+k-r)!}{(k-r)!} \cdot \frac{k!}{(i+j+k)!}.$$

This is the coefficient of x^i in

$$F(x) = \frac{k!}{(i+j+k)!}(1+x)^j x^{i-k}\left(\frac{d}{dx}\right)^{i+j}[(1+x)^i x^{j+k}].$$

Use Leibniz's rule to evaluate the derivative. Then

$$F(x) = \frac{k!}{(i+j+k)!}\sum_{r=0}^{i}\binom{i+j}{r}\frac{(j+k)!}{(k-i+r)!}\frac{i!}{(i-r)!}x^r(1+x)^{i+j-r},$$

---

† There was a misprint in the original version of this identity.

in which the coefficient of x^i is

$$\frac{(i+j)!(j+k)!}{(i+j+k)!j!} \sum_{r=0}^{i} \binom{i}{r}\binom{k}{i-r}.$$

But $\sum_{r=0}^{i} \binom{i}{r}\binom{k}{i-r}$ is the coefficient of x^i in $(1+x)^{i+k}$, namely $\binom{k+i}{i}$. Therefore

$$\sum_{r=0}^{\min(i,j,k)} \binom{i}{r}\binom{j}{r}\binom{k}{r}\binom{i+j+k}{r}^{-1} = \frac{(i+j)!(j+k)!(k+i)!}{i!j!k!(i+j+k)!}.$$

Problem 75-4*, *A Combinatorial Identity*, by P. BARRUCAND (Université Paris IV, France).

Let

$$A(n) = \sum_{i+j+k=n} \frac{n!^2}{i!^2 j!^2 k!^2},$$

where i, j, k are integers ≥ 0, and let

$$B(n) = \sum_{m=0}^{n} \binom{n}{m}^3,$$

so that $A(n)$ is the sum of the squares of the trinomial coefficients of rank n and $B(n)$ is the sum of the cubes of binomial coefficients of rank n ($A(n) = 1, 3, 15, 93, 639, \cdots$, $B(n) = 1, 2, 10, 56, 346, \cdots$).

Prove that

$$A(n) = \sum_{m=0}^{n} \binom{n}{m} B(m).$$

Editorial note. Equivalently, one has to prove that

$$\sum_{m=0}^{n} \sum_{r=0}^{m} \binom{n}{m}\binom{m}{r}^3 = \sum_{m=0}^{n} \binom{2m}{m}\binom{n}{m}^2.$$

For other properties, e.g., recurrences, integral representations, etc., the proposer refers to his papers in Comptes Rendus Acad. Sci. Paris, 258 (1964), pp. 5318–5320 and 260 (1965), pp. 5439–5541. He also notes that his solution is a tedious indirect one. [M.S.K.]

By equating the constant term in the binomial expansion on both sides of the identity

$$(1) \quad \{1 + (1 + x)(1 + y/x)(1 + 1/y)\}^n = \left\{1 + \frac{1+x}{y}\right\}^n \left\{1 + y\left(1 + \frac{1}{x}\right)\right\}^n$$

or an equivalent identity, the desired result was obtained by D. R. BREACH (University of Canterbury, Christchurch, New Zealand), D. McCARTHY (University of Waterloo), D. MONK (University of Edinburgh, Scotland) and P. E. O'NEIL (University of Massachusetts, Boston).

Editorial note.

With a little more work, one can obtain a more general identity. Expanding (1) yields

$$\sum_{r=0}^{n} \sum_{i=0}^{r} \sum_{j=0}^{r} \sum_{k=0}^{r} \binom{n}{r}\binom{r}{i}\binom{r}{j}\binom{r}{k} x^{i-j} y^{j-k}$$

$$= \sum_{r=0}^{n} \sum_{s=0}^{n} \binom{n}{r}\binom{n}{s} y^{s-r} \sum_{i=0}^{r} \sum_{j=0}^{s} \binom{r}{i}\binom{s}{j} x^{i-j}.$$

Equating the coefficients of $x^u y^v$ on both sides gives

$$\sum_{r=0}^{n} \sum_{k=0}^{r} \binom{n}{r}\binom{r}{k}\binom{r}{v+k}\binom{r}{u+v+k}$$

$$= \sum_{r=0}^{n} \binom{n}{r}\binom{n}{r+v} \sum_{j=0}^{r+v} \binom{r}{u+j}\binom{r+v}{j} = \sum_{r=0}^{n} \binom{n}{r}\binom{n}{r+v}\binom{2r+v}{r-u}.$$

Using differential equations, the proposer in a second solution establishes the equivalent identity

$$\sum \frac{x^n A(n)}{n!} = e^x \sum \frac{x^n B(n)}{n!}.$$ [M.S.K.]

A Multinomial Summation

Problem 87-2*, *by* DONALD RICHARDS *and* STAMATIS CAMBANIS (University of North Carolina).

Evaluate in closed form the sum of the squared multinomial coefficients

$$S(n,k) = \sum \left[\frac{k!}{k_1! k_2! \cdots k_n!} \right]^2$$

where the sum is over all nonnegative integers $k_1, \cdots, k_n$ such that $k_1 + \cdots + k_n = k$.

Remarks.

1. For $n = 2$, the sum is well known; it becomes

$$\sum_{j=0}^{k} \binom{k}{j}^2 = \binom{2k}{k}.$$

But for $n \geq 3$, nothing seems to be available.

2. An equivalent formulation is the problem of evaluating the integral

$$(2\pi)^{-n} \int_0^{2\pi} \cdots \int_0^{2\pi} |e^{i\theta_1} + \cdots + e^{i\theta_n}|^{2k} \, d\theta_1 \cdots d\theta_n$$

for complex k. (By Carlson's theorem on analytic functions, it is sufficient to assume that k is a nonnegative integer.) This integral appeared in work on stochastic processes (cf. S. Cambanis et al., *Ergodic properties of stationary stable processes*, Stochastic Process. Appl., 1987, to appear) and also in the area of harmonic analysis (cf. R. E. Edwards and K. A. Ross, *Helgason's number and lacunarity constants*, Bull. Austral. Math. Soc., 9 (1973), pp. 187–218). Using generating functions, it can also be shown that

$$S(n, k) = k! \frac{\partial^k}{\partial t^k} [{}_0F_1(1, t)]_{t=0}^n;$$

however, for $n \geq 3$ this representation seems just as intractable as the previous ones.

3. From some remarks made by Edwards and Ross, the integral may be interpreted as the expected value $E(D^{2k})$, where D is the distance from the origin of a unit-step Pearson random walker after n steps. (The Pearson random walk consists of starting at the origin, walking in the plane for a distance c_1 at random angle θ_1, then proceeding for a distance c_2 at a random angle θ_2, etc.). In the case when $c_1 = c_2 = \cdots = 1$, the probability density function of D appears (in K. V. Mardia's well-known book on directional data) in terms of an integral of products of modified Bessel functions. Since Mardia has no explicit results when $n = 3$, it seems that his integrals are just as difficult as ours.

Several comments and conjectures were submitted. PIERRE BARRUCAND (University of Paris) referred to his papers [1], [2] and a previous problem [3], which provide a generating function, integral representations and recurrence relations for $S(n, k)$, in particular,

(*)
$$[I_0(2\sqrt{x})]^n = \sum_{k=0}^{\infty} \frac{S(n, k)x^k}{(k!)^2}$$

and its consequence

(**)
$$S(n, k) = \sum_{j=0}^{k} \binom{k}{j}^2 S(n-m, j) S(m, k-j).$$

O. G. RUEHR (Michigan Technological University) uses (**) to get

$$S(3, k) = \sum_{j=0}^{k} \binom{k}{j}^2 \binom{2j}{j} \quad \text{and} \quad S(4, k) = \sum_{j=0}^{k} \binom{k}{j}^2 \binom{2j}{j} \binom{2k-2j}{k-j}$$

and conjectures that

$$\lim_{k \to \infty} S(n, k) = \frac{n^{2k+n/2}}{(4k\pi)^{(n-1)/2}}$$

based upon numerical computations.

GEORGE E. ANDREWS (Pennsylvania State University) computes $S(n, k)$ and notes the occurrence of large primes which implies implausibility of a closed form finite product representation. He conjectures that $S(n, k) \equiv 0(\mathrm{mod}\ n)$.

LOUIS W. KOLITSCH (University of Tennessee at Martin) proves Andrew's conjecture as follows.

Let

$$S_n(k) = \sum \binom{k}{k_1 k_2 \cdots k_n}^2$$

where the sum is over all compositions of k into exactly n nonnegative parts. Define an equivalence relation on the compositions of k into exactly n nonnegative parts as follows. Two compositions are equivalent if one can be obtained from the other through a cyclic permutation of the parts. This equivalence relation divides the compositions of k into exactly n nonnegative parts into classes where the number of elements in a particular class is the cyclic order of any element in the class.

Therefore

$$S_n(k) = \sum d(\pi) \binom{k}{k_1 k_2 \cdots k_n}^2$$

where the sum is over one representative π from each equivalence class and $d(\pi)$ is the cyclic order of π. We can then rewrite this sum as

$$S_n(k) = \sum d(\pi) \left(\frac{((n/d)\lambda)!}{(\lambda_1! \lambda_2! \cdots \lambda_d!)^{n/d}} \right)^2 \quad \text{where } \lambda = \lambda_1 + \lambda_2 + \cdots + \lambda_d.$$

Suppose p^a is the highest power of a prime p dividing n/d. Let $p^j, j \geq 0$, be the highest power of p dividing gcd $(\lambda_1, \cdots, \lambda_d)$. Then p^j divides λ since $\lambda_1 + \cdots + \lambda_d = \lambda$. We can easily see that p^a divides $((n/d)\lambda)!/(\lambda_1! \lambda_2! \cdots \lambda_d!)n/d$ since the multinomial coefficients satisfy

$$\binom{x}{x_1 x_2 \cdots x_m} \equiv 0 \left(\mathrm{mod}\ \frac{x}{(x, x_1, \cdots, x_m)} \right).$$

Therefore $S_n(k) \equiv 0(\mathrm{mod}\ n)$.

(*Note.* the same proof shows that

$$\sum \binom{k}{k_1 \cdots k_n}^n \equiv 0(\mathrm{mod}\ n)$$

for positive integral n.)

Note added in proof. In a communication from GEORGE H. WEISS (Silver Spring, Maryland) to the proposers, Weiss notes that a formal solution of the Pearson Random Walk is given in his joint paper, J. E. Kiefer, G. H. Weiss, "The Pearson Random Walk," A.I.P. Conference Proceedings, 109 (American Institute of Physics), 1984, pp. 11–32.

REFERENCES

[1] P. Barrucand, *Sur la somme des puissances des coefficients multinomiaux et les puissances successives d'une fonction de Bessel*, C. R. Acad. Sci. Paris, 253 (1964), pp. 5318–5320.
[2] ——, *Quelques inte'grales relatives aux fonctions de Bessel et aux sommes de carré des coefficients multinomiaux*, C. R. Acad. Sci. Paris, 260 (1965), pp. 5439–5441.
[3] ——, *Problem 75-4*, A combinatorial identity*, this Review, 17 (1975), p. 168.

Supplementary References
Combinatorics

[1] I. Anderson, *First Course in Combinatorial Mathematics*, Clarendon Press, Oxford,
[2] C. Berge, *Principles of Combinatorics*, Academic Press, N.Y., 1971.
[3] G. Berman and K. Fryer, *Introduction to Combinatorics*, Academic Press, N.Y., 1972.
[4] K. P. Bogart, *Introductory Combinatorics*, Pitman, Boston, 1983.
[5] R. A. Brualdi, *Introductory Combinatorics*, Elsevier North-Holland, N.Y. 1978.
[6] F. N. David and D. E. Barton, *Combinatorial Chance*, Griffin, London, 1962.
[7] D. I. A. Cohen, *Basic Techniques of Combinatorial Theory*, Wiley, N.Y., 1978.
[8] L. Comptet, *Advanced Combinatorics*, Reidel, Dortrecht, 1974.
[9] H. W. Gould, *Combinatorial Identities*, West Virginia University, Morgantown, 1972.
[10] I. P. Goulden and D. M. Jackson, *Combinatorial Enumeration*, Wiley, N.Y.,1983.
[11] R. L. Graham, D. E. Knuth, and O.Patashnik, *Concrete Mathematics*, Addison-Wesley, Reading, 1989.
[12] R. L. Graham, B. L. Rothschild, and J. H. Spencer, *Ramsey Theory*, Wiley, N.Y., 1980.
[13] M. Hall, *Combinatorial Theory*, Blaidell, Waltham, 1967.
[14] N. L. Johnson, and S. Kotz, *Urn Models and Their Applications*, Wiley, N.Y., 1977.
[15] D. E. Knuth, *The Art of Computer Programming, III, Sorting and Searching*, Addison-Wesley, Reading, 1973.
[16] C. L. Liu, *Introduction to Combinatorial Mathematics*, McGraw-Hill, N.Y., 1968.
[17] L. Lovasz, *Combinatorial Problems and Exercises*, Elsevier North-Holland, N.Y., 1979.
[18] P. A. MacMahon, *Combinatorial Analysis*, Chelsea, N.Y., 1960.
[19] E. B. McBride, *Obtaining Generating Functions*, Springer-Verlag, Berlin, 1971.
[20] S. G. Mohanty, *Lattice Path Counting and Applications*, Academic Press, N.Y., 1979.
[21] T. V. Narayana, *Lattice Path Combinatorics with Statistical Applications*, University of Toronto Press, Toronto, 1979.
[22] I. Niven, *Mathematics of Choice*, MAA, Washington, D.C., 1965.
[23] J. Riordan, *An Introduction to Combinatorial Analysis*, Wiley, N.Y., 1958.
[24] J. Riordan, *Combinatorial Identities*, Wiley, N.Y., 1968.
[25] F. Roberts, *Applied Combinatorics*, Prentice-Hall, N.J., 1984.
[26] G-C. Rota, ed., *Studies in Combinatorics*, Washington, D.C., 1978.
[27] G-C. Rota et al, *Finite Operator Calculus*, Academic Press, N.Y., 1975.
[28] H. Ryser, *Combinatorial Mathematics*, MAA, Washington, D.C., 1963.
[29] L. Takas, *Combinatorial Methods in the Theory of Stochastic Processes*, Wiley, N.Y., 1967.
[30] I. Tomescu, *Introduction to Combinatorics*, Collet's, London, 1975.
[31] I. Tomescu, *Problems in Combinatorics and Graph Theory*, Wiley, N.Y., 1985.
[32] A. Tucker, *Applied Combinatorics*, Wiley, N.Y., 1980.
[33] N. Y. Vilenkin, *Combinatorics*, Academic Press, N.Y., 1971.
[34] W. A. Whitworth, *Choice or Chance*, Hafner, N.Y., 1948.
[35] H. S. Wilf, *Generating Functionology*, Academic, San Diego, 1990.

5. SERIES

Problem 75-3, *A Power Series Expansion*, by U. G. HAUSSMANN (University of British Columbia).

Last year, a former engineering student of ours wrote to the mathematics department concerning a problem encountered in the electrical design for the appurtenant structures of the Mica Dam on the Columbia River in British Columbia. These structures include a spillway, low level outlets, intermediate level outlets, auxiliary service buildings and a power intake structure.

The engineers obtained a function

(1) $$f(u) = [\exp(u + nu) + \exp(-nu)]/[1 + \exp u],$$

where

(2) $$\cosh u = 1 + x/2$$

and where x is a ratio of resistances. Moreover, they suspected that if $y = f[u(x)]$, then

(3) $$y(x) = \sum_{k=0}^{n} \binom{n+k}{2k} x^k.$$

Show that (3) is valid.

Solution by M. E. H. ISMAIL (University of Toronto).

To exhibit the dependence of $f(u)$ and $y(x)$ on n, let us use $f(n, u)$ and $y(n, x)$ instead of $f(u)$ and $y(x)$, respectively. It is clear that $y(0, x) = f(0, u)$ and $y(1, x) = f(1, u)$. Direct and easy manipulations yield

$$y(n + 2, x) - 2y(n + 1, x) + y(n, x) = xy(n + 1, x),$$

that is,

$$y(n + 2, x) + y(n, x) = 2y(n + 1, x) \cosh u.$$

On the other hand, $f(n, u)$ satisfies the same recurrence since

$$f(n + 2, u) + f(n, u) = \frac{(e^u + e^{-u}) \exp(u + (n + 1)u) + (e^{-u} + e^u) \exp(-(n + 1)u)}{1 + \exp u}$$

$$= 2f(n + 1, u) \cosh u.$$

Therefore $y(n, x) = f(n, u)$.

Editorial note. Lossers gives the explicit sum

$$\sum_{k=0}^{n} \binom{n+k}{2k} x^k = \frac{2(2+x-t)^n}{4+x+t} + \frac{2(2+x+t)^n}{4+x-t},$$

where $t^2 = x^2 + 4x$. Byrd notes that the latter sum occurs in his paper, *Expansion of analytic functions in polynomials associated with Fibonacci numbers*, Fibonacci Quart., 1 (1963), p. 17, and that for $x = 1$, the sum reduces to the odd Fibonacci numbers F_{2n+1}. Carlitz shows more generally that if

$$f(\lambda, u) = [e^{(\lambda+1)u} - e^{-\lambda u}]/[e^u + 1],$$

where λ is an arbitrary complex number and $2 \cosh u = 2 + t^2$, then

$$f(\lambda, u) = \sum_{k=0}^{\infty} \binom{\lambda+k}{2k} t^{2k}$$

in some region about $t = 0$.

The other solutions used induction, generating functions, differential equations and the known expansions of

$$\frac{\cos(2n+1)\theta}{\cos\theta} \quad \text{or} \quad \frac{\sin n\theta}{\sin\theta}.$$

Power Series of an Elliptic Function

Problem 81-14, by M. L. GLASSER (Clarkson College) (corrected).

Show that

$$4K^2\left\{\sqrt{\frac{1-\sqrt{1-2z}}{2}}\right\} = \pi^2 \sum_{k=0}^{\infty} \frac{[(2k-1)!!]^3 z^k}{[k!]^2 [(4k)!!]},$$

where $(-1)!! = 1$, $(4k)!! = 4\cdot8 \cdots (4k)$, etc.

Solution by N. T. SHAWAGFEH (Clarkson College)

$$\text{Let } I = K^2\left\{\sqrt{\frac{1-\sqrt{1-2z}}{2}}\right\} = \frac{\pi^2}{4}\left[{}_2F_1\left[\frac{1}{2},\frac{1}{2}; 1; \frac{1-\sqrt{1-2z}}{2}\right]\right]^2.$$

Using Kummer's identity [1] gives

$${}_2F_1[2\alpha, 2\beta; \alpha+\beta+\tfrac{1}{2}; \tfrac{1}{2}\{1-(1-t)^{1/2}\}] = {}_2F_1[\alpha, \beta; \alpha+\beta+\tfrac{1}{2}; t].$$

With $t = 2z$, $\alpha = \tfrac{1}{4}$, $\beta = \tfrac{1}{4}$ we can write I as

$$I = \frac{\pi^2}{4}\left[{}_2F_1\left[\frac{1}{4},\frac{1}{4}; 1; 2z\right]\right]^2.$$

But Clausen's theorem states [2]

$${}_3F_2[2\alpha, 2\beta, \alpha+\beta; 2(\alpha+\beta), \alpha+\beta+\tfrac{1}{2}; t] = [{}_2F_1[\alpha, \beta; \alpha+\beta+\tfrac{1}{2}; t]]^2.$$

Hence

$$I = \frac{\pi^2}{4}{}_3F_2\left[\frac{1}{2},\frac{1}{2},\frac{1}{2}; 1, 1; 2z\right] = \frac{\pi^2}{4}\sum_{k=0}^{\infty} \frac{[(\tfrac{1}{2})_k]^3}{[k!]^3}(2z)^k = \frac{\pi^2}{4}\sum_{k=0}^{\infty} \frac{[(2k-1)!!]^3}{[k!]^2[(4k)!!]}z^k.$$

where

$$(2k - 1)!! = 2^k \left(\tfrac{1}{2}\right)_k, \qquad (4k)!! = 4^k k!.$$

REFERENCES

[1] A. ERDÉLYI et al., *Higher Transcendental Functions*, vol. I, McGraw-Hill, New York, 1953.
[2] T. CLAUSEN, J. Reine Angew. Math. 3, 89 (1828).

Comment by W. B. JORDAN (Scotia, NY).
The formula is equivalent to

$$\left(\frac{2K}{\pi}\right)^2 = 1 + \left(\frac{1}{2}\right)^3 (2kk')^2 + \left(\frac{1 \cdot 3}{2 \cdot 4}\right)^3 (2kk')^4 + \cdots \quad \left(k < \frac{1}{\sqrt{2}}\right),$$

which was published by Ramanujan in 1913, prefaced by "It is well known that ..."
(G. H. Hardy, ed., *Collected Papers of S. Ramanujan* (Chelsea, New York 1962, p. 36).

Limit of a Sum

Problem 81-7, *by* D. J. NEWMAN (Temple University).
Prove that

$$\lim_{n \to \infty} \sum_{k=0}^{n} (-1)^k \sqrt{\binom{n}{k}} = 0.$$

Solution by W. B. JORDAN (Scotia, NY).
Let

$$S_n = \sum_{k=0}^{m} (-1)^k \binom{n}{k}^{1/2} = \sum_{k=0}^{m} (-1)^k \left[\frac{\Gamma(n+1)}{\Gamma(n+1)-k)\Gamma(k+1)}\right]^{1/2}.$$

From the symmetry of the binomial coefficients around $k = \tfrac{1}{2}n$, $S_n = 0$ for n odd, and we need consider only even n. Consider the integral of

$$Q = \left[\frac{\Gamma(n+1)}{\Gamma(n+1-z)\Gamma(z+1)}\right]^{1/2} \csc \pi z$$

around a rectangle whose corners are $-\tfrac{1}{2} \pm ib$ and $n + \tfrac{1}{2} \pm ib$. Inside the rectangle the gamma functions have no zeros or poles, so the only singularities of Q are simple poles at $z = 0, 1, \cdots, n$; the residues of $\csc \pi z$ at these points are $(-1)^z/\pi$, and so

$$S_n = \frac{1}{2i} \int Q \, dz.$$

Now

$$\csc \pi z = \frac{\Gamma(1-z)\Gamma(z)}{\pi} = \left[\csc \pi z \frac{\Gamma(1-z)\Gamma(z)}{\pi}\right]^{1/2},$$

so

$$Q^2 = \frac{\Gamma(n+1)\csc \pi z \Gamma(1-z)}{\pi z \Gamma(n+1-z)} = \frac{\Gamma(n+1)\csc \pi z}{\pi z(n-z)(n-1-z)\cdots(1-z)}.$$

When $z = x + ib, |m - z| \geq b$,

$$|\csc \pi z| = \frac{2}{|e^{\pi(ix-b)} - e^{\pi(b-ix)}|} \sim 2e^{-\pi b},$$

so $|Q^2| < 2\Gamma(n+1)e^{-\pi b}/\pi b^{n+1}$, which goes to 0 as b goes to ∞. The integrals on the horizontal sides of the rectangle are 0, leaving

$$S_n = \frac{1}{2i}\left[\int_{n+1/2-i\infty}^{n+1/2+i\infty} + \int_{-1/2+i\infty}^{-1/2-i\infty}\right] Q \, dz.$$

In the first integral put $z = n + \frac{1}{2} + it$, $dz = i\, dt$. Since n is even,

$$\csc \pi z = \csc \pi(\tfrac{1}{2} + it) = [\pi^{-1}\csc \pi(\tfrac{1}{2} + it)\Gamma(\tfrac{1}{2} - it)\Gamma(\tfrac{1}{2} + it)]^{1/2},$$

$$Q = \left[\pi^{-1}\frac{\Gamma(n+1)\Gamma(\tfrac{1}{2} + it)}{\Gamma(n + \tfrac{3}{2} + it)\cosh \pi t}\right]^{1/2}$$

In the second integral put $z = -\frac{1}{2} - it$ and find $Q = -$(the above Q). The integrals on the vertical sides double up, giving

$$S_n = \int_{-\infty}^{\infty}\left[\frac{\pi^{-1}\Gamma(n+1)\Gamma(\tfrac{1}{2} + it)}{\Gamma(n + \tfrac{3}{2} + it)\cosh \pi t}\right]^{1/2} dt.$$

Now

$$\left|\frac{\Gamma(n + \tfrac{3}{2} + it)}{\Gamma(\tfrac{1}{2} + it)}\right| = |(n + \tfrac{1}{2} + it)(n - \tfrac{1}{2} + it)\cdots(\tfrac{1}{2} + it)|$$

$$\geq (n + \tfrac{1}{2})(n - \tfrac{1}{2})\cdots(\tfrac{1}{2}) = \frac{\Gamma(n + \tfrac{3}{2})}{\sqrt{\pi}},$$

so

$$|S_n| < \left[\frac{\Gamma(n+1)}{\sqrt{\pi}\,\Gamma(n + \tfrac{3}{2})}\right]^{1/2}\int_{-\infty}^{\infty} \mathrm{sech}^{1/2}\pi t\, dt,$$

whose limit as n approaches ∞ is 0, completing the proof.

Also solved by J. GRZESIK (TRW Defense & Space Systems Group) *and the proposer,* who both indicate that the result still holds if the exponent $\frac{1}{2}$ in the sum is changed to p where $0 < p < 1$.

Editorial note: A. MEIR (University of Alberta) raised the question of which other sequences besides $\binom{n}{k}$ have the same limiting property. The proposer casually mentions that the results hold for other sequences. Meir also raised the question of the asymptotic sum where the exponent is $p = 1 + \varepsilon$. In this regard, for exponents $p = 2$ or 3, the sums are clearly unbounded since we have the explicit summations

$$\sum_{k=0}^{2n}(-1)^k\binom{2n}{k}^2 = (-1)^n\binom{2n}{n}, \qquad \sum_{k=0}^{2n}(-1)^k\binom{2n}{k}^3 = (-1)^n\frac{(3n)!}{(n!)^3}.$$

[M.S.K.]

Problem 69-14*, *Sum of Inverse Powers of Cosines*, by L. A. GARDNER, JR. (Sperry Rand Research Center).

The vth cumulant of a certain quadratic form in n independent standardized normal variates involves the sum

$$S_{n,v} = \left(\frac{\pi}{2n}\right)^{2v} \sum_{k=1}^{n-1} \left\{\cos\frac{k\pi}{2n}\right\}^{-2v},$$

where v and n are positive integers. It can be shown that

$$\lim_{n \to \infty} S_{n,v} = \zeta(2v)$$

for every fixed $v \geq 1$ where ζ is the Riemann zeta function. Can one find a "simpler" closed form expression for $S_{v,n}$ for all v and n?

Editorial note. A related problem (Amer. Math. Monthly, 1967, Problem 5486) is to show that

$$\sum_{k=0}^{n-1} \csc^2\left\{\frac{\pi(2k+1)}{2n}\right\} = n^2. \qquad\qquad \text{[H.E.F.]}$$

Solution by MICHAEL E. FISHER (Cornell University).
We shall show how to evaluate the general sums

(1) $$Q_{n,v}(\delta) = \sum_{r=1}^{n-1}\left[\sin\left(\frac{r\pi + \delta}{n}\right)\right]^{-2v},$$

where n and v are positive integers and $|\delta| \leq \frac{1}{2}\pi$. (The empty sums $Q_{1,v}$ will be interpreted as zero.) To see the relation to the sums

(2) $$S_{m,v} = \left(\frac{\pi}{2m}\right)^{2v} \sum_{k=1}^{m-1}\left[\cos\frac{k\pi}{2m}\right]^{-2v} = \left(\frac{\pi}{2m}\right)^{2v} \sum_{r=1}^{m-1}\left[\sin\frac{r\pi}{2m}\right]^{-2v},$$

defined in the original problem, take $\delta = 0$ and $n = 2m$ in (1). On separating off the term for $r = m$, namely $[\sin(\pi/2)]^{-2v} = 1$, and combining the remaining sums into the form (2), we find

(3) $$S_{m,v} = \tfrac{1}{2}(\pi/2m)^{2v}[Q_{2m,v}^0 - 1],$$

where, here and below, the superscript zero denotes $\delta = 0$. We shall, incidentally, also evaluate the sum

(4) $$L_n(\delta) = \sum_{r=1}^{n-1} \ln\sin\left(\frac{r\pi + \delta}{n}\right).$$

In order to construct a generating function for the $Q_{n,v}(\delta)$, consider the function

(5) $$F_n(z) = (z^n - \rho^n e^{2i\delta})(z^n - \rho^n e^{-2i\delta})$$

which is readily factorized in terms of the nth roots of unity. On setting

(6) $$z = \rho^{-1}, \qquad \rho = e^{\gamma(\zeta)} = (1 + \zeta^2)^{1/2} + \zeta,$$

so that $\zeta = \sinh\gamma$, the factorization yields the identity

(7) $G_n(\delta;\zeta) = 2[\cosh 2n\gamma(\zeta) - \cos 2\delta] = 2^n \prod_{r=0}^{n-1} \left[\sin^2 \left(\frac{r\pi + \delta}{n} \right) + \zeta^2 \right].$

Hence, with the definition

(8) $H_n(\delta;\zeta) = \ln \{G_n(\delta;\zeta)/[\sin^2 (\delta/n) + \zeta^2]\}$

and, in particular for $\delta = 0$,

(9) $H_n^0(\zeta) = 2 \ln [2 \sinh n\gamma(\zeta)/\zeta],$

we have

(10) $H_n(\delta;\zeta) - n \ln 2 = \sum_{v=0}^{\infty} h_{n,v}(\delta)\zeta^{2v} = \sum_{r=1}^{n-1} \ln \left[\sin^2 \left(\frac{r\pi + \delta}{n} \right) + \zeta^2 \right].$

By expanding the logarithms in powers of ζ^2 (either purely formally or, for $|\zeta| < \sin(\pi/2n)$, in an absolutely convergent series), we see that $H_n(\delta;\zeta)$ is a generating function for the sums $Q_{n,v}(\delta)$. Explicitly, we have (see (1) and (4))

(11) $L_n(\delta) = \tfrac{1}{2}h_{n,0}(\delta) = \tfrac{1}{2}H_n(\delta;0) - \tfrac{1}{2}n \ln 2,$

which yields the result

(12) $L_n(\delta) = \ln [2 \sin \delta/\sin (\delta/n)] - \tfrac{1}{2}n \ln 2,$

and, for $v \geq 1$,

(13) $Q_{n,v}(\delta) = (-)^{v+1}vh_{n,v}(\delta) = -(v/v!)[(-d/d\zeta^2)^v H_n(\delta;\zeta)]_{\zeta=0}.$

By performing the indicated differentiations, we may obtain

(14) $Q_{n,1}(\delta) = n^2(\sin \delta)^{-2} - [\sin (\delta/n)]^{-2}$

and

(15) $Q_{n,2}(\delta) = n^4(\sin \delta)^{-4} - \tfrac{2}{3}n^2(n^2 - 1)(\sin \delta)^{-2} - [\sin (\delta/n)]^{-4}.$

Having found the result (14), however, it is easier to find the sums for higher v by using the recurrence relation

(16) $(2v + 1)Q_{n,v+1} = 2vQ_{n,v} + n^2(d/d\delta)^2[Q_{n,v}/2v],$

which is readily proved by differentiating the definition (1) twice with respect to δ. (With the convention $[Q_{n,v}/2v]_{v=0} = -L_n(\delta)$, this relation holds even for $v = 0$.) Iteration of (16) establishes

(17) $Q_{n,v}(\delta) = \sum_{t=1}^{v} A_{v,t}(n)(\sin \delta)^{-2t} - [\sin (\delta/n)]^{-2v},$

where the $A_{v,t}(n)$ are polynomials in n^2 of degree v which can be calculated recursively from

(18) $v(2v + 1)A_{v+1,t} = 2(v^2 - t^2n^2)A_{v,t} + (t - 1)(2t - 1)n^2A_{v,t-1},$

with

(19) $A_{1,1} = n^2$ and $A_{v,t} = 0$ for $t \geq v + 1.$

The extreme coefficients are easily seen to be

(20) $A_{v,v}(n) = n^{2v}, \quad A_{v,1}(n) = \dfrac{2^{v-1}n^2}{(v-1)!}\displaystyle\prod_{u=1}^{v-1}\dfrac{(u^2-n^2)}{2u-1}.$

Alternatively, let us note that by the definition (1) the expression (17) must remain bounded when $\delta \to 0$. Consequently, the v polynomials $A_{v,t}(n)$ can be determined for any fixed v merely by noting that the inverse powers of δ^2 generated by the Laurent expansion of $[\sin(\delta/n)]^{-2v}$ in (17) must be cancelled identically by corresponding terms from the expansion of the summands $A_{v,t}(n)(\sin\delta)^{-2t}$. In this way one can also show that

(21) $A_{v,v-1}(n) = -\tfrac{1}{3}vn^{2(v-1)}(n^2-1), \qquad\qquad v \geq 2,$

and

(22) $A_{v,v-2}(n) = \tfrac{1}{90}vn^{2(v-2)}(n^2-1)[(5v-11)n^2-5v-1], \qquad v \geq 3.$

We now specialize to the case $\delta = 0$. Either directly from (9) or by letting $\delta \to 0$ in (14) and (15), we find

(23) $Q^0_{n,1} = \tfrac{1}{3}(n^2-1), \qquad Q^0_{n,2} = \tfrac{1}{45}(n^2-1)(n^2+11),$

and hence, via (3),

(24) $S_{m,1} = \dfrac{\pi^2}{6}\left(1 - \dfrac{1}{m^2}\right), \qquad S_{m,2} = \dfrac{\pi^4}{90}\left(1 + \dfrac{5}{2m^2} - \dfrac{7}{2m^4}\right).$

More generally it follows directly from the result (17) that $Q^0_{n,v}$ is a polynomial of degree v in n^2 with rational coefficients; that is,

(25) $Q^0_{n,v} = \displaystyle\sum_{t=0}^{v} q_{n,v,t}n^{2(v-t)}.$

For a given v the $v+1$ coefficients $q_{n,v,t}$ can be determined from the $v+1$ values

(26) $\begin{aligned} &Q^0_{1,v} = 0, \quad Q^0_{2,v} = 1, \quad Q^0_{3,v} = 2^{2v+1}/3^v, \\ &Q^0_{4,v} = 1 + 2^{v+1}, \quad \cdots, \quad Q^0_{v,v}, \quad Q^0_{v+1,v}, \end{aligned}$

by simply imposing the equality for $n = 1, 2, \cdots, v+1$. With the aid of the Lagrange polynomial interpolation formula this yields the explicit expression

(27) $Q^0_{n,v} = \displaystyle\sum_{t=1}^{v+1} Q^0_{t,v}\prod_{\substack{s=1\\s\neq t}}^{v+1}\dfrac{(n^2-s^2)}{(t^2-s^2)}.$

Equally, the sums $S_{m,v}$ must be polynomials of degree v in $1/m^2$, and, hence, a formula similar to (27) can be written in terms of the values $S_{1,v} = 0, S_{2,v}, S_{3,v}, \cdots, S_{v+1,v}$.

Insofar as the formula (27) is a "simple closed form expression" for $Q^0_{n,v}$ (and thereby for $S_{m,v}$), it solves the problem. For large n, however, an asymptotic formula is more desirable. In fact, the limit

(28) $\displaystyle\lim_{n\to\infty} n^{-2v}Q^0_{n,v} = 2\zeta(2v)/\pi^{2v} = 2^{2v}B_v/(2v)!$

is easily established directly from the definition (1). (Here $\zeta(\cdot)$ is the Riemann zeta function and B_l is the lth Bernoulli number.) It follows from this that the leading coefficient in (25) is

(29) $$q_{n,v,0} = 2\zeta(2v)/\pi^{2v}.$$

In order to obtain expressions for the higher order coefficients we write $\zeta = \xi/n$ and formally expand the generating function (9) in inverse powers of n. This yields

(30)
$$H_n^0\left(\frac{\xi}{n}\right) = 2\ln n + 2\ln\left(\frac{2\sinh\xi}{\xi}\right) - \tfrac{1}{3}n^{-2}\xi^3 \coth\xi$$
$$+ \frac{1}{180}n^{-4}\xi^5(27\coth\xi - 5\xi\,\mathrm{cosech}^2\,\xi) - \cdots.$$

On expanding in powers of ξ^2 with the aid of Bernoulli's numbers, we rederive (29) and obtain in addition, for $v \geq 2$,

(31) $$q_{n,v,1} = \tfrac{1}{3}vq_{n,v-1,0} = \tfrac{2}{3}v\zeta(2v-2)/\pi^{2v-2}.$$

For $v = 1$ one obtains $q_{n,1,1} = -\tfrac{1}{3}$ which checks with (23) as does the value of $q_{n,2,1}$ given by (31). The second order coefficient $q_{n,v,2}$ derives from the last term presented in (30). It can be expressed as a quadratic form in the Bernoulli numbers B_1 to B_v, but a more compact expression is not apparent.

    Finally, we employ these results to supplement the formulas (24) by

(32) $$S_{m,v} = \zeta(2v) + \tfrac{1}{12}v\zeta(2v-2)/m^2 + O(m^{-4}).$$

A Hyperbolic Series Identity

Problem 79-8, *by* CHIH-BING LING (Virginia Polytechnic Institute and State University).

    Show that, for $a > 0$,

$$\sum_{n=0}^{\infty} \frac{1}{\cosh(2n+1)a} = \sum_{n=0}^{\infty} \frac{(-1)^n}{\sinh(2n+1)a}.$$

Solution by L. CARLITZ (Duke University).

More generally, for $r, s \geq 1$, we have

$$\sum_{n=0}^{\infty} \frac{2^s\binom{n+s-1}{n}}{\cosh^r(2n+s)a} = 2^{r+s}\sum_{n=0}^{\infty} \frac{\binom{n+s-1}{n}e^{-ar(2n+s)}}{(1+e^{-2a(2n+s)})^r}$$

$$= 2^{r+s}\sum_{m,n=0}^{\infty} (-1)^m\binom{n+s-1}{n}\binom{m+r-1}{m}e^{-a(2m+r)(2n+s)}$$

$$= 2^{r+s}\sum_{m=0}^{\infty} \frac{(-1)^m\binom{m+r-1}{m}e^{-as(2m+r)}}{(1+e^{-2a(2m+r)})^s} = \sum_{m=0}^{\infty} \frac{(-1)^m 2^r\binom{m+r-1}{m}}{\sinh^s(2m+r)a}.$$

    H. E. FETTIS (Mountain View, California) and O. P. LOSSERS (Technological University, Eindhoven, the Netherlands) each showed by using the known Fourier series for the Jacobian elliptic functions $\mathrm{sn}\,u$ and $\mathrm{cn}\,u$ [P. F. Byrd, M. D. Friedman,

Handbook of Elliptic Integrals for Engineers and Physicists, Springer-Verlag, Heidelberg, 1971, pp. 304–305] that each of the given sums equals kK/π, where $a = \pi K'/2K$.

W. A. AL-SALAM (University of Alberta) in addition to proving the given identity, establishes several q analogues, e.g.,

$$\sum_{n=0}^{\infty} \frac{(-1)^n x^n q^n}{(1-xq)(1-xq^2)\cdots(1-xq^{n+1})} = \sum_{n=0}^{\infty}\sum_{m=0}^{\infty} (-1)^n (xq)^{n+m}\begin{bmatrix} n+m \\ m \end{bmatrix}$$

$$= \sum_{n=0}^{\infty}\sum_{m=0}^{\infty} (-1)^m (xq)^{n+m}\begin{bmatrix} n+m \\ m \end{bmatrix}$$

$$= \sum_{n=0}^{\infty} \frac{x^n q^n}{(1+xq)(1+xq^2)\cdots(1+xq^{n+1})}.$$

Problem 76-2, *An Infinite Sum*, by MURRAY GELLER (Jet Propulsion Laboratory, Pasadena, California).

In a recent paper by Chih-Bing Ling[1], the summations $\sum_{n=1}^{\infty} (\cosh^{2s} n\pi c)^{-1}$ were analytically evaluated for $c = 1, \sqrt{3}, 1/\sqrt{3}$ and the results were explicitly given for $s = 1, 2, 3$. Show that

$$\sum_{n=1}^{\infty} (\cosh^6 n\pi\sqrt{2})^{-1} = \frac{2\sqrt{2}}{15\pi} + \frac{\sqrt{2}}{15}A + \frac{(\sqrt{2}-1)}{12}A^2 + \frac{(5-3\sqrt{2})}{120}A^3 - \frac{1}{2},$$

where $A = \sqrt{2}\Gamma^4(\tfrac{1}{8})/16\pi^2\Gamma^2(\tfrac{1}{4})$.

REFERENCE

[1] CHIH-BING LING, *On summation of series of hyperbolic functions*, SIAM J. Math. Anal., 5 (1974), pp. 551–562.

Solution by W. B. JORDAN (Scotia, N.Y.).

We begin with the formula [1, p. 489]

$$\frac{\theta_2'(z)}{\theta_2(z)} = -\tan z - \sin 2z \sum_{n=1}^{\infty} \frac{4q^{2n}}{1+2q^{2n}\cos 2z + q^{4n}}.$$

Put $q = e^{-u}$ and $S_j(u) = \sum_{n=1}^{\infty} \text{sech}^j nu$. Then for Re $(u) > 0$

$$\frac{\theta_2'(z)}{\theta_2(z)} + \tan z = -\sin 2z \sum_{n=1}^{\infty} \frac{1}{\cosh^2 nu - \sin^2 z}$$

$$= -\sin 2z \sum_{n=1}^{\infty} (\text{sech}^2 nu + \sin^2 z \, \text{sech}^4 nu + \sin^4 z \, \text{sech}^6 nu + \cdots)$$

$$= -\sin 2z \sum_{j=1}^{\infty} S_{2j}(u) \sin^{2j-2} z.$$

Values of $S_{2j}(u)$ can be obtained from this formula by differentiating a sufficient number of times with respect to z, and putting $z = 0$. To simplify the results, write $f(z) = -\theta_2'(z)/\theta_2(z)$ and get

$$2S_2(u) = f'(0) - 1,$$

$$12S_4(u) = f'''(0) + 4f'(0) - 6,$$

$$240S_6(u) = f^{(v)}(0) + 20f'''(0) + 64f'(0) - 120.$$

Since $\theta_2'(0) = \theta_2'''(0) = \theta_2^{(v)}(0) = 0$, the derivatives of f are

$$f'(0) = -\theta_2''/\theta_2,$$

$$f'''(0) = \frac{3\theta_2''^2}{\theta_2^2} - \frac{\theta_2^{(iv)}}{\theta_2},$$

$$f^{(v)}(0) = -\frac{30\theta_2''^3}{\theta_2^3} + \frac{15\theta_2''\theta_2^{(iv)}}{\theta_2^2} - \frac{\theta_2^{(vi)}}{\theta_2},$$

all evaluated at $z = 0$.

When u is large q is small and both the sech series and the θ_2 series converge rapidly; either is suitable for calculation. When u is small it may be more convenient to use formulas in terms of complete elliptic integrals. We use k and k' to denote the complementary moduli, and K and E the complete Jacobi integrals, for the nome q. The conversion formulas [1, p. 470] are

$$\theta_2'' = -4q\,\partial\theta_2/\partial q,$$

$$\theta_2^{(iv)} = 16q\left(\frac{\partial\theta_2}{\partial q} + q\frac{\partial^2\theta_2}{\partial q^2}\right),$$

$$\theta_2^{(vi)} = -64q\left(\frac{\partial\theta_2}{\partial q} + 3q\frac{\partial^2\theta_2}{\partial q^2} + q^2\frac{\partial^3\theta_2}{\partial q^3}\right),$$

$$\theta_2^2 = 2kK/\pi, \qquad \log q = -\pi K'/K,$$

$$dq/q\,dk = \pi^2/(2kk'^2K^2), \qquad d\theta_2/\theta_2 dq = EK/(\pi^2 q),$$

and so $\theta_2''/\theta_2 = -4EK/\pi^2$.

Similarly,

$$\theta_2^{(iv)}/\theta_2 = 32K^2(\tfrac{3}{2}E^2 - k'^2K^2)/\pi^4,$$

$$\theta_2^{(vi)}/\theta_2 = -64K^3[15E^3 - 30k'^2EK^2 + 8k'^2K^3(1 + k'^2)]/\pi^6,$$

and so

$$S_2(u) = \frac{2}{\pi^2}EK - \frac{1}{2},$$

$$S_4(u) = \frac{4K}{3\pi^2}\left(E + \frac{2}{\pi^2}k'^2K^3\right) - \frac{1}{2},$$

$$S_6(u) = \frac{32k'^2K^6}{15\pi^6}(1 + k'^2) + \frac{8k'^2K^4}{3\pi^4} + \frac{16EK}{15\pi^2} - \frac{1}{2},$$

$$= \frac{A^3}{240}k'^2(1 + k'^2) + \frac{A^2k'^2}{24} + \frac{2A}{15}\cdot\frac{E}{K} - \frac{1}{2},$$

where $A = 8K^2/\pi^2$. We note in passing that

$$S_1(u) = \sum_{n=1}^{\infty} \frac{2q^n}{1+q^{2n}} = \frac{K}{\pi} - \frac{1}{2}.$$

For certain values of u, such as π, $\pi\sqrt{3}$ and $\pi/\sqrt{3}$, the values of k', K, E are known in terms of radicals and Gamma functions. In particular [2], for $u = \pi\sqrt{2}$,

$$k = \sqrt{2} - 1, \qquad k'^2 = 2k,$$

$$K = \Gamma^2(\tfrac{1}{8})/2^{1/4} \cdot 8\Gamma(\tfrac{1}{4}), \qquad \frac{E}{K} = \frac{1}{\sqrt{2}}\left(1 + \frac{\pi}{4K^2}\right),$$

which leads to the proposer's result.

Comment. For this value of u the first term of the sech^6 series is 1.7×10^{-10}, and the ratio of consecutive terms is about $\exp(-6\pi\sqrt{2}) = 2.6 \times 10^{-12}$. Thus the original series is eminently suitable for calculation, whereas the elliptic integral formula suffers from the cancellation of its first ten significant digits.

REFERENCES

[1] E. T. WHITTAKER AND G. N. WATSON, A Course in Modern Analysis, Cambridge University Press, London, 1952.
[2] J. W. L. GLAISHER, Mess. Math., 24(1895), pp. 24–47.

Also solved by the proposer, who notes that many more results of a similar nature are available in an unpublished manuscript on the application of theta functions to the evaluation of Gaussian and hyperbolic summations by the proposer and M. Saffren.

Problem 77-18, An Infinite Summation, by A. M. LIEBETRAU (Johns Hopkins University).
    Show that

(A)
$$\sum_{j=1}^{\infty} \alpha_j^{-6} \left[\frac{\sin \alpha_j - \sinh \alpha_j}{\cos \alpha_j + \cosh \alpha_j}\right]^2 = \frac{1}{80},$$

where the α_j's are the positive solutions to the equation

$$(\cos \alpha)(\cosh \alpha) + 1 = 0.$$

This identity follows from a problem in statistics, that of finding the distribution of a certain functional of a Gaussian process $\eta(t)$ with covariance kernel

$$E[\eta(t)\eta(u)] = K(t, u) = \begin{cases} \frac{2}{3}(3t^2u - t^3), & 0 \le t \le u \le 1, \\ \frac{2}{3}(3u^2t - u^3), & 0 \le u \le t \le 1. \end{cases}$$

Solution by the proposer.
    In order to obtain the distribution of a certain random functional of a Poisson process [2], it became necessary to solve the following eigenvalue problem: Express the positive symmetric function

(1)
$$K(t, u) = \begin{cases} \frac{2}{3}(3t^2u - t^3), & 0 \le t \le u \le 1, \\ \frac{2}{3}(3u^2t - u^3), & 0 \le u \le t \le 1, \end{cases}$$

in the form

(2)
$$\sum_{j=1}^{\infty} \lambda_j^{-1} f_j(t) f_j(u),$$

where λ_j is an eigenvalue and $f_j(t)$ is the corresponding normalized eigenfunction of the system

(3)
$$f(t) = \lambda \int_0^1 K(t, u) f(u)\, du, \qquad \int_0^1 f_i(t) f_k(t) = \delta_{jk}.$$

In (3), δ_{jk} is the Kronecker delta.

Substitution of (1) into the first equation of (3) yields

(4)
$$f(t) = \frac{2}{3} \lambda \left\{ \int_0^t (3u^2 t - u^3) f(u)\, du + \int_t^1 (3t^2 u - t^3) f(u)\, du \right\}.$$

Successive differentiation of (4) with respect to t yields

(5)
$$f^{(4)}(t) = 4\lambda f(t) \equiv \alpha^4 f(t),$$

which is easily seen to have the solution

(6)
$$f(t) = c_1 e^{-\alpha t} + c_2 e^{\alpha t} + c_3 \cos(\alpha t) + c_4 \sin(\alpha t)$$

for suitable constants c_1, c_2, c_3, c_4.

Boundary conditions for determining the c_j's are obtained from considering $f(0)$, $f'(0)$, $f''(0)$ and $f'''(0)$. Substitution of the appropriate derivatives of (6) into (4) produces the following system of equations:

$$c_1 + c_2 + c_3 = 0,$$

$$c_1 - c_2 - c_4 = 0,$$

(7)
$$(\alpha + 1) e^{-\alpha} c_1 + (1 - \alpha) e^{\alpha} c_2 - (\alpha \sin \alpha + \cos \alpha) c_3 + (\alpha \cos \alpha - \sin \alpha) c_4 = 0,$$

$$e^{-\alpha} c_1 - e^{\alpha} c_2 - (\sin \alpha) c_3 + (\cos \alpha) c_4 = 0.$$

Elimination of c_1 and c_2 from (7) yields, after some manipulation:

(8)
$$(\cosh \alpha + \cos \alpha) c_3 + (\sinh \alpha + \sin \alpha) c_4 = 0,$$
$$(\sinh \alpha - \sin \alpha) c_3 + (\cosh \alpha + \cos \alpha) c_4 = 0.$$

The equations (8) have nontrivial solution if, and only if,

(9)
$$(\cosh \alpha)(\cos \alpha) + 1 = 0.$$

Being a covariance function (1) is positive definite, so it is necessary to consider only positive solutions to (9); let α_j denote the jth smallest, so that $\lambda_j = \frac{1}{4}\alpha_j^4$, $j = 1, 2, \cdots$. From (7) and (9), it follows that

$$c_3 = \frac{\cos \alpha_j + \cosh \alpha_j}{\sin \alpha_j - \sinh \alpha_j} c_4 \equiv \kappa_j c_4,$$

$$c_2 = -\tfrac{1}{2}(c_3 + c_4) = -\tfrac{1}{2}(\kappa_j + 1) c_4,$$

$$c_1 = \tfrac{1}{2}(c_4 - c_3) = -\tfrac{1}{2}(\kappa_j - 1) c_4.$$

Finally, c_4 is chosen so that $\int_0^1 f_j^2(t)\,dt = 1$. A lengthy, but elementary, calculation yields

(10)
$$c_4 = \kappa_j^{-1} = \frac{\sin \alpha_j - \sinh \alpha_j}{\cos \alpha_j + \cosh \alpha_j};$$

hence

$$f_j(t) = -\tfrac{1}{2}(1 - \kappa_j^{-1})\,e^{-\alpha_j t} - \tfrac{1}{2}(1 + \kappa_j^{-1})\,e^{\alpha_j t} + \cos(\alpha_j t) + \kappa_j^{-1}\sin(\alpha_j t).$$

Now, by Mercer's theorem on positive definite kernels, it follows (see Churchill [1] or Riesz and Nagy [3], for example) that the series (2) converges absolutely and uniformly to $K(t, u)$ in the unit square. Thus,

$$\int_0^1 \int_0^1 K(t, u)\,dt\,du = \frac{4}{3}\int_0^1 \int_0^u (3t^2 u - t^3)\,dt\,du = \frac{1}{5}$$

(11)
$$= \int_0^1 \int_0^1 \sum_{j=1}^\infty \lambda_j^{-1} f_j(t) f_j(u)\,dt\,du = \sum_{j=1}^\infty \lambda_j^{-1} \int_0^1 \int_0^1 f_j(t) f_j(u)\,dt\,du$$

$$= \sum_{j=1}^\infty \lambda_j^{-1}\left[\int_0^1 f_j(t)\,dt\right]^2 = \sum_{j=1}^\infty \lambda_j^{-1}(2\lambda_j^{-1/2}\kappa_j^{-2})$$

$$= 2\sum_{j=1}^\infty (\tfrac{1}{4}\alpha_j^4)^{-3/2}\kappa_j^{-2} = 16\sum_{j=1}^\infty \alpha_j^{-6}\kappa_j^{-2}.$$

We conclude from (11) that

$$\sum_{j=1}^\infty \alpha_j^{-6}\kappa_j^{-2} = \sum_{j=1}^\infty \alpha_j^{-6}\left[\frac{\sin \alpha_j - \sinh \alpha_j}{\cos \alpha_j + \cosh \alpha_j}\right]^2 = \frac{1}{80},$$

where $\{\alpha_j\}_1^\infty$ are the positive solutions to (9) and κ_j^{-1} is given by (10). Moreover, since (9) is an even function of α,

$$\sum_{j=-\infty}^\infty \alpha_j^{-6}\kappa_j^{-2} = \sum_{j=-\infty}^\infty \alpha_j^{-6}\left[\frac{\sin \alpha_j - \sinh \alpha_j}{\cos \alpha_j + \cosh \alpha_j}\right]^2 = \frac{1}{40},$$

where $\alpha_{-j} = -\alpha_j$.

REFERENCES

[1] R. V. CHURCHILL, *Fourier Series and Boundary Value Problems*, McGraw-Hill, New York, 1963.
[2] A. M. LIEBETRAU, *Some tests of randomness based upon the second-order properties of the Poisson process*, Math. Sci. Tech. Report 249, The Johns Hopkins University, Baltimore, 1976.
[3] F. REISZ AND B. NAGY, *Functional Analysis*, Ungar, New York, 1955.

Editorial note. The previous proof can be simplified by noting from (4) that $f(0) = f'(0) = 0 = f''(1) = f'''(1)$. This leads to a simpler set of equations for (7) and (8). M. L. GLASSER (Clarkson College of Technology) sketches a proof by contour integrals for the related identities:

$$\sum_{j=1}^{\infty} \alpha_j^{-3} \left\{ \frac{\sin \alpha_j - \sinh \alpha_j}{\cos \alpha_j + \cosh \alpha_j} \right\} = 0,$$

$$\sum_{j=1}^{\infty} (\alpha_j^4 - \lambda^4)^{-1} = \left\{ \frac{\sin \lambda \cosh \lambda - \sinh \lambda \cos \lambda}{1 + \cos \lambda \cosh \lambda} \right\} \bigg/ 4\lambda^3, \qquad |\lambda| < \alpha_1.$$

He also indicates that (A) can be derived this way but since the poles involved are second order, the calculations get messy.   [M.S.K.]

An Infinite Sum

Problem 85-9, by M. L. GLASSER (Clarkson University).
   Evaluate the sum

$$S_0 = \sum_{k=2}^{\infty} k \coth^{-1} k (4k^2 - 3).$$

The solutions by C. GEORGHIOU (University of Patras, Greece) *and* I. P. E. KINNMARK (University of Notre Dame) were essentially the same. This solution is as follows:
   We show that

$$S_0 = \frac{1}{2} \log \frac{9}{2\pi}.$$

For $|x| > 1$, we have

$$\coth^{-1} x = \frac{1}{2} \log \frac{x+1}{x-1}$$

and since $\coth^{-1} x = O(x^{-1})$ as $x \to +\infty$, it is clear that S_0 exists. Then it is easy to see that

$$e^{2S_0} = \prod_{k=2}^{\infty} \left(\frac{2k-1}{2k+1} \right)^{2k} \left(\frac{k+1}{k-1} \right)^k.$$

Finally, using Stirling's formula gives

$$\prod_{k=2}^{n} \left(\frac{2k-1}{2k+1} \right)^{2k} \left(\frac{k+1}{k-1} \right)^k = \frac{9n}{2^{2n+1}} \frac{(2n!)^2}{(n!)^4} \left(\frac{n^2+n}{4n^2+4n+1} \right) \to \frac{9}{2\pi}$$

as $n \to +\infty$, from which the above result follows.

   Editorial note: The proposer in his solution obtains as a by-product the identity

$$n! = \left(\frac{n}{e} \right)^n \sqrt{2\pi n} \prod_{k=n}^{\infty} \left\{ (1 + 1/k)^{k+1/2} / e \right\}$$

and notes that an elementary derivation by induction was given by D. MERMIN in a private communication.

Problem 66–14*, *A Finite Series*, by J. P. CHURCH (E. I. du Pont de Nemours and Co.).

The following series has arisen in a study of neutron transport theory in which an extension to higher harmonics was being considered for the transport kernel:

$$S_m = \binom{2m}{m}^{-1} \sum_{r=0}^{[m/2]} \frac{(-1)^r (2m - 2r)!}{r!(m - r)!(m - 2r)!(m + 1 + a - 2r)},$$

where $a = 1$ for m odd and $a = 0$ for m even. It has been verified that S_m is zero for $m = 2$ through 9. Also, the series has been computed in double precision arithmetic on an IBM 7040 for values of $m = 2$ through 100. Roundoff error was persistent, but in no case did the value of the series exceed $\pm.2 \times 10^{-9}$, and in most cases was far smaller. Consequently, it is conjectured that the series is identically zero for all $m > 1$.

The solutions by D. E. AMOS (University of Missouri), H. W. GOULD (West Virginia University), J. M. HORNER (University of Alabama in Huntsville), and A. MILEWSKI (IBM France) were the same and are given as follows.

Since the Legendre polynomial of degree m is given by

$$P_m(x) = 2^{-m} \sum_{r=0}^{[m/2]} \frac{(-1)^r (2m - 2r)! x^{m-2r}}{r!(m - r)!(m - 2r)!},$$

it follows that

$$2^{-m} \binom{2m}{m} S_m = \int_0^1 x^a P_m(x) \, dx = \frac{1}{2} \int_{-1}^1 x^a P_m(x) \, dx.$$

Since the integral is zero for $m > 1$, by the orthogonality property of the Legendre polynomials, the conjecture is valid.

Both R. BOJANIC (Ohio State University) and D. L. LANSING (NASA, Langley Research Center), in their solutions, generalize the previous result by using the known integral

$$I_m^r = \int_0^1 x^r P_m(x) \, dx = \frac{\sqrt{\pi}\,\Gamma(1 + r)}{2^{r+1}\Gamma\left(\dfrac{2 + r - m}{2}\right)\Gamma\left(\dfrac{3 + r + m}{2}\right)},$$

$$\text{Re } r > -1 \quad \text{and} \quad m = 0, 1, 2, 3, \cdots.$$

Still further generalizations are pointed out by Lansing; in particular, the one obtained by starting out with the integral

$$\int_0^1 x^r (1 - x^2)^{n/2} P_m^n(x) \, dx,$$

where $P_m^n(x)$ is an associated Legendre polynomial.

Problem 70-7, Summation of a Series, by NORMAN D. MALMUTH (North American Rockwell Science Center).

The following identity establishes an area rule for the hypersonic aerodynamic efficiency of delta wing-slender body combinations [1]. We consider

---

[1] N. MALMUTH, *Aerodynamic characteristics of hypersonic wing-body combinations according to rotational linear theory*, Rep. NA68-466, North American Rockwell Corporation, Los Angeles Division, 1968.

$x \in R \equiv \{x | 0 < x < \infty\}$. If

$$y \equiv e^x, \quad a \equiv \frac{3y^{-2} - 1}{3y^2 - 1}, \quad b \equiv \frac{1 - a}{1 + a},$$

$$c_n \equiv \cosh nx, \quad n = 0, \pm 1, \pm 2, \cdots,$$

$$-\phi_n \equiv \frac{3c_{2(n+3)} - c_{2(n+2)}}{3c_{2(n+1)} - c_{2(n+2)}} = \frac{1 + b \tanh 2(n + 2)x}{1 - b \tanh 2(n + 2)x},$$

then

$$(1) \quad U_n^\infty \equiv \sum_n^\infty c_{2k+5} \Big/ \prod_n^k \phi_m \equiv \frac{c_{2n+5}}{\phi_n} + \frac{c_{2n+7}}{\phi_n \phi_{n+1}} + \frac{c_{2n+9}}{\phi_n \phi_{n+1} \phi_{n+2}} + \cdots = U_n^F,$$

where

$$U_n^F \equiv \frac{c_{2(n+2)} - 3c_{2(n+1)}}{4c_1}$$

and U_n^∞ converges absolutely and uniformly in R by D'Alembert's ratio test.

(a) Prove (1).

(b)*Obtain U_n^F directly as a particular solution to the difference equation:

$$\phi_n U_n = U_{n+1} + c_{2n+5}.$$

Solution by the proposer.

The ascending continued fraction representation

$$U_n^\infty = \frac{c_{2n+5} + \dfrac{c_{2n+7} + \dfrac{c_{2n+9}}{\phi_{n+2}} +}{\phi_{n+1}}}{\phi_n}$$

is generated by the first order linear difference equation

$$(2) \qquad\qquad \phi_n U_n = U_{n+1} + c_{2n+5}.$$

Substitution and use of the identity $2c_1 c_n = c_{n+1} + c_{n-1}$ verifies that U_n^F is a particular solution of (2). Since U_n^∞ is also a particular solution, it differs from U_n^F by at most a solution of the homogeneous version of (2); i.e.,

$$(3) \qquad\qquad (U_n^\infty - U_n^F) \Big/ \prod_k^{n-1} \phi_m = C,$$

where C and the lower index k on the product are constants independent of n. The constant C may be determined by studying (3) for a fixed arbitrary $x \in R$ as $n \to \infty$. For this purpose, we note that

$$|U_n^\infty| < \sum_n^\infty y^{2k+5} \Big/ \left| \prod_n^k \phi_m \right|.$$

There are two cases to consider: (i) $x \in R_1$ if $0 < x \le \frac{1}{2} \ln 3$, and (ii) $x \in R_2$ if $\frac{1}{2} \ln 3 \le x < \infty$. ($R_1$ is associated with the range of gas specific heat ratios γ with $1 < \gamma \le 2$.)

(i) For $x \in R_1$, we have $1 > a \ge 0, 0 < b \le 1 \Rightarrow \{|\phi_n|\} \uparrow$.
Thus,

$$\left|\prod_{n}^{k} \phi_m\right| > |\phi_n|^{k-n+1}.$$

We observe that for any $x \in R$, $\lim_{n\to\infty} |\phi_n| = |a|^{-1}$ and that for any $x > 0$ (there exists a $\mu > 0$), $|a|^{-1} - y^2 \equiv \mu$.

Consider a number p with $0 < p \leq 1$. For arbitrary fixed p, and $x \in R_1$ (an n_0 exists) for all $n > n_0$, $p\mu < |\phi_n| - y^2 \leq \mu$. The last inequality insures the convergence of the majorizing geometric series $\sum_{n}^{\infty} y^{2k+5} |\phi_n|^{n-k-1}$ to $y^{2n+5}/(|\phi_n| - y^2)$. Thus for $x \in R_1$,

$$\sum_{n}^{\infty} y^{2k+5} \Big/ \left|\prod_{n}^{k} \phi_m\right| < y^{2n+5}/(p\mu).$$

(ii) For an arbitrary fixed $x \in R_2$, $0 \geq a > -1$, $1 \leq b < \infty$, and hence (an n_0 exists) for all $n > n_0$, $1 - b \tanh 2(n+2)x < 0 \Rightarrow \{|\phi_n|\}\downarrow$. Hence, $|\phi_n| \geq |a|^{-1}$, and

$$\left|\prod_{n}^{k} \phi_m\right| > |a|^{n-k-1},$$

which implies that

$$\sum_{.n}^{\infty} y^{2k+5} \Big/ \left|\prod_{n}^{k} \phi_m\right| < y^{2n+5}/\mu.$$

Accordingly, $U_n^{\infty} = O(y^{2n})$ uniformly for $x \in R$ as $n \to \infty$. From its definition, U_n^F has the same order in this limit. Since $\prod_{k}^{n-1} \phi_m = O(|a|^{-n})$, the left-hand side of (3) is $O((y^2|a|)^n)$, which is $o(1)$ uniformly for $x \in R$. Thus $C \equiv 0$, proving the identity.

For $x = 0$,
$$U_n^{\infty} = -1 + 1 - 1 + 1 - \cdots = -\tfrac{1}{2},$$

i.e., the series is Euler summable.

For $x \to \infty$,
$$U_n^{\infty} \to y^{2n+3}/4.$$

A Fourier Sine Series

Problem 86-10 *by* M. L. GLASSER (Clarkson College) *and* H. E. FETTIS (Former Collaborating Problem Editor).

Find a closed form expression for the sum of the Fourier series

$$\sum_{k=1}^{\infty} (-1)^k (k^2 - \alpha^2)^{-1} \sin kx$$

in terms of the hypergeometric function. Also, when α is rational, show that the sum can be expressed in elementary terms.

Solution by A. A. JAGERS (Technische Hogeschool Twente, Enschede, the Netherlands).

Assume $\alpha \geq 0$. Clearly this is no restriction. For $|z| \leq 1$ define

$$M(z)=\sum_{k=1}^{\infty}(k^2-\alpha^2)^{-1}z^k.$$

Then Im $M(-e^{ix})$ coincides with the given Fourier sine series. Let $m\in Z$, $m\geq0$. Then, for $m<\alpha<m+1$ and $z\neq1$,

$$M(z)=\frac{z}{2\alpha}\left\{\frac{F(1,1-\alpha;2-\alpha;z)}{1-\alpha}-\frac{F(1,1+\alpha;2+\alpha;z)}{1+\alpha}\right\}$$

$$=\frac{z}{2\alpha}\left\{\sum_{n=1}^{m}(n-\alpha)^{-1}z^{n-1}+\frac{z^mF(1,1+m-\alpha;2+m-\alpha;z)}{1+m-\alpha}-\frac{F(1,1+\alpha;2+\alpha;z)}{1+\alpha}\right\},$$

by Gauss' definition of the hypergeometric series (note that $2\alpha(k^2-\alpha^2)^{-1}=(k-\alpha)^{-1}-(k+\alpha)^{-1}$). For $\alpha=0$ we get a dilogarithm:

$$M(z)=\sum_{k=1}^{\infty}k^{-2}z^k=-\int_0^z\log(1-t)\frac{dt}{t}=-z\int_0^1\frac{\log t}{1-tz}dt.$$

By extending the hypergeometric functions involved the function M may be extended to an analytic function $\tilde{M}$ on $C\backslash(-\infty,-1]$ as follows:

$$\tilde{M}(z)=\frac{1}{2\alpha}\sum_{n=1}^{m}(n-\alpha)^{-1}z^n+\frac{z}{2\alpha}\int_0^1\frac{z^mt^{m-\alpha}-t^\alpha}{1-tz}dt$$

(for $\alpha\downarrow0$ we regain the dilogarithm above). This integral representation, related to a Stieltjes transform, can be easily verified by expansion of $(1-tz)^{-1}$ in a geometric series for $|z|\leq1$ and subsequent termwise integration.

Now assume that $\alpha\in Q$. Then, to show that $M(z)$ can be expressed in elementary terms, it suffices to do so for

$$\int_0^1\frac{t^\beta}{1-tz}dt$$

where $\beta=p/q\in Q$, $\beta>-1$, p, $q\in Z$, $q>0$. This is easily accomplished by the substitution $t=u^q$, which gives

$$q\int_0^1\frac{u^{p+q-1}}{1-zu^q}du$$

with $p+q-1\geq0$, followed by a partial fraction expansion.   □

Remark. For $m=0$ the function $(1-\alpha^2)M(z)/z$ may be viewed as the moment generating function of a probability distribution on $(0,1)$ with density $(1-\alpha^2)(t^{-\alpha}-t^\alpha)/2\alpha$ for $0<t<1$. Hence this function admits a continued fraction expansion of "chain sequence type" (cf., e.g., H. S. Wall, *Analytic Theory of Continued Fractions*, Chelsea, New York, 1967).

Editorial note. N. Ortner finds several integral representations for the sum and obtains the following for $x=m\pi/n$:

$$-\frac{\pi \sin (m\pi\alpha/n)}{2\alpha \sin \pi\alpha}+\frac{1}{2n\alpha}\sum_{k=1}^{n-1}\sin\frac{km\pi}{n}\left[\Psi\left(\frac{n+\alpha\pm k}{2n}\right)-\Psi\left(\frac{\alpha+k}{n}\right)\right]$$

$(+$ if m odd, $-$ if m even$)$.

For rational α, the proposers found that

$$\sum_{k=1}^{\infty}\frac{(-1)^k \sin kx}{n^2k^2-m^2}$$

$$=\frac{1}{mn}\sum_{k=1}^{n-1}\sin\frac{m}{n}(x+k\pi)\frac{\sin (k\pi m/n)}{\sin (\pi m/n)}\cdot\log\left|\frac{\sin (1/2n)(x+(2k+1)\pi)}{\sin (1/2n)(x+(2k-1)\pi)}\right|$$

$$(|x|<\pi, (m,n)=1, n\geq 2).$$

Problem 73-10*, *A Summation Identity*, by H. M. SRIVASTAVA (University of Victoria, Victoria, British Columbia, Canada).

Let

$$F(t) = t^3 + at^2 + bt + c$$

and

$$G(p,q) = (2p+q+a-1)+\frac{(2p+q+a+1)F(p)}{F(-p-q-a)}$$

$$+\frac{(2p+q+a+3)F(p)F(p+1)}{F(-p-q-a)F(-p-q-a-1)}+\cdots,$$

where a, b, c, p and q are constants. Prove or disprove that

$$G(p,q) = G(q,p).$$

Solution by OTTO G. RUEHR (Michigan Technological University).

Denote the zeros of F by x, y and z. In hypergeometric notation, we have

$$G(p,q) = A \, _5F_4\left[\begin{array}{cccccc} p-x, & p-y, & p-z, & 1+\dfrac{A}{2}, & 1; \\ & & & & \\ r+x, & r+y, & r+z, & \dfrac{A}{2}; \end{array} -1\right],$$

where $A = 2p+q+a-1$ and $r = p+q+a$. In order to simplify the expression for G and display the symmetry, we employ the following result [1, p. 28]:

$$_6F_5\left[\begin{array}{cccccc} A, & 1+\dfrac{A}{2}, & B, & C, & D, & E; \\ & & & & & \\ & \dfrac{A}{2}, & 1+A-B, & 1+A-C, & 1+A-D, & 1+A-E; \end{array} -1\right]$$

$$= \frac{\Gamma(1 + A - D)\Gamma(1 + A - E)}{\Gamma(1 + A)\Gamma(1 + A - D - E)} {}_3F_2 \left[\begin{array}{ccc} 1 + A - B - C, & D, & E; \\ & 1 + A - B, & 1 + A - C; \end{array} \; 1 \right].$$

Making the substitutions $A = 2p + q + a - 1, B = p - x, C = p - y, D = p - z$ and $E = 1$, we obtain

$$G(p, q) = (p + q + a + z - 1) {}_3F_2 \left[\begin{array}{ccc} q - z, & p - z, & 1; \\ & p + q + a + x, & p + q + a + y; \end{array} \; 1 \right],$$

from which the symmetry is obvious.

REFERENCE

[1] W. N. BAILEY, *Generalized Hypergeometric Series*, Stechert-Hafner, New York and London, 1964.

Two Infinite Sums

Problem 79-12, *by* P. J. DE DOELDER (Eindhoven University of Technology, Eindhoven, The Netherlands).

(1) Evaluate in closed form:

$$S(p, q, x) := \sum_{n=1}^{\infty} \frac{J_p(nx)J_q(nx)}{n^{2m}};$$

$J_p(x)$ and $J_q(x)$ are Bessel functions of order p and q; $p + q = 2l$; $p - q = 2s$; $l = 0, 1, 2, \cdots$; $s = 0, 1, 2, \cdots$; $0 \leq x \leq \pi$.
In particular, for $p + q > 2m$, show that (1) is given by

$$\sum_{n=1}^{\infty} \frac{J_p(nx)J_q(nx)}{n^{2m}} = \frac{1}{2} \frac{\Gamma(2m)\Gamma(l - m + \frac{1}{2})}{\Gamma(m + s + \frac{1}{2})\Gamma(m - s + \frac{1}{2})\Gamma(m + l + \frac{1}{2})} \left(\frac{x}{2}\right)^{2m-1}$$

(2) Evaluate in closed form:

$$T(p, q, x) := \sum_{n=1}^{\infty} \frac{J_p(nx)J_q(nx)}{n^{2m+1}};$$

$p + q = 2l + 1$; $p - q = 2s + 1$; $l = 0, 1, 2, \cdots$; $s = 0, 1, 2, \cdots$; $m = 0, 1, 2, \cdots$; $0 \leq x \leq \pi$.
In particular, for $p + q > 2m + 1$, show that (2) is given by

$$\sum_{n=1}^{\infty} \frac{J_p(nx)J_q(nx)}{n^{2m+1}} = \frac{1}{2} \frac{\Gamma(2m + 1)\Gamma(l - m + \frac{1}{2})}{\Gamma(m + l + \frac{3}{2})\Gamma(m + s + \frac{3}{2})\Gamma(m - s + \frac{1}{2})} \left(\frac{x^2}{4}\right)^{m}$$

The above two problems arose from telecommunication theory.

Solution by H. E. FETTIS (Mountain View, CA).
We consider the more general expression

(1) $$\bar{S}_{2k}(p, q, x, y) = \sum_{n=1}^{\infty} \frac{J_p(nx)J_q(ny)}{(nx/2)^p (ny/2)^q} n^{2k},$$

where k is an integer. Then

(2)
$$S_m(p, q, x) = \left(\frac{x}{2}\right)^{p+q} \bar{S}_{2k}(p, q, x, x)$$

with

(3)
$$m = \frac{p+q}{2} - k, \qquad p+q \text{ even}$$

and

(4)
$$T_m(p, q, x) = \left(\frac{x}{2}\right)^{p+q} \bar{S}_{2k}(p, q, x, x)$$

with

(5)
$$m = \frac{p+q-1}{2} - k, \qquad p+q \text{ odd.}$$

By differentiation, we find that

(6)
$$\frac{\partial}{\partial x} \bar{S}_{2k}(p-1, q, x, y) = -\frac{x}{2} \bar{S}_{2k+2}(p, q, x, y).$$

By using the known result [1, p. 103, #53], [2, p. 123, §4E, #11],

(7)
$$\bar{S}_0(p, q, x, y) = -\frac{1}{2} \frac{1}{\Gamma(p+1)\Gamma(q+1)} + \frac{\Gamma(\frac{1}{2})_2F_1(\frac{1}{2}-q, \frac{1}{2}, p+1; (x/y)^2)}{\Gamma(p+1)\Gamma(q+\frac{1}{2})y},$$
$$p, q > -\frac{1}{2}, \quad 0 < x < y < \pi,$$

and the fact that

(8)
$$\frac{d}{dz} {}_2F_1(a, b, c; z) = \frac{ab}{c} {}_2F_1(a+1, b+1, c+1; z),$$

we find by induction that for $k > 0$,

(9)
$$\bar{S}_{2k}(p, q, x, y) = \frac{1}{2}\left(\frac{2}{y}\right)^{2k+1} \frac{\Gamma(k+\frac{1}{2})_2F_1(\frac{1}{2}-q+k, k+\frac{1}{2}, p+1; (x/y)^2)}{\Gamma(p+1)\Gamma(\frac{1}{2}+q-k)}$$

Setting $x = y$ and noting that

(10)
$$_2F_1(a, b, c; 1) = \frac{\Gamma(c)\Gamma(c-a-b)}{\Gamma(c-a)\Gamma(c-b)},$$

we obtain

(11)
$$\bar{S}_{2k}(p, q, x, x) = \frac{1}{2} \frac{\Gamma(k+\frac{1}{2})\Gamma(p+q-2k)}{\Gamma(p+\frac{1}{2}-k)\Gamma(q+\frac{1}{2}-k)\Gamma(p+q+\frac{1}{2}-k)}\left(\frac{x}{2}\right)^{-1-2k}$$

If $p+q$ is even, (11) may be brought into agreement with the proposer's result for relation (a) by setting

(12)
$$m = \frac{p+q}{2} - k, \qquad p = l+s, \quad q = l-s, \quad p+q > 2m,$$

and in agreement with the proposer's result for (b)[1] if $p+q$ is odd by taking

(13) $\qquad m = \dfrac{p+q-1}{2}-k, \qquad p=l+s+1, \quad q=l-s, \quad p+q>2m+1.$

Note that (11) is more general that either (a) or (b), since it includes both as special cases, and requires only that $p, q > -\frac{1}{2}$ and $0 < x < \pi$.

For negative $k(p+q<2m)$, (6) is used in the backward sense:

(14) $\qquad \bar{S}_{-2k}(p, q, x, y) = \bar{S}_{-2k}(p, q, 0, y) - \dfrac{1}{2}\displaystyle\int_0^x \bar{S}_{-2k+2}(p+1, q, t, y)t\, dt,$

with S_0 given by (7).

The term $S_{-2k}(p, q, 0, y)$ is (for $k>0$) a polynomial of degree $2k$, and obeys the same differential-recurrence relation (6) with respect to y and p as $\bar{S}(p, q, x, y)$ does with respect to x and q. From this property, we obtain the relationship

(15) $\qquad \bar{S}_{-2k}(p, q, 0, y) = \bar{S}_{-2k}(p, q, 0, 0) - \dfrac{1}{2}\displaystyle\int_0^y \bar{S}_{-2k+2}(p, q+1, 0, t)t\, dt$

with

(16) $\qquad\qquad S_{-2k}(p, q, 0, 0) = \dfrac{\zeta(2k)}{\Gamma(p+1)\Gamma'(q+1)},$

where $\zeta(2k)$ is the Riemann zeta-function:

$$\zeta(z) = \sum_{n=1}^{\infty} n^{-z}.$$

Again, by induction we find the general form of $\bar{S}_{-2k}$ to be

(17) $\qquad S_{-2k}(p, q, 0, y) = P_{-2k}(y) + \dfrac{1}{2}\dfrac{\Gamma(\frac{1}{2}-k)}{\Gamma(1+p)\Gamma(\frac{1}{2}+q+k)}\left(\dfrac{2}{y}\right)^{1-2k}$

where $P_{-2k}(y)$ is a polynomial in y^2 of degree k:

(18) $\qquad P_{-2k}(y) = \dfrac{1}{\Gamma(p+1)\Gamma(q+1)}\bigg\{\zeta(2k) - \dfrac{\zeta(2k-2)(\frac{1}{2}y)^2}{(q+1)1!} + \dfrac{\zeta(2k-4)(\frac{1}{2}y)^4}{(q+1)(q+2)2!}$

$\qquad\qquad\qquad\qquad\qquad -\cdots - \dfrac{1}{2}\dfrac{(-)^k(\frac{1}{2}y)^{2k}}{(q+1)(q+2)\cdots(q+k)k!}\bigg\},$

$\qquad P_0(y) = -\dfrac{1}{2}\dfrac{1}{\Gamma(p+1)\Gamma(q+1)}.$

Finally, by successive applications of (14) we obtain for $\bar{S}_{-2k}(p, q, x, y)$ the following expression:

$\qquad S_{-2k}(p, q, x, y) = \bigg\{ P_{-2k}(y) - \dfrac{P_{-2k+2}(y)}{(p+1)1!}(\frac{1}{2}x)^2 + \dfrac{P_{-2k+4}(y)}{(p+1)(p+2)2!}$

$\qquad\qquad\qquad\qquad\qquad -\cdots - \dfrac{1}{2}\dfrac{(-)^k(\frac{1}{2}x)^{2k}}{(p+1)(p+2)\cdots(p+k)k!}\bigg\}$

---

[1] In the proposer's result for T_m, the negative sign in the last factor on the right-hand side should be deleted.

$$+\frac{1}{2}\frac{\Gamma(\frac{1}{2}-k)(2/y)^{1-2k}}{\Gamma(1+p)\Gamma(\frac{1}{2}+q+k)}{}_2F_1\left(\frac{1}{2}-q-k,\frac{1}{2}-k,1+p;\left(\frac{x}{y}\right)^2\right).$$

By inserting the appropriate expressions for the $P_{-2k}(y)$, a more symmetrical result is obtained. For $k = 1, 2$, e.g., we get:

$$S_{-2}(p,q,x,y)=\frac{1}{\Gamma(p+1)\Gamma(q+1)}\left\{\zeta(2)+\frac{1}{8}\left[\frac{x^2}{(p+1)}+\frac{y^2}{(q+1)}\right]\right\}$$

$$-\frac{\Gamma(\frac{1}{2})y}{2\Gamma(1+p)\Gamma(\frac{1}{2}+q)}{}_2F_1\left(-\frac{1}{2}-q,-\frac{1}{2},1+p,\left(\frac{x}{y}\right)^2\right),$$

$$S_{-4}(p,q,x,y)=\frac{1}{\Gamma(p+1)\Gamma(q+1)}\left\{\zeta(4)-\frac{\zeta(2)}{4}\left[\frac{x^2}{(p+1)}+\frac{y^2}{(q+1)}\right]\right.$$

$$-\frac{x^2y^2}{32(p+1)(q+1)}-\frac{1}{64}\left[\frac{x^4}{(p+1)(p+2)}+\frac{y^4}{(q+1)(q+2)}\right]\right\}$$

$$+\frac{\Gamma(\frac{1}{2})y^3}{12\Gamma(1+p)\Gamma(\frac{5}{2}+q)}{}_2F_1\left(-\frac{3}{2}-q,-\frac{3}{2},1+p;\left(\frac{x}{y}\right)^2\right).$$

REFERENCES

[1] A. ERDÉLYI ET AL., *Higher Transcendental Functions*, vol. 2, McGraw-Hill, New York, 1953.
[2] V. MANGULUS, *Handbook of Series*, Academic Press, New York, 1965.

Infinite Sums of Bessel Functions

Problem 85-14, *by* D. M. PETKOVIĆ (University of Nis, Nis, Yugoslavia).

Let N be the set of all natural numbers and let k, m, $n \in N$. Prove that for (and only for) $k > m$ the following sums are valid:

(i) $\displaystyle\sum_{n=1}^{\infty}\frac{J_{2k-1}(nx)}{n^{2m-1}}=\frac{(2k-2m-1)!!}{(2k+2m-3)!!}x^{2m-2},\qquad \begin{cases} m=1, & 0\leq x<2\pi, \\ m>1, & 0\leq x\leq 2\pi, \end{cases}$

(ii) $\displaystyle\sum_{n=1}^{\infty}\frac{J_{2k}(nx)}{n^{2m}}=\frac{(2k-2m-1)!!}{(2k+2m-1)!!}x^{2m-1},\qquad \begin{cases} m=1, & 0\leq x<2\pi, \\ m>1, & 0\leq x\leq 2\pi, \end{cases}$

where $J_k(x)$ are Bessel functions of the first kind and of integral order.

The above sums are special cases, $k > m$, of series arising in the solutions or accelerating the convergence of solutions of certain electrostatic problems.

Note that $(-1)!! = 1$, $(0)!! = 1$.

Solution by M. L. GLASSER (Clarkson University).

Let

$$S(x)=\sum_{n=1}^{\infty}\frac{J_\nu(nx)}{n^\alpha},\qquad \nu>-\frac{1}{2},\ \alpha>0.$$

The convergence of Schlomilch series is extensively discussed in Chapter 19 of Watson's treatise on Bessel functions. If

$$C_0 \equiv \max\{1-\sigma, \ -\nu\} < \frac{3}{2},$$

then the Mellin transform of $S(x)$ can be computed by termwise integration and the series can then be summed. Hence, by the Mellin inversion formula

$$\text{(1)} \qquad S(x) = \frac{1}{4\pi i} \int_{c-i\infty}^{c+i\infty} (x/2)^{-s} \frac{\Gamma\left(\dfrac{s+\nu}{2}\right)}{\Gamma\left(\dfrac{\nu-s}{2}+1\right)} \zeta(\alpha+s)\, ds, \qquad c_0 < c < \frac{3}{2}.$$

For $0 < x < 2\pi$ the contour can be closed in the left-hand plane. For simplicity assume that $\alpha - \nu - 1$ is not an even integer so the integrand has no double poles. Then by residues

$$\text{(2)} \qquad S(x) = \frac{1}{2} \frac{\Gamma\left(\dfrac{\nu+1-\alpha}{2}\right)}{\Gamma\left(\dfrac{\nu+\alpha+1}{2}\right)} (x/2)^{\alpha-1} + \sum_{l=0}^{\infty} \frac{(-1)^l}{l!} \frac{\zeta(\alpha-2l-\nu)}{\Gamma(\nu+1+l)} (x/2)^{2l+\nu}.$$

If $\alpha - \nu$ is an even integer, due to the vanishing of the zeta function the series terminates. The validity of (2) is trivial for $x=0$; for $x \geq 2\pi$ one must examine carefully the decay of the integrand of (1) into the left-hand plane. For example, with $\nu = 2k-1$, $\alpha = 2m-1$ the series in (2) disappears for $k > m$ and

$$S(x) = \frac{1}{2} \frac{\Gamma\left(k-m+\tfrac{1}{2}\right)}{\Gamma\left(k+m-\tfrac{1}{2}\right)} (x/2)^{2m-2} = \frac{(2k-2m-1)!!}{(2k+2m-3)!!} x^{2m-2}.$$

In the same way one can treat the series

$$\sum_{n=1}^{\infty} \frac{x(n)}{n^\alpha} J_\nu(nx),$$

where $x(n)$ is an elementary Dirichlet character.

[See B. Berndt, Pub. Elek. Fak. Univ. Beograd, 386 (1972) and M. L. Glasser, Math. Comp., 37 (1981), p. 499.]

A Summation of Bessel Functions

Problem 84-18, by HENRY E. FETTIS[1] (Mountain View, California).
    In [1, formula 57, 32.2] we have the sum

$$\sum_{k=1}^{\infty} (-1)^{k-1} \frac{k^{2m-\nu}}{k^2 - a^2} J_\nu(kx) = \frac{\pi a^{2m-\nu-1}}{2a \sin a\pi} J_\nu(ax)$$

for $0 < x < \pi$, $\mathrm{Re}(\nu) > 2m - \tfrac{5}{2}$, and $m = 1, 2, 3, \cdots$.
    It is known that this result is also valid in the closed interval $0 \leq x \leq \pi$ when $\nu = 2m$ (see Problem 79-18) and when $m=0$ and $\mathrm{Re}(\nu) > -\tfrac{3}{2}$, with the addition of the term

$$-\left(\frac{x}{2a}\right)^\nu \{2a^2 \Gamma(\nu+1)\}^{-1}$$

---

[1] Deceased December 15, 1984.

(see [2, §4D, part III]).

Find a closed form for the sum when $m = -1, -2, -3$ and $0 \leq x \leq \pi$.

REFERENCES

[1] E. R. HANSEN, *A Table of Series and Products*, Prentice-Hall, Englewood Cliffs, NJ, 1975.
[2] V. MANGULIS, *Handbook of Series for Scientists and Engineers*, Academic Press, New York, 1965.

Solution by ELDON HANSEN (Lockheed Missiles & Space Company).

Using partial fractions,

$$\frac{1}{k^{2m}(k^2-a^2)} = \frac{1}{a^{2m}(k^2-a^2)} - \sum_{s=1}^{m} \frac{1}{k^{2s}a^{2m-2s+2}},$$

we obtain

(1)
$$\sum_{k=1}^{\infty} \frac{(-1)^k}{k^{2m}(k^2-a^2)} \cos(kx\cos y)$$

$$= \frac{1}{a^{2m}} \sum_{k=1}^{\infty} \frac{(-1)^k}{k^2-a^2} \cos(kx\cos y)$$

$$- \sum_{s=1}^{m} \frac{1}{a^{2m-2s+2}} \sum_{k=1}^{\infty} \frac{(-1)^k}{k^{2s}} \cos(kx\cos y).$$

Equations (17.3.10) and (17.4.7) of [1] can be written as

(2)
$$\sum_{k=1}^{\infty} \frac{(-1)^k}{k^2-a^2} \cos kz = \frac{1}{2a^2} - \frac{\pi}{2a} \cos az \csc \pi a$$

and

(3)
$$\sum_{k=1}^{\infty} \frac{(-1)^k}{k^{2s}} \cos kz = \frac{(-1)^{s-1}}{2(2s)!} z^{2s} \sum_{k=0}^{2s} \binom{2s}{k} \left(\frac{\pi}{z}\right)^k (2-2^k) B_k.$$

Letting $z = x\cos y$ and substituting (2) and (3) into (1), we obtain

$$\sum_{k=1}^{\infty} \frac{(-1)^k}{k^{2m}(k^2-a^2)} \cos(kx\cos y)$$

$$= \frac{1}{2a^{2m+2}} - \frac{\pi}{2a^{2m+1}} \cos(ax\cos y) \csc \pi a$$

$$+ \sum_{s=1}^{m} \frac{(-1)^s}{a^{2m-2s+2}(2s)!} \sum_{k=0}^{2s} \binom{2s}{k} \pi^k (1-2^{k-1})(x\cos y)^{2s-k} B_k.$$

We now multiply this equation by $\sin^{2\nu} y$ and integrate from 0 to $\pi/2$. Equations 8.411(4) and 3.621(5) of [2] can be written

(4) $$\int_0^{\pi/2} \sin^{2\nu} y \cos(z \cos y)\, dy = \frac{1}{2}\left(\frac{2}{z}\right)^\nu \Gamma\left(\frac{1}{2}\right)\Gamma\left(\nu+\frac{1}{2}\right) J_\nu(z)$$

and

$$\int_0^{\pi/2} \sin^{2\nu} y \cos^{2s-k} y\, dy = \frac{\Gamma(\nu+1/2)\Gamma(s-k/2+1/2)}{2\Gamma(\nu+s+1-k/2)}.$$

Using these relations, the integration yields the desired result

(5) $$\sum_{k=1}^\infty \frac{(-1)^k}{k^{2m+\nu}(k^2-a^2)} J_\nu(kx)$$

$$= -\frac{\pi}{2a^{2m+\nu+1}} \csc \pi a\, J_\nu(ax)$$

$$+ \sum_{s=0}^m \sum_{n=0}^s \frac{(-1)^s \pi^{2n} a^{2s-2m-2}(x/2)^{\nu+2s-2n}}{(2n)!\,\Gamma(s-n+1)\Gamma(s+\nu-n+1)} (1-2^{2n-1}) B_{2n}.$$

We have simplified the result by noting that $B_k = 0$ for $k = 3, 5, 7, \cdots$ and using the relation

$$\Gamma(2z) = \pi^{-1/2} 2^{2z-1} \Gamma(z)\Gamma(z+\tfrac{1}{2}).$$

Equations (2) and (3) impose the condition $-\pi < x < \pi$ and equation (4) is valid only if $\mathrm{Re}(\nu) > -\frac{1}{2}$. The partial fraction expansion used to obtain equation (1) is valid only for $m = 0, 1, 2, \cdots$. However, our result is valid for $m = 0, \pm 1, \pm 2, \cdots$ provided we interpret the sum in the right member of (5) as zero when $m < 0$.

Note that the initial equation in the statement of problem 84-18* is incorrectly copied from [1]. The additive term is also incorrect and is correctly given in equation (57.32.2) of [1]. Equation (5) gives the correct results.

Editorial Note. L. Lorch (York University) *and* P. Szegö (San Jose, California) just sketch out a solution and note that the closed formed expressions can also be gotten for analogous series in which the Bessel function is replaced by a Struve function.

Series Involving Bessel Functions

Problem 85-25, *by* ELDON HANSEN (Lockheed Missiles & Space Co.).
    Assume that a closed form is known for the seriés

$$S_m(a) = \sum_{k=1}^\infty c_k k^{m-a} J_a(kx)$$

where J_a is the Bessel function of the first kind and where the coefficients c_k are independent of a. Sum the series $S_{m+2n}(a)$ $(n = 0, 1, 2, \cdots)$ by expressing it as a finite sum of terms involving $S_m(b)$ for various values of b.

The result can be used to sum the series in Problem 84-18* [1] which was proposed by Henry E. Fettis.

REFERENCE

[1] HENRY E. FETTIS, *Problem 84-18*, A summation of Bessel functions*, this Review, 26 (1984), pp. 430–431.

The solutions by O. P. LOSSERS (Eindhoven University of Technology, Eindhoven, the Netherlands) *and the proposer were the same and are as follows:*

From G. N. Watson, *Theory of Bessel Functions*, 1966, p. 45, equation (1) we borrow the formula

$$2aJ_a(kx) = kx[J_{a-1}(kx) + J_{a+1}(kx)],$$

so that

$$S_{m+2}(a) = \sum_{k=1}^{\infty} c_k k^{m-(a-2)} J_a(kx)$$

$$= \sum_{k=1}^{\infty} c_k k^{m-(a-2)} \left[\frac{2(a-1)}{kx} J_{a-1}(kx) - J_{a-2}(kx) \right]$$

$$= \frac{2(a-1)}{x} S_m(a-1) - S_m(a-2).$$

Using the latter formula repeatedly, it follows by induction that

$$S_{m+2n}(a) = \sum_{k=0}^{n} (-1)^k \left(\frac{x}{2} \right)^{k-n} \binom{n}{k} \frac{\Gamma(a-k)}{\Gamma(a-n)} S_m(a-n-k).$$

Problem 76–11, A Bessel Function Summation, by B. C. BERNDT (University of Illinois).

A.   Let $j_{\nu,n}$ denote the nth positive zero of the ordinary Bessel function $J_\nu(z)$, where $\nu > -1$. If $ai \neq 0$, $\pm j_{\nu,n}$, $1 \leq n < \infty$, show that

(1)
$$\sum_{n=1}^{\infty} \frac{1}{j_{\nu,n}^2 + a^2} = \frac{1}{2ai} \frac{J_{\nu+1}(ai)}{J_\nu(ai)}.$$

It is to be noted that for $\nu = \frac{1}{2}$, (1) reduces to the known result

$$\sum_{n=1}^{\infty} \frac{1}{n^2 + a^2} = \frac{\pi}{2a} \coth(\pi a) - \frac{1}{2a^2}.$$

B.   State and prove a general theorem on the summation of rational functions of zeros of Bessel functions for which (1) is the special case corresponding to the rational function $1/(z^2 + a^2)$.

Solution by P. LEVY (Brooklyn, NY)

THEOREM. *Let $F(z) = p(z)/q(z)$ be an even, rational function whose only poles are simple ones at $z_1, z_2, \cdots, z_m$. Assume further that* (i) $z_k \neq 0, \pm j_{v,n}$, $k = 1, 2, \cdots, m$, $n = 1, 2, \cdots$, *and that* (ii) $\deg q(z) \geq \deg p(z) + 2$. *Then if $v > -1$,*

$$\sum_{n=1}^{\infty} F(j_{v,n}) = \frac{1}{2} \sum_{k=1}^{m} \frac{J_{v+1}(z_k)}{J_v(z_k)} \cdot \frac{p(z_k)}{q'(z_k)}.$$

Proof. Consider the integral of $F(z) J_{v+1}(z)/J_v(z)$ around a rectangle with vertices at $\pm A \pm iB$, where A and B are chosen so that the rectangle contains all the z_k's and does not pass through any of the $j_{v,n}$'s. It is known that for $v > -1$, $J_{v+1}(z)/J_v(z)$ is bounded on the rectangle [1, p. 498]. Thus, (ii) insures that the integral goes to zero as A and B go to infinity.

Condition (i) guarantees that the integrand is analytic except for simple poles at $\pm j_{v,n}$ and z_k. Application of the residue theorem together with the recurrence formula,

$$z J_v'(z) = v J_v(z) - z J_{v+1}(z),$$

yields the desired result.

As required, (1) follows as a special case of the theorem with $F(z) = 1/(z^2 + a^2)$. This result is also given in the above cited reference.

In addition, note that (i) and the restriction that the poles of $F(z)$ be simple may be omitted provided that one takes into account the resulting pole at the origin and the presence of higher order poles in the integrand. For example, if $F(z) = z^{-4}$, an analysis similar to that above would yield

$$\sum_{n=1}^{\infty} j_{v,n}^{-4} = \frac{1}{16(v+1)^2(v+2)}.$$

REFERENCE

[1] G. N. WATSON, *A Treatise on the Theory of Bessel Functions*, 2nd ed., Cambridge University Press, Cambridge, England, 1944.

Problem 68-20, A Hypergeometric Series, by JERRY L. FIELDS and YUDELL L. LUKE (Midwest Research Institute).

If n is a positive integer or zero, evaluate the hypergeometric series

(1) $$A_n = {}_4F_3\left(\begin{matrix} -n, f+1, 1, \dfrac{f(n+e-1)+z(e-1+f)}{n+1} \\ e, f, 1-z \end{matrix}\middle| 1\right)$$

and consider special cases such as $e = z + \lambda n$, $\lim_{n \to \infty} A_n$, etc.

Solution by the proposers,

We prove that

(2) $$A_n = \frac{z(n+f)(e-1)}{f(z-n)(n+e-1)}, \qquad f(z-n)(n+e-1) \neq 0.$$

For particular values of the parameters in (1) and (2), limiting forms of (1) must be taken. The same is true in what follows. Consider

$$(3) \qquad V(z) = (-z)^{-1} \, {}_4F_3\!\left(\begin{matrix} -n, b, c, d + hz \\ e, f, 1 - z \end{matrix}\, \middle|\, 1\right).$$

Clearly $V(z)$ is a rational function of z with poles at $z = 0, 1, \cdots, n$. To develop the partial fraction decomposition of $V(z)$, we proceed as follows: If

$$y_k(z) = \frac{(d + hz)_k}{-z(1 - z)_k}, \qquad k = 0, 1, \cdots, n,$$

then by residue theory

$$y_k(z) = \sum_{r=0}^{k} \frac{(-1)^r (d + hr)_k}{(r - z)r!(k - r)!},$$

so that

$$V(z) = \sum_{k=0}^{n} \frac{(-n)_k (b)_k (c)_k}{(e)_k (f)_k k!} \sum_{r=0}^{k} \frac{(-1)^r (d + hr)_k}{(r - z)r!(k - r)!}.$$

Upon interchanging the order of summations, we find

$$V(z) = \sum_{r=0}^{n} \frac{(-1)^r (-n)_r (b)_r (c)_r (d + hr)_r}{(r - z)(e)_r (f)_r (r!)^2} q_r,$$

$$(4) \qquad q_r = {}_4F_3\!\left(\begin{matrix} r - n, r + b, r + c, d + (h + 1)r \\ r + e, r + f, r + 1 \end{matrix}\, \middle|\, 1\right).$$

Next we specialize the parameters so as to obtain a particularly simple form for q_r. Thus let

$$c = 1, \qquad b = f + 1.$$

Then q_r can be expressed as the sum of two ${}_2F_1$'s each of unit argument which can be summed by Gauss' theorem. We find

$$q_r = \frac{\Gamma(r + e)\Gamma(n + e - d - r(h + 1))}{\Gamma(n + e)\Gamma(e - d - rh)}\left[1 - \frac{(r - n)(d + r(h + 1))}{(r + f)(-n + d + 1 - e + r(h + 1))}\right].$$

Now set

$$d + r(h + 1) = \rho(r + f), \qquad -n + d + 1 - e + r(h + 1) = \rho(r - n),$$

$$\rho = h + 1, \qquad r = 1, 2, \cdots, n - 1.$$

Thus with

$$h = \frac{e - f - 1}{n + f}, \qquad d = \frac{f(n + e - 1)}{n + f},$$

we have

$$q_r = 0, \quad r = 0, 1, \cdots, n - 1, \quad q_n = 1,$$

and (2) readily follows.

The following special cases are of interest.

(5) $\quad {}_3F_2\left(\begin{matrix} -n, f+1, 1 \\ f, 1-z \end{matrix} \middle| \begin{matrix} z+f \\ n+f \end{matrix}\right) = \dfrac{z(n+f)}{f(z-n)}, \qquad n = 0, 1, 2, \cdots, \quad f(z-n) \neq 0.$

(6) $\quad {}_3F_2\left(\begin{matrix} -n, n+e-1-z, 1 \\ e, 1-z \end{matrix} \middle| 1\right) = \dfrac{z(e-1)}{(z-n)(n+e-1)},$

$$\qquad\qquad n = 0, 1, 2, \cdots, \quad (z-n)(n+e-1) \neq 0.$$

Equations (5), (6) are the limiting forms of (1), (2) with $e \to \infty$ and $f \to \infty$, respectively. Note that the limiting form $z \to \infty$ gives nothing new as (1) is symmetric in the parameters e and $1 - z$. Also (6) is a special case of Saalschutz's theorem. If in (1), (2) we put $e = x + \lambda n$ and let $n \to \infty$, it can be easily shown that

(7) $\quad {}_3F_2\left(\begin{matrix} f+1, 1, (\lambda+1)f + \lambda z \\ f, 1-z \end{matrix} \middle| \begin{matrix} -1 \\ \lambda \end{matrix}\right) = \dfrac{-\lambda z}{f(\lambda+1)}, \qquad f(\lambda+1) \neq 0,$

$$|\lambda| > 1 \quad \text{or} \quad |\lambda| = 1 \quad \text{and} \quad R(\{(\lambda+1)(f+z)\}) < -1.$$

Note that (5) is (7) with $\lambda = -(n+f)/(z+f)$. If in (5) we let $n \to \infty$, we get the curious result

(8) $\qquad\qquad {}_2F_2\left(\begin{matrix} f+1, 1 \\ f, 1-z \end{matrix} \middle| -(z+f)\right) = -\dfrac{z}{f}, \qquad f \neq 0,$

and with $f = 1$,

(9) $\quad {}_1F_1\left(\begin{matrix} 2 \\ 1-z \end{matrix} \middle| -(1+z)\right) = -z \quad \text{or} \quad {}_1F_1\left(\begin{matrix} -1-z \\ 1-z \end{matrix} \middle| 1+z\right) = -ze^{1+z}.$

Problem 63-17, *On a Double Summation,* by PERRY SCHEINOK (Hahnemann Medical College).

Evaluate the following double summation which has arisen in a probability problem:

$$\sum_{\substack{m=1}}^{\infty} \sum_{\substack{n=1 \\ (m,n)=1}}^{\infty} \frac{x^{m-1} y^{n-1}}{1 - x^m y^n}.$$

Solved by W. L. NICHOLSON (Hanford Laboratories).

Suppose $\sum_{i=1}^{\infty} f_i$ and $\sum_{j=1}^{\infty} g_j$ are absolutely convergent series. The resulting double series of their product can be summed in any order. In particular, partitioning the positive integer pairs (i, j) into equivalence classes $\{(km, kn) \mid (m, n) = 1\}$ and summing first within classes gives

$$\left(\sum_{i=1}^{\infty} f_i\right)\left(\sum_{j=1}^{\infty} g_j\right) = \sum_{i,j=1}^{\infty} f_i g_j = \sum_{(m,n)=1}^{\infty} \left(\sum_{k=1}^{\infty} f_{km} g_{kn}\right).$$

The present problem is the special case $f_i = x^{i-1}$ and $g_j = y^{j-1}$ with $|x|, |y| < 1$. Using the geometric series sum formula for each of the series in parentheses gives

$$\left(\frac{1}{1-x}\right) \cdot \left(\frac{1}{1-y}\right) = \sum_{(m,n)=1} \frac{x^{m-1} y^{n-1}}{1 - x^m y^n}.$$

A different version of this problem with $f_i = g_i = 2^{-i}$ appears as an elementary problem in *The American Mathematical Monthly*, October, 1963, p. 893, Problem E 1550.

Double Summations

Problem 87-10 by DANNY SUMMERS (Memorial University).

Given that

$$f(n,N) = \sum_{j=0}^{N-1} (-1)^j \cos^{n-1} \frac{\pi}{N}\left(j+\frac{1}{2}\right) \sin \frac{\pi}{N}\left(j+\frac{1}{2}\right)$$

where n and N are integers such that $n \geq N \geq 2$, find simple closed-form expressions for

$$\sum_{n=N}^{\infty} n^\alpha f(n,N)$$

in each of the cases $\alpha = 0, 1, 2,$ and 3.

These sums arise in the classical, symmetric, one-dimensional random walk problem with one reflecting barrier and one absorbing barrier.

Solution by MICHAEL E. FISHER (Cornell University).

The generating function here is

$$(1) \qquad\qquad S_N(z) = \sum_{n=N}^{\infty} z^n f(n,N)$$

where

$$f(n,N) = \sum_{j=0}^{N-1} f_j(n,N),$$

$$f_j(n,N) \equiv (-1)^j \cos^{n-1} \frac{\pi}{N}\left(j+\frac{1}{2}\right) \sin \frac{\pi}{N}\left(j+\frac{1}{2}\right).$$

Then,

$$(2) \qquad\qquad S_N^l \equiv \sum_{n=N}^{\infty} n^l f(n,N), \qquad l = 0, 1, 2, \cdots$$

$$= \left(z\frac{\partial}{\partial z}\right)^l S_N(z)\Bigg|_{z=1} = \left(-\frac{\partial}{\partial \theta}\right)^l S_N(e^{-\theta})\Bigg|_{\theta=0}.$$

Now note that $f_{-j-1} = f_j$ so that

$$(3) \qquad f(n,N) = -\frac{1}{2n} \sum_{j=0}^{N-1} (-1)^j \frac{\partial}{\partial \alpha}\left\{\cos\left[\frac{\pi}{N}\left(j+\frac{1}{2}\right)+\alpha\right]\right\}^n \Bigg|_{\alpha=0}.$$

Expanding $\cos^n \theta = (e^{i\theta} + e^{-i\theta})^n/2^n$ by the binomial theorem and performing the sums on j, we get

$$(4) \qquad f(n,N) = -N\frac{\partial}{\partial \alpha} \sum_{r=0}^{n} \frac{1}{2^n}\binom{n}{r} e^{i[(\pi/2N)+\alpha](n-2r)} \Delta_{N,n-2r}\Bigg|_{\alpha=0}$$

where

(5) $\qquad \Delta_{N,n-2r}=1 \quad \text{if } n-2r=\pm(2k-1)N \quad \text{for } k=0,1,2,\cdots,$

$\qquad\qquad\qquad =0 \quad \text{otherwise.}$

Substituting in (1) and rearranging the double sum on n and r into sums on $k=1,2,$ $\cdots$ and q, with $n=(2k-1)N+2q$ and $q=0,1,2,\cdots$ (a graphical representation of the nonvanishing terms is helpful) yields two sets of terms, one set with $r=q$ and $n-2r=(2k-1)N$ and another with $r=n-q$ and $n-2r=-(2k-1)N$. Performing the α differentiation then yields

(6) $\qquad S_N(z)=2N \sum\limits_{k=1}^{\infty} \sum\limits_{q=0}^{\infty} \dfrac{(-1)^{k+1}(2k-1)N}{(2k-1)N+2q}\binom{(2k-1)N+2q}{q}\left(\dfrac{z}{2}\right)^{(2k-1)N+2q}$

Our task now is to sum a binomial series on t with coefficients involving $\binom{n+(p+1)t}{t}$, where here $p=1$ and $t=q$. This can be done by considering the function $w(z)$ defined by the solution of

(7) $\qquad w(1+w^p)=z \quad \text{for } p=0,1,2,\cdots \quad \text{with } \dfrac{w}{z}\to 1 \quad \text{as } z\to 0.$

If we expand $[w(z)]^n$ in powers of z using Cauchy's theorem and change the integration variable from z to w, we obtain the identity

(8) $\qquad [w(z)]^n= \sum\limits_{t=0}^{\infty} \dfrac{(-1)^t n}{n+(p+1)r}\binom{n+(p+1)t}{t}z^{pt+n}$

for sufficiently small z. Evidently, this is just what is required in (6). We find

(9) $\qquad S_N(z)=2N \sum\limits_{k=1}^{\infty} (-1)^{k+1}\left(\dfrac{-2w(z)}{z}\right)^{(2k-1)N}=\dfrac{2N}{(-2w/z)^N+(-2w/z)^{-N}}.$

The required sums will now follow from (2). A somewhat more convenient form of solution is provided by setting $z=e^{-\theta}$ and $-2w/z=e^{\phi}$, yielding

(10) $\qquad S_N(e^{-\theta})=\dfrac{N}{\cosh N\phi(\theta)}$

in which $\phi(\theta)$ is a solution of

(11) $\qquad\qquad\qquad \cosh\phi=e^{\theta}$

which behaves like

$$\phi^2=2\theta+\dfrac{2}{3}\theta^2+\dfrac{4}{45}\theta^3+\cdots$$

(found readily by expanding (11) in powers of θ and ϕ and solving recursively). Finally, we obtain

(12) $\qquad S_N(e^{-\theta})=N+N^3(-\theta)+\left(\dfrac{5}{3}N^5-\dfrac{2}{3}N^3\right)\dfrac{(-\theta)^2}{2!}$

$\qquad\qquad\qquad +\dfrac{61N^7-50N^5+4N^2}{15}\dfrac{(-\theta)^3}{3!}+\cdots$

in which the coefficients are the required sums, i.e.,

$$S_N^0=N, \qquad S_N^1=N^3,$$

Editorial note. The proposer notes that $p(n, N) = f(n, N)/N$ is the probability distribution function for the number of steps n taken by a particle to exit the symmetric, one-dimensional random walk system in which N is the number of steps separating the particle source from the sink. Also, the above infinite sums, which represent moments of this distributuion, have applications in theoretical space physics.

Problem 73-26, A Double Summation, by B. C. BERNDT (University of Illinois).

Let h and k be positive integers such that $(h, k) = 1$. Evaluate in closed form the nonabsolutely convergent double series

$$S(h, k) = \sum_{\substack{n=1 \\ mh \neq nk}}^{\infty} \sum_{m=1}^{\infty} \frac{1}{m^2 h^2 - n^2 k^2}.$$

Then deduce that

$$S(h, k) + S(k, h) = \pi^2/4hk.$$

Solution by the proposer.

Let

$$((x)) = \begin{cases} x - [x] - \frac{1}{2} & \text{if } x \text{ is not an integer}, \\ 0 & \text{otherwise}. \end{cases}$$

Then the classical Dedekind sum $s(h, k)$ is defined by

$$s(h, k) = \sum_{n \bmod k} ((n/k))((hn/k)).$$

If h and k are coprime positive integers, then

$$(1) \qquad s(h, k) + s(k, h) = -\frac{1}{4} + \frac{1}{12}\left(\frac{h}{k} + \frac{k}{h} + \frac{1}{hk}\right),$$

which is the famous reciprocity theorem for Dedekind sums [1, p. 4].

We shall employ the Poisson summation formula. If f is of bounded variation on $[a, b]$, then

$$(2) \qquad \frac{1}{2}\sum_{n=a}^{b}{}' \{f(n + 0) + f(n - 0)\} = \int_a^b f(x)\, dx + 2 \sum_{n=1}^{\infty} \int_a^b f(x) \cos(2\pi nx)\, dx,$$

where the prime on the summation sign indicates that when $n = a$ and when $n = b$, only the terms $f(a + 0)$ and $f(b - 0)$, respectively, are counted. In (2), put $a = 0$, $b = k$, and $f(x) = ((x/k))((hx/k))$. Observe that $f(0 + 0) = f(k - 0) = 1/4$. Also [1, p. 23],

$$\int_0^k ((x/k))((hx/k))\, dx = \frac{k}{12h}.$$

Since $(h, k) = 1$, (2) then yields

(3) $$\frac{1}{4} + s(h, k) = \frac{1}{12h} + 2 \sum_{n=1}^{\infty} \int_0^k ((x/k))((hx/k)) \cos(2\pi nx)\, dx.$$

In order to calculate the integrals on the right-hand side of (3), we let $x = ky$, replace $((hy))$ by its Fourier series, and then invert the order of summation and integration which is justified by the bounded convergence of the Fourier series of $((hy))$ on $0 \le y \le 1$. Accordingly, we obtain

$$\int_0^k ((x/k))((hx/k)) \cos(2\pi nx)\, dx$$

(4)
$$= -\frac{k}{\pi} \sum_{m=1}^{\infty} \frac{1}{m} \int_0^1 (y - \tfrac{1}{2}) \sin(2\pi mhy) \cos(2\pi nky)\, dy$$

$$= -\frac{k}{2\pi} \sum_{m=1}^{\infty} \frac{1}{m} \int_0^1 y\{\sin(2\pi y(mh + nk)) + \sin(2\pi y(mh - nk))\}\, dy.$$

A routine integration by parts gives

$$\int_0^1 y\{\sin(2\pi y(mh + nk)) + \sin(2\pi y(mh - nk))\}\, dy$$

(5)
$$= \begin{cases} -\dfrac{mh}{\pi(m^2h^2 - n^2k^2)}, & mh - nk \ne 0, \\[3mm] -\dfrac{1}{2\pi(mh + nk)}, & mh - nk = 0. \end{cases}$$

Substituting (5) into (4) and then (4) into (3), we obtain

(6) $$\frac{1}{4} + s(h, k) = \frac{k}{12h} + \frac{hk}{\pi^2} \sum_{\substack{n=1 \\ mh \ne nk}}^{\infty} \sum_{m=1}^{\infty} \frac{1}{m^2h^2 - n^2k^2} + \frac{k}{2\pi^2} \sum_{\substack{n=1 \\ mh = nk}}^{\infty} \sum_{m=1}^{\infty} \frac{1}{m(mh + nk)}.$$

Since $(h, k) = 1$, $mh = nk$ if and only if $n = rh$, $1 \le r < \infty$. Thus

(7) $$\frac{k}{2\pi^2} \sum_{\substack{n=1 \\ mh = nk}}^{\infty} \sum_{m=1}^{\infty} \frac{1}{m(mh + nk)} = \frac{1}{4\pi^2 hk} \sum_{r=1}^{\infty} \frac{1}{r^2} = \frac{1}{24hk}.$$

After substituting (7) into (6) and rearranging, we find that

(8) $$S(h, k) = \frac{\pi^2}{hk} \left\{ \frac{1}{4} - \frac{k}{12h} - \frac{1}{24hk} + s(h, k) \right\}.$$

By symmetry,

(9) $$S(k, h) = \frac{\pi^2}{hk}\left\{\frac{1}{4} - \frac{h}{12k} - \frac{1}{24hk} + s(k, h)\right\}.$$

Add (8) and (9) and then use (1) to obtain

(10) $$S(h, k) + S(k, h) = \pi^2/4hk.$$

As the referee has remarked, (10) holds in the more general case when $(h, k) = d$. This is easily seen by using the more general reciprocity theorem for Dedekind sums,

$$s(h, k) + s(k, h) = -\frac{1}{4} + \frac{1}{12}\left(\frac{h}{k} + \frac{k}{h} + \frac{d^2}{hk}\right).$$

REFERENCE

[1] HANS RADEMACHER AND EMIL GROSSWALD, *Dedekind sums*, Carus Mathematical Monograph no. 16, Mathematical Association of America, Washington, D.C., 1972.

An Infinite Triple Summation

Problem 80-13, *by* M. L. GLASSER (Clarkson College).
   Show that

$$S = \sum_{i,j,k} (\operatorname{sgn} i)(\operatorname{sgn} j)(\operatorname{sgn} k)(\operatorname{sgn} (i+j-k))/i^2 j^2 = \pi^2 \ln 2,$$

where the sums are over all positive and negative odd integers, and each is understood to be a sum from $-N$ to N with the limit $N \to \infty$ taken at the end.
   This result is needed in calculating the free energy of superfluid helium [1], [2].

REFERENCES

[1] D. RAINER AND J. W. SERENE, Phys. Rev., B13 (1976), pp. 4745–4748.
[2] J. C. RAINWATER, Phys. Rev., B18 (1978), pp. 3728–3729.

Solution by the proposer.
   In [2] it is shown that

(1) $$S = -8 \int_0^1 \frac{\tanh^{-1} x \ln x}{x(1-x^2)} \, dx - \frac{7}{2}\zeta(3).$$

By noting that $2 \tanh^{-1} x = \ln(1+x) - \ln(1-x)$ and using partial fractions, the integral in (1) can be broken up into the six integrals over the interval $(0, 1)$:

$$I_1 = \int [\ln x \ln (1+x)/(1-x)] \, dx, \qquad I_4 = \int [\ln x \ln (1-x)/(1-x)] \, dx,$$

$$I_2 = \int [\ln x \ln (1+x)/x] \, dx, \qquad I_5 = \int [\ln x \ln (1-x)/x] \, dx,$$

$$I_3 = \int [\ln x \ln (1+x)/(1+x)] \, dx, \qquad I_6 = \int [\ln x \ln (1-x)/(1+x)] \, dx.$$

By elementary means it can be shown that

$$I_4 = I_5 = -\frac{4}{3}I_2 = -8I_3,$$

and, from [1, p. 567],

$$I_2 = -\frac{3}{4}\zeta(3).$$

From the identity

$$\ln x \ln (1-x) = \frac{1}{2}\left[\ln^2 (1-x) + \ln^2 x - \ln^2 \left(\frac{1-x}{x}\right)\right],$$

we have

$$2I_6 = \int_0^1 [\ln^2 (1-x)/(1+x)]\, dx + \int_0^1 [\ln^2 x/(1+x)]\, dx - \int_0^1 \ln^2 \left(\frac{1-x}{x}\right)\frac{dx}{1+x}.$$

The second integral on the RHS is known [1, p. 540], and the first and third can be expressed in terms of Euler's dilogarithm,

$$\phi(x) = -\int_0^x \frac{\ln (1-t)}{t}\, dt,$$

for which $\phi(1) = \zeta(2)$, $\phi(\frac{1}{2}) = \pi^2/12 - \frac{1}{2}\ln^2 2$. Treating I_4 similarly, we find

$$I_1 = \zeta(3) - \frac{\pi^2}{4}\ln 2, \qquad I_6 = \frac{13}{8}\zeta(3) - \frac{\pi^2}{4}\ln 2.$$

Therefore, we have

$$\int_0^1 \frac{\ln x \tanh^{-1} x}{x(1-x^2)}\, dx = -\frac{7}{16}\zeta(3) - \frac{\pi^2}{8}\ln 2.$$

Also solved by O. G. RUEHR (Michigan Technological University), who writes $S = \frac{7}{2}\zeta(3) + S'$, where

$$S' = 8 \sum_{n=0}^{\infty} \frac{1}{2n+1} \sum_{m=0}^{\infty} \frac{1}{(2n+2m+3)^2} = 2\,{}_4F_3\left[\begin{matrix}1,1,1,1\\2,2,\frac{3}{2}\end{matrix};1\right]$$

$$= 2\int_0^{\infty} x e^{-x}{}_2F_1\left[\begin{matrix}1,1\\\frac{3}{2}\end{matrix};e^{-x}\right] dx = -8\int_0^{\pi/2} \theta \sin \theta\, d\theta.$$

He concludes using known integrals from [1] and the following:

$$\int_0^{\pi/2} \frac{\phi(\pi - \phi)\, d\phi}{\sin \phi} = \frac{7}{2}\zeta(3).$$

REFERENCE

[1] I. S. GRADSTEYN AND I. M. RYZHIK, *Tables of Integrals, Series and Products*, Academic Press, New York, 1965.

A Triple Sum

Problem 82-7, by P. E. MERILEES (International Meteorological Institute, Stockholm).
Given that $I_{mnp} = 1$ if $m = n + p$, $= 0$ otherwise, show that

$$T_{ljk} \equiv \sum_{m=-M}^{M} \sum_{n=-M}^{M} \sum_{p=-M}^{M} \exp\{i(m\lambda_l - n\lambda_j - p\lambda_k)\} I_{mnp}$$

$$= \frac{\sin(2M+1)\theta' \sin(2M+1)\theta''}{\sin\theta' \sin\theta''} - Q_{ljk},$$

where

$$Q_{ljk} = \frac{1}{\sin\theta}\left\{ \frac{\sin[(2M+1)\theta + M\theta']\sin(M+1)\theta'}{\sin\theta'} \right.$$

$$\left. - \frac{\sin[(2M+1)\theta + M\theta'']\sin(M+1)\theta''}{\sin\theta''} \right\}$$

and $2\theta = \lambda_j - \lambda_k$, $2\theta' = \lambda_l - \lambda_k$, $2\theta'' = \lambda_l - \lambda_j$. Show further that

$$T_{ljk} = \frac{1}{2\pi} \int_0^{2\pi} \frac{\sin(2M+1)\left(\frac{\lambda - \lambda_l}{2}\right)\sin(2M+1)\left(\frac{\lambda - \lambda_j}{2}\right)\sin(2M+1)\left(\frac{\lambda - \lambda_k}{2}\right) d\lambda}{\sin\left(\frac{\lambda - \lambda_l}{2}\right)\sin\left(\frac{\lambda - \lambda_j}{2}\right)\sin\left(\frac{\lambda - \lambda_k}{2}\right)}.$$

The problem arose in the theory of spectral models of atmospheric flow.

Solution by W. B. JORDAN (Scotia, NY).
In the interests of greater symmetry we change the signs of θ' and m, getting

$$2\theta' = \lambda_k - \lambda_l, \quad \theta + \theta' + \theta'' = 0, \quad I_{mnp} = 0 \quad \text{unless } m + n + p = 0.$$

Now $\sin(2M+1)\phi/\sin\phi = \sum_{q=-M}^{M} \exp(2iq\phi)$, so the integral is

$$\frac{1}{2\pi} \int_0^{2\pi} \sum_{m=-M}^{M} \sum_{n=-M}^{M} \sum_{p=-M}^{M} \exp i[(m + n + p)\lambda - (m\lambda_l + n\lambda_j + p\lambda_k)] \, d\lambda.$$

But, for q an integer,

$$\frac{1}{2\pi} \int_0^{2\pi} \exp(iq\lambda) \, d\lambda = \begin{cases} 1 & \text{if } q = 0, \\ 0 & \text{otherwise.} \end{cases}$$

and the integral is equal to the given triple summation.

First solution: Reduction of the summation. Note that all terms for which $n \neq -m - p$ vanish, so the n-summation may be dropped, provided we delete terms where $|m + p| > M$. For brevity, put

$$u = \exp(i\theta) = \exp \tfrac{1}{2}i(\lambda_j - \lambda_k), \quad v = \exp(i\theta') = \exp \tfrac{1}{2}i(\lambda_k - \lambda_l),$$
$$w = \exp(i\theta'') = \exp \tfrac{1}{2}i(\lambda_l - \lambda_j), \quad \text{so } uvw = 1,$$
$$k = 2M + 1.$$

Then

$$T = \sum\sum \exp i[m(\lambda_l - \lambda_j) - p(\lambda_j - \lambda_k)]$$

$$= \left(\sum_{m=-M}^{M} \sum_{p=-M}^{M} - \sum_{m=-M}^{-1} \sum_{p=-M}^{-M-m-1} - \sum_{m=1}^{M} \sum_{p=-M+1-m}^{M} \right) w^{2m} u^{-2p}$$

$$= \frac{w^{-2M} - w^{2M+2}}{1 - w^2} \cdot \frac{u^{2M} - u^{-2M-2}}{1 - u^{-2}} - \sum_{m=-M}^{-1} w^{2m} \frac{u^{2M} - u^{2M+2m}}{1 - u^{-2}}$$

$$- \sum_{m=1}^{M} w^{2m} \frac{u^{-2M-2+2m} - u^{-2M-2}}{1 - u^{-2}},$$

$$(u - u^{-1})T = \frac{w^k - w^{-k}}{w - w^{-1}}(u^k - u^{-k}) - u^k \left[\frac{w^{-2M} - 1}{1 - w^2} - \frac{(uw)^{1-k} - 1}{1 - u^2 w^2} \right]$$

$$- u^{-k} \left[\frac{w^2 u^2 - (wu)^{k+1}}{1 - w^2 u^2} - \frac{w^2 - w^{k+1}}{1 - w^2} \right],$$

$$(u - u^{-1})(w - w^{-1})(u^{-1}w^{-1} - uw)\,T$$

$$= (w - w^{-1})[u^k((uw)^{-k} - (uw)^{-1}) - u^{-k}(uw - (uw)^k)]$$

$$+ (u^{-1}w^{-1} - uw)[(w^k - w^{-k})(u^k - u^{-k}) - u^k(w^{-1} - w^{-k}) - u^{-k}(w - w^k)]$$

$$= (w - w^{-1})(w^{-k} + w^k - w^{-1}u^{k-1} - wu^{1-k}) + (v - v^{-1})(v^k + v^{-k} - w^{-1}u^k - wu^{-k})$$

$$= (u - u^{-1})(u^k + u^{-k}) + (v - v^{-1})(v^k + v^{-k}) + (w - w^{-1})(w^k + w^{-k})$$

or, dividing out $4i$,

$$-2 \sin \theta \sin \theta' \sin \theta'' \cdot T = \sin \theta \cos k\theta + \sin \theta' \cos k\theta' + \sin \theta'' \cos k\theta'',$$

where $k = 2M + 1$. The proposer's solution does not reconcile with this.

Second solution: Reduction of the integral. Since the integrand is expressible as a polynomial in $\cos (\lambda - \lambda_n)$, the integral from 2π to 4π is equal to that from 0 to 2π, so $T = (1/4\pi) \int_0^{4\pi}$. For brevity, put

$$a_n = \exp(-i\lambda_n/2) \quad \text{for } n = 1, 2, 3, \qquad k = 2M + 1.$$

Let $z = \exp(i\lambda/2)$, $d\lambda = 2dz/iz$.

The path of integration for z is the unit circle. Since the λ-integrand is finite for λ real, the z-integrand has no singularities on the path.

$$T = \frac{1}{4\pi} \int \frac{[(za_1)^k - (za_1)^{-k}][(za_2)^k - (za_2)^{-k}][(za_3)^k - (za_3)^{-k}]}{[za_1 - (za_1)^{-1}][za_2 - (za_2)^{-1}][za_3 - (za_3)^{-1}]} \cdot \frac{2dz}{iz}$$

$$= \frac{(a_1 a_2 a_3)^{1-k}}{2\pi i} \int \frac{[1 - (za_1)^{2k}][1 - (za_2)^{2k}][1 - (za_3)^{2k}]}{[1 - (za_1)^2][1 - (za_2)^2][1 - (za_3)^2]} \cdot z^{2-3k}\, dz.$$

The only singularity inside the path is a pole at $z = 0$. We evaluate its residue by expanding in Maclaurin series. Splitting into partial fractions we have

$$\frac{(a_1^2 - a_2^2)(a_2^2 - a_3^2)(a_3^2 - a_1^2)}{\text{Denominator}} = \frac{a_1^4(a_3^2 - a_2^2)}{1 - (za_1)^2} + \text{cycle},$$

where "cycle" means two more terms, formed by cycling the subscripts.

$$\text{Numerator} = 1 - (a_1^{2k} + a_2^{2k} + a_3^{2k})z^{2k} + O(z^{4k})$$

so

$$(a_1 a_2 a_3)^{k-1}(a_1^2 - a_1^2)(a_2^2 - a_3^2)(a_3^2 - a_1^2)\,T$$

= coef of z^{3k-3} in

$$a_1^4(a_3^2 - a_2^2)(1 + z^2 a_1^2 + z^4 a_1^4 + \cdots)[1 - (a_1^{2k} + a_2^{2k} + a_3^{2k})z^{2k}] + \text{cycle}$$

$$= a_1^4(a_3^{2k} - a_2^2)[a_1^{3k-3} - (a_1^{2k} + a_2^{2k} + a_3^{2k})a_1^{k-3}] + \text{cycle}$$

$$= (a_2^2 - a_3^2)a_1^{k+1}(a_2^{2k} + a_3^{2k}) + \text{cycle}.$$

Divide by $4i(a_1 a_2 a_3)^{k+1}$, noting that

$$(a_2/a_3) - (a_3/a_2) = 2i \sin \theta, \qquad (a_2/a_3)^k + (a_3/a_2)^k = 2 \cos k\theta.$$

and get the same result as in the first solution.

Editorial note: As indicated by Jordan, the solution given in the problem statement is incorrect. Both the proposer's solution and that of Glasser agreed, upon editorial modification, with those given by Jordan.

Problem 72-27, *A Triple Sum,* by P. G. KIRMSER and K. K. HU (Kansas State University).

Determine the sum

$$S = \sum_{i=1}^{N-1} \sum_{j=1}^{N-1} \sum_{k=1}^{N-1} \frac{(\sin kj\pi/N)(\sin ij\pi/N)}{(1 - \cos j\pi/N)}, \qquad N \geq 2.$$

The problem arose in determining the total amount of dirt removed in a grading operation by a road grader passing over an irregular surface repeatedly.

Solution by W. B. JORDAN (Scotia, New York).

The k and i summations are well known; i.e.,

$$\sum_{k=1}^{N-1} \sin kj\pi/N = \sum_{i=1}^{N-1} \sin ij\pi/N = \begin{cases} \cot j\pi/2N, & j \text{ odd}, \\ 0, & j \text{ even}, \end{cases}$$

and so

$$S = \sum_{j=\text{odd}}^{N-1} (\cot^2 j\pi/2N)/(2 \sin^2 j\pi/2N).$$

Consider the integral of $Q = \tan Nz \cot^2 z/(2 \sin^2 z)$ around a rectangle in the complex plane. The vertical sides of the rectangle are $x = -b, x = \pi - b$; the horizontal sides are $y = \pm iR$, with b small and R large. Since Q has period π, the integrals on the vertical sides cancel each other. As R tends to infinity, Q tends to zero, and the integrals on the horizontal sides vanish, leaving 0 for the integral. Within the rectangle, Q has a triple pole at $z = 0$ with residue $N(N^2 - 1)/6$ and simple poles at $z = j\pi/2N, j = 1, 3, \cdots, 2N - 1$, with residue $-(\cot^2 j\pi/2N)$ $/(2N \sin^2 j\pi/2N)$. When N is odd, the point $z = \pi/2$ is not a pole, and is to be omitted from the list of residues. Therefore, whether N is even or odd, there are an even number of these simple poles, and the residue at $2N - j$ is the same as at j. The sum of all these residues is then $-2S/N$, and therefore

$$S = N^2(N^2 - 1)/12.$$

Editorial Note. Carty refers to related results in the solution of Problem 69-14 (Jan. 1971). The proposers' solution was an indirect one. They showed that the triple sum and the answer satisfied the same linear equation.

Problem 60-10*, *Some Multiple Summations,* by W. L. Bade (AVCO Research and Advanced Development Division).

By applying the Born-von Kármán method to the electronic motions, using a model which represents each molecule as an isotropic harmonic dipole-oscillator, the London-van der Waals cohesive energy of a linear lattice is calculated in the dipole-dipole approximation.[1] One problem that arises is to calculate the sum

$$I = \sum_{n=1}^{N} \left\{ \Phi\left(\frac{n}{N}\right) \right\}^{s}$$

where

$$\Phi(x) = \sum_{r=1}^{\infty} r^{-3} \cos 2\pi xr.$$

A good approximation to I can be gotten in the form

$$I \cong 2^{-s}N \sum_{t=0}^{s} \binom{s}{t} S_t^{(s)} \qquad\qquad s \ll N$$

where

$$S_t^{(s)} = \sum_{r_1=1}^{\infty} \cdots \sum_{r_s=1}^{\infty} \frac{\delta(-r_1 - r_2 \cdots - r_t + r_{t+1} + \cdots + r_s)}{r_1^3 r_2^3 \cdots r_s^3}.$$

Here $\delta(x)$ is a function defined by $\delta(0) = 1$, $\delta(x) = 0$, $x \neq 0$. It follows that $S_1^{(2)} = \sum_{r=1}^{\infty} r^{-6} = \zeta(6)$. Can any other of the sums $S_t^{(s)}$ be found in closed form?

[1] W. L. Bade and J. G. Kirkwood, *Drude-Model Calculation of Dispersion Forces II. The Linear Lattice,* Jour. of Chem. Physics, Vol. 27, No. 6 (Dec., 1957), pp. 1284–1288.

Supplementary References
Series

[1] W.N.Bailey, *Generalized Hypergeometric Series*, Cambridge University Press, 1935.

[2] N.K.Bary, *A Treatise on Trigonometric Series, I,II*, MacMillan, N.Y., 1964.

[3] T.J.I'A.Bromwich, *An Introduction to the Theory of Infinite Series*, MacMillan, London, 1947.

[4] W.E.Byerly, *Fourier Series and Spherical, Cylindrical and Ellipsoidal Harmonics*, Dover, N.Y.,

[5] H.S.Carslaw, *Introduction to the Theory of Fourier Series and Integrals*, Dover, N.Y., 1930.

[6] P.Dienes, *The Taylor Series*, Dover, N.Y., 1957.

[7] I.S.Gradshteyn and I.M.Ryzhik, *Table of Integrals, Series and Products*, Academic Press, N.Y.,
     1980.

[8] E.R.Hansen, *A Table of Series and Products*, Prentice-Hall, N.J., 1975.

[9] G.H.Hardy and W.W.Rogosinski, *Fourier Series*, MacMillan, N.Y., 1944.

[10] G.H.Hardy, *Divergent Series*, Clarendon Press, Oxford, 1949.

[11] I.I.Hirschman,Jr., *Infinite Series*, Holt, Rinehart and Winston, N.Y., 1962.

[12] D.Jackson, *Fourier Series and Orthogonal Polynomials*, MAA, Washington, D.C., 1941.

[13] L.B.W.Jolley, *Summation of Series*, Dover, N.Y., 1961.

[14] K.Knopp, *Theory and Application of Infinite Series,*, Blackie, London, 1949.

[15] V.Mangulis, *Handbook of Series for Scientists and Engineers*, Academic Press, 1965.

[16] A.P.Prudnikov, Y.A.Brychkov and O.I.Marichev, *I,II, Integrals and Series*, Gordon and Breach,
     N.Y., 1986.

[17] I.J.Schwatt, *An Introduction to the Operations with Series*, Chelsea, N.Y., 1924.

[18] G.N. Watson, *a Treatise on the Theory of Bessel Functions*, Macmillan, N.Y., 1948.

[19] A.D.Wheelon, *Tables of Summable Series and Integrals Involving Bessel Functions*, Holden-
     Day, San Francisco, 1968.

[20] H.S. Wilf, *Generating Functionology*, Academic, San Diego, 1990.

[21] A. Zygmund, *Trigonometric Series, I, II*, Cambridge University Press, Cambridge, 1959.

6. SPECIAL FUNCTIONS

A Conjectured Property of Legendre Functions

Problem 79-14*, by A. K. RAINA *and* V. SINGH (Tata Institute of Fundamental Research, Bombay, India).

The following conjecture arises in extending various asymptotic high energy theorems on scattering amplitudes in particle reactions to finite energies:

Let the successive maxima of $|P_\nu(\cos\theta)|$, considered as a function of $\nu(\nu \geq 0)$ for a fixed $\theta(\pi/2 \geq \theta > 0)$, be denoted by $m_0, m_1, m_2, \cdots$. Then, $m_0 > m_1 > m_2 > \cdots$.

Numerical evidence for the conjecture is quite strong.

Solution by W. B. JORDAN (Scotia, NY).

We give two proofs of the conjecture, both based on approximating the Legendre function. In the first, the approximation is accurate for θ near 90°, but deteriorates as θ decreases until, at around 15°, the validity of the proof becomes dubious. A by-product, however, is a proof for all θ for sufficiently large n. The other approximation is accurate for θ near zero but deteriorates as θ increases; the proof based on it is probably valid to at least 45°.

When $\theta = 90°$ the first peak of P_n is negative and occurs at $n = 1.92326$ (this happens to be the value of n where $\Gamma((n+1)/2)$ has its minimum). As θ decreases the first peak remains negative and occurs at greater and greater n. Consequently, we are interested only in values of n greater than about 2.

Equation numbers are from the *Handbook of Mathematical Functions* [1]. For our first approximation we start with (8.10.4) with $\mu = 0$ and $z = \cos\theta$:

$$P_n(\cos\theta) = (2\pi \sin\theta)^{-1/2} e^{-i\pi/4} \Gamma(n+1)/\Gamma(n+\tfrac{3}{2})$$
$$\cdot [e^{(n+\frac{1}{2})i\theta} F(\tfrac{1}{2}, \tfrac{1}{2}; n+\tfrac{3}{2}; z) + e^{i\pi/2} e^{-(n+\frac{1}{2})i\theta} F(\tfrac{1}{2}, \tfrac{1}{2}; n+\tfrac{3}{2}; \bar{z})],$$

where

$$z = \frac{e^{i(\theta - \pi/2)}}{2\sin\theta} = \frac{1}{2} - \frac{1}{2} i \cot\theta$$

and $\bar{z}$ is its complex conjugate. Let

$$\beta = \left(n + \frac{1}{2}\right)\theta - \frac{\pi}{4}.$$

Then,

$$P_n(\cos\theta) = (2\pi \sin\theta)^{-1/2} \frac{\Gamma(n+1)}{\Gamma(n+\frac{3}{2})} [e^{i\beta} F(\tfrac{1}{2}, \tfrac{1}{2}; n+\tfrac{3}{2}; z) + \text{conj}].$$

For n large, $\Gamma(n+1)/\Gamma(n+\tfrac{3}{2})$ can be approximated by $(n+\tfrac{3}{4})^{-1/2}$; the error is 0.20% at $n = 2$ and 0.07% at $n = 4$. The hypergeometric series converges for $|z| < 1$, i.e., for $\theta > 30°$, but it is an asymptotic representation for all $\theta > 0$ and its leading term is 1.

194

Then, for n large,

$$P_n(\cos \theta) \cong (2\pi \sin \theta)^{-1/2}(n +\tfrac{3}{4})^{-1/2}\, 2 \cos \beta.$$

Peaks occur for β a multiple of π (very nearly), i.e., at intervals of π/θ in n; and their magnitude decreases like $n^{-1/2}$. This proves the conjecture when n is large.

For n not so large, the approximation worsens as θ approaches 0. To improve on it, we take one more term of the F-series:

$$F \cong 1 + \frac{z}{4n +6} = 1 + \frac{1 - i \cot \theta}{8n + 12}$$

$$\cong \left(1 + \frac{1}{8n + 12}\right) e^{-i\phi}, \qquad \phi = \frac{\cot \theta}{8n + 12},$$

giving $P_n(\cos \theta) \cong r \cos (\beta - \phi)$ where $r = 2(2\pi \sin \theta)^{-1/2}(n +\tfrac{3}{4})^{-1/2}(1 + (1/(8n + 12)))$. Peaks occur when

$$0 = dP_n/dn$$

$$= r' \cos (\beta - \phi) - r(\beta' - \phi') \sin (\beta - \phi)$$

or, since $\beta' = \theta$,

$$\tan (\beta - \phi) = r'/r(\theta - \phi'),$$

and the peaks lie on the curve

$$\text{Peak } P_n = \pm[1 + (r'/r(\theta - \phi'))^2]^{-1/2} r.$$

To a rather good approximation, $r'/r = -1/2(n +\tfrac{3}{4})$ and ϕ' is negligible compared with θ, so

$$\pm \text{Peak } P_n = 2(2\pi \sin \theta)^{-1/2}\left(1 + \frac{1}{8n + 12}\right)\left[n + \frac{3}{4} + \frac{1}{4\theta^2(n +\tfrac{3}{4})}\right]^{-1/2}.$$

At the kth peak, β is roughly $k\pi$, so $\theta \cong (k +\tfrac{1}{4})\pi/(n +\tfrac{1}{2})$, and $1/4\theta^2(n +\tfrac{3}{4})$ is about $(n +\tfrac{1}{4})/(4\pi^2(k +\tfrac{1}{4})^2)$, which is independent of θ. Thus, in spite of appearances, the correction does not increase as θ decreases and the formula should give the magnitude of the peaks accurately until θ is quite small. Since for fixed θ, it is a decreasing function of n, the conjecture is proved for all but small θ.

For our second proof, we expand P_n in a Neumann series of Bessel functions. We start with Mehler's integral:

$$P_n(\cos \theta) = \frac{2}{\pi} \int_0^\theta [2(\cos t - \cos \theta)]^{-1/2} \cos \left(n + \frac{1}{2}\right) t\, dt.$$

Change the variable from t to u and v by means of

$$1 - \cos t = (1 - \cos \theta) \sin^2 u, \qquad t = \theta \sin v,$$

so

$$\sin \tfrac{1}{2}t = \sin \tfrac{1}{2}\theta \sin u$$

and

$$\cos \tfrac{1}{2}t\, dt = 2 \sin \tfrac{1}{2}\theta \cos u\, du = \theta \cos \tfrac{1}{2}t \cos v\, dv.$$

Then,

$$P_n(\cos \theta) = \frac{2}{\pi} \int_0^{\pi/2} \sec \tfrac{1}{2}t \cos \left(n + \frac{1}{2}\right) t \, du.$$

Let $x = (n + \tfrac{1}{2})\theta$. Then, by (9.1.42)

$$\cos (n + \tfrac{1}{2})t = \cos (x \sin v)$$

$$= J_0(x) + 2 \sum_{k=1}^{\infty} J_{2k}(x) \cos 2kv$$

and

$$\sec \frac{1}{2}t \, du = \frac{\theta \cos v \, dv}{2 \sin \tfrac{1}{2}\theta \cos u},$$

so

$$P_n(\cos \theta) = \frac{\theta}{\pi \sin \tfrac{1}{2}\theta} \int_0^{\pi/2} [J_0(x) + 2 \sum] M \, dv$$

where $M = \cos v / \cos u$. Now,

$$\frac{\theta}{\pi \sin \tfrac{1}{2}\theta} \int_0^{\pi/2} M \, dv = \frac{2}{\pi} \int_0^{\pi/2} \sec \frac{1}{2}t \, du$$

$$= \frac{2}{\pi} \int_0^{\pi/2} \left(1 - \sin^2 \frac{1}{2}\theta \sin^2 u\right)^{-1/2} du = \frac{2}{\pi} K\left(\sin \frac{1}{2}\theta\right),$$

a complete elliptic integral.

When θ is small, M is a slowly-varying function of v, its Fourier harmonics are small, and the series for P_n converges rapidly. We expand M in a power series in θ:

$$\sin^2 \tfrac{1}{2}\theta \cos^2 u = \sin^2 \tfrac{1}{2}\theta - \sin^2 (\tfrac{1}{2}\theta \sin v)$$

$$= \sin[\tfrac{1}{2}\theta(1 - \sin v)] \sin [\tfrac{1}{2}\theta(1 + \sin v)].$$

Let $v = 90° - 2w$, $p = \theta \sin^2 w$, $q = \theta \cos^2 w$. Then,

$$M^2 = \frac{\sin^2 2w \sin^2 \tfrac{1}{2}\theta}{\sin p \sin q} = \frac{4pq \sin^2 \tfrac{1}{2}\theta}{\theta^2 \sin p \sin q}.$$

Now,

$$\frac{\sin p \sin q}{pq} = \left(1 - \frac{\theta^2}{6} \sin^4 w + \cdots\right)\left(1 - \frac{\theta^2}{6} \cos^4 w + \cdots\right)$$

$$= 1 - \frac{\theta^2}{24}(3 + \cos 4w) + \cdots,$$

so

$$M = \frac{2}{\theta} \sin \frac{1}{2} \theta \left[1 + \frac{\theta^2}{48}(3 - \cos 2v) + O(\theta^4)\right]$$

Through θ^2 there are therefore no terms with $2k > 2$, while the coefficient of J_2 is

$$\frac{4}{\pi} \int_0^{\pi/2} -\frac{\theta^2}{48} \cos^2 2v \, dv = -\frac{\theta^2}{48},$$

and our approximation is

$$P_n(\cos\theta) \cong \frac{2}{\pi} K\left(\sin\frac{1}{2}\theta\right) J_0(x) - \frac{\theta^2}{48} J_2(x).$$

The functions J_0 and $-J_2$ are roughly in phase with each other, and they both have peaks that decrease with increasing x, or n. This proves the conjecture so long as θ is small enough for the approximation to be valid. But calculation shows that, at $\theta = 45°$, the two Bessel terms give about 0.1% error in the first few peaks and in the peaks at $n = 80.5$ and $n = 4000.5$. The hypergeometric formula says successive peaks are $\pi/\theta = 4$ apart in n, so their amplitude ratio is $(n/(n+4))^{1/2}$. For this to be as small as even twice our 0.1%, n must be over 1000 and the "large n" proof is applicable. We conclude the conjecture is true in the range $0 < \theta < 45°$, and therefore for all θ.

REFERENCE

[1] M. ABRAMOWITZ AND I. STEGUN, *Handbook of Mathematical Functions*, U.S. Government Printing Office, National Bureau of Standards, 1965.

An Identity

Problem 79-15, by J. D. LOVE (Australian National University, Canberra, Australia).

In an analysis of the electrostatic potential for two charged dielectric spheres, the following identity arises:

$$\operatorname{csch} x = P_n(\cosh x)Q_n(\cosh x) + Q_n(\cosh x)\sum_{m=0}^{n-1} P_m(\cosh x)\, e^{(n-m)x}$$

$$+ P_n(\cosh x)\sum_{m=n+1}^{\infty} Q_m(\cosh x)\, e^{(n-m)x}$$

where $P_n(\cosh x)$ and $Q_n(\cosh x)$ are modified Legendre functions of the first and second kinds, respectively. Prove the identity for real $x > 0$ and nonnegative integers n.

Solution by O. P. LOSSERS (Eindhoven University of Technology, Eindhoven, The Netherlands).

Let the right-hand side of the identity be denoted by S. Then by use of the integral [1, p. 154]

$$P_n(\cosh x)Q_m(\cosh x) = \frac{1}{2}\int_{-1}^{1} \frac{P_n(t)P_m(t)}{\cosh x - t}\, dt, \qquad n \leqq m,$$

S can be reduced to

$$S = \frac{1}{2} e^{nx} \int_{-1}^{1} \frac{P_n(t)}{\cosh x - t}\left(\sum_{m=0}^{\infty} P_m(t)\, e^{-mx}\right) dt.$$

The infinite series can be evaluated by means of the generating function for Legendre polynomials [1, p. 154], viz.,

$$\sum_{m=0}^{\infty} P_m(t)\, e^{-mx} = (1 - 2\, e^{-x}t + e^{-2x})^{-1/2} = 2^{-1/2}\, e^{1/2x}(\cosh x - t)^{-1/2},$$

thus leading to

$$S = 2^{-3/2} e^{(n+1/2)x} \int_{-1}^{1} \frac{P_n(t)}{(\cosh x - t)^{3/2}} \, dt.$$

Because of [2, p. 822, 7.225 (4)],

$$\int_{-1}^{1} \frac{P_n(t)}{(\cosh x - t)^{1/2}} \, dt = \frac{2\sqrt{2}}{2n+1} e^{-(2n+1)(x/2)}.$$

It follows by differentiation with respect to x that

$$\int_{-1}^{1} \frac{P_n(t)}{(\cosh x - t)^{3/2}} \, dt = \frac{2\sqrt{2}}{\sinh x} e^{-(n+1/2)x},$$

which shows that

$$S = \operatorname{csch} x.$$

REFERENCES

[1] A. ERDÉLYI, W. MAGNUS, F. OBERHETTINGER AND F. G. TRICOMI, *Higher Transcendental Functions*, Vol. I, McGraw-Hill, New York, 1953.
[2] I. S. GRADSTEYN AND I. M. RYZHIK, *Tables of Integrals, etc.*, Academic Press, New York, 1965.

A Generalized Hypergeometric-Type Identity

Problem 88-11, *by* C. C. GROSJEAN (State University of Ghent, Belgium).
  Let the rational function $f(a, b, c, d; k)$ be defined as follows:

$$f(a, b, c, d; k) = \sum_{m=0}^{2k+1} \sum_{n=0}^{2k+1-m} \frac{(-2k-1)_{m+n}}{(a+c-k)_{m+n}} \frac{(2a)_m (b)_m}{m!(2b)_m} \frac{(2c)_n (d)_n}{n!(2d)_n}$$

with the usual notation

$$(z)_0 = 1, \quad (z)_j = z(z+1) \cdots (z+j-1) \quad \forall z \in \mathbb{C}, \quad \forall j \in \mathbb{N}_0,$$

whereby $k \in \mathbb{N}$ and $a, b, c,$ and d represent arbitrary real or complex numbers, except that b and $d \notin \{0, -\frac{1}{2}, -1, \cdots, -k\}$ as well as $a + c \notin \{0, \pm 1, \pm 2, \cdots, \pm k\}$, solely in order to avoid the appearance of zero denominators leading to indeterminacies in the above-mentioned definition.
  Prove that $f(a, b, c, d; k)$ is identically equal to zero.
  This problem arose from some work on generalized hypergeometric functions which I was led to from a solution of a problem in electrostatics.

Solution by O. P. LOSSERS (Eindhoven University of Technology, Eindhoven, the Netherlands).
  We define

$$F^{h,k,l}_{r,s,t}\left(\begin{matrix}(a_h);(b_k);(c_l)\\(d_r);(e_s);(f_t)\end{matrix}\middle|\, x,y\right) = \sum_{p=0}^{\infty} \sum_{q=0}^{\infty} \frac{[(a_h)]_{p+q}[(b_k)]_p[(c_l)]_q \, x^p y^q}{[(d_r)]_{p+q}[(e_s)]_q[(f_t)]_q \, p! \, q!}$$

where (a_h) represents the array of symbols $a_1, a_2, \cdots, a_h$, and $[(a_h)]_{p+q}$ means the product $(a_1)_{p+q} \cdots (a_h)_{p+q}$. Then

$$f(a,b,c,d;k)=F^{1;2;2}_{1;1;1}\left(\begin{matrix}-2k-1;b,2a;d,2c\\a+c-k;2b;2d\end{matrix}\middle|1,1\right).$$

Choosing in formula (23) of [1]

$$\lambda_1=b,\quad \gamma_1=2a,\quad \lambda_2=d,\quad \gamma_2=2c,$$

we obtain

$$\tfrac{1}{2}(\gamma_1+\gamma_2-2k)=\tfrac{1}{2}(2a+2e-2h)=a+c-k,\quad 2\lambda_1=2b,\quad 2\lambda_2=2d.$$

We find that $f(a, b, c, d; k) = 0$.

REFERENCE

[1] C. C. GROSJEAN AND R. K. SHARMA, *Transformation formulae for hypergeometric series in two variables* II, Simon Stevin, 62 (1988), pp. 97–125.

Problem 78-5, *Evaluation of Weierstrass Zeta Functions*, by CHIH-BING LING (Virginian Polytechnic Institute and State University).
Show that

(i) $\zeta(\tfrac{1}{2}|1, i)= \pi/2,$

(ii) $\zeta(\tfrac{1}{2}|1, e^{\pi i/3})= \pi/\sqrt{3},$

(iii) $\zeta(\tfrac{1}{2}|1, e^{\pi i/6}/\sqrt{3})= \pi\sqrt{3},$

where $\zeta(z|2\omega_1, 2\omega_2)$ is a Weierstrass zeta function of z with double pseudo-periods $2\omega_1$ and $2\omega_2$.
Solution by the proposer.
Numerical evidence of the three relations was obtained by the proposer some time ago [1]. To prove them analytically, we make use of the following two Legendre relations [2]:

(1)
$$\omega_2\zeta(\omega_1|2\omega_1, 2\omega_2)-\omega_1\zeta(\omega_2|2\omega_1, 2\omega_2)=\tfrac{1}{2}\pi i,$$
$$\omega_3\zeta(\omega_2|2\omega_1, 2\omega_2)-\omega_2\zeta(\omega_3|2\omega_1, 2\omega_2)=\tfrac{1}{2}\pi i,$$

where

(2) $\omega_1+\omega_2+\omega_3 = 0.$

Firstly, let $2\omega_1 = 1$ and $2\omega_2 = i$. The first Legendre relation gives

(3) $\dfrac{i}{2}\zeta(\tfrac{1}{2}|1, i)-\tfrac{1}{2}\zeta(\tfrac{1}{2}i|1, i)=\tfrac{1}{2}\pi i.$

Since

$$\zeta(\tfrac{1}{2}i|1, i) = -i\zeta(\tfrac{1}{2}|-i, 1) = -i\zeta(\tfrac{1}{2}|1, i),$$

the result in (i) follows immediately.

Secondly, let $2\omega_1 = e^{-\pi i/6}/\sqrt{3}$ and $2\omega_2 = e^{\pi i/6}/\sqrt{3}$. Substitute the values into the first Legendre relation and further make use of the relations:

(4) $\zeta(e^{\pm\pi i/6}/2\sqrt{3}|e^{-\pi i/6}/\sqrt{3}, e^{\pi i/6}/\sqrt{3}) = \sqrt{3}\, e^{\mp\pi i/6}\zeta(\tfrac{1}{2}|1, e^{\pi i/3}).$

The result in (ii) follows.

Thirdly, let $2\omega_1$ and $2\omega_2$ take the same values as in the second case. Here $2\omega_3 = -1$. The second Legendre relation gives

(5) $-\tfrac{1}{2}\zeta(e^{\pi i/6}/2\sqrt{3}|2\omega_1, 2\omega_2) - \dfrac{1}{2\sqrt{3}}e^{\pi i/6}\zeta(-\tfrac{1}{2}|2\omega_1, 2\omega_2) = \tfrac{1}{2}\pi i.$

By (4) and the result in (ii), the first Weierstrass zeta function involved is

$$\text{1st function} = \sqrt{3}\, e^{-\pi i/6}\zeta(\tfrac{1}{2}|1, e^{\pi i/3}) = \pi\, e^{-\pi i/6}$$

By choosing an equivalent pair of pseudo-periods, the second function involved is

$$\text{2nd function} = -\zeta(\tfrac{1}{2}|1, e^{\pi i/6}/\sqrt{3}).$$

The result in (iii) is readily deduced.

REFERENCES

[1] C. B. LING, *Doctoral thesis*, London University, England, 1937.
[2] E. T. COPSON, *Theory of Functions of a Complex Variable*, Oxford University Press, London, 1935.

Problem 74-22*, Fourier Coefficients of a Function Involving Elliptic Integrals of the First Kind, by E. O. SCHULZ-DUBOIS (IBM Zurich Research Laboratory, Switzerland).

Consider Fourier coefficients

$$a_n = \frac{1}{\pi}\int_0^{2\pi} f(x)\cos nx\, dx$$

of the function $f(x)$, which is even about $x = 0$ ($f(x) = f(-x)$) and odd about $x = \pi/2$ ($f(\pi - x) = -f(x)$), hence periodic of period 2π, and defined for $0 < x < \pi/2$ by

$$f(x) = \begin{cases} K(k), & 0 \le x \le x_k, \\ F(k, \sin^{-1}((1/k)\cos x)), & x_k \le x \le \pi/2, \end{cases}$$

with $0 < k \le 1$, $x_k = \cos^{-1} k$, where K and F are the Legendre complete and incomplete elliptical integrals of the first kind, respectively. Obviously, even-indexed Fourier coefficients vanish ($a_{2n} = 0$).

Show that for the particular choice $k = 1/\sqrt{2}$, every second odd-indexed Fourier coefficient also vanishes, i.e., $a_{4n-1} = 0$. This fact first emerged with limited accuracy for a finite number of coefficients by numerical computation.

The problem arose in connection with interdigital transducers for acoustic surface waves, where $f(x)$ is the electrostatic potential along a surface on which there are alternatingly biased thin metal strips. Vanishing of a_n indicates the inability to excite the nth harmonic surface wave.

Solution by R. N. HILL (University of Delaware).

The vanishing of a_{4n-1} for $k = 1/\sqrt{2}$ follows immediately from the fact that

$$(1) \qquad a_{2n+1} = [2/(2n+1)](-1)^n P_n(1 - 2k^2),$$

where P_n is the Legendre polynomial. The result (1) can be established by noting first that

$$(2) \qquad a_{2n+1} = \frac{4}{\pi} \operatorname{Re} \int_0^{\pi/2} F(k, \sin^{-1}((1/k)\cos x)) \cos((2n+1)x)\, dx.$$

An integration by parts which differentiates F and integrates the cosine changes (2) into

$$(3) \qquad a_{2n+1} = \frac{4}{(2n+1)\pi k} \int_{x_k}^{\pi/2} U_{2n}(\cos x)(1 - k^{-2}\cos^2 x)^{-1/2} \sin x\, dx,$$

where U is the Chebyshev polynomial of the second kind, defined by

$$(4) \qquad U_l(\cos x) = \sin((l+1)x)/\sin x.$$

The explicit formula,

$$(5) \qquad U_{2n}(ky) = \sum_{m=0}^{n} (-1)^m \binom{2n-m}{m} (2ky)^{2(n-m)},$$

and a change to the variable $y = k^{-1}\cos x$ can be used to produce from (3) a finite series which can be integrated term by term with the aid of the beta function. The new series which results from this term by term integration can be recognized as a polynomial case of the hypergeometric series. Hence

$$(6) \qquad a_{2n+1} = [2/(2n+1)](-1)^n {}_2F_1(-n, n+1; 1; k^2).$$

The result (1) now follows from an identity between the Legendre polynomial and the hypergeometric function.

On the Roots of a Bessel Function Equation

Problem 86-8 by ANTHONY D. RAWLINS (Brunel, The University of West London).
Show that the roots of the equation

$$F(z) J_\nu(z) + z J_\nu'(z) = 0$$

where $\operatorname{Im} F(z) < 0$ and $\nu > -1$ (or $\nu = n$ an integer) can only lie in the second and fourth quadrant of the z-plane.

This problem arose from an investigation of a sound field radiated from an acoustically lined circular duct. It was necessary to show that only certain waves could propagate along the duct. These waves corresponded to the roots of the above equation when F took a particular form.

Solution by A. A. JAGERS (Technische Hogeschool Twente, Enschede, the Netherlands).

Let $G(z) = zJ'_\nu(z)/J_\nu(z)$. Then

$$G(z) + \nu = \frac{zJ_{\nu-1}(z)}{J_\nu(z)} = 2\nu + \frac{-z^2|}{|2(\nu+1)} \frac{-z^2|}{|2(\nu+2)} \frac{-z^2|}{|2(\nu+3)} \cdots,$$

a continued fraction corresponding to the recurrence relation

$$J_{\nu-1}(z) + J_{\nu+1}(z) = 2\nu z^{-1}J_\nu(z).$$

Let $A_n(z, \nu)$ denote the nth approximant of this continued fraction. Then $A_0(z, \nu) = 2\nu$, $A_1(z, \nu) = 2\nu - z^2/(2\nu + 2)$ and

$$A_{n+2}(z,\nu) = 2\nu - 1/\{(2\nu+2)z^{-2} - A_n(z, \nu+2)^{-1}\}$$

for $n \geq 0$. By induction on n, it follows that $\operatorname{Im} A_n(z, \nu) \leq 0$ if $\nu > -1$ and $\operatorname{Im} z^2 \geq 0$. Hence, $\operatorname{Im} G(z) \leq 0$ if $\nu > -1$ and z does not lie inside the second or fourth quadrant of the complex plane. To complete the proof of the required result, use the fact that $J_{-n}(z) = (-1)^n J_n(z)$ for integer n and that $J_\nu(z)$ and $zJ'_\nu(z)$ have no zeros in common except possibly $z = 0$.

REFERENCE

[1]   G. N. WATSON, *Theory of Bessel Functions*, Cambridge University Press, Cambridge, 1962.

Monotonicity of Bessel Functions

Problem 87-11*, by S. W. RIENSTRA (Katholieke Universiteit, Nijmegen, the Netherlands).

The eigenvalue equation related to a problem of sound propagation in hard-walled annular ducts is given by

$$x^2\{J'_n(x)Y'_n(xh) - Y'_n(x)J'_n(xh)\} = 0$$

with $n = 0, 1, 2, \cdots, 0 < h < 1$, where J'_n and Y'_n denote derivatives of Bessel functions, while h is the ratio between the inner and outer duct radii. We are interested in the solutions $x = \alpha(h)$ as a function of h. In particular, we want to show that

$$\frac{d\alpha}{dh} = \frac{\alpha f_n(\alpha)}{h\{f_n(\alpha h) - f_n(\alpha)\}} \quad \text{where} \quad f_n(x) = \frac{J'_n(x)^2 + Y'_n(x)^2}{1 - n^2/x^2},$$

is always finite or $f_n(\alpha h) - f_n(\alpha) \neq 0$.

If $n = 0$, the latter is true since $f_0(x) = J_1(x)^2 + Y_1(x)^2$ is a decreasing function [1]. It is also true for $n \geq 1$ if $\alpha h < n$ and $\alpha > n$ ($\alpha \leq n$ does not occur). If $\alpha h = n$, $d\alpha/dh = 0$. Finally, it will also be true for the case $\alpha h > n$ if $f_n(x)$ is decreasing for $x > n$. In view of numerical evidence that this is so for $n = 0, 1, \cdots, 100$, it is conjectured to be true. Prove or disprove.

REFERENCE

[1]   G. N. WATSON, *A Treatise on the Theory of Bessel Functions*, Cambridge Univ. Press, New York, 1948, p. 446.

Solution by W. B. JORDAN (Scotia, New York).

We write y for hx. The given eigenvalue equation is

(1) $$J_n'(x)Y_n'(y) - J_n'(y)Y_n'(x) = 0.$$

Its derivative with respect to x is

$$J_n''(x)Y_n'(y) + y'J_n'(x)Y_n''(y) - y'J_n''(y)Y_n'(x) - J_n'(y)Y_n''(x) = 0.$$

We simplify this expression by using (1), the Wronskian

$$J_{n+1}(t)Y_n(t) - J_n(t)Y_{n+1}(t) = 2/\pi t$$

and the recurrence formulas

$$C_n'(t) = C_{n-1}(t) - \frac{n}{t}C_n(t) = -C_{n+1}(t) + \frac{n}{t}C_n(t)$$

in which C may be either J or Y. We get

$$(a) + \frac{n}{x^2}(b) + y'\left((c) + \frac{n}{y^2}(d)\right) = 0$$

where

$$(a) = J_n'(y)Y_{n+1}'(x) - J_{n+1}'(x)Y_n'(y)$$

$$= \frac{2}{\pi x}\left[\frac{n(n+1)}{x^2} - 1\right]\frac{J_n'(y)}{J_n'(x)},$$

$$(b) = J_n'(y)Y_n(x) - J_n(x)Y_n'(y) = -\frac{2}{\pi x}\frac{J_n'(y)}{J_n'(x)},$$

$$(c) = J_{n+1}'(y)Y_n'(x) - J_n'(x)Y_{n+1}'(y)$$

$$= -\frac{2}{\pi y}\left[\frac{n(n+1)}{y^2} - 1\right]\frac{J_n'(x)}{J_n'(y)},$$

$$(d) = J_n(y)Y_n'(x) - J_n'(x)Y_n(y) = \frac{2}{\pi y}\frac{J_n'(x)}{J_n'(y)}$$

so

$$\frac{1}{x}\left(\frac{n^2}{x^2} - 1\right)J_n'^2(y) - \frac{y'}{y}\left(\frac{n^2}{y^2} - 1\right)J_n'^2(x) = 0.$$

Now $y' = h + xh'$, so for dx/dh to be ∞ it is necessary that $h' = 0$, which can occur only if

(2) $$\frac{x^2}{x^2 - n^2}J_n'^2(x) = \frac{y^2}{y^2 - n^2}J_n'^2(y).$$

We are to show that (1) and (2) cannot both be true. In the notation of [1, §9.2], namely,

$$J_n'(t) = N_n(t)\cos\phi_n(t), \qquad Y_n'(t) = N_n(t)\sin\phi_n(t),$$

(1) gives

$$\sin\left[\phi_n(y)-\phi_n(x)\right]=0, \qquad \phi_n(x)=\phi_n(y)+k\pi,$$

k being some integer; and (2) becomes

$$f_n(x)=f_n(y)$$

where

$$f_n(t)=t^2 N_n^2(t)/(t^2-n^2)=2/\pi t\phi_n'(t)$$

on using (9.2.21). For t large, $N_n^2(t)=2/\pi t$, so $\phi_n'(t)=1-n^2/t^2$; thus $\phi_n'(n)=0$ and $\phi_n'(\infty)=1$. For $n\geq 2$ and $t\geq n$ the formula $1-n^2/t^2$ is a decent first approximation to $\phi_n'(t)$, so ϕ_n' is monotone increasing and f_n is monotone decreasing. It follows that $f_n(x)$ cannot equal $f_n(y)$ if $x\neq y$.

REFERENCE

[1] M. ABRAMOWITZ AND I. STEGUN, Handbook of Mathematical Functions, National Bureau of Standards, Washington, D.C., 1965.

Also solved by J. A. COCHRAN (Washington State University) who refers to results in his paper "The analyticity of cross-product Bessel function zeros," *Proc. Cambridge Philos. Soc.*, 62 (1966), pp. 215–256.

Problem 77-20, *When is the Modified Bessel Function Equal to its Derivative?*, by I. NASELL (Royal Institute of Technology, Stockholm, Sweden).
Prove that the equation

$$I_\nu(x)=I_\nu'(x)$$

has exactly one positive solution $x=\xi(v)$ for each $\nu>0$. Investigate the properties of the function ξ.

Solution by M. L. GLASSER (Clarkson College of Technology).
We note that

$$I_\nu(x)=c_\nu x^\nu \prod_{n=1}^{\infty}\left(1+\frac{x^2}{j_{\nu,n}^2}\right)$$

so that

$$\phi_\nu(x)\equiv\left[\ln I_\nu(x)\right]'=\frac{\nu}{x}+2x\sum_{n=1}^{\infty}(x^2+j_{\nu,n}^2)^{-1}=\frac{I_\nu'(x)}{I_\nu(x)}.$$

It is clear that for $\nu>0$, $\phi_\nu(x)$ is positive, continuous and monotonically decreasing. For small x, $\phi_\nu(x)>1$; for large x, we have from the known asymptotic estimate $I_\nu(x)=e^x(2\pi x)^{-1}\{1+O(1/x)\}$ that $\phi_\nu(x)\sim 1-1/2x+O(1/x^2)<1$. Therefore $\phi_\nu(x)=1$ has a unique positive solution $x=\xi(\nu)$.

Also solved by the proposer, who shows additionally that

$$\nu+\nu^2>\xi(\nu)>\max(\nu,\nu^2), \qquad \nu>0$$

$$\xi(\nu)\sim\nu \quad \text{as} \quad \nu\to 0, \qquad \xi(\nu)\sim\nu^2 \quad \text{as} \quad \nu\to\infty.$$

He also conjectures that ξ has a positive derivative and refers to his paper, *Rational bounds for ratios of modified Bessel functions*, to appear in SIAM J. Math. Anal.

In a late solution, J. BOERSMA and P. J. DE DOELDER (Technological University, Eindhoven, the Netherlands) also show that

$$\xi(\nu) = \nu + \frac{\nu^2}{2} + \frac{\nu^4}{16} - \frac{\nu^5}{32} + O(\nu^6) \qquad \text{for } \nu \to 0,$$

$$\xi(\nu) = \nu^2 + \frac{1}{2} + \frac{\nu^{-2}}{4} + \frac{\nu^{-4}}{8} + O(\nu^{-6}) \quad \text{for } \nu \to \infty,$$

$$\xi(\nu) \approx \nu^4/(\nu^2 - \tfrac{1}{2}) \quad \text{(correct to four decimal places when } \nu \geq 4),$$

$$\xi'(\nu) > 0.$$

Gamma Function Expansions

Problem 83-8, by W. B. JORDAN (Scotia, NY).

Prove that

(a) $\quad \log \dfrac{\Gamma(z + 1/2)}{\sqrt{z}\,\Gamma(z)} = -\sum_{r=1}^{k} \dfrac{(1 - 2^{-2r}) B_{2r}}{r(2r - 1) z^{2r-1}} + O(z^{-2k+1/2}),$

(b) $\quad \log \dfrac{\Gamma(z + 3/4)}{\sqrt{z}\,\Gamma(z + 1/4)} = -\sum_{r=1}^{k} \dfrac{E_{2r}}{4r(4z)^{2r}} + O(z^{-2k-1/2}),$

(c)* $\quad \dfrac{1}{z} \left[\dfrac{\Gamma(z + 3/4)}{\Gamma(z + 1/4)} \right]^2 = 1 + \dfrac{2u}{1+} \dfrac{9u}{1+} \dfrac{25u}{1+} \dfrac{49u}{1+} \cdots,$

where B_{2r} and E_{2r} are Bernoulli and Euler numbers, and $u = 1/64z^2$.

Solution by O. P. LOSSERS (Eindhoven University of Technology, Eindhoven, the Netherlands).

(a), (b) The results immediately follow from Barnes' asymptotic expansion [1, form. 1.18(12)]

$$\log \Gamma(z + \alpha) \sim (z + \alpha - \tfrac{1}{2}) \log z - z + \tfrac{1}{2} \log (2\pi)$$

$$+ \sum_{n=1}^{\infty} \frac{(-1)^{n+1} B_{n+1}(\alpha)}{n(n + 1)} z^{-n}, \qquad z \to \infty, \quad |\arg z| < \pi,$$

in which $B_{n+1}(\alpha)$ is the Bernoulli polynomial, and the known values [1, form. 1.14(7)]

$$B_{n+1}(\tfrac{1}{2}) - B_{n+1}(0) = -2^{-n}(2^{n+1} - 1) B_{n+1},$$

$$B_{n+1}(\tfrac{3}{4}) - B_{n+1}(\tfrac{1}{4}) = 2^{-2n-1}(n + 1) E_n;$$

furthermore, recall that B_{n+1} vanishes for even n and E_n vanishes for odd n. The remainder terms may in fact be replaced by $O(z^{-2k-1})$ in case (a), and by $O(z^{-2k-2})$ in case (b).

In the same manner it can be shown that

$$\log \frac{\Gamma(z + \alpha + \tfrac{1}{2})}{\sqrt{z}\,\Gamma(z + \alpha)} \sim \sum_{n=1}^{\infty} \frac{(-1)^{n+1} E_n(2\alpha)}{2^{n+1} n} z^{-n}, \qquad z \to \infty, \quad |\arg z| < \pi,$$

where $E_n(2\alpha)$ is the Euler polynomial. This expansion includes the results (a) and (b) as special cases.

(c)* The continued fraction is known from Perron [2, p. 36, eq. (24)].

Also solved by V. BELEVITCH (Philips Research Laboratory, Brussels), BRUCE BERNDT (University of Illinois at Urbana-Champaign), M. L. GLASSER (Clarkson College), OTTO G. RUEHR (Michigan Technological University), ROBERT E. SHAFER (Berkeley, CA), JAMES A. WILSON (Iowa State University) *and the proposer.*

In their discussions of part (c), Belevitch mentioned Perron [2]; Berndt referred to Bauer [3], Ramanujan [4], and Stieltjes [5]; Ruehr appealed to Wall [7]; and Glasser referred to Belevitch [6].

Wilson employed the theory of orthogonal polynomials to prove (c) and obtained the generalization

$$\frac{1}{z}\frac{\Gamma(\tfrac{3}{4}+p+z)}{\Gamma(\tfrac{1}{4}+p+z)}\frac{\Gamma(\tfrac{3}{4}-p+z)}{\Gamma(\tfrac{1}{4}-p+z)} = 1 + \frac{(1-16p^2)/32z^2}{1} + \frac{c_1/z^2}{1} + \frac{c_2/z^2}{1} + \cdots$$

for Re $z > 0$, with $c_n = [(2n+1)^2 - 16p^2]/64$.

Shafer also referred to Ramanujan [4] and derived the following generalization:

$$\frac{\Gamma(z+\alpha+\beta+\tfrac{3}{4})\Gamma(z+\alpha-\beta+\tfrac{3}{4})\Gamma(z-\alpha+\beta+\tfrac{3}{4})\Gamma(z-\alpha-\beta+\tfrac{3}{4})}{\Gamma(z+\alpha+\beta+\tfrac{1}{4})\Gamma(z+\alpha-\beta+\tfrac{1}{4})\Gamma(z-\alpha+\beta+\tfrac{1}{4})\Gamma(z-\alpha-\beta+\tfrac{1}{4})}$$

$$= z^2 + \tfrac{1}{16} - \alpha^2 - \beta^2 - \cfrac{2(\tfrac{1}{16}-\alpha^2)(\tfrac{1}{16}-\beta^2)}{z^2 + \tfrac{1}{16} + 1^2 - \alpha^2 - \beta^2 - \cfrac{(\tfrac{9}{16}-\alpha^2)(\tfrac{9}{16}-\beta^2)}{z^2 + \tfrac{1}{16} + 2^2/2 - \alpha^2 - \beta^2 - \cdots}}$$

REFERENCES

[1] A. ERDÉLYI, W. MAGNUS, F. OBERHETTINGER AND F. G. TRICOMI, *Higher Transcendental Functions, Vol. I,* McGraw-Hill, New York, 1953.
[2] O. PERRON, *Die Lehre von den Kettenbrüchen, Band* II, Teubner, Stuttgart, 1957.
[3] F. BAUER, *Von einem Kettenbruche Euler's und einem Theorem von Wallis,* Abh. Bayer. Akad. Wiss., 11 (1872), pp. 96–116.
[4] S. RAMANUJAN, *Collected Papers,* Chelsea, New York, 1962, p. xxvii.
[5] T. J. STIELTJES, *Note sur quelques fractions continues,* Quart. J. Math., 25 (1891), pp. 198–200.
[6] V. BELEVITCH, *The Gauss hypergeometric ratio as a positive real function,* SIAM J. Math. Anal., 13 (1982), pp. 1024–1040.
[7] H. S. WALL, *Analytic Theory of Continued Fractions,* Van Nostrand, New York, 1948, p. 371, l. 94.9.

Evaluation of a Hypergeometric Function

Problem 84-11, by OTTO G. RUEHR (Michigan Technological University).

Evaluate I_k for integral $k \geq 2$:

$$I_k = \binom{k^2+k}{2k} {}_kF_{k-1}\left(\begin{matrix} k-k^2, k+2, k+3, \cdots, 2k \\ 2, 3, \cdots, k \end{matrix} \Big| 1\right).$$

The binomial coefficient is included to ensure that I_k is an integer.

The problem arose in an attempt to solve Problem E3004 of the American Mathematical Monthly, 90 (1983), p. 400.

Solution by GEORGE GASPER (Northwestern University).

Let $(a)_n = a(a+1)\cdots(a+n-1) = \Gamma(a+n)/\Gamma(a)$ and let $m_1,\cdots,m_p$ be nonnegative integers. From Minton's formula [2, (14)]

(1)

$$
{}_{p+1}F_p\left(\begin{array}{c} -(m_1+\cdots+m_p), b_1+m_1,\cdots,b_p+m_p \\ b_1,\cdots,b_p \end{array}; 1\right)
$$

$$
= \frac{(-1)^{m_1+\cdots+m_p}(m_1+\cdots+m_p)!}{(b_1)_{m_1}\cdots(b_p)_{m_p}}
$$

which is easily proved by induction, it immediately follows for integral $k \geq 2$ that

$$
I_k = \binom{k^2+k}{2k}\frac{(k^2-k)!}{(2)_k(3)_k\cdots(k)_k} = \frac{1!2!3!\cdots(k-1)!(k^2+k)!}{(k+1)!(k+2)!\cdots(2k)!}.
$$

For basic hypergeometric analogues and extensions of (1), see [1].

REFERENCES

[1] G. GASPER, *Summation formulas for basic hypergeometric series*, SIAM J. Math. Anal., 12 (1981), pp. 196–200.

[2] B. M. MINTON, *Generalized hypergeometric function of unit argument*, J. Math. Phys., 11 (1970), pp. 1375–1376.

Problem 77-2, A Gaussian Hypergeometric Function Identity, by P. W. KARLSSON (The Technical University, Lyngby, Denmark).
Establish the identity,

$$
\frac{1-x}{a+c}{}_2F_1(a, 1-b; a+c+1; 1-x){}_2F_1(b, 1-a; b+c; x)
$$

$$
+\frac{x}{b+c}{}_2F_1(a, 1-b; a+c; 1-x){}_2F_1(b, 1-a; b+c+1; x)
$$

$$
= \frac{\Gamma(a+c)\Gamma(b+c)}{\Gamma(c+1)\Gamma(a+b+c)}.
$$

Solution by M. E. H. ISMAIL (McMaster University, Hamilton, Ontario, Canada).
We have to assume $|x| \leq 1$ and $|1-x| \leq 1$. We also assume Re $(c+a) > 1$ and Re $(c+b) > 1$. The identity holds if $x(1-x) = 0$, by Gauss' theorem. Next consider the case $|x| \in (0, 1)$ and denote the left hand side of the identity by $G(x)$. Apply Euler's transformation for hypergeometric function to all the functions appearing in $G(x)$ to get

$$
(1-x)^{-c-a}x^{-c-b}G(x)
$$

$$
= \frac{1}{a+c}{}_2F_1(c+1, c+a+b; a+c+1; 1-x){}_2F_1(c, b+c+a-1; b+c, x) +
$$

$$\frac{1}{b+c}{}_2F_1(c,a+c+b-1;a+c;1-x){}_2F_1(c+1,b+c+a;b+c+1;x).$$

Thus,

$$c(c+a+b-1)(1-x)^{-c-a}x^{-c-b}G(x)$$

$$={}_2F_1(c,c+a+b-1;a+c;1-x)\frac{d}{dx}{}_2F_1(c,b+c+a-1;b+c;x)$$

$$-{}_2F_1(c,c+a+b-1;b+c;x)\frac{d}{dx}{}_2F_1(c,c+a+b-1;a+c;1-x)$$

$$=\text{Wronskian of two solutions of same hypergeometric differential equation}$$

$$=A\,\exp\left\{-\int^x\frac{b+c-(2c+b+a)t}{t(1-t)}\,dt\right\}=Ax^{-c-b}(1-x)^{-c-a},$$

where A is a constant. Therefore $G(x)$ is a constant for $|x|\in(0,1)$. Letting $x\to0$ one can evaluate the constant by Gauss' theorem.

Also solved by O.P. Lossers (University of Technology, Eindhover, the Netherlands) who specialized and reduced a known hypergeometric identity (cf. Erdelyi et al., Higher Transcendental Functions, vol. I, p. 85 (13)).

An Identity for Double Hypergeometric Series

Problem 85-24*, by H. M. SRIVASTAVA (University of Victoria).

Let (a_p) abbreviate the array of p parameters $a_1,\cdots,a_p$, with similar interpretations for (b_q), et cetera, and define a general double hypergeometric series by

$$F_{q:s;v}^{p:r;u}\left[\begin{array}{c}(a_p):(\alpha_r);(\gamma_u);\\(b_q):(\beta_s);(\delta_v);\end{array}x,y\right]$$

$$=\sum_{l,m=0}^{\infty}\frac{\Pi_{j=1}^p(a_j)_{l+m}\Pi_{j=1}^r(\alpha_j)_l\Pi_{j=1}^u(\gamma_j)_m}{\Pi_{j=1}^q(b_j)_{l+m}\Pi_{j=1}^s(\beta_j)_l\Pi_{j=1}^v(\delta_j)_m}\frac{x^l}{l!}\frac{y^m}{m!},$$

where $(\lambda)_l=\Gamma(\lambda+l)/\Gamma(\lambda)$.

Prove (or disprove) that, for all integers $n\ge0$,

$$(*)\quad F_{q:1;0}^{p:2;1}\left[\begin{array}{c}(a_p);\,\alpha,\beta;\,\gamma-\alpha-\beta;\\(b_q):\gamma+n;\,-;\end{array}z,z\right]$$

$$=\frac{1}{(\gamma-+\alpha)_n(\gamma-\beta)_n}\sum_{k=0}^n\binom{n}{k}(\alpha)_k(\beta)_k(\gamma+k)_{n-k}(\gamma-\alpha-\beta)_{n-k}$$

$$\cdot{}_{p+2}F_{q+1}\left[(a_p),\gamma-\alpha,\gamma-\beta;(b_q),\gamma+k;z\right],$$

provided that each side exists.

Solution by W. A. AL-SALAM *and* J. L. FIELDS (University of Alberta).

We first remark that (∗) maybe considered as an identity in formal power series in which case it can be "generalized" as

$$(**) \quad \sum_{l,m=0}^{\infty} w_{l+m} \frac{(\alpha)_l (\beta)_l (\gamma-\alpha-\beta)_m}{(\gamma+n)_l \, l! \, m!} z^{l+m}$$

$$= \frac{(\gamma)_n}{(\gamma-\alpha)_n (\gamma-\beta)_n} \sum_{k=0}^{n} \binom{n}{k} \frac{(\alpha)_k (\beta)_k (\gamma-\alpha-\beta)_k}{(\gamma)_k} \sum_{j=0}^{\infty} \frac{(\gamma-\alpha)_j (\gamma-\beta)_j}{j! \, (\gamma+k)_j} w_j z^j$$

for any sequence $\{w_\nu\}$. This is also equivalent to the "transformation formula," valid for nonnegative integers n, k,

$$(1) \quad \frac{(\gamma)_n (\gamma-\alpha-\beta)_n}{(\gamma-\alpha)_n (\gamma-\beta)_n} \, {}_3F_2 \left[\begin{matrix} -n, \, \alpha, \, \beta; \\ \gamma+k, \, 1-n-\gamma+\alpha+\beta; \end{matrix} \, 1 \right]$$

$$= \frac{(\gamma)_k (\gamma-\alpha-\beta)_k}{(\gamma-\alpha)_k (\gamma-\beta)_k} \, {}_3F_2 \left[\begin{matrix} -k, \, \alpha, \, \beta; \\ \gamma+n, \, 1-k-\gamma+\alpha+\beta; \end{matrix} \, 1 \right].$$

(∗∗) is also equivalent to

$$(2) \quad (1-z)^{\alpha+\beta-\gamma} \, {}_2F_1 \left[\begin{matrix} \alpha, \, \beta; \\ \gamma+n; \end{matrix} \, z \right] = \frac{(\gamma)_n}{(\gamma-\alpha)_n (\gamma-\beta)_n} \sum_{k=0}^{n} \binom{n}{k}$$

$$\cdot \frac{(\alpha)_k (\beta)_k (\gamma-\alpha-\beta)_{n-k}}{(\gamma)_k} \, {}_2F_1 \left[\begin{matrix} \gamma-\alpha, \, \gamma-\beta; \\ \gamma+k; \end{matrix} \, z \right].$$

The equivalence of these formal identities can be seen easily if we recall that

$$\{a_k\} = \{b_k\} \Leftrightarrow \sum a_k z^k = \sum b_k z^k \Leftrightarrow \sum a_k w_k z^k = \sum b_k w_k z^k;$$

all series are taken in a formal sense.

Now although the proposer likes to think of (∗) as a generalization of Euler's transformation (quoted in the proposal) it is however a consequence of it (i.e., it follows from the Euler transformation). To see this, let $D_z = d/dz$. We get

$$D_z^n \left\{ (1-z)^{\alpha+\beta-\gamma} \, {}_2F_1 \left[\begin{matrix} \alpha, \, \beta; \\ \gamma; \end{matrix} \, z \right] \right\} = \frac{(\gamma-\alpha)_n (\gamma-\beta)_n}{(\gamma)_n} \, {}_2F_1 \left[\begin{matrix} \gamma-\alpha+n, \, \gamma-\beta+n; \\ \gamma+n; \end{matrix} \, z \right]$$

$$= (1-z)^{\alpha+\beta-\gamma-n} \, {}_2F_1 \left[\begin{matrix} \alpha, \, \beta; \\ \gamma+n; \end{matrix} \, z \right] \frac{(\gamma-\alpha)_n (\gamma-\beta)}{(\gamma)_n}.$$

Now using the Leibnitz formula on the left-hand side and Euler's transformation once more we get (2). Replacing z^k by $w_k z^k$ in (2) we get (∗∗).

In a much similar way, taking the nth q-derivative of the q-analogue of the Euler transformation

$$ {}_2\phi_1 \left[\begin{matrix} a, \, b; \\ c; \end{matrix} \, z \right] = \frac{(zab/c; \, q)_\infty}{(z; \, q)_\infty} \, {}_2\phi_1 \left[\begin{matrix} c/a, \, c/b; \, q; \\ c; \end{matrix} \, \frac{zab}{c} \right] $$

we get a q-analogue of $(**)$, namely,

$$\sum_{l,m=0}^{\alpha} \frac{(a; q)_l(b; q)_l(c/ab; q)_m}{(cq^n; q)_l(q; q)_l(q; q)_m} w_{l+m} z^{l+m}$$

$$= \frac{(c; q)_n c^n}{(c/a; q)_n(c/b; q)_n a^n b^n} \sum_{k=0}^{n} \begin{bmatrix} n \\ k \end{bmatrix} q^{(1/2)(n-k)(n-k-1)} \frac{(a; q)_k(b; q)_k}{(c; q)_k}$$

$$\cdot \left(\frac{ab}{c} q^{1-n+k}; q\right)_{n-k} \sum_{j=0}^{\alpha} \frac{(c/a; q)_j(c/b; q)_j}{(q; q)_j(cq^k; q)_j} \left(\frac{abq^k z}{c}\right)^j w_j$$

and the analogue of (1) is

$$\frac{(c; q)_m(c/ab; q)_m}{(c/a; q)_m(c/b; q)_m} {}_3\phi_2\left[\begin{matrix} q^{-m}, a, b; \\ cq^n, (ab/c)q^{1-m}; \end{matrix} q^{1+n}\right]$$

$$= \frac{(c; q)_n(c/ab; q)_n}{(c/a; q)_n(c/b; q)_n} {}_3\phi_2\left[\begin{matrix} q^{-n}, a, b; \\ cq^m, (ab/c)q^{1-n}; \end{matrix} q^{1+m}\right].$$

Comment by R. G. BUSCHMAN (University of Wyoming).

If we apply [2, (4.8.3.12)] to the ${}_{p+2}F_{q+1}$ functions and do some simplifications, we can write the sum in the form

$$\frac{(\gamma)_n(\gamma-\alpha-\beta)_n}{(\gamma-\alpha)_n(\gamma-\beta)_n} \frac{\Gamma(\gamma)}{\Gamma(\alpha)\Gamma(\gamma-\alpha)} \int_0^1 x^{\gamma-\alpha-1}(1-x)^{\alpha-1} {}_2F_1\left(\begin{matrix} -n, \beta; \\ 1-\gamma+\alpha+\beta-n; \end{matrix} 1-x\right)$$

$$\cdot {}_{p+1}F_q\left(\begin{matrix} \gamma-\beta, (a_p); \\ (b_q); \end{matrix} zx\right) dx.$$

The formula [1, (A.1.2.6)] allows us to rewrite this as

$$\frac{(\gamma)_n(\gamma-\alpha-\beta)_n}{(\gamma-\alpha)_n(\gamma-\beta)_n} F\begin{matrix} 0: p+2; 3 \\ 1: q; \qquad 1 \end{matrix}\left(\begin{matrix} -: \gamma-\alpha, \gamma-\beta, (a_p); -n, \beta; \\ \gamma: (b_q); 1-\gamma+\alpha+\beta-n; \end{matrix} z, 1\right).$$

REFERENCES

[1] HAROLD EXTON, *Handbook of Hypergeometric Integrals*, Ellis Horwood Ltd., 1978.
[2] L. J. SLATER, *Generalized Hypergeometric Functions*, Cambridge University Press, Cambridge–London, 1966.

Also solved by the proposer, who employed his generalization of Saalschütz theorem,

(†) $\quad {}_3F_2\left[\begin{matrix} a, b, -N; \\ c+n, a+b-c-N+1; \end{matrix} 1\right]$

$$= \frac{(c-a+n)_{N-n}(c-b+n)_{N-n}}{(c-n)_{N-n}(c-a-b)_N} \sum_{k=0}^{n} \binom{n}{k} \frac{(a)_k(b)_k(c-a-b)_{n-k}}{(c+N)_k},$$

and who also obtained the generalization $(**)$ above.

Supplementary References
Special Functions

[1]   M.Abramowitz and I.Stegun, *Handbook of Mathematical Functions*, N.B.S., Washington, D.C., 1964.
[2]   E.Artin, *The Gamma Function*, Holt, N.Y., 1964.
[3]   R.A. Askey, *Orthogonal Polynomials and Special Functions*, SIAM, Philadelphia, 1975.
[4]   R.A. Askey, ed., *Theory and Application of Special Functions*, Academic Press, N.Y., 1975.
[5]   W.N.Bailey, *Generalized Hypergeometric Series*, Cambridge University Press, Cambridge, 1935.
[6]   W.W.Bell, *Special Functions for Scientists and Engineers*, Van Nostrand, London, 1968.
[7]   R.Bellman, *A Brief Introduction to Theta Functions*, Holt, Rinehart and Winston, N.Y., 1961.
[8]   F.Bowman, *Introduction to Elliptic Functions with Applications*, Dover, N.Y., 1961.
[9]   P.F.Byrd and M.D.Friedman, *Handbook of Elliptic Integrals for Engineers and Scientists*, Springer-Verlag, Berlin, 1971.
[10] B.C.Carlson, *Special Functions of Applied Mathematics*, Academic Press, N.Y., 1977.
[11] A.Cayley, *Elementary Treatise on Elliptic Functions*, Dover, N.Y., 1961.
[12] H.M. Edwards, *Riemann's Zeta Function*, Academic Press, N.Y., 1974.
[13] A.Erdelyi, W.Magnus, F.Oberhettinger, F.G.Tricomi, *Higher Transcendental Functions, I,II,III*, Mcgraw-Hill, N.Y., 1953, 1955.
[14] A.Gray, *Bessel Functions and their Applications to Physics*, MacMillan, London, 1952.
[15] A.G.Greenhill, *The Applications of Elliptic Functions*, Dover, N.Y., 1959.
[16] E.W.Hobson, *The Theory of Spherical and Ellipsoidal Harmonics*, Chelsea, N.Y., 1955.
[17] D.F.Lawden, *Elliptic Functions and Applications*, Springer-Verlag, N.Y.,1989.
[18] 'N.N.Lebedev, *Special Functions and their Applications*, Dover, N.Y., 1972.
[19] L.Lewin, *Dilogarithms and Associated Functions*, MacDonald, London, 1958.
[20] Y.L.Luke, *The Special Functions and their Approximations, I, II*, Academic Press, N.Y., 1969.
[21] T.M.MacRobert, *Spherical Harmonics*, Dover, N.Y., 1948.
[22] W.Magnus and F.Oberhettinger, *Formulas and Theorems for the Special Functions of Mathematical Physics*, Chelsea, N.Y., 1949.
[23] W.Miller, *Lie Theory and Special Functions*, Academic Press, N.Y., 1968.
[24] F.J.Olver, *Asymptotics and Special Functions*, Academic Press, N.Y., 1974.
[25] E.D.Rainville, *Special Functions*, Chelsea, N.Y., 1971.
[26] F.E.Relton, *Applied Bessel Functions*, Dover, N.Y., 1965.
[27] G.Sansone, *Orthogonal Functions*, Interscience, N.Y., 1956.
[28] L.J.Slater, *Confluent Hypergeometric Functions*, Cambridge University Press, London, 1960.
[29] L.J.Slater, *Generalized Hypergeometric Functions*, Cambridge University Press, London, 1966.
[30] I.N.Sneddon, *Special Functions of Mathematical Physics*, Longman, London, 1980.
[31] H.M.Srivastava, *Special Functions in Queuing Theory*, Academic Press, N.Y., 1982.
[32] E.C.Titchmarsh, *The Zeta-Function of Riemann*, Cambridge University Press, Cambridge, 1930.
[33] G.N.Watson, *A Treatise on the Theory of Bessel Functions*, MacMillan, N.Y., 1948.
[34] E.T.Whitaker, G.N.Watson, *A Course of Modern Analysis*, Macmillan, N.Y., 1946.

7. ORDINARY DIFFERENTIAL EQUATIONS

Problem 77-16, *A First Order Nonlinear Differential Equation*, by I. RUBINSTEIN (Weizmann Institute of Science, Rehovot, Israel).

Solve the differential equation

$$\frac{dr}{dt} + t^{-1}\sqrt{r^2 + a^2} = b, \qquad t > 1,$$

where $r(1) = r_0$ and a, b are constants.

The equation arose in a problem of charge transfer by a dissociated water within a layer of a strong 1,1 valent electrolyte solution.

Solution by W. W. MEYER (General Motors Research Laboratories).

We consider the more general equation

(1)
$$\frac{dr}{dt} + \frac{1}{f}\frac{df}{dt}\sqrt{r^2 + a^2} = b\frac{df}{dt}$$

where f is a given function of t, nonvanishing and differentiable but otherwise arbitrary. We can assume that $f(1) = 1$ without loss of generality. Since (1) can be rewritten as

(2)
$$\sqrt{r^2 + a^2}\,\frac{d}{dt}\ln\left[(\sqrt{r^2 + a^2} + r)f/a\right] = b\frac{df}{dt},$$

we let

(3)
$$(\sqrt{r^2 + a^2} + r)f/a = z^{-1}.$$

Then

$$\sqrt{r^2 + a^2} = \frac{a}{2}(zf + z^{-1}f^{-1})$$

and (2) becomes

$$-\frac{a}{2}(zf + z^{-1}f^{-1})z^{-1}\frac{df}{dt} = b\frac{df}{dt}$$

or, if $b \neq 0$,

(4)
$$2f\frac{df}{dz} + \alpha(f^2 + z^{-2}) = 0$$

where $\alpha = a/b$. With f^2 as dependent variable, (4) is linear inhomogeneous; its solution is

(5)
$$f^2 = e^{\alpha(c-z)} + \alpha\int_z^c e^{\alpha(\zeta - z)}\zeta^{-2}\,d\zeta$$

where c is a constant, the value of z corresponding to $t = 1$. From (3) and the conditions $f(1) = 1$, $r(1) = r_0$ it follows that

(6) $$c = a/(\sqrt{r_0^2 + a^2} + r_0) = (\sqrt{r_0^2 + a^2} - r_0)/a.$$

Our solution is parametric. For any value of z we can determine an (f, r) pair, f from (5) and, subsequently, r from

(7) $$r = \frac{a}{2}(z^{-1}f^{-1} - zf).$$

If the value obtained for f is in the range of $f(t)$, then the inverse mapping $f \to t$ gives one or more (t, r) pairs. When $b = 0$, the (f, r) pairs come directly from (7) with $z = c$.

For the problem as stated, $f(t) = t$. Because of the restriction $t > 1$, we take the range of z to be $(0, c)$ if $\alpha > 0$ and (c, ∞) if $\alpha < 0$.

A Differential Equation

Problem 79-20*, by J. D. LOVE (Australian National University, Canberra, Australia).
Derive a bounded solution of the equation

$$\frac{dy}{dt} = y(\lambda t), \qquad y(0) = 1,$$

where λ is a constant > 1.

The proposer notes that he heard about the problem in New Zealand and gathers that the problem and solution originated in Queensland.

Solution by P. O. FREDERICKSON (Los Alamos Scientific Laboratory).

It is shown in [1] that this equation has real almost-periodic solutions for any $\lambda > 1$. Observe first that the Dirichlet series

$$y_\beta(t) = \sum_n c_n \exp(\beta\lambda^n t)$$

converges uniformly on the closed halfplane Re $(\beta t) \leq 0$, if $\beta \neq 0$ and

$$c_n = c_0\beta^{-n}\lambda^{-n(n+1)/2}.$$

Moreover, $y_\beta(0)$ is defined by a Laurent series which converges everywhere but $\beta = 0$, and therefore has only isolated zeros. Everywhere else we may choose $c_0 = c_0(\beta)$ so that $y_\beta(0) = 1$, as requested. If Re $(\beta) = 0$ the function $y_\beta(t)$ is almost-periodic on the real line, and hence bounded, while the function $(y_\beta + y_{-\beta})/2$ is real almost-periodic. Finally, it is only necessary to match terms in the series for $y'_\beta(t)$ and $y_\beta(\lambda t)$ to see that $y_\beta(t)$ is a solution.

REFERENCE

[1] P. O. FREDERICKSON, *Dirichlet series solutions for certain functional differential equations*, in the proceedings of the Japan-United States seminar on ordinary differential and functional equations, Lecture Notes in Mathematics 243, Springer-Verlag, New York, 1971.

Editorial note. A.J.E.M. Janssen and W.A.J. Luxemburg (California Institute of Technology) constructed integral representations for a general class of solutions all of which, they showed, were asymptotic to the formal series

$$\sum_{n=0}^{\infty} \frac{\lambda^{n(n-1)/2}t^n}{n!}.$$

Comment by L. Fox (Oxford University Computing Laboratory, Oxford, England). The equation $dy/dt = y(\lambda t)$, in a more general vector form $dy/dt = Ay(\lambda t) + By(t)$, occurred in the formulation of a problem submitted in 1970 to the Oxford University–Industry Study Group. A discussion of the formulation was given in the paper

> J. R. OCKENDON AND A. B. TAYLER. *The dynamics of a current collection system for an electric locomotive.* Proc. Roy. Soc., A322 (1971), pp. 447–468.

Analytical and numerical solutions for both $\lambda < 1$ and $\lambda > 1$ appear in

> L. FOX, D. F. MAYERS, J. R. OCKENDON AND A. B. TAYLER. *On a functional differential equation.* J. Inst. Math. Applics., 8 (1971), pp. 271–307,

and a detailed theoretical treatment is given in

> T. KATO AND J. B. MCLEOD, *The functional-differential equation $y'(x) = ay(\lambda x) + by(x)$.* Bull. Amer. Math. Soc., 77 (1971), pp. 891–937.

P. DIAMOND AND V. G. HART (University of Queensland, Queensland, Australia) noted that a bounded solution was given by E. W. BOWEN (Armidale, New South Wales) and also refer to the latter two references of Fox.

Problem 72-8, A Nonlinear Differential Equation, by O. G. RUEHR (Michigan Technological University).

If $x(t)$ is analytic at $t = 0$ and satisfies

$$\dot{x} \exp\{t\dot{x}^2\} = 1, \qquad x(0) = 0, \qquad \dot{x} = dx/dt,$$

determine its power series expansion. The problem arose as an approximation to a nonlinear heat flow equation.

Solution by C. C. ROUSSEAU (Memphis State University).

Clearly, if $\dot{x}(t)$ satisfies $\dot{x} \exp\{t\dot{x}^2\} = 1$, then $\dot{x}(0) = 1$ and for all t, $x(t) > 0$. Taking the logarithm of both sides and letting $\dot{x}(t) = \exp\{u(t)\}$, we obtain

(1) $$u = -t \exp\{2u\}.$$

If $|t| < 1/(2e)$ and $C = \{u|\ |u| = 1/2\}$, then for every point on C, $|t \exp\{2u\}| < |u|$. Thus, by Rouché's theorem, we insure that (1), viewed as an equation in u, has a unique root in the interior of C. Then, by Lagrange's theorem (cf. [E. T. WHITTAKER and G. N. WATSON, *A Course in Modern Analysis,* 4th ed., Cambridge Univ. Press, Cambridge, 1927, § 7.32]), we obtain the power series expansion

(2) $$\dot{x}(t) = 1 + \sum_{n=1}^{\infty} (-1)^n[(2n + 1)^{n-1}/n!]t^n.$$

Integrating term-by-term and using the initial condition $x(0) = 0$, we obtain

(3) $$x(t) = \sum_{n=0}^{\infty} (-1)^n[(2n + 1)^{n-1}/(n + 1)!]t^{n+1}.$$

The final series converges for $|t| < 1/(2e)$, corresponding to the fact that $t = -1/(2e)$ is a branch point of (1).

Editorial note. J. BENDALL (Royal Armament Research and Development Establishment, Fort Halstead, Seven Oaks, Kent, England) gives a solution which came directly out of a computer as a tape which is printed out on a specially constructed typewriter-punch in the form shown below. The program, MIRMADE and the language MIRFAC are due to H. J. GAWLIK and F. J. BERRY of the same establishment and it enables one to program his problem in standard notation. (Interested persons may contact H. J. GAWLIK.) The given differential equation is replaced by the pair of first order equations

$$\dot{y} = -y^3 z, \qquad \dot{z} = 2y^2 z^2 + 4ty^4 z^3,$$

where

$$y = \dot{x}, \qquad 1/z = 1 + 2ty^2 \quad \text{and} \quad y = 1, \quad z = 1 \quad \text{at } t = 0.$$

Up to 15 terms of the series were requested and product overflow occurred in computing the term in t^9.

Printout of MIRMADE data tape.

variables
t, y
initial
$y = 1, \quad z = 1$
terms
15
equations
$y' = -y^3 z$
$z' = -2y^2 z^2 + 4 + y^4 z^3$
Solve.
Solution: $y = 1 - t + 5t^2/2! - 49t^3/3! - 729t^4/4! - 14641t^5/5!$
$\qquad\qquad + 371293t^6/6! - 11390625t^7/7! + 410338673t^8/8!$

(For accent, the terms $t^n/n!$ are printed in red.)

A computer solution is also given for the more general equation $\dot{x} \exp(t\dot{x}^p) = 1$ and which is guessed correctly as

$$x = \sum_{n=0}^{\infty} (-1)^n (pn + 1)^{n-1} t^{n+1}/(n+1)!.$$

Incidentally, a parametric solution for the latter differential equation can be gotten as follows: Let $\dot{x} = e^{-\lambda}$ giving $t = \lambda e^{p\lambda}$. Then,

$$dt = (1 + p\lambda) e^{p\lambda} d\lambda,$$

$$dx = (1 + p\lambda) e^{(p-1)\lambda} d\lambda.$$

Integrating, we obtain

$$x = \{1 + (p\lambda - 1) e^{(p-1)\lambda}\}/(p-1).$$

The proposer obtains the latter parametric solution for $p = 2$ and then obtains the power series expansion from it. [M.S.K.]

A Nonlinear Second Order Differential Equation

Problem 81-17*, by E. Y. RODIN (Washington University).

Find the general solution of the differential equation

$$\frac{d^2y}{dx^2} + x^2 e^y = 0, \qquad x > 0.$$

The problem arose in a study of the nonisothermal flow of a Newtonian fluid between parallel plates.

Editorial note. The equation can at least be reduced to a first order one. [M.S.K.]

The Comments by H. E. FETTIS (Mountain View, CA) *and* G. WANNER (Université de Genève, Genève, Switzerland) are essentially the same.

The change of variables $x^2 = \xi$, $y(x) = \eta(\xi)$ transforms the given equation into

(1)
$$\xi\eta'' + \frac{\eta'}{2} + \frac{\xi e^\eta}{4} = 0,$$

which is a special case of the equation

(2)
$$\xi\eta'' + a\eta' + b\xi e^\eta = 0$$

and is treated in some detail in Kamke, *Differentialgleichung,* Vol. 1 (Chelsea, New York, 1948 & 1971, #6.76). According to the discussion there, the general solution is only known in closed form when $a = 0$ and $a = 1$. For the latter case, cf. problem #79.11. The case $a = 2$ is known as "Emden's equation"; see R. Emden, *Gaskugeln,* (Leipzig & Berlin, 1907), and H. Lemke (J. für Math., 152 (1913), pp. 118–137).

A particular solution (but not the general one) when $b(a - 1) > 0$ is

$$\eta = \ln \frac{2(a - 1)}{b\xi^2}.$$

As indicated in the Kamke book, (1) can be transformed by $u(t) = \xi\eta'$, $t = \xi^2 e^\eta$ into

(3)
$$t(2 + u)u' - \tfrac{1}{2}u + \tfrac{1}{4}t = 0 \qquad \text{(eq. 1.237).}$$

A further transformation $v(u) = (-u/2 + t/4)^{-1}$ yields

(4)
$$v' = u\left(\frac{u}{2} + 1\right)v^3 + \left(u + \frac{5}{2}\right)v^2,$$

which is an Abel differential equation.

A Nonlinear Differential Equation

Problem 84-3, by M. A. ABDELKADER (Alexandria, Egypt).

The general solution of the differential equation

(1)
$$v'' + x^2 e^v = 0, \qquad x > 0$$

of Problem 81-17 (October, 1981, p. 524) apparently could not be found (see October, 1982, pp. 480-481). Find that of

(2)
$$w'' + x^2 e^w = 2x^{-2}, \qquad x > 0,$$

which might approximate that of (1) for large values of x.

Solution by A. D. OSBORNE (University of Keele, England).

The equation

(1)
$$w'' + x^n e^w = nx^{-2}, \qquad x > 0,$$

for any number n, is as easy to solve as the equation proposed. Letting $v(x) = -x^n e^w$,

(1) reduces to

(2) $$vv'' = v'^2 + v^3 \Rightarrow v'^2 = 2v^3 + kv^2$$

where k is an arbitrary constant. The solution of (2) is obtained by direct integration, and the solution of (1) is

$$e^w = \begin{cases} \dfrac{k}{2x^n} \operatorname{sech}^2 \left(\dfrac{k^{1/2}}{2} x + K \right), & k \neq 0, \\[3mm] \dfrac{-4}{x^n (K - \sqrt{2}\, x)^2}, & \end{cases}$$

where K is another constant. This method of solution can be generalised to equations of the type,

(3) $$w'' + f(x)e^w = g(x).$$

It is easily shown that (3) reduces to (2) under the transformation $v(x) = -f(x)e^w$ if and only if $f^2 g = f'^2 - ff''$.

Editorial note. Both Habashy and Oser note that w is not a good approximation to v.

A Nonlinear Differential Equation

Problem 79-11, by D. K. Ross (La Trobe University, Victoria, Australia).
    Find the general solution of the ordinary nonlinear differential equation

$$\frac{1}{x} \frac{d}{dx} \left(x \frac{dy}{dx} \right) = e^{-\varepsilon y} \quad \text{with } x > 0,$$

and where $\varepsilon = 1$ or -1.
    Problems of this kind arise in the study of the behavior of an electro-chemical flow in a packed cylindrical bed electrode; see, for example, the work of Ayre [1].

REFERENCE

[1] P. J. AYRE, *The performance of packed bed electrodes*, Ph.D. thesis, University of Newcastle-upon-Tyne, 1973.

    Editorial Comment: Many of the solutions noted that there was no need to restrict ε to ± 1. A smaller subset of solvers (indicated by an asterisk after their names) solved the more general equation

(1) $$\Psi_{xx} + \frac{1}{x} \Psi_x + Kx^n e^{\Psi} = 0.$$

    D. A. KEARSLEY* (National Bureau of Standards) refers to his paper (J. Res. Nat. Bur. Standards, 67B (1963), pp. 245–247), which contains the following solutions for

$x > 0$:

$$\Psi = \log\left\{\frac{2m^2}{Kx^{n+2}\cosh^2(m\log x + a)}\right\}, \quad K > 0,$$

$$\Psi = \log\left\{\frac{-2m^2}{Kx^{n+2}\sinh^2(m\log x + a)}\right\}, \quad K < 0,$$

$$\Psi = \log\left\{\frac{-2m^2}{Kx^{n+2}\sin^2(m\log x + a)}\right\}, \quad K < 0,$$

$$\Psi = \log\left\{\frac{-2}{Kx^{n+2}(\log x + a)^2}\right\}, \quad K < 0,$$

$$\Psi = m\log x + a, \quad K = 0.$$

He also notes that (1) arises in the steady flow of viscous fluids with a viscosity depending exponentially on temperature and in calculating the temperature distribution in a dielectric in an alternating field.

Equation (1) can be reduced to

$$\frac{d^2r}{ds^2} + K e^r = 0$$

by means of the substitutions $x = e^s$, $\Psi = r - (n+2)s$ and then can be solved by quadratures.

W. N. SELANDER (Chalk River Nuclear Laboratories) also refers to the paper of Kearsley, and W. SQUIRE (West Virginia University) refers to a paper of H. Lemke (J. Reine Angew. Math., 142 (1913), pp. 118–138) on the density distribution of gaseous bodies.

Problem 73-12, *A Nonlinear Differential Equation*, by OTTO G. RUEHR (Michigan Technological University).

Determine the general solution of

$$\left\{\frac{d^2f}{dx^2} + 2\right\}f = \left\{\frac{df}{dx} + 2x\right\}\left\{\frac{df}{dx} + x\right\}.$$

The problem arose in modeling the Helmholtz equation in two dimensions.

Solution by the proposer.

Assuming that $f' + 2x \neq 0$, we have

$$\frac{f'' + 2}{f' + 2x} = \frac{f'}{f} + \frac{x}{f}.$$

Integration yields

$$\log|f' + 2x| = \log|f| + \log|c_1| + \int \frac{x}{f}\,dx,$$

and hence

$$\frac{f'}{f} + \frac{2x}{f} = c_1 \exp\left\{\int \frac{x}{f}\,dx\right\}.$$

Letting $x/f = u''/u'$ and integrating, we obtain

$$f(u')^2 = c_2 \exp\{c_1 u\}.$$

Substituting $f = xu'/u''$ and noting the identity

$$(d^2u/dx^2)/(du/dx)^3 = -d^2x/du^2,$$

we obtain the linear differential equation

$$\frac{d^2x}{du^2} + \frac{\exp\{-c_1 u\}}{c_2} x = 0.$$

Let

$$\frac{-2\exp\{-c_1 u/2\}}{c_1\sqrt{c_2}} = t + c,$$

where c is a constant. Then

$$f = c_2 \exp\{c_1 u\}(dx/du)^2 = (dx/dt)^2,$$

and the equation for x becomes

$$\frac{d^2x}{dt^2} + \frac{1}{t+c}\frac{dx}{dt} + x = 0.$$

Hence, the general solution in parametric form is

$$x = AJ_0(t + c) + BY_0(t + c),$$

$$f = [AJ_1(t + c) + BY_1(t + c)]^2,$$

where A and B are arbitrary constants. Note that the limit $c \to \infty$ corresponds to the solution $f(x) = C - x^2$.

Editorial note. It is instructive to consider the initial value problem for this differential equation. Letting $t = 0$ correspond to the initial point $(x_0, f(x_0), f'(x_0))$, one finds that c is given by

$$c = \frac{2\sqrt{f(x_0)}}{f'(x_0) + 2x_0},$$

and then A and B are determined by the equations

$$AJ_0(c) + BY_0(c) = x_0,$$

$$AJ_1(c) + BY_1(c) = \sqrt{f(x_0)}.$$

The solution holds in an interval $[x_1, x_2]$ containing x_0, where x_1 and x_2 are determined by the condition $dx/dt = 0$. [C.C.R.]

Problem 64–10, A Boundary Value Problem, by M. S. KLAMKIN (Ford Scientific Laboratory).

The Thomas-Fermi equation

(1)
$$\frac{d^2y}{dx^2} = \sqrt{y^3/x},$$

subject to the boundary conditions

$$y(0) = 1, \qquad y(\infty) = 0,$$

arises in the problem of determining the effective nuclear charge in heavy atoms [1]. Transform this boundary-value problem into an initial value one.

Solution by the proposer.

If we let $x = 1/t$, then (1) is transformed into

$$\{t^4 D^2 + 2t^3 D\}y = \sqrt{y^3 t},$$

subject to the boundary conditions

$$y(0) = 0, \qquad y(\infty) = 1.$$

Now let $y'(0) = \lambda$ (to be determined) and [2]

(2) $$y(t) = \lambda^{3/2} F(\lambda^{-1/2} t).$$

It follows that $F(x)$ satisfies the initial value problem

$$\{x^4 D^2 + 2x^3 D\}F = \sqrt{F^3 x},$$

$$F(0) = 0, \qquad F'(0) = 1.$$

Then by letting $t \to \infty$ in (2), we have

$$\lambda = F(\infty)^{-2/3}.$$

Consequently, the initial boundary value problem has been transformed into two similar initial-value problems. This avoids interpolation techniques for numerically determining λ.

REFERENCES

[1] H. T. DAVIS, *Introduction to Nonlinear Differential and Integral Equations*, U. S. Atomic Energy Commission, 1960, pp. 405–407.
[2] M. S. KLAMKIN, *On the transformation of a class of boundary value problems into initial value problems for ordinary differential equations*, this Journal, 4 (1962), pp. 43–47.

Also solved by Newman Fisher (San Francisco State College) using variational methods. He shows that $-\frac{6}{7}y'(0)$ is given by the stationary value of

$$I[y] = \int_0^\infty \left\{ (y')^2 + \frac{4}{5} \sqrt{y^5/x} \right\} dx.$$

A Linear Differential Equation

Problem 83-10, *by* J. GRUENDLER (North Carolina A&T State University).

Find the general solution to

$$\frac{d^2 x}{dt^2} + 2c \frac{dx}{dt} - (1 - 2 \operatorname{sech}^2 t)x = 0.$$

This equation arose as one uncoupled portion of the variational equation along a homoclinic orbit for a dynamical system in R^4.

Solution by C. GEORGHIOU (University of Patras, Patras, Greece).

(a) Let $c = 0$. Then $x_1 = \operatorname{sech} t$ is a solution of the given ODE. A second (linearly independent) solution is readily found to be $x_2 = t \operatorname{sech} t + \sinh t$, and the general solution in this case is

$$x = A \operatorname{sech} t + B(t \operatorname{sech} t + \sinh t).$$

(b) When $c \neq 0$, the substitution $x = y e^{-ct}$ gives

(1) $$\frac{d^2y}{dt^2} - (1 + c^2 - 2 \operatorname{sech}^2 t)y = 0.$$

The solution of (1) is found by using the following proposition due to Darboux [1] and cited in [2, p. 132].

PROPOSITION. *Let $u = u(t)$ be the general solution of the equation*

(2) $$\frac{d^2u}{dt^2} = [g(t) + h]u$$

and let $f = f(t)$ be any particular solution of (2) for the particular value $h = h_1$. Then the general solution of the equation

(3) $$\frac{d^2y}{dt^2} = \left[f(t) \frac{d^2}{dt^2} \left(\frac{1}{f(t)} \right) + h - h_1 \right] y$$

for $h \neq h_1$ is

(4) $$y = u'(t) - u(t) \frac{f'(t)}{f(t)}.$$

By taking $g(t) \equiv 0$, $h_1 = 1$, $f(t) = \cosh t$ and $h = 1 + c^2 \equiv b^2$, (3) becomes identical to (1), $u = Ae^{bt} - Be^{-bt}$ and the general solution to the original equation is by (4),

$$x = Ae^{(b-c)t}[b - \tanh t] + Be^{-(b+c)t}[b + \tanh t].$$

REFERENCES

[1] G. DARBOUX, C. R. Acad. Sci. Paris, 94 (1882), p. 1456.
[2] E. L. INCE, *Ordinary Differential Equations,* Dover, New York, 1956.

Problem 65-4, Duality in Differential Equations, by MOSTAFA A. ABDELKADER (Giza, Egypt).

(A) Show that the solutions of the differential equations

(1) $$\xi'' + \sigma \xi'^n = 0 \qquad\qquad \left(\xi' \equiv \frac{d\xi}{d\sigma} \right)$$

and

(2) $$\xi \xi'' + \theta \xi'^2 + \sigma = 0$$

lead to the solutions of the equations

$$\xi'' + \sigma \xi'^m = 0 \quad \text{and} \quad \xi \xi'' + \lambda \xi'^2 + \sigma = 0,$$

where

$$m = \frac{-5 - 3n}{3 + n} \quad \text{and} \quad \lambda = \frac{-\theta}{1 + 3\theta}.$$

(B) For the values $n = -1$ and $\theta = 0$ (which are the roots of $m = n$ and $\lambda = \theta$), (1) and (2) reduce to

(3) $\xi\xi'' + \sigma = 0.$

Show that (3) provides a connection between the Langmuir-Blodgett space-charge equation for cylinders,

(4) $$\frac{d^2y}{d\xi^2} + \frac{1}{\xi}\frac{dy}{d\xi} - \frac{1}{\xi\sqrt{y}} = 0,$$

and the Blasius equation of hydrodynamics,

(5) $$\frac{d^3f}{d\eta^3} + f\frac{d^2f}{d\eta^2} = 0.$$

   Solution by the proposer.
   (A) Differentiating (1) and substituting from (1) yields

(6) $\xi\xi''' - n\xi'\xi'' + \xi^{n+1} = 0.$

Making the transformation

$$x = \xi'\xi^{-(n+2)/3}, \qquad y = x\xi\frac{dx}{d\xi},$$

(6) is transformed (after considerable manipulation) to the first-order equation

(7) $$\frac{dy}{dx} = -\frac{(n+5)x}{3} + \frac{1}{9y}\{(n+2)(n-1)x^3 - 9\}.$$

The value of $n = -5$ makes (7) separable (it is also a root of $m = n$). Thus we can exclude it and make the substitutions

$$x = -\left(\frac{3}{n+5}\right)^{2/3} X, \qquad y = -\left(\frac{3}{n+5}\right)^{1/3} Y,$$

which changes (7) to

(8) $$\frac{dY}{dX} = X + \frac{1}{Y}\{g_1(n)X^3 + 1\},$$

where

$$g_1(n) = \frac{(n+2)(n-1)}{(n+5)^2}.$$

We now need the following lemma:
   If

$$g(n) = \frac{a + bn + cn^2}{\alpha + \beta n + \gamma n^2}, \qquad (g(n) \neq \text{const.}),$$

and

$$m = \frac{\begin{vmatrix} a & b \\ \alpha & \beta \end{vmatrix} - \begin{vmatrix} c & a \\ \gamma & \alpha \end{vmatrix} n}{\begin{vmatrix} c & a \\ \gamma & \alpha \end{vmatrix} + \begin{vmatrix} c & b \\ \gamma & \beta \end{vmatrix} n},$$

then

$$g(m) \equiv g(n).$$

It now follows that $g_1(m) \equiv g_1(n)$ when $m = -(5 + 3n)/(3 + n)$. Thus the solution of (8) leads to that of (1), and conversely, with n replaced by m.

Using the above, one can immediately integrate (1) for the (n, m) values of $(-5, -5)$, $(0, -\frac{5}{3})$ and $(1, -2)$. Also can one find solutions for values corresponding to $(-\frac{1}{2}, -\frac{7}{5})$ and $(-4, -7)$.

Differentiating (2) yields

$$(9) \qquad\qquad \xi\xi''' + (1 + 2\theta)\xi'\xi'' + 1 = 0.$$

Making the transformation

$$x = \xi'\xi^{-1/3}, \qquad y = x\xi \frac{dx}{d\xi},$$

(9) is transformed into

$$(10) \qquad\qquad \frac{dy}{dx} = -(2\theta + \tfrac{4}{3})x - \frac{1}{9y}\{(6\theta + 2)x^3 + 9\}.$$

Since the case $2\theta + \tfrac{4}{3} = 0$ makes (10) separable, we can exclude it and make the substitution

$$x = (2\theta + \tfrac{4}{3})^{-2/3}X, \qquad y = (2\theta + \tfrac{4}{3})^{-1/3}Y,$$

which changes (10) to

$$(11) \qquad\qquad \frac{dY}{dX} = -X - \frac{1}{Y}\{g_2(\theta)X^3 + 1\},$$

where

$$g_2(\theta) = \frac{1 + 3\theta}{2(2 + 3\theta)^2}.$$

It follows from the lemma that $g_2(\lambda) \equiv g_2(\theta)$ when $\lambda = -\theta/(1 + 3\theta)$.

(B) In (4) let $\xi \equiv \xi(s)$ where s is a new variable such that

$$y^{1/2} = -2^{-1/3}\frac{d\xi}{ds}.$$

We then get

$$(11) \qquad\qquad \xi\xi_3 + \xi_1\xi_2 + 1 = 0, \quad \text{where} \quad \xi_r = \frac{d^r\xi}{ds^r}.$$

Integrating (11) yields

$$\xi\xi_2 + s = c \quad (\text{const.}),$$

or

$$\xi\xi'' + \sigma = 0,$$

where

$$\sigma = s - c.$$

For the Blasius equation (5), we set

$$\frac{df}{d\eta} = \sigma \quad \text{and} \quad \frac{d^2f}{d\eta^2} = \xi$$

and obtain

(12) $$\xi' + f = 0.$$

Differentiating (12), we obtain

$$\xi'' + \frac{df}{d\sigma} = 0 = \xi\xi'' + \sigma.$$

Editorial note. Since (2) is immediately integrable for $\theta = 1$, it is also integrable for $\theta = -\frac{1}{4}$.

It would be of interest to find a general class of differential equations, say, $F(y'', y', y, x, \lambda) = 0$, where λ is a parameter, and a class of transformations which take the differential equation into the dual equation $F(\xi'', \xi', \xi, \sigma, \lambda') = 0$ of the same form.

Problem 60-4*, Vorticity Interaction, by SIN-I CHENG (Princeton University)

The shock wave over the blunt nose of a slender body in a hypersonic stream leaves a large vorticity in the downstream flow field. This vorticity interacts with the displacement flow of the boundary layer over the body to produce an induced pressure gradient. For hypersonic flow at very high altitudes, this self-induced pressure gradient along the body surface is so large as to govern the development of the boundary layer itself. This is distinctly different from ordinary boundary layers for which the pressure gradient acting on the boundary layer is known and essentially is independent of the downstream development of the boundary layer.

The mathematical analysis of one such an interacting boundary layer leads to the following proposed problem:

Determine α and L such that

$$\lim_{x\to\infty} F'(x) = L,$$

where $F(x)$ satisfies the differential equation

(1) $$\{D^3 + x^2D^2 - xD + 1\}F(x) = \alpha,$$

and boundary conditions

$$F(0) = 0,$$
$$F'(0) = 0,$$
$$F''(0) = 1.$$

Solution by Yudell L. Luke (Midwest Research Institute, Kansas City, Missouri*).

It is easy to see that $F(x) = \alpha$ is the particular solution of (1) and that $F(x) = x$ is a complementary solution. By the usual power series approach, the other two complementary solutions of (1) may be expressed in hypergeometric

---

* This solution was obtained using results of work supported by the Applied Mathematics Laboratory of the David W. Taylor Model Basin.

form [1] as

(2)
$$F_1(x) = {}_2F_2(-\tfrac{1}{3}, -\tfrac{1}{3}; \tfrac{1}{3}, \tfrac{2}{3}; -\xi),$$

(3)
$$F_2(x) = \tfrac{1}{2}x^2\, {}_2F_2(\tfrac{1}{3}, \tfrac{1}{3}; \tfrac{4}{3}, \tfrac{5}{3}; -\xi).\,\xi = \frac{x^3}{3}.$$

Using the boundary conditions at the origin, it follows that

(4)
$$F(x) = \alpha - \alpha F_1(x) + F_2(x).$$

To determine the behavior of $F(x)$ for x large, we appeal to the asymptotic theory of hypergeometric functions. For the case at hand, we follow Meijer [2] who shows that

(5)
$$\frac{\Gamma(a_1)\Gamma(a_2)}{\Gamma(1+b_1)\Gamma(1+b_2)}\, {}_2F_2\left(\begin{matrix} a_1, a_2 \\ 1+b_1, 1+b_2 \end{matrix}; -z\right) \frown L_{2,2}(z),$$

$$|z| \to \infty, |\arg z| < \frac{\pi}{2}$$

where

(6)
$$L_{2,2}(z) = \frac{z^{-a_1}\Gamma(a_1)\Gamma(a_2-a_1)}{\Gamma(1+b_1-a_1)\Gamma(1+b_2-a_1)}\, {}_3F_1\left(\begin{matrix} a_1, a_1-b_1, a_1-b_2 \\ 1+a_1-a_2 \end{matrix}; z^{-1}\right)$$

$$+ \text{ a like expression with } a_1 \text{ and } a_2 \text{ interchanged.}$$

Note that for both (2) and (3) $a_1 = a_2 = b_1$ and the expression for $L_{2,2}(z)$ must be found by a limiting process. By L'Hospital's rule,

(7)
$$L_{2,2}(z) = \frac{z^{-1-a_1}\Gamma(a_1+1)}{\Gamma(b_2-a_1)}\, {}_4F_2\left(\begin{matrix} 1, 1, a_1+1, a_1+1-b_2 \\ 2, 2 \end{matrix}; z^{-1}\right)$$

$$+ \frac{z^{-a_1}\Gamma(a_1)}{\Gamma(1+b_2-a_1)}\{\ln z - \psi(a_1) - \psi(1+b_2-a_1) + \psi(1)\},$$

where $\psi(z)$ is the logarithmic derivative of the gamma function.

As a remark aside, (7) is not given by Meijer. In some unpublished notes, we study a more general function notated $L_{p,q}(z)$ for the case where two numerator parameters of a given ${}_pF_q$ differ by an integer or zero.

Collecting (2–7), we get

(8)
$$F(x) \frown \alpha + \frac{\alpha}{27}\frac{\Gamma(\tfrac{1}{3})}{\Gamma(\tfrac{2}{3})}\xi^{-2/3}G(x) + \frac{3^{-1/3}}{9}\frac{\Gamma(\tfrac{2}{3})}{\Gamma(\tfrac{1}{3})}\xi^{-2/3}G(x)$$

$$+ \frac{\alpha}{3}\frac{\Gamma(\tfrac{1}{3})}{\Gamma(\tfrac{2}{3})}\xi^{1/3}\left[\ln \xi - \psi\left(-\frac{1}{3}\right) - \psi\left(\frac{2}{3}\right) + \psi(1)\right]$$

$$+ 3^{-1/3}\frac{\Gamma(\tfrac{2}{3})}{\Gamma(\tfrac{1}{3})}\xi^{1/3}\left[\ln \xi - \psi\left(\frac{1}{3}\right) - \psi\left(\frac{4}{3}\right) + \psi(1)\right],$$

$$|\xi| \to \infty, \qquad |\arg \xi| < \frac{\pi}{2},$$

where

(9)
$$G(x) = {}_4F_2\left(\begin{matrix} 1, 1, \tfrac{2}{3}, \tfrac{4}{3} \\ 2, 2 \end{matrix}; \xi^{-1}\right).$$

If $\lim_{x \to \infty} F'(x) = L$, a constant, then the coefficient of $\ln \xi$ in (8) must vanish. This determines

$$
(10) \qquad\qquad \alpha = -\left\{ \frac{3^{1/3}\Gamma(\frac{2}{3})}{\Gamma(\frac{1}{3})} \right\}^2
$$

A straightforward calculation gives

$$
(11) \qquad L = \frac{3^{-2/3}\Gamma(\frac{2}{3})}{\Gamma(\frac{1}{3})} \left[3 + 2\psi\left(\frac{5}{3}\right) - 2\psi\left(\frac{4}{3}\right) \right] = 0.88152.
$$

REFERENCES

1. A. ERDÉLYI, et al., *Higher Transcendental Functions*, v. 1, McGraw-Hill Book Company, Inc., 1953.
2. C. S. MEIJER, *"On the G-Function, VIII,"* Nederlandsche Akademie Van Wetenschappen, Proceedings, pp. 1165–1175, 1946.

Editorial note. An equivalent expression for L is $\{\Gamma(\frac{2}{3}/3^{1/3})\}^2$. J. Ernest Wilkins notes the following after his solution:

These results are consistent with those asserted by Lu Ting, *Boundary Layer over a Flat Plate in Presence of Shear Flow.* The Physics of Fluids, vol 3 (1960), pp 78–81, who claims that, for the solution $f(\eta)$ of the system

$$
3f''' + \eta^2 f'' - \eta f' + f = -\beta, f(0) = f'(0) = 0,
$$

$f'(\infty) = 1, f''(0) = 0.7866, \beta = 0.8695.$

If we set $f(\eta) = 3^{1/3}L^{-1}F(x)$, $x = 3^{-1/3}\eta$, $\beta = -3^{1/3}L^{-1}\alpha$, then $F(x)$ is the solution of the system discussed above and so we would get $\beta = 3^{5/3}/\Gamma^2(\frac{1}{3}) = 0.8695, f''(0) = 3^{-1/3}L = 3^{1/3}/\Gamma^2(\frac{2}{3}) = 0.7866.$

Problem 65-2, *A Third Order Differential Equation*, by DONALD E. AMOS (University of Missouri).

The differential equation

$$
[D^3 + t^2 D + 3t]y = 0
$$

arises in a problem describing the motion of a particle in a magnetic field.
  (1) Identify the power series solutions in terms of special functions,
  (2) evaluate the associated integral

$$
\int_0^t xy(x)\, dx,
$$

and
  (3) find asymptotic expressions for large t in (1) and (2).

Solution by SIDNEY SPITAL (California State Polytechnic College).
  The power series solution of the differential equation is straightforward and yields

$$
y(t) = \sum_{n=0}^{\infty} \left\{ A' \frac{(-1)^n}{(2n)!} \left(\frac{t^2}{2}\right)^{2n} + B' \frac{(-1)^n t^{4n+1}}{(4n+1)!} + C' \frac{(-1)^n}{(2n+1)!} \left(\frac{t^2}{2}\right)^{2n+1} \right\}.
$$

In terms of the Fresnel integrals[†]

$$C\left(\frac{t}{\sqrt{\pi}}\right) = \frac{1}{\sqrt{\pi}} \int_0^t \cos\left(\frac{x^2}{2}\right) dx, \quad S\left(\frac{t}{\sqrt{\pi}}\right) = \frac{1}{\sqrt{\pi}} \int_0^t \sin\left(\frac{x^2}{2}\right) dx,$$

$$y(t) = A' \cos\frac{t^2}{2} + B' \sin\frac{t^2}{2} + \sqrt{\pi} C'\left\{\cos\left(\frac{t^2}{2}\right) C\left(\frac{t}{\sqrt{\pi}}\right) + \sin\left(\frac{t^2}{2}\right) S\left(\frac{t}{\sqrt{\pi}}\right)\right\}.$$

y can be put in more compact form by means of the Fresnel auxiliary functions[†]

$$f\left(\frac{t}{\sqrt{\pi}}\right) = \left[\frac{1}{2} - S\left(\frac{t}{\sqrt{\pi}}\right)\right] \cos\frac{t^2}{2} - \left[\frac{1}{2} - C\left(\frac{t}{\sqrt{\pi}}\right)\right] \sin\frac{t^2}{2},$$

$$g\left(\frac{t}{\sqrt{\pi}}\right) = \left[\frac{1}{2} - C\left(\frac{t}{\sqrt{\pi}}\right)\right] \cos\frac{t^2}{2} + \left[\frac{1}{2} - S\left(\frac{t}{\sqrt{\pi}}\right)\right] \sin\frac{t^2}{2}.$$

Thus,

$$y(t) = A \cos\frac{t^2}{2} + B \sin\frac{t^2}{2} + Cg\left(\frac{t}{\sqrt{\pi}}\right).$$

Since

$$\int_0^t xg\left(\frac{x}{\sqrt{\pi}}\right) dx = \frac{1}{2} - f\left(\frac{t}{\sqrt{\pi}}\right),$$

it follows that

$$\int_0^t xy(x) \, dx = A \sin\frac{t^2}{2} + B\left[1 - \cos\frac{t^2}{2}\right] + C\left[\frac{1}{2} - f\left(\frac{t}{\sqrt{\pi}}\right)\right].$$

Since also,[†]

$$f\left(\frac{t}{\sqrt{\pi}}\right) \sim \frac{1}{t\sqrt{\pi}} + O\left(\frac{1}{t^5}\right),$$

$$g\left(\frac{t}{\sqrt{\pi}}\right) \sim \frac{1}{t^3\sqrt{\pi}} + O\left(\frac{1}{t^7}\right),$$

we finally have

$$y(t) \sim A \cos\frac{t^2}{2} + B \sin\frac{t^2}{2} + \frac{C}{t^3\sqrt{\pi}} + O\left(\frac{1}{t^7}\right),$$

$$\int_0^t xy(x) \, dx \sim A \sin\frac{t^2}{2} + B\left(1 - \cos\frac{t^2}{2}\right) + C\left(\frac{1}{2} - \frac{1}{t\sqrt{\pi}}\right) + O\left(\frac{1}{t^5}\right).$$

D. L. LANSING (NASA, Langley Research Center) reduces the differential equation to an inhomogeneous Bessel equation and obtains the complete asymptotic expansions

$$y(t) \sim A \cos\frac{t^2}{2} + B \sin\frac{t^2}{2} + \frac{2C}{\pi t^3} \sum_{n=0}^{\infty} \Gamma\left(2n + \frac{3}{2}\right)\left(\frac{-4}{t^4}\right)^n,$$

[†] W. GAUTSCHI, *Error Function and Fresnel Integrals*, Handbook of Mathematic Functions, National Bureau of Standards, Washington, D. C., 1964, pp. 300–302.

$$\int_0^t xy(x)\,dx \sim A \sin \frac{t^2}{2} + B\left(1 - \cos \frac{t^2}{2}\right)$$

$$+ C\left\{\frac{1}{2} - \frac{1}{\pi t}\sum_{n=0}^{\infty} \Gamma\left(2n + \frac{1}{2}\right)\left(\frac{-4}{t^4}\right)^n\right\}.$$

H. E. FETTIS (Wright-Patterson AFB, Ohio) first reduces the differential equation to

$$2xu' + 3u = 0,$$

where $x = t^2/2$, $u = y'' + y$. This then leads to the solution as given before and the complete asymptotic expansions.

Problem 76-6, *An n-th Order Linear Differential Equation*, by M. S. KLAMKIN (University of Waterloo).

Solve the differential equation

$$[x^{2n}(D - a/x)^n - k^n]y = 0.$$

Editorial note. Most of the solutions reduced the equation simply to $[x^{2n}D^n - k^n]u = 0$ and then referred to Kamke's *Differentialgleichungen*. N. Ortner also solved the dual equation $[x^n(D - a/x)^{2n} - k^n]y = 0$ in a similar fashion. More generally, it is just as easy to solve the pair of equations

$$[x^{2n}(D + \phi'(x))^n - k^n]y = 0, \qquad [x^n(D + \phi'(x))^{2n} - k^n]y = 0,$$

for by the exponential shift theorem, they reduce to

$$[x^{2n}D^n - k^n]y\,e^\phi = 0, \qquad [x^nD^{2n} - k^n]y\,e^\phi = 0.$$

The latter pair can be solved in terms of solutions of first order equations by using the known dual operational identities

$$x^{2n}D^n \equiv [x^2D + (1-n)x]^n, \qquad x^nD^{2n} \equiv [xD^2 + (1-n)D]^n. \quad [\text{M.S.K.}]$$

A Nonlinear System of Differential Equations

Problem 80-18, *by* G. N. LEWIS *and* O. G. RUEHR (Michigan Technological University).

Find the general solution of the following system of differential equations:

$$\frac{dY_1}{dt} = r(Y_1 \sin t - Y_2 \cos t)(r \cos t - Y_1),$$

$$\frac{dY_2}{dt} = r(Y_1 \sin t - Y_2 \cos t)(r \sin t - Y_2).$$

The problem arose in the study of trailer-truck jackknifing.

Solution by DEBORAH FRANK LOCKHART (Michigan Technological University).

Let $Y_1 = u_1 \sin t + u_2 \cos t$, $Y_2 = -u_1 \cos t + u_2 \sin t$. Then

$$\frac{du_1}{dt} = u_2 - ru_1^2, \qquad \frac{du_2}{dt} = (r^2 - ru_2 - 1)u_1.$$

If $r^2 = 1$, let $u_2 = 1/(rz)$. Then $z'' = 1$ and $u_1 = z'/(rz)$. Thus $u_1 = (B+t)/T$, $u_2 = 1/T$, where $T = r(A + Bt + \frac{1}{2}t^2)$.

If $r^2 \neq 1$, let $u_2 = (r^2 - 1)z/(1 + rz)$. Then $z'' + (1 - r^2)z = 0$ and $u_1 = z'/(1 + rz)$. Thus $z = (A/\mu) \sinh \tau$ and

$$u_1 = \frac{\mu A \cosh \tau}{\mu + rA \sinh \tau}, \, u_2 = \frac{u^2 A \sinh \tau}{\mu + rA \sinh \tau},$$

where $\mu = \sqrt{r^2 - 1}$ and $\tau = \mu(t + B)$. This solution is *real-valued for $r^2 \neq 1$.*

(*Note.* As $A \to \infty$, $u_1 \to (\mu/r) \coth \tau$, $u_2 \to \mu^2/r$, which is a solution. As $t \to \infty$, $(u_1, u_2) \to (\mu/r, \mu^2/r)$, which is a critical point (when defined) of the system for u_1 and u_2.)

Solution by W. WESTON MEYER (General Motors Research Laboratories). With

(1) $$Y_1 + iY_2 = Ze^{it} \to \text{Im}(Z) = -Y_1 \sin t + Y_2 \cos t,$$

we compress the system to a single equation

(2) $$\frac{d}{dt}(Ze^{it}) = -r(re^{it} - Ze^{it})\, \text{Im}(Z),$$

or, in fluxional notation, after cancelling e^{it},

(3) $$\dot{Z} + iZ + r(r - Z)\, \text{Im}(Z) = 0.$$

This is Riccati-like; under the change of variable

(4) $$Z = -r^{-1}\dot{W}/\text{Im}(W)$$

it transforms, after cancellation of $-1/\text{Im}(W)$, to

(5) $$\frac{d}{dt}(r^{-1}\dot{W}) + ir^{-1}\dot{W} + r\, \text{Im}(\dot{W}) = 0$$

if r is real and nonzero. Let u and $-v$ be the real and imaginary parts of $r^{-1}W$:

(6) $$r^{-1}\dot{W} = u - iv \to \text{Im}(W) = -c - \int_a^t r(s)v(s)\, ds.$$

Then (5) implies that $\dot{u} + v - r^2 v = -\dot{v} + u = 0$, hence that $u = \dot{v}$ and

(7) $$\ddot{v} + (1 - r^2)v = 0.$$

Generally, for r defined throughout an interval (a, b),

(8) $$Y_1(t) + iY_2(t) = [\dot{v}(t) - iv(t)]\left[c + \int_a^t r(s)v(s)\, ds\right]^{-1} e^{it}, \qquad a \leq t \leq b,$$

where v is an arbitrary real solution of (7) and c is an arbitrary real constant. Formula (8) does not actually require r to be zero-free; it encompasses even the singular case $r = 0$ if we allow v to take on a complex value wherever r vanishes.

Problem 77-17, *A System of Second Order Differential Equations*, by L. CARLITZ (Duke University).

Solve the following system of differential equations:

(1) $$F''(x) = F(x)^3 + F(x)G(x)^2,$$

(2) $$G''(x) = 2G(x)F(x)^2$$

where $F(0) = G'(0) = 1$, $F'(0) = G(0) = 0$.

Solution by C. GIVENS (Michigan Technological University).

Observe that in addition to the given initial conditions, one also has $F''(0) = 1$, $G''(0) = 0$ from the system of differential equations. If one sets $H = F^2 - G^2$, then $H(0) = 1$, $H'(0) = H''(0) = 0$. Continuing to compute, one obtains $H'''(x) = 6F(x)^2 H'(x)$. Thus, $H^{(n)}(0) = 0$ for $n \geq 3$ also and consequently $H \equiv 1$.

From $G'' = 2F^2 G$, it follows that $G'' = 2G(1 + G^2)$ and, upon multiplication by $2G'$, that $[(G')^2]' = [(G^2 + 1)^2]'$. Thus, $G' = G^2 + 1$ and so $G = \tan x$. But then $F = \sec x$ from $F^2 = G^2 + H$.

Editorial note. B. Margolis, in her solution, first noted that it was easy to find a solution $F(x) = \sec x$, $G(x) = \tan x$ and then establishes uniqueness. The proposer obtained the above system of equations by considering generating functions associated with *up-down and down-up* permutations of $\{1, 2, \cdots, n\}$. [M.S.K.]

A System of Ordinary Differential Equations

Problem 84-15, *by* GENGZHE CHANG (University of Science and Technology of China, Hefei, Anhui, China).

Find the solution of the system of ordinary differential equations

(1) $$tX'(t) = AX(t), \text{ with } X(1) \text{ given,}$$

where

$$X(t) = \begin{bmatrix} x_1(t) \\ \vdots \\ x_n(t) \end{bmatrix}$$

and A is an $n \times n$ constant matrix with eigenvalues $1, 2, \cdots, n$.

Solution by JOHN A. CROW (Student, California State University, Fullerton).

Consider the general system of ordinary differential equations

(1) $$X'(t) - f'(t)AX(t) = 0$$

where X is a column vector and A is an $n \times n$ constant matrix with distinct nonzero eigenvalues $\lambda_1, \cdots, \lambda_n$. As may be verified by direct substitution, the general solution is:

(2) $$X(t) = \exp\{Af(t)\} \cdot \alpha$$

where α is an arbitrary constant column vector. Now suppose $X(t_0)$ is specified. Then using the fact that $\{\exp(At)\}^{-1} = \exp(-At)$, it follows that

(3) $$X(t) = \exp\{A[f(t) - f(t_0)]\} \cdot X(t_0).$$

Without loss of generality, assume $f(t_0) = 0$. Since the eigenvalues are distinct and nonzero, then there is a similarity transformation P such that $B = P^{-1}AP = \text{diag}\{\lambda_1, \cdots, \lambda_n\}$. Using the relation $\exp(At) = P \exp(Bt) P^{-1}$, and the fact that

$$\exp(Bt) = \begin{bmatrix} e^{\lambda_1 t} & & 0 \\ & \ddots & \\ 0 & & e^{\lambda_n t} \end{bmatrix}$$

it follows that

(4) $$X(t) = P \begin{bmatrix} e^{\lambda_1 f(t)} & & 0 \\ & \ddots & \\ 0 & & e^{\lambda_n f(t)} \end{bmatrix} P^{-1} X(t_0).$$

In the special case where $f'(t) = 1/t$, $t_0 = 1$, and $\lambda_j = j$, then (4) reduces to

$$X(t) = P \begin{bmatrix} t^1 & & & 0 \\ & t^2 & & \\ & & \ddots & \\ 0 & & & t^n \end{bmatrix} P^{-1} X(1).$$

These results can be extended to nonhomogeneous systems of ODE's.

NANCY WALLER *and the proposer* give the following explicit solution by using the Cayley–Hamilton theorem:

$$X(t) = X(1) + \sum_{k=1}^{n} \frac{A(A-I)(A-2I) \cdots (A-(k-1)I) X(1)(t-1)^k}{k!}.$$

A. S. FERNÁNDEZ (E. T. S. Ingenieros Industriales de Madrid) uses the Lagrange interpolating polynomial to obtain

$$X(t) = \sum_{i=1}^{n} \frac{(-1)^{i-1} t^i}{(n-i)!(i-1)!} \prod_{j \neq i} (A - jI) X(1).$$

J. ROPPERT (Wirtschaftsuniversität Wien) also gives a solution if A has multiple eigenvalues by means of Jordan decomposition. Z. J. KABALA and I. P. E. KINNMARK (Princeton University) show how to solve (1) as above as well as when A has multiple eigenvalues. M. LATINA (Community College of Rhode Island) in his solution notes that the method can be extended to more general Euler–Cauchy systems

$$\sum_{k=1}^{m} t^k A_k X^{(k)}(t) = 0.$$

Problem 61–5*, Flame Propagation, by WILLIAM SQUIRE (West Virginia University).

The thermal theory of flame propagation is based on the partial differential equation

$$\frac{\partial T}{\partial t} = \alpha \nabla^2 T + Q(T),$$

where $Q(T)$ is the rate of heat liberation. For the case of a one-dimensional deflagration wave, this can be reduced to a second order ordinary differential equation by assuming

$$T = f(x - vt),$$

where v is the velocity of the wave.

By letting the temperature become the independent variable and the temperature gradient the dependent variable, it can be further reduced to a first order equation.

The flame speed can then be computed from the eigenvalue m of

$$(1) \qquad\qquad\qquad y \frac{dy}{dz} + my = Q(z)$$

subject to $y(0) = y(1) = 0$, where y is a dimensionless temperature gradient and z a dimensionless temperature. The smallest eigenvalue only is physically significant here.

ZELDOVICH ("Theory of Flame Propagation", NACA TM 1282, June 1951) has proven the existence of an eigenvalue m if

$$Q(z) > 0 \qquad \text{for} \qquad 0 < z < 1 - \epsilon,$$

$$Q(z) = 0 \qquad \text{for} \qquad 1 - \epsilon < z < 1.$$

I. Can the restrictions on $Q(z)$ be relaxed?

Zeldovich also states that for $Q(z) = z(1 - z)$ (corresponding to a cold flame supported by an autocatalytic reaction), solutions exist for all $m \geq 2$. It then can be verified that

$$y = \frac{2}{\sqrt{6}} \{(1 - z) - (1 - z)^{3/2}\}$$

is a solution corresponding to

$$m = \frac{5}{\sqrt{6}} > 2.$$

II. Can other particular solutions or a general solution be found? Alternatively, can the minimum value of m be estimated without exhibiting the solution?

It is to be noted that this eigenvalue problem also occurs in certain other biological problems relating to the spread of epidemics or mutations, and that the proposer has solved part I.

Solution by JOSEPH E. WARREN (Gulf Research and Development Company).

Let
$$y = mp,$$
$$z = 1 - \theta,$$
$$Q(z) = Q(1 - \theta) = \Phi(\theta) \int_0^1 Q(1 - \theta) \, d\theta,$$
$$m = \left[\frac{1}{\lambda} \int_0^1 Q(1 - \theta) \, d\theta \right]^{1/2}.$$

Then,

(2)
$$p \frac{dp}{d\theta} - p = -\lambda\Phi(\theta); \qquad p(0) = p(1) = 0,$$

where
$$\int_0^1 \Phi(\theta) \, d\theta = 1$$

I. The existence of solutions to Equation (2) has been investigated rather extensively. The existence conditions derived by RICHARDSON ("*Existence and Stability of One-Dimensional, Steady-State Combustion Waves*", Fourth International Symposium on Combustion, 182, Williams and Wilkins, Baltimore, 1953) are the following:

$$-\infty < \Phi(\theta) < +\infty \text{ for } 0 < \theta < 1$$

(3)
$$\Phi(0) = 0; -\infty < \frac{d\Phi}{d\theta}(0) < +\infty$$

$$\Phi(1) = 0; -\infty < \frac{d\Phi}{d\theta}(1) \leq 0$$

While these conditions are sufficient, they are not necessary for the problem being considered.

If Equation (2) is integrated over the interval from 0 to θ, the form obtained is that which was studied by KLEIN ("*A Contribution to Flame Theory*", Philosophical Transactions of the Royal Society, A*249*, 389, 1957); i.e.,

(4)*
$$p = \left[2 \int_0^\theta p \, d\theta - 2\lambda \int_0^\theta \Phi(\theta) \, d\theta \right]^{1/2}; \qquad p(1) = 0$$

In the solution of Equation (4), Picard's method of successive approximations will converge and will yield a unique continuous result if the Cauchy-Lipschitz conditions are satisfied.

For the completely exothermic problem that has been posed, the existence conditions are as follows:

$$0 \leq \Phi(\theta) < +\infty \text{ for } 0 < \theta \leq 1$$

(5)
$$\Phi(0) = 0; 0 \leq \frac{d\Phi}{d\theta}(0) < +\infty$$

From the physical problem, it is apparent that "p" must be continuous, single-valued and non-negative; furthermore, evaluating (4) at $\theta = 1$ gives

---

* Using iterative methods with the starting approximation $p = \theta(1 - \theta)$, λ can be determined numerically with a convergence rate of one significant figure per two iterations for an Arrhenius-type reaction rate function.

(6)
$$\lambda = \int_0^1 p \, d\theta$$

The point, $p = 0$ and $\theta = 1$, represents a saddle-point when λ has the value indicated by Equation (6).

II. SPALDING ("*One-Dimensional Laminar Flame Theory for Temperature Explicit Reaction Rates, Part II*", Combustion and Flame, *1*, no. 2, 296, 1957) has tabulated particular solutions to Equation (2) for the following reaction-rate functions:

(7)
$$\Phi(\theta) = \left\{ 2n \, \frac{(n+1)}{(n-1)} \right\} \theta^n (1 - \theta^{n-1})$$

(8)
$$\Phi(\theta) = \left(\frac{2}{n^2}\right)(n+1)(n+2)[2(n+1)\theta^{n+1} - (6n+5)\theta^{2n+1}$$
$$+ 2(3n+2)\theta^{3n+1} - (2n+1)\theta^{4n+1}].$$

(9)
$$\Phi(\theta) = \begin{cases} 0, 0 \leq \theta \leq \theta_i, \\ \dfrac{1}{(1-\theta_i)}, \theta_i < \theta < 1. \end{cases}$$

(10)
$$\Phi(\theta) = \begin{cases} 0, 0 \leq \theta < \theta_i \text{ and } \theta_i + h < \theta \leq 1, \\ \dfrac{1}{h}, \theta_i \leq \theta \leq \theta_i + h. \end{cases}$$

WARREN, REED and PRICE ("*Theoretical Considerations of Reverse Combustion in Tar Sands*", AIME Transactions, *219*, 109, 1960) have presented an additional particular solution for an artificial reaction rate function.

(11)
$$\Phi(\theta) = \begin{cases} 0, 0 \leq \theta \leq \theta_i \\ \dfrac{2\theta}{(1-\theta_i)}, \theta_i < \theta < 1 \end{cases}$$

Under the restriction $\Phi(0) = 0$, Equation (2) clearly indicates that $p \leq \theta$; therefore, from Equation (6), the following upper limit on the eigenvalue can be obtained:

(12)
$$\lambda \leq \tfrac{1}{2}$$

ADLER ("*The Limits of the Eigenvalue of the Laminar Flame Equation in Terms of the Reaction Rate-Temperature Centroid*", Combustion and Flame, *3*, no. 3, 389, 1959) extended Spalding's centroid concept to obtain both upper and lower limits on λ. The functional approach of Adler was modified by Warren (*Reverse Combustion*, Ph.D. Thesis, Pennsylvania State University, 1960) to obtain a sharper result; the upper limit, λ_u, and the lower limit, λ_L, are defined by the following equations:

(13)
$$\theta_c = \begin{cases} \tfrac{4}{3}(\lambda_u)^{1/2}; 0 \leq \lambda_u \leq \tfrac{1}{4}, \\ \dfrac{1}{\lambda_u}\left\{\dfrac{1}{2} - \dfrac{2}{3}\left(\dfrac{1}{2} - \lambda_u\right)\left[3 - 2\left(\dfrac{1}{2} - \lambda_u\right)^{1/2}\right]\right\}; \end{cases} \qquad \tfrac{1}{4} \leq \lambda_u \leq \tfrac{1}{2}$$

(14)
$$\theta_c = (2\lambda_L)^{1/2}; \qquad 0 \leq \lambda_L \leq \tfrac{1}{2},$$

where
$$\theta_c = \int_0^1 \theta \Phi(\theta) \, d\theta.$$

Problem 63-15, On a Periodic Solution of a Differential Equation, by G. W. VELTKAMP (Technological University, Eindhoven).

a) Consider the differential equation

(1)
$$\frac{dy}{dt} + f(y) = p(t),$$

where
  (i) p is continuous and periodic with period 1,
  (ii) f is continuously differentiable for all y,
  (iii) $f(y_2) > f(y_1)$ whenever $y_2 > y_1$,
  (iv) $\lim_{y \to \pm \infty} f(y) = \pm \infty$.
Prove that (1) admits exactly one solution with period 1.
  b) Suppose that, besides (i) to (iv) also
  (v) $p(t + \frac{1}{2}) = -p(t)$ for all t,
  (vi) $f(-y) = -f(y)$ for all y.
Prove that the periodic solution $y = z(t)$ of (1) satisfies

$$z(t + \tfrac{1}{2}) = -z(t) \qquad\qquad \text{for all } t.$$

  c) Suppose that, besides (i) to (vi) also
(vii) $p(t) > 0$ for $0 < t < \frac{1}{2}$.
Let t_1 be defined by

$$0 < t_1 < \frac{1}{2}, \qquad \int_{t_1}^{t_1+1/2} p(t)\, dt = 0.$$

Prove that there exists a number t_0 such that

$$0 < t_0 < t_1$$

and

$$z(t) > 0 \quad \text{for} \quad t_0 < t < t_0 + \tfrac{1}{2}.$$

Solution by I. I. KOLODNER (Carnegie Institute of Technology).

1a (Existence). From (i) it follows that p is bounded, i.e., $|p| \leq a$. From (ii)–(iv), we deduce the existence of y and $\bar{y}$ such that $f(\bar{y}) = a + 1$ and $f(y) = -(a + 1)$. Thus $p(t) - f(\bar{y}) \leq -1$ and $p(t) - f(y) \geq 1$, implying that in the strip $S = R \times [y, \bar{y}]$ of the (t, y) plane, the direction field of the differential equation is confining. The proof may now be completed by a standard argument. [From the confining property it follows that for every $c \in [y, \bar{y}]$, (1) has a unique solution $z(t, c)$ defined *on* [0, 1], with the initial value c. Since $y \leq z(t, c) \leq \bar{y}$, it follows from the continuity of f that the function $c \to z(1, c)$ maps $[y, \bar{y}]$ continuously into itself and hence has a fixed point c_0. The function $z(t) = z(t, c_0)$ defined first on [0, 1] and extended periodically to R, is obviously the desired solution.]

1b (Uniqueness). Let z be a periodic solution and u any solution of (1). Then $v = u - z$ is a solution of the equation

(2)
$$y' = g(t, y),$$

where

$$g(t, y) = p(t) - z'(t) - f(z(t) + y)).$$

In view of (iii), sgn $g(t, y) = -$sgn y, and thus, in view of boundedness of $g(t, y)$ for any fixed y, the solution $v = 0$ of (2) is a forward attractor. It follows that it is the only periodic solution of (2).

2. Define $u(t) = -z(t + \frac{1}{2})$. Then

$$u'(t) + f(u(t)) = -z'(t + \frac{1}{2}) - f(z(t + \frac{1}{2})) = -p(t + \frac{1}{2}) = p(t).$$

Since

$$u(t + 1) = -z(t + \frac{3}{2}) = -z(t + \frac{1}{2}) = u(t),$$

$u(t) = z(t)$ by uniqueness of periodic solutions of (1).

3. First we show that $z(0) < 0$. From (vi) and (vii) it follows that $z \neq 0$ on $[0, \frac{1}{2}]$. In view of (ii), (iii) and (vi),

$$f(y) = yg(y),$$

where $g(y) > 0$ if $y \neq 0$ and g is an even function. Thus z is a solution of the equation

$$y' + g(z(t))y = p(t),$$

whence we get the representation

$$z(t) = z(0) \exp (h(t)) + \int_0^t p(\tau) \exp (h(t) - h(\tau)) \, d\tau,$$

where $h(t) = -\int_0^t g(z(\tau)) \, d\tau$. Since g is even it follows in view of part 2 that $g(z(t))$ has period $\frac{1}{2}$ and one deduces that $h(t + \frac{1}{2}) = h(t) + h(\frac{1}{2}), h(1) = 2h(\frac{1}{2})$. The periodicity condition $z(1) = z(0)$ together with (v) yields now, after some simplifications,

$$z(0) = -\left[1 + \exp \left(-h \left(\frac{1}{2}\right)\right)\right]^{-1} \int_0^{1/2} p(\tau) \exp (-h(\tau)) \, d\tau.$$

Thus, $z(0) < 0$ in view of (vii).

Since $z(\frac{1}{2}) = -z(0) > 0$, z has a zero $t_0 \in (0, \frac{1}{2})$. Since at any zero $\tau \in (0, \frac{1}{2})$ of z, $z'(\tau) = p(\tau) > 0$, this zero is unique, and thus sgn $z(t) = $ sgn $(t - t_0)$ on $[0, \frac{1}{2}]$. Hence $z(t) > 0$ on $(t_0, \frac{1}{2}]$. For any $t \in [\frac{1}{2}, t_0 + \frac{1}{2})$, $z(t) = -z(t - \frac{1}{2}) > 0$, since $(t - \frac{1}{2}) \in [0, t_0)$. Thus $z(t) > 0$ for $t_0 < t < t_0 + \frac{1}{2}$.

Finally we show that $t_0 < t_1$. For $t \in [0, \frac{1}{2}]$, define

$$k(t) = \int_t^{t+1/2} p(\sigma) \, d\sigma.$$

Then $k(0) > 0$, $k(\frac{1}{2}) < 0$ and since k is strictly decreasing, there exists a unique $t_1 \in (0, \frac{1}{2})$ such that sgn $k(t) = -$sgn $(t - t_1)$. Suppose that $t_0 \geq t_1$. Integrating (1) from t_0 to $t_0 + \frac{1}{2}$ we get then

$$z \left(t_0 + \frac{1}{2}\right) - z(t_0) + \int_{t_0}^{t_0+1/2} f(z(\tau)) \, d\tau = k(t_0) \leq 0.$$

This is a contradiction since

$$z(t_0 + \frac{1}{2}) = -z(t_0) = 0,$$

while

$$\int_{t_0}^{t_0+1/2} f(z(\tau))\, d\tau > 0.$$

Problem 67-1, A Nonlinear Eigenvalue Problem, by C. J. Bouwkamp (Philips Research Laboratories, Eindhoven, The Netherlands).

Let $0 < a < 1$, and let $f(x)$ be sufficiently smooth for $0 \leq x \leq 1$ such that $f(0) = 0$ and $f(x) > 0$ for $0 < x < 1$. Further, let k be the smallest positive value such that

$$\varphi''(r) + r^{-1}\varphi'(r) + k^2 f(\varphi(r)) = 0, \qquad 0 \leq r \leq 1,$$

with $\varphi(0) = a$, $\varphi'(0) = 0$, $\varphi(1) = 0$.

Show that $k = 2\sqrt{u(0)}$, where $u(x)$, for $0 \leq x \leq a$, is the solution of the following initial value problem:

$$u'(x) = v(x),$$

$$v'(x) = \begin{cases} \dfrac{v^2(x)}{u(x)}[1 + f(x)v(x)] & \text{for} \quad x \neq a, \\[2ex] -\dfrac{1}{2}\dfrac{d}{dx}\left(\dfrac{1}{f(x)}\right)_{x=a} & \text{for} \quad x = a, \end{cases}$$

with $u(a) = 0$, $v(a) = -1/f(a)$.

Solution by the proposer.

The parameter k can be eliminated from the differential equation by setting $x = kr$, $\varphi(r) = \psi(x)$. Thus k is the smallest positive zero of $\psi(x)$, where

$$\psi''(x) + x^{-1}\psi'(x) + f(\psi(x)) = 0, \qquad x \geq 0,$$

and $\psi(0) = a$, $\psi'(0) = 0$.

It is easy to show that this zero exists and that $\psi(x)$ is monotonically decreasing from a to zero as x increases from 0 to k. Setting $t = \frac{1}{4}x^2$, $T = \frac{1}{4}k^2$, $\psi(x) = \chi(t)$, we find

$$t\chi'' + \chi' + f(\chi) = 0, \qquad 0 \leq t \leq T,$$

with $\chi(0) = a$, $\chi'(0) = -f(a)$, $\chi(T) = 0$.

The function $\chi(t)$ has a unique inverse, $t = t(\chi)$, and the latter satisfies

$$\frac{d^2 t}{d\chi^2} = \frac{1}{t}\left(\frac{dt}{d\chi}\right)^2\left[1 + f(\chi)\frac{dt}{d\chi}\right], \qquad 0 \leq \chi \leq a,$$

with $t(a) = 0$, $t'(a) = -1/f(a)$, and $t(0) = T$. The right-hand side has to be properly defined at $t = a$ since it is of the undetermined form $0/0$. The final result is equivalent to the system (u, v) as indicated.

Remark. The special case $f(x) = x(1 - x)$ was numerically solved by the proposer in Nederl. Akad. Wetensch. Proc. Ser. A, 68 (1965), pp. 539–547, by a different method, involving interpolation. He has since solved the problem for $f(x) = [x(1 - x)]^2[x(1 - x) + b]^{-1}$ for about 1000 combinations of a and b in the intervals $0 < a < 1$ and $0 < b < 3$, using the present direct method.

Problem 64-17, A Property of Real Solutions to Bessel's Equation, by G. W.
VELTKAMP (Technological University, Eindhoven, Netherlands).

Let $\Phi(x)$ and $\Psi(x)$ be independent real solutions of the Bessel equation

$$x \frac{d}{dx}\left(x \frac{dy}{dx}\right) + (x^2 - \nu^2)\dot{y} = 0.$$

Also, let x_1 and x_2 be two roots of $\Phi(x)$, i.e.,

$$\Phi(x_1) = \Phi(x_2) = 0.$$

Show that if $x_2 > x_1 > 0$, then

$$|\Psi(x_1)| > |\Psi(x_2)| > 0.$$

Solution by J. J. M. BRANDS (Technological University, Eindhoven, Netherlands).

The problem can be generalized to the following:
Let $\Phi(x)$ and $\Psi(x)$ be independent real solutions of the differential equation

(1) $$\frac{d}{dx}\left(p(x) \frac{dy}{dx}\right) + q(x)y = 0,$$

where $p(x)$ and $q(x)$ are real continuous functions for $x > 0$ such that

$$p(x) > 0 \quad \text{and} \quad \frac{d}{dx}\{p(x)q(x)\} > 0.$$

Then if $\Phi(x_1) = \Phi(x_2) = 0$, $x_2 > x_1 > 0$,

$$|\Psi(x_1)| > |\Psi(x_2)| > 0.$$

By Abel's formula,

$$p(x)[\Phi(x)\Psi'(x) - \Phi'(x)\Psi(x)] = C \text{ (nonzero constant)}.$$

Thus it follows that neither of $\Psi(x)$ or $\Phi'(x)$ can vanish at x_1 or x_2. Whence,

$$|\Psi(x_1)| = \frac{|C|}{p(x_1)|\Phi'(x_1)|},$$

(2)

$$|\Psi(x_2)| = \frac{|C|}{p(x_2)|\Phi'(x_2)|}.$$

Multiplying (1) (taking $y = \Phi$) by $p(x)\Phi'(x)$ and integrating between x_1 and x_2
gives

$$[p(x_2)\Phi'(x_2)]^2 - [p(x_1)\Phi'(x_1)]^2 = \int_{x_1}^{x_2} \Phi(x)^2 D\{p(x)q(x)\} \, dx > 0.$$

Hence,

(3) $$p(x_2)|\Phi'(x_2)| > p(x_1)|\Phi'(x_1)|.$$

The desired inequality immediately follows now from (2) and (3).

A Nonoscillation Result

Problem 80-8, by D. K. ROSS (La Trobe University, Victoria, Australia).
Consider the second-order linear differential equation

$$y'' + p(x)y' + q(x)y = 0$$

in the interval $a < x < b$ where $p(x)$ and $q(x)$ are continuous and $q(x) < 0$. If $\phi(x)$ is a nontrivial solution of the above equation, prove that the function $W(x) = \phi'(x)\phi(x)$ has at most one zero in the interval.

Solution by R. R. BURNSIDE (Paisley College of Technology, Scotland).

Since ϕ is nontrivial and p, q are continuous, $\phi(x_1)$ and $\phi'(x_1)$ cannot both be zero for any $x_1 \in (a, b)$. Also, W is continuous and

$$W'(x) + p(x)W(x) = \{\phi'(x)\}^2 - q(x)\{\phi(x)\}^2 > 0, \qquad a < x < b.$$

Hence, if $W(x_1) = 0$ for some $x_1 \in (a, b)$, then $W'(x_1) > 0$ and W is locally increasing at x_1. Since W is continuous, it cannot have more than one zero in (a, b). A similar result clearly holds in the case of the second-order nonlinear equation $y'' + p(x)y' + q(x)y^n = 0$ where n is odd.

Supplementary References
Ordinary Differential Equations

[1]   H.Bateman, *Differential Equations*, Chelsea, N.Y., 1966.

[2]   R.Bellman, *Stability Theory of Differential Equations*, McGraw-Hill, N.Y., 1953.

[3]   R.Bellman and K.L.Cooke, *Modern Elementary Differential Equations*, Addison-Wesley, Reading, 1971.

[4]   G.Birkoff and G-C.Rota, *Ordinary Differential Equations*, Xerox, 1969.

[5]   G.Boole, *A Treatise on Differential Equations*, 5th ed., Chelsea, N.Y.

[6]   W.E.Boyce and R.C.DiPrima, *Elementary Differential Equations and Boundary Value Problems*, Wiley, 1977.

[7]   G.F.Carrier and C.E.Pearson, *Ordinary Differential Equations*, Blaisdell, Waltham, 1968.

[8]   L.Cesari, *Asymptotic Behavior and Stability Problems in Ordinary Differential Equations*, Academic Press, 1963.

[9]   E.A.Coddington and N.Levinson, *Theory of Ordinary Differential Equations*, McGraw-Hill, N.Y., 1955.

[10] A.R.Forsyth, *Differential Equations*, MacMillan, London, 1948.

[11] A.Halanay, *Differential Equations: Stability, Oscillations, Time Lags*, Academic Press, 1966.

[12] P.Hartman, *Ordinary Differential Equations*, Wiley, N.Y., 1964.

[13] E.I.Ince, *Ordinary Differential Equations*, Dover, N.Y., 1953.

[14] E.Kamke, *Differentialgleichungen: Losungsmethoden und Losungen*, Akademische Verlagagesellschaft, Leipsig, 1943.

[15] S.Lefschetz, *Differential Equations, Geometric Theory*, Wiley, N.Y., 1963.

[16] N.W.McLachlan, *Ordinary Non-Linear Differential Equations in Engineering and Physical Sciences*, Clarendon Press, Oxford, 1950.

[17] G.M.Murphy, *Ordinary Differential Equations and their Solutions*, Van Nostrand, Princeton, 1960.

[18] O.Plaat, *Ordinary Differential Equations*, Holden-Day, San Francisco, 1971.

[19] T.L.Saaty and J.Bram, *Nonlinear Mathematics*, McGraw-Hill, N.Y., 1964.

[20] G.Sansone and R.Conti, *Non-Linear Differential Equations*, Pergamon, N.Y., 1952.

[21] G.F.Simmons, *Differential Equations with Applications and Historical Notes*, McGraw-Hill, N.Y., 1972.

[22] J.J.Stoker, *Nonlinear Vibrations in Mechanical and Electrical Systems*, Wiley, N.Y., 1950.

[23] W.Wasow, *Asymptotic Expansions of Ordinary Differential Equations*, Wiley, N.Y., 1966.

8. PARTIAL DIFFERENTIAL EQUATIONS

Problem 77-4, Solutions to Linear Partial Differential Equations Involving Arbitrary Functions, by A. UNGAR (National Institute for Mathematical Sciences, Pretoria, Africa).

Let $x = (x_1, x_2, \cdots, x_n)$ be a set of n real variables and let L be the linear differential operator

$$L\{f(x)\} = \sum_{i=1}^{N} a_{p_i}(x) \frac{\partial^{p_i} f(x)}{\partial x^{p_i}}.$$

Here p_i are multi-indices of order n. For a multi-index p of order n, $p = (p_1, p_2, \cdots, p_n)$, where the entries are integers, $|p| = p_1 + p_2 + \cdots + p_n$ and

$$\frac{\partial^p}{\partial x^p} = \frac{\partial^{|p|}}{\partial x_1^{p_1} \partial x_2^{p_2} \cdots \partial x_n^{p_n}}.$$

The coefficients $a_{p_i}(x)$ are functions of x and N is an integer.

Prove that

$$f(x) = S_1(x) A [S_2(x)],$$

where $S_1(x)$ and $S_2(x)$ are specified functions and A is an arbitrary suitably differentiable function of $S_2(x)$, satisfies the linear partial differential equation

(1) $$L\{F(x)\} = 0$$

in a domain, if and only if

$$g(x) = S_1(x) e^{\alpha S_2(x)}$$

is a particular solution of (1) in that domain, for every real α in some interval.

Examples. $e^{\alpha(x+iy)}$ and $A(x+iy)$ satisfy the Laplace equation and $(1/R) e^{\alpha(R-ct)}$ and $(1/R) A(R-ct)$, $R^2 = x^2 + y^2 + z^2$, satisfy the wave equation for an arbitrary suitably differentiable function A.

Solution by W. ALLEGRETTO (University of Alberta, Edmonton, Alberta, Canada).

The first implication is immediate. For the other, assume $g(x) = S_1(x) \exp(\alpha S_2(x))$ solves (1) for some domain in the x-space and for α in some interval. Now consider $L\{S_1 A(S_2)\}$ for arbitrary smooth A. Expanding out, we find that for some q_i (functions of x, S_1, S_2 and their derivatives but not A) and some integer m that

$$L\{S_1 A(S_2)\} = \sum_{i=0}^{m} q_i A^{(i)}(S_2)$$

where $A^{(i)} = d^i A/dt^i$. On letting $A(t) = e^{\alpha t}$, we obtain

$$0 = L\{S_1 e^{\alpha S_2}\} = e^{\alpha S_2}\left\{\sum_{i=0}^{m} q_i \alpha^i\right\}.$$

Thus for any fixed x in the domain,

(2) $$\sum_{i=0}^{m} q_i \alpha^i = 0.$$

But since (2) is a polynomial in α with infinitely many zeros as α varies over the interval, it must identically vanish, i.e., $q_i \equiv 0$ in the domain. Whence, also

$$L\{S_1 A(S_2)\} = 0.$$

The result is also valid if $S_1 e^{\alpha S_2}$ solves the equation for only $m+1$ values of α.

Problem 60-1, *Steady-State Diffusion-Convection,* by G. H. F. GARDNER (Gulf Research & Development Company).

When a homogeneous fluid flows through a porous material, such as sandstone or packings of small particles, its molecules are scattered by the combined action of molecular diffusion and convective mixing. Thus a sphere of tagged fluid particles expands as it moves and its deformation may be resolved into the longitudinal and transverse component. The transverse mixing, which is many times less than the longitudinal mixing, has been investigated at turbulent rates of flow but has received little attention when the flow rate is low as for subterranean fluid movement. The following experiment was set up to investigate transverse mixing at low flow rates.

A rectangular porous block with impermeable sides was mounted with one side horizontal. An impermeable horizontal barrier AB divided the block into equal parts for about one-third of its length. One fluid was pumped at a constant rate into the block above AB and another was pumped at an equal rate below AB. After passing B the fluids mingled and a steady-state distribution was attained.

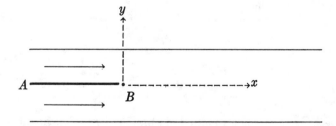

The fluids have approximately equal densities and the heavier was flowed under AB so that the equilibrium would be stable.

The steady-state distribution of the fluids is assumed to be given by

(1) $$\frac{\partial S}{\partial x} = \alpha \frac{\partial^2 S}{\partial x^2} + \beta \frac{\partial^2 S}{\partial y^2},$$

where $S(x, y)$ denotes the fractional amount of the lower fluid present at the point (x, y). The equation was obtained under the following assumptions (for an analogous problem, see H. Bateman, *Partial Differential Equations*, p. 343):

(1) Streamline motion and molecular diffusion cause the dispersion.

(2) The coefficient α reflects the dispersion of flow caused by velocity variations along each streamline and molecular diffusion in the direction of flow.

(3) The coefficient β reflects the dispersion perpendicular to the direction of flow caused by diffusion between steamlines.

The boundary conditions to be satisfied are

(a) $\dfrac{\partial S}{\partial y} = 0$ on the impermeable boundaries,

(b) $S \to 0$ for $y > 0$, $x \to -\infty$, and

(c) $S \to 1$ for $y < 0$, $x \to -\infty$.

Solve equation (1) for an infinite medium. Here boundary condition (a) is replaced by

(a') $\dfrac{\partial S}{\partial y} = 0$ for $y = 0$, $x < 0$.

Solution by the proposer.

The use of parabolic coordinates (ξ, η) instead of rectangular coordinates (x, y) is more appropriate because the semi-infinite boundary $y = 0$, $x < 0$ is simply given by $\eta = 0$. Thus, writing

(2) $$\frac{x}{\alpha} = \frac{1}{2}(\eta^2 - \xi^2), \quad \frac{y}{\sqrt{\alpha\beta}} = \xi\eta,$$

the differential equation is transformed to

(3) $$\frac{\partial^2 S}{\partial \xi^2} + \frac{\partial^2 S}{\partial \eta^2} + \xi \frac{\partial S}{\partial \xi} - \eta \frac{\partial S}{\partial \eta} = 0,$$

and the boundary conditions become

(a'') $\dfrac{\partial S}{\partial \eta} = 0, \eta = 0,$

(b') $S \to 0, \xi \to \infty,$

(c') $S \to 1, \xi \to -\infty.$

If we now assume that S is a function of ξ only and is independent of η, equation (3) reduces to

(4) $$\frac{d^2 S}{d\xi^2} + \xi \frac{dS}{d\xi} = 0.$$

The general solution of equation (4) may be written

(5) $$S = A \int_0^{\xi/\sqrt{2}} e^{-u^2}\, du + B.$$

Boundary condition (a″) is obviously satisfied. Conditions (b′) and (c′) can be satisfied by choosing the constants A and B appropriately. Hence the required solution may be written,

(6)
$$S = \tfrac{1}{2}[1 - erf\ \xi/\sqrt{2}].$$

The curves of constant concentration are given by constant values of ξ and therefore are parabolas confocal with the end of the barrier.

One simple result is perhaps noteworthy. The vertical concentration gradient at points on the x-axis is given by

(7)
$$\left(\frac{\partial S}{\partial y}\right)_{y=0} = \frac{1}{\sqrt{4\pi\beta x}}.$$

It is independent of α and hence gives a convenient way of measuring β experimentally.

Generalization of the solution for an infinite medium, given by equation (6), to certain bounded regions is easily accomplished by use of the method of images.

Problem 60-6*, Steady-State Plasma Arc, by Jerry Yos (AVCO Research and Advanced Development Division).

In studying the positive column of an electric arc, a model is considered in which the arc strikes between two plane electrodes in an infinite channel. The sides of the channel are held at a fixed temperature $T = 0$ and are perfect electrical insulators. The electrodes are held at fixed potentials and are perfect thermal insulators (Fig. 3). The steady-state distributions of temperature and electrical potential in the arc are then determined by the equations for the conservation of current and for the energy balance between the electrical heating of the gas and the cooling due to thermal conduction to the walls, i.e.

(1)
$$\nabla \cdot (\sigma \nabla \phi) = 0,$$

(2)
$$\sigma(\nabla \phi)^2 = -\nabla \cdot (k \nabla T).$$

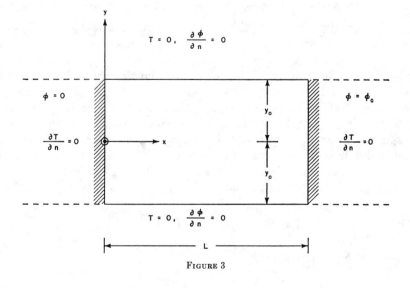

FIGURE 3

Here,

$$\sigma = \sigma(T), \qquad k = k(T)$$

are the electrical and thermal conductivities of the gas, respectively. The boundary conditions at the electrodes are

(3)
$$\phi = 0, \qquad \frac{\partial T}{\partial n} = 0, \quad \text{for} \quad x = 0,$$

$$\phi = \phi_0, \qquad \frac{\partial T}{\partial n} = 0, \quad \text{for} \quad x = L.$$

The boundary conditions at the walls of the channel are

(4)
$$T = 0, \qquad \frac{\partial \phi}{\partial n} = 0, \quad \text{for} \quad y = \pm Y_o .$$

One solution of this problem can be readily found in the form

$$\phi = \frac{\phi_0 x}{L}, \qquad T = T(y),$$

where $T(y)$ is determined implicitly from

(5)
$$\frac{\phi_0 y}{L} = \int_T^{T_m} \frac{k \, dT}{\sqrt{2 \int_T^{T_m} k\sigma \, dT}}$$

and where the maximum temperature T_m is given by

(6)
$$\frac{\phi_0 Y_o}{L} = \int_0^{T_m} \frac{k \, dT}{\sqrt{2 \int_T^{T_m} k\sigma \, dT}} .$$

Is this solution unique?

Problem 60-7*, *A Free Boundary Problem*, by Chris Sherman (AVCO Research and Advanced Development Division).

In analyzing the phenomena occurring in the column of an electric arc operating with forced convection, a set of coupled partial differential equations involving the dependent variables **J**, **E** (electric current density and field strength), **U**, p (velocity and pressure of the gas stream), and T (gas temperature) arises (see Problem 60-6). This set may be reduced to a single equation by the following drastic simplifying assumptions: **U** is taken to be a constant directed along the x-axis of a rectangular coordinate system, **E** a constant directed along the z-axis, and T is assumed to be a function of x and y only. **J** is related to **E** by **J** $= \sigma(T)$**E** where $\sigma(T)$, the electrical conductivity, is a known function of temperature. The equation which results is the conservation of energy equation

(1)
$$\frac{\partial^2 T}{\partial x^2} + \frac{\partial^2 T}{\partial y^2} - 2a \frac{\partial T}{\partial x} + b^2 \sigma(T) = 0,$$

where a and b are constants. If

$$\sigma(T) = cT \qquad (T \geq T_0) \quad \text{in region 1,}$$

$$\sigma(T) = 0 \qquad (T \leq T_0) \quad \text{in region 2,}$$

where c and T_0 are known constants, a solution of equation (1) which includes the determination of the shape of the boundary separating regions 1 and 2 is sought. The boundary conditions to be satisfied are

1. the separating boundary B passes through the point $(x_0, 0)$,
2. T and $\partial T/\partial n$ are continuous across B,
3. T bounded,
4. $\lim_{\sqrt{x^2+y^2} \to \infty} T = 0$.

The following questions are of particular interest:

(a) Does a stable solution exist?
(b) If it does, is the solution unique?
(c) Is the separating boundary B closed?

A similar alternative but simpler problem in which

$$\sigma(T) = c, \qquad (T \geq T_0) \quad \text{in region 1,}$$

$$\sigma(T) = 0, \qquad (T \leq T_0) \quad \text{in region 2,}$$

is also of interest.

Problem 69-11, *Ohmic Heating,* by J. A. LEWIS (Bell Telephone Laboratories).

Consider an isotropic, homogeneous, conducting body, ohmically heated by the passage of direct current between perfectly conducting electrodes on its surface, the rest of the surface being electrically and thermally insulated. Show that the maximum temperature in the body depends only on the potential difference between the electrodes, the electrode temperature, and the electrical and thermal conductivities, in general functions of the temperature, and is independent of the size and shape of the body and of the electrode configuration.

Solution by the proposer.

The potential V and temperature T in a body with electrical and thermal conductivities $\sigma(T)$, $k(T)$ satisfy the equations

(1) $$\nabla \cdot (\sigma \nabla V) = 0, \qquad \nabla \cdot (k \nabla T) + \sigma |\nabla V|^2 = 0$$

and the boundary conditions on the electrodes

$$V = V_0 \quad \text{on } S_0, \qquad V = V_1 \quad \text{on } S_1, \qquad T = T_0 \quad \text{on } S_0 + S_1,$$

while on the insulated surface,

$$\partial V/\partial n = \partial T/\partial n = 0.$$

One may verify by direct substitution that the pair of functions V, $T(V)$ satisfy the equations and boundary conditions provided that $T(V)$ is given by

(2) $$\int_{T_0}^{T(V)} \frac{k(T)}{\sigma(T)} dT = \frac{1}{2}(V - V_0)(V_1 - V).$$

The maximum temperature, obtained by differentiation with respect to V, is given by

$$\int_{T_0}^{T_{max}} \frac{k(T)}{\sigma(T)} dT = \frac{1}{8}(V_1 - V_0)^2,$$

which is the desired result.

   Editorial note. The result is also valid if the boundary condition $T = T_0$ on $S_0 + S_1$ is extended to

$$T = T_0 \quad \text{on } S_0 \quad \text{and} \quad T = T_1 \quad \text{on } S_1.$$

Assuming that $V_1 > V_0$, $T_1 > T_0$, we replace (2) by

(2)' $$\int_{T_0}^{T(V)} \frac{k(T)}{\sigma(T)} dT = \frac{1}{2}(V - V_0)(V_1 - V) + \frac{V - V_0}{V_1 - V_0} \int_{T_0}^{T_1} \frac{k(T)}{\sigma(T)} dT.$$

It still follows easily that the pair of functions V, $T(V)$ satisfy (1) and the boundary conditions.

   The maximum temperature, obtained by differentiation of the right-hand side of (2)' (or by completing the square), occurs when $V = a + (V_0 + V_1)/2$ and is now given by

$$\int_{T_0}^{T_{max}} \frac{k(T)}{\sigma(T)} dT = \frac{1}{8}(V_1 - V_0 + 2a)^2,$$

where

$$a = \frac{1}{V_1 - V_0} \int_{T_0}^{T_1} \frac{k(T)}{\sigma(T)} dT,$$

again giving the desired result.

   Since $V_0 \leq V \leq V_1$, the previous equation for T_{max} is only valid if $2a < V_1 - V_0$. If $2a \geq V_1 - V_0$, then $T_{max} = T_1$ and occurs on the boundary S_1. A similar argument applies for the case $T_1 < T_0$. [M.S.K.]

Postscript. H.J. Wintle gives an application of the above results in his paper, *A note on electrothermal breakdown in insulation*, IEEE Trans. Elec. Insulation, 24(1989) 139-141. He also mistakenly notes a correction for the above that if the condition $|2a| < |V_1 - V_0|$ is violated, then the maximum temperature occurs adjacent to one electrode, and is the bigger of T_0, T_1.

   Problem 62–1*, A Steady-State Temperature, by ALAN L. TRITTER (Data Processing Inc.) and A. I. MLAVSKY (Tyco, Inc.).

   Consider the steady-state temperature ($T(r, z)$) distribution boundary-value problem for an infinite solid bounded by two parallel planes:

(1) $$\frac{\partial^2 T}{\partial r^2} + \frac{1}{r} \frac{\partial T}{\partial r} + \frac{\partial^2 T}{\partial z^2} = 0, \quad 0 < z < H, \quad r \geq 0,$$

$$\left\{ -k \frac{\partial T}{\partial z} = \begin{matrix} Q, r < R \\ 0, r > R \end{matrix} \right\}_{z=0},$$

$$\{T = 0\}_{z=H},$$

$$|T| < M \text{ (boundedness condition)},$$

(all the parameters involved are constants). Determine the temperature at the point $r = z = 0$.

The solutions by E. DEUTSCH (Institute of Mathematics, Bucharest, Rumania), THOMAS ROGGE (Iowa State University), J. ERNEST WILKINS JR. (General Dynamics Corporation) and M. S. KLAMKIN (University of Buffalo) were essentially the same and are given by the following:

Letting

$$\phi(\lambda, z) = \int_0^\infty r J_0(\lambda r) T(r, z) \, dr,$$

it follows by integration by parts that the Hankel transform of Eq. (1) is

$$\{D^2 - \lambda^2\}\phi = 0,$$

subject to the boundary conditions

$$k \left. \frac{\partial \phi}{\partial z} \right]_{z=0} = \int_0^R Q r J_0(\lambda r) \, dr = \frac{QR J_1(\lambda R)}{\lambda},$$

$$\{\phi = 0\}_{z=H}.$$

Consequently,

$$\phi(\lambda, z) = \frac{QR}{k} \frac{J_1(\lambda R)}{\lambda^2} \frac{\sinh \lambda(H - z)}{\cosh H}.$$

Inverting the latter transform:

$$(2) \qquad T(r, z) = \frac{QR}{k} \int_0^\infty \frac{\sinh \lambda(H - z)}{\lambda \cosh \lambda H} J_0(\lambda r) J_1(\lambda R) \, d\lambda.$$

On letting $H \to \infty$, we obtain

$$\lim_{H \to \infty} T(r, z) = \frac{QR}{k} \int_0^\infty e^{-\lambda z} J_0(\lambda r) J_1(\lambda R) \frac{d\lambda}{\lambda}$$

which corresponds to a result given in Carslaw and Jaeger, Conduction of Heat in Solids, Oxford University Press, London, 1959, p. 215.

In particular, the temperature at $r = 0$, $z = 0$, is given by

$$(3) \qquad T(0, 0) = \frac{QR}{k} \int_0^\infty \lambda^{-1} J_1(\lambda R) \tanh \lambda H \, d\lambda.$$

The series expansion

$$(4) \qquad T(0, 0) = \frac{QR}{k} \left\{ 1 - \frac{R}{H} \sum_{m=1}^\infty \frac{(-1)^{m+1}}{m + \sqrt{m^2 + R^2/4H^2}} \right\}$$

is obtained by expanding $\tanh \lambda H$ into the exponential series

$$\tanh \lambda H = 1 - 2 \sum_{m=1}^\infty (-1)^{m+1} e^{-2m\lambda H}$$

and employing the integral

$$\int_0^\infty \lambda^{-1} e^{-a\lambda} J_1(\lambda R) \, d\lambda = (\sqrt{a^2 + R^2} - a)/R$$

(Watson, Theory of Bessel Functions, Cambridge University Press, London, 1952, p. 386).

Deutsch also obtains the alternate series expansion

$$(5) \qquad T(0, 0) = \frac{QH}{k} \left\{ 1 - \frac{4R}{\pi H} \sum_{n=1,3,5,\cdots} \frac{1}{n} K_1 \left(\frac{\pi n R}{2H} \right) \right\}.$$

This latter expansion is suitable for numerical calculations for large values of the parameter R/H. It shows that $\lim_{H \to 0} T(0, 0) = 0$ which is what one would expect physically. It also implies that

$$\sum_{m=1}^{\infty} \frac{(-1)^{m+1}}{m + \sqrt{m^2 + \lambda^2}} = \frac{1}{2\lambda} - \frac{1}{4\lambda^2} \quad \text{as} \quad \lambda \to \infty.$$

For small values of R/H, Wilkins expanded

$$\sum_{m=1}^{\infty} \frac{(-1)^{m+1}}{m + \sqrt{m^2 + \lambda^2}} = \sum_{m=1}^{\infty} \frac{(-1)^{m+1}}{\lambda^2} \left\{ \sqrt{m^2 + \lambda^2} - m \right\}$$

into powers of λ using the binomial theorem. The coefficients depend on the Zeta function $\zeta(2m + 1)$. In particular,

$$\lim_{H \to \infty} T(0, 0) = \frac{QR}{k}.$$

By a superposition integral, we can also find the temperature distribution from (2) for the more general flux condition

$$k \frac{\partial T}{\partial z} \bigg]_{z=0} = Q(r), \qquad r \geq 0.$$

Amos also gives the following series expansion for $T(r, z)$:

$$T(r, z) = \begin{cases} \dfrac{Q}{k} (H - z) - \dfrac{4QR}{\pi k} \displaystyle\sum_{n=1}^{\infty} \dfrac{(-1)^{n+1}}{2n - 1} I_0(\lambda_n r) K_1(\lambda_n R) \sin \lambda_n(H - z), \\ \hfill r < R, \\ \dfrac{4QR}{\pi k} \displaystyle\sum_{n=1}^{\infty} \dfrac{(-1)^{n+1}}{2n - 1} K_0(\lambda_n r) I_1(\lambda_n R) \sin \lambda_n(H - z), \hfill r \geq R, \end{cases}$$

where $\lambda_n = (2n - 1)\pi/2H$. This reduces to (5) for $r = z = 0$.

Warren also gives series expansions for both small and large H/R.

For extensions of this problem to the unsteady-state in finite or infinite cylinders see *Unsteady Heat Transfer into a Cylinder Subject to a Space- and Time-Varying Surface Flux*, by M. S. Klamkin, Tr-2-58-5, AVCO Research and Advanced Development Division, May, 1958.

Problem 70-23. A Heat Transfer Problem, by J. ERNEST WILKINS, JR. (Howard University).

Solve the partial differential equation

$$k\nabla^2 T \equiv k\left\{\frac{1}{r}\frac{\partial}{\partial r}\left(r\frac{\partial T}{\partial r}\right) + \frac{1}{r^2}\frac{\partial^2 T}{\partial\theta^2}\right\} = -q, \qquad 0 < r < a, \quad 0 \le \theta \le 2\pi,$$

subject to the boundary condition

$$-k\frac{\partial T}{\partial r} = h(\theta)T, \qquad r = a, \quad 0 \le \theta \le 2\pi,$$

which describes the temperature distribution T in an infinitely long, uniformly heated rod of radius a and thermal conductivity k, when the heat transfer coefficient h varies around its mean value h_0 over the circumference in such a manner that

$$h(\theta) = h_0(1 + \varepsilon \cos n\theta)$$

for some nonzero integer n and some number ε such that $0 < \varepsilon < 1$.

Solution by G. W. VELTKAMP (Technological University, Eindhoven, Netherlands).

Since $T = -qr^2/(4k)$ is a solution of the inhomogeneous differential equation, we assume that the full solution can be represented as

(1)
$$T = \frac{qa^2}{4k}\left[1 - \frac{r^2}{a^2} + \sum_{j=0}^{\infty}{}' c_j\left(\frac{r}{a}\right)^{jn}\cos jn\theta\right]$$

(where the $'$ indicates that the term with $j = 0$ should be taken half). This "ansatz" is motivated by separation of variables together with the regularity condition at $r = 0$ and the observation that the mode-coupling mechanism constituted by the boundary condition will not excitate other modes than those present in (1).

We write the boundary condition as

(2)
$$-\frac{a}{n}\frac{\partial T}{\partial r} = (v + \lambda \cos n\theta)T \quad \text{at } r = a,$$

where

$$v = h_0 a/(nk), \qquad \lambda = \varepsilon v.$$

Substitution of (1) into (2) gives

$$\frac{2}{n} - \sum_{j=1}^{\infty} jc_j \cos jn\theta = (v + \lambda \cos n\theta) \sum_{j=0}^{\infty}{}' c_j \cos jn\theta$$

$$= v\sum_{j=0}^{\infty}{}' c_j \cos jn\theta + \tfrac{1}{2}\lambda\left[c_1 + \sum_{j=1}^{\infty}(c_{j+1} + c_{j-1})\cos jn\theta\right].$$

Equating coefficients of $\cos jn\theta$ gives

(3)
$$vc_0 + \lambda c_1 = 4/n,$$

(4)
$$(v + j)c_j + \tfrac{1}{2}\lambda(c_{j+1} + c_{j-1}) = 0.$$

Comparing the difference equation (4) with the recurrence relation for the Bessel functions, and observing that convergence of (1) at $r = a$ implies boundedness of the c_j, we find

$$c_j = A(-1)^j J_{j+v}(\lambda).$$

Substitution into (3) gives

$$\frac{4}{n} = A(\nu J_\nu(\lambda) - \lambda J_{\nu+1}(\lambda)) = A\lambda J_\nu'(\lambda).$$

It is well known that $J_\nu'(z) \neq 0$ for $0 < |z| < \nu$. Hence the condition $0 < \varepsilon < 1$ ensures that $J_\nu'(\lambda) = J_\nu'(\varepsilon\nu)$ is different from zero. Therefore the solution is

$$T = \frac{qa^2}{4k}\left[1 - \frac{r^2}{a^2} + \frac{4}{n}\sum_{j=0}^{\infty}{}' (-1)^j \frac{J_{j+\nu}(\varepsilon\nu)}{\varepsilon\nu J_\nu'(\varepsilon\nu)}\left(\frac{r}{a}\right)^{jn} \cos jn\theta\right],$$

where $\nu = h_0 a/(nk)$.

Problem 64–5*, *A Physical Characterization of a Sphere*, by M. S. KLAMKIN (University of Buffalo).

Consider the heat conduction problem for a solid:

$$\frac{\partial T}{\partial t} = \nabla^2 T.$$

Initially, $T = 0$.
On the boundary, $T = 1$.

The solution to this problem is well known for a sphere and, as to be expected, it is radially symmetric. Consequently, the equipotential (isothermal) surfaces do not vary with the time (the temperature on them, of course, varies). It is conjectured for the boundary value problem above, that the sphere is the only bounded solid having the property of invariant equipotential surfaces. If we allow unbounded solids, then another solution is the infinite right circular cylinder which corresponds to the spherical solution in two-dimensions.

Problem 73–24*, *An Inverse Drum Problem*, by L. FLATTO and D. J. NEWMAN (Yeshiva University).

Consider the vibration of a drum. Here a thin elastic membrane of uniform areal density is stretched to a uniform tension and is held fixed at its boundary R, which is a simple closed plane curve. It is conjectured that if there is a solution of the wave equation

$$\frac{\partial^2 z}{\partial t^2} = c^2\left\{\frac{\partial^2 z}{\partial x^2} + \frac{\partial^2 z}{\partial y^2}\right\}$$

such that ∇z vanishes along some simple closed curve within or on R, then the drum is a circular one.

Problem 62–15,* *A Property of Harmonic Functions*, by M.S. Klamkin (University of Buffalo).

A. For what functions F do there exist harmonic functions ϕ satisfying

$$\left(\frac{\partial\phi}{\partial x}\right)^2 + \left(\frac{\partial\phi}{\partial y}\right)^2 + \left(\frac{\partial\phi}{\partial z}\right)^2 = F(\phi).$$

B. Give a physical interpretation for (A).

Supplementary References
Partial Differential Equations

[1] W.F.Ames, *Nonlinear Partial Differential Equations in Engineering, I, II*, Academic Press, N.Y., 1965, 1972.

[2] H.Bateman, *Partial Differential Equations of Mathematical Physics*, Dover, N.Y., 1944.

[3] J.R.Cannon, *The One-Dimensional Heat Equation*, Addison-Wesley, Reading, 1984.

[4] H.S.Carslaw and J.C.Jaeger, *Conduction of Heat in Solids*, Clarendon Press, Oxford, 1959.

[5] C.R.Chester, *Techniques in Partial Differential Equations*, McGraw-Hill, N.Y., 1971.

[6] R.V.Churchill, *Fourier Series and Boundary Value Problems*, McGraw-Hill, N.Y., 1963.

[7] E.T.Copson, *Partial Differential Equations*, Cambridge University Press, London, 1975.

[8] R.Courant and D.Hilbert, *Methods of Mathematical Physics, I,II*, Interscience, N.Y., 1953, 1962.

[9] J.Crank, *Mathematics of Diffusion*, Oxford University Press, London,1956.

[10] J.W.Dettman, *Mathematical Methods in Physics and Engineering*, McGraw- Hill, N.Y., 1966.

[11] G.F.D.Duff, *Partial Differential Equations*, University of Toronto Press, Toronto, 1956.

[12] B.Epstein, *Partial Differential Equations*, McGraw-Hill, N.Y., 1962.

[13] A.Friedman, *Partial Differential Equations of Parabolic Type*, Prentice-Hall, N.J., 1964.

[14] A.R.Forsyth, *Differential Equations*, MacMillan, London, 1948.

[15] P.R.Garabedian, *Partial Differential Equations*, Wiley, N.Y., 1964.

[16] E.Goursat, *A Course in Mathematical Analysis, II*, Dover, N.Y., 1945.

[17] G.Hellwig, *Partial Differential Equations*, Blaisdell, N.Y., 1964.

[18] J.Jeans, *Mathematical Theory of Electricity and Magnetism*, Cambridge University Press, Cambridge, 1948.

[19] F.John, *Plane Waves and Spherical Means*, Interscience, N.Y. 1955.

[20] H.Lamb, *Hydrodynamics*, Dover, N.Y., 1945.

[21] L.M.Milne-Thompson, *Theoretical Hydrodynamics*, MacMillan, N.Y., 1950.

[22] I.G.Petrofsky, *Lectures on Partial Differential Equations*, Interscience, N.Y., 1954.

[23] M.H.Protter, H.F.Weinberger, *Maximum Principles in Differential Equations*, Blaisdell, N.Y., 1967.

[24] Lord Rayleigh, *The Theory of Sound*, Dover, N.Y., 1945.

[25] W.R.Smythe, *Static and Dynamic Electricity*, McGraw-Hill, N.Y.,1939.

[26] I.N.Sneddon, *Elements of Partial Differential Equations*, McGraw-Hill, N.Y.,1957.

[27] I.N.Sneddon, *Mixed Boundary Value Problems in Potential Theory*, InterScience, N.Y., 1966.

[28] A.Sommerfeld, *Partial Differential Equations in Physics*, Academic Press, N.Y., 1949.

[29] I.Stakgold, *Boundary Value Problems of Mathematical Physics, I,II*, MacMillan, N.Y., 1967, 1968.

[30] J.A.Stratton, *Electromagnetic Theory*, McGraw-Hill, 1941.

[31] A.N.Tychonov and A.A.Samarski, *Partial Differential Equations of Mathematical Physics*, Holden-Day, San Francisco, 1964.

[32] H.F.Weinberger, *A First Course in Partial Differential Equations*, Blaidell, N.Y., 1965.

9. DEFINITE INTEGRALS

Problem 77-3, *A Definite Integral of N. Bohr*, by P. J. SCHWEITZER (IBM Research Center).

N. Bohr [1] investigated the integral

$$K = \int_0^\infty F(x)(F'(x) - \ln x) \, dx$$

where

$$F(x) = \int_{-\infty}^\infty \frac{\cos (xy) \, dy}{(1 + y^2)^{3/2}}$$

is related to a modified Bessel function [2] and he numerically obtained the rough approximate result $K \approx -0.540$. Find an exact expression for K.

REFERENCES

[1] NIELS BOHR, *Collected Works*, vol. I, L. Rosenfeld, ed., North-Holland, Amsterdam, 1972, pp. 554–557.

[2] M. ABRAMOWITZ AND I. STEGUN, EDS., *Handbook of Mathematical Functions*, Applied Mathematics Series 55, U.S. Government Printing Office, National Bureau of Standards, 1964, Eq. 9.6.25.

Solution by D. E. AMOS (Sandia Laboratories).

Express K in the form

$$K = \left[\frac{F^2(x)}{2}\right]_0^\infty - \int_0^\infty \ln x F(x) \, dx$$

where

$$F(x) = 2 \int_0^\infty \frac{\cos (xy)}{(1 + y^2)^{3/2}} \, dy = 2x K_1(x).$$

Then, with $\lim_{x \to 0} x K_1(x) = 1$, we have

$$K = -2 - 2 \int_0^\infty x (\ln x) K_1(x) \, dx.$$

Consider the integral [2, p. 486]

$$G(a) = \int_0^\infty x^a K_1(x) \, dx = 2^{a-1} \Gamma\left(\frac{a}{2} + 1\right) \Gamma\left(\frac{a}{2}\right) = 2^{a-2} a \Gamma^2\left(\frac{a}{2}\right).$$

Then

252

$$G'(1) = \int_0^\infty x(\ln x) K_1(x)\, dx.$$

and

$$K = -2 - 2G'(1).$$

Logarithmic differentiation of $G(a)$ produces

$$G'(a) = G(a)\left[\ln 2 + \frac{1}{a} + \psi\left(\frac{a}{2}\right)\right] = 2^{a-2} a \Gamma^2\left(\frac{a}{2}\right)\left[\ln 2 + \frac{1}{a} + \psi\left(\frac{a}{2}\right)\right]$$

and

$$G'(1) = \frac{\pi}{2}[1 - \gamma - \ln 2],$$

where γ is Euler's constant. Hence

$$K = -2 - \pi[1 - \gamma - \ln 2] \approx -1.15063.$$

Editorial note. Bohr's comments on the evaluation of K are contained in a letter which he wrote to his brother, Harald Bohr. Bohr first derives a series expansion for F based on the fact that F satisfies the differential equation $F'' - (1/x)F' - F = 0$. He also derives an asymptotic expansion for F. Then, evidently, Bohr employs the two series representations in appropriate intervals and uses numerical integration techniques to evaluate K. After what he describes as "some days of numerical drudgery", he obtains $K \approx -0.540$. The exact source of error in Bohr's result is, perhaps, a subject for historical speculation. Amos notes the interesting numerical fact that $(4/\pi)G'(1) \approx -0.54073$, which fosters speculation to the effect that the numerical value quoted by Bohr refers to only part of the integral defining K. In any case, it is worth noting that Bohr's basic approach is viable enough. With the aid of a computer, it is a relatively easy matter to implement Bohr's program and so obtain $K \approx -1.15063$. [C.C.R.]

Problem 77-8, A Definite Integral, by M. L. GLASSER (Clarkson College of Technology).
　　Prove that

$$\int_0^\infty \frac{\log |J_0(x)|}{x^2}\, dx = -\frac{\pi}{2}.$$

Editorial note. A class of such integrals has been treated by B. Berndt and the author by complex integration (Aequationes Math., to appear).
　　Solution by the proposer.
　　The trigonometric integral

(1)
$$PV \int_0^\infty \frac{\tan u}{u}\, du = \frac{\pi}{2}$$

can be traced back to Euler and can be derived by elementary means. On the other hand the analogous Bessel function integral

(2)
$$PV \int_0^\infty \frac{J_1(u)}{u J_0(u)}\, du$$

appears to be unknown and several attempts to evaluate it based on analogy with (1) have been unsuccessful. In this solution, we describe a method by which (1) and (2) can be evaluated along with a host of related results all of which appear to be new.

Our procedure is based on Wiener's generalized Tauberian theorems [1], which can be rephrased as follows:

THEOREM. *Let K_0 be such that*

(i)
$$\sum_{k=-\infty}^{\infty} \max_{k \leq x \leq k+1} |K_0(x)| < \infty,$$

(ii)
$$k_0(u) \equiv \int_{-\infty}^{\infty} dx\, e^{iux} K_0(x) \neq 0, \quad \text{for real } u,$$

and let $dg(x)$ be a finite measure, where $g(x)$ is of bounded variation. Then, if

$$\lim_{x \to \infty} \int_{-\infty}^{\infty} K_0(x-y)\, dg(y) = A k_0(0),$$

then

$$\lim_{x \to \infty} \int_{-\infty}^{\infty} K(x-y)\, dg(y) = A k(0)$$

for any function K which satisfies (i) *and where $k(u)$ is defined in analogy with* (ii).

As an example, let $K_0(x) = e^{-x} \ln (1+e^{2x})$ and let $dg_0(x) = e^{-x} dv_0(e^x)$, where $v_0(x)$ is the number of positive zeros of $J_0(x)$ less than or equal to x. It is clear that the hypotheses of the theorem are satisfied, where $k_0(u) = \pi(1-iu)^{-1} \operatorname{sech}(\pi u/2)$. Now consider

$$F_0(x) = \int_{-\infty}^{\infty} K_0(x-y)\, dg_0(y) = e^{-x} \int_{-\infty}^{\infty} \ln (1+e^{2(x-y)})\, dv_0(e^y).$$

If $s = e^x$, $t = e^y$, then

$$F_0(x) = \frac{1}{s} \int_0^{\infty} \ln \left(1 + \frac{s^2}{t^2}\right) dv_0(t) = \frac{1}{s} \sum_{n=1}^{\infty} \ln \left(1 + \frac{s^2}{j_{0,n}^2}\right) = \frac{1}{s} \ln I_0(s).$$

Thus,

$$\lim_{x \to \infty} F_0(x) = \lim_{x \to \infty} e^{-x} \ln I_0(e^x) = 1$$

since $I_0(u) \sim e^u (2\pi u)^{-1/2}$. Because $k_0(0) = \pi$, we see that $A = 1/\pi$. Therefore, for any function K such that $\sum_{-\infty}^{\infty} \max_{k \leq x \leq k+1} |K(x)| < \infty$ we have

$$\lim_{x \to \infty} \int_{-\infty}^{\infty} K(x-y)\, dg_0(y) = \frac{1}{\pi} \int_{-\infty}^{\infty} K(y)\, dy.$$

In particular, let

$$K(x) = \int_{-\infty}^{x} e^{-u} \ln |1 - e^{2u}|\, du,$$

which decays exponentially for $|x| \to \infty$. Then

$$F(x) = \int_{-\infty}^{\infty} K(x-y)\, dg_0(y) = \frac{1}{2} \int_0^{\infty} dv_0(t) \int_s^{\infty} du \ln \left[\left(1 - \frac{1}{u^2 t^2}\right)^2\right]$$

where $s = e^{-x}$, $t = e^y$. By changing the order of integration, as allowed by Fubini's theorem, we find

$$\lim_{x \to \infty} F(x) = \frac{1}{2} \int_0^\infty du \ln J_0^2 \left(\frac{1}{u} \right) = \frac{1}{\pi} \int_{-\infty}^\infty \int_{-\infty}^x e^{-u} \ln |1 - e^{2u}| \, du$$

$$= -\frac{1}{2\pi} \int_{-\infty}^\infty du \ln \left(\frac{1+e^u}{1-e^u} \right)^2 = -\frac{\pi}{2}.$$

After an integration by parts we have

$$PV \int_0^\infty \frac{du}{u} \frac{J_1(u)}{J_0(u)} = \frac{\pi}{2}.$$

Proceeding in the same way, but replacing $J_0(x)$ by $\Gamma(\nu+1)x^{-\nu}J_\nu(x)$, we obtain

$$\int_0^\infty \frac{du}{u^2} \{ \nu \ln u - \ln 2^\nu \Gamma(\nu+1) - \ln |J_\nu(u)| \} = \frac{\pi}{2}, \qquad \nu > -\frac{1}{2}.$$

By a slight change in the choice of K and use of the formula [2]

$$ber^2 \frac{1}{x} + bei^2 \frac{1}{x} = \prod_{n=1}^\infty [1 + (xj_{0,n})^{-4}],$$

we obtain

$$\int_0^\infty du \ln \left\{ ber^2 \frac{1}{u} + bei^2 \frac{1}{u} \right\} = \left(\frac{\pi}{8} - 1 \right) \sqrt{2}.$$

REFERENCES

[1] N. WIENER, *The Fourier Integral and Certain of its Applications*, Dover, New York, p. 74.
[2] M. L. GLASSER AND M. S. KLAMKIN, unpublished.

A Conjectured Definite Integral

Problem 85-16*, by A. H. NUTTALL (Naval Underwater Systems Center, New London, CT).

It is conjectured that

$$\int_0^\pi \left\{ \frac{\sin x}{x} \exp \left(x \cot x \right) \right\}^\nu dx = \frac{\pi \nu^\nu}{\Gamma(1+\nu)} \qquad \text{for } \nu \geq 0.$$

Prove or disprove.

The integral arose in a study of cross correlators. The conjectured result was discovered numerically first from the result for $\nu = \frac{1}{2}$ for which the computer output for the integral was recognized as $\sqrt{2\pi}$. The above result has been confirmed numerically to 15 decimal places for numerous values of ν in the range $[0, 150]$.

Solution by C. J. BOUWKAMP (Technische Hogeschool Eindhoven, Eindhoven, the Netherlands).

Nuttall's conjecture is true. Further, the integral is a special case of a general class of integral representations. Let $r(\beta) > 0$ be continuously differentiable on $[0, \pi)$ with

$r(\beta) \to \infty$ as $\beta \to \pi$; define

$$f(\nu,\beta):= \cos\{\nu\{r\sin\beta - \beta)\}\} + (r'/r)\sin\{\nu(r\sin\beta - \beta)\}.$$

Then

(1) $$\int_0^\pi r^{-\nu} \exp(\nu r \cos\beta) f(\nu,\beta)\, d\beta = \frac{\pi\nu^\nu}{\Gamma(\nu+1)}.$$

The special (and most simple) choice for r is $r(\beta)=\beta/\sin\beta$, which makes $f=1$. Then replacing β by x gives the required formula.

The proof of (1) goes via a Hankel-type integral,

$$\frac{i}{2}\int_\infty^{(0^+)} (-s)^{-\nu-1} e^{-\nu s}\, ds = \frac{\pi\nu^\nu}{\Gamma(\nu+1)},$$

the integration-path being parametrized through polar coordinates:

$$s=-r(\beta)\exp(i\beta), \qquad ds=-ir(1-ir'/r)\exp(i\beta)\, d\beta,$$

and assuming that the path is symmetric with respect to the real axis.

Editorial note. Most solvers employed contour integration in some fashion using a Hankel-type integral as above. Jones wrote

$$\lim_{x\to 0} \pi \left(\frac{2}{x}\right)^\nu J_\nu(\nu x) = \frac{\pi\nu^\nu}{\nu!}$$

where

$$J_\nu(\nu x) = \frac{1}{\pi}\int_0^\pi \exp\{-\nu F(\theta,x)\}\, d\theta,$$

and

$$F(\theta,x) = -\cot\theta (\theta^2 - x^2\sin^2\theta)^{1/2} + \ln\{\theta + (\theta^2 - x^2\sin^2\theta)^{1/2}\}/x\sin\theta,$$

Mallows specialized his previous result [1] which was proved independently [2]. Cosgrove and Glasser also used Mallow's result which they generalized and will describe elsewhere.

Hornor and Rousseau point out the connection between this problem and the analytic continuation of the series $\sum_{n=1}^\infty (n^{n-1}/n!)w^n$ which is a well-known Lagrange–Bürman expansion for the solution of $w=ze^{-z}$.

REFERENCES

[1] C. L. MALLOWS, *Problem 6245*, posed December, 1978, solution in Amer. Math. Monthly, 83 (1980), p. 584.
[2] R. EVANS, M. E. H. ISMAIL AND D. STANTON, *Coefficients in expansions of certain rational functions*, Canad. J. Math., 34 (1982), pp. 1011–1024.

A Normalization Constant

Problem 86-9 by JEROLD R. BOTTIGER (U. S. Army Chemical Research and Development Center) *and* DAVID K. COHOON (Temple University).

We wish to predict the orientation of a small axisymmetric particle carrying a fixed electrical dipole moment $\vec{\mu}$ which is suspended in a gas with temperature T and is in an electric field whose electric vector $\vec{E}$ is assumed to be in the direction of the positive z-axis of the laboratory coordinate system. We let θ be the angle between the electric field vector $\vec{E}$ and a vector $\vec{R}$ which is parallel to the axis of symmetry of the particle and we let α be the fixed angle between $\vec{R}$ and the dipole moment vector $\vec{\mu}$ which is fixed in a moving coordinate system whose axes are defined by poles and other points on the tumbling particle. The probability that the angle θ between the axis of symmetry of the particle and the electric vector $\vec{E}$ satisfies $\theta_1 < \theta < \theta_2$ is

$$\int_{\theta_1}^{\theta_2} P_a(\theta)\, d\theta = c \int_{\theta_1}^{\theta_2} \exp\left(b\cos(\alpha)\cos(\theta)\right) I_0(b\sin(\alpha)\sin(\theta))\sin(\theta)\, d\theta$$

where c is a normalization constant, b is directly proportional to the product of the lengths of the dipole moment and electric field vectors and is inversely proportional to temperature T of the gas, and I_0 is the zeroth order modified Bessel function. Find c by evaluating the right side of the equation

$$\frac{1}{c} = \int_0^{\pi} \exp\left(b\cos(\alpha)\cos(\theta)\right) I_0(b\sin(\alpha)\sin(\theta))\sin(\theta)\, d\theta,$$

thereby determining the probability distribution $P_a(\theta)$.

Solution by M. L. GLASSER (Clarkson University).

In spherical coordinates the average of $f(z)$ over the unit sphere

$$\frac{1}{2\pi}\int_0^{\pi}\sin\theta\, d\theta \int_{-\pi}^{\pi} d\phi f(\cos\theta) = \int_{-1}^{1} f(x)\, dx$$

is clearly invariant under rotations of the sphere. In particular, under a rotation about the x-axis by an angle α

$$z \to \cos\alpha\cos\theta - \sin\alpha\sin\theta\sin\phi.$$

Hence for any function f

(*) $$\frac{1}{2\pi}\int_{-\pi}^{\pi} d\phi \int_0^{\pi}\sin\theta\, d\theta\, f(\cos\alpha\cos\theta - \sin\alpha\sin\theta\sin\phi) = \int_{-1}^{1} f(x)\, dx.$$

Since

$$I_0(x) = \frac{1}{2\pi}\int_{-\pi}^{\pi} e^{-x\sin\phi}\, d\phi,$$

by choosing $f(x) = e^{bx}$ we obtain $c = \pi b\, \mathrm{sech}\, b$.

In many cases at least one of the integrations in (*) is known, leading to an integral identity. For example,

(i) $f(x) = J_0(xy)$, $\alpha = \pi/2$ gives ($x = \cos\theta$)

$$\int_0^1 \frac{xJ_0^2(xy)}{\sqrt{1-x^2}}\, dx = J_0(2y) + \frac{\pi}{2}[H_0(2y)J_1(2y) - H_1(2y)J_0(2y)];$$

(ii) $f(x) = K_0(|xy|)$, $\alpha = \pi/2$ gives

$$\int_0^1 dx \frac{xI_0(xy)K_0(xy)}{\sqrt{1-x^2}} = K_0(2y) + \frac{\pi}{2}[L_1(2y)K_0(2y) - L_0(2y)K_1(2y)].$$

Neither of these appears to have been tabulated.

Problem 68-17, *A Definite Integral*, by B. F. LOGAN, C. L. MALLOWS and L. A. SHEPP (Bell Telephone Laboratories).

Evaluate the integral

$$I = \int_0^\infty e^{-w/2}\sqrt{w}\, du,$$

where

$$w = \frac{u}{1 - e^{-u}}.$$

The integral arose in a probability problem.

Solution by WILLIAM B. JORDAN (G. E. Knolls Atomic Power Laboratory).

Letting $w = u + v$, it follows that

$$e^u = w/v, \qquad\qquad we^{-w} = ve^{-v},$$
$$du = dw/w - dv/v, \qquad u]_0^\infty = w]_1^\infty = v]_1^0.$$

Whence,

$$I = \int_{u=0}^\infty \{we^{-w}\}^{1/2}\left\{\frac{dw}{w} - \frac{dv}{v}\right\}$$

or

$$I = \int_1^\infty \{we^{-w}\}^{1/2}\frac{dw}{w} - \int_1^0 \{ve^{-v}\}^{1/2}\frac{dv}{v}$$

$$= \int_0^\infty t^{-1/2}e^{-t/2}\, dt = \sqrt{2\pi}.$$

M. L. GLASSER (Battelle Memorial Institute) obtains the following generalization by means of the generating function for the generalized Laguerre polynomials:

$$\int_0^\infty \{ye^{-y}\}^v\, dx = \int_0^\infty \{ye^{-y}\}^v\, dy, \qquad v > 0,$$

where $y = x/(e^x - 1)$ or $x/(1 - e^{-x})$.

In the solution by the proposers, it was noted that Logan and Shepp obtained the result $(2\pi)^{-1/2}I = 1$ as the expression of the probabilistic fact that a standard Wiener process is almost certain to meet a boundary curve. Additionally, Mallows obtains the still further generalization:

Suppose (i) $h(x)$ and its derivative $h'(x)$ are positive and monotone for $-\infty < x < \infty$, with $h(x) = h(-x) + x$ and $h(-\infty) = 0$. Also suppose (ii) $g(w)$ (defined for $0 \leq w < \infty$) satisfies $g(h(x)) \equiv g(h(-x))$ for $-\infty < x < \infty$. Then $\int_0^\infty g(h(x))\, dx$ $= \int_0^\infty g(w)\, dw$ (with slight additional generality, for any f we have $\int_0^\infty f(g(h(x)))\, dx$ $= \int_0^\infty f(g(w))\, dw$). The proof is straightforward. In the present case we have $g(w) = (we^{-w})^{1/2}, h(x) = x/(1 - e^{-x})$.

In general, if the function $g(h(x))$ is given, it may not be easy to see how to choose h so that (i) and (ii) are satisfied. If h is given satisfying (i) (there are many such functions; it is only necessary that $h'(x) - \frac{1}{2}$ is odd and monotone, with $h(-\infty) = 0$), then it is easy to construct many functions $g(w)$ satisfying (ii). Any function of $|h^{-1}(w)|$ will do; alternatively we may write $g(w) = G(w, w - h^{-1}(w))$ where $G(w, w')$ is any symmetric function. (Then $g(w) = G(h(x), h(-x))$, where $w = h(x)$.) In the present case $G(w, w') = (|w - w'|\min(w, w')/ww')^{1/2}$.

Problem 61-9*, A Definite Integral, by W. L. BADE (Avco Research and Advanced Development Division).

Evaluate the integral

$$Q = \int_0^\infty xe^{-2x}\{\psi(x)\}^2\, dx,$$

where

$$\psi(x) = \int_0^{\pi/2} \{1 - e^{x(1-\csc\theta)}\} \sec^2\theta\, d\theta.$$

Physically Q is proportional to the cross-section corresponding to the exponential repulsive potential $\phi = Ae^{-r/r_0}$ in the limit of high relative velocity.

A numerical integration for Q gives a value of 0.3333. Consequently, it is conjectured that $Q = \frac{1}{3}$.

Solution: Most of the solutions were based on the explicit evaluation of $\psi(x)$ which can be accomplished by a variety of substitutions, i.e., $y = \csc\theta$, $\cosh u = \csc\theta$, etc. This yields

$$\psi(x) = xe^x \int_0^\infty e^{-x\cosh u}\, du = xe^x \int_1^\infty \frac{e^{-xy}}{\sqrt{y^2 - 1}}\, dy = xe^x K_0(x),$$

and thus

$$Q = \int_0^\infty x^3 K_0^2(x)\, dx.$$

R.E. Simmons, by using the known integral

$$\int_0^\infty t^{u-1} K_\nu(t)\, dt = 2^{u-2} \Gamma\left(\frac{u - \nu}{2}\right) \Gamma\left(\frac{u + \nu}{2}\right),$$

(Watson, p. 388) also obtains the following generalization of Q:

$$Q(\lambda) = \int_0^\infty x^{\lambda-3} e^{-2x}\{\psi(x)\}^2\, dx = 2^{\lambda-3} \frac{\Gamma^4(\lambda/2)}{\Gamma(\lambda)}.$$

Another Definite Integral

Problem 84-9, by M. L. GLASSER (Clarkson College).
Evaluate

$$\int_0^\infty \frac{C_0(|\alpha x - \beta/x|)}{x^3+1}\, dx$$

where $C_0(z)=aJ_0(z)+bY_0(z)+cK_0(z)$ and $\alpha, \beta \geq 0$.

Solution by the proposer.
Let $u = \alpha x - \beta/x$, so

(*) $$x^2 - \left(\frac{u}{\alpha}\right)x - \frac{\beta}{\alpha} = 0.$$

Then, if F is any function,

$$\int_{-\infty}^\infty \frac{F(u)}{x^2+1}\, dx = \int_{-\infty}^\infty F(u)\left[\frac{x_1'(u)}{x_1^2+1} + \frac{x_2'(u)}{x_2^2+1}\right] du,$$

where x_1, x_2 are the roots of (*) and the prime denotes the derivative with respect to u. But the quantity in square brackets is

$$\frac{d}{du}\left[\tan^{-1}x_1 + \tan^{-1}x_2\right] = \frac{d}{du}\tan^{-1}\left(\frac{x_1+x_2}{1-x_1x_2}\right) = \frac{\alpha+\beta}{u^2+(\alpha+\beta)^2},$$

since by Newton's identities $x_1 + x_2 = u/\alpha$ and $x_1 x_2 = -\beta/\alpha$. Hence, for any even function F,

$$\int_0^\infty \frac{F(u)}{x^2+1}\, dx = (\alpha+\beta)\int_0^\infty \frac{F(x)}{x^2+(\alpha+\beta)^2}\, dx.$$

Therefore, from known results [1, eqs. 2, 4, 5, p. 330],

$$\int_0^\infty \frac{J_0(\alpha x - \beta/x)}{x^2+1}\, dx = \frac{\pi}{2}\left[I_0(\alpha+\beta) - \mathbf{L}_0(\alpha+\beta)\right],$$

$$\int_0^\infty \frac{Y_0(|\alpha x - \beta/x|)}{x^2+1}\, dx = -K_0(\alpha+\beta),$$

$$\int_0^\infty \frac{K_0(|\alpha x - \beta/x|)}{x^2+1}\, dx = \frac{\pi^2}{4}\left[\mathbf{H}_0(\alpha+\beta) - Y_0(\alpha+\beta)\right].$$

REFERENCES

[1] Y. L. LUKE, *Integrals of Bessel Functions*, McGraw-Hill, New York, 1962.

Problem 63-10, *A Definite Integral*, by J. L. BROWN, JR. (Pennsylvania State University).

Evaluate the definite integral

$$I_m = \int_0^\infty \frac{\sin mx}{x} J_0{}^m(x) \, dx,$$

where m is an arbitrary positive integer.

Solution by WILLIAM F. TRENCH (Radio Corporation of America).

Starting with the known integral representation

$$J_0(x) = \frac{1}{2\pi} \int_{-\pi}^\pi e^{ix\cos\theta} \, d\theta,$$

it follows that

$$I_m = \frac{1}{2(2\pi)^m} \int_{\theta_1=-\pi}^\pi \cdots \int_{\theta_m=-\pi}^\pi \left\{ \int_{-\infty}^\infty \frac{\sin mx}{x} e^{ix(\cos\theta_1+\ldots+\cos\theta_m)} \, dx \right\} d\theta_1 \cdots d\theta_m \, .$$

Since

$$\int_{-\infty}^\infty \frac{\sin tx}{x} e^{ixy} \, dx = \pi$$

for $t > 0$ and $|y| < t$, it follows that $I_m = \pi/2$. Also, in a similar fashion one can show that

$$\int_0^\infty \frac{\sin tx}{x} J_0{}^m(x) \, dx = \frac{\pi}{2},$$

where m is any positive integer $\leq t$, and that

$$\int_0^\infty \frac{\sin tx}{x} J_n{}^m(x) \, dx = 0$$

if m and n are positive integers, at least one of which is even, and $t \geq m$.

A Definite Integral of Bessel Functions

Problem 87-18, *by* M. L. GLASSER (Clarkson University).

Various integrals of the function $F(x) = I_0(x) - L_0(x)$ (where $L_0(x)$ is a modified Struve function) arise in the study of the diffusion of a swarm of charged particles normal to an applied electric field [1]. Let

$$J_n = \int_0^\infty \{F(x)\}^n \, dx.$$

It is known that J_1 is divergent. Show that $J_2 = 1$ and $\lim nJ_n = \pi/2$. Can any other values of J_n be obtained in closed form?

REFERENCE

[1] M. L. GLASSER, *Line shape analysis for ion mobility spectroscopy*, Analytical Chemistry, submitted.

Solution by A. A. JAGERS (Universiteit Twente, Enschede, the Netherlands).

Since

$$F(x) = \frac{2}{\pi} \int_0^\infty \sin (xt)(1+t^2)^{-1/2} \, dt \qquad (x>0),$$

we have

$$J_2 = \int_0^\infty \{F(x)\}^2 \, dx = \frac{2}{\pi} \int_0^\infty (1+t^2)^{-1} \, dt = 1$$

by Parseval's identity for the Fourier sine transform. From a second integral representation

$$F(x) = \frac{2}{\pi} \int_0^{\pi/2} e^{-x\cos\theta} \, d\theta,$$

it follows that F is a positive decreasing function on $[0, \infty)$ with $F(0) = 1$. Then $\lim nJ_n = \lim (n+1)J_n = \lim (n+1) \int_0^\delta F(x)^n \, dx$ for all $\delta > 0$, and hence $\lim_{n\to\infty} nJ_n = \lim_{\delta\downarrow 0} \lim_{n\to\infty} \{F(\delta)^{n+1} - 1\}/F'(0) = -1/F'(0) = \pi/2$.

Editor's note. A. Adams and O.G. Ruehr computed J_n as follows:

$J_3 = .58757\,01401$, $\quad J_4 = .42369\,89107$, $\quad J_5 = .33257\,14107$,
$J_6 = .27402\,74648$, $\quad J_7 = .23312\,00349$, $\quad J_8 = .20288\,55443$, $\quad J_9 = .17961\,51120$.

They noted the formal asymptotic expansion

$$\int_0^\infty [u(x)]^n \, dx = \frac{1}{a_1 n} + \frac{2a_2 - a_1^2}{2a_1^3 n^2} + \frac{a_1^4 - 6a_1^2 a_2 + 12a_2^2 - 6a_1 a_3}{6a_1^5 n^3} + \cdots$$

where a_j are the Maclaurin coefficients of $u(x)$, and reported that it was numerically effective in this case.

Problem 69-12, *An Elliptic Integral*, by A.W. Gillies (The City University. London, England).

Show that

$$J = \int_0^1 \frac{2D(k) - K(k)}{\sqrt{1-k^2}} \frac{dk}{k} = \frac{\pi}{4},$$

where $k^2 D(k) = K(k) - E(k)$ and $K(k)$, $E(k)$ are the complete elliptic functions of the first and second kind, respectively.

The integral arose in a problem in electromagnetism.

Solution by J. K. M. JANSEN (Technological University, Eindhoven, The Netherlands).

Let

$$I(\varphi) = \int_0^1 \frac{(2 - k^2)F(\varphi, k) - 2E(\varphi, k) + \frac{1}{2}k^2 \sin 2\varphi}{k^3 \sqrt{1-k^2}} \, dk,$$

where $F(\varphi, k)$ and $E(\varphi, k)$ are incomplete elliptic functions of the first and second kind, respectively. Then, $J = I(\pi/2)$. Since

$$\frac{dI(\varphi)}{d\varphi} = \int_0^1 \frac{[\sqrt{1 - k^2 \sin^2 \varphi} - 1]\cos 2\varphi}{k\sqrt{1 - k^2}\sqrt{1 - k^2 \sin^2 \varphi}} \, dk = (\cos 2\varphi) \log |\cos \varphi|,$$

$$I\left(\frac{\pi}{2}\right) = I(0) + \int_0^{\pi/2} (\cos 2\varphi) \log \cos \varphi \, d\varphi = \frac{\pi}{4}.$$

Editorial note. Fettis notes that

$$2D(k) - K(k) = \frac{1}{k} Q_{1/2}\left(\frac{2}{k^2} - 1\right),$$

where $Q_{1/2}$ is an associated Legendre function, and also obtains the generalization

$$\int_0^1 \frac{Q_{n-1/2}((2/k^2) - 1)}{k^2\sqrt{1 - k^2}} \, dk = \frac{\pi}{4n}, \qquad n \geq 1. \quad \text{[M.S.K.]}$$

Problem 71-15, *An Elliptic Integral*, by M. L. GLASSER (Battelle Memorial Institute).

Show that

$$\int_0^1 \frac{K(u) \, du}{\sqrt{1 - u}} = 2\sqrt{2a}K(a)K(a'),$$

where $K(k)$ is the complete elliptic integral of the first kind, $a = \sqrt{2} - 1$, $a' = \sqrt{1 - a^2}$.

Editorial note. The proposer comments that he obtained the result as a special case of a more complicated integral (to be published) and hopes that in iew of its apparent simplicity, a more direct evaluation can be found. [M.S.K.]

Solution by H. E. FETTIS (Wright-Patterson AFB).

Starting with the known result [1, No. 531.07]

$$I = \int_0^{\pi/2} \int_0^{\pi/2} \frac{dx \, dy}{(1 - k_1^2 \sin^2 x - k_2^2 \sin^2 y)^{1/2}} = \frac{2}{1 + k_2'} K(k_3)K'(k_4),$$

where

$$k_3 = \frac{k_1' - (1 - k_1^2 - k_2^2)^{1/2}}{1 + k_2'}, \qquad k_4 = \frac{k_1' + (1 - k_1^2 - k_2^2)^{1/2}}{1 + k_2'},$$

we set $k_1^2 = k_2^2 = k^2$, so that $k_3 = k_4 = k'/(1 + k)$. Next we make the transformation

$$\sin^2 x = \frac{p^2}{1 + p^2}, \qquad \sin^2 y = \frac{q^2}{1 + q^2},$$

giving

$$I = \int_0^\infty \int_0^\infty \frac{dp\,dq}{\{(1 + p^2)(1 + q^2)(1 + k'^2 p^2 + k^2 q^2)\}^{1/2}}.$$

On letting $k'p = r \cos \theta$, $kq = r \sin \theta$, we obtain

$$I = \frac{1}{kk'} \int_0^\infty \int_0^{\pi/2} \frac{r\,dr\,d\theta}{\{(1 + r^2)(1 + r^2 k^{-2} \sin^2 \theta)(1 + r^2 k'^{-2} \cos^2 \theta)\}^{1/2}}.$$

Integrating the latter with respect to θ (see [1, No. 284.00]) and setting $r^2 = t$ gives

$$\int_0^\infty \frac{K(F(t))\,dt}{\{(1 + t)(t^2 + t + \lambda)\}^{1/2}} = \frac{4}{1 + k} K\left(\frac{k'}{1 + k}\right) K'\left(\frac{k'}{1 + k}\right),$$

where $F(t)^2 = (t^2 + t)/(t^2 + t + \lambda^2)$ and $\lambda = kk'$.

The proposed problem is a special case of the latter and follows by letting $\lambda = \frac{1}{2}$, applying Gauss' transformation and letting $u = t/(1 + t)$.

REFERENCE

[1] P. F. BYRD AND M. D. FRIEDMAN, *Handbook of Elliptic Integrals for Engineers and Scientists*, Springer-Verlag, Heidelberg, 1971.

Also solved by O.G. Ruehr (Michigan Technological University) and the proposer. pp.294-296

Editorial note. All three solvers also showed that

$$K(a) = \frac{\pi^{3/2}(1 + a^2)^{-1/2}}{2\Gamma(\frac{5}{8})\Gamma(\frac{7}{8})}, \qquad K(a') = \frac{\pi^{3/2}(1 + a^2)^{-1/2}}{\sqrt{2}\Gamma(\frac{5}{8})\Gamma(\frac{7}{8})}.$$

The latter is not new since it is known that

$$\int_0^1 \frac{dx}{\sqrt{1 - x^8}} = \frac{K(\sqrt{2} - 1)}{\sqrt{2}} = \frac{1}{8} \frac{\Gamma(\frac{1}{8})\Gamma(\frac{1}{2})}{\Gamma(\frac{5}{8})}. \qquad \text{[M.S.K.]}$$

Problem 71-27, *An Infinite Integral,* by HEDAYAT YASAIMAIBODI (George Washington University).

Show that

$$\int_{-\infty}^\infty \{x'x - (x - a)'A^{-1}(x - a)\} \exp\{-(x - a)'A^{-1}(x - a)\}\,dx$$

$$= \frac{(k - 1 + 2a'a)\pi^{k/2}}{2},$$

where a and x are $k \times 1$ matrices and

$$A^{-1} = \begin{bmatrix} 1 & 1 & 1 & 1 & \cdots & 1 \\ 1 & 2 & 2 & 2 & \cdots & 2 \\ 1 & 2 & 3 & 3 & \cdots & 3 \\ \vdots & & & & & \vdots \\ 1 & 2 & 3 & 4 & \cdots & k \end{bmatrix}.$$

Editorial note. The proposer notes that the above result has some applications in stochastic processes and information theory.

Solution by INGRAM OLKIN (Stanford University).

We prove the following more general result, which is somewhat standard in multivariate analysis.

LEMMA. *Let Y, U and M be $k \times t$ matrices $(t \leq k)$, and let Σ be a $k \times k$ positive definite matrix. If*

$$I_{k,t}(M, B; U, \Sigma)$$

(1)
$$= \int_{\Omega(Y)} \frac{|\Sigma|^{-k/2}}{(2\pi)^{kt/2}} [\operatorname{tr}(Y + M)'B^{-1}(Y + M)] \exp[-(1/2)\operatorname{tr}(Y + U)'\Sigma^{-1}(Y + U)] dY,$$

where $\Omega(Y) = \{Y: -\infty < y_{ij} < \infty,\ \text{all } i,j\}$, then

$$I_{k,t}(M, B; U, \Sigma) = \operatorname{tr} B^{-1}[t\Sigma + (U - M)(U - M)'].$$

Proof. A simplification in (1) is achieved by letting $Z = \Sigma^{-1/2}(Y + U)$. The Jacobian of this transformation is $|\Sigma|^{k/2}$, and (1) becomes

$$I_{k,t}(M, B; U, \Sigma)$$

$$= \int_{\Omega(Z)} \frac{1}{(2\pi)^{kt/2}} [\operatorname{tr}(\Sigma^{1/2}Z - U + M)'B^{-1}(\Sigma^{1/2}Z - U + M)] \exp[-(1/2)\operatorname{tr} Z'Z] dZ.$$

$$= \int_{\Omega(Z)} \frac{1}{(2\pi)^{kt/2}} [\operatorname{tr} \Sigma^{1/2}B^{-1}\Sigma^{1/2}Z'Z - 2\operatorname{tr} Z'\Sigma^{1/2}B^{-1}(U - M)$$

$$+ \operatorname{tr}(U - M)'B^{-1}(U - M)] \exp[-(1/2)\operatorname{tr} ZZ'] dZ.$$

By symmetry, the second term is zero. If we let

$$w_{ij} = \int_{\Omega(Z)} (Z'Z)_{ij} \exp[-(1/2)\operatorname{tr} Z'Z] dZ \quad \text{and} \quad W = (w_{ij}),$$

a direct computation yields $w_{ij} = 0$, $i \neq j$, and $w_{ii} = t$, so that $W = tI$. By linearity, the first term is equal to $t \operatorname{tr} \Sigma^{1/2}B^{-1}\Sigma^{1/2}$, and the result follows from the fact that

$$\int_{\Omega(Z)} \frac{1}{(2\pi)^{kt/2}} \exp[-(1/2)\operatorname{tr} ZZ'] dZ = 1.$$

In Problem 71-27, $t = 1$, and the answer is given by

$$(2\pi)^{k/2}|A/2|^{k/2}\{I_{k,1}(0,1;a,(1/2)A) - I_{k,1}(a,A;a,(1/2)A)]\}$$
$$= \pi^{k/2}|A|^{k/2}\{\text{tr } [aa' + A/2] - \text{tr } A^{-1}A/2\}$$
$$= \{a'a + (1/2) \text{tr } A - (1/2)k\}\pi^{k/2}|A|^{k/2}.$$

The matrix A^{-1} is well known since it is the inverse of the tridiagonal matrix

$$A = \begin{bmatrix} 2 & -1 & 0 & 0\cdots & 0 & 0 \\ -1 & 2 & -1 & 0\cdots & 0 & 0 \\ 0 & -1 & 2 & -1\cdots & 0 & 0 \\ & & & \cdots & & \\ 0 & 0 & 0 & 0\cdots +2 & -1 \\ 0 & 0 & 0 & 0\cdots -1 & 1 \end{bmatrix}.$$

and $|A| = 1$. Thus tr $A = 2k - 1$, giving the desired result.

A Definite Integral

Problem 84-8*, by J. A. MORRISON (Bell Laboratories, Murray Hill, NJ).
   Prove directly that

$$I \equiv \int_0^1 \left|\frac{2}{\pi} \tan^{-1}\left[\frac{2}{\pi} \tan^{-1}\left(\frac{1}{x}\right)\right] + \frac{1}{\pi} \log\left(\frac{1+x}{1-x}\right)\right| - \frac{1}{2}\left|\frac{dx}{x} = \frac{1}{2} \log\left(\frac{\pi}{2\sqrt{2}}\right).$$

This result was obtained indirectly from the consideration of a Wiener–Hopf integral equation with kernel

$$k(z) = \sin^{-1}(e^{-|z|}) = \sum_{n=0}^{\infty} \frac{(2n)!}{(2^n n!)^2 (2n+1)} e^{-(2n+1)|z|},$$

which was investigated by the proposer (SIAM J. Appl. Math., 42 (1982), pp. 588–607). Let $k_N(z)$ denote the truncated kernel, in which the sum is from $n = 0$ to N, and let $\pm i\tau_m(N)$, $m = 1, \cdots, N$, and $\pm i\sigma_m$, $m = 1, 2, \cdots$, denote the zeros of the Fourier transforms of $k_N(z)$ and $k(z)$, respectively. Then

$$\lim_{N\to\infty} \prod_{m=1}^{N} \left[\frac{\tau_m(N)}{\sigma_m}\right]^2 = \frac{\pi}{2\sqrt{2}},$$

(although $\lim_{N\to\infty} \tau_m(N) = \sigma_m$, for fixed m), and the asymptotic evaluation of this product leads to a Riemann sum, and to the above value of I. Numerical evaluation of I showed that e^{2I} agrees with $\pi/(2\sqrt{2})$ to 11 significant figures, the difference being consistent with the error estimate for the integral.

Principal Value of an Integral

Problem 81-5*, by H. E. FETTIS (Mountain View, California).

It is known that

$$\text{PV} \int_0^{\pi} \frac{\cos \lambda \theta \; d\theta}{\cos \phi - \cos \theta} = \frac{\pi \sin \lambda \phi}{\sin \phi}$$

when λ is an integer (PV denotes the Cauchy principal value of the integral). Evaluate the integral for nonintegral λ.

Problem 66–18, An Incomplete Laplace Transform, by P. J. SHORT (White Sands Missile Range).

Evaluate in closed form:

$$I = \int_0^x e^{-t} I_0(2\sqrt{xt}) \; dt.$$

Here $I_0(z)$ is the modified Bessel function of the first kind of order zero, and $x \geq 0$.

The integral occurs in propagation studies, heat transfer and ionic exchange of columns.

Solution by S. STEIN (Sylvania Communications Systems Laboratories). Letting $t = u^2/2$, one obtains

$$I = \int_0^{\sqrt{2x}} \exp\left(-u^2/2\right) I_0(u\sqrt{2x}) \; u \; du.$$

In this form, the integral can be recognized in terms of Marcum's Q-function (see [1]), which is defined as

$$Q(a, b) = \int_b^{\infty} \exp\left(\frac{u^2 + a^2}{2}\right) I_0(au) \; u \; du.$$

For the special case $b = 0$, $Q(a, 0) = 1$. Its complement, $R(a, b) = 1 - Q(a, b)$, is familiar in operations research as the circular coverage function. A summary of results involving the Q-function was given by this solver in Appendix A of [2].

By using the known "symmetry" property of the Q-function,

(1) $$Q(a, b) + Q(b, a) = 1 + \exp\left(-\frac{a^2 + b^2}{2}\right) I_0(ab),$$

and noting that

$$I = e^x \{1 - Q(\sqrt{2x}, \sqrt{2x})\},$$

it follows that

$$I = \tfrac{1}{2} e^x \{1 - e^{-2x} I_0(2x)\}.$$

Also, since

$$I_2 = \int_0^{\infty} e^{-t} I_0(2\sqrt{xt}) \; dt = e^x Q(\sqrt{2x}, 0) = e^x,$$

it follows that

$$\lim_{x \to \infty} \frac{I}{I_2} = \frac{1}{2}.$$

REFERENCES

[1] J. I. MARCUM, *A statistical theory of target detection by pulsed radar*, RM-753, The Rand Corporation, Santa Monica, California, 1948.
[2] M. SCHWARTZ, W. R. BENNETT AND S. STEIN, *Communication Systems and Techniques*, McGraw-Hill, New York, 1966.

An Inverse Laplace Transform

Problem 88-20*, *by* O. G. RUEHR (Michigan Technological University) *and* W. R. YOUNG (Scripps Institute of Oceanography).
   Find the inverse Laplace transform of

$$e^{-xp^\alpha}/p^\gamma, \quad \mathrm{Re}\,x > 0, \quad \mathrm{Re}\,\alpha < 1.$$

What happens as $\alpha \to 1^-$? This problem has arisen in studying the dispersion of random walks along a line subject to trapping or arrest for long periods of time. Special cases have occurred earlier in electromagnetic theory [1] and heat conduction [2].

REFERENCES

[1] F. M. RAGAB, *The inverse Laplace transform of an exponential function*, Comm. Pure Appl. Math., 11 (1958), pp. 115–127.
[2] O. G. RUEHR, *A class of functions possessing elementary integral transforms*, presented at annual meeting, Society for Industrial and Applied Mathematics, 1969.

   Solutions were received from C.C. GROSJEAN (State University of Ghent) *and the proposers and a remark from* J.S. LEW (IBM Thomas J. Watson Research Center); they are outlined below.
   The proposers write $L^{-1}(\exp(-xs^\alpha)/s^\gamma) = t^{\gamma-1}\,R(x/t^\alpha)$ so that, after scaling

$$(*) \qquad\qquad e^{-y} = \int_0^\infty e^{-z}\,z^{\gamma-1}\,R(y/z^\alpha)\,dz$$

   Multiplying (*) by y^{p-1} and integrating from zero to infinity, i.e. taking the Mellin transform, they obtain, using Fubini's theorem

$$\int_0^\infty z^{p-1}\,R(z)\,dz = \Gamma(p)/\Gamma(\gamma + \alpha p)$$

Calculating the Mellin inversion by residues using the poles $p = -n$ of the gamma function yields

$$R(\alpha,\gamma;x) = \sum_{n=0}^\infty \frac{(-1)^n x^n}{n!\,\Gamma(\gamma - \alpha n)}$$

Note that R is entire in x for Re $\alpha < 1$. Thus

$$L^{-1}(\exp(-xs^\alpha)/s^\gamma) = \sum_{n=0}^\infty \frac{(-1)^n x^n t^{\gamma-\alpha n-1}}{n!\Gamma(\gamma - \alpha n)}$$

The proposers calculated $R(1 - \epsilon, 0; x)$ for small positive ϵ and found the expected sharp peaks near $x = 1$. They also point out that

$$\lim_{\alpha \to 1^-} \int_0^\infty R(\alpha, 0; x)dx = 1$$

from the Mellin transform given above.

Lew remarks that the solution of the problem should be related to Faxen's integral [1]

$$F_1(\alpha, \beta; y) = \int_0^\infty \exp(-t + y\tau^\alpha)\tau^{\beta-1}dt$$

and supplies additional references [2], ..., [7].

Indeed, comparing the Maclaurin expansions of R and F_i we see, with the aid of the gamma function reflection formula, that

$$R(\alpha, \gamma; x) = \frac{1}{2\pi i}\left\{ e^{i\pi(1-\gamma)}F_i(\alpha, 1 - \gamma; -xe^{i\pi\alpha}) - e^{i\pi(1-\gamma)}F_i(\alpha, 1 - \gamma; -xe^{-i\pi\alpha})\right\}$$

which gives us directly the representation (suitably restricted)

$$R(\alpha, \gamma; x) = \frac{1}{\pi} \int_0^\infty y^{-\gamma} \exp(-y - xy^\alpha \cos \pi\alpha) \sin[xy^\alpha \sin \pi\alpha + \pi\gamma]dy$$

Both the proposers and Grosjean found similar integral representations.

In addition, Professor Grosjean gives a lengthy, careful analysis to show the continuity of the process as $\alpha \to 1^-$ and the consequent δ-like behavior and notes that x must be real when $\alpha = 1$.

References

[1] F. W. J. Olver, *Asymptotics and Special Functions*, Academic Press, New York and London, 1974, p. 332.

[2] G. H. Hardy, *Notes on the asymptotic value of a definite integral and the coefficients in a power series*, Messenger of Mathematics, Vol. 46, 1917, pp. 70-73.

[3] William Feller, *An Introduction to Probability Theory and Its Applications*, Wiley, New York, p. 548.

[4] N. Bleistein, R. A. Handelsman, and J. S. Lew, *Functions whose Fourier coefficients decay at infinity: an extension of the Riemann-Lebesgue Lemma*, SIAM J. Math Anal., Vol. 3, No. 3, August, 1972, p. 489.

[5] E. J. Riekstiņš, *Asymptotic Expansions of Integrals*, Vol. 2 (Russian), Izdat. "Zinatne," Riga, 1977, Section 18.3 (Mathmatical Reviews 57 # 3715).

[6] H. Faxén (1921), *Expansion in series of the integral $\int_0^\infty exp[-x(t \pm t^{-n})]t^2 \, dt$*, Ark. Mat. Fys., 15, no. 13, pp. 1-57.

[7] Abramowitz and Stegun, *Handbook of Mathematical Functions*, p. 1002.

A Laplace Transform

Problem 88-13, *by* M. L. GLASSER (Clarkson University, Potsdam, New York).
The Laplace transform

$$\phi(p) = \int_0^\infty t^{-1/2} e^{-pt} e^{-t^2/4\alpha} \, dt$$

which is needed extensively in field theory and statistical mechanics, is tabulated in [1], for example. However, this (and the value listed elsewhere) is only valid for real $p \geq 0$. Show that for p real and Re $\alpha \geq 0$

$$\phi(p) = \pi \sqrt{\frac{|p|\,\alpha}{2}} e^{\alpha p^2/2} \left\{ (1 - \operatorname{sgn} p) I_{1/4}(\alpha p^2/2) + \frac{\sqrt{2}}{\pi} K_{1/4}(\alpha p^2/2) \right\}.$$

Since this is not the analytic continuation of the result for $p > 0$, it is possible that erroneous results may have appeared due to the uncritical use of the tabulated formula.

REFERENCE

[1] A. ERDELYI, ED., *Tables of Integral Transforms*, Vol. 1, McGraw-Hill, New York, 1954, p. 146.

Solution by NORBERT ORTNER *and* PETER WAGNER (University of Innsbruck).
   The Laplace transform of the function $f(t) = t^{-1/2} e^{-t^2/4\alpha}$ is given for Re $\alpha > 0$ in [1, p. 43, 5.44] by

$$\Phi(p) := \sqrt{\alpha p}\, e^{\alpha p^2/2} K_{1/4}(\alpha p^2/2), \quad p > 0.$$

By [2, p. 125, Thm. 13.8.6] the absolutely convergent integral $\Psi(p) = \int_0^\infty f(t) e^{-pt} \, dt$ is a holomorphic function of p in the whole complex plane. On the other hand, the function $\Phi(p)$ can be analytically continued to the Riemannian surface S of $\log p$ since the same holds true for the square root and the function $K_\nu(z)$. Taking into account [3, Form. 8.475.5] we obtain $K_{1/4}((e^{2\pi i}z)^2) = -K_{1/4}(z^2)$, interpreting $e^{2\pi i}z$ as the point on S which lies above z on the next leaf. Since $(e^{2\pi i}z)^{1/2} = -z^{1/2}$, we can infer that $\Phi(p)$ is actually an analytic function in $\mathbb{C}\backslash 0$ and, by continuity, even in $\mathbb{C}$. Using the uniqueness of the analytic continuation we conclude that Φ and Ψ must coincide everywhere, i.e.,

$$\int_0^\infty t^{-1/2} e^{-pt} e^{-t^2/4\alpha} \, dt = \sqrt{\alpha\,|p|}\, \exp\left[(i \arg p + \alpha p^2)/2 \right] K_{1/4}(\alpha\,|p|^2 e^{2i \arg p}/2)$$

for all complex p different from 0. Finally, we want to point out that, in contrast to the assertion of the proposer, the value of $\Psi(p)$ for negative real p can be deduced by the method of analytic continuation. Indeed, applying again [3, Form. 8.475.5], i.e., $K_{1/4}(e^{m\pi i}z) = e^{-m\pi i/4} K_{1/4}(z) - \sqrt{2}\pi i \sin(m\pi/4) I_{1/4}(z)$, with $z = \alpha p^2/2$, $p = e^{\pi i}|p| < 0$, we obtain

$$K_{1/4}(e^{2\pi i}\alpha p^2/2) = -i K_{1/4}(\alpha p^2/2) - \sqrt{2}\pi i I_{1/4}(\alpha p^2/2)$$

which together with $\sqrt{p} = i\sqrt{|p|}$ yields the desired result if Re $\alpha > 0$. In the case $\alpha \neq 0$, Re $\alpha = 0$, the given integral diverges for negative real p.

REFERENCES

[1] L. BADII AND F. OBERHETTINGER, *Tables of Laplace Transforms*, Springer-Verlag, Berlin, New York, 1973.
[2] J. DIEUDONNÉ, *Treatise on Analysis*, Vol. II, Academic Press, New York, 1970.
[3] I. S. GRADSHTEYN AND I. M. RYZHIK, *Table of Integrals, Series and Products*, Academic Press, New York, 1980.

A Moment Generating Function

Problem 85-22, *by* E. O. GEORGE *and* C. C. ROUSSEAU (Memphis State University).

The moment generating function for the mid-range of a random n-sample from the logistic distribution is given by

$$M_n(s) = n(n-1) \int_0^1 du \int_0^u dv \left[uv/((1-u)(1-v)) \right]^{s/2} (u-v)^{n-2}.$$

Evaluate this integral in terms of the gamma function.

Solution by R. G. BUSCHMAN (University of Wyoming).

Let $v = uw$ in the inner integral and convert it to an Euler integral [1, (1.1.6)] so that we have

$$M_n(s) = \left[\Gamma(n+1)\Gamma(1+s/2)/\Gamma(n+s/2) \right]$$
$$\cdot \int_0^1 u^{s+n-1}(1-u)^{-s/2} {}_2F_1\left(\begin{matrix} s/2, 1+s/2; \\ n+s/2; \end{matrix} u \right) du.$$

This integral can be evaluated [1, (4.8.3.12)] to obtain

$$M_n(s) = \left[\Gamma(n+1)\Gamma(1+s/2)\Gamma(1-s/2)\Gamma(n+s)/(\Gamma(n+s/2)\Gamma(1+n+s/2)) \right]$$
$$\cdot {}_3F_2\left(\begin{matrix} n+s, s/2, 1+s/2; \\ 1+n+s/2, n+s/2; \end{matrix} 1 \right).$$

If we now apply Dixon's theorem [1, (2.3.3.5)] and simplify, we have

$$M_n(s) = \Gamma(1+s/2)\Gamma(1-s/2)\Gamma(n/2+s/2)\Gamma(n/2-s/2)/\Gamma^2(n/2).$$

REFERENCE

[1] L. J. SLATER, *Generalized Hypergeometric Functions*, Cambridge University Press, London–Cambridge, 1966.

Problem 65-5, *A Double Integral*, by N. MULLINEUX (College of Advanced Technology, Birmingham, England) and J. R. REED (G. E. C. Engineering Co., Birmingham, England).

The following double integral arose in an electromagnetic problem where it was necessary to show that $I_0 = \pi/\alpha^2$ in order to check that the total eddy-current in a thick sheet of iron was equal and opposite to the line current carried in a conductor above the iron:

$$I_0 = \int_{-\infty}^{\infty} dx \int_0^{\infty} \frac{e^{-h\lambda} \cos x\lambda \, d\lambda}{\mu\lambda\sqrt{\lambda^2 + \alpha^2} + \lambda^2 + \alpha^2}.$$

Establish the value of I_0 and generalize the result.

Solution by D. L. LANSING (NASA, Langley Research Center).

A generalization of the integral I_0 when $\mu = 1$ occurs in the following problem which is associated with Laplace's equation in two dimensions and hence has an appropriate interpretation in electrostatics, hydrodynamics, and steady-state diffusion.

The field $\varphi(x, y)$ produced at any point of the (x, y)-plane by a distribution of line sources with strength $2\pi\alpha^{-n}J_n(\alpha y)$ per unit length along the positive y-axis is

$$\varphi(x, y) = \frac{1}{2\alpha^n} \int_0^{\infty} \log [x^2 + (y - \eta)^2] J_n(\alpha\eta) \, d\eta.$$

The flux density on the line $y = -h$ is

$$(1) \qquad -\frac{\partial\varphi(x, y)}{\partial y}\bigg]_{y=-h} = \frac{1}{\alpha^n} \int_0^{\infty} \frac{h + \eta}{x^2 + (h + \eta)^2} J_n(\alpha\eta) \, d\eta.$$

By substituting into (1) the integral representation

$$\frac{h + \eta}{x^2 + (h + \eta)^2} = \int_0^{\infty} e^{-(h+\eta)\lambda} \cos x\lambda \, d\lambda,$$

interchanging the order of integration and carrying out the η integration, one obtains the following for the flux density:

$$\int_0^{\infty} \frac{e^{-h\lambda} \cos \lambda x}{\sqrt{\lambda^2 + \alpha^2}[\lambda + \sqrt{\lambda^2 + \alpha^2}]^n} \, d\lambda.$$

The total flux F across this line is then

$$(2) \qquad F = \int_{-\infty}^{\infty} \int_0^{\infty} \frac{e^{-h\lambda} \cos \lambda x}{\sqrt{\lambda^2 + \alpha^2}[\lambda + \sqrt{\lambda^2 + \alpha^2}]^n} \, d\lambda \, dx,$$

which is a generalization of I_0 when $\mu = 1$.

But since source flow is symmetric and the field is "incompressible," F must equal one-half of the total strength of the sources producing the field. Hence,

$$(3) \qquad F = \frac{1}{2} \int_0^{\infty} \frac{2\pi J_n(\alpha\eta)}{\alpha^n} \, d\eta = \frac{\pi}{\alpha^{n+1}}.$$

Using the Dirac delta function, one can obtain this result directly from (2) by interchanging the order of integration and using the properties

$$\int_{-\infty}^{\infty} \cos x\lambda \, dx = 2\pi\delta(\lambda) \quad \text{and} \quad \int_0^{\infty} \delta(\lambda)f(\lambda) \, d\lambda = \frac{1}{2}f(0).$$

If this procedure is applied to

$$\int_{-\infty}^{\infty} \int_{0}^{\infty} \frac{e^{-h\lambda} \cos \lambda x}{\sqrt{\lambda^2 + \alpha^2}[\mu\lambda + \sqrt{\lambda^2 + \alpha^2}]^n} \, d\lambda \, dx,$$

which is a more general extension of I_0 than (2), one again obtains the value π/α^{n+1}.

Problem 67-20, A Quadruple Integral, by L. CARLITZ (Duke University).

Evaluate the quadruple integral:

$$\int_0^1 \int_0^1 \int_0^1 \int_0^1 \frac{(1 - x^2 y^2 z^2 t^2) \, dx \, dy \, dz \, dt}{\sqrt{(1 - x^2)(1 - y^2)(1 - z^2)(1 - t^2)(1 + x^2 y^2 z^2 t^2)^3}}.$$

Solution by JAMES S. PETTY (Wright-Patterson Air Force Base).

By using the Taylor series expansion

$$\frac{1 - \xi}{(1 - \xi)^{3/2}} = \sum_{n=0}^{\infty} \frac{\Gamma(n + \frac{1}{2})}{n! \, \Gamma(\frac{1}{2})} \frac{\Gamma(n + \frac{5}{4})}{\Gamma(\frac{5}{4})} \frac{\Gamma(\frac{1}{4})}{\Gamma(n + \frac{1}{4})} (-\xi)^n$$

$$= {}_2F_1\left(\begin{matrix} \frac{1}{2}, \frac{5}{4}; -\xi \\ \frac{1}{4} \end{matrix}\right),$$

where ${}_pF_q$ is a generalized hypergeometric function [4], [5], the integral may be expressed as

$$I = \sum_{n=0}^{\infty} \frac{\Gamma(n + \frac{1}{2})}{n! \, \Gamma(\frac{1}{2})} \frac{\Gamma(n + \frac{5}{4})}{\Gamma(\frac{5}{4})} \frac{\Gamma(\frac{1}{4})}{\Gamma(n + \frac{1}{4})} (-1)^n \left[\int_0^1 \frac{x^{2n} \, dx}{(1 - x^2)^{1/2}}\right]^4.$$

Since

$$\int_0^1 \frac{x^{2n} \, dx}{(1 - x^2)^{1/2}} = \frac{\pi}{2} \frac{\Gamma(n + \frac{1}{2})}{n! \, \Gamma(\frac{1}{2})},$$

then

$$I = \left(\frac{\pi}{2}\right)^4 \sum_{n=0}^{\infty} \left[\frac{\Gamma(n + \frac{1}{2})}{n! \, \Gamma(\frac{1}{2})}\right]^5 \frac{\Gamma(n + \frac{5}{4})\Gamma(\frac{1}{4})}{\Gamma(\frac{5}{4})\Gamma(n + \frac{1}{4})} (-1)^n$$

$$= \left(\frac{\pi}{2}\right)^4 {}_6F_5\left(\begin{matrix} \frac{1}{2}, \frac{5}{4}, \frac{1}{2}, \frac{1}{2}, \frac{1}{2}, \frac{1}{2}; -1 \\ \frac{1}{4}, 1, 1, 1, 1 \end{matrix}\right).$$

Whipple found the transformation [5]

$${}_6F_5\left(\begin{matrix} a, 1 + \frac{1}{2}a, b, c, d, e; -1 \\ \frac{1}{2}a, 1 + a - b, 1 + a - c, 1 + a - d, 1 + a - e \end{matrix}\right)$$

$$= \frac{\Gamma(1 + a - d)\Gamma(1 + a - e)}{\Gamma(1 + a)\Gamma(1 + a - d - e)} {}_3F_2\left(\begin{matrix} 1 + a - b - c, d, e; 1 \\ 1 + a - b, 1 + a - c \end{matrix}\right),$$

from which, with $a = b = c = d = e = \frac{1}{2}$,

$$I = \left(\frac{\pi}{2}\right)^3 {}_3F_2\left(\begin{matrix} \frac{1}{2}, \frac{1}{2}, \frac{1}{2}; 1 \\ 1, 1 \end{matrix}\right).$$

Dixon's theorem [4], [5],

$$_3F_2\left(\begin{matrix} a, b, c;\ 1 \\ 1+a-b, 1+a-c \end{matrix}\right)$$

$$= \frac{\Gamma(1+\tfrac{1}{2})\Gamma(1+a-b)\Gamma(1+a-c)\Gamma(1+\tfrac{1}{2}a-b-c)}{\Gamma(1+a)\Gamma(1+a-b-c)\Gamma(1+\tfrac{1}{2}a-b)\Gamma(1+\tfrac{1}{2}a-c)},$$

then gives

$$I = 4\left(\frac{\pi}{2}\right)^2 \left[\frac{\Gamma(\tfrac{5}{4})}{\Gamma(\tfrac{3}{4})}\right]^2,$$

which, by using Legendre's duplication formula for gamma functions [4], reduces to

$$I = \tfrac{1}{32}\Gamma^4(\tfrac{1}{4}).$$

REFERENCES

[4] A. Erdélyi, W. Magnus, F. Oberhettinger and F. G. Tricomi, *Higher Transcendental Functions*, vol. 1, McGraw-Hill, New York, 1953.
[5] W. N. Bailey, *Generalized Hypergeometric Functions*, Cambridge University Press, London, 1935.

Also solved by the proposer who obtains the summation of the $_6F_5$ function by means of formula III 27 of L. J. Slater, *Generalized Hypergeometric Functions*, Cambridge University Press, New York, 1967, p. 245.

An n-Space Integral

Problem 88-9*, by SEUNG-JIN BANG (Seoul, Korea).
   Find a closed-form expression for the integral

$$f(\vec{x}t) = \int d\vec{r}\, \exp(i\vec{x}\cdot\vec{r})\frac{\sin\left(t(r^2+m^2)^{1/2}\right)}{(r^2+m^2)^{1/2}}$$

where $\vec{r}$ and $\vec{x}$ are n-dimensional vectors and the integration is over all of n-space.
   This integral represents the spatial part of the causal Green function for the Klein–Gordon operator

$$K = \partial_t^2 - \nabla^2 + m^2$$

which describes the relativistic dynamics of a meson field [1], [2].

REFERENCES

[1] F. Treves, *Basic Linear Partial Differential Equations*, Academic Press, New York, 1975, §7.
[2] S. Schweber, *Introduction to the Relativistic Quantum Theory of Fields*, Academic Press, New York, 1961.

Solution by M. L. GLASSER (Clarkson University).
   Let θ denote the polar angle of $\vec{r}$ with respect to $\vec{x}$. Then, due to cylindrical symmetry, for arbitrary positive n

$$f(\vec{x},t) = \frac{2\pi^{(n-1)/2}}{\Gamma[(n-1)/2]}\int_0^\infty r^{n-1}\,dr \int_0^\pi d\theta\, \sin^{n-2}\theta\, \exp(ikr\cos\theta)\frac{\sin t\sqrt{r^2+m^2}}{\sqrt{r^2+m^2}}.$$

However, the angular integral is

$$\pi^{1/2}(2/kr)^{(n/2)-1}\Gamma\left(\frac{n-1}{2}\right)J_{(n/2)-1}(kr),$$

so

$$f(\vec{x},t)=\frac{(2\pi)^{n/2}}{k^{(n-1)/2}}\int_0^\infty dr\,\sqrt{kr}J_{(n/2)-1}(kr)r^{(n-1)/2}\frac{\sin t\sqrt{r^2+m^2}}{\sqrt{r^2+m^2}}.$$

This integral converges for $0<n<3$, where it is a tabulated Hankel transform. Thus for $0<n<3$,

(*) $f(\vec{x},t)=(\pi/2)^{1/2}(2\pi)^{n/2}m^{(n-1)/2}(t^2-k^2)^{(1-n)/4}J_{(1-n)/2}[m\sqrt{t^2-k^2}]H(t-k),$

where H denotes the unit step function. It is interesting that the zero dimensional limit, often considered in statistical mechanics, does exist and is

$$m^{-1}\sin\left(m\sqrt{t^2-k^2}\right)H(t-k).$$

Also solved by CARL C. GROSJEAN (State University of Ghent, Ghent, Belgium) who showed that (*) holds generally, provided that the divergent integrals (for $n\geq2$) are appropriately handled (e.g., in the Cesàro sense), and that this f satisfies the Klein–Gordan equation

An Integral on $SO(3)$

Problem 84-1*, by B. E. EICHINGER (University of Washington).
     The following problem is encountered in the classical statistical mechanics of flexible bodies. As usual, let $SO(n)$ denote the special orthogonal group consisting of $n\times n$ orthogonal matrices with determinant $+1$. Let a and b be diagonal matrices,

$$a=\text{diag}(a_1,a_2,\cdots,a_n),\qquad -\infty<a_1<a_2<\cdots<a_n<\infty,$$
$$b=\text{diag}(b_1,b_2\cdots b_n),\qquad 0<b_1<b_2<\cdots b_n<\infty$$

and let $F_n(a,b)$ be the integral

$$F_n(a,b)=\int_{SO(n)}\exp\left[i\,\text{Tr}(ahbh^{-1})\right]dh,$$

where dh is the left-invariant measure on $SO(n)$. For $n=2$, the evaluation of $F_n(a,b)$ reduces to

$$F_2(a,b)=\int_0^{2s}\exp\left[i\,\text{Tr}\{ah(\theta)bh^{-1}(\theta)\}\right]d\theta,$$

where

$$h(\theta)=\begin{bmatrix}\cos(\theta)&\sin(\theta)\\-\sin(\theta)&\cos(\theta)\end{bmatrix}.$$

Thus one easily finds $F_2(a,b)=2\pi J_0(\Delta a\Delta b/2)\exp[i\,\text{Tr}(a)\,\text{Tr}(b)/2]$ where $\Delta a=a_2-a_1$ and $\Delta b=b_2-b_1$.
     Evaluate $F_3(a,b)$. A nice account of integration on $SO(3)$ is given in chapter 4 of

Fourier Series and integrals by H. Dym and H. P. McKean (Academic Press, New York, 1972).

Solution by DONALD RICHARDS (University of Wyoming and University of North Carolina).

We shall compute the integral

$$F_n(a,b) = \int_{SO(n)} \exp\left[i\operatorname{Tr}(ahbh^{-1})\right] dh_S,$$

where $SO(n)$ is the special orthogonal group, $a = \operatorname{diag}(a_1, a_2, \cdots, a_n)$, $b = \operatorname{diag}(b_1, b_2, \cdots, b_n)$ with a_i, $b_j \in \mathbb{R}$, and dh_S is the normalized Haar measure on $SO(n)$. Actually, if $O(n)$ is the full orthogonal group and dh_0 is the corresponding normalized Haar measure, the function

$$G_n(a,b) = \int_{O(n)} \exp\left[i\operatorname{Tr}(ahbh^{-1})\right] dh_0$$

is important for multivariate statistical theory. It follows from the work of A. T. James and others (cf. Muirhead [1]) that

$$G_n(a,b) = \sum_{k=0}^{\infty} \sum_{\kappa} \frac{i^k C_K(a) C_K(b)}{k! C_K(I_n)}.$$

Here I_n is the $n \times n$ identity matrix; $K = (k_1, \cdots, k_n)$ is a partition of k i.e. $k_1 \geq k_2 \geq \cdots \geq k_n \geq 0$ are nonnegative integers, and $k_1 + \cdots k_n = k$. Also, $C_K(\cdot)$ is the zonal polynomial corresponding to K; when $n = 2$, $C_K(\cdot)$ is expressible in Legendre polynomials. We prove that if n is odd then $F_n(a,b) = G_n(a,b)$. To this end, write $O(n)$ as the disjoint union of $SO(n)$ and a coset $C = h_1 SO(n)$, where $\det(h_1) = -1$. Then

$$G_n(a,b) = \left(\int_{SO(n)} + \int_C \right) \exp\left[i\operatorname{Tr}(ahbh^{-1})\right] dh_0$$

$$= \frac{1}{2} F_n(a,b) + H_n(a,b),$$

where we have used the relation $dh_0|_{SO(n)} = \frac{1}{2} dh_S$, and $H_n(a,b)$ denotes the integral over C. To evaluate $H_n(a,b)$, replace h by $-h$. This maps C bijectively onto $SO(n)$, leaves dh_0 invariant and shows that $H_n(a,b) = \frac{1}{2} F_n(a,b)$.

When n is even, $H_n(a,b)$ is evaluated by replacing h by ph, where p is a permutation matrix chosen to interchange the first two rows of h. It follows from our techniques that in this case

$$G_n(a,b) = \frac{1}{2} \left[F_n(a,b) + F_n(p^{-1}ap, b) \right].$$

The problem of evaluating $F_n(a,b)$ was earlier posed by I. Satake [2].

REFERENCES

[1] R. J. MUIRHEAD, *Aspects of Multivariate Statistical Theory*, John Wiley, New York, 1983.
[2] I. SATAKE, *Problem 2 in the Problem List from the Special Session on Analysis on Symmetric Complex Domains*, Notices of the AMS, 27 (1980), p. 442.

Problem 74-6*, *Three Multiple Integrals*, by M. L. MEHTA (Centre d'Etudes Nucleaires de Saclay, Gif-sur-Yvette, France).

Define $Q(\mathbf{r}_1) \equiv 1$ and

$$Q(\mathbf{r}_1, \mathbf{r}_2, \cdots, \mathbf{r}_n) \equiv \prod_{j>k}^{n} |\mathbf{r}_j - \mathbf{r}_k|^2$$

for $n > 1$, where $\mathbf{r}_1, \mathbf{r}_2, \cdots, \mathbf{r}_n$ are vectors in s-dimensional Euclidean space. Evaluate the multiple integrals

$$F_1(n, s) = \int \cdots \int Q(\mathbf{r}_1, \cdots, \mathbf{r}_n) \prod_{i=1}^{n} \exp\{-|\mathbf{r}_i|^2\} \, d\mathbf{r}_i,$$

$$F_2(n, s-1) = \int \cdots \int_{|\mathbf{r}_i|=1} Q(\mathbf{r}_1, \cdots, \mathbf{r}_n) \prod_{i=1}^{n} d\Omega_i,$$

and

$$F_3(n, s) = \int \cdots \int_{|\mathbf{r}_i| \leq 1} Q(\mathbf{r}_1, \cdots, \mathbf{r}_n) \prod_{i=1}^{n} d\mathbf{r}_i$$

for $s = 1$, 2 and 3. In integrals F_1, F_2 and F_3, the region of integration for each vector is the whole of E_s, the surface of the unit sphere in E_s, and the interior of the unit sphere in E_s, respectively. Note the $(s-1)$ in F_2 since the integrations are only over the angles.

Editorial note. The proposer offers the following comments concerning motivations for the study of the integrals F_1, F_2 and F_3. "These integrals may be considered to be the average electrostatic energies or the partition functions at a particular temperature of a certain Coulomb gas. The charges are restricted to remain inside the unit sphere (integral F_3), or restricted to move on the surface of the unit sphere (integral F_2), or they are free to move in the entire space. In the last case, one applies a central harmonic attraction to keep the charges from dispersing to infinity (integral F_1). However, this was not the original motivation to study these multiple integrals. They turn up in many a priori unrelated branches of applied mathematics. For example, $F_1(n, 1)$ and $F_2(n, 1)$ are the normalization constants for the unitary Gaussian or circular ensembles of random matrices [1], [3], and $F_3(n, 1)$ is encountered in the study of the distribution of zeros of stationary random functions [2]. One encounters $F_1(n, 2)$ in the theory of random matrices [1], [3] and in the statistical theory of coherent radiation. The integrals $F_1(n, 3)$ and $F_2(n, 2)$ are related to the theory of non-crossing random walks, and for this application it is especially important to have results which are valid for large n." The proposer notes that $F_1(n, s)$ and $F_3(n, s)$ are known for $s = 1$ and $s = 2$, and that $F_2(n, s-1)$ is known for $s = 2$. However, $F_2(n, 2)$ is not known. [C.C.R.]

REFERENCES

[1] J. GINIBRE, *Statistical ensembles of complex, quaternion, and real matrices*, J. Mathematical Phys., 6 (1965), pp. 440–449.
[2] M. S. LONGUET-HIGGINS, *The distribution of intervals between zeros of a stationary random function*, Philos. Trans. Roy. Soc. London Ser. A, 254 (1961–1962), pp. 557–599.
[3] M. L. MEHTA, *Random Matrices and the Statistical Theory of Energy Levels*, Academic Press, New York, 1967.

Problem 69-4, Limit of an Integral, by W. A. J. LUXEMBURG (California Institute
of Technology).

Determine

$$\lim_{n\to\infty} \sqrt{n} \int_0^\infty \left\{ \frac{J_1(2x)}{x} \right\}^n dx.$$

Solution by NELSON M. BLACHMAN (Sylvania Electronic Systems).

The given problem is a special case of

$$\lim_{n\to\infty} \sqrt{n} \int_0^\infty f^n(x)\, dx,$$

where $f(0) = 1$ is the unique global maximum of $f(x)$ and there is an a such that
$f(x) = 1 - ax^2 + o(x^2)$ as $x \to 0$. Hence, $f(x) = e^{-ax^2 + o(x^2)}$. When raised to a
high power this expression is significant only for very small x, and $o(x)^2$ can therefore be neglected. The integrand is thus essentially Gaussian, and the limit is
readily found to be $\frac{1}{2}\sqrt{\pi/a}$. For the case of $f(x) = J_1(2x)/x$, we have $a = \frac{1}{2}$, and
the result is $\sqrt{\frac{1}{2}\pi}$.

The Gaussian property suggests that the problem may have a probabilistic
interpretation, since the foregoing solution is related to the derivation of the
central limit theorem for independent, identically distributed random variables
(iidrv's). This is indeed the case whenever $f(x)$ is a characteristic function, and
$J_1(2x)/x$ in fact happens to be the characteristic function of the distribution with
density $\rho(y) = \pi^{-1}(1 - y^2/4)^{1/2}$ for $-2 < y < 2$ and $\rho(y) = 0$ elsewhere [*Bateman Project Tables of Integral Transforms*, vol. 1, p. 43]. Thus, $f^n(x)$ is the characteristic function of the sum of n iidrv's, and its Fourier transform is the probability
density function of this sum. In particular, $\int_{-\infty}^\infty f^n(x)\, dx/(2\pi)$ is the value of this
density at the origin, and the limit therefore gives an asymptotic expression
$1/\sqrt{4\pi an}$ for this density.

The second moment of the distribution with characteristic function $f(x)$
is $2a$, and the central limit theorem thus gives us the still more general result

$$\frac{1}{\pi}\int_0^\infty f^n(x)\cos xy\, dx \sim \frac{1}{2\sqrt{\pi an}} e^{-y^2/(4an)}.$$

Also solved by ANTON SUHADOLC (University of Wisconsin) who shows that

$$\lim_{n\to\infty} \sqrt{n} \int_0^\infty \left\{ \frac{J_p(2x)p!}{x^p} \right\}^n dx = \sqrt{\pi(1 + p)}/2, \qquad p > 0,$$

and the proposer.

Editorial note. Suhadolc's solution is equivalent to the method of Laplace
for the asymptotic expansion of an integral of the form

$$\int_a^b \Phi(x)F(x)^n\, dx,$$

where Φ and F satisfy certain conditions. See G. PÓLYA and G. SZEGÖ, *Aufgaben and Lehrsätze aus der Analysis*, vol. I, Dover, N.Y., 1945, pp. 78, 244, or N. G. DE BRUIJN, *Asymptotic Methods in Analysis*, Noordhoff, Amsterdam, 1961, pp. 63–65. [M.S.K.]

Problem 72-11, *Limit of an Integral*, by N. MULLINEUX and J. R. REED (University of Aston, Birmingham, England).

Determine

$$I = \lim_{\varepsilon \to 0} \int_0^{\varepsilon} \log \left\{ \frac{|\sin t - \varepsilon/2|}{\sin \varepsilon/2} \right\} \frac{dt}{\sin t}.$$

The problem arose in the numerical inversion of singular integral equations such as those which occur in aerodynamics and problems associated with plane transmission lines.

Editorial note. Most solvers (see later discussion) noted that the integral as given above does not exist since for fixed $\varepsilon \neq 0$, the integrand behaves like $t^{-1} \log \{(\varepsilon/2) \sin \varepsilon/2\}$ as $t \to 0$. The error is a typographical one since in the original proposal the term $|\sin t - t/2|$ appears as $|\sin (t - \varepsilon/2)|$. In the following solution of O. P. LOSSERS (Technological University of Eindhoven, Eindhoven, the Netherlands) the latter expression is used.

Replacing ε by 2ε we get

$$\int_0^{2\varepsilon} \log \left\{ \frac{|\sin (t - \varepsilon)|}{\sin \varepsilon} \right\} \frac{dt}{\sin t} = \int_0^{2\varepsilon} \log |\cos t - \cot \varepsilon \cdot \sin t| \frac{dt}{\sin t}$$

$$= \int_0^{2\varepsilon} \log \cos t \frac{dt}{\sin t} + \int_0^{2\varepsilon} \log |1 - \cot \varepsilon \cdot \tan t| \frac{dt}{\sin t}.$$

Since $(\log \cos t)/\sin t = -t/2 + O(t^3)$ $(t \to 0)$, we have that the first term is $O(\varepsilon^2), (\varepsilon \to 0)$. By substitution of $v = \cot \varepsilon \cdot \tan t$ we get

$$I = \lim_{\varepsilon \to 0} \left\{ O(\varepsilon^2) + \int_0^{\cot \varepsilon \cdot \tan 2\varepsilon} \frac{\log |1 - v|}{v} (1 + v^2 \tan^2 \varepsilon)^{-1/2} \, dv \right\}$$

$$= \int_0^2 \frac{\log |1 - v|}{v} \, dv = -\frac{\pi^2}{4}.$$

Problem 75–20, *Limit of an Integral*, by M. L. GLASSER (University of Waterloo, Ontario, Canada).

Show that

$$\lim_{n \to \infty} n \int_0^{\infty} I_n(x) J_n(x) K_n(x) \, dx = 8^{-1/2},$$

where, as usual, I_n, J_n, K_n are Bessel functions.

Solution by J. BOERSMA (Technical University, Eindhoven, the Netherlands).

Let the integral be denoted by A_n. Then it is shown in two different ways that $\lim_{n\to\infty} A_n = 2^{-3/2}$.

Firstly, replace the integration variable x by nx, yielding

$$(1) \qquad A_n = n^2 \int_0^\infty I_n(nx)J_n(nx)K_n(nx)\, dx.$$

In the latter integral we insert the uniform asymptotic expansion

$$(2) \qquad I_n(nx)K_n(nx) = \frac{1}{2n(1+x^2)^{1/2}}[1+O(n^{-1})], \qquad n\to\infty, \quad x\geq 0,$$

obtainable from Olver [1, Chap. 10, § 7.4]; the expansion (2) is uniformly valid for $x\geq 0$. Then the leading term of A_n becomes, by means of [2, formula 8.5 (11)] and (2),

$$(3) \qquad \frac{n}{2}\int_0^\infty (1+x^2)^{-1/2}J_n(nx)\, dx = \tfrac{1}{2}nI_{n/2}(\tfrac{1}{2}n)K_{n/2}(\tfrac{1}{2}n) = 2^{-3/2}[1+O(n^{-1})],$$
$$n\to\infty.$$

The contribution of the remainder term in (2) can be estimated by means of the inequality (cf. Watson [3, formulas 8.5(9), 13.74(1)])

$$(4) \qquad |J_n(nx)| \leq \left(\frac{2}{\pi n}\right)^{1/2}|x^2-1|^{-1/4}, \qquad\qquad x\geq 0,$$

thus yielding

$$(5) \qquad \int_0^\infty (1+x^2)^{-1/2}|J_n(nx)|\, dx = O(n^{-1/2}), \qquad n\to\infty.$$

Then it is found from (3) and (5) that

$$(6) \qquad A_n = 2^{-3/2}+O(n^{-1/2}), \qquad \lim_{n\to\infty} A_n = 2^{-3/2}.$$

Secondly, by means of [2, formula 8.13(20)], A_n can be expressed in terms of Legendre functions, viz.,

$$(7) \qquad A_n = e^{n\pi i}\frac{n\Gamma(\tfrac{3}{2}n+\tfrac{1}{2})}{\Gamma(-\tfrac{1}{2}n+\tfrac{1}{2})}P_{n/2-1/2}^{-n}(5^{1/2})Q_{n/2-1/2}^{-n}(5^{1/2}).$$

Finally, Mr. J. K. M. JANSEN of our University determined the numerical value of A_n for $n = 5, 10, 20$ from (7):

$$A_5 = 0.350412, \quad A_{10} = 0.352736, \quad A_{20} = 0.353347,$$

correct to six decimal places. These values do agree very well with (10); e.g., for $n = 10$ the error turns out to be 10^{-5}. By extrapolation from A_5, A_{10}, A_{20}, Jansen found the "numerical" limit to be 0.353553, which is identical to $2^{-3/2}$ in all six decimal places.

REFERENCES

[1] F. W. J. OLVER, *Asymptotics and Special Functions*, Academic Press, New York, 1974.
[2] A. ERDÉLYI, W. MAGNUS, F. OBERHETTINGER AND F. G. TRICOMI, *Tables of Integral Transforms*, vol. II, McGraw-Hill, New York, 1954.
[3] G. N. WATSON, *A Treatise on the Theory of Bessel Functions*, 2nd Ed., Cambridge University Press, Cambridge, 1958.
[4] A. ERDÉLYI, W. MAGNUS, F. OBERHETTINGER AND F. G. TRICOMI, *Higher Transcendental Functions*, vol. I, McGraw-Hill, New York, 1953.
[5] R. C. THORNE, *The asymptotic expansion of Legendre functions of large degree and order*, Philos. Trans. Roy. Soc. London Ser. A, 249(1957), pp. 597–620.

Also solved by D. E. AMOS (Sandia Laboratories) and A. G. GIBBS (Battelle Memorial Institute), who both showed that

$$\lim_{n \to \infty} n \int_0^\infty I_n(x) J_n(ax) K_n(bx)\, dx = \tfrac{1}{2}(1 + a^2)^{-1/2}.$$

Editorial note. Procedures for obtaining the asymptotic behavior of the latter integral and more general ones are to appear in a paper by F.W.J. Oliver and the proposer (Utilitas Math. 12(1977) 225–239).

A Conjectured Limit

Problem 80-7* (corrected), *by* H. P. ROBINSON (Lafayette, California).
  In calculating the two expressions

$$I = \int_0^\infty \{\zeta(x) - 1\}\, dx,$$

$$L = \lim_{n \to \infty} \left\{ \sum_{k=2}^n \frac{1}{\ln k} - \int_0^n \frac{dx}{\ln x} \right\},$$

it was found that $I = L$ to at least 43 decimal places. It is conjectured that $I = L$. The integrals are Cauchy principal values.

  Solution by D. J. NEWMAN (Temple University) *and* D. V. WIDDER (Harvard University).
  For $x > 1$, we have

$$\zeta(x) - 1 = \sum_{k=2}^\infty \frac{1}{k^x} = \int_2^\infty \frac{dt}{t^x} + \sum_{k=2}^\infty \left(\frac{1}{k^x} - \int_k^{k+1} \frac{dt}{t^x} \right) = \frac{2^{1-x}}{x-1} + \sum_{k=2}^\infty \left(\frac{1}{k^x} - \int_k^{k+1} \frac{dt}{t^x} \right):$$

However, since these terms are bounded by $(x+1)/k^{x+1}$, this series converges throughout $\mathrm{Re}\, x > 0$, and so the equality persists there by analytic continuation. Integration therefore gives

$$\int_0^\infty \{\zeta(x) - 1\}\, dx = \int_0^\infty \frac{2^{1-x}\, dx}{x-1} + \sum_{k=2}^\infty \left(\frac{1}{\ln k} - \int_k^{k+1} \frac{dt}{\ln t} \right),$$

the integrals being principal values and the term by term integration (and interchange) being justified by the fact that said terms (and integrand) are all positive. Finally, changing variables by $x = 1 - \ln t / \ln 2$ yields

$$\int_0^\infty \frac{2^{1-x}}{x-1} \, dx = -\int_0^2 \frac{dt}{\ln t},$$

which completes the proof.

Editorial note. Most solvers used the Euler-Maclaurin sum formula or referred to Titchmarsh, *The Theory of the Riemann Zeta-Function*, (Oxford, 1951). The proposer's calculation yields

$$I = L = -.24323\ 83428\ 90980\ 75541\ 50591\ 35465\ 46230\ 71783\ 049.$$

Supplementary References
Definite Integrals

[1]   A.Erdelyi, W.Magnus, F.Oberhettinger, and F.G.Tricomi, *Tables of Integral Transforms, I,II*, McGraw-Hill, N.Y., 1953, 1954.

[2]   I.S.Gradshteyn and I.M.Ryzhik, *Tables of Integrals, Series and Products*, Academic Press, N.Y., 1980.

[3]   I.I.Hirschman and D.V.Widder, *The Convolution Transform*, Princeton University Press, Princeton, 1955.

[4]   Y.L.Luke, *Integrals of Bessel Functions*, McGraw-Hill, N.Y., 1962.

[5]   F.Oberhettinger, *Tables of Bessel Transforms*, Springer-Verlag, N.Y., 1972.

[6]   F.Oberhettinger, *Tables of Mellin Transforms*, Springer-Verlag, N.Y., 1974

[7]   F.Oberhettinger and L.Badii, *Tables of Laplace Transforms*, Springer- Verlag, N.Y., 1973.

[8]   A.P.Prudnikov, Y.A.Brychkov, and Q.I.Marichev, *I,II, Integrals and Series*, Gordon and Breach, N.Y., 1986.

[9]   G.E.Roberts and H.Kaufman, *Table of Laplace Transforms*, Saunders, Philadelphia, 1966.

[10] I.N.Sneddon, *Fourier Transforms*, McGraw-Hill, N.Y., 1951.

[11] E.C.Titchmarsh, *Introduction to the Theory of Fourier Integrals*, Clarendon Press, Oxford, 1948.

[12] A.D.Wheelon, *Tables of Summable Series and Integrals Involving Bessel Functions*, Holden-Day, San Francisco, 1968.

[13] D.V.Widder, *The Laplace Transform*, Princeton University Press, Princeton, 1946.

10. INTEGRAL EQUATIONS

Problem 74–8, An Integral Equation for Crystal Growth, by J. S. Lew and R. Ghez
(IBM T. J. Watson Research Center).

A mathematical model of crystal growth yields the following equation,
with $a, b > 0$:

$$f(t) = e^{-t}\left\{1 - e^{bt} - a \int_0^t \frac{f(u)\,du}{\sqrt{\pi(t-u)}}\right\}.$$

The limiting thickness of deposited material is observable in experiments and is
proportional to

$$c_0(a, b) = \int_0^\infty f(t)\,dt.$$

Establish the following relations:

(i) $$c_0(a, b) = b \sum_{n=0}^\infty \frac{(-a)^n}{(n+1)(n+1+b)\sqrt{n!}},$$

(ii) $$\frac{e^{-t} + \int_0^t f(u)\,du}{1 + a e^{-t}/\sqrt{\pi t}} = c_0(a, b) + O(e^{-t}/t^{3/2}) \quad \text{as } t \to \infty,$$

(iii) $$c_0(a, b) = \left\{\frac{4 \log a}{\pi a^2}\right\}^{1/2} \cdot [1 + O(1/\log a)] \quad \text{as } a \to \infty.$$

Solution by the proposers.

The integral equation can be written as

(1) $$f = g - Af,$$

where g and A are given by

(2) $$g(t) = e^{-t}\{1 - e^{-bt}\},$$

$$Af(t) = a e^{-t} \int_0^t [\pi(t-u)]^{-1/2} f(u)\,du.$$

For complex t, the integral defining $Af(t)$ is taken along the line segment from 0 to
t. By induction, we shall show that for $n = 0, 1, \cdots$,

(3) $$|A^n g(t)/b| \le \frac{a^n |t|^{(n+2)/2}}{\Gamma((n+4)/2)} \quad \text{in Re } (t) \ge 0.$$

283

Indeed, if $n = 0$ and $t = x + iy$, then the square of (3) can be rewritten as

(4) $(1 - e^{-bx})^2 + 4 e^{-bx} \sin^2 (by/2) \leq b^2 (x^2 + y^2)$ for $x \geq 0$,

and proved by elementary arguments. Moreover, if we assume (3) for any nonnegative n and put $v = |t|$, $w = |u|$ in the next calculations, then we verify

$$|A^{n+1}g(t)/b| \leq \frac{a}{\sqrt{\pi}} \int_0^t |t - u|^{-1/2} |A^n g(u)/b| \, |du|$$

(5)
$$\leq \frac{a^{n+1}}{\Gamma\left(\frac{1}{2}\right)\Gamma\left(\frac{n+4}{2}\right)} \int_0^v (v - w)^{-1/2} w^{(n+2)/2} \, dw$$

$$= \frac{a^{n+1} v^{(n+3)/2}}{\Gamma\left(\frac{n+5}{2}\right)}.$$

The Picard iterates of (1) converge uniformly by (3) to the solution $f(t)$ on each compact subset of Re $(t) \geq 0$. Indeed,

(6) $|f(t)| \leq \sum_{n=0}^{\infty} |A^n g(t)| \leq ba^{-2} [1 + a\sqrt{|t|}][\exp(a^2 |t|) - 1],$

so that $f(t)$ is continuous on Re $(t) \geq 0$ and analytic on Re $(t) > 0$.

The inequality (6) as $|t| \to \infty$ implies that the Laplace transform,

(7) $$Lf(s) = \int_0^{\infty} e^{-st} f(t) \, dt,$$

is absolutely convergent for Re $(s) > a^2$ and analytically continuable [1, p. 367] into $|s| > a^2$, $|\arg s| < \pi$. The inequality (6) as $|t| \to 0$ implies that this transform is $O(s^{-2})$ as $|s| \to \infty$ [1, p. 473], uniformly in $|\arg s| \leq \theta$ for each $\theta < \pi$. Also the transforms of the Picard iterates converge to the transform $Lf(s)$, at least for Re $(s) > a^2$, by the inequality (6) and the dominated convergence theorem. Thus we can transform the given (1) to produce

(8) $Lf(s) = b/(s+1)(s+1+b) - aLf(s+1)/\sqrt{s+1},$

and iterate this difference equation to obtain

(9)

$$Lf(s) = \frac{b}{(s+1)(s+1+b)} + b \sum_{n=1}^{\infty} \frac{(-a)^n}{(s+n+1)(s+n+1+b)\sqrt{(s+1)\cdots(s+n)}}.$$

For all complex s except on the half-line $(-\infty, -1]$, the series (9) converges uniformly on compact sets, hence providing an analytic continuation of $Lf(s)$.

Clearly $Lf(s)$ has a Taylor series

(10) $$Lf(s) = \sum_{n=0}^{\infty} c_n(a, b)s^n \quad \text{for } |s| < 1,$$

and $c_0(a, b)$ has the form

(11) $c_0(a, b) = Lf(0) = \int_0^\infty f(t)\, dt,$

in agreement with our previous definition. By (9) and (11), as required,

(12) $c_0(a, b) = b \sum_{n=0}^\infty \dfrac{(-a)^n}{(n+1)(n+1+b)\sqrt{n!}}.$

By (8) and (10), in addition,

(13) $\dfrac{Lf(s)}{s} = \dfrac{b}{s(s+1)(s+1+b)} + a \sum_{n=0}^\infty (s+1)^{n-1/2} \sum_{m=0}^n c_m(a, b)$

for $|s+1| < 1$. The decay of $Lf(s)$ near ∞ permits us to express $\int_0^t f(u)\, du$ via the standard inversion integral, and deform the contour into the left half-plane. Then an inverse Abelian theorem of Doetsch [2, p. 159; 4, Thm. 3] permits us to note the residue $Lf(0)$ at $s = 0$, invert (13) term by term at $s = -1$, and obtain the asymptotic expansion for $t \to +\infty$:

(14) $\int_0^t f(u)\, du \sim c_0(a, b) - e^{-t} + a\, e^{-t} \sum_{n=0}^\infty \dfrac{t^{-n-1/2}}{\Gamma(-n+1/2)} \sum_{m=0}^n c_m(a, b).$

Truncation of (14) yields the statement, as required,

(15) $c_0(a, b)[1 + a\, e^{-t}/\sqrt{\pi t}] + O(e^{-t}/t^{3/2}) = e^{-t} + \int_0^t f(u)\, du.$

Finally, we may rewrite (12) as a contour integral

(16) $c_0(a, b) = \dfrac{b}{2\pi i} \int_C \dfrac{\exp(z \log a)}{a(b+z)\sqrt{z}\,\Gamma(z+1)} \pi \csc(\pi z)\, dz.$

Here the contour C runs in from $+\infty$ below the positive real axis, crosses into the first quadrant between 0 and 1 and runs out to $+\infty$ above the positive real axis. However, the given contour can be deformed into the left half-plane, and the previously cited theorem can be extended to this case [2, p. 159; 5]. Moreover, near $z = 0$,

(17) $\dfrac{b\pi \csc(\pi z)}{a(b+z)\sqrt{z}\,\Gamma(z+1)} = a^{-1} z^{-3/2}\left[1 + \sum_{n=1}^\infty d_n z^n\right],$

and hence, as $a \to +\infty$,

(18) $c_0(a, b) \sim \left\{\dfrac{4 \log a}{\pi a^2}\right\}^{1/2}\left[1 + \dfrac{\sqrt{\pi}}{2} \sum_{n=1}^\infty \dfrac{d_n (\log a)^{-n}}{\Gamma(-n+3/2)}\right].$

Further properties of this equation, and a description of its use, are contained in a recent paper of the authors [3].

REFERENCES

[1] G. DOETSCH, *Handbuch der Laplace Transformation, Vol. I: Theorie der Laplace Transformation*, Birkhäuser Verlag, Basel, 1950.
[2] ———, *Handbuch der Laplace Transformation, Vol. II: Anwendungen der Laplace Transformation*, Birkhäuser Verlag, Basel, 1955.

[3] R. Ghez and J. S. Lew, *Interface kinetics and crystal growth under conditions of constant cooling rate, I: Constant diffusion coefficient*, J. Crystal Growth, 20 (1973), pp. 273–282.

[4] R. A. Handelsman and J. S. Lew, *Asymptotic expansion of Laplace convolutions for large argument and tail densities for certain sums of random variables*, SIAM J. Math. Anal., 5 (1974), pp. 425–451.

[5] V. Riekstina, *Asymptotic expansions of some integrals and the sums of power series*, Latvian Math. Yearbook, 8 (1970), pp. 223–239.

Problem 72-2*, A Family of Fredholm Integral Equations, by J. L. Casti (Systems Control, Inc.).

Consider the family of Fredholm integral equations indexed by the upper limit of integration,

$$(1) \qquad u(t, x) = 1 + \int_0^x e^{-|t-y|} u(y, x)\, dy, \qquad 0 \leq t \leq x < \infty.$$

It is a simple matter to show that a closed form solution is given by $u(t, x) = (\sin t + \cos t)/\cos x$. Without making use of this closed form solution, prove that a unique continuous solution to (1) exists except for a countable number of values of x, the odd multiples of $\pi/2$.

This problem arises in a simplified one-dimensional transport process [1], [2]. The solution method for this simplified version will presumably shed light on the more realistic problems for which no closed form solutions are available.

REFERENCES

[1] R. Bellman and R. Kalaba, *Transport theory and invariant imbedding*, Nuclear Reactor Theory, G. Birkhoff and E. Wigner, eds., Amer. Math. Soc., Providence, R.I., 1961, pp. 206–208.

[2] J. Casti, *Invariant imbedding and the solution of Fredholm integral equations with displacement kernels*, Rep. R-639-PR, The RAND Corporation, Santa Monica, Calif., 1970.

Solution by Otto G. Ruehr (Michigan Technological University).

The stated closed form solution is incorrect. It should be

$$u(t, x) = -1 + (\sec x)\{(\sin x - \cos x + 1)\sin t + (\sin x + \cos x + 1)\cos t\}.$$

To attack the problem from a more general point of view, consider the associated homogeneous equation

$$(2) \qquad v(t, x) = \lambda \int_0^x e^{-|t-y|} v(y, x)\, dy.$$

The Fredholm theory guarantees that (1) has a unique continuous solution for those values of x for which $\lambda = 1$ is not an eigenvalue of (2). By splitting the integral into two parts to remove the absolute value and by successive differentiation, we obtain the Sturm–Liouville problem with periodic boundary conditions

$$\frac{d^2 v(t, x)}{dt^2} + (2\lambda - 1) v(t, x) = 0,$$

$$(3)$$

$$v(0, x) = \frac{dv(t, x)}{dt}\bigg]_{t=0}, \qquad v(x, x) = -\frac{dv(t, x)}{dt}\bigg]_{t=x}.$$

A nontrivial solution of (3) is

$$v(t, x) = \sin t\sqrt{2\lambda - 1} + \sqrt{2\lambda - 1}\cos t\sqrt{2\lambda - 1},$$

provided that λ satisfies

$$(\lambda - 1)\sin x\sqrt{2\lambda - 1} = \sqrt{2\lambda - 1}\cos x\sqrt{2\lambda - 1}.$$

Hence, $\lambda = 1$ is an eigenvalue only if x is an odd multiple of $\pi/2$. In this case, the integral equation (1) has a continuous solution only in the case that the "forcing function," $F(t) = 1$, is orthogonal to all the eigensolutions. In particular, this would require that

$$0 = \int_0^x (\sin t + \cos t)\, dt = 1 - \cos x + \sin x.$$

The above holds if $x = x_j = (4j - 1)\pi/2$, $j = 1, 2, \cdots$, and does not hold if $x = x_k = (4k - 3)\pi/2$, $k = 1, 2, \cdots$. In fact, as predicted by the Fredholm theory, we have infinitely many solutions for $x = x_j$:

$$u(t, x_j) = -1 + A\sin t + (A + 2)\cos t, \qquad A \text{ arbitrary and real.}$$

One of these ($A = -1$) can be obtained from the corrected closed form solution by taking the limit as $x \to x_j$.

Problem 67-16, *An Integral Equation*, by C. J. BOUWKAMP (Philips Research Laboratories, Eindhoven, The Netherlands).

In a paper[1] on the study of granularity in photographic plates, the following integral equation had arisen:

$$g(a) = \frac{8}{\pi a^2}\int_0^{2a} \rho r(\rho)\, d\rho \int_0^{\cos^{-1}(\rho/2a)} \sin^2\theta\, d\theta.$$

Here, $g(a)$ denotes the "granularity" defined for $a \geq 0$, and $r(\rho)$ denotes the "autocorrelation" defined for $0 \leq \rho \leq 2a$.

Given $g(a)$, determine $r(\rho)$ assuming appropriate conditions of smoothness.
Solution by DAVID GOOTKIND (Avco Corporation).
First let $s = 2a$, $g(s/2) = h(s)$, $\cos\theta = x/s$ to give

$$h(s) = \frac{32}{\pi s^4}\int_0^s \rho r(\rho)\, d\rho \int_\rho^s \sqrt{s^2 - x^2}\, dx.$$

Interchanging the order of integration, letting $\rho^2 = u$, $x^2 = v$, $s^2 = t$, $F(v) = \int_0^v r(\sqrt{u})\, du$, and then differentiating with respect to t gives

$$\frac{\pi}{4}\frac{d}{dt}\{t^2 h(\sqrt{t})\} = \int_0^t \frac{F(v)}{\sqrt{v}}\frac{dv}{\sqrt{t - v}}.$$

The solution of this Abel equation is

[1] A. Marriage and E. Pitts, *Relation between Granularity and Auto-correlation*, J. Opt. Soc. Amer., 46 (1956), pp. 1019–1027. The authors state: "A solution of the integral equation in closed form cannot be easily found."

$$F(x) = \frac{\sqrt{x}}{4} \frac{d}{dx} \int_0^x \frac{d(t^2 h \sqrt{t})}{\sqrt{x - t}}.$$

Finally, differentiating both sides with respect to x and letting $\sqrt{x} = \rho$, we obtain

$$r(\rho) = \frac{1}{16\rho} \frac{d^2}{d\rho^2} \int_0^\rho \frac{d(\lambda^4 g(\lambda/2))}{\sqrt{\rho^2 - \lambda^2}}.$$

Editorial note. The proposer, in addition to giving a solution similar to that above, gives a solution in the following alternate form:

Let

$$\lambda_n = \frac{n + 2}{2^{n+3}} \frac{\Gamma\left(\frac{1}{2}\right) \Gamma\left(3 + \frac{n}{2}\right)}{\Gamma\left(\frac{3}{2} + \frac{n}{2}\right)} \rho^n, \qquad\qquad n \geq 0.$$

Then if $g(a) = \sum_n g_n a^n$,

$$r(\rho) = \sum_n \lambda_n g_n \rho^n.$$

A proof follows by direct substitution.

F. Downton (University of Birmingham) in his solution notes that the given integral equation is a variant of one arising in geophysics (see H. Jefferys, *The Earth*, 3rd ed., Cambridge University Press, 1952, pp. 43–44).

Problem 65-7*, *Solution to an Integral Equation,*† by A. J. Strecok (Argonne National Laboratory).

Show that a solution of the equation

(1) $$1 = e^{-x^2} + \int_0^x 2te^{-t^2} \psi(x, t)\, dt$$

is given by

(2) $$\psi(x, t) = \frac{2}{\pi} (x^2 - t^2)^{-1/2} \int_0^t e^{u^2}\, du,$$

and give conditions on $\psi(x, t)$ which make this a unique solution.

Solution by C. J. Bouwkamp (Technological University, Eindhoven, Netherlands).

The first part of the problem is simple in view of the solution given of *Problem* 65-6 if applied to $s = -1$, viz.,

$$\int_0^x 2te^{-t^2} \psi(x, t)\, dt = \frac{4}{\pi} I(-1, x) = 1 - e^{-x^2}, \qquad\qquad x > 0.$$

The second part of the problem is not well posed. Equation (1) is not a proper integral equation. The functional equation (1) has an abundance of solutions,

---

† Work performed under the auspices of the United States Atomic Energy Commission.

and conditions on $\psi(x, t)$ that will make the particular solution (2) the only solution of (1) seem to be quite artificial. This is substantiated by noting it is easy to construct solutions of the types $f(t)g(x)$ and $f(t) + g(x)$, where $f(t)$ is wholly arbitrary. Thus,

$$\psi_1(x, t) = f(t)(1 - e^{-x^2}) \left(\int_0^x 2te^{-t^2} f(t) \, dt \right)^{-1}$$

and

$$\psi_2(x, t) = 1 + f(t) - (1 - e^{-x^2})^{-1} \int_0^x 2te^{-t^2} f(t) \, dt$$

are solutions of (1) for any $f(t)$ making the right-hand side meaningful. Apparently, the proposer is interested in solutions that are singular at $x = t$. However, there is a whole class of such solutions. This is shown as follows.

Assume that (1) is meant to hold for $x > 0$, and that $\psi(x, t)$ is defined and continuous for $0 \leq t < x$. It is somewhat easier to discuss the problem in transformed variables. To that end, set

$$x^2 = \xi, \quad t^2 = \tau, \quad \chi(\xi, \tau) = \psi(x, t).$$

Then (1) becomes

(3) $$1 = e^{-\xi} + \int_0^\xi e^{-\tau} \chi(\xi, \tau) \, d\tau,$$

and the function (2) corresponds to

(4) $$\chi(\xi, \tau) = \frac{1}{\pi} (\xi - \tau)^{-1/2} \int_0^\tau e^u u^{-1/2} \, du,$$

this being a particular slution of (3). In fact, by invoking the theory of Abel's integral equation, this can be proved independently of the solution to *Problem* 65-6.

Now assume that

$$\chi(\xi, \tau) = (\xi - \tau)^{-\alpha} e^\tau g(\tau),$$

where $g(t)$ is an unknown function and $0 < \alpha < 1$. Then (4) transforms into the generalized Abel equation

$$\int_0^\xi \frac{g(\tau)}{(\xi - \tau)^\alpha} \, d\tau = 1 - e^{-\xi},$$

which is solved by

$$g(\tau) = \frac{e^{-\tau} \sin(\alpha\pi)}{\pi} \int_0^\tau e^u u^{\alpha-1} \, du.$$

That is to say, in terms of the original variables, the function

(5) $$\psi_3(x, t) = \frac{2 \sin(\alpha\pi)}{\pi(x^2 - t^2)^\alpha} \int_0^t e^{u^2} u^{2\alpha-1} \, du$$

is a particular solution of (1) for any α with $0 < \alpha < 1$. The proposer's solution (2) is obtained for $\alpha = 1/2$.

Of course, there exist still other solutions of (1). Thus, if the proposer writes his unknown function $\psi(x, t)$ as a product of the kernel function $(x^2 - t^2)^{-1/2}$ and an unknown function $h(t)$, the solution indicated for $h(t)$ is unique, by invoking the existing theory of Abel's integral equation. But then the kernel function should have been prescribed.

Problem 75-9, *A Singular Integral Equation*, by M. L. GLASSER (University of Waterloo, Ontario, Canada).

In studying the effects of boundaries on the thermodynamics of a one-dimensional spin system with singular interaction, it was found that the partition function satisfies the singular integral equation

(1) $$Z(s) = e^{-q} \int_0^1 dt\, Z(t)/(1-st)^p.$$

(a) Show that in the case $p = 1$, $q = \ln \pi$, the exact solution of (1) is

$$Z(s) = A(1-s)^{-1/2} \mathbf{K}(s^{1/2}),$$

where $\mathbf{K}(k)$ denotes the complete elliptic integral of the first kind with modulus k and A is an arbitrary constant.

(b)* Are there any other exactly solvable cases?

Solution by B. C. CARLSON (Ames Laboratory–ERDA, Iowa State University).

If $p = 1$ and $e^{-q} = \pi^{-1} \sin \pi a$, where $0 < \operatorname{Re} a < 1$, a solution (which can be multiplied by an arbitrary constant) is

$$Z(s) = (1-s)^{a-1} {}_2F_1(a, a; 1; s), \qquad |\arg (1-s)| < \pi.$$

Because of the relation [1, 2.1(23)],

$$(1-s)^{a-1} {}_2F_1(a, a; 1; s) = (1-s)^{-a} {}_2F_1(1-a, 1-a; 1; s),$$

the values a and $1 - a$ correspond to the same solution, which is real for real s if a is real or if $\operatorname{Re} a = \frac{1}{2}$ (i.e., if $\bar{a} = 1 - a$). The proposer's solution is the case $a = \frac{1}{2}$. No solutions were found for $p \ne 1$.

The complex eigenvalues and continuous spectrum illustrate strikingly the peculiarities of singular kernels. A real symmetric kernel that is nonsingular can have only real denumerable eigenvalues. The kernel $(1 - st)^{-1}$ is real (for real s and t) and symmetric but singular, because it is not square integrable on the region $0 \le s \le 1$, $0 \le t \le 1$. Its spectrum (the set of admissible values of e^{-q}) covers the open right half-plane.

In order to prove that

$$\int_0^1 (1-st)^{-1}(1-t)^{a-1} {}_2F_1(a, a; 1; t)\, dt = \frac{\pi}{\sin \pi a}(1-s)^{a-1} {}_2F_1(a, a; 1; s),$$

we may assume that $\operatorname{Re} s < \frac{1}{2}$ and $0 < \operatorname{Re} a < \frac{1}{2}$, for the permanence of functional relations will insure that the equality continues to hold on the region $|\arg (1 - s)| < \pi$, $0 < \operatorname{Re} a < 1$, where both sides are analytic. Under the more restrictive conditions, the series expansions,

$$(1-st)^{-1} = (1-s)^{-1}\left[1-\frac{s}{s-1}(1-t)\right]^{-1} = (1-s)^{-1}\sum_{m=0}^{\infty}\left(\frac{s}{s-1}\right)^{m}(1-t)^{m},$$

$$_2F_1(a, a; 1; t) = \sum_{n=0}^{\infty}\frac{(a)_n(a)_n}{n!n!}t^n,$$

are absolutely and uniformly convergent for $0 \le t \le 1$. Integration term by term transforms the integral into an absolutely convergent double series,

$$(1-s)^{-1}\sum_{m=0}^{\infty}\sum_{n=0}^{\infty}\left(\frac{s}{s-1}\right)^{m}\frac{(a)_n(a)_n}{n!n!}\int_0^1 t^n(1-t)^{a-1+m}\,dt$$

$$= (1-s)^{-1}\sum_{m=0}^{\infty}\sum_{n=0}^{\infty}\left(\frac{s}{s-1}\right)^{m}\frac{(a)_n(a)_n\Gamma(a+m)}{n!\Gamma(1+a+m+n)}$$

$$= (1-s)^{-1}\sum_{m=0}^{\infty}\left(\frac{s}{s-1}\right)^{m}\frac{\Gamma(a+m)}{\Gamma(1+a+m)}\sum_{n=0}^{\infty}\frac{(a)_n(a)_n}{(1+a+m)_n n!}$$

$$= (1-s)^{-1}\sum_{m=0}^{\infty}\left(\frac{s}{s-1}\right)^{m}\frac{\Gamma(a+m)\Gamma(1-a+m)}{m!m!}$$

$$= (1-s)^{-1}\Gamma(a)\Gamma(1-a)\sum_{m=0}^{\infty}\frac{(a)_m(1-a)_m}{m!m!}\left(\frac{s}{s-1}\right)^{m}$$

$$= (1-s)^{-1}\frac{\pi}{\sin \pi a}{}_2F_1\left(a, 1-a; 1; \frac{s}{s-1}\right)$$

$$= \frac{\pi}{\sin \pi a}(1-s)^{a-1}{}_2F_1(a, a; 1; s).$$

We have used [1, 1.5(1), 1.5(5)] in the first equality, [1, 2.1(14)] in the third, [1, 1.2(6)] in the fifth, and [1, 2.1(22)] in the last.

REFERENCE

[1] A. ERDÉLYI, W. MAGNUS, F. OBERHETTINGER AND F. G. TRICOMI, *Higher Transcendental Functions*, McGraw-Hill, New York, 1953.

Problem 68-13, *A Volterra Integral Equation*, by B. A. FUSARO (Queens College, North Carolina).

Solve the integral equation

$$D^{m-2}F(t) = \int_0^t z^{m-1}(\sinh z)G(\lambda)\,d\lambda$$

for $G(\lambda)$ which is assumed to be twice differentiable. Here $z = c\sqrt{2(t-\lambda)}$, $t > 0$, and m is an integer ≥ 2.

Solution by the proposer.

Put $K(t-\lambda) = z^{m-1}\sinh z$ and then write the integral equation in operator form, expressing the right side as a convolution,

(1) $$D^{m-2}F = K*G.$$

Equation (1) has a unique solution of exponential order given by

(2) $$G = L^{-1}(\phi) \qquad \text{with } \phi = L(D^{m-2}F)/L(K),$$

where L denotes the Laplace transform. This solution results after noting that K and $D^{m-2}F$ are of exponential order, and applying the convolution theorem to (1) to get

$$L(D^{m-2}F) = L(K*G) = L(K) \cdot L(G).$$

The kernel K is positive-definite so that the term $L(K)$ never vanishes, and hence (2) is a solution of the given equation. The uniqueness of (2) can be shown, e.g., by assuming there exists a second solution H so that $K*(G - H) = 0$; the positive-definiteness of K, and standard continuity arguments, yield $G - H = 0$.

   Remark. The integral equation arose out of the work of E. C. Young, *Riemann's method and the characteristic value and Cauchy problem for the damped wave equation* (to appear).

Problem 75–18, *A Nonlinear Integral Equation*, by O. G. RUEHR (Michigan Technological University).

   Find a continuous $F(t)$ for $t \geq 0$ satisfying

$$t\{F(t)\}^2 = \int_0^1 \frac{F(ts) - F(t-ts)}{1-2s}\, ds$$

and $F(0) = C > 0$.

   Solution by P. G. KIRMSER (Kansas State University).

   Differentiating the given equation n times with respect to t and then setting $t = 0$ leads to the recursion formula

$$n \sum_{j=0}^{n-1} \binom{n-1}{j} F^{(j)}(0) F^{(n-1-j)}(0) = F^{(n)}(0) \int_0^1 \frac{s^n - (1-s)^n}{1-2s}\, ds$$

for the nth derivative at zero in terms of derivatives of lower orders.

   The integral on the right side can be expressed as

$$-\int_0^1 \frac{s^n - (1-s)^n}{s - (1-s)}\, ds = -\sum_{j=1}^n \int_0^1 s^{n-j}(1-s)^{j-1}\, ds = -\sum_{j=1}^n \frac{(n-j)!(j-1)!}{n!}.$$

With this evaluation, the recursion formula becomes

$$F^{(n)}(0) = \frac{n \sum_{j=0}^{n-1} \binom{n-1}{j} F^{(j)}(0) F^{(n-1-j)}(0)}{-\sum_{j=1}^n [(n-j)!(j-1)!/n!]}$$

For $F(0) = C$, this formula is satisfied by

$$F^{(n)}(0) = (-1)^n C^{n+1}(n!)^2.$$

Thus, the Maclaurin's series for $F(t)$ is

$$F(t) = C \sum_{j=0}^{\infty} (-1)^j (j!) C^n t^n.$$

This divergent series was summed by Euler, as shown in Hardy [1, p. 26], to yield

$$F(t) = C \int_0^\infty \frac{e^{-w}\,dw}{1 + Ctw}$$

as the continuous solution of the problem posed.

REFERENCE

[1] G. H. HARDY, *Divergent Series*, Oxford University Press, London, 1949.

Supplementary References
Integral Equations

[1] C.T.H.Baker, *The numerical treatment of integral equations*, Clarendon Press, Oxford, 1977.
[2] H.T.Davis, *Introduction to Nonlinear Differential and Integral Equations*, U.S. Atomic Energy Commission, Washington, D.C., 1960.
[3] F.D.Gakhov, *Boundary Value Problems*, Addison-Wesley, Reading, 1966.
[4] C.D.Green, *Integral Equation Methods*, Barnes and Noble, N.Y., 1969.
[5] H.Hochstadt, *Integral Equations*, Wiley, N.Y., 1973.
[6] W.V.Lovitt, *Linear Integral Equations*, Dover, 1950.
[7] M.I.Imanaliev, *Integral Equations*, Noordhoff, Groningen, 1975.
[8] S.G.Miklin, *Integral Equations*, Pergamon, London, 1957.
[9] N.I.Muskhelishvili, *Singular Integral Equations*, Noordhoff, Groningen, 1953.
[10] N.I.Muskhelishvili, *Integral Equations*, Noordhoff, Leyden, 1977.
[11] W.A.Pogorzelski, *Integral Equations and their Applications*, Pergamon, N.Y., 1966.
[12] F.G.Tricomi, *Integral Equations*, Interscience, N.Y., 1957.
[13] K.Yoshida, *Lectures on Differential and Integral Equations*, Interscience, N.Y., 1960.

11. MATRICES AND DETERMINANTS

Problem 65-1, A Least Squares Estimate of Satellite Attitude, by GRACE WAHBA (IBM—Federal Systems Division).

Given two sets of n points $\{\mathbf{v}_1, \mathbf{v}_2, \cdots, \mathbf{v}_n\}$, and $\{\mathbf{v}_1^*, \mathbf{v}_2^*, \cdots, \mathbf{v}_n^*\}$, where $n \geq 2$, find the rotation matrix M (i.e., the orthogonal matrix with determinant $+1$) which brings the first set into the best least squares coincidence with the second. That is, find M which minimizes

$$\sum_{j=1}^{n} \| \mathbf{v}_j^* - M\mathbf{v}_j \|^2.$$

This problem has arisen in the estimation of the attitude of a satellite by using direction cosines $\{\mathbf{v}_k^*\}$ of objects as observed in a satellite fixed frame of reference and direction cosines $\{\mathbf{v}_k\}$ of the same objects in a known frame of reference. M is then a least squares estimate of the rotation matrix which carries the known frame of reference into the satellite fixed frame of reference.

Solution by J. L. FARRELL and J. C. STUELPNAGEL (Westinghouse Defense and Space Center).

Let k denote the dimension of the column vectors $\mathbf{v}_1, \cdots, \mathbf{v}_n, \mathbf{v}_1^*, \cdots, \mathbf{v}_n^*$ and let V and V^* denote the two $k \times n$ matrices obtained by juxtaposing $\mathbf{v}_1, \cdots, \mathbf{v}_n$ and $\mathbf{v}_1^*, \cdots, \mathbf{v}_n^*$, respectively.

For any orthogonal matrix M, define $Q(M)$ as the sum of squares to be minimized, so

$$Q(M) = \sum_{j=1}^{n} \| \mathbf{v}_j^* - M\mathbf{v}_j \|^2 = \operatorname{tr} (V^* - MV)^T (V^* - MV),$$

where tr denotes the trace function and a superscript T denotes transposition.

$Q(M)$ may be rewritten as

$$Q(M) = \operatorname{tr} (V^{*T} - V^T M^T)(V^* - MV) = \operatorname{tr} V^{*T}V^* + \operatorname{tr} V^T V - 2 \operatorname{tr} V^T M^T V^*.$$

Since the first two terms are independent of M, $Q(M)$ is minimized by maximizing $F(M) = \operatorname{tr} V^T M^T V^*$, which may be written as

$$F(M) = \operatorname{tr} M^T V^* V^T.$$

It is a well-known fact that an arbitrary real square matrix A can be written as a product UP, where U is orthogonal and P is symmetric and positive semi-definite. Furthermore, if A is nonsingular, U is uniquely defined and P is positive definite. If A is singular, U is not unique, but it may be taken to have determinant $+1$. (The corresponding statement of the first result above for complex A may be found in [1, §2.8] and the result for real A follows from it.) Applying this result to $A = V^* V^T$, we have $F(M) = \operatorname{tr} M^T UP$. Since P is symmetric, there is an orthogonal matrix N such that NPN^T is a diagonal matrix D, whose diagonal elements $d_1, \cdots, d_k$ are arranged in decreasing order. All d_i are nonnegative, since P is positive semidefinite. Now, letting $X = NM^T \cdot UN^T$, we obtain

$$F(M) = \operatorname{tr} M^T U N^T D N = \operatorname{tr} N M^T U N^T D = \operatorname{tr} XD = \sum_{i=1}^{k} d_i x_{ii}.$$

Since $F(M)$ is a linear function of the nonnegative numbers $d_1, \cdots, d_k$, its maximum is attained when the diagonal elements of X attain their maximum values. Because X is an orthogonal matrix, all elements of X are between -1 and 1, so $F(M)$ is maximized when $x_{ii} = 1$, $x_{ij} = 0$, $i \neq j$.

Because $\det M$ is required to be $+1$, $\det X = \det (NM^T U N^T) = (\det N)^2 \cdot \det M \det U = \det U$. If $\det U = -1$, then it is required that $\det X = -1$, and it is not hard to see that

$$X = \begin{pmatrix} I_{k-1} & 0 \\ 0 & -1 \end{pmatrix}$$

is a solution (since $d_1 \geq d_2 \geq \cdots \geq d_k$). Letting X_0 be the matrix which maximizes $F(M)$ ($X_0 = I$ or $X_0 = \begin{pmatrix} I_{k-1} & 0 \\ 0 & -1 \end{pmatrix}$, according as $\det U = +1$ or -1), $X_0 = NM_0^T U N^T$, or $M_0 = U N^T X_0^T N$ is a rotation matrix which minimizes the sum of squares $Q(M)$. If $V^* V^T$ is nonsingular, it is the unique rotation matrix which does so.

REFERENCE

[1] M. Marcus, *Basic Theorems in Matrix Theory*, Nat. Bureau of Stds., Appl. Math. Series No. 57, 1960.

R. H. Wessner (Hughes Aircraft Company) in his solution points out that if $\det A \neq 0$, then $V^* V^T = A = UP$,

$$U = (A^T)^{-1}(A^T A)^{1/2}, \quad P = (A^T A)^{1/2},$$

where $(A^T A)^{1/2}$ is the symmetric square root of $A^T A$ with positive eigenvalues, and, hence, for $\det A > 0$,

$$M_0 = (VV^{*T})^{-1}(VV^{*T}V^* V^T)^{1/2}.$$

J. R. Velman (Hughes Aircraft Company) in his solution demonstrates that in the case $\det A < 0$, $M_0 = U(I - 2G)$ where G is any one-dimensional projection satisfying $GE_1 = G$, where E_1 is the eigenspace of the smallest eigenvalue of P, hence Farrel and Stuelpnagle's solution in this case is unique if the smallest eigenvalue of P has multiplicity one.

J. E. Brock (U. S. Naval Postgraduate School) solved the problem for $\det V^* V^T \geq 0$ by differentiating

$$\alpha = -\operatorname{tr} [V^T M^{-1} V^* + V^{*T} MV]$$

with respect to each of the 9 elements of M and setting the results equal to 0. The resulting equations turn out to be

$$M^T A M^T = A^T,$$

which implies that $M^T A$ is symmetric, $(M^T A)(M^T A) = A^T A$, $M^T A$ is any symmetric square root $(A^T A)^{1/2}$ of $A^T A$, and $M = (A^T)^{-1}(A^T A)^{1/2}$. He then gives an example in which the actual residual sum of squares is minimized by taking the positive definite symmetric square root.

Problem 63-7, On Commutative Rotations, by JOEL BRENNER (Stanford Research Institute).

Show that if two nontrivial proper rotations in E_3 are commutative, then they are rotations about the same axis or else rotations through 180° about two mutually perpendicular axes.

Solution by THEODORE KATSANIS (NASA, Lewis Research Center).

Let A and B denote two nontrivial commutative rotations, and let x_A and x_B denote their respective axes. Let P be a plane containing x_A and parallel to x_B. If x_A is parallel to x_B, let O be any point on x_A; otherwise, let O be the intersection of the projection on P of x_B with x_A. In either case $A(O) = O$, so that $B(O) = BA(O) = AB(O)$. Since $B(O)$ is left fixed by A, it must lie on x_A. But if x_B does not intersect x_A, $B(O)$ cannot lie on x_A. Hence x_B intersects x_A and $B(O) = O$.

Consider now a point S on x_A, $S \neq O$. Then $A(S) = S$, so that $B(S) = BA(S) = AB(S)$, and $B(S)$ is left fixed by A. Hence $B(S)$ lies on x_A. This means that either $x_A = x_B$, or else x_B is perpendicular to x_A, in which case B must also be a 180° rotation.

Solution by K. A. POST (Technological University, Eindhoven, Netherlands).

Suppose A has the matrix

$$\begin{pmatrix} 1 & 0 & 0 \\ 0 & c & s \\ 0 & -s & c \end{pmatrix}$$

where c and s stand for $\cos \phi$ and $\sin \phi$ respectively, i.e., A represents a rotation about the x-axis through an angle ϕ. Let $B = (b_{ij})$ represent a rotation such that $AB = BA$. Equating both sides we get six equations

$$\begin{cases} (1 - c)b_{12} + sb_{13} = 0, \\ sb_{12} + (c - 1)b_{13} = 0, \end{cases} \quad \begin{cases} sb_{31} + (c - 1)b_{21} = 0, \\ (1 - c)b_{31} + sb_{21} = 0, \end{cases}$$
$$s(b_{32} + b_{23}) = 0, \qquad s(b_{22} - b_{33}) = 0.$$

As $s^2 + (c - 1)^2 = 2 - 2c \neq 0$ (A is nontrivial), we obtain $b_{12} = b_{13} = b_{21} = b_{31} = 0$. Now there are two cases.

(1) $\qquad b_{22} = b_{33} = c', \qquad b_{32} = -b_{23} = s', \qquad b_{11} = 1.$

(Rotations about the same axis).

(2) $\quad s = 0, c = -1, \qquad b_{11} = -1, \qquad b_{22} = -b_{33} = c', \qquad b_{32} = b_{23} = s'.$

(Rotations through 180° about mutually perpendicular axes).

Commuting Matrix Exponentials

Problem 88-1*, by DENNIS S. BERNSTEIN (Harris Corporation, Melbourne, Florida).

In feedback control theory for sampled-data systems, the equivalent discrete-time dynamics matrix is given by e^{Ah}, where h is the sample interval and A is the dynamics matrix for the original continuous-time system. When A is perturbed by A_0 (possibly due to some modeling uncertainty), then it is necessary to consider $e^{(A+A_0)h}$. For robust control system design it thus may be of interest to know when $e^{(A+A_0)h}$ can be decomposed into a nominal part involving A and a perturbed part involving A_0. Analogous questions arise in the study of bilinear systems of the form

$$\dot{x} = Ax + uBx,$$

where u is a scalar control. In this case the Lie group generated by A and B plays a central role. Again, it is of interest to know how e^{A+B} is related to e^A and e^B, the principal result being the Baker–Campbell–Hausdorff formula. Of course, it is well known that when

$$(1) \qquad\qquad\qquad AB = BA,$$

where A, B are real $n \times n$ matrices, then both

$$(2) \qquad\qquad\qquad e^A e^B = e^B e^A$$

and

$$(3) \qquad\qquad\qquad e^A e^B = e^{A+B}$$

hold. It is less well known that the converse is not true. Specifically, examples are given in [1] showing that (2) may hold while (1) and (3) are violated, and that (3) may hold while (1) is violated. Interestingly, it is stated without proof in [1] that (3) implies (2). Prove this claim or find a counterexample. A copy of [1] is available from the proposer.

REFERENCE

[1] M. FRÉCHET, *Les solutions non commutables de l'équation matricielle $e^X e^Y = e^{X+Y}$*, Rend. Circ. Mat. Palermo (2), (1952), pp. 11–27.

Solution by EDGAR M. E. WERMUTH (Zentralinstitut für Angewandte Mathematik, Jülich, Federal Republic of Germany).

The claim is false. An explicit counterexample is given by

$$A = \begin{pmatrix} 0 & 0 & 0 & 0 \\ 0 & 0 & 0 & 0 \\ 0 & 0 & a & -b \\ 0 & 0 & b & a \end{pmatrix}, \qquad B = \begin{pmatrix} 0 & 0 & 1 & 0 \\ 0 & 0 & 0 & 1 \\ 0 & 0 & 0 & 0 \\ 0 & 0 & 0 & 0 \end{pmatrix},$$

where $z = a + ib$ is a solution of $e^z - z = 1$, e.g., $a = 2.088843 \cdots$ and $b = 7.461489 \cdots$. We get

$$e^A e^B = e^{A+B} = \begin{pmatrix} 1 & 0 & 1 & 0 \\ 0 & 1 & 0 & 1 \\ 0 & 0 & a+1 & -b \\ 0 & 0 & b & a+1 \end{pmatrix}, \qquad e^B e^A = \begin{pmatrix} 1 & 0 & a+1 & -b \\ 0 & 1 & b & a+1 \\ 0 & 0 & a+1 & -b \\ 0 & 0 & b & a+1 \end{pmatrix}.$$

Moreover, we can prove the following.[1]

THEOREM. *If A and B are square matrices with elements, then $e^A e^B = e^B e^A$ if and only if $AB = BA$.*

Editorial note. Solvers pointed out several relevant papers, including a correction and counterexample published by Fréchet in [2].   [C.C.R.]

REFERENCES

[1] M. FRÉCHET, *Les solutions non commutables de l'équation matricielle* $e^X e^Y = e^{X+Y}$, Rend. Circ. Mat. Palermo (2), (1952), pp. 11–27.

[2] ――, *Rectification*, Rend. Circ. Mat. Palermo (2), (1953), pp. 71–72.

[3] W. GIVENS, *Review of* Les solutions non commutables de liquation matricielle $e^X e^Y = e^{X+Y}$, M. FRÉCHET, Math. Reviews, 14 (1953), p. 237.

[4] C. W. HUFF, *On pairs of matrices (of order two) A, B satisfying the condition* $e^A e^B = e^{A+B} \neq e^B e^A$, Rend. Circ. Mat. Palermo (2), (1953), pp. 326–330.

[5] A. G. KAKAR, *Non-commuting solutions of the matrix equation* $\exp(X+Y) = \exp X \exp Y$, Rend. Circ. Mat. Palermo (2), (1953), pp. 331–345.

Problem 73-3, Commuting Matrices, by ANDREW ACKER (Universität Karlsruhe, Germany).

Given that $K(t)$ denotes a differentiable matrix function of a real variable t and that $\dot{K}(t)$ denotes its derivative.

(1) Show that: If, on the interval I, the eigenvalues of $K(t)$ are distinct, then a necessary and sufficient condition that $K(t_1)$ commutes with $K(t_2)$ (abbreviated: $K(t_1) \sim K(t_2)$) for all t_1 and t_2 in I is that $K(t) \sim \dot{K}(t)$ for each t in I.

(2)   Give a counterexample of the necessary and sufficient condition in (1) when the eigenvalues of $K(t)$ are not distinct at all points in I.

(3) Show that: If the eigenvalues of $K(t)$ fail to be distinct only at $t_0 \in I$, if $K(t)$ is p times continuously differentiable near t_0 and $\sum_{i=0}^{p} \alpha_i K^{(i)}(t_0)$ has distinct eigenvalues for some arbitrary choice of $\alpha_0, \alpha_1, \cdots, \alpha_p$, then the necessary and sufficient condition in (1) holds.

The above problems arose in a study of a factorization theorem for operator-value functions.

Solution by the proposer.

Actually, one must assume that $K(t)$ is continuously differentiable (c.d.) or, more generally, that $K(t)$ is absolutely continuous. The first assumption will be used since.

(1) The proof of (1) uses two well-known facts:

(a) If $K(t)$ is a c.d. square matrix function and $K(t_0)$ has distinct eigenvalues, then there is an $\varepsilon > 0$ such that the eigenvalues and normalized eigenvectors of $K(t)$ can be chosen as c.d. functions of t in the interval $I_\varepsilon(t_0) = (t_0 - \varepsilon, t_0 + \varepsilon)$.

(b) Assume the eigenvalues $\lambda_1, \lambda_2, \cdots, \lambda_n$ of K are distinct and that $K\phi_i = \lambda_i \phi_i$ and $K^+ \psi_i = \lambda_i^* \psi_i$ for $i = 1, \cdots, n$ (where K^+ is the adjoint, and the vectors ϕ_i and ψ_i are nontrivial). Then $\langle \phi_i, \psi_i \rangle \neq 0$ for $i = 1, \cdots, n$.

We first prove (1) on the interval $I_\varepsilon(t_0)$:

Let $\lambda(t)$ be one of the c.d. eigenvalue functions of $K(t)$ and let $K(t) - \lambda(t)I$

$= A(t)$. Then $A(t)\phi(t) = 0$ and $A^+(t)\psi(t) = 0$ on $I_\varepsilon(t_0)$, where $\phi(t)$ and $\psi(t)$ are the c.d. normalized eigenvector functions of $K(t)$ at $\lambda(t)$ and $K^+(t)$ at $\lambda^*(t)$. Therefore $\dot{A}(t)\phi(t) + A(t)\dot{\phi}(t) = 0$ on $I_\varepsilon(t_0)$. The condition $A(t) \sim \dot{A}(t)$ implies (since the eigenvalues of $A(t)$ are distinct) that $\phi(t)$ is an eigenvector of $\dot{A}(t)$, i.e., that $\dot{A}(t)\phi(t) = \alpha(t)\phi(t)$ for some function $\alpha(t)$. The relations $\dot{A}\phi + A\dot{\phi} = 0$ and $\dot{A}\phi = \alpha\phi$ together imply that $A\dot{\phi} + \alpha\phi = 0$. Whence

$$\langle A^+\psi, \dot{\phi}\rangle + \alpha\langle\psi, \phi\rangle = \langle\psi, A\dot{\phi}\rangle + \alpha\langle\psi, \phi\rangle = \langle\psi, A\dot{\phi} + \alpha\phi\rangle = 0.$$

Since $A^+\psi = 0$ and $\langle\psi, \phi\rangle \neq 0$, it follows that $\alpha(t) = 0$ on $I_\varepsilon(t_0)$, and therefore that $A(t)\phi(t) = 0$ and $A(t)\dot{\phi}(t) = 0$ both hold. Now $A(t)$ has (up to a complex factor) only one eigenvector at the eigenvalue 0. Whence there is a function $C(t)$ such that $\dot{\phi}(t) = C(t)\phi(t)$. $C(t)$ is continuous since $C(t) = \langle\phi(t), \dot{\phi}(t)\rangle$. Therefore $\phi(t) = \phi(t_0)\exp\int_{t_0}^t C(t')\,dt'$ for t in $I_\varepsilon(t_0)$. Thus at each t in $I_\varepsilon(t_0)$, $\phi(t)$ is an eigenvector of $K(t)$ if and only if $\phi(t_0)$ is. The eigenvectors of $K(t)$ can be assumed to be constant. Finally, $K(t)$ is diagonalized by a fixed invertible transformation $X = (\phi_1(t_0), \cdots, \phi_n(t_0))$ throughout $I_\varepsilon(t_0)$, and $K(t_1) \sim K(t_2)$ for all t_1 and t_2 in $I_\varepsilon(t_0)$.

Now we prove (1): Let $I = (a, b)$, t_0 be arbitrary in (a, b) and

$$t_1 = \sup \{t | \phi_1(t_0), \cdots, \phi_n(t_0) \text{ are eigenvectors of } K(t) \text{ on } [t_0, t]\}.$$

By continuity, $\phi_1(t_0), \cdots, \phi_n(t_0)$ are also eigenvectors of $K(t_1)$. If one assumes $t_1 < b$, a contradiction is obtained as follows: $K(t_1)$ has distinct eigenvalues, so there is an interval $I_\varepsilon(t_1)$ on which $\phi_1(t_0), \cdots, \phi_n(t_0)$ are eigenvectors of $K(t)$, contradicting the definition of t_1. Therefore, by the argument used above, $K(t_1) \sim K(t_2)$ for $t_1, t_2 \in (a, b)$.

(2) Let

$$K(t) = \begin{cases} A_+ t^2, & t \geq 0, \\ A_- t^2, & t \leq 0, \end{cases}$$

where A_+ and A_- are noncommuting constant matricies. Then $K(t) \sim \dot{K}(t)$ for all t, but $K(t_1) \nsim K(t_2)$ for $t_1 < 0 < t_2$.

(3) Let $t_0 = 0$. $K(t)$ has constant eigenvectors $\phi_1^-, \cdots, \phi_n^-$ for $t < 0$ and $\phi_1^+, \cdots, \phi_n^+$ for $t > 0$. These are also eigenvectors of $S(t) = \sum_{i=0}^p \alpha_i K^{(i)}(t)$ for $t < 0$ and $t > 0$, respectively. By continuity of $S(t)$, $\{\phi_1^-, \cdots, \phi_n^-\}, \{\phi_1^+, \cdots, \phi_n^+\}$ are two sets of eigenvectors of $S(0)$. The eigenvalues of $S(0)$ are distinct, so the second set must be merely a reordering (with complex factors) of the first. Thus $\phi_1^-, \cdots, \phi_n^-$ are eigenvectors of $K(t)$ for all $t \in I$. Therefore $K(t_1) \sim K(t_2)$ for $t_1, t_2 \in I$ as in (1).

Editorial note. The proposer also notes, with proof, the following similar result for matrix functions of a complex variable.

Let $K(z)$ be a square matrix function which is analytic throughout a region R of the complex plane. If there is at least one point z_0 in R such that the eigenvalues of $K(z_0)$ are distinct and if $K(z) \sim \dot{K}(z)$ in a neighborhood of z_0, then $\{K(z)|z \in R\}$ is a commuting set.

For related results, see J. F. P. Martin, *Some results on matrices which commute with their derivatives*, SIAM J. Appl. Math., 15 (1967), pp. 1171–1183 and A. Acker, *Absolute continuity of eigenvectors of time-varying operators*, Proc. Amer. Math. Soc., 42 (1974), pp. 198–201.

On a Nonlinear Matrix Differential Equation

Problem 85-8, *by* P. SCHWEITZER (University of Rochester).

Let $f(z) = \sum_{K=1}^{\infty} c_K z^K$ be an analytic square matrix function of z, with c_1 and c_2 given, that satisfies the matrix differential equation

(1)
$$\frac{df(z)}{dz} = \frac{f(z) A f(z)}{z^2}$$

for some given square matrix A. Show this is possible if and only if

(2) $c_1 A c_1 = c_1,$

(3) $c_1 A c_2 = c_2 A c_1 = c_2.$

If these conditions are met, show that the solution is unique and give closed form expressions for $\{c_K, K \geq 3\}$ and $f(z)$ in terms of c_1 and c_2.

Solution by GILBERT N. LEWIS (Michigan Technological University).

Substituting the power series expression for $f(z)$ into (1) and equating coefficients of different powers of z yield

(4)
$$K c_K = \sum_{j=1}^{K} c_j A c_{K-j+1}, \qquad K = 1, 2, 3, \cdots.$$

When $K = 1$, we have (2). The equation for $K = 2$ is

(5) $2 c_2 = c_1 A c_2 + c_2 A c_1.$

Premultiplication of (5) by $c_1 A$, postmultiplication of (5) by $A c_1$, and use of (2) yield

$$c_1 A c_2 = c_1 A c_2 A c_1 = c_2 A c_1 = c_2,$$

which is (3). For $K > 2$, we again pre- and postmultiply in (4) by $c_1 A$ and $A c_1$, respectively, and use the equations derived from lower values of K to obtain

$$c_K = c_j A c_{K-j+1} \quad \text{for any } j = 1, 2, \cdots, K.$$

By induction, we find that

$$c_K = (c_2 A)^{k-2} c_2 \quad \text{for } K = 2, 3, 4, \cdots.$$

Thus,

$$f(z) = c_1 z + \sum_{K=2}^{\infty} (c_2 A z)^{K-2} c_2 z^2.$$

If $|z|$ is small enough, this last series converges to

$$f(z) = c_1 z + (I - c_2 A z)^{-1} c_2 z^2,$$

the unique solution of equation (1).

Problem 77-14*, *A Matrix Convergence Problem*, by G. K. KRISTIANSEN (Research Establishment RISØ, Roskilde, Denmark).

Let $P = \{p_{rs}\}$ be a symmetric matrix having (i) $p_{rs} = 0$ for $|r - s| > 1$ and $p_{rs} > 0$ otherwise, (ii) spectral radius 1, and (iii) $p_{s-1,s} + p_{s+1,s} \leq 1$ for all s. Denote by e^T the $1 \times n$ matrix with all entries 1, and let $I = \{\delta_{rs}\}$ be the $n \times n$ unit matrix. Let c be a nonnegative $n \times 1$ matrix with $e^T c = 1$. Prove or disprove that the matrix

$$F = (I - c\,e^T)P$$

has spectral radius at most equal to 1. If a counterexample is found, try to minimize the order n. The problem arose in an investigation of methods for solution of the neutron diffusion equation in reactor physics.

Partial solution by C. GIVENS (Michigan Technological University).

1. If uv' is a rank one perturbation of a matrix A, i.e. $B = A + uv'$, then

$$\det B = \det A + v'\,\text{adj}A\,u.$$

The proof is simple in the invertible case and the result follows in general by continuity.

2. If the result of part 1 is applied to

$$s1 - F = sl - P + c(Pe)',$$

then the characteristic polynomial $c_F(s)$ of F is given by

$$c_F(s) = c_P(s) + e'P\,\text{adj}\,(s1 - P)c.$$

But,

$$P\,\text{adj}(s1 - P) = [s1 - (s1 - P)]\,\text{adj}\,(s1 - P) = s\,\text{adj}\,(s1 - P) - c_P(s)1.$$

Thus

(*) $$c_F(s) = s[e'\,\text{adj}\,(s1 - P)c].$$

3. Since P is real symmetric, irreducible, nonnegative and imprimitive, there is a unique real eigenvalue (namely, $s = 1$) with modulus one. Consequently,

$$\text{adj}\,(s1 - P) > 0 \quad \text{for } s \geq 1.$$

(cf. Marcus–Minc, *A Survey of Matrix Theory and Matrix Inequalities*, p. 125). Therefore, $e > 0$, $c \geq 0$ imply $c_F(s) \neq 0$ for $s \geq 1$; i.e. *F can have no real eigenvalue in* $[1, \infty)$.

4. As (*) indicates, F always has a zero eigenvalue. The corresponding eigenvector is $z_0 = \text{adj}\,Pc$. Indeed, if s is an eigenvalue of F, then the corresponding eigenvector is $z_s = \text{adj}\,(s1 - P)c$. Thus,

$$c_F(s) = s(e'z_s)$$

and the eigenvalues of F are those values that make members of the one parameter family z_s lie in the hyperplane perpendicular to e. This shows that no eigenvector of F can have all of its components of one sign.

5. The tridiagonal $n \times n$ matrix P with constant diagonal $a = 1 - \cos \theta_n$, $\theta_n = \pi/(n+1)$ and with nonzero off-diagonal elements $b = \frac{1}{2}$ satisfies the hypotheses of the problem. Moreover, its smallest eigenvalue $s = 1 - 2\cos \theta_n$ approaches minus one as n goes to infinity. It was hoped that the perturbation could then create an eigenvalue of F less than minus one for n sufficiently large. However, computer calculation for a variety of c's showed an eigenvalue of F moving towards minus one for $n \leq 20$ and then retreating for $20 < n \leq 35$.

6. A subset of the class of stochastic (hence doubly stochastic, by symmetry) matrices satisfies the hypotheses on P. In several different ways, it can be shown in this case that

$$c_F(s) = s(s-1)^{-1}c_P(s).$$

Therefore, in the stochastic case, the spectral radius of F is less than one.

Problem 73-20, *A Special Class of Matrices*, by J. F. FOLEY (Baltimore Gas and Electric Company).

Let Q be an $n \times n$ stochastic matrix with no two rows or columns alike (here the row sums are unity).

Find conditions on Q sufficient to insure that the matrix $T = Q^2$ is a matrix in which all rows are identical and

$$0 \leqq t_{ij} = t_j < 1, \qquad \sum_{k=1}^{n} t_k = 1, \qquad i,j = 1,2,\cdots,n.$$

These matrices arise from the study of Markov processes with a discrete time sequence in which equilibrium is reached after the first time transaction.

Solution by JOHN Z. HEARON (National Institutes of Health).

Since a row stochastic matrix is of rank one if and only if it has all rows identical, the problem can be restated as follows. Characterize the set of all n-square matrices Q which are row stochastic of rank greater than one and such that Q^2 is of rank one.

If Q is any n-square matrix whatever of rank $r > 1$, range $R(Q)$ and nullspace $N(Q)$, it is clear that for Q^2 to be of rank one, it is necessary and sufficient for there to be $r - 1$ linearly independent vectors in the intersection $R(Q) \cap N(Q)$. This condition is possible only if $\dim(N(Q)) = n - r \geqq r - 1$. Thus the set we seek is the intersection of the set of all n-square row stochastic matrices and the set of all n-square matrices of rank r such that $\dim(R(Q) \cap N(Q)) = r - 1$. This last condition can be exploited in many ways. We give two.

If $\{p_1, p_2, \cdots, p_{n-r}\}$ is a basis for $N(Q)$, we must have vectors v_i such that $Qv_i = p_i$ for $i = 1, 2, \cdots, r - 1$. It follows that $\lambda = 0$ is a characteristic value of (algebraic) multiplicity $n - 1$. The remaining characteristic value must be $\lambda = 1$, since Q is stochastic. Thus the Jordan normal form is

(1) $\operatorname{diag}(1, N, N, \cdots, N, 0, 0, \cdots, 0),$

where there are $r - 1$ matrices

$$N = \begin{bmatrix} 0 & 1 \\ 0 & 0 \end{bmatrix}$$

and $n - 2r + 1$ zeros. Thus the set we seek is the set of all row stochastic matrices with Jordan form (1).

We now give a construction of the desired matrices. Fix n and r such that $n - r \geqq r - 1$. Let u be the vector $[1, 1, \cdots, 1]^T$. Select an arbitrary nonnegative vector y_1, normalized so that $u^T y_1 = 1$, and any $r - 1$ independent vectors

$y_2, y_3, \cdots, y_r$ orthogonal to u. It is easily verified that $y_1, y_2, \cdots, y_r$ are linearly independent. Now select any $r - 1$ independent vectors $x_2, x_3, \cdots, x_r$ from the orthogonal complement of the set $\{y_1, y_2, \cdots, y_r\}$. It is easily verified that $u = x_1$, $x_2, \cdots, x_r$ are linearly independent.

Then $Q = \sum_{i=1}^{r} x_i y_i^T$ satisfies $Qu = u$ and $Q^2 = x_1 y_1^T$. Since $x_1 y_1^T \geq 0$ by construction, we can have $Q \geq 0$ by suitably scaling the $x_2, x_3, \cdots, x_r$. Then Q is row stochastic, of rank r and such that Q^2 is of rank one. Here $x_2, x_3, \cdots, x_r$ are the $r - 1$ vectors in $R(Q) \cap N(Q)$, since $N(Q)$ is spanned by the orthogonal complement of $\{y_1, y_2, \cdots, y_r\}$. The orthogonality of u to each of $y_2, \cdots, y_r$ is necessary and sufficient for $Qu = u$. The choice $y_1 = u/n$ makes Q doubly stochastic.

Conversely, if Q is of rank r, we may write $Q = \sum_{i=1}^{r} x_i y_i^T$, where $\{x_1, x_2, \cdots, x_r\}$ is a basis for $R(Q)$ and $\{y_1, y_2, \cdots, y_r\}$ is a basis for $R(Q^T)$. If Q is stochastic, then $u \in R(Q)$ and we are at liberty to take $x_1 = u$. The requirement that $\operatorname{rank}(Q^2) = 1$ implies that $r - 1$ of the vectors $x_1, x_2, \cdots, x_r$ are in $N(Q)$. Thus $x_2, x_3, \cdots, x_r$ are in the orthogonal complement of $\{y_1, y_2, \cdots, y_r\}$, since $x_1 = u$ obviously is not. Finally, $Qu = u$ requires that u be orthogonal to each of $y_2, y_3, \cdots, y_r$. Thus the structure given above is necessary and sufficient.

A Matrix Problem

Problem 82-6, by C. G. BROYDEN (University of Essex, Colchester, England).

Let X be an $m \times n$ matrix with no null columns and with $q_i \geq 1$ nonzero elements in its ith row, $1 \leq i \leq m$. Let $Q = \operatorname{diag}(q_i)$, and D be the diagonal matrix whose diagonal elements are equal to the corresponding diagonal elements of $X^T X$. Determine the conditions under which the matrix $D - X^T Q^{-1} X$ is

(a) positive definite,

(b) positive semidefinite.

This problem arose in connection with an algorithm for scaling examination marks.

Solution by the proposer.

Denote the ith row of X by x_i^T and define the row vector s_i^T to be that vector for which $s_{ij} = 0$ if $x_{ij} = 0$ and $s_{ij} = 1$ if $x_{ij} \neq 0$. Let Y be a diagonal matrix of order n and define the row vector u_i^T by

(1) $$u_i^T = x_i^T Y (I - q_i^{-1} s_i s_i^T).$$

Note that $I - q_i^{-1} s_i s_i^T$ is idempotent. It follows that

(2) $$u_i^T u_i = x_i^T Y (I - q_i^{-1} s_i s_i^T) Y x_i = x_i^T Y^2 x_i - q_i^{-1} (x_i^T Y s_i)^2.$$

Since $x_{ij} = 0$ whenever $s_{ij} = 0$, it follows that

(3) $$x_i^T Y s_i = x_i^T Y e,$$

where $e = (1, 1, \cdots, 1)^T$. Letting $y = Ye$, a straightforward calculation yields

(4) $$\sum_{i=1}^{m} u_i^T u_i = \sum_{i=1}^{m} \{x_i^T Y^2 x_i - q_i^{-1} (x_i^T y)^2\} = y^T (D - X^T Q^{-1} X) y.$$

Consequently, $D - X^T Q^{-1} X$ is positive semidefinite for all X.

Let $(a_1, a_2, \cdots, a_m)^T$ and $(b_1, b_2, \cdots, b_n)^T$ be vectors having no zero components and choose each x_{ij} to be either 0 or else $a_i b_j$. Then setting $Y = \operatorname{diag}(b_j^{-1})$, we find

(5) $$x_i^T Y = a_i s_i^T,$$

so that (1) yields

(6) $$u_i^T = 0^T, \qquad i = 1, \cdots, m.$$

Thus, in this case $D - X^T Q^{-1} X$ is not positive definite. In general, $D - X^T Q^{-1} X$ is positive definite unless there exists a nonzero diagonal matrix Y and a sequence of scalars $a_1, a_2, \cdots, a_m$ such that (5) holds for $i = 1, 2, \cdots, m$.

Problem 74–16, A Matrix Problem, by C. R. CRAWFORD (Erindale College, Ontario, Canada).

If A denotes a given nonsingular square matrix (real or complex) of order $n = rs$, prove that there exist nonzero column vectors $x = (x_1, x_2, \cdots, x_n)^T$ and $b = (b_1, b_2, \cdots, b_n)^T$ such that

(1) $$Ax = b,$$

(2) $$\sum_{i=1}^{r} x_i = \sum_{i=r+1}^{2r} x_i = \cdots = \sum_{i=n-r+1}^{n} x_i,$$

(3) $$b_{kr+1} = b_{kr+2} = \cdots = b_{kr+r} \qquad (k = 0, 1, \cdots, s-1).$$

The problem arose in power system engineering in the study of mutual impedance in a coil with relatively high voltage and current but with constant frequency.

Solution by THOMAS FOREGGER (Bell Laboratories, Murray Hill, New Jersey).

The condition (2) can be rewritten as a system of $(s-1)$ homogeneous equations in the n variables $x_1, \cdots, x_n$. The condition (3) can be written as a system of $s(r-1)$ homogeneous equations in the n variables $x_1, \cdots, x_n$. Thus we have a homogeneous system of $s - 1 + s(r-1) = n - 1$ equations in n variables. The system must have a nonzero solution x, and since A is nonsingular, $b = Ax$ is also nonzero.

Problem 71-25*, Balancing Chemical Equations, by MICHAEL JONES (University of Dallas).

Suppose that A and B are $p \times n$ and $p \times m$ matrices, respectively, with non-negative entries. Give necessary and sufficient conditions such that the matrix-vector equation $Ax = By$ has positive solutions. If possible, find explicit expressions for x and y. This problem had arisen in the attempt to formulate a general method of balancing chemical equations (see R. Crocker, *Application of Diophantine equations to problems in chemistry*, J. Chem. Educ., 45 (1968), pp. 731–733).

Problem 64-19, Nth Power of a Matrix, by JAMES F. FOLEY (Baltimore Gas and Electric Company).

If the matrix M is given by

$$M = \begin{Vmatrix} 1 - \lambda + \lambda^2 & 1 - \lambda \\ \lambda - \lambda^2 & \lambda \end{Vmatrix},$$

determine M^n.

This problem has arisen in determining the result of repeated partial mixing of two completely miscible fluids.

Solution by F. W. PONTING (University of Aberdeen, Scotland).

The characteristic values of M satisfy $\det(xI - M) = (x - 1)(x - \lambda^2) = 0$. Thus M can be expressed in the form $M = TDT^{-1}$, where

$$T = \frac{1}{\sqrt{1 + \lambda}} \begin{Vmatrix} 1 & -1 \\ \lambda & 1 \end{Vmatrix}, \qquad T^{-1} = \frac{1}{\sqrt{1 + \lambda}} \begin{Vmatrix} 1 & 1 \\ -\lambda & 1 \end{Vmatrix}, \qquad \lambda \neq -1,$$

and

$$D = \begin{Vmatrix} 1 & 0 \\ 0 & \lambda^2 \end{Vmatrix}.$$

Whence, $M^n = TD^nT^{-1}$, or

$$M^n = \frac{1}{1 + \lambda} \begin{Vmatrix} 1 + \lambda^{2n+1} & 1 - \lambda^{2n} \\ \lambda - \lambda^{2n+1} & \lambda + \lambda^{2n} \end{Vmatrix}.$$

For $\lambda = -1$, we find by continuity or induction that

$$M^n = \begin{Vmatrix} 2n + 1 & 2n \\ -2n & 1 - 2n \end{Vmatrix}.$$

Solution by R. F. RINEHART (U. S. Naval Postgraduate School, Monterey).

The problem can be regarded as one of calculating the matrix functional value $f(M)$, where $f(z) = z^n$. As is well known [F. R. Gantmacher, *Theory of Matrices*, Chelsea, New York, 1959] $f(M) = L(M)$, where $L(z)$ is the Lagrange-Hermite polynomial "interpolating" $f(z)$ at the characteristic values λ_1, λ_2 of M:

$$L(z) = \begin{cases} \dfrac{z - \lambda_2}{\lambda_1 - \lambda_2} f(\lambda_1) + \dfrac{z - \lambda_1}{\lambda_2 - \lambda_1} f(\lambda_2) & \text{if } \lambda_1 \neq \lambda_2, \\ (z - \lambda_1) f'(\lambda_1) + f(\lambda_1) & \text{if } \lambda_1 = \lambda_2. \end{cases}$$

Thus,

$$M^n = \frac{M - \lambda^2 I}{1 - \lambda^2} + \frac{M - I}{\lambda^2 - 1} \lambda^{2n}, \qquad \qquad \lambda^2 \neq 1,$$

which simplifies to the result given above. For $\lambda^2 = 1$,

$$M^n = (M - I)n + I.$$

For $\lambda = 1$, $M^n = I$, while for $\lambda = -1$ we obtain the result given above.

Editorial note. Other solutions were obtained by means of the Cayley-Hamilton theorem, Sylvester's theorem, induction, or from a difference equation.

V. O. MOWERY (Bell Telephone Laboratories) in his solution notes that the problem occurs in many different applications and in his paper [*On hypergeometric functions in iterated networks*, IEEE Trans. Circuit Theory, CT-11 (1964), pp. 232–247] gives references to a list of papers on the problem which have appeared in circuit theory literature. He also shows that if

$$T = \begin{Vmatrix} A & B \\ C & D \end{Vmatrix},$$

then

$$T^n = \Delta^{(n-1)/2} U_{n-1}(x)\, T - \Delta^{n/2} U_{n-2}(x)\, I,$$

where $x = (A + D)/2\sqrt{\Delta}$, $\Delta = AD - BC \neq 0$, and $U_n(x)$ is the Chebyshev polynomial of the second kind.

J. ASTIN (University College of Wales, England) in his solution refers to a forthcoming paper of his to be published in the Mathematical Gazette concerning the nth power of a square matrix of general order.

T. KRISHNAN and T. J. RAO (Indian Statistical Institute, Calcutta) in their solution note that the sum of the elements of the matrix M is invariant under exponentiation. Motivated by this, they determined the necessary and sufficient conditions for a 2×2 matrix to have this invariant sum property. These conditions involve what is called the Lucas polynomial† of the matrix.

For a 2×2 matrix A, $A^N = u_N(p, q)A - qu_{N-1}(p, q)I$, where $p =$ trace of A, and $q =$ determinant of A. $u_N(p, q)$ is called a Lucas polynomial of order N. Let us define two more quantities associated with a 2×2 matrix $A = (a_{ij})$. Let $2r = (a_{11} + a_{22}) - (a_{12} + a_{21})$, $s = a_{11} + a_{12} + a_{21} + a_{22}$. Suffixes will denote the matrix with which these quantities are associated. For the matrix M, $p = 1 + \lambda^2$, $q = \lambda^2$, and $u_N = (1 - \lambda^{2N})/(1 - \lambda^2)$.

THEOREM. *A necessary and sufficient condition for a* 2×2 *matrix to have the sum of elements* s *invariant under exponentiation is that the Lucas polynomial of the matrix is* $s/2$ *times the Lucas polynomial of the matrix* M, *with* $\lambda = \sqrt{r}$.

Proof. Barakat† has shown that

(1)
$$u_N = pu_{N-1} - qu_{N-2}, \quad \text{and} \quad A^N = u_N A - qu_{N-1}I.$$

$$\text{Hence,} \quad s_{A^N} = u_N s_A - 2qu_{N-1}.$$

If $s_{A^N} = s_A = s$ for all N, then $s = 2qu_{N-1}/(u_N - 1)$, for all N, which together with (1) reduces to $u_N = ru_{N-1} + (s/2)$. Solving this difference equation, we finally have $u_N = (s/2)(1 - r^N)/(1 - r)$.

Problem 76–9*, A Matrix Stability Problem, by S. VENIT (California State University at Los Angeles).

Let

$$P = \begin{bmatrix} B & C \\ I & 0 \end{bmatrix}$$

be a real, square matrix of order $2n$, partitioned into four $n \times n$ blocks. Assume that I and 0 are the identity and null matrices (of order n), respectively, and that the only nonzero elements of B and C are given by $b_{ij} = 2r_j/(1 + 2r_j)$ when $|i - j| = 1$, and $c_{ij} = (1 - 2r_j)/(1 + 2r_j)$ when $i = j$ $(i, j = 1, 2, \cdots, n)$, where the r_j are arbitrary positive numbers.

Show either that the spectral radius of P is less than 1 for all positive integers n, or find a counterexample.

This problem arose in considering the matrix stability of a DuFort–Frankel-type difference scheme. See *Numerical Methods for Partial Differential Equations,*

---

† R. BARAKAT, *The matrix operator* e^x *and Lucas polynomials*, J. Math. Phys., **43** (1964), pp. 333–335.

by W. F. Ames, Barnes and Noble, New York, 1969. In this reference, the author shows that the spectral radius is less than 1 in the special case when all r_j are equal.

Solution by A. R. CURTIS (Atomic Energy Research Establishment, Harwell, Oxfordshire, England).

Writing $s_j = 2r_j/(1+2r_j)$, $j = 1$ to n, we note that $0 < s_j < 1$ and that the partitioned matrix P has the form

$$P = \begin{bmatrix} AD & I - 2D \\ I & 0 \end{bmatrix},$$

where D is the nonsingular matrix diag $\{s_1, s_2, \cdots, s_n\}$ and A is the symmetric $n \times n$ matrix whose elements are zero except for those immediately above and below the main diagonal, which have the value unity. The eigenvalues of A are known to be $2\cos(\pi k/(n+1))$, $k = 1, 2, \cdots, n$, so its Rayleigh quotients lie in the range $\pm 2\cos(\pi/(n+1))$.

Let λ be a nonzero eigenvalue of P. The lower blocks of P show that the corresponding eigenvalue x can be partitioned in the form

$$x = \begin{bmatrix} \lambda v \\ v \end{bmatrix},$$

where v is an n-vector. Let $u = Dv$ and let u^* be the complex conjugate transpose of u. We pre-multiply the upper blocks by u^*, obtaining the real quadratic equation for λ

(1) $$f(\lambda) \equiv \beta \lambda^2 - 2c\lambda + 2 - \beta = 0,$$

where

$$2c = u^*Au/u^*u, \qquad |c| \leqq \cos(\pi/(n+1)) < 1,$$

and

$$\beta = u^*D^{-1}u/u^*u > 1.$$

For $\lambda = e^{i\theta}$ we find that

(2) $$f(\lambda) = 2e^{i\theta}[(\cos\theta - c) + i(\beta - 1)\sin\theta].$$

Because of the above bounds on c and β, as arg (λ) increases by 2π, arg $(f(\lambda))$ increases by 4π; the easiest way to see this is to consider the graphs of real and imaginary parts of the expression in square brackets in (2). Therefore, by the principle of the argument, $f(\lambda)$ has two zeros inside the unit circle, and since it is quadratic it has no others. It follows that the spectral radius of P is less than 1.

Problem 72-3*, *A Conjectured Positive Definite Matrix*, by K. W. SCHMIDT (University of Manitoba, Canada).

It is conjectured that the matrix

$$A_k = [a_{ij}(k)],$$

where A_0 is the 0-matrix and A_1 is the I-matrix, is positive definite if

$$a_{i0}(r) = a_{i(n+1)}(r) = 0,$$

$$4a_{ij}(r) - a_{ij}(r-1) - a_{ij}(r+1) - a_{i(j-1)}(r) - a_{i(j+1)}(r) = 0,$$

$$1 \le i \le n, \qquad 1 \le j \le n, \qquad 2 \le r+1 \le k.$$

The problem arose in efforts to find a direct solution for the rectangular Poisson difference system.

Solution by MELVYN CIMENT and CLEVE MOLER (University of Michigan).

Let J be the $n \times n$ "shift" matrix with 1's on the superdiagonal and 0's elsewhere. Then the matrices A_r defined by the problem satisfy the three term recurrence

$$A_0 = 0,$$

$$A_1 = I,$$

$$A_{r+1} = A_r(4I - J - J^T) - A_{r-1}, \qquad r = 1, 2, \cdots.$$

Note that A_2 is a matrix with 4's on the diagonal and -1's on the super- and subdiagonals. Furthermore,

$$A_r = p_r(A_2),$$

where $p_r(x)$ is the polynomial defined by

$$p_1(x) = 1,$$

$$p_2(x) = x,$$

$$p_{r+1}(x) = x \cdot p_r(x) - p_{r-1}(x), \qquad r = 2, 3, \cdots.$$

Since A_2 is symmetric, so are all the A_r's.

Let $\lambda_{i,r}$, $i = 1, \cdots, n$, denote the eigenvalues of A_r. The spectral mapping theorem or any easy induction argument shows that

$$\lambda_{i,r} = p_r(\lambda_{i,2}).$$

Clearly $\lambda_{i,1} = 1$ and Gerschgorin's theorem shows that

$$\lambda_{i,2} > 2.$$

Hence,

$$\lambda_{i,r+1} > 2\lambda_{i,r} - \lambda_{i,r-1},$$

and consequently,

$$\lambda_{i,r} > r \quad \text{for all } i.$$

This certainly proves that A_r is positive definite.

It can be shown that

$$\lambda_{i,2} = 2 + 4\sin^2 \frac{i\pi}{2n+2}, \qquad i = 1, \cdots, n.$$

Hence for each fixed i the above inequalities become equalities as $n \to \infty$. It can also be shown that $p_r(x)$ is essentially the Chebyshev polynomial of the second kind, that $p_r(x)$ is monotonic increasing in x for $x \geq 2$, and hence the relation $\lambda_{i,r} = p_r(\lambda_{i,2})$ actually preserves the ordering of the eigenvalues. Finally, by using the fact that $\lambda_{i,2} < 6$ and solving the difference equation $p_1 = 1$, $p_2 = 6$, $p_{r+1} = 6p_r - p_{r-1}$, it can be shown that the largest eigenvalue of A_r grows at most like $(3 + \sqrt{8})^r$.

Solution by OTTO G. RUEHR (Michigan Technological University).

Let

$$a_{ij}(k) = \frac{2}{n+1} \sum_{s=1}^{n} \frac{\sin(i\pi s/(n+1)) \sin(j\pi s/(n+1)) \sinh(k\psi_s)}{\sinh \psi_s},$$

where

$$\cosh \psi_s = 2 - \cos\left(\frac{\pi s}{n+1}\right).$$

Each term in the finite sum satisfies the given recurrence relations. For $k = 0$, we obtain the zero matrix and since

$$\frac{2}{n+1} \sum_{s=1}^{n} \sin\left(\frac{i\pi s}{n+1}\right) \sin\left(\frac{j\pi s}{n+1}\right) = \delta_{ij},$$

we have the unit matrix for $k = 1$. The above equation also shows that we have in effect a spectral resolution of A_k from which we can determine the eigenvalues

$$\lambda_s = \frac{\sinh k\psi_s}{\sinh \psi_s}.$$

Since $\lambda_s > 0$ for $k \geq 1$ and A_k is symmetric, we conclude that the matrix is positive definite. Finally we note that the components of the inverse can be written

$$a_{ij}^{-1}(k) = \frac{2}{n+1} \sum_{s=1}^{n} \frac{\sin(i\pi s/(n+1)) \sin(j\pi s/(n+1)) \sinh \psi_s}{\sinh k\psi_s}.$$

Problem 76–8, A Matrix Inequality, by W. ANDERSON, JR. and G. TRAPP (West Virginia University).

Let A and B be Hermitian positive definite (HD) matrices. Write $A \geq B$ if $A - B$ is HD. Show that

$$A^{-1} + B^{-1} \geq 4(A + B)^{-1}.$$

Solution by M. H. MOORE (Vector Research Inc.).

The matrix inverse function is "matrix convex" on the class of all HD matrices: for any HD matrices A and B, and for any scalar $0 \le \lambda \le 1$,

$$[\lambda A + (1-\lambda)B]^{-1} \le \lambda A^{-1} + (1-\lambda)B^{-1}.$$

Proofs of the latter fact, most of which involve simultaneously diagonalizing A and B, may be found in any of [1]–[5]. The result claimed is just the special case $\lambda = \frac{1}{2}$.

REFERENCES

[1] T. CACOULLOS AND I. OLKIN, *On the bias of functions of the characteristic roots of a random matrix*, Biometrika, 52 (1965), pp. 87–94.
[2] T. GROVES AND T. ROTHENBERG, *A note on the expected value of the inverse of a matrix*, Ibid., 56 (1969), p. 690.
[3] M. H. MOORE, *A convex matrix function*, Amer. Math. Monthly, 80 (1973), pp. 408–409.
[4] V. K. SRIVASTAVA, *On the expectation of the inverse of a matrix*, Sankhyā Ser. A, 32 (1970), p. 336.
[5] P. WHITTLE, *A multivariate generalization of Tchebichev's inequality*, Quart. J. Math. Oxford Ser., 9 (1958), pp. 232–240.

Solution by E. H. LIEB (Princeton University).

This inequality is a special case of a more general Schwarz inequality for matrices [1].

THEOREM. *If* $\{C_1, \cdots, C_k\}$ *are* HD *matrices and* $\{D_1, \cdots, D_k\}$ *are any matrices, then*

$$\left(\sum_{i=1}^{k} D_i^* \right) \left(\sum_{i=1}^{k} C_i \right)^{-1} \left(\sum_{i=1}^{k} D_i \right) \le \sum_{i=1}^{k} D_i^* C_i^{-1} D_i.$$

Proof. Let $X_i = C_i^{-1/2} D_i - C_i^{1/2} L$, with $L \equiv (\sum_{i=1}^{k} C_i)^{-1}(\sum_{i=1}^{k} D_i)$. Then

$$0 \le \sum_{i=1}^{k} X_i^* X_i = \sum_{i=1}^{k} D_i^* C_i^{-1} D_i - \left(\sum_{i=1}^{k} D_i^* \right) \left(\sum_{i=1}^{k} C_i \right)^{-1} \left(\sum_{i=1}^{k} D_i \right).$$

To solve the problem take $k = 2$, $C_1 = A$, $C_2 = B$ and $D_1 = D_2 = I$.

REFERENCE

[1] E. H. LIEB AND M. B. RUSKAI, *Some operator inequalities of the Schwarz type*, Advances in Math., 12 (1974), pp. 269–273.

Problem 76-20*, On the Extreme Eigenvalues of an $n \times n$ Matrix, by L. B. BUSHARD (Babcock and Wilcox Research Center).

Find estimates, as functions of n, on the largest and smallest eigenvalues of the $n \times n$ matrix $A_n = (a_{ij})$: $a_{ij} = 1/(1+|i-j|)$, $i, j = 1, \cdots, n$. The equation $y' = -A_n y$ arises as a potential test problem for stiff ordinary differential equation solvers.

Solution by DENNIS C. JESPERSEN (Mathematics Research Center, University of Wisconsin).

Since the matrix A_n is symmetric, all of its eigenvalues are real and we may find one-sided bounds for $\lambda_{\max}(A_n)$ and $\lambda_{\min}(A_n)$ by evaluating the Rayleigh quotient $x^T A_n x/(x^T x)$ for specific choices of x. Let $x = (1, 1, \cdots, 1)^T$; then $x^T A_n x/(x^T x) = 2 \log n + O(1)$, which is thus a lower bound for $\lambda_{\max}(A_n)$. Let $x = (-1, 1, -1, \cdots, (-1)^n)^T$; then $\lambda_{\min}(A_n) \le x^T A_n x/(x^T x) = \log 4 - 1 + O(1/n)$.

An upper bound for $\lambda_{\max}(A_n)$ is easily obtained from Gershgorin's theorem, which yields $\lambda_{\max}(A_n) \leq 1 + 2(1/2 + 1/3 + \cdots + 1/[n/2]) + O(1/n) = 2\log n + O(1)$. Hence, $\lambda_{\max}(A_n) = 2\log n + O(1)$.

A lower bound for $\lambda_{\min}(A_n)$ may be found as follows. Define

$$f(\theta) = \sum_{k=-\infty}^{\infty} \frac{1}{1+|k|} e^{ik\theta}, \qquad\qquad 0 < \theta < 2\pi.$$

Now

$$f(\theta) = -1 + (\pi - \theta)\sin\theta - (\cos\theta)\log(2 - 2\cos\theta)$$

(cf. [1, pp. 188–189] or [2, p. 239]) and one can readily verify that $f(\theta) \geq f(\pi) = \log 4 - 1$ for $0 < \theta < 2\pi$. It follows that for any x,

$$x^T A_n x = \sum_{j,k=1}^{n} x_j x_k \left\{ \frac{1}{2\pi} \int_0^{2\pi} f(\theta) e^{-i(j-k)\theta}\, d\theta \right\}$$

$$= \frac{1}{2\pi} \int_0^{2\pi} f(\theta) \left| \sum_{k=1}^{n} x_k e^{ik\theta} \right|^2 d\theta$$

$$\geq (\log 4 - 1) x^T x.$$

Hence $\lambda_{\min}(A_n) \geq \log 4 - 1$, so we have $\log 4 - 1 \leq \lambda_{\min}(A_n) \leq \log 4 - 1 + O(1/n)$. Computer calculation for $n = 30$ gave $\lambda_{\min}(A_{30}) = .38690$ (correct to the number of digits shown), which is to be compared with $\log 4 - 1 \doteq .38629$.

Since the spread of the eigenvalues of A_n is relatively small, we might say that the system $y' = -A_n y$ is not very stiff.

Thanks are due to R. Askey for pointing out reference [2] and to P. G. Nevai for his help in bounding $x^T A_n x / (x^T x)$ from below.

REFERENCES

[1] T. J. I. BROMWICH, *An Introduction to the Theory of Infinite Series*, 2nd ed., Macmillan, New York, 1942.
[2] E. R. HANSEN, *A Table of Series and Products*, Prentice-Hall, Englewood Cliffs, NJ, 1975.

Problem 72-18, Eigenvalues of a Reflected Tridiagonal Matrix, by T. S. CHOW (Boeing Computer Services).

Let $J = (j_{ik})$ be the $n \times n$ tridiagonal matrix: $j_{ik} = x$ $(i = k)$, $j_{ik} = a$ $(|i - k| = 1)$, and $j_{ik} = 0$ (otherwise). Construct the matrix Q by arranging the rows of J in the reverse order with alternating algebraic sign changes:

$$Q = \begin{bmatrix} & & & & a & x \\ & & & -a & -x & -a \\ & & a & x & a & \\ & & \cdots & & & \\ (-)^{n-1}x & (-)^{n-1}a & & & & \end{bmatrix}.$$

Determine the eigenvalues of Q in terms of the eigenvalues λ_i of J.

Solution by O. G. RUEHR (Michigan Technological University).

Let $S = (s_{jk})$ be the $n \times n$ matrix such that $s_{jk} = (-1)^{j-1}$ if $k + j = n + 1$ and $s_{jk} = 0$ otherwise:

$$S = \begin{bmatrix} & & & & & & 1 \\ & & & & & -1 & \\ & & & & 1 & & \\ & & & -1 & & & \\ & & \cdots & & & & \\ (-1)^{n-1} & & & & & & \end{bmatrix}.$$

Note that premultiplication by S reverses the order of rows and changes the signs of even rows, while postmultiplication by $(-1)^n S$ reverses the order of columns and changes the signs of odd columns. Thus,

$$SQ = (-1)^{n-1}J$$

and

$$(-1)^n QS = J - 2xI,$$

where I denotes the identity matrix. Since $S^2 = (-1)^{n-1}I$, we have

$$(-1)^n QSSQ = -Q^2 = (-1)^{n-1}(J^2 - 2xJ).$$

Hence, if λ is an eigenvalue of J, there is an eigenvalue μ of Q such that

$$\mu^2 = (-1)^n(\lambda^2 - 2x\lambda) = (-1)^n[(\lambda - x)^2 - x^2].$$

Since the eigenvalues of J,

$$\lambda_j = x - 2a \cos \frac{j\pi}{n+1}, \qquad\qquad j = 1, 2, \cdots, n,$$

are well known, we see that for $a \neq 0$ there is an ambiguity in the sign of the square root only if $\lambda = x$ (n odd). As λ runs through its other values both square roots should be taken. If $\lambda = x$, $\mu = (-1)^{(n-1)/2}x$.

Problem 61–5′, On the Eigenvalues of a Matrix, by CARL E. SEALANDER (Boeing Scientific Research Laboratories).

Determine the characteristic roots of the matrix $A = \| a_{ij} \|$ where

(1) $\qquad a_{ij} = \sin \frac{i\pi}{n} \sin \frac{j\pi}{n} \cos \frac{(i-j)\pi}{n}, \qquad\qquad (i, j = 1, 2, \cdots, n-1).$

This problem has arisen in connection with finding an orthogonal transformation to reduce a certain quadratic form to a sum of squares. The form arises from considering a body subject to a system of n central forces of arbitrary magnitude, and directed radially outwards along the rays $\theta = 2\pi i/n, i = 0, 1, 2, \cdots, (n-1)$, in a polar coordinate system. One might, for example, think of the resultant centrifugal force due to a system of concentrated masses on the vertices of a regular n-gon.

Solution by A. W. SAENZ (U. S. Naval Research Laboratory).

In the trival case $n = 2$, (1) implies that the only characteristic value of A is 1. From now on, we shall only deal with the cases $n \geq 3$.

The characteristic equation for A can be written in the form

$$(2) \qquad \lambda^{n-1} + \sum_{m=1}^{n-1} (-1)^m c_m \lambda^{n-m-1} = 0,$$

where c_m equals the sum of all the principal minors of A of order m.

We shall now prove that

$$(3) \qquad c_m = 0, \qquad m \geq 3, \qquad n \geq 4.$$

Introducing the matrix $B = \| b_{rs} \|$ where

$$(4) \qquad b_{rs} = \sin x_r \sin y_s \cos (x_r - y_s), \qquad (r, s = 1, 2, 3)$$

it follows that

$$(5) \qquad \det B = 0,$$

for all x_r and y_s.

But (1), (4), and (5) imply for $n \geq 4$ that any 3-rowed minor of A vanishes, which clearly implies that (3) holds.

Employing (3) and the well known expressions for c_1 and c_2, (2) becomes

$$(6) \qquad \lambda^{n-3}\{\lambda^2 - (\operatorname{tr} A) + \tfrac{1}{2}[(\operatorname{tr} A)^2 - \operatorname{tr} A^2]\} = 0, \qquad (n \geq 3),$$

where $\operatorname{tr} A$ denotes the trace of A.

By means of simple trigonometric identities and elementary summation formulas, one finds that

$$(7) \qquad \operatorname{tr} A = \frac{n}{2}, \qquad \operatorname{tr} A^2 = \frac{10 n^2}{64}, \qquad (n \geq 2).$$

For $n \geq 3$, it follows from (6) and (7) that A has $(n - 3)$ characteristic values equal to zero, and the two non-zero values of

$$\lambda_1 = \frac{n}{8}, \qquad \lambda_2 = \frac{3n}{8}.$$

Problem 68-10, *Rank and Eigenvalues of a Matrix*, by SYLVAN KATZ (Aeronutronic Division, Philco-Ford Corporation) and M. S. KLAMKIN (Ford Scientific Laboratory).

Determine the rank and eigenvalues of the $n \times n$ $(n \geq 3)$ matrix $\| A_{rs} \|$ where $A_{rs} = \cos (r - s)\theta$ and $\theta = 2\pi/n$. This problem arose in a study of electromagnetic wave propagation.

Solution by G. J. FOSCHINI (Bell Telephone Laboratories, Holmdel, New Jersey).

From elementary complex algebra it follows that if k is an integer then

$$(1) \qquad \sum_{j=1}^{n} e^{i\theta k j} = \begin{cases} 0, & k \not\equiv 0 \,(\mathrm{mod}\, n), \\ n, & k \equiv 0 \,(\mathrm{mod}\, n). \end{cases}$$

Using (1) we see that the Vandermonde $\|n^{-1/2} e^{i\theta rs}\|$ has inverse $\|n^{-1/2} e^{-i\theta rs}\|$ and furthermore that

$$\|n^{-1/2} e^{i\theta rs}\| \cdot \|A_{rs}\| \cdot \|n^{-1/2} e^{-i\theta rs}\|$$

$$= \|n^{-1/2} e^{i\theta rs}\| \cdot \left\| \frac{e^{i\theta(r-s)}}{2} + \frac{e^{-i\theta(r-s)}}{2} \right\| \cdot \|n^{-1/2} e^{-i\theta rs}\|$$

has zero entries except in the $(1, 1)$ and $(n-1, n-1)$ positions where $1/2$ appears. Thus the rank of $\|A_{rs}\|$ is 2 and its eigenvalues are $1/2$ and 0 with multiplicities 2 and $n-2$ respectively.

Additionally, the same similarity transformation of $\|\sin (r-s)\theta\|$ yields a matrix with zero entries except in the $(1, n-1)$ and $(n-1, 1)$ positions where $(1/2i)$ appears. Thus the rank of $\|\sin (r-s)\theta\|$ is 2 and its eigenvalues are 0 (multiplicity $n-2$), $i/2$ and $-i/2$.

Solution by CARLENE ARTHUR and CECIL ROUSSEAU (Baylor University).

It is well known that the characteristic equation can be found by computing $\mathrm{Tr}(A^k)$, $k = 1, 2, \cdots, n$. To this end, we compute

$$[A^2]_{rs} = \sum_{t=1}^{n} [\cos (r-t)\theta][\cos (s-t)\theta] = \tfrac{1}{2} \sum_{t=1}^{n} \{\cos (r-s)\theta + \cos (r+s-2t)\theta\}.$$

Since

$$\sum_{t=1}^{n} \cos (r+s-2t)\theta = \mathrm{Re} \left\{ e^{i(r+s)\theta} \sum_{t=1}^{n} e^{-2it\theta} \right\} = 0, \qquad n \geq 3,$$

$$A^2 = \frac{n}{2} A$$

and, more generally,

$$A^k = \left(\frac{n}{2}\right)^{k-1} A.$$

Thus,

$$\mathrm{Tr}(A) = n, \qquad \mathrm{Tr}(A^k) = 2\left(\frac{n}{2}\right)^k.$$

Since the sum of the kth powers of the eigenvalues of $A = \mathrm{Tr}(A^k)$, it immediately follows that the eigenvalues are $n/2, n/2, 0, 0, \cdots, 0$ and the rank of the matrix is 2.

In a similar fashion, it can be shown that the eigenvalues of the $n \times n$ matrix $\|B_{rs}\|$ where $B_{rs} = \cos (r+s)\theta$ and $\theta = \pi/(n+1)$ are $(n-1)/2$, $(-n-1)/2$, $0, 0, \cdots, 0$.

Generalization by HARRY APPELGATE (City College of New York).

Let $t_1, t_2, \cdots, t_n$ be n real numbers $(n \geq 2)$ such that at least one difference $t_i - t_j$ is not a multiple of π. Then the matrix A with entries $a_{ij} = \cos (t_i - t_j)$ has rank 2.

Proof. Define vectors

$$
c = \begin{pmatrix} \cos t_1 \\ \cos t_2 \\ \vdots \\ \cos t_n \end{pmatrix}, \qquad
s = \begin{pmatrix} \sin t_1 \\ \sin t_2 \\ \vdots \\ \sin t_n \end{pmatrix}.
$$

It is easy to see that $A = cc^T + ss^T$, where T means transpose. The condition that some difference $t_i - t_j$ is not a multiple of π shows that c and s are linearly independent. If x is an arbitrary vector, $Ax = (c, x)c + (s, x)s$ where $(\,,\,)$ is the usual scalar product. Hence $Ax = 0$ if and only if $(c, x) = (s, x) = 0$. This shows that the kernel of A is the orthogonal complement of the 2-dimensional subspace generated by c and s. Hence dim (ker A) $= n - 2$ which implies rank (A) $= 2$.

If $n \geq 3$ and $t_i = 2\pi i/n$ we get the problem as stated.

Remark. We get a similar result if $a_{ij} = \sin(t_i - t_j)$.

Editorial note. Carlitz refers to some related results in the paper *Some Cyclotomic Matrices*, Acta Arith., 5 (1959), pp. 293–308. The proposers' solution and generalization that the rank is 2 for all $\theta \neq m\pi$ $(n \geq 3)$ appears in a publication preprint of the Ford Scientific Laboratory. [H.E.F.]

Problem 67–12, *Convergence of an Iteration*, by JOEL L. BRENNER (Stanford Research Institute).

The following problem arose from a consideration of the allocation of some production facilities.

Given a mixture of operations $(1, 2, \cdots, n)$, an operations researcher wishes to calculate the unique proportions $x = (x_1, x_2, \cdots, x_n)$ that satisfy $Ax = x$, where A is a known linear operation. Find for him sufficient conditions for the convergence of the following iteration: Let y_0 be arbitrary, $y_1 = Ay_0$, $z_1 = \frac{1}{2}(y_0 + y_1)$, $y_{n+1} = Az_n$, $a_{n+1} = \frac{1}{2}(z_n + y_{n+1})$, $n = 1, 2, \cdots$.

Solution.[1]

Let $z_0 = y_0$. Then

$$
z_1 = \tfrac{1}{2}(z_0 + Az_0) = \tfrac{1}{2}(I + A)z_0,
$$

where I is the identity operator $(Ix = x)$. For $n = 1, 2, \cdots$, we have $z_{n+1} = \frac{1}{2}(z_n + Az_n) = \frac{1}{2}(I + A)z_n$. Thus the iteration is reduced to: Let z_0 be arbitrary and

$$
z_{n+1} = \tfrac{1}{2}(I + A)z_n, \qquad\qquad n = 0, 1, 2, \cdots
$$

This is simply the power method to find the largest eigenvalue and corresponding eigenvector of the linear operator $\frac{1}{2}(I + A)$. The iterations "converge" if the largest eigenvalue is simple and all other eigenvalues are smaller in magnitude (see D. K. Faddeev and V. N. Faddeeva, *Computational Methods of Linear Algebra*, W. H. Freeman, San Francisco, 1963, p. 291 ff.).

If λ is an eigenvalue of A with corresponding eigenvector x, then $\frac{1}{2}(1 + \lambda)$ is an eigenvalue of $\frac{1}{2}(I + A)$ with corresponding eigenvector x. We desire that the

---

[1] As no name appeared with this solution, there is no acknowledgment of the solver.

iterations converge to an eigenvector $x_0 = Ax_0$, i.e., x_0 corresponds to the eigen-value $\lambda_0 = 1$. Since $\frac{1}{2}(1 + \lambda_0) = 1$, the iterations will converge to x_0 if and only if:

(a) $\lambda_0 = 1$ is a simple eigenvalue of A;
(b) there is an $\epsilon > 0$ such that, if λ is any eigenvalue of A other than λ_0, then

$$\tfrac{1}{2}|1 + \lambda| \leq 1 - \epsilon.$$

Condition (b) is satisfied if each eigenvalue of A other than λ_0 lies inside the circle in the complex plane centered at -1 with radius 2. The rate of convergence of the iterations depends on the value of ϵ, being faster if ϵ is larger.

Depending on the location of the eigenvalues of A, the rate of convergence may be improved by the following modification of the iterations. Let a be any real number, $a \neq -1$, and z_0 be arbitrary. Let

$$z_{n+1} = (aI + A)z_n/(a + 1), \qquad n = 0, 1, \cdots.$$

Then the iterations will converge if and only if:
(a) $\lambda_0 = 1$ is a simple eigenvalue of A;
(b) there is an $\epsilon > 0$ such that, if λ is any eigenvalue of A other than λ_0, then

$$|a + \lambda| \leq (1 - \epsilon)|a + 1|.$$

Generally, there will be an optimum value for a which depends on the eigenvalues of A.

D. K. Faddeev and V. N. Faddeeva (loc. cit.) consider other techniques for improving the rate of convergence.

Problem 68-6, A Continued Fraction Representation of Eigenvalues, by THOMAS E. PHIPPS, JR. (U.S. Naval Ordnance Laboratory).

The equation satisfied by Laguerre polynomials

$$\left\{D^2 + \frac{2}{x}D + \frac{\lambda}{x} - \frac{1}{4}\right\}y = 0,$$

$$\lambda > 0, \qquad 0 \leq x < \infty,$$

is commonly solved by a change of variable, $y = v \exp(-x/2)$. If instead one substitutes directly

$$y = \sum_{n=0}^{\infty} C_n x^n,$$

one obtains the three-term recurrence relation

$$C_{n+1} + b_n C_n - a_n C_{n-1} = 0, \qquad n = 1, 2, \cdots,$$

where

$$a_n \equiv \frac{1}{4(n + 1)(n + 2)}, \qquad b_n \equiv \frac{\lambda}{(n + 1)(n + 2)}.$$

Following a method familiar from elementary treatments of the Mathieu equation, with the use of

$$\frac{C_n}{C_{n-1}} = \frac{a_n}{b_n + C_{n+1}/C_n},$$

one may develop the "indicial equation" in the form

$$-\frac{\lambda}{2} = \frac{C_1}{C_0} = \frac{a_1}{b_1} + \frac{a_2}{b_2} + \frac{a_3}{b_3} + \cdots .$$

Solutions of this equation for λ should yield the well-known positive integral eigenvalues of the nonrelativistic Schroedinger hydrogen atom, $\lambda = 1, 2, \cdots$. Unfortunately, something has gone wrong, because for $\lambda > 0$ every partial quotient and convergent of the expression on the right is positive, while the left-hand side is plainly negative. Hence there are no positive eigenvalues and the hydrogen atom blows up. How would you propose to save the world in this extremity?

Solution by ROBERT SINGLETON (Wesleyan University).

The major error is in proceeding to the infinite continued fraction. The propriety of an equation such as that on which the proposer hangs his case is treated in [1, Section 57]. The formulation in the problem differs slightly from Perron's but the conversion is easily made.

Perron's discussion is adequate to locate the error but it does not well illuminate what is going on. For any initial conditions C_0, C_1 the recursion system has a unique solution. When the continued fraction converges (as it does in this problem) it specifies a particular set of initial conditions having a distinguished solution with the property, loosely, that its terms are of minimum order in the index. See [1, Section 57, Theorem 46C] for a proper statement. A second interesting property comes from Theorem 46A. When all coefficients in the recursion system are positive then every solution, except perhaps the distinguished solution, has some negative terms. The continued fraction is irrelevant to a solution arising from another set of initial conditions, as is the case in this problem.

It seems unnecessary to force a determination of the positive integral eigenvalues through continued fractions since the standard substitution does elegantly. But as a *tour de force* this can be accomplished properly by converting the differential equation directly into a continued fraction. The direct method of [1, Section 80] does not work well since the coefficients of the differential equation are not polynomials. But the equation may be rearranged as

$$y = \frac{2-x}{1-\lambda}\left(\frac{1}{2}y + y'\right) + \frac{x}{1-\lambda}\left(\frac{1}{4}y + y' + y''\right).$$

If now we define a sequence of functions by

$$f_0(x) = y(x),$$

$$f_n(x) = \tfrac{1}{2}f_{n-1}(x) + f'_{n-1}(x), \qquad\qquad n \geq 0,$$

we find by successive differentiation and substitution that

$$f_n = \frac{(n+2)-x}{(n+1)-\lambda}f_{n+1} + \frac{x}{(n+1)-\lambda}f_{n+2}, \qquad\qquad n \geq 0.$$

By substitution and transformation we have, formally, the continued fraction

$$(1 - \lambda)\frac{f_0}{f_1} = 2 - x + \frac{(2 - \lambda)x}{3 - x} + \frac{(3 - \lambda)x}{4 - x} + \frac{(4 - \lambda)x}{5 - x} + \cdots .$$

In general, one has again the question of when the continued fraction represents the function on the left. But in particular, when λ is any positive integer exceeding 1, the continued fraction is finite and does represent a rational function which may be taken as f_0/f_1. This can, in fact, be integrated to provide a solution of the original differential equation. Thus the eigenvalues appear naturally.

Of course, this really is a *tour de farce* (sic) since the standard substitution shows how to define the functions f_i so as to eliminate the exponential and obtain a rational function.

Also solved by the proposer.

REFERENCE

[1] O. PERRON, *Die Lehre von den Kettenbrüchen*, Chelsea, New York, 1950.

Further editorial comment. The usual criterion for determining the eigenvalues of the Schroedinger equation is that the function be square-integrable. This point is not made evident in either solution.

The original equation is satisfied by $e^{-x/2}$ times the Laguerre polynomial of order $\lambda - 1$ and degree 1, viz:

$$y = e^{-x/2}L^1_{\lambda-1}(x) = \lambda e^{-x/2}{}_1F_1(1 - \lambda, 2, x),$$

Singleton's solution (which is a special case of a more general continued fraction for the logarithmic derivative of the confluent hypergeometric function (see E. L. Ince, *Ordinary Differential Equations*, Dover, New York, 1944, p. 180)) can be put in a form which also exhibits the eigenvalue $\lambda = 1$; by rewriting it in terms of the original variable y one has

$$\frac{y'}{y} = -\frac{1}{2} + \frac{1 - \lambda}{2 - x} + \frac{2 - \lambda}{3 - x} + \cdots + .$$

For more information on the question of continued fractions as related to linear difference equations, the reader is referred to recent works of Wimp(ARL 69-0186, Nov. 1969) and Gautschi(loc. cit., bibliography). [H.E.F.]

Problem 74–14, A Generalization of the Vandermonde Determinant, by S. VENIT (California State University at Los Angeles).

Evaluate the $n \times n$ determinant $D(m, n) = |A_{rs}|$, where $n > m \geq 0$ and

$$A_{rs} = \frac{(n - r)!x_1^{n+1-r-s}}{(n + 1 - r - s)!}, \qquad 1 \leq s \leq m + 1, \quad 1 \leq r \leq n + 1 - s$$

$$A_{rs} = x_{s-m}^{n-r}, \qquad m + 1 < s \leq n, \quad 1 \leq r \leq n,$$

$$A_{rs} = 0, \qquad 1 \leq s \leq m + 1, \quad n + 1 - s < r \leq n.$$

Solution by O. P. LOSSERS (Technological University, Eindhoven, the Netherlands).

Since the jth column, $1 \le j \le m+1$, is the $(j-1)$th derivative of the first column, it follows that

$$D(m, n) = \prod_{t=1}^{m+1} \left(\frac{\partial}{\partial y_t}\right)^{t-1} VM(y_1, \cdots, y_{m+1}, x_2, \cdots, x_{n-m}) \bigg|_{y_1 = y_2 = \cdots = y_{m+1} = x_1},$$

where $VM(y_1, \cdots, y_{m+1}, x_2, \cdots, x_{n-m})$ is the Vandermonde determinant, i.e.,

$$VM(y_1, \cdots, y_{m+1}, x_2, \cdots, x_{n-m})$$

$$= \prod_{1 \le i < j \le m+1} (y_i - y_j) \cdot \prod_{\substack{1 \le k \le m+1 \\ 2 \le l \le n-m}} (y_k - x_l) \cdot \prod_{2 \le p < q \le n-m} (x_p - x_q).$$

The only nonzero contributions to $D(m, n)$ come from applying all derivatives to the first of the three factors. Furthermore,

$$\prod_{t=1}^{m+1} \left(\frac{\partial}{\partial y_t}\right)^{t-1} \prod_{1 \le i < j \le m+1} (y_i - y_j) \bigg|_{y_1 = y_2 = \cdots = y_{m+1} = x_1} = (-1)^{m(m+1)/2} \prod_{t=1}^{m+1} (t-1)!.$$

Hence we obtain the result

$$D(m, n) = (-1)^{m(m+1)/2} \prod_{t=1}^{m+1} (t-1)! \cdot \prod_{2 \le l \le n-m} (x_1 - x_l)^{m+1} \cdot \prod_{2 \le p < q \le n-m} (x_p - x_q)$$

$$= (-1)^{m(m+1)/2} \prod_{t=1}^{m+1} (t-1)! \cdot \prod_{2 \le l \le n-m} (x_1 - x_l)^m \cdot \prod_{1 \le p < q \le n-m} (x_p - x_q).$$

Two Equal Determinants

Problem 78-15, *by* R. SHANTARAM (Indian Statistical Institute, New Delhi, India).

Define $m[2n] = \binom{2n}{n}$, $n = 0, 1, 2, \cdots$. Let $T(n)$ be the $n \times n$ matrix whose (i, j) element is $m[2(i+j-1)]$, $i, j = 1, 2, \cdots, n$ and $S(n)$ be the $(n+1) \times (n+1)$ matrix whose (i, j) element is $m[2(i+j)]$, $i, j = 0, 1, \cdots, n$. Prove that

$$\det T(n) = \det S(n) = 2^n.$$

Solution by A. A. JAGERS (Technische Hogeschool Twente, Enschede, the Netherlands).

Let $G_n = [g_{ij}]$ be a $(n+1) \times (n+1)$ matrix with $g_{ij} = c_{i+j}$, where $c_{2k+1} = 0$, $c_{2k} = \binom{2k}{k}$ for $k = 0, 1, 2, \cdots$. Then G_n may be viewed as a Gram matrix with respect to the inner product on $C[0, \pi]$ defined by $(f, g) = \pi^{-1} \int_0^\pi f(x) g(x) \, dx$. In fact, $g_{ij} = (f_i, f_j)$, if $f_k(x) = 2^k \cos^k x$. Hence G_n can be written as $G_n = P^T P$, where the kth column of P contains the representation of f_k with respect to the orthonormal system $1, \sqrt{2} \cos x$, $\sqrt{2} \cos(2x), \cdots$. Clearly P is a triangular matrix with $1, \sqrt{2}, \sqrt{2}, \cdots$ on the diagonal. Thus $\det G_n = (\det P)^2 = 2^n$. Now $\det S_0 = 1$, and, for $m > 0$, $\det G_{2m} = \det S_m \cdot \det T_m$, $\det G_{2m-1} = \det S_{m-1} \cdot \det T_m$. (Consider the principal submatrix of G_m formed by the even numbered rows and columns (respectively odd numbered rows and columns).) Hence $\det S_n = \det T_n = 2^n$.

Problem 79-3*, *A Determinant*, by A. E. BARKAUSKAS and D. W. BANGE (University of Wisconsin—La Crosse).

Find either a closed form solution or a simple recurrence to evaluate the $n \times n$ determinant $|a_{ij}|$ where $a_{ij} = a_{ji}$, $a_{ii} = c + 1$ (c an integer > 1), $a_{12} = 1$, $a_{i,ci+k} = 1$ for $k = 1$ to c and $ci + k \leq n$; all other $a_{ij} = 0$.

The problem arose in counting the spanning trees of a certain class of outerplanar graphs.

Problem 69-3, *A Determinant*, by J. PRASAD (University of Khartoum, Khartoum, Sudan).

Evaluate the $(n + 1)$st order determinant $|A_{rs}|$, where

$$A_{rs} = \sum_{i=0}^{n} \frac{(a_{r-1} + b_i)^m}{a_i + a_{s-1}}, \qquad r, s, = 1, 2, \cdots, n + 1,$$

m integral.

Solution by A. S. HOUSEHOLDER (University of Tennessee).

The determinant to be evaluated is easily seen to be the product of two determinants and these can be evaluated independently. This being the case, it is just as easy to consider a slightly more general problem, and for convenience the notation will be modified mildly. However, it will be assumed that $m \geq -1$. With that restriction on m, the problem to be considered is the evaluation of the determinant whose elements are

$$\sum_{j=1}^{n} (r_i + s_j)^m (a_j + b_k)^{-1},$$

which is the product of a determinant whose elements are

$$(r_i + s_j)^m$$

and one whose elements are

$$(a_i + b_j)^{-1}.$$

But the second determinant was evaluated by Cauchy, and it is equal to

$$\frac{\zeta^{1/2}(a_1, a_2, \cdots, a_n)\zeta^{1/2}(b_1, b_2, \cdots, b_n)}{\prod_{i,j=1}^{n} (a_i + b_j)},$$

where $\zeta^{1/2}$ is the Vandermonde. Hence only the first needs to be considered. But if $m = -1$, that is also Cauchy's determinant. Hence, suppose $m \geq 0$.

It is clear that this is a polynomial in the r_j and s_k, of degree m in each r_j and in each s_k, that it is an alternant in the r_j and also in the s_k, hence contains as a divisor the product of the Vandermondes

$$\zeta^{1/2}(r_1, r_2, \cdots, r_n)\zeta^{1/2}(s_1, s_2, \cdots, s_n),$$

hence that the remaining factor is a polynomial of degree $m - (n - 1)$ in each r_j and in each s_k that is symmetric in each of the two sets, and further that it is symmetric in the two sets themselves (i.e., if each r_j is interchanged with s_j, its value is unchanged). Hence, it vanishes identically if $0 \leq m < n - 1$, and is a constant if $m = n - 1$.

The evaluation can be made by means of a theorem due to Garbieri (T. Muir, *The Theory of Determinants*, vol. III, Dover, New York, 1960, pp. 163–165):

Let

$$x^n + c_1 x^{n-1} + \cdots + c_n = (x - r_1)(x - r_2) \cdots (x - r_n),$$

$$x^n + d_1 x^{n-1} + \cdots + d_n = (x - s_1)(x - s_2) \cdots (x - s_n),$$

and let

$$f(x, y) = \sum_{i=0}^{n+h} \sum_{j=0}^{n+h} a_{ij} x^i y^i.$$

Then

$$|f(r_1, s_1) f(r_2, s_2), \cdots, f(r_n, s_n)|$$

$$= (-1)^{h+1} \zeta^{1/2}(r_1, r_2, \cdots, r_n) \zeta^{1/2}(s_1, s_2, \cdots, s_n)$$

$$\begin{vmatrix}
a_{00} & a_{01} & \cdots & a_{0,n+h} & d_n & 0 & \cdots \\
a_{10} & a_{11} & \cdots & a_{1,n+h} & d_{n-1} & d_n & \cdots \\
\cdot & \cdot & \cdots & & \cdot & \cdot & \cdots \\
a_{n+h,0} & a_{n+h,1} & \cdots & a_{n+h,n+h} & 0 & 0 & \cdots \\
c_n & c_{n-1} & \cdots & 0 & 0 & 0 & \cdots \\
0 & c_n & \cdots & 0 & 0 & 0 & \cdots \\
\cdot & \cdot & \cdots & \cdot & \cdot & \cdot & \cdots
\end{vmatrix},$$

where the determinant on the left is represented by its diagonal element, and that on the right is of order $n + 2h + 2$.

The matrix of the determinant on the right can be partitioned in an obvious manner as $\begin{pmatrix} A & D \\ C & 0 \end{pmatrix}$.

If

$$f(x, y) = (x + y)^{n-1},$$

then C is a row vector, D a column vector, and in A the only nonzero elements are those immediately above and to the left of the secondary diagonal, and these are the binomial coefficients. Hence the determinant has the value

$$(-1)^{(n+1)(n+2)/2} \prod_{j=0}^{n-1} \binom{n-1}{j},$$

and this is the determinantal multiplier of the product of the Vandermondes.

If $h = m - n \geq 0$, the elements along the secondary diagonal of A are the binomial coefficients $\binom{m}{j}$, and all others are zero. The evaluation can be made by means of the identity (a somewhat special case of a more general one)

$$\begin{vmatrix} A & D \\ C & 0 \end{vmatrix} = |A| \ |-CA^{-1}D|.$$

Evidently

$$|A| = (-1)^{m(m+1)/2} \prod_{j=0}^{m} \binom{m}{j},$$

and the matrix A^{-1} has nonzero elements only along the secondary diagonal where they are the reciprocals of the binomial coefficients. The second of the two determinants on the right above is of order $h + 1$; hence

$$|-CA^{-1}D| = (-1)^{h+1}|CA^{-1}D|.$$

By the Binet–Cauchy theorem, this is equal to a sum of $\binom{m+1}{h+1}$ terms, each of which is the product of a determinant formed from columns of C and one from rows of D, divided by the product of certain binomial coefficients. An explicit representation would be rather complicated and will not be attempted.

Editorial note. For related results on the double alternant determinant see T. Muir and W. H. Metzler, *A Treatise on the Theory of Determinants*, Dover, New York, 1960, pp. 348–352, 358–361. [M.S.K.]

Zeros of a Determinant

Problem 82·8, by NANCY J. BOYNTON (Michigan Technological University).

Show that if $\rho < \rho_k$, $f_k(z)$ has exactly $k + 2$ zeros inside the open unit circle, and if $\rho \geq \rho_k$, $f(z)$ has at most $k + 1$ zeros inside the open unit circle. Here,

$$f_k(z) = \begin{vmatrix} R_z & 0 & 0 & \cdots & 0 & 0 & z^2 \\ z^2 & R_z & 0 & \cdots & 0 & 0 & 0 \\ 0 & z^2 & R_z & \cdots & 0 & 0 & 0 \\ \cdot & \cdot & \cdot & \cdots & \cdot & \cdot & \cdot \\ \cdot & \cdot & \cdot & \cdots & \cdot & \cdot & \cdot \\ 0 & 0 & 0 & \cdots & z^2 & R_z & 0 \\ z^2 & z^2 & z^2 & \cdots & z^2 & 2z^2 & R_z + z \end{vmatrix}, \quad (k+2) \times (k+2)$$

where $R_z = [-2(\rho + 1)z + 2\rho]$ and $\rho_k = (3 \cdot 2^k - 1)/(4 \cdot 2^k - 1)$.

This problem arose in the study of a single-service Markovian queuing system with two servers in series and a buffer of k waiting spaces between the two servers.

Solution by A. A. JAGERS (Technische Hogeschool Twente, Enschede, the Netherlands).

We assume $\rho > 0$. For $\rho \leq 0$ our arguments fail and, more important, for $\rho \leq 0$ the statement of the problem is no longer true. For example, if $\rho = -1$ and $k \geq 2$, then $R_z \equiv -2$ and by formula (2) below and Rouché's theorem the number of zeros of $f_k(z)$ inside the unit circle $C: = \{z; |z| = 1\}$ is equal to the number of zeros of $(z - 2\rho)R_z^{k+2}$ inside C, that is, is equal to 0.

First we will calculate $f_k(z)$. We use some auxiliary determinants denoted by $\tilde{f}_k(z)$, $g_k(z; a)$, and $c_k(z)$, which all three differ from $f_k(z)$ in the last row only. Let the last row of $\tilde{f}_k(z)$, $g_k(z; a)$ and $c_k(z)$ be given as:

$$[z^2, z^2, \cdots, z^2, 2z^2, R_z + z^2], \quad [a, a, \cdots, a], \quad [0, 0, \cdots, 0, z^2, R_z] \quad \text{respectively.}$$

Then clearly $g_k(z; a) = a\, g_k(z; 1)$ and $c_k(z) = R_z^{k+2} + (-1)^{k+1}(z^2)^{k+2}$. Moreover, since the determinant function is linear in its last column as well as in its last row, we have $f_k(z) = \tilde{f}_k(z) + (z - z^2) R_z^{k+1}$ and $\tilde{f}_k(z) = g_k(z; z^2) + c_k(z)$ respectively. On the other hand, if we add all rows of $\tilde{f}_k(z)$, we see that $\tilde{f}_k(z) = g_k(z; R_z + 2z^2)$. By combining these results we obtain $\tilde{f}_k(z) = (R_z + 2z^2)(R_z^{k+2} - (-z^2)^{k+2})/(R_z - (-z^2))$, a sum of a geometric series. Hence

(1)
$$f_k(z) = R_z^{k+2} + z R_z^{k+1} - \left\{ \sum_{n=0}^{k} (-z^2)^{n+2} R_z^{k-n} \right\} - (-z^2)^{k+2}.$$

Also, since $R_z + 2z^2 = 2(z - 1)(z - \rho)$,

(2)
$$f_k(z) = (z - 1) \frac{(z - 2\rho) R_z^{k+2} - z^3 R_z^{k+1} - 2(z - \rho)(-z^2)^{k+2}}{R_z + z^2}.$$

Since $|R_z| \geq 2$ for z on C ($|z| = 1$), it follows by the triangle inequality applied to the series in (1) that

(3)
$$|R_z|^{k+2} \geq \frac{|R_z|^{k+2} - 1}{|R_z| - 1} + 1 \geq |f_k(z) - R_z^{k+2}|,$$

with equality if and only if $|R_z| = 2$ and (hence) $z = 1$. In particular 1 is the only zero of $f_k(z)$ on C. Furthermore

(4)
$$\lim_{z \to 1} \frac{f_k(z)}{z - 1} = 2(-1)^{k+2}\{2^k(4\rho - 3) + 1 - \rho\}.$$

Thus 1 is a multiple zero of $f_k(z)$ if $\rho = \rho_k$, in which case $2^k(4\rho - 3) + 1 - \rho = 0$, and only a simple zero if $\rho \neq \rho_k$. Actually, for $z = 1$ and $\rho = \rho_k$ (so that $\frac{2}{3} \leq \rho \leq \frac{3}{4}$) the derivative of $(R_z + z^2) f_k(z)/(z - 1)$ is nonzero; so 1 is a double zero of $f_k(z)$ if $\rho = \rho_k$. Let N be the number of zeros of $f_k(z)$ inside C, where $\rho \neq \rho_k$, then

(5)
$$\left(N + \frac{1}{2}\right) 2\pi i = \oint_C \frac{f_k'(z)}{f_k(z)}\, dz,$$

a Cauchy principal value. This value may be determined by a refinement of the proof of Rouché's theorem. In fact, in view of (3) we may write $f_k(z)$ as a product $g(z) h(z)$, where $g(z) = R_z^{k+2}$, $h(1) = 0$, and $|h(z) - 1| < 1$ for z on C, $z \neq 1$. Since $\rho > 0$, the unique zero of R_z, which is $\rho/(\rho + 1)$, lies inside C. Hence by the "principle of the argument" the increase in $\text{Arg}(g(z))$ as z traverses C (in the positive sense) is equal to $(k + 2)2\pi$. Moreover it follows from (4) that for $\varepsilon \in \mathbb{R}$, $|\varepsilon|$ small, $2^{k+1} h(1 + \varepsilon i) \approx \varepsilon i\{2^k(4\rho - 3) + 1 - \rho\}$, an imaginary number with argument $\pi/2$ or $-\pi/2$ depending on the sign of ε and on whether $\rho > \rho_k$ or $\rho < \rho_k$.

Since $h(z)$ remains inside a circle of radius 1 around 1, if z traverses C, $z \neq 1$, the increase of $\text{Arg}(h(z))$ is π for $\rho < \rho_k$ and $-\pi$ for $\rho > \rho_k$. The corresponding increase of $\text{Arg}(f_k(z)) = \text{Arg}(g(z)) + \text{Arg}(h(z))$ is $(2k + 5)\pi$ or $(2k + 3)\pi$. Hence $N = k + 2$ if $\rho < \rho_k$ and $N = k + 1$ if $\rho > \rho_k$ and since the degree of $f_k(z)$ is $2k + 4$ all entries in the following table are known.

# of zeros of $f_k(z)$	$z = 1$	$\|z\| = 1, z \neq 1$	$\|z\| < 1$	$\|z\| > 1$
$0 < \rho < \rho_k$	1	0	$k + 2$	$k + 1$
$\rho = \rho_k$	2	0	$k + 1$	$k + 1$
$\rho > \rho_k$	1	0	$k + 1$	$k + 2$

Problem 72-12, *Construction of a Maximal* $(0, 1)$ *Determinant*, by J. R. VENTURA, JR. (Naval Underwater Systems Center).

It is known [1] that if $n = 2^m - 1$ for some positive integer m, then the determinant of the $n \times n$ matrix V_n having a maximal value over all $n \times n$ matrices having 0's and 1's as entries is given by

$$f(n) = |V_n| = 2^{-n}(n + 1)^{(n + 1)/2}.$$

Construct a $(2n + 1) \times (2n + 1)$ maximal $(0, 1)$ determinant.

REFERENCE

[1] J. COHN, *On the value of determinants*, Proc. Amer. Math. Soc., 14 (1963), pp. 581–588.

Solution by the proposer.

Let U_n be the $n \times n$ matrix consisting of all 1's. Let V_{2n+1} be given by

$$V_{2n+1} = \begin{bmatrix} 0 \cdots 0 & 1 \cdots 1 & 1 \\ V_n & U_n - V_n & 1 \\ & & \vdots \\ & & 1 \\ V_n & V_n & 0 \\ & & \vdots \\ & & 0 \end{bmatrix}.$$

Then,

$$|V_{2n+1}| = \begin{vmatrix} V_n & -V_n \\ V_n & V_n \end{vmatrix} = 2^n |V_n|^2$$

$$= \frac{(2n + 2)^{n+1}}{2^{2n+1}} = F(2n + 1).$$

Problem 69–8, *Bounds for Maximal Two Element Determinants*, K. W. SCHMIDT (University of Manitoba, Canada).

If $F_n(a, b)$ denotes the maximum absolute value of an nth order determinant, all of whose elements are a and b (real), show that

$$|A - aB| \leq F_n(a, b) \leq \max \{|A|, |bB|\},$$

where
$$A = (a + b)|a - b|^{n-1}F_n(0, 1),$$
$$B = |a - b|^{n-1}F_{n-1}(0, 1)$$

and
$$0 \leq |a| \leq |b|.$$

Solution by the proposer.

It is sufficient to prove the case $a = x$, $b = 1$, $|x| < 1$. It is known that[1]

(1) $S(x, 1) = (x - 1)^{n-1}L(x),$

where $S(a, b) =$ determinant of a matrix of order n in which each element is either a or b,

$$L(x) = kx + h,$$

with $k = S(1, 0)$, and $h = (-1)^{n-1}S(0, 1)$.

Consequently,

(2) $\max L(x) = \max |L(x)|$, $\max L(-1) = F_{n-1}$, $\max L(0) = F_n$,

where F_{n-1} and F_n stand for $F_{n-1}(0, 1)$ and $F_n(0, 1)$.

Since there exist always the two pairs k, h where

$$k = 0, \quad h = (-1)^{n-1}F_{n-1},$$
$$k = R, \quad h = (-1)^{n-1}F_n \quad \text{with } F_n - F_{n-1} \leq R \leq F_n,$$

then lower bounds for $\max L(x)$ are given by F_{n-1} and

$$\min \{(x + 1)F_n, (x + 1)F_n - xF_{n-1}\}.$$

Hence, because of (1),

(3) $F_n(x, 1) \geq |x - 1|^{n-1} \max \{F_{n-1}, \min \{(x + 1)F_n, (x + 1)F_n - xF_{n-1}\}\}.$

For $0 \leq x < 1$ it follows, since $|k| \leq F_n$ and $|h| \leq F_n$, that

$$\max L(x) \leq (x + 1)F_n,$$

and because of (1),

(4) $F_n(x, 1) \leq |x - 1|^{n-1}(x + 1)F_n.$

For $-1 \leq x < 0$ it will now be proved by contradiction that

(5) $\max L(x) \leq (x + 1)F_n - xF_{n-1},$

and because of (1),

(6) $F_n(x, 1) \leq |x - 1|^{n-1}|(x + 1)F_n - xF_{n-1}|.$

Assuming some $x = r$, for which $L(x) = k_r x + h_r$ does not satisfy (5), it follows that

$$L(r) = (r + 1)F_n - rF_{n-1} + e, \quad e > 0,$$

$$L(x) = L(r) - k_r(r - x),$$

$$L(x) = (F_n - F_{n-1} - k_r)(r - x) + (x + 1)F_n - xF_{n-1} + e,$$

(7) $L(-1) \geq F_{n-1} + e$ for $F_n - F_{n-1} \geq k_r$,

(8) $L(0) \geq F_n + e$ for $F_n - F_{n-1} \leq k_r$.

Since, because of (2) neither (7) nor (8) can hold, (5) is true. Finally, substitution of $x = a/b$ in the statements (3), (4) and (6) yields the desired result.

A Nonnegative Determinant

Problem 83-11, *by* G. N. LEWIS (Michigan Technological University).

An $m \times m$ matrix A has positive diagonal elements, negative subdiagonal elements, and zeros elsewhere. Matrix B is obtained from A by replacing the jth column with positive elements.

(a) Prove that det $B > 0$.

(b) If the magnitudes of the nonzero elements of A are all greater than one, find a lower bound on det B in terms of the positive column elements.

This problem arose in the study of the numerical solution of large deflections of a horizontal cantilever beam with nonlinear material behavior under concentrated point loads.

Solution by JACQUES FARGES (Student, University of Victoria).

Regarding B as a partitioned matrix and expanding about the jth column gives

$$\det B = \prod_{i=j+1}^{m} a_i \cdot \sum_{k=1}^{j} \left[(-1)^{j-k} c_k \cdot \prod_{i=1}^{k-1} a_i \cdot \prod_{i=k}^{j-1} b_i \right]$$

$$= \prod_{i=j+1}^{m} a_i \cdot \sum_{k=1}^{j} \left[c_k \cdot \prod_{i=1}^{k-1} a_i \cdot \prod_{i=k}^{j-1} |b_i| \right] > 0.$$

If $a_i > 1$ for $1 \leq i \leq m$ and $|b_i| > 1$ for $1 \leq i \leq m - 1$, then $\det B > \sum_{k-1}^{j} c_k$.

In fact, if A is a lower triangular matrix with positive diagonal elements, negative elements below the diagonal, and B is obtained as previously, a similar proof shows $\det B > 0$. Specifically (using matrix notation)

$$\det B = \left[\prod_{i-j+1}^{m} a_{i,i} \right] \sum_{k-1}^{j} \left[(-1)^{j-k} c_k D_k \prod_{i-1}^{k-1} a_{i,i} \right],$$

where $D_k = 1$ for $k = j$, and for $k \neq j$

$$D_k = \begin{vmatrix} a_{k+1,k} & a_{k+1,k+1} & & 0 \cdots\cdots\cdots 0 \\ a_{k+2,k} & a_{k+2,k+1} & a_{k+2,k+2} & \\ \vdots & & & \ddots & 0 \\ a_{j-1,k} & \cdots\cdots\cdots\cdots\cdots\cdots & a_{j-1,j-1} \\ a_{j,k} & \cdots\cdots\cdots\cdots\cdots\cdots & a_{j,j-1} \end{vmatrix}.$$

Using cofactor expansion from the right, there are two choices for each minor encountered: 1) $a_{i,i}$ which is made negative by the cofactor sign pattern, or 2) some $a_{j,i}$ (where $j > i$) which is negative and remains so, again by the cofactor sign pattern. Therefore,

$$D_k = \sum_R \left[\prod_{q-k}^{j-1} - |a_{p_q, q}| \right],$$

where $R: \{p_q\}_{q-k}^{j-1}, k + 1 \leq p_q \leq j, p_q \geq q, p_s \equiv p_t$ iff $s = t$. Taking out the factor $(-1)^{j-k}$ from D_k, and with this, cancelling the identical factor in $\det B$ implies that $\det B > 0$.

If $|a_{j,i}| > 1$ when $a_{j,i} \neq 0$, then

$$\det B > \sum_{k-1}^{j-2} c_k (2)^{j-1-k} + c_{j-1} + c_j.$$

Problem 73-17, A Hadamard-Type Bound on the Coefficients of a Determinant of Polynomials, by A. J. GOLDSTEIN and R. L. GRAHAM (Bell Telephone Laboratories).

If $A = (a_{ij})$ is an $n \times n$ matrix, a classical inequality of Hadamard [2, p. 253] asserts that

$$|\det A| \leq \left(\prod_{i=1}^{n} \sum_{j=1}^{n} |a_{ij}|^2 \right)^{1/2} \equiv H(A).$$

In recent studies on coefficient growth in greatest common divisor algorithms for polynomials, W. S. Brown [1] was led to inquire about possible analogues of this inequality for the case in which the entries of the matrix are *polynomials*.

Let $A(x) = (A_{ij}(x))$ be a matrix whose elements are polynomials and let $a_0, a_1, \cdots$ be the coefficients of the polynomial representation of $\det A(x)$. If $W = (w_{ij})$, where w_{ij} denotes the sum of the absolute values of the coefficients of $A_{ij}(x)$, then show that

$$\left(\sum |a_k|^2 \right)^{1/2} \leq H(W).$$

REFERENCES

[1] W. S. Brown, *On Euclid's algorithm and the computation of polynomial greatest common divisors*, J. Assoc. Comput. Mach., 18 (1971), pp. 478–504.

[2] F. R. Gantmacher, *The Theory of Matrices*, vol. 1, Chelsea, New York, 1960.

Solution by O. P. Lossers (Technological University, Eindhoven, the Netherlands).

Since $|A_{kl}(e^{it})| \leq w_{kl}$, it follows from Hadamard's inequality that

$$|\det A(e^{it})|^2 \leq \prod_{k=1}^{n} \sum_{l=1}^{n} |A_{kl}(e^{it})|^2 \leq \prod_{k=1}^{n} \sum_{l=1}^{n} w_{kl}^2 = (H(W))^2.$$

However,

$$\frac{1}{2\pi} \int_0^{2\pi} |\det A(e^{it})|^2 \, dt = \frac{1}{2\pi} \int_0^{2\pi} \left(\left(\sum a_k \, e^{ikt} \right) \left(\sum \bar{a}_l \, e^{-ilt} \right) \right) dt = \sum |a_k|^2.$$

Hence

$$\sum |a_k|^2 = \frac{1}{2\pi} \int_0^{2\pi} |\det A(e^{it})|^2 \, dt \leq \frac{1}{2\pi} \int_0^{2\pi} (H(W))^2 \, dt = (H(W))^2.$$

Supplementary References
Matrices and Determinants

[1] A.C.Aitken, *Determinants and Matrices*, Oliver and Boyd, Edinburgh, 1976.

[2] R.Bellman, *Introduction to Matrix Analysis*, McGraw-Hill, N.Y., 1960.

[3] R.Cooke, *Infinite Matrices and Sequence Spaces*, MacMillan, London, 1950.

[4] P.J.Davis, *The Mathematics of Matrices*, Blaisdell, N.Y., 1965.

[5] P.J.Davis, *Circulant Matrices*, Wiley, N.Y., 1979.

[6] W.L.Ferrar, *Algebra*, Oxford University Press, London, 1951.

[7] R.A.Frazer, W.J.Duncan and A.R.Collar, *Elementary Matrices and Some Applications to Dynamics and Differential Equations*, Cambridge University Press, London.

[8] F.R.Gantmacher, *The Theory of Matrices, I,II*, Chelsea, N.Y., 1959.

[9] A.S.Householder, *The Theory of Matrices in Numerical Analysis*, Blaisdell, N.Y., 1964.

[10] M.Marcus and H.Minc, *A Survey of Matrix Theory and Matrix Inequalities*, Allyn and Bacon, Boston, 1964.

[11] L.Mirsky, *An Introduction to Linear Algebra*, Clarendon Press, Oxford, 1972.

[12] T.Muir, *The Theory of Determinants in the Historical Order of Development, I,II*, Dover, 1960.

[13] T.Muir and W.H.Metzler, *A Treatise on the Theory of Detrminants*, Dover, N.Y., 1960.

[14] S.Perlis, *Theory of Matrices*, Addison-Wesley, Reading, 1952.

[15] L.A.Pipes, *Matrix Methods for Engineering*, Prentice-Hall, N.J., 1963.

[16] R.M.Pringle, *Generalized Inverse Matrices with Applications to Statistics*, Griffin, London, 1971.

[17] N.J.Pullman, *Matrix Theory and its Applications*, Dekker, N.Y., 1976.

[18] H.Schwerdtfeger, *Introduction to Linear Algebra and the Theory of Matrices*, Noordhoff, Groningen, 1961.

[19] G.E.Shilov, *Linear Algebra*, Dover, N.Y., 1977.

[20] R.F.Scott, *A Treatise on the Theory of Determinants*, Cambridge University Press, Cambridge, 1880.

[21] R.M.Thrall, *Vector Spaces and Matrices*, Wiley, N.Y., 1957.

[22] H.W.Turnbull, *The Theory of Determinants, Matrices, and Invariants*, Blackie, London, 1928.

[22] R.S.Varga, *Matrix Iterative Analysis*, Prentice-Hall, N.J., 1962.

[23] J.H.Wilkinson, *The Algebraic Eigenvalue Problem*, Clarendon Press, Oxford, 1965.

12. NUMERICAL APPROXIMATIONS AND ASYMPTOTIC EXPANSIONS

The Attenuated Abel Transform

Problem 87-1*, by MOSHE DEUTSCH (Bar-Ilan University, Ramat-Gan, Israel).

The Radon transform relates the line-of-sight radiance of an extended source of radiation to its emission coefficient distribution. It forms the mathematical basis of both emission and absorption computed tomography, flame and plasma diagnostics and other related techniques. For a cylindrically symmetric and optically thin source, where no absorption occurs, this transform reduces to the Abel transform. Its well-known analytic inversion formulae can be employed, in this case, to obtain the physically important emission coefficient distribution from the measured radiance data.

If, however, the cylindrically symmetric source has a finite (radially) constant absorption, then the Radon transform reduces to the *attenuated* Abel transform, the analytic inversion of which is considerably more complicated. For a measured radiance function represented by a spline function, the following integral occurs in the inversion formula:

$$I(a,b) = \int_a^b (x^2 + A^2)^{1/2} \cos x \, dx$$

where a, b and A are constants related to the absorption in the source and the knots of the spline. This finite Fourier cosine transform is not listed in any of the widely used transform tables.

Can this integral be expressed in a closed form? If not, can it be expressed as a rapidly convergent series for arbitrary a, b and A?

Solution by W. B. JORDAN (Scotia, New York).

If some standard numerical integration formula is used to compute I there are two potential sources of trouble. First, if A is small and the interval a to b spans or comes close to $x = 0$, the radical can have rapid fluctuation (its derivative is discontinuous when $A = 0$) and no polynomial can fit it very well. Second, if $b - a$ is large, the integration covers several cycles of the cosine and again many integration points are needed. We give several formulas, which among them will cover the waterfront.

(1) If $b - a$ is not too large and not too close to $x = 0$, Gaussian integration is an easy solution.

(2) It is possible to construct a Gaussian-type formula using $\cos x$ as the weight function. However, this weight is one-signed only over intervals of length π, from $(n - \frac{1}{2})\pi$ to $(n + \frac{1}{2})\pi$; and if a and/or b does not happen to coincide with one of these points, special formulas are needed. The method was deemed impractical.

(3) We can obtain an integral with a nonoscillatory integrand by noting that $I(0, b)$ satisfies the Bessel equation

$$\frac{d^2 I}{dA^2} - \frac{1}{A}\frac{dI}{dA} - I = -Q$$

where $Q(A) = (b^2 + A^2)^{1/2} \sin b + (b^2 + A^2)^{-1/2} b \cos b$ and so

$$I = AK_1(A)\left(b \sin b + \cos b - 1 + \int_0^A Q(x)I_1(x)\,dx\right) + AI_1(A)\int_A^\infty Q(x)K_1(x)\,dx.$$

(4) When b is large and is greater than A we compute $I(0, b)$ by computing to ∞ (which involves a K_1-Bessel function) and then subtracting off the tail (which involves the cosine-integral function). To get an integral that converges at ∞ we integrate by parts twice:

$$I = Q(A) - R \quad (Q \text{ as defined above})$$

where

$$R = A^2 \int_0^b (x^2 + A^2)^{-3/2} \cos x\,dx = AK_1(A) - A^2 T,$$

$$T = \int_b^\infty (x^2 + A^2)^{-3/2} \cos x\,dx.$$

Expand the radical in powers of A/x and integrate term by term:

$$T = \int_b^\infty \left(1 - \frac{3A^2}{2x^2} + \frac{15A^4}{8x^4} - \cdots\right)x^{-3}\cos x\,dx \quad (b > A)$$

$$= U_3 - \frac{3}{2}A^2 U_5 + \frac{15}{8}A^4 U_7 - \frac{35}{16}A^6 U_9 + \cdots.$$

Integration by parts gives the recurrence formula

$$U_{n+2} = \frac{1}{n(n+1)}(U_n + b^{-n}\sin b - nb^{-n-1}\cos b)$$

with $U_1 = -Ci(b)$.

(5) When A is small put $x = At$ and expand the cosine:

$$I(0, b) = A^2 \int_0^h (1 + t^2)^{1/2}\cos At\,dt \quad \left(h = \frac{b}{A}\right)$$

$$= A^2 \sum_{n=0}^\infty (-1)^n A^{2n} H_{2n}/(2n)!$$

where $H_k = \int_0^h (1 + t^2)^{1/2} t^k\,dt$

$$H_0 = \tfrac{1}{2}h(1 + h^2)^{1/2} + \tfrac{1}{2}\log(h + (1 + h^2)^{1/2}),$$

$$H_k = (h^{k-1}(1 + h^2)^{3/2} - (k-1)H_{k-2})/(k+2) \quad (k > 1).$$

(6) For the intermediate case $A > b$ expand the square root in powers of x and integrate term by term:

$$I(0,b) = A \int_0^b \left[1 + \frac{1}{2}\left(\frac{x}{A}\right)^2 - \frac{1}{8}\left(\frac{x}{A}\right)^4 + \frac{1}{16}\left(\frac{x}{A}\right)^6 - \cdots \right] \cos x \, dx$$

$$= A\left(F_0 + \frac{F_2}{2A^2} - \frac{F_4}{8A^4} + \frac{F_6}{16A^6} - \cdots \right) \qquad (A > b),$$

$$F_0 = \sin b,$$

$$F_k = b^k \sin b + k b^{k-1} \cos b - k(k-1)F_{k-2} \qquad (k > 1).$$

(7) When A and b are both large we can get an asymptotic series by successive integrations by parts; terms of the series are expressible as Legendre polynomials. Put $u = (A^2 + x^2)^{1/2}$ and get

$$\int u \cos x \, dx = u \sin x + u' \cos x - u'' \sin x - u''' \cos x + \cdots$$

$$= (u - u'' + u^{iv} - \cdots) \sin x + (u' - u''' + u^v - \cdots) \cos x \big|_0^b.$$

Put $x = A \cot \theta$, so $u = A \csc \theta$, $v = 1/u = 1/A \sin \theta$. Then

$$u' = \frac{du}{dx} = \cos \theta = xv,$$

$$u'' = v + xv', \quad u''' = 2v' + xv'', \quad u^{(n+1)} = nv^{(n-1)} + xv^{(n)}.$$

But [1]

$$v^{(n)} = \left(\frac{d}{d(A \cot \theta)} \right)^n \frac{1}{A} \sin \theta = \frac{(-1)^n}{A^{n+1}} n! \sin^{n+1}\theta \, P_n(\cos \theta)$$

so

$$u^{(n+1)} = \frac{(-1)^{n-1}}{A^n} n! \sin^n\theta \, (P_{n-1}(\cos \theta) - P_n(\cos \theta) \cos \theta)$$

$$= \frac{(-1)^{n-1}A^2(n-1)!}{u^{n+2}} P'_n(z), \quad z = \frac{x}{u} \quad (n > 0).$$

When n is odd $u^{(n)}(x=0) = 0$, so u and its derivatives are to be evaluated only at $x = b$, and so

$$I(0,b) = \left[u + \frac{A^2}{u^2}\left(-\frac{P'_1}{u} + \frac{2!P'_3}{u^3} - \frac{4!P'_5}{u^5} + \cdots \right) \right] \sin b$$

$$+ \left[\frac{b}{u} + \frac{A^2}{u^2}\left(\frac{P'_2}{u^2} - \frac{3!P'_4}{u^4} + \frac{5!P'_6}{u^6} - \cdots \right) \right] \cos b \quad \text{asymptotically}$$

in which $u = (b^2 + A^2)^{1/2}$ and the argument of P' is b/u.

REFERENCE

[1] E. T. WHITTAKER AND G. N. WATSON, A Course of Modern Analysis, Cambridge Univ. Press, Cambridge, U.K., 1927.

Problem 77-13*, *A Property of the First Erlang Function,* by ILIA KAUFMAN (Bell Canada, Ottawa, Ontario, Canada).

Let

$$f_c(x) = (x+c)[B(x+c|x-1) - B(x+c|x)], \qquad x \geq 1, \quad c \geq 0,$$

where

$$B(y|x) = \frac{e^{-y}y^x}{\int_y^\infty e^{-t}t^x\, dt}.$$

The function B, or its restriction to integral values of x,

$$B(y|n) = \frac{y^n/n!}{\sum_{k=0}^n y^k/k!},$$

is called the *first Erlang function*. It is easy to prove that for any fixed value of c,

$$\lim_{x\to\infty} f_c(x) = 2/\pi.$$

Determine or numerically estimate

$$\Delta = \inf_{c \geq 0} \sup_{x \geq 1} |f_c(x) - 2/\pi|.$$

Our own numerical results suggest that for $c = 0.75$, $\sup_{x \geq 1} |f_c(x) - 2/\pi|$ is sufficiently small that it may be neglected in some practical calculations. This problem arises in queuing theory and has applications in telephone plant engineering.

Solution by W. B. JORDAN (Scotia, New York).

We approximate the solution algebraically and refine it numerically. Formula numbers are from the reference below. Let

$$G(p, q) = \frac{1}{pB(p|q)} = p^{-q-1}\, e^p \int_p^\infty e^{-t}t^q\, dt = \int_0^\infty e^{-pu}(1+u)^q\, du.$$

Besides being called "Erlang", this function is called a confluent hypergeometric function (13.2.5), an incomplete gamma function (6.5.3), and a chi-square probability function. The connection with the latter is (26.4.2)

$$G(p, q) = p^{-q-1}\, e^p \Gamma(q+1) Q(2p|2q+2).$$

Let $L(x, c) = G(x+c, x-1)$ and $M(x, c) = G(x+c, x)$ so

$$f_c(x) = \frac{1}{L} - \frac{1}{M}.$$

From the recurrence formula (26.4.8) for Q, we get

$$pG(p, q) = 1 + qG(p, q-1),$$

so that

$$(x+c)M = 1 + xL, \qquad f_c(x) = \frac{1-cL}{(1+xL)L}.$$

We give three formulas for computing L. If x is a (not too large) integer, $(1+u)^q$ can be expanded and integrated term by term to give

$$L = \frac{1}{x+c} + \frac{x-1}{(x+c)^2} + \frac{(x-1)(x-2)}{(x+c)^3} + \frac{(x-1)(x-2)(x-3)}{(x+c)^4} + \cdots.$$

This tests nicely for efficient calculation. In particular, for $x = 1$:

$$L(1, c) = \frac{1}{1+c}, \qquad f_c(1) = \frac{1+c}{2+c}.$$

A convenient formula for any x comes from the continued fraction expansion (24.4.10) of Q. (Caution: The printed formula is truncated too soon to show the law of formation. The numerators of $x^2/2$ should be $1, 1, 2, 3, 4, \cdots$.) Recursion formulas for this fraction are

$$P_0 = 1, \quad P_1 = \frac{x+c+1}{c+2}, \quad Q_0 = Q_1 = 1,$$

$$r_n = (n-1)(x-n)/(2n+c)(2n-2+c),$$

$$P_n = P_{n-1} + r_n P_{n-2}, \qquad Q_n = Q_{n-1} + r_n Q_{n-2},$$

$$L = P/(x+c)Q.$$

Convergence is good.

For x large, we have the asymptotic formula

$$L = \int_0^\infty e^{-x[u-\log(1+u)]} e^{-cu} (1+u)^{-1} du.$$

The coefficient of x is a maximum at $u = 0$, so that an appropriate change of variable is

$$\tfrac{1}{2}t^2 = u - \log(1+u), \qquad u = t + t^2/3 + t^3/36 - t^4/270 + \cdots,$$

which produces

$$L = \tfrac{1}{2}\pi z (1 + a_1 z + a_2 z^2 + \cdots),$$

where

$$z = (2/\pi x)^{1/2}, \quad a_1 = -\left(c + \frac{1}{3}\right), \quad a_2 = \left(c^2 + \frac{1}{6}\right)\frac{\pi}{4}.$$

Then,

$$f_c(x) = \frac{2}{\pi}(1 + b_1 z + b_2 z^2 + \cdots),$$

where

$$b_1 = \left(2 - \frac{\pi}{2}\right)c - \frac{1}{3}, \qquad b_2 = \left(\frac{\pi}{3} - 1\right)(c - 3c^2) + \frac{1}{3} - \frac{\pi}{12}.$$

Unfortunately, these asymptotic series do not give adequate accuracy in the region of interest.

Consider the family of curves f vs. x, with c as parameter. All curves have $f = 2/\pi$ as their asymptote. If $c > .7766$, $b_1 > 0$; $b_2 > 0$ in this vicinity also, so the curve is (probably) entirely above its asymptote. If $c = .7519$, $f_c(1) = 2/\pi$ and the curve is (probably) entirely below it. The value of c for the inf sup therefore lies between these values, probably closer to the latter. The desired curve starts slightly above $2/\pi$, dips to a minimum at an equal amount below it, and then rises to its asymptote.

The asymptotic expansion implies that f has a minimum of $(2/\pi)(1-b_1^2/(4b_2))$ when $z = -b_1/(2b_2)$. If we set $(2/\pi)-(\text{this min})=f_c(1)-2/\pi$ and solve for c we find $c = .7556$. Numerical calculation in this vicinity leads to

$$c = .7562, \qquad \inf \sup = .00056 \pm 10\%$$

with $x = 14$ (roughly) at the min.

REFERENCE

M. ABRAMOWITZ AND I. STEGUN, *Handbook of Mathematical Functions*, National Bureau of Standards, U.S. Government Printing Office, Washington, DC, 1965.

Problem 61–14 *, *On a Zero of an Implicit Function*, by M. A. MEDICK (Michigan State University).

The integral

$$I(t, a) = \int_0^\infty \frac{\lambda e^{-\lambda}}{\sqrt{1 + a^2\lambda^2}} \sin\left\{ \frac{t}{\sqrt{1 + a^2\lambda^2}} \right\} d\lambda$$

is associated with the bending moment response of a spherical shell in an acoustic medium to a step pressure wave. The first positive non-trivial zero $t_0(a)$ of $I(t, a) = 0$ for fixed "a" is useful in assessing the relative importance of bending and membrane stresses in the shell. It is of interest to obtain approximations for $t_0(a)$ for the ranges
(a) $a \ll 1$,
(b) $a \gg 1$,
(c) $a \approx 1$.

Solution by DAVID ROTHMAN (Rocketdyne) and KENNETH S. DAVIS (San Jose State College.)

(a) $a \ll 1$. Choose R such that $1 \ll R \ll 1/a$. Then,

$$I(t, a) \approx \int_0^R \lambda e^{-\lambda}(1 - a^2\lambda^2/2) \sin t(1 - a^2\lambda^2/2 + 3a^4\lambda^4/8) \, d\lambda$$

$$\approx (\sin t) \int_0^R \lambda e^{-\lambda}(1 - a^2\lambda^2/2) \, d\lambda - t(\cos t) \int_0^R \lambda e^{-\lambda}(a^2\lambda^2/2 - 5a^4\lambda^4/8) \, d\lambda$$

$$\approx (1 - 3a^2) \sin t - (3a^2 - 75a^4)t \cos t.$$

At a zero,

$$\frac{\tan t_0}{t_0} \approx 3a^2 - 66a^4.$$

At the first zero, $t_0 = \pi + x$, where $x \ll 1$. Whence,

$$t_0 \approx \pi(1 + 3a^2 - 57a^4).$$

(b) $a \gg 1$. Choose ϵ such that $1/a \ll \epsilon \ll 1$, and let

$$Z_i = \int_0^\infty \lambda e^{-\lambda}(1 + a^2\lambda^2)^{-i} \, d\lambda.$$

Then,

$$Z_1 \approx \int_0^\epsilon \lambda (1 + a^2\lambda^2)^{-1}\, d\lambda + a^{-2} \int_\epsilon^\infty \lambda^{-1} e^{-\lambda}\, d\lambda$$

$$\approx (\ln a + \gamma)/a^2, \qquad\qquad \text{where} \quad \gamma = .5772 \cdots.$$

Since

$$\frac{\partial}{\partial a} \{a^{2i} Z_i\} = 2i\{a^{2i+2} Z_{i+1}\}/a^3,$$

it follows that

$$Z_{i+1} \approx 1/2ia^2, \qquad\qquad i = 1, 2, \cdots.$$

Also, since

$$I(t, a) = \sum_{i=0}^\infty \frac{(-1)^i t^{2i+1} Z_{i+1}}{(2i + 1)!},$$

we have

$$I(t, a) \approx t(\ln a + \gamma)/a^2 + \sum_{i=1}^\infty \frac{(-1)^i t^{2i+1}}{(2i + 1)!(2ia^2)}.$$

At the first positive zero,

$$\ln a + \gamma \approx \int_0^{t_0} \frac{x - \sin x}{x^2}\, dx$$

$$\approx -1 + \int_0^{t_0} \frac{1 - \cos x}{x}\, dx \approx -1 + \ln t_0 + \gamma.$$

Thus,

$$t_0 \sim ae.$$

R. E. WALTERS and W. SQUIRE (West Virginia University) determine t_0 numerically by using a generalized Gauss-Laguerre quadrature [1], i.e.,

$$\int_0^\infty \lambda^n e^{-\lambda} F(\lambda)\, d\lambda = \sum_{i=1}^N W_i F(\lambda_i).$$

TABLE 1

First positive values of $t_0(a)$ for $I(t, a) = 0$

a	N				
	1	4	8	12	16
0	3.14159	3.14159	3.14159	3.14159	3.14159
0.1	3.20379	3.22251	2.22253	3.22253	3.22253
0.2	3.38359	3.39011	3.38957	3.38960	3.38956
0.4	4.02312	3.76761	3.76192	3.76271	3.76240
0.6	4.90732	4.08196	4.12990	4.12633	4.12730
0.8	5.92755	4.33067	4.50977	4.48240	4.48544
1.0	7.0248	4.5756	4.9193	4.8562	5.3114
1.5	9.9347	5.3340	10.0773	8.6537	6.0990
2.0	12.953	6.277	11.660	9.935	10.010
3.0	19.110	8.443	14.938	12.160	11.012
5.0	31.57	13.20	22.06	16.96	14.54
10.0	62.91	25.62	41.27	30.25	24.80

Tables for the values of the weights W_i and the corresponding values of the abscissas λ_i are also given in [1].

The roots t_0 for specific a's can be computed by any of the usual root finding methods with the aid of a computer.

In order to compare the convergence of the values of t_0, computations were performed for various values of N; $N = 1, 4, 8, 12$ and 16 were used. The results are shown in Table 1. It is to be noted that the accuracy in the values of $t_0(a)$ decreases with increasing a.

REFERENCE

[1] P. Rabinowitz and G. Weiss, *Tables of abscissas and weights for numerical evaluation of integrals of the form* $\int_0^\infty e^{-x}x^n f(x)\, dx$, MTAC, 13 (1959), pp. 285–294.

Editorial Note: On comparing the first solution with the values in the table of the second solution, one sees that there is a rather good agreement.

Problem 76-13*, *An Average Relative Speed Approximation*, by L. K. Arnold (D. H. Wagner, Associates), L. Dodson (Naval Underwater Systems Center) and L. Rosen (Center for Naval Analyses).

Two ships A and B are cruising along straight line paths in a planar ocean at constant speed u and v, respectively. If B's direction is a random variable uniformly distributed over $(0, 2\pi]$, then the expected speed of B relative to A is given by

$$z = \int_0^{2\pi} \{u^2 + v^2 - 2uv \cos \theta\}^{1/2} \, d\theta/(2\pi).$$

We have found that

$$\bar z = x + .27 y^2/x,$$

where $x = \max(u, v)$, $y = \min(u, v)$, is a fair approximation to z. More, precisely, $|z - \bar z| \leq .25$ knots for $u, v = 0(.25)40$. Prove or disprove the latter error bound for all u, v between 0 and 40.

Solution by W. B. Jordan (Scotia, New York).

Choose v to be the larger of u and v. Put $u/v = k$, so $0 \leq k \leq 1$. Then

$$z = \frac{v}{\pi} \int_0^\pi (1 + k^2 - 2k \cos \theta)^{1/2} \, d\theta.$$

We take $\bar z$ in the form $\bar z = v(a + bk^2)$ and we solve two problems:

(A) For $a = 1$ and $b = .27$, what is $\max |\bar z - z|$?
(B) What values of a and b minimize $\max |\bar z - z|$?

The work is simplified by first performing a Landen transformation on z. Put

$$\theta = \phi - \sin^{-1}(k \sin \phi), \qquad W^2 = 1 - k^2 \sin^2 \phi,$$

to give

$$z = \frac{v}{\pi} \int_0^\pi (W - 2k \cos \phi + W^{-1}k^2 \cos^2 \phi) \, d\phi = \frac{2v}{\pi}(E + k^2 B)$$

where E is the complete elliptic integral of the second kind, and

$$B = \int_0^{\pi/2} W^{-1} \cos^2 \phi \, d\phi = (E - k'^2 K)/k^2.$$

Derivatives are

$$dz/d(k^2) = vB/\pi, \qquad d\bar{z}/d(k^2) = vb.$$

The function B/π increases from .25 at $k = 0$ to $1/\pi = .31831$ at $k = 1$. That the increase is monotone can be seen from the integral defining B. The equation $d(\bar{z} - z)/d(k^2) = 0$ therefore has exactly one root if b lies between these limits. For $b = .27$ this root is $k^2 = .506$ (great precision is not needed here, because the curves of z and $\bar{z}$ are parallel at this point.) At this k^2,

$\bar{z} = 1.136620 \, v, \qquad z = 1.131142 \, v,$

$\bar{z} - z = .005478 \, v = .219$ knots when $v = 40$ knots.

It is also necessary to check the endpoints. At $k = 0$ the error is zero. At $k = 1$,

$\bar{z} = 1.27 \, v, \qquad z = 4v/\pi = 1.273240 \, v,$

$\bar{z} - z = -.003240 \, v = -.130$ knots when $v = 40$ knots.

Therefore the worst error in problem (A) is .219 knots.

The max error is minimized by taking by $\bar{z}$ parallel to the line joining the endpoints of z, and halfway from this line to the parallel tangent. Then $b = 4/\pi - 1 = .273240$. This is the value of B/π when $k^2 = .567$. Here

$$z = 1.147709 \, v, \qquad \bar{z} = (a + .154927)v, \qquad \bar{z} - z = (a - .992782)v.$$

The error at $k = 0$ is $(a - 1)v$. For these two errors to be equal and opposite, $a = .996391$, and the wrorst error is .003609 $v = .144$ knots. That is: The formula

$$\bar{z} = .99639x + .27324 \, y^2/x$$

reduces the worst error by about a third.

Note. A $4D$ table of the B-function, for $k^2 = 0(.01)1$, is printed in Chapter 5B of E. Jahnke and F. Emde, *Tables of Functions*, Dover, New York, 1945.

Problem 64-18, *An Expansion for the Absorptivity of a Gas*, by J. ERNEST WILKINS, JR. (General Dynamics/General Atomic).

The apparent absorptivity of an infinite cylinder of gas has been expressed by Nusselt (Z. Ver. Deut. Ing., 70 (1926), p. 763) in the form

$$\alpha = 1 - \frac{4}{\pi} \int_0^{\pi/2} \int_0^{\pi/2} e^{-2\mu \cos y \sec x} \cos^2 x \cos y \, dx \, dy,$$

where μ is a physical property of the gas (logarithmic decrement of radiation) multiplied by the radius of the cylinder. Find series expansions for α for both small and large positive values of μ.

Solution by HENRY E. FETTIS (Wright-Patterson Air Force Base, Ohio).

For large μ, the integral in y can be expressed in terms of modified Bessel functions of order 1, i.e.,

$$\int_0^{\pi/2} e^{-2\mu \cos y \sec x} \cos y \, dy = 1 - \frac{\pi}{2} \{ I_1(2\mu \sec x) - \mathbf{L}_1(2\mu \sec x) \}.$$

The known asymptotic expansion for the right-hand side is

$$\frac{1}{\pi} \sum_{n=1}^{\infty} \frac{(-1)^{n+1} \Gamma(\frac{1}{2} + n) \cos^{2n} x}{\mu^{2n} \Gamma(\frac{3}{2} - n)}.$$

Whence,

$$\alpha \sim 1 - \frac{1}{\pi^{3/2}} \sum_{n=1}^{\infty} \frac{(2n+1)\Gamma(n+\frac{1}{2})^3}{\mu^{2n}(2n-1)(n+1)!}.$$

For small μ, differentiating with respect to μ, we get $\alpha(0) = 0$, $\alpha'(0) = 2$, $\alpha''(0) = -16/3$, and

$$\alpha'''(\mu) = \frac{32}{\pi} \int_0^{\pi/2} \int_0^{\pi/2} e^{-2\mu \cos y \sec x} \sec x \cos^4 y \, dy.$$

Setting $\cosh \theta = \sec x$,

$$\alpha'''(\mu) = \frac{32}{\pi} \int_0^{\infty} d\theta \int_0^{\pi/2} e^{-2\mu \cos y \cosh \theta} \cos^4 y \, dy = \frac{32}{\pi} \int_0^{\pi/2} K_0(2\mu \cos y) \cos^4 y \, dy.$$

One can show that the last integral can be integrated to

$$\alpha'''(\mu) = 2I_2(\mu)K_2(\mu) - 8I_1(\mu)K_1(\mu) + 6I_0(\mu)K_0(\mu),$$

but it will be simpler to use the power series for $K_0(x)$,

$$K_0(x) = \sum_{n=0}^{\infty} (x/2)^{2n} \{ \psi(n+1) - \log x/2 \}/(n!)^2,$$

and the definite integrals

$$\int_0^{\pi/2} \cos^{2n+4} y \, dy = \frac{\sqrt{\pi}}{2} \frac{\Gamma(n+\frac{5}{2})}{\Gamma(n+3)},$$

$$\int_0^{\pi/2} \{\cos^{2n+4} y\} \log \cos y \, dy = \frac{\sqrt{\pi}}{4} \frac{\Gamma(n+\frac{5}{2})}{\Gamma(n+3)} \left\{ \psi\left(n+\frac{5}{2}\right) - \psi(n+3) \right\}.$$

On carrying out the indicated operations, we obtain

$$\alpha = 2\mu - \frac{8\mu^2}{3} + \frac{1}{\sqrt{\pi}} \sum_{n=0}^{\infty} \frac{\mu^{2n+3}\Gamma(n+\frac{1}{2})A_n}{n!\,(n+1)!\,(n+2)!},$$

where

$$A_n = 2\psi(n+1) + \psi(n+3) - \psi(n+\tfrac{5}{2}) - 2\log \mu.$$

Radius of Convergence of a Power Series

Problem 85-18*, *by* J. E. WILKINS (Chicago, IL).

The power series

(1)
$$v = \sum_{n=0}^{\infty} A_n u^n$$

is a formal solution of the differential system

(2)
$$d(u\,dv/du)/du = v^4, \qquad v(0) = 1,$$

if the coefficients A_n are defined so that

(3)
$$A_0 = 1, \quad B_m = \sum_{k=0}^{m} A_k A_{m-k}, \quad A_{n+1} = (n+1)^{-2} \sum_{m=0}^{n} B_m B_{n-m}.$$

What is the radius of convergence ρ of the power series (1)? The differential system occurs in the study of straight triangular fins that radiate heat to surroundings at absolute zero [1]. It was asserted in [1] that $\rho \geq \frac{1}{2}$.

REFERENCE

[1] J. E. WILKINS, *Minimizing the mass of thin radiating fins*, J. Aero/Space Sci., 37 (1960), pp. 145–146.

Editorial note. Although the proposer has not determined ρ, he has decent bounds for it as well as conjectured very close bounds for it.

Solution by W. B. JORDAN (Scotia, NY).
First solution:
We use the D'Alembert ratio test. We used the proposer's recurrence relations to compute the first 60 coefficients. For n between 40 and 60 their ratios can be approximated by

$$A_{n-1}/A_n = .79493729 + .2648880n^{-1} + .118004n^{-2}$$

correct to 8 decimal places, implying ρ is very close to .7949373.
Second solution:
Translate the origin to an arbitrary point h by means of $u = h(1-x)$. The differential equation becomes

$$\frac{d}{dx}\left((1-x)\frac{dv}{dx} \right) = hv^4$$

which has the solution

$$v_1 = \left(\frac{10}{9hx^2} \right)^{1/3} \left(1 - \frac{2}{21}x - \frac{11}{882}x^2 + \cdots \right),$$

so the equation has a movable singularity. If h is determined so that $v = 1$ when $x = 1$, then the point $u = h$ will be a singularity of the proposer's series; it may well be the one that limits the convergence, so $\rho = h$. However, v_1 converges slowly (if at all) at $x = 1$ so this calculation is inconvenient. The leading factor suggests the singularity can be removed by putting $v = w^{-2/3}$, which promises to give a rapidly converging series. The equation becomes

$$\frac{5u}{3}\left(\frac{dw}{du} \right)^2 - w\frac{d}{du}\left(u\frac{dw}{du} \right) = \frac{3}{2}, \qquad w(0) = 1.$$

When $u = \rho$: $x = 0$, $v = \infty$ and $w = 0$, therefore ρ is computed as a root of $w(u) = 0$. The expansion of w is

$$w = 1 - \frac{3}{2}u + \frac{3}{8}u^2 - \frac{5}{48}u^3 + \cdots = \sum_{n=0}^{\infty} a_n u^n.$$

The recurrence for the coefficients is

$$a_k = \left(\frac{1}{3k^2}\right) \sum_{j=1}^{k-1} j(5k - 8j) a_j a_{k-j}, \qquad k > 1, \quad a_1 = -1.5.$$

The promise was not fulfilled. The first eight coefficients alternate in sign and decrease rapidly, but all thereafter are negative. They decrease (numerically) slowly to a_{24}, after which they grow. A rough calculation using the D'Alembert ratio gives the radius of convergence at about 0.795, so we still have a singularity.

We computed the first 100 coefficients, and we solved the equation $w(u) = 0$ using 40, 50, 60 and 100 terms of the series. To 8 decimal places all four solutions were .79493668. The next 2 digits were 40, 20, 14, and 09, implying

$$\rho = .7949366807.$$

Since this is quite close to the first solution it is highly probable we have found the right singularity.

Also solved by M. ORLOWSKI *and* M. PACHTER (National Research Institute for Mathematical Sciences, CSIR, Pretoria, South Africa) *and by the proposer* who showed that $.794714 < \rho < .794942$ and presented evidence that $.794932 < \rho < .794944$.

Editorial note. Consider the more general problem $uv'' + v' = v^{\alpha}$, $v(0) = 1$, $\alpha \geq 0$. Let $v(u)$ have Maclaurin series coefficients $A_n(\alpha)$ and define $\lambda(\alpha) = 1/\rho(\alpha)$. Numerical experiments suggest that there is a critical value, $\bar{\alpha} \cong 2.1$, so that for $\alpha > \bar{\alpha}$, $\lambda = F(\alpha) \cong \sum_{h=0}^{\infty} B_h \alpha^{1-n}$, while for $\alpha < \bar{\alpha}$, $\lambda = G(\alpha) \cong \sum_{h=0}^{\infty} b_n \alpha^n$. (Exceptions: $\alpha = 0, 1, 2$, see below). Coefficient estimates are as follows:

$$B_0 = .5 \qquad\qquad b_0 = 1.$$
$$B_1 = -.78986\ 81352 \qquad b_1 = -.3550614058$$
$$B_2 = .17005\ 82583 \qquad b_2 = .04699\ 84101$$
$$B_3 = .07878\ 19504 \qquad b_3 = .00146\ 95230$$
$$B_4 = .02277\ 05725 \qquad b_4 = -.00272\ 05557$$
$$B_5 = .00767\ 34705 \qquad b_5 = .00114\ 53208.$$
$$B_6 = .00584\ 97841$$

For $0 < \alpha < \bar{\alpha}$, (except for $\alpha = 1, 2$) $A_n(\alpha)$ is positive for a number of terms, then alternates in sign; for $\alpha > \bar{\alpha}$, $A_n(\alpha) > 0$. The number of positive terms increases without bound as α approaches $\bar{\alpha}$ from below. $\lambda(\alpha)$ is *not continuous* at $\alpha = 0, 1, 2$. For $\alpha = 0$, clearly $v_0(u) = 1 + u$, which is entire, so $\lambda(0) = 0$. However, $\lim_{\alpha \to 0^+} \lambda(\alpha) = G(0) = 1$. Similarly $v_1(u) = I_0(2\sqrt{u})$, so $\lambda(1) = 0$, yet $\lim_{\alpha \to 1} \lambda(\alpha) = .6916602761 = 4/\delta^2$, where δ is the smallest zero of J_0. (I_0 and J_0 are, of course, Bessel functions.) Elementary perturbation analysis explains these phenomena. It would be interesting to have an explanation for the discontinuity of λ at $\alpha = 2$. Finally, we record the exact values of a few of the coefficients given above. $B_0 = \frac{1}{2}$, $B_1 = 5/2 - \pi^2/3$, $b_0 = 1.$, $b_1 = \pi^2/6 - 2$, $b_2 = 1 - \pi^2/6 + 7\pi^4/360 - \zeta(3)$. [O. G. R.]

Asymptotic Behavior of a Melting Interface

Problem 89-3*, *by* O. G. RUEHR (Michigan Technological University).

The unidimensional one-phase Stefan problem is to find $u(x, t)$ and $X(t)$ satisfying the following:

(i) $u_{xx} = u_t,$ $0 < x < X(t),$

(ii) $u[X(t), t] = 0,$ $-u_x[X, t] = \dfrac{dX}{dt},$ $X(0) = 0,$

(iii) $u(0, t) = h(t)$ or $-u_x(0, t) = g(t).$

This describes the melting of a solid $(x \geq 0)$ initially at the melting temperature with prescribed temperature, $h(t)$, or heat flux, $g(t)$, at the fixed boundary $(x = 0)$. The liquid-solid interface is at $x = X(t)$. Physical constants have been removed by similarity transformations (see [1] for details). An example solution is $u = e^{t-x} - 1$, $X = t$.

(iv) $u = \displaystyle\sum_{n=1}^{\infty} \frac{(d^n/dt^s)[X(t)-x]^{2n}}{(2n)!}.$

It has long been known and easily verified from (iv) that if h is constant, then $x(t)$ is proportional to $\sqrt{t}$. We seek the asymptotic behavior of X for $g(t) \equiv 1$ and conjecture that

(v) $\displaystyle\lim_{t \to \infty} \frac{x^2(t)}{t \log (t)} = 2.$

Prove or disprove this conjecture (v).

REFERENCES

[1] G. W. EVANS II, E. ISAACSON, AND J. K. L. MACDONALD, *Stefan-like problems*, Quart. Appl. Math., 9 (1951), pp. 185–193.
[2] O. G. RUEHR, *The one-phase Stefan problem*, Internal Memorandum, Radiation Laboratory, University of Michigan, Ann Arbor, MI, 1962.

Solution by G. F. NEWELL (University of California, Berkeley, CA).

The proposer's conjecture is correct, but the proof can be inferred more easily from the physics of the problem than the formal mathematics.

The formulation has been made nondimensional so that if $g(t) \cong 1$ for $t \leq 1$, the interface $X(t)$ with $X(0) = 0$ at first absorbs most of the heat injected at rate $g(t) = 1$ at $x = 0$ and travels at a speed $dX(t)/dt = 1$. As the interface moves away from the heat source, however, most of the heat injected by time t will not have reached $X(t)$ by time t and the speed of the interface slows down. Indeed, in a time t, the heat will travel a distance of order only $t^{1/2}$.

If there were no interface, each unit of heat injected at $x = 0$ and time τ will generate a heat distribution (temperature) at time $t > \tau$ of

$$[\pi(t-\tau)]^{-1/2} \exp (-x^2/4(t-\tau)),$$

the "fundamental" solution of the heat equation. A continuous source of $g(\tau)$ at $x = 0$ will generate a temperature

$$u_0(x, t) = \int_0^t [\pi(t - \tau)]^{-1/2} \exp(-x^2/4(t - \tau))g(\tau) \, d\tau.$$

For (moderately) large values of $x^2/4t$, we can expand $t - \tau$ in powers of τ/t and approximate this by

$$u_0(x, t) = \frac{\exp(-x^2/4t)}{(\pi t)^{1/2}} \int_0^t \exp\left(-\frac{x^2}{4t}\frac{\tau}{t}\right)g(\tau) \, d\tau$$

$$\cong \frac{\exp(-x^2/4t)4t^2}{(\pi t)^{1/2}x^2} \quad \text{for } g(\tau) = 1.$$

This shows, first, that the temperature drops very rapidly for moderately large values of $x^2/4t$, and, second, that this "wavefront" receives a contribution from a range of τ of order $t(4t/x^2)$, a small fraction of the time t (provided $g(\tau)$ does not grow too rapidly) but a time range large compared with 1 for sufficiently large t.

For $t \gg 1$, the interface $X(t)$ obviously must ride the front of this wave in some way. If it got too far ahead of the wave, it would not absorb any heat and would be slowed down, but, if it lagged behind the front, it would travel fast enough to catch up. The only question is, *where* in the steep wavefront does $X(t)$ ride?

The solution $u(x, t)$ with an interface $X(t)$ can, in principle, be constructed by successive "reflections" off the boundaries $X(t)$ and $x = 0$. Any heat injected at time τ will take a time of order at least $X^2(\tau)$ to reach the interface, even if the interface stays still after time τ. The heat distribution from the source will not "feel" the interface $X(t)$ until it can reach the interface. When some heat does reach the interface, it is absorbed and cannot continue to spread forward and backward. To correct for this, we can add to the u_0 a negative distributed source along the interface $X(t)$ to cancel the u_0 at $X(t)$ but add nothing to the gradient $\partial u_0/\partial x$. This "reflected" heat will eventually reach $x = 0$ and be reflected from this boundary, etc.

It is not necessary to write this formal solution. Except in the early stages $t \lesssim 1$ when $X(t)$ is of order 1, these successive reflections obviously will have a negligible effect on the temperature distribution as compared with the primary source, because the amount of heat absorbed at $X(t)$ by time t is only a small fraction of that generated at $x = 0$. The conclusion is that the motion of $X(t)$ is determined mainly by the heat absorbed from u_0, i.e.,

$$(1) \qquad \frac{dX}{dt} \cong -\frac{\partial}{\partial x} u_0(x, t)\bigg|_{x = X} = \int_0^t \frac{X}{\pi^{1/2}} \frac{\exp(-X^2/4(t - \tau))}{2(t - \tau)^{3/2}} g(t) \, d\tau.$$

The errors resulting from the approximations that have been made so far are negligible compared with those we will make now in an attempt to solve this equation analytically for $X(t)$ with $t \gg 1$.

If we change the integration variable to u with

$$u^2 = x^2/4(t - \tau),$$

(1) becomes

$$(2) \qquad \frac{dX}{dt} = \frac{2}{\sqrt{\pi}} \int_{X/2\sqrt{t}}^\infty \exp(-u^2)g\left(t - \frac{X^2}{4u^2}\right) du$$

$$= \text{erfc}(X/2\sqrt{t}) \quad \text{for } g(\tau) = 1.$$

It is more convenient to solve this equation for t as a function of $Z = X^2/4t$ than for X as a function of t. For this purpose, we can write (2) in the form

$$t^{1/2} = \frac{\sqrt{Z}\left[1 + \dfrac{t/Z}{dt/dZ}\right]}{\text{erfc}(\sqrt{Z})}$$

or

(3) $$\frac{1}{2}\log t = -\log(\text{erfc}(\sqrt{Z})) + \frac{1}{2}\log Z + \log\left[1 + \frac{t/Z}{dt/dZ}\right].$$

For sufficiently large Z, $-\log(\text{erfc}(\sqrt{Z})) \cong Z$, and, as a very crude approximation, we can take $\frac{1}{2}\log t \sim Z$, which gives the conjectured result

(4) $$\lim_{t\to\infty}\frac{X^2(t)}{t\log t} = \lim_{t\to\infty}\frac{4Z(t)}{\log t} = 2.$$

Actually for large Z we can obtain a much better approximation by using the asymptotic expansion $\text{erfc}(\sqrt{Z}) \cong e^{-Z}/\sqrt{\pi}Z$ and taking as a first approximation

$$\tfrac{1}{2}\log t_1 = Z + \log(\sqrt{\pi}Z),$$

neglecting the last term of (3). We can then obtain a jth approximation t_j by substituting t_{j-1} in the last term of (3). Thus, as a second approximation

$$\tfrac{1}{2}\log t_2 = Z + \log(\sqrt{\pi}Z) + \log[1 + 1/2(Z+1)].$$

Although the conjectured limit (4) is correct, it does not give a very accurate description of what is happening for (reasonably) large but finite t. Equation (4) suggests that to reach $Z = 2$ would require a time of about $\log t = 4$, $t \sim 54$. The approximations t_1, t_2, t_3, however, give $\log t_1 = 6.53$, $\log t_2 = 6.84$, and $\log t_3 = 6.85$, and thus $t \sim 944$. Even this is too low. If we use more accurate bounds for $\text{erfc}(\sqrt{Z})$, we obtain $7.11 \lesssim \log t \lesssim 7.22$, i.e., $t \sim 1300$. By this time, about 1300 units of heat have been inserted at $x = 0$ but only about $X = 2(tZ)^{1/2} \sim 100$ units have been absorbed at X.

Even for $Z = 1$, we find that $4.1 \lesssim \log t \lesssim 4.3$, i.e., $t \sim 65$.

For $g(t) = 1$, the temperature $u(0, t)$ at $x = 0$ increases proportionally to $t^{1/2}$ and $Z(t)$ increases more or less like $\log t^{1/2}$. More generally, we should find that $Z(t)$ grows more or less like $\log u(0, t)$. Thus for $u(0, t) = \text{constant}$, $Z(t) \sim \text{constant}$ and $X(t)$ is proportional to $t^{1/2}$, but, if $u(0, t)$ increases exponentially, $Z(t)$ is proportional to t as is $X(t)$.

Remark by the proposer. The Maclaurin series for $X(t)$ has radius of convergence equal to zero. A good approximation for $X(t)$ is given parametrically by

$$x = (2\lambda - 1)e^\lambda + 1, \qquad t = \lambda e^{2\lambda}.$$

Problem 64–7, An Asymptotic Series, by N. G. DE BRUIJN (Technological University, Eindhoven, Netherlands).

Let $\phi(x)$ be infinitely often differentiable for $x \geq 0$, and let

$$\int_0^\infty |\phi^{(k)}(x)|\,dx$$

be convergent for each $k = 0, 1, 2, \cdots$. Define

$$F(t) = \sum_{n=1}^{\infty} n^{-1}\phi(nt), \qquad\qquad t > 0.$$

Show that $F(t) + \phi(0) \log t$ has an asymptotic development in the form of an asymptotic series $\sum_{n=0}^{\infty} c_n t^n$ if $t > 0$, $t \to 0$.

Solution by the proposer.

Introducing a positive constant λ, we put $\phi_1(x) = \phi(x) - \phi(0)e^{-\lambda x}$. Then ϕ_1 still has the properties attributed to ϕ, and moreover $\phi_1(0) = 0$. We put $x^{-1}\phi_1(x) = \eta(x)$, and we apply the Euler-Maclaurin sum formula to $\sum_0^{\infty} \eta(nt)$ (we can apply it to the infinite series since $\eta^{(k)}(x) \to 0 (x \to \infty)$ for each k, and $\int_0^{\infty} |\eta^{(k)}(x)| \cdot dx \to \infty$):

$$\sum_{1}^{\infty} \eta(nt) = t^{-1}\int_0^{\infty} \eta(x) - \frac{1}{2}\eta(0) - \sum_{k=1}^{m} \frac{B_{2k}\, t^{2k-1}\eta^{(2k-1)}(0)}{(2k)!}$$

$$- t^{2m}\int_0^{\infty} \frac{\eta^{(2m)}(x)B_{2m}(tx - [tx])}{(2m)!}\, dx.$$

Thus we obtain the following asymptotic series for $F(t) + \phi(0) \log t$:

$$\int_0^{\infty} x^{-1}(\phi(x) - \phi(0)e^{-\lambda x})\, dx - \phi(0) \log \frac{1 - e^{-\lambda t}}{\lambda t}$$

$$- \frac{1}{2} t\eta(0) - \sum_{k=1}^{\infty} \frac{B_{2k}\, t^{2k}\eta^{2k-1}(0)}{(2k)!}.$$

We finally want to get rid of λ. We note that $\int_0^{\infty} x^{-1}(e^{-x} - e^{-\lambda x})\, dx = \log \lambda$ and that the coefficients of the asymptotic development of $F(t) + \phi(0) \log t$ should not depend on λ. We now evaluate these coefficients by taking $\lambda = 0$. Then we have $\eta^{(k)}(0) = (k + 1)^{-1}\phi^{(k+1)}(0)$, and we obtain for the asymptotic series

$$\int_0^{\infty} x^{-1}(\phi(x) - \phi(0)e^{-x})\, dx - \frac{1}{2} t\phi'(0) - \sum_{k=1}^{\infty} \frac{B_{2k}\, t^{2k}\phi^{(2k)}(0)}{(2k)(2k)!}.$$

We remark that there is strict equality if $\phi(x) = e^{-\lambda x}$ $(\lambda > 0)$.

Problem 64-1, An Asymptotic Expansion, by H. O. Pollak and L. Shepp (Bell Telephone Laboratories).

Show that

$$e^{-x} \sum_{n=1}^{\infty} \frac{x^n}{n!} \log n = \log x - \frac{1}{2x} + O(x^{-2}).$$

This problem has arisen in studying the entropy of the Poisson distribution

$$H(x) = \sum_{n=1}^{\infty} p_n(x) \log p_n(x),$$

where

$$p_n(x) = \frac{x^n e^{-x}}{n!}.$$

Solution by J. H. VAN LINT (Technological University, Eindhoven, Netherlands).

We use the known formula for Euler's constant

(1) $$S_n = \sum_{r=1}^{n} r^{-1} \sim \log n + \gamma + \frac{1}{2(n+1)} + O\left(\frac{1}{n^2}\right).$$

Note that

(2) $$e^{-x} \sum_{1}^{\infty} \frac{x^n}{n!\, n^2} = O\left\{\frac{e^{-x}}{x^2} \sum_{1}^{\infty} \frac{x^{n+2}}{(n+2)!}\right\} = O(x^{-2}).$$

It now follows from (1) and (2) that

$$e^{-x} \sum_{1}^{\infty} \frac{x^n \log n}{n!}$$

$$= e^{-x} \sum_{1}^{\infty} \frac{S_n x^n}{n!} - \gamma(1 - e^{-x}) - \frac{e^{-x}}{2} \sum_{n=1}^{\infty} \frac{x^n}{(n+1)!} + O(x^{-2})$$

$$= \gamma + \log x - \text{Ei}\,(-x) - \gamma(1 - e^{-x}) - \frac{e^{-x}}{2x}(e^x - 1 - x) + O(x^{-2})$$

$$= \log x - \frac{1}{2x} + O(x^{-2}).$$

To obtain more terms of the asymptotic expansion, we just use more terms in the expansion (1).

Problem 63-11, An Asymptotic Result on a Zero of a Function, by R. L. GRAHAM and L. SHEPP (Bell Telephone Laboratories).

If x_n, $n \geq 2$, is the unique positive solution to

$$x^{-n} = \sum_{r=1}^{\infty} (x + r)^{-n},$$

show that

$$\frac{x_n}{n} \to \frac{1}{\log 2}.$$

Solution by R. BOJANIC (The Ohio State University).

We shall consider here a slightly more general problem.

Let $\{a_n\}$ and $\{b_n\}$ be two sequences of positive real numbers such that

(1) $$a_n = \sum_{\nu=1}^{\infty} \left(1 + \frac{\nu}{n} b_n\right)^{-n}, \qquad n = 2, 3, \cdots.$$

It is easy to see that if one sequence is given, the other is uniquely determined by (1). We shall prove here that

(i) $$\lim_{n \to \infty} b_n = b, (b > 0) \Rightarrow \lim_{n \to \infty} a_n = (e^b - 1)^{-1},$$

(ii) $$\lim_{n \to \infty} a_n = a, \, (a > 0) \Rightarrow \lim_{n \to \infty} b_n = \log \left(1 + \frac{1}{a}\right).$$

In particular, if $a_n = 1$, and $b_n = n/x_n$, $n = 2, 3, \cdots$, we obtain from (ii) the required asymptotic result.

The proofs of (i) and (ii) follow from the inequalities

(2) $$a_n \geq (e^{b_n} - 1)^{-1}, \quad \text{i.e.,} \quad b_n \geq \log \left(1 + \frac{1}{a_n}\right), \qquad\qquad n \geq 2,$$

(3) $$a_n b_n \leq 2, \qquad\qquad n \geq 2,$$

and the following lemma.

LEMMA. *If $\{b_n\}$ is bounded, then for any $\delta > 0$ there exists N_δ such that*

(4) $$n \geq N_\delta \Rightarrow a_n \leq (e^{b_n/(1+2\delta)} - 1)^{-1} + \frac{2}{b_n} (1 + \delta)^{-n+1}.$$

To prove (i), assume that $\lim b_n = b$, $b > 0$. Then from the first of the inequalities (2) it follows that

(5) $$\lim \inf a_n \geq (e^b - 1)^{-1}.$$

On the other hand, since $\lim b_n = b$, $b > 0$, using (4), for any $\delta > 0$ and $n \geq N_\delta$ we get

(6) $$\lim \sup a_n \leq (e^{b/(1+2\delta)} - 1)^{-1},$$

and (i) follows from (5) and (6) since δ can be chosen arbitrarily close to zero.

To prove (ii), assume $\lim a_n = a$, $a > 0$. Then from the second of the inequalities (2) we get

(7) $$\lim \sup b_n \geq \log \left(1 + \frac{1}{a}\right).$$

Next, from $a_n \geq \frac{1}{2}a > 0$, $n \geq N_a$, and (3) it follows that $\{b_n\}$ is bounded. For any $\delta > 0$ and $n \geq N_\delta$, we have by (4),

$$a_n \leq (e^{b_n/(1+2\delta)} - 1)^{-1} + \frac{2}{b_n} (1 + \delta)^{-n+1}.$$

Since $\log (1 + t) \geq t/(1 + t)$ for $t \geq 0$, we have by (2),

$$b_n \geq \log \left(1 + \frac{1}{a_n}\right) \geq (1 + a_n)^{-1}, \qquad\qquad n \geq 2,$$

and the preceding inequality becomes

$$a_n \leq (e^{b_n/(1+2\delta)} - 1)^{-1} + 2(1 + a_n)(1 + \delta)^{-n+1}.$$

From this inequality and (3), we obtain

$$b_n \leq (1 + 2\delta) \log \left(1 + \frac{1}{a_n} + 2 \left(1 + \frac{1}{a_n}\right)(1 + \delta)^{-n+1} e^{2/b_n(1+2\delta)}\right), \qquad\qquad n \geq N_\delta.$$

Since $\lim a_n = a$, $a > 0$, it follows that

(8) $$\lim \sup b_n \leq (1 + 2\delta) \log \left(1 + \frac{1}{a}\right),$$

and (ii) follows from (7) and (8) since δ can be chosen arbitrarily close to zero.
We have still to prove the inequalities (2), (3), and the lemma.

The inequalities (2) follow from

$$\left(1 + \frac{\nu}{n} b_n\right)^n \leq e^{\nu b_n}.$$

We have, namely,

$$a_n = \sum_{\nu=1}^{\infty} \left(1 + \frac{\nu}{n} b_n\right)^{-n} \geq \sum_{\nu=1}^{\infty} e^{-\nu b_n} = (e^{b_n} - 1)^{-1}.$$

On the other hand,

$$a_n = \sum_{\nu=1}^{\infty} \left(1 + \frac{\nu}{n} b_n\right)^{-n} \leq \int_0^{\infty} \left(1 + \frac{t}{n} b_n\right)^{-n} dt = \frac{n}{(n-1)b_n} \leq \frac{2}{b_n}, \qquad n \geq 2,$$

which proves (3).

Proof of the lemma. For any $\delta > 0$, let

$$a_n = \left(\sum_{\nu=1}^{p_n} + \sum_{\nu=p_n+1}^{\infty}\right)\left(1 + \frac{\nu}{n} b_n\right)^{-n} = S_n + T_n,$$

where $p_n = [\delta_n/b_n] + 1$. To estimate S_n, we observe that for $1 \leq \nu \leq p_n$ we have

$$\frac{\nu}{n} b_n \leq \frac{p_n}{n} b_n \leq \left(\frac{\delta_n}{b_n} + 1\right)\frac{b_n}{n} = \delta + \frac{b_n}{n} \leq 2\delta, \qquad n \geq N_\delta,$$

since $\{b_n\}$ is bounded. Using this inequality and $\log(1+t) \geq t/(1+t)$, $t \geq 0$, we find that

$$n \log\left(1 + \frac{\nu}{n} b_n\right) \geq \frac{b_n}{1 + \frac{\nu}{n} b_n} \geq \frac{\nu b_n}{1 + 2\delta},$$

i.e.,

$$\left(1 + \frac{\nu}{n} b_n\right)^{-n} \leq e^{-b_n/(1+2\delta)}$$

for all $1 \leq \nu \leq p_n$ and $n \geq N_\delta$. Thus,

$$S_n = \sum_{\nu=1}^{p_n} \left(1 + \frac{\nu}{n} b_n\right)^{-n} \leq \sum_{\nu=1}^{p_n} e^{-\nu b_n/(1+2\delta)} \leq \sum_{\nu=1}^{\infty} e^{-\nu b_n/(1+2\delta)},$$

i.e.,

(9) $$S_n \leq (e^{b_n/(1+2\delta)} - 1)^{-1}, \qquad n \geq N_\delta.$$

On the other hand,

$$T_n = \sum_{\nu=p_n+1}^{\infty} \left(1 + \frac{\nu}{n} b_n\right)^{-n} \leq \int_{p_n}^{\infty} \left(1 + \frac{t}{n} b_n\right)^{-n} dt \leq \int_{\delta n/b_n}^{\infty} \left(1 + \frac{t}{n} b_n\right)^{-n} dt,$$

i.e.,

(10) $$T_n \leq \frac{n}{(n-1)b_n} (1 + \delta)^{-n+1} \leq \frac{2}{b_n} (1 + \delta)^{-n+1}, \qquad n \geq 2.$$

Finally, (4) follows from (9) and (10).

The proposers show, by using the implicit function theorem, that x_n has the asymptotic expansion

$$x_n = \frac{n}{\log 2} + c_0 + \frac{c_1}{n} + \cdots,$$

where $c_0 = -\frac{3}{2}$, $c_1 = \frac{25}{15} \log 2$.

Problem 71-20*, *An Asymptotic Expansion*, by P. BARRUCAND (Université de Paris, France).

If

$$\frac{xe^{-kh(x)}}{(1 + x) \log (1 + x)} = \sum_{n=0}^{\infty} P_n(k)x^n,$$

where

$$h(x) = \frac{1}{\log (1 + x)} - \frac{1}{2} - \frac{1}{x},$$

find an asymptotic expansion of $P_n(k)$ for $k \to \infty$.

Solution by OTTO G. RUEHR (Michigan Technological University).[3]

Consider the more general expansion problem

$$a(x)F[kb(x)] = \sum_{n=0}^{\infty} P_n(k)x^n,$$

where

$$a(x) = \sum_{i=0}^{\infty} a_i x^i, \quad F(y) = \sum_{m=0}^{\infty} F_m y^m \quad \text{and} \quad b(x) = \sum_{n=1}^{\infty} b_n x^n.$$

Employing the series for the integral powers of $b(x)$:

$$[b(x)]^m = \sum_{n=m}^{\infty} b_{m,n} x^n, \qquad\qquad b_{1,n} = b_n,$$

we obtain

$$a(x)F[kb(x)] = \sum_{i=0}^{\infty} a_i x^i \sum_{m=0}^{\infty} k^m F_m \sum_{n=m}^{\infty} b_{m,n} x^n$$

$$= \sum_{i=0}^{\infty} a_i x^i \sum_{n=0}^{\infty} x^n \sum_{m=0}^{n} k^m F_m b_{m,n}$$

$$= \sum_{n=0}^{\infty} x^n \sum_{i=0}^{n} a_{n-i} \sum_{m=0}^{i} k^m F_m b_{m,n}$$

$$= \sum_{n=0}^{\infty} x^n \sum_{m=0}^{n} F_m k^m \sum_{i=m}^{n} a_{n-i} b_{m,i}.$$

---

[3] A previous solution appears in the July, 1972 issue.

Comparing coefficients of powers of x shows that $P_n(k)$ is the polynomial

$$P_n(k) = \sum_{m=0}^{n} F_m k^m \sum_{i=m}^{n} a_{n-i} b_{m,i} = \sum_{j=0}^{n} k^{n-j} F_{n-j} \sum_{i=n-j}^{n} a_{n-i} b_{n-j,i}$$

$$= \sum_{j=0}^{n-1} k^{n-j} F_{n-j} \sum_{s=0}^{j} a_{j-s} b_{n-j,n-j+s} + a_n.$$

The determination of the behavior of $P_n(k)$ for large k requires the calculation of $b_{n,n+p}$ for small values of p and arbitrary n. To this end, we use the following lemma which can be established by induction.

LEMMA. *Let $b_{m,n}$ be defined as above. Then*

$$b_{n,n+p} = \sum_{j=1}^{p} \binom{n}{j} b_1^{n-j} R_{j,p},$$

where

$$R_{1,p} = b_{p+1} = b_{1,p+1} \quad and \quad R_{j,p} = \sum_{s=j-1}^{p-1} R_{j-1,s} R_{1,p-s}, \qquad p \ge j \ge 2,$$

i.e.,

$$[R(x)]^j = \sum_{p=j}^{\infty} R_{j,p} x^p,$$

where

$$R(x) = \frac{b(x) - b_1 x}{x}.$$

Using the lemma, we obtain

$$P_n(k) = k^n \sum_{j=0}^{n-1} \frac{F_{n-j}}{k^j} \left\{ a_j b_1^{n-j} + \sum_{r=1}^{j} \binom{n-j}{r} b_1^{n-j-r} Q_{j,r} \right\} + F_0 a_n,$$

where

$$Q_{j,r} = \sum_{s=r}^{j} a_{j-s} R_{r,s}.$$

It can be shown that the Q's are generated by

$$a(x)[R(x)]^r = a(x)\left[\frac{b(x) - b_1 x}{x}\right]^r = \sum_{j=r}^{\infty} Q_{j,r} x^j.$$

Returning to the special case of the problem, we find that

$$F_n = \frac{(-1)^n}{n!}, \quad a_0 = 1,$$

$$a_i = (-1)^i + \sum_{j=1}^{i} \frac{(-1)^{j-1} a_{i-j}}{j+1}, \qquad i \ge 1,$$

$$b_i = a_i + a_{i+1}, \quad i \ge 1,$$

$$b_1 = -\tfrac{1}{12},$$

$$P_n(k) \simeq \frac{k^n}{(12)^n n!}\left[1 - \frac{6n^2}{k} + \frac{6n(n-1)}{5k^2}[50 + 68(n-2) + 15(n-2)(n-3)]\right.$$

$$-\frac{36n(n-1)(n-2)}{5k^3}[90 + 142(n-3) + 53(n-3)(n-4)$$

$$\left. + 5(n-3)(n-4)(n-5)]\cdots\right].$$

Asymptotic Behavior of a Sequence

Problem 79-5 *by* L. ERLEBACH *and* O. RUEHR (Michigan Technological University).
A sequence $\{a_n\}$ is defined as follows:

$$a_n = n(n-1)a_{n-1} + \tfrac{1}{2}n(n-1)^2 a_{n-2}, \qquad n \geq 3; \quad a_1 = 0, \quad a_2 = 1.$$

Determine how a_n behaves for large n. The problem arose in counting Hamiltonian cycles for bipartite graphs.

Solution by DONALD E. KNUTH (Stanford University).
Let $b_n = 2a_{n+2}/(n+2)!(n+1)!$, so that the recurrence is transformed into $b_n = b_{n-1} + b_{n-2}/2n$ for $n \geq 2$, $b_0 = b_1 = 1$. The generating function $B(z) = \sum_{n\geq 0} b_n z^n$ satisfies $2B'(z) = 2zB'(z) + (2+z)B(z)$; hence we find $B(z) = e^{-z/2}(1-z)^{-3/2}$. Writing $e^{-z/2} = e^{-1/2}\sum_{m\geq 0}(1-z)^m/2^m m!$, we have therefore $b_n = \sum_{m\geq 0} e^{-1/2}f_m(n)/2^m$ where $f_m(n)$ is the coefficient of z^n in $(1-z)^{m-3/2}$. Now

$$f_m(n)/f_{m+1}(n) = \binom{m-3/2}{n}\Big/\binom{m-1/2}{n} = (2n+1-2m)/(1-2m);$$

so $f_m(n) = O(n^{-m}f_0(n))$ for fixed m as $n \to \infty$, and we have found an asymptotic series for b_n. In particular, $f_0(n) = (2n+1)f_1(n)$ and

$$f_1(n) = \binom{2n}{n}2^{-2n} = (\pi n)^{-1/2}(1 - \tfrac{1}{8}n^{-1} + O(n^{-2})),$$

hence $b_n = (\pi e n)^{-1/2}(2n + \tfrac{5}{4} + O(n^{-1}))$. The original numbers satisfy $a_n \sim n!^2\sqrt{\pi e n} \sim 2\sqrt{\pi}\, n^{2n+1/2}e^{-2n-1/2}$.

M. KLAWE (University of Toronto) *and* W. PULLEYBLANK *and* W. VOLLMER-HAUS (University of Calgary) established asymptotic bounds for the more general problem $s_1 = 0$, $s_2 = \alpha > 0$; $s_n = f(n)s_{n-1} + g(n)s_{n-2}$, $n \geq 3$ where f and g are positive real-valued functions.

The proposers' solution uses a generating function to get the terminating series form

$$a_n = \frac{n!(n-1)\Gamma(n-\tfrac{1}{2})}{\Gamma(\tfrac{1}{2})}\,{}_1F_1[2-n;\tfrac{3}{2}-n;-\tfrac{1}{2}]$$

from which, by means of Kummer's formula, the series developed above by Knuth is obtained:

$$a_n = \frac{(2n)!(2n-2)}{2^{2n}e^{1/2}(2n-1)}\,{}_1F_1[-\tfrac{1}{2};\tfrac{3}{2}-n;\tfrac{1}{2}].$$

This representation is exact, asymptotic, and convergent.

Asymptotic Behavior of an n-fold Integral

Problem 82-20, *by* A. M. ODLYZKO, L. A. SHEPP *and* D. SLEPIAN (Bell Laboratories, Murray Hill, NJ).

Determine the asymptotic behavior of the n-fold integral

$$I_n = \int_0^x \cdots \int_0^x \exp\left(-\sum_{j=1}^n x_j^2 + \frac{1}{2n}\left(\sum_{j=1}^n x_j\right)^2\right) dx_1 \cdots dx_n$$

as $n \to \infty$.

Solution by J. GRZESIK (TRW, Energy Research Center, Redondo Beach, CA).

The identity

(1) $$\exp(x^2) = \pi^{-1/2} \int_{-\infty}^\infty dk \exp[-(k^2 + 2kx)],$$

applied to the linear sum in the exponent of I_n, allows one to reduce that latter into the form:

(2) $$I_n = \pi^{-1/2}\left(\frac{\pi^{1/2}}{2}\right)^n (2n)^{1/2}[F_+(n) + F_-(n)],$$

with

(3) $$F_\pm(n) = \int_0^\infty dk \exp(-nk^2)[1 \pm \mathrm{erf}(k)]^n.$$

For large n, the contribution of $F_-(n)$ becomes negligible, whereas that of $F_+(n)$ is estimated by saddle-point integration. The saddle point per se is found numerically to lie at $k_{sp} = 0.3578345 \cdots$, and the net outcome of such approximation yields the diverging result

(4) $$I_n \underset{n\to\infty}{\sim} \alpha \exp(\beta n),$$

with $\alpha = 1.1500399 \cdots$, $\beta = 0.0784462 \cdots$. The growth with n which (4) attributes to I_n is consistent with an a priori upper bound

(5) $$I_n \leqq \exp(\gamma n),$$

$\gamma = 0.2257913 \cdots$, set by use of the Cauchy–Schwarz inequality.

Pannatoni and Ruehr considered the more general integral

$$I(n, \sigma) = \int_0^\infty \cdots \int_0^\infty \exp\left\{-\sum_{j=1}^n x_j^2 + \frac{\sigma^2}{n}\left(\sum_{j=1}^n x_j\right)^2\right\} dx_1 \cdots dx_n.$$

If $0 < \sigma < 1$, then

$$I(n, \sigma) \sim \left(\frac{-2}{h''(u_0, \sigma)}\right)^{1/2} \exp\{n\, h(u_0, \sigma)\} \quad \text{as } n \to \infty,$$

where

$$h(u, \sigma) = -(1 - \sigma^2)\, u^2 + \ln\left\{\frac{\sqrt{\pi}}{2}\, \text{erfc}\,(-\sigma u)\right\},$$

$$h''(u, \sigma) = \left(\frac{\partial^2}{\partial u^2}\right) h(u, \sigma).$$

The parameter u_0 depends implicitly on σ; it is given by

$$u_0 = \frac{z_0}{\sigma},$$

where z_0 is the unique real solution of the equation

$$z_0 \exp(z_0^2)\, \text{erfc}\,(-z_0) = \frac{\sigma^2}{(1 - \sigma^2)\, \sqrt{\pi}}.$$

The problem posed by Odlyzko, Shepp and Slepian corresponds to the case $\sigma = 1/\sqrt{2}$. Ruehr calculates $\bar{\sigma}$ so that $h(u_0, \bar{\sigma}) = 0$ and hence $0 < I(\infty, \bar{\sigma}) < \infty$.

Remark by the proposers. Let $X_1, \cdots, X_n$ be a zero mean Gaussian vector with covariance $EX_i^2 = 1$, $EX_iX_j = \rho$, $i \neq j$, where to obtain positive definiteness, $-1/(n-1) \leq \rho \leq 1$, and let $P_n(\rho) = P(X_i > 0, i = 1, \cdots, n)$. It is easy to see that $I_n = 2^{1/2}\pi^{n/2} P_n(1/(n + 1))$. The identity used to express I_n as a single integral comes alternately from the observation that if $\eta_0, \cdots, \eta_n$ are standard normal, then $X_i = \sqrt{1 - \rho}\,\eta_i + \sqrt{\rho}\,\eta_0$, $i = 1, \cdots, n$, realize the above vector, for $\rho \geq 0$. The present problem is concerned with $\rho = 1/(n + 1) > 0$. For $\rho < 0$, however, $\sqrt{\rho}$ is not real and $P_n(\rho)$ must be found by analytic continuation from $\rho > 0$. The corresponding problem for a general covariance is widely believed to be intractable, i.e., not reducible to a simple formula or even a single integral.

The Complexity of the Standard Form of an Integer

Problem 89-11*, *by* K. V. LEVER (Brunel University, Uxbridge, United Kingdom, and G. E. C. Hirst Research Centre, Wembley, Middlesex, U.K.).

Adaptive equalization of a time-varying linear communication channel requires the implementation of the approximate inverse of a discrete convolution operator. Under certain conditions (the single-precursor channel model) it can be shown [1], [2] that the complexity of the deconvolution operator of span n (with equalizer impulse response coefficients $\{e_{n-1}, \cdots, e_2, e_1, 1\}$) may be minimized by means of a "divide-and-conquer" principle based on the structure of the subgroup lattice of the cyclic group $\mathbb{C}_n$. When n is composite with standard form (product of powers of primes)

$$n = p_1^{m_1} p_2^{m_2} \cdots p_k^{m_k}$$

the equalizer deconvolution can be decomposed into a cascade of M subequalizers in series, where

$$M = m_1 + m_2 + \cdots + m_k.$$

This canonical form corresponds to a convolutional factorization of the original impulse response given above, and consists of m_1 subsystems with impulse responses having $p_1 - 1$ nonunity coefficients, m_2 with $p_2 - 1$ coefficients, and so on. The *complexity*, defined as the total number of nonunity coefficients, is therefore

$$C(n) = m_1(p_1 - 1) + m_2(p_2 - 1) + \cdots + m_k(p_k - 1).$$

The total number of different equalizer architectures (given by the structure of the K-module induced by the subgroup algebra) can be shown to be the multinomial coefficient

$$D(n) = M!/(m_1!m_2! \cdots m_k!).$$

Linearity allows the permutation of the M subsystems into an arbitrary order: the total number of equivalent minimal configurations is therefore

$$E(n) = M!^2/(m_1!m_2! \cdots m_k!).$$

These number-theoretic functions vary too erratically to permit the estimation of their asymptotic behaviour by elementary techniques, but in [3] it is shown that both M and k are of *average* order $\log(\log(n))$. What are the average (or normal) orders of $C(n)$, $D(n)$, and $E(n)$?

REFERENCES

[1] K. V. Lever, F. M. Clayton, and G. J. Janacek, *Minimisation of architectural complexity in the design of adaptive equalisers for digital radio systems*, in Mathematical Modelling for Information Technology, A. O. Moscardini and E. H. Robson, eds., Ellis–Horwood, London, 1988, pp. 60–78.

[2] M. T. Dudek, M. A. Kitching, S. A. Baigent, F. M. Clayton, M. L. Fielding, and K. V. Lever, *The series equalizer: a new architecture for adaptive equalization in digital radio systems*, in Proc. International Conference on Data Communications Technology, National Institute of Higher Education, Limerick, Ireland, September 1988, pp. 127–136.

[3] G. H. Hardy and E. M. Wright, *An Introduction to the Theory of Numbers*, Fourth edition, Oxford University Press, Oxford, London, 1975, pp. 354–358.

Supplementary References
Numerical Analysis

[1] F. S. Acton, *Analysis of Straight-Line Data*, Dover, N.Y., 1959.

[2] ———, *Numerical Methods that Work*, Harper and Row, N.Y., 1970.

[3] J. H. Ahlberg, E. N. Nilson and J. L. Walsh, *The Theory of Splines and their Applications*, Academic, N.Y., 1967.

[4] W. F. Ames, *Nonlinear Partial Differential Equations in Engineering, II*, Academic, N.Y., 1972.

[5] ———, *Numerical Methods for Partial Differential Equations*, Barnes & Noble, N.Y., 1969.

[6] L. Auslander and R. Tolimieri, *"Is computing with the finite Fourier transform pure or applied mathematics?"* Bull. AMS (1979) pp. 847-897.

[7] P. B. Bailey, L. F. Shampine and D. E. Waltman, *Nonlinear Two Point Boundary Value Problems*, Academic, N.Y., 1968.

[8] R. Bellman, K. L. Cooke and J. A. Lockette, *Algorithms, Graphs, and Computers*, Academic, N.Y., 1970.

[9] I. S. Berezin and N. P. Zhidkov, *Computing Methods, I, II*, Pergamon, Oxford, 1965.

[10] G. H. Brown, Jr., *"On Halley's variation of Newton's method,"* AMM (1977) pp. 726-728.

[11] R. L. Burden, J. D. Faires and A. C. Reynolds, *Numerical Analysis*, Prindle, Weber & Schmidt, Boston, 1978.

[12] E. W. Cheney, *Introduction to Approximation Theory*, McGraw-Hill, N.Y., 1966.

[13] A. M. Cohen, ed., *Numerical Analysis*, McGraw-Hill, N.Y., 1973.

[14] L. Collatz, *Numerical Treatment of Differential Equations*, Springer-Verlag, Berlin, 1960.

[15] E. T. Copson, *Asymptotic Expansions*, Cambridge University Press, Cambridge, 1965.

[16] G. Dahlquist and A. Bjorck, *Numerical Methods*, Prentice-Hall, N.J., 1974.

[17] P. J. Davis, *Interpolation and Approximation*, Blaisdell, Waltham, 1963.

[18] P. J. Davis and P. Rabinowitz, *Numerical Integration*, Blaisdell, Waltham, 1967.

[19] N. G. De Bruijn, *Asymptotic Methods in Analysis*, Interscience, N.Y., 1961.

[20] F. De Kok, *"On the method of stationary phase for multiple integrals,"* SIAM J. Math. Anal. (1971) pp. 76-104.

[21] R. B. DINGLE, *Asymptotic Expansions: Their Derivation and Interpretation*, Academic, N.Y., 1973.

[22] A. ERDELYI, *Asymptotic Expansions*, Dover, N.Y., 1956.

[23] A. FINBOW, *"The bisection method: A best case analysis,"* AMM (1985) pp. 285–286.

[24] G. E. FORSYTHE, *"Generation and use of orthogonal polynomials for data fitting with a digital computer,"* SIAM J. Appl. Math. (1957) pp. 74–88.

[25] G. E. FORSYTHE, M. A. MALCOLM AND C. B. MOLER, *Computer Methods for Mathematical Problems*, Prentice-Hall, N.J., 1977.

[26] G. E. FORSYTHE AND W. R. WASOW, *Finite Difference Methods for Partial Differential Equations*, Wiley, N.Y., 1960.

[27] M. R. GAREY AND D. S. JOHNSON, *"Approximation algorithms for combinatorial problems: An annotated bibliography,"* in J. F. Traub, ed., *Algorithms and Complexity: New Directions and Recent Results*, Academic, N.Y., 1976, pp. 41–52.

[28] ———, *Computers and Intractability: A Guide to the Theory of NP-Completeness*, Freeman, San Francisco, 1979.

[29] J. M. HAMMERSLEY, *"Monte Carlo methods for solving multivariable problems,"* Ann. N.Y. Acad. Sci. (1960) pp. 844–874.

[30] R. W. HAMMING, *Numerical Methods for Scientists and Engineers*, McGraw-Hill, N.Y., 1973.

[31] D. R. HARTREE, *Numerical Analysis*, Clarendon Press, Oxford, 1958.

[32] P. HENRICI, *Elements of Numerical Analysis*, Wiley, N.Y., 1964.

[33] ———, *Essentials of Numerical Analysis with Pocket Calculator Demonstration*, Wiley, N.Y., 1982.

[34] F. B. HILDEBRAND, *Introduction to Numerical Analysis*, McGraw-Hill, N.Y., 1974.

[35] A. S. HOUSEHOLDER, *Principles of Numerical Analysis*, McGraw-Hill, N.Y., 1953.

[36] E. ISAACSON AND H. B. KELLER, *Analysis of Numerical Methods*, Wiley, N.Y., 1966.

[37] L. V. KANTOROVICH AND V. I. KRYLOV, *Approximate Methods of Higher Analysis*, Noordhoff, Groningen, 1958.

[38] S. KARLIN AND W. J. STUDDEN, *Tchebycheff Systems: With Applications in Analysis and Statistics*, Interscience, N.Y., 1966.

[39] D. KATZ, *"Optimal quadrature points for approximating integrals when function values are observed with error,"* MM (1984) pp. 284–290.

[40] H. B. KELLER, *Numerical Methods for Two-Point Boundary Value Problems*, Blaisdell, Waltham, 1968.

[41] M. S. KLAMKIN, *"Transformation of boundary value problems into initial value problems,"* J. Math. Anal. Appl. (1970) pp. 308–330.

[42] D. E. KNUTH, *"Algorithms,"* Sci. Amer. (1977) pp. 63–80, 148.

[43] ———, *The Art of Computer Programming I, II, III*, Addison-Wesley, Reading, 1973.

[44] Z. KOPAL, *Numerical Analysis*, Chapman & Hall, London, 1961.

[45] C. LANCZOS, *Applied Analysis*, Prentice-Hall, N.J., 1956.

[46] G. G. LORENTZ, *Approximation of Functions*, Holt, Rinehart and Winston, N.Y., 1966.

[47] J. E. MCKENNA, *"Computers and experimentation in mathematics,"* AMM (1972) pp. 294–295.

[48] G. MEINHARDUS, *Approximations of Functions: Theory and Numerical Methods*, Springer-Verlag, N.Y., 1967.

[49] G. H. MEYER, *Initial Value Methods for Boundary Value Problems*, Academic, N.Y., 1973.

[50] G. MIEL, *"Calculator calculus and roundoff error,"* AMM (1980) pp. 243–351.

[51] J. J. H. MILLER, ed., *Topics in Numerical Analysis I, II, III*, Academic, N.Y., 1972, 1974 and 1976.

[52] T. Y. NA, *Computational Methods in Engineering Boundary Value Problems*, Academic, N.Y., 1979.

[53] T. R. F. NONWEILER, *Computational Mathematics: An Introduction to Numerical Approximation*, Horwood, Chichester, 1984.

[54] F. W. J. OLVER, *Asymptotics and Special Functions*, Academic, N.Y., 1974.

[55] J. M. ORTEGA AND W. C. RHEINBOLDT, *Iterative Solutions of Nonlinear Equations in Several Variables*, Academic, N.Y., 1970.

[56] A. M. OSTROWSKI, *Solution of Equations and Systems of Equations*, Academic, N.Y., 1966.

[57] M. J. D. POWELL, *Approximation Theory and Methods*, Cambridge University Press, Cambridge, 1981.

[58] R. D. RICHTMYER AND K. W. MORTON, *Difference Methods for Initial Value Problems*, Interscience, N.Y., 1964.

[59] T. R. RIVLIN, *An Introduction to the Approximation of Functions*, Dover, N.Y., 1969.

[60] S. M. ROBERTS AND J. G. SHIPMAN, *Two-Point Boundary Value Problems: Shooting Methods*, Elsevier, N.Y., 1972.

[61] P. ROZSA, ed., *Numerical Methods*, North-Holland, Amsterdam, 1980.

[62] A. SARD AND S. WEINTRAUB, *A Book of Splines*, Wiley, N.Y., 1971.

[63] L. F. SHAMPINE AND M. K. GORDON, *Computer Solutions of Ordinary Differential Equations: The Initial Value Problem*, Freeman, San Francisco, 1975.

[64] O. D. SMITH, *Numerical Solution of Partial Differential Equations: Finite Difference Methods*, Clarendon Press, Oxford, 1978.

[65] S. K. STEIN, *"The error of the trapezoidal method for a concave curve,"* AMM (1976) pp. 643–645.

[66] F. STUMMEL, *"Rounding error analysis of elementary numerical algorithms,"* Computing, Suppl. (1980) pp. 169–195.

[67] J. L. SYNGE, *The Hypercircle Method in Mathematical Physics: A Method for the Approximate Solution of Boundary Value Problems*, Cambridge University Press, Cambridge, 1957.

[68] F. SZIDAROVSZKY AND S. YAKOWITZ, *Principles and Processes of Numerical Analysis*, Plenum Press, N.Y., 1976.

[69] J. TODD, ed., *A Survey of Numerical Analysis*, McGraw-Hill, N.Y., 1962

[70] H. S. WALL, *"A modification of Newton's method,"* AMM (1948) pp. 90–94.

[71] D. M. YOUNG AND R. D. GREGORY, *A Survey of Numerical Mathematics I, II*, Addison-Wesley, Reading, 1973.

13. INEQUALITIES

Problem 60–5, A Resistor Network Inequality, by ALFRED LEHMAN (University of Wisconsin).

Consider the $m \times n$ series-parallel resistor network of Figure 1. The driving-point resistance between the terminals A and B is given by

$$R_{AB} = \left\{ \sum_{i=1}^{m} \left[\sum_{j=1}^{m} r_{ij} \right]^{-1} \right\}^{-1}.$$

The addition of short-circuits to the network of Figure 1 produces Figure 2. The driving-point resistance between terminals C and D is given by

$$R_{CD} = \sum_{q=1}^{n} \left\{ \sum_{p=1}^{m} [r_{pq}]^{-1} \right\}^{-1}.$$

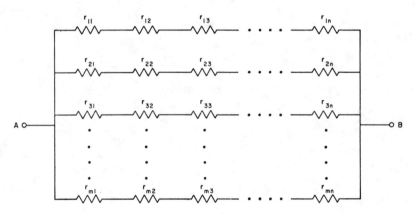

FIGURE 1

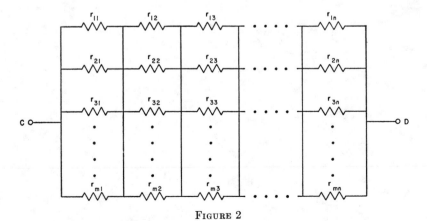

FIGURE 2

Since short-circuits in a resistor network cannot increase resistance,

$$R_{AB} \geqq R_{CD},$$

or equivalently

$$\sum_{i=1}^{m} \sum_{q=1}^{n} \left\{ \sum_{j=1}^{n} \sum_{p=1}^{m} \frac{r_{ij}}{r_{pq}} \right\}^{-1} \leqq 1.$$

Give a direct proof of the latter inequality where $r_{ij} \geqq 0$.

Solution by FAZLOLLAH REZA (Syracuse University).

We shall first prove the inequality $R_{AB} \geqq R_{CD}$ for $n = m = 2$, and then extend it by induction to the general case. The proof is based on the concavity of the function $\phi(x, y) = (x^{-1} + y^{-1})^{-1}$, $(x, y \geqq 0)$. That ϕ is concave follows from

$$Q = \phi_{xx} u^2 + 2\phi_{xy} uv + \phi_{yy} v^2 = -2(uy - vx)^2 / (x + y)^3.$$

Whence,

$$2\phi \left(\frac{x_1 + x_2}{2}, \frac{y_1 + y_2}{2} \right) \geqq \phi(x_1, y_1) + \phi(x_2, y_2)$$

and

$$2 \left\{ \frac{2}{r_{11} + r_{12}} + \frac{2}{r_{21} + r_{22}} \right\}^{-1} \geqq \left\{ \frac{1}{r_{11}} + \frac{1}{r_{21}} \right\}^{-1} + \left\{ \frac{1}{r_{12}} + \frac{1}{r_{22}} \right\}^{-1},$$

(note that $\phi(kx, ky) = k\phi(x, y)$). This establishes the inequality for $n = m = 2$. To extend the proof, it is first shown that the inequality is true for any positive integer $n > 2$. The non-negative numbers r_{ij} can be written in the matrix form

$$\left\| \begin{array}{cccc} r_{11} & r_{12} & \cdots & r_{1n} \\ r_{21} & r_{22} & \cdots & r_{2n} \end{array} \right\|.$$

To show that the inequality is valid for $m = 2, n = 3$, we note that the inequality is already valid for

$$\left\| \begin{array}{cc} r_{11} + r_{12} & r_{13} \\ r_{21} + r_{22} & r_{23} \end{array} \right\|.$$

On the other hand, it was established that

$$\phi(r_{11} + r_{12}, r_{21} + r_{22}) \geqq \phi(r_{11}, r_{21}) + \phi(r_{12}, r_{22}).$$

Thus the theorem holds for $m = 2$, $n = 3$ and by induction for all $n \geqq 2$.

The theorem holds for the $2 \times n$ array

$$\left\| \begin{array}{cccc} (r_{11}^{-1} + r_{21}^{-1})^{-1} & (r_{12}^{-1} + r_{22}^{-1})^{-1} & \cdots & (r_{1n}^{-1} + r_{2n}^{-1})^{-1} \\ r_{31} & r_{32} & \cdots & r_{3n} \end{array} \right\|.$$

Due to the concavity of ϕ, the first row of the latter array can be replaced by the following suitable $2 \times n$ array without violating the inequality. That is, the theorem holds for

$$\left\| \begin{array}{cccc} r_{11} & r_{12} & \cdots & r_{1n} \\ r_{21} & r_{22} & \cdots & r_{2n} \\ r_{31} & r_{32} & \cdots & r_{3n} \end{array} \right\|.$$

Continuing by induction, the inequality holds for arbitrary positive integers m and n.

Solution by the proposer.

The inequality

$$\sum_{i=1}^{m} \sum_{q=1}^{n} \left\{ \sum_{j=1}^{n} \sum_{p=1}^{m} \frac{r_{ij}}{r_{pq}} \right\}^{-1} \leqq 1, \qquad\qquad (r_{ij} > 0)$$

is a special case $(\alpha = -1)$ of the Minkowski inequality

$$\left[\sum_{i=1}^{m} \left(\sum_{j=1}^{n} r_{ij} \right)^{\alpha} \right]^{1/\alpha} \underset{\geqq}{\leqq} \sum_{q=1}^{n} \left[\sum_{p=1}^{m} (r_{pq})^{\alpha} \right]^{1/\alpha}, \qquad \alpha \underset{<}{\overset{>}{}} 1 \quad \text{and} \quad r_{ij} \geqq 0.$$

This latter inequality and several proofs are to be found in (3) (see particularly 2.11.4 and 2.11.5 on page 31).

It is the purpose of this solution to discuss the relation between the Minkowski inequality and the concept of a short-circuit in a network of non-linear resistors, each of voltage-current characteristic

$$E_k = (\text{sign } I_k)(r_k |I_k|^{-1/\alpha}), \qquad\qquad r_k > 0$$

(E_k denotes the voltage, I_k the current and r_k the resistance associated with the kth resistor. The constant α is to be the same for all resistors of the network.)

In the linear case $(\alpha = -1)$ where each resistor obeys Ohm's law the result is essentially given by Jeans (4) on pages 320 to 324. Assume a passive network of linear resistors (i.e. obeying Ohm's law $E_k = r_k I_k$) excited by a single unit-current source. Consider the dissipated power $(\sum_k r_k I_k^2)$ resulting from any current distribution obeying Kirchoff's current (continuity) law.* The unique current distribution which also satisfies Kirchoff's voltage law† is the unique distribution which minimizes the dissipated power (Jeans Theorem 357, page 322). Hence consider two such networks, the second being derived from the first by a sequence of short-circuits. The actual current distribution of the first network also satisfies the conditions of Kirchoff's current law for second network. Hence the actual power dissipation in the second network cannot exceed that of the first. Since the networks are excited by a unit current source, the equivalent resistance between the excited terminals is equal to the dissipated power. Hence the terminal-to-terminal resistance of the second network cannot exceed that of the first (essentially Jeans Theorem 359, page 324). The inequality of problem 60–5 then results from a computation of the resistance of the two given series-parallel networks by means of the series and parallel combination rules $a + b$ and $(a^{-1} + b^{-1})^{-1}$.

This result can be generalized to networks of resistors each having a voltage current characteristic of the form $E_k = (\text{sign } I_k)(r_k |I_k|^{-1/\alpha})$ where $r_k > 0$ and $\alpha < 0$, α being the same for all resistors of the network. By Theorems 1 and 3 of Duffin (2) and the remark on page 438 of Birkhoff and Diaz (1) it follows that the unique current distribution which satisfies Kirchoff's current law and minimizes the dissipated power $\sum_k r_k |I_k|^{[1-(1/\alpha)]}$ is the unique current distribution satisfying both of Kirchoff's laws. Since the resulting terminal-to-

---

* The algebraic sum of currents entering any vertex is 0.

† The algebraic sum of resistor voltages around any closed circuit is 0.

terminal behavior of the network is easily shown to have the form $E = (\text{sign } I)(r|I|^{-(1/\alpha)})$, E and I being the applied voltage and current, the network has an equivalent resistance r and this resistance cannot be increased by short-circuits. The Minkowski inequality for $\alpha < 0$ then follows, as in the linear case, from the computation of the resistance of the two networks of problem 60–5 by means of the appropriate series and parallel combination rules $a + b$ and $(a^\alpha + b^\alpha)^{1/\alpha}$.

The previous argument holds where E_k is an increasing function of I_k, that is where $\alpha < 0$. For $\alpha > 0$ it is possible to formulate a theory of *series-parallel* resistor networks which yield the Minkowski inequality by power minimization for $0 < \alpha < 1$ and by power maximization for $\alpha > 1$. The electrical significance of such networks is open to question. Consider, for example, the networks a and b of figure 3. For either, the equivalent resistance between the * marked terminals, calculated by the series and parallel formulas, is 8. This corresponds to equality in the 2×2 case of the Minkowski inequality for $\alpha = \frac{1}{2}$. The result can also be obtained from the Kirchoff's law current distribution $I_k = \frac{1}{2}$, $k = 1, 2, 3, 4$. If non-series-parallel networks are considered the Kirchoff's law current distribution need not be unique. For example, in the network c of figure 3, the distribution $I_1 = I_4 = I_5 = \frac{1}{3}$,[a] $I_2 = I_3 = \frac{2}{3}$ satisfies both of Kirchoff's laws and yields a resistance of 45/4 between the * marked terminals. But by symmetry the same can be said for the current distribution $I_2 = I_3 = I_5 = \frac{1}{3}$,[a] $I_1 = I_4 = \frac{2}{3}$. Furthermore the I's may be chosen so as to satisfy Kirchoff's current law and have arbitrarily large magnitudes. Hence the dissipated power may be made arbitrarily small. In the case of series-parallel networks however the restriction of current flows to a single direction insures that a unique current distribution appropriately maximizes or minimizes the dissipated power.

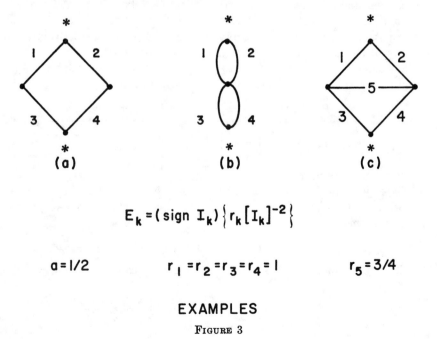

$$E_k = (\text{sign } I_k)\left\{r_k[I_k]^{-2}\right\}$$

$$a = 1/2 \qquad\qquad r_1 = r_2 = r_3 = r_4 = 1 \qquad\qquad r_5 = 3/4$$

EXAMPLES

FIGURE 3

---

[a] With the proper orientation for I_5.

(1) G. BIRKHOFF and J. B. DIAZ, *Non-linear network problems*, Quart. of Appl. Math., vol. 13 (1955) pp. 431–443.

(2) R. J. DUFFIN, *Non-linear networks II A*, Bull. Amer. Math. Soc., vol. 53 (1947) pp. 963–971.

(3) G. H. HARDY, J. E. LITTLEWOOD, and G. PÓLYA, *Inequalities*, Cambridge, 1934.

(4) J. JEANS, *The mathematical theory of electricity and magnetism*, Cambridge, fifth ed. 1925.

Editorial note. It is to be noted that the Reza solution yields a straight-forward easily remembered proof of the Minkowski inequality. Also by starting with the function

$$\phi(x, y) = \{x^\alpha + y^\alpha\}^{1/\alpha}$$

with general α, the inductive proof of Reza (or the similar inductive proof of reference 3, pp. 38–39) then yields the Minkowski inequality for all values of the exponent.

Problem 68–1, A Network Inequality, by J. C. TURNER and V. CONWAY (Huddersfield College of Technology, England).

If $p + q = 1, 0 < p < 1$, and m, n are positive integers > 1, give a direct analytic proof of the inequality

$$(1 - p^m)^n + (1 - q^n)^m > 1.$$

The problem had arisen in the comparison of the reliabilities R_A and R_B of the two systems in Fig. 4. Here,

$$R_A = 1 - (1 - p^m)^n \quad \text{and} \quad R_B = (1 - q^n)^m,$$

where p denotes the probability of any component operating and it is assumed that all the components operate independently. Reliability is defined as the probability that the system operates successfully under the given conditions. A system is said to operate successfully if at least one path is made from input to output, when all the components are activated simultaneously.

Since the set of all possible paths through the network A is a subset of the set for network B, it is clear that the above inequality must hold.

Editorial note. Particular nice special cases of the above inequality are

$$\left\{1 - \left(\frac{1}{2}\right)^n\right\}^m + \left\{1 - \left(\frac{1}{2}\right)^m\right\}^n > 1, \qquad \frac{1}{2^m} + \frac{1}{2^{1/m}} < 1,$$

the latter inequality being known.

The original inequality can also be shown to hold for nonintegral $m, n > 1$.

Other related inequalities can be obtained by introducing one line of short-circuits at a time, e.g.,

$$1 - (1 - p^m)^n \leq \{1 - (1 - p^r)^n\}\{1 - (1 - p^{m-r})^n\} \leq (1 - q^n)^m.$$

Also, if the probability of each component operating can be different, we obtain

$$\prod_{i=1}^{n}\left\{1 - \prod_{j=1}^{m} p_{ij}\right\} + \prod_{j=1}^{m}\left\{1 - \prod_{i=1}^{n} q_{ij}\right\} > 1.$$

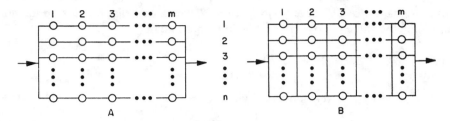

FIG. 4

For a similar derivation of another network inequality, see Problem 60–5, this Review, 4 (1962), pp. 150–155. [M.S.K.]

Solution by L. CARLITZ (Duke University).

We shall show that, for $n > 1, m > 1$,

(1)
$$\prod_{i=1}^{n}\left\{1 - \prod_{j=1}^{m} p_{ij}\right\} + \prod_{j=1}^{m}\left\{1 - \prod_{i=1}^{n} q_{ij}\right\} > 1,$$

where

$$p_{ij} + q_{ij} = 1, \qquad 0 < p_{ij} < 1, \qquad i = 1, \cdots, n, \quad j = 1, \cdots, m.$$

To begin with, we show that

(2)
$$(a + b - ab)(c + d - cd) > ac + bd - abcd,$$

where

$$0 < a < 1, \quad 0 < b < 1, \quad 0 < c < 1, \quad 0 < d < 1.$$

Indeed,

$$(a + b - ab)(c + d - cd) - (ac + bd - abcd)$$
$$= ad + bc - ab(c + d) - cd(a + b) + 2abcd$$
$$= ad(1 - b - c + bc) + bc(1 - a - d + ad)$$
$$= ad(1 - b)(1 - c) + bc(1 - a)(1 - d) > 0.$$

This proves (2).

If in (2) we replace a, b, c, d by $1 - a, 1 - b, 1 - c, 1 - d$, respectively, we obtain

$$(1 - ab)(1 - cd) + (1 - a'c')(1 - b'd') > 1,$$

where

$$a' = 1 - a, \qquad b' = 1 - b, \qquad \text{etc.}$$

This evidently proves (2) when $n = m = 2$.

We now show that, for $n \geq 2$,

(3)
$$\prod_{i=1}^{n}(1 - p_{i1}p_{i2}) + \left\{1 - \prod_{i=1}^{n} q_{i1}\right\}\left\{1 - \prod_{i=1}^{n} q_{i2}\right\} > 1.$$

Since we have already proved (3) when $n = 2$, we assume that it holds up to and including the value n. To carry out the induction, we put

(4)
$$\bar{q}_{ij} = \begin{cases} q_{ij}, & 1 \le i < n, \\ q_{nj}q_{n+1,j}, & i = n, \end{cases}$$

and $\bar{p}_{ij} = 1 - \bar{q}_{ij}$. By the inductive hypothesis,

$$\prod_{i=1}^{n} (1 - \bar{p}_{i1}\bar{p}_{i2}) + \left\{1 - \prod_{i=1}^{n} \bar{q}_{i1}\right\}\left\{1 - \prod_{i=1}^{n} \bar{q}_{i2}\right\} > 1,$$

that is,

$$\prod_{i=1}^{n-1} (1 - p_{i1}p_{i2}) \cdot (1 - \bar{p}_{n1}\bar{p}_{n2}) + \left\{1 - \prod_{i=1}^{n+1} q_{i1}\right\}\left\{1 - \prod_{i=1}^{n+1} q_{i2}\right\} > 1.$$

To complete the induction, we need only show that

(5)
$$(1 - p_{n1}p_{n2})(1 - p_{n+1,1}p_{n+1,2}) > 1 - \bar{p}_{n1}\bar{p}_{n2}.$$

Since

(6)
$$\bar{p}_{nj} = 1 - \bar{q}_{nj} = 1 - q_{nj}q_{n+1,j},$$

(5) is equivalent to

$$(1 - p_{n1}p_{n2})(1 - p_{n+1,1}p_{n+1,2}) + (1 - q_{n1}q_{n+1,1})(1 - q_{n2}q_{n+1,2}) > 1,$$

which has already been proved.

We have therefore proved (1) for $m = 2$ and all n: or equivalently for $n = 2$ and all m. We now assume that (1) holds for all m and for all $i = 1, 2, \cdots, n$. We again use the notation (4). By the inductive hypothesis,

$$\prod_{i=1}^{n} \left\{1 - \prod_{j=1}^{m} \bar{p}_{ij}\right\} + \prod_{j=1}^{m} \left\{1 - \prod_{i=1}^{n} \bar{q}_{ij}\right\} > 1,$$

that is,

$$\prod_{i=1}^{n-1} \left\{1 - \prod_{j=1}^{m} p_{ij}\right\} \cdot \left\{1 - \prod_{j=1}^{m} \bar{p}_{nj}\right\} + \prod_{j=1}^{m} \left\{1 - \prod_{i=1}^{n+1} q_{ij}\right\} > 1.$$

To complete the induction, we need only show that

$$\left\{1 - \prod_{j=1}^{m} p_{nj}\right\}\left\{1 - \prod_{j=1}^{m} p_{n+1,j}\right\} > 1 - \prod_{j=1}^{m} \bar{p}_{nj}:$$

in view of (6), this is the same as

$$\left\{1 - \prod_{j=1}^{m} p_{nj}\right\}\left\{1 - \prod_{j=1}^{m} p_{n+1,j}\right\} + \prod_{j=1}^{m} (1 - q_{nj}q_{n+1,j}) > 1.$$

But this has already been proved.

Editorial note. JOEL BRENNER (University of Arizona) in his solution also establishes that for

$$0 < p, q, r < 1, \qquad p + q + r = 1, \qquad m, n, t > 1, \qquad s = mnt,$$

$$\{1 - p^{s/m}\}^m + \{1 - q^{s/n}\}^n + \{1 - r^{s/t}\}^t > 2,$$

or in homogeneous form

$$\{(a + b + c)^{s/m} - a^{s/m}\}^m + \{(a + b + c)^{s/n} - b^{s/n}\}^n$$
$$+ \{(a + b + c)^{s/t} - c^{s/t}\}^t > 2(a + b + c)^s$$

with $a, b, c > 0$.

Correspondingly, the proposed inequality in homogeneous form is

$$\{(a + b)^m - a^m\}^n + \{(a + b)^n - b^n\}^m > (a + b)^{mn},$$

$$m, n > 1, \qquad a, b > 0.$$

Also he notes that the inequalities are reversed if $0 < m, n, t < 1$.

Postscript. See J.L. Brenner, *Some inequalities from switching theory*, J. Comb. Theory 7(1969) 197-205.

Subadditivity of Teletraffic Special Functions

Problem 80-19*, *by* E. ARTHURS AND B. W. STUCK (Bell Laboratories, Murray Hill, NJ).

I. Show that the following inequality holds:

$$\sum_{i=1}^{N} G(S_i, A_i) \geq G\left(\sum_{i=1}^{N} S_i, \sum_{i=1}^{N} A_i\right),$$

where

$$0 \leq A_i < S_i, \qquad S_i = 1, 2, \cdots, \qquad i = 1, \cdots, N,$$

$$G(S, A) = AB(S, A), \qquad B(S, A) = \frac{A^S/S!}{\sum_{k=0}^{S} A^k/k!}.$$

$B(S, A)$ is called the Erlang blocking function in teletraffic theory.

II. Show that the following inequality holds:

$$\sum_{i=1}^{N} H(S_i, A_i) \geq H\left(\sum_{i=1}^{N} S_i, \sum_{i=1}^{N} A_i\right),$$

where

$$0 \leq A_i < S_i, \qquad S_i = 1, 2, \cdots, \qquad i = 1, \cdots, N,$$

$$H(S, A) = \frac{A}{S - A} C(S, A), \qquad C(S, A) = \frac{A^S/(S-1)! \, (S-A)}{\sum_{k=0}^{S-1} A^k/k! + A^S/(S-1)! \, (S-A)}.$$

$C(S, A)$ is called the Erlang delay function in teletraffic theory.

These problems are related, because H and G are related by the identity

$$H(S, A) = \frac{S}{S - A} \frac{G(S, A)}{S - A + G(S, A)},$$

while Erlang's blocking and delay functions are related by

$$C(S, A) = \frac{SB(S, A)}{S - A + AB(S, A)}.$$

The mathematical property we wish to show is that the functions $G(S, A)$ and $H(S, A)$ are subadditive in the above sense.

The physical interpretation of this property for $G(S, A)$ is as follows. Telephone calls arrive to a set of links. We can partition the calls into N groups, intended for N sets of links. Given the average amount of traffic within each group, A_i, and the number of links for each set, S_i, we adopt the following measure of goodness: if a call arrives and finds all links within its group occupied, it is rejected or blocked, and we wish to minimize the average amount of rejection. Although there is a great deal of numerical evidence that one should pool links rather than dedicate links to different call types to minimize total average blocking, there does not appear in the literature any analytic proof of this property. Furthermore, we have checked with numerous colleagues who have worked extensively in the teletraffic area, and none of them have been able to demonstrate analytically that subadditivity holds!

The physical interpretation of the second problem concerning $H(S, A)$ is that one is given a set of processors and work to be executed on these processors. The work can be partitioned into N subsets, and the average amount of work per subset is A_i. $H(S, A)$ is proportional to the mean delay experienced by a cluster with S processors that on the average must handle A units of work. There exists an overwhelming body of evidence that the total mean delay is less when there is a single partition of work versus multiple partitions. We have checked with numerous colleagues who are quite knowledgeable on queueing theory, and to date no one has been able to produce a satisfactory analytic proof of this property!

The interested reader is referred to the literature (e.g., see D. Jagerman, *Some properties of the Erlang loss function*, Bell System Technical Journal, 53 (1974), pp. 525–551, and the references therein).

Editorial note. The two inequalities I, II are proved together with stronger stochastic comparison results by D. R. SMITH and W. WHITT, *Resource sharing for efficiency in traffic systems*, Bell System Tech. J., 60 (1981), pp. 39–55.

Problem 68-15*, A Set of Inequalities, by JAMES D. RILEY (Rolling Hills Estates, California).

If $\{a_i\}$ denotes a set of n positive numbers, determine for which n the inequality

$$\left\{ \sum_{i=1}^{n} a_i^{-2} \right\}^{1/2} \leq \sum_{i=1}^{n} a_i^{-1} - \sum_{i=1}^{n-1} \sum_{j=i+1}^{n} (a_i + a_j)^{-1}$$

$$(1) \qquad + \sum_{i=1}^{n-2} \sum_{j=i+1}^{n-1} \sum_{k=j+1}^{n} (a_i + a_j + a_k)^{-1} - \cdots + (-1)^{n+1} \left\{ \sum_{i=1}^{n} a_i \right\}^{-1}$$

is valid.

For example, if $n = 2$, the inequality is

$$\sqrt{1/a^2 + 1/b^2} \leq 1/a + 1/b - 1/(a + b),$$

which is valid.

Physically, if the a_i denote failure rates for n components in parallel, then the right-hand side of the inequality is the mean time to failure of the system. The left side provides a useful lower bound for those n in which the inequality is valid.

Editorial note. The proposer has determined an n for which the inequality is invalid and notes that it is also valid for $n = 3$ by means of tedious algebraic manipulations.

Partial solution[1] by ANTHONY J. STRECOK (Argonne National Laboratory).

Let $S_n(a_1, a_2, \cdots, a_n)$ denote the right side of (1). Since

$$S_n(1, 1, \cdots, 1) = \sum_{j=1}^{n} (-1)^{j-1} {}_nC_j/j = \int_0^1 \{1 - (1 - x)^n\}/x \, dx$$

$$= \int_0^1 (1 - y^n)/(1 - y) \, dy = \sum_{j=1}^{n} 1/j,$$

inequality (1) is invalid when

(2) $$n^{1/2} > \sum_{j=1}^{n} 1/j.$$

$n = 7$ is the smallest value for which (2) is true. Because $(n + 1)^{1/2} - n^{1/2} > 1/(n + 1)$ for $n \geq 7$, equation (2) also holds for $n + 1$. Consequently (1) cannot be generally valid for $n > 6$.

Let $f_n(a_1, a_2, \cdots, a_n) = [S_n^2(a_1, a_2, \cdots, a_n) - \sum_{i=1}^{n} a_i^{-2}](\sum_{j=1}^{n} a_j)^2$. Inequality (1) is valid whenever $0 \leq f_n(a_1, a_2, \cdots, a_n)$ and $0 \leq S_n(a_1, a_2, \cdots, a_n)$. It is possible to set $a_1 = 1$ without losing generality because both members of inequality (1) can be multiplied by any positive constant. A direct calculation shows that $f_2(1, a_2) = 1$ and $0 < S_n(1, a_2)$.

For $3 \leq n \leq 6$, $f_n(1, a_2, \cdots, a_n)$ attains a minimum at $a_2 = \cdots = a_n = 1$. This appears to be the only minimum for $a_j > 0$, $j = 2, \cdots, n$, though this conjecture does not seem easy to prove rigorously. If this is a unique minimum, then $S_n(1, 1, \cdots, 1) > n^{1/2}$, $f_n(1, a_2, \cdots, a_n) > 0$, and inequality (1) is generally valid when $n \leq 6$.

Subsequent further comment by F.W. Steutel (University of Texas).

Denoting by $M_n = M_n(a_1, a_2, \cdots, a_n)$ the expectation of $\max(X_1, X_2, \cdots, X_n)$ where the X_j are independent and exponentially distributed random variables with means a_j^{-1}, we have

$$M_n = \int_0^\infty x \, dF_n(x) = \int_0^\infty (1 - F_n(x)) \, dx = \int_0^\infty \left\{1 - \prod_{j=1}^{n} (1 - e^{-a_j x})\right\} dx$$

$$= \int_0^\infty \{\sum e^{-a_j x} - \sum\sum e^{-(a_j + a_k)x} + \cdots\} dx,$$

where F_n denotes the distribution function of $\max(X_1, X_2, \cdots, X_n)$. It now follows that the required inequality reads

[1] This work was performed under the auspices of the U.S. Atomic Energy Commission.

(1)
$$M_n \geq \left\{ \sum_{j=1}^{n} a_j^{-2} \right\}^{1/2}.$$

Clearly (1) does not hold generally for large n (not for any $n \geq 7$, in fact). For in the case that $a_j = a$ for all j, we have

$$M_n = \int_0^\infty \{1 - (1 - e^{-ax})^n\} \, dx = \frac{1}{a} \int_0^1 \frac{1 - u^n}{1 - u} \, du = \frac{1}{a} \sum_1^n \frac{1}{j} \sim a^{-1} \log n$$

whereas

$$\left\{ \sum_1^n a_j^{-2} \right\}^{1/2} = a^{-1} \sqrt{n}.$$

It is easy to show that

$$(\min a_j)^{-1} \leq M_n \leq \sum_1^n a_j^{-1},$$

(2)
$$(\max a_j)^{-1} \sum_1^n \frac{1}{k} \leq M_n \leq (\min a_j)^{-1} \sum_1^n \frac{1}{k}.$$

From (2) it follows that (1) cannot hold for large n if the a_j's are bounded away from zero and infinity. Another bound can be obtained as follows:

$$M_{n+1} = \int_0^\infty \left\{ 1 - \prod_1^n (1 - e^{-a_j x}) \right\} dx + \int_0^\infty e^{-a_{n+1} x} \prod_1^n (1 - e^{-a_j x}) \, dx$$

$$= M_n + \int_0^\infty e^{-a_{n+1} x} F_n(x) \, dx.$$

By Fubini's theorem,

$$\int_0^\infty e^{-a_{n+1} x} F_n(x) \, dx = \int_0^\infty e^{-a_{n+1} x} \, dx \int_0^x dF_n(y)$$

$$= \int_0^\infty dF_n(y) \int_y^\infty e^{-a_{n+1} x} \, dx = a_{n+1}^{-1} \int_0^\infty e^{-a_{n+1} y} \, dF_n(y).$$

Therefore,

$$M_{n+1} = M_n + a_{n+1}^{-1} E \exp\left(-a_{n+1} \max(X_1, X_2, \cdots, X_n)\right),$$

where E denotes expectation. Then by Jensen's inequality,

$$M_{n+1} \geq M_n + a_{n+1}^{-1} e^{-a_{n+1} M_n}.$$

Since $x + a^{-1} e^{-ax}$ is increasing for $x > 0$, we obtain the lower bounds $M_n > M_n^*$ ($n = 1, 2, \cdots$) by solving

$$M_1^* = M_1 = a_1^{-1}, \qquad M_{n+1}^* = M_n^* + a_{n+1}^{-1} e^{-a_{n+1} M_n^*}.$$

The best bounds seem to be obtained if

$$a_1 \geq a_2 \geq \cdots \geq a_n.$$

Problem 71-28, *An Inequality,* by M. S. KLAMKIN (Ford Motor Company) and D. J. NEWMAN (Yeshiva University).

Determine the largest value of the constant k such that

$$a^3 + b^3 + c^3 \geq 3abc + k(a - b)(b - c)(c - a)$$

for all nonnegative a, b, c.

Solution by C. C. ROUSSEAU (Memphis State University).

The arithmetic-geometric mean inequality gives

$$a^3 + b^3 + c^3 \geq 3abc$$

for all nonnegative a, b, c. Hence, taking k to be positive, we need only consider the case where $(a - b)(b - c)(c - a) > 0$. Without loss of generality, we can require that $a < b < c$. The desired value of k is given by

$$k = \min_{\substack{a,b,c \geq 0 \\ a < b < c}} \frac{a^3 + b^3 + c^3 - 3abc}{(a - b)(b - c)(c - a)}.$$

Using the identity,

$$a^3 + b^3 + c^3 - 3abc = \tfrac{1}{2}(a + b + c)[(a - b)^2 + (b - c)^2 + (c - a)^2],$$

we write

$$k = \min_{\substack{a \geq 0 \\ s > t > 0}} \frac{(3a + s + t)[s^2 + t^2 + (s - t)^2]}{2st(s - t)},$$

where $s = c - a$ and $t = b - a$. From the above equation, it is clear that $a = 0$ for the minimum. Hence, letting $c = bx$, we have simply

$$k = \min_{x > 1} \frac{x^3 + 1}{x(x - 1)}.$$

By elementary calculus, the desired value of x is the real root, greater than 1, of the quartic equation,

$$x^4 - 2x^3 - 2x + 1 = 0.$$

The quartic equation can be factored to yield

$$[x^2 + (\sqrt{3} - 1)x + 1][x^2 - (\sqrt{3} + 1)x + 1] = 0.$$

It follows that the desired value of k is given by

$$k = \frac{\alpha^3 + 1}{\alpha(\alpha - 1)}$$

where $\alpha = \tfrac{1}{2}(1 + \sqrt{3} + \sqrt[4]{12})$. An approximate numerical value is

$$k \doteq 4.403669475.$$

Editorial note. A simpler expression for k, as was noted by the proposers and several of the solvers, is $k = \sqrt{9 + 6\sqrt{3}}$. D. Shanks and the proposers also noted that it follows symmetrically that the smallest permissible value of the constant is

$-\sqrt{9 + 6\sqrt{3}}.$

If a,b,c denote the sides of an arbitrary triangle, then the largest and smallest permissible values of k are $\pm 2\sqrt{9 + 6\sqrt{3}}$. This follows from a duality relation established by the first proposer (*Duality in triangle inequalities*, Ford Motor Company Preprint, July 1971. Also, see Notices Amer. Math. Soc., August 1971, p. 782); i.e., if a,b,c are the sides of a triangle, then there exist three nonnegative numbers x,y,z such that

$$x = s - a, \quad y = s - b, \quad z = s - c, \qquad 2s = a + b + c,$$

and, conversely, for any three nonnegative numbers x,y,z there exist sides of a triangle a,b,c and here

$$a = y + z, \quad b = z + x, \quad c = x + y.$$

Then, corresponding to any inequality in x,y,z, we have a corresponding inequality in a,b,c and conversely, i.e.,

$$F(x,y,z) \geq 0 \Rightarrow F(s - a, s - b, s - c) \geq 0,$$

$$G(a,b,c) \geq 0 \Rightarrow G(y + z, z + x, x + y) \geq 0. \qquad [\text{M.S.K.}]$$

Problem 77-12*, *Conjectured Inequalities*, by PETER FLOR (University of Cologne, West Germany).

Establish or disprove the following inequalities where all the variables are positive:

(1) $\quad a^3 + b^3 + c^3 + 3abc \geq a^2(b + c) + b^2(c + a) + c^2(a + b);$

(2)
$$39a^3 + 15a(b^2 + c^2) + 20ad^2 + 5bc(b + c + d)$$
$$\geq 10a^2(b + c) + 43a^2d + 39abc + ad(b + c);$$

(3)
$$5(a^4 + b^4 + c^4 + d^4) + 6(a^2c^2 + b^2d^2) + 12(a^2 + c^2)bd + 12(b^2 + d^2)ac$$
$$\geq 2(a^3 + b^3 + c^3 + d^3)(a + b + c + d) + 4(a + c)(b + d)(ac + bd)$$
$$+ 2(a^2 + c^2)(b^2 + d^2) + 8abcd.$$

Note that obviously, (1) and (2) do not hold for all *real* values of the variables. The situation for (3) is the same: consider $a = b = 1$, $c = d = -1$. Further note that on equating the variables, equality is obtained in all cases.

All three are particular cases of a general inequality which I published as a conjecture ten years ago and on which no progress has been reported so far (see Bull. Amer. Math. Soc., vol. 72, Research Problem 1, p. 30). Their proof might indicate a method for attacking this old conjecture.

Editorial note. The proposer does have a proof of (1).

Solution by O. G. RUEHR (Michigan Technological University).

(1) By symmetry, we can assume $a \geq b \geq c$. Then write

$$a(a - b)^2 + (a - b)^2(b - c) + c(a - c)(b - c) \geq 0.$$

(2) First express the inequality as

$$3a(b-a)^2+3a(c-a)^2+15a(b-c)^2+16a(a-d)^2+2a(2a-c-d)^2$$
$$+2a(2a-b-d)^2+(a-b)(a-c)(5b+5c+5d+a)\geqq 0.$$

If $b=c$, we are done. Then by symmetry assume $b>c$. Since now the relation holds otherwise, we can also take $b>a>c$. Define $X=b-a$, $Y=a-c$, $U=a-d$ and write

$$8aX^2+11aU^2+18aY^2+14aXY+5XY^2+2a(U+Y)^2+2a(U-X)^2+5cU^2$$
$$+10cX^2+5/2[X^2+U^2+(X+U)^2]\geqq 0.$$

This holds since X and Y are positive by assumption.

(3) The inequality can be written as a sum of non-negative terms as follows:

$$2ab(a-b)^2+2ac(a-c)^2+2ad(a-d)^2+2bc(b-c)^2+2bd(b-d)^2+2cd(c-d)^2$$
$$+\tfrac{3}{2}(a-b+d-c)^2(a-d+b-c)^2+\tfrac{1}{2}(a-b+d-c)^2(b-c+d-a)^2$$
$$+\tfrac{1}{2}(a-b+c-d)^2(a-d+b-c)^2+\tfrac{1}{2}(a-b+c-d)^2(b-c+d-a)^2\geqq 0.$$

This decomposition was motivated in part by considering the special case $c=b$, $d=b$ which reduces to $3(a-b)^2(a^2+b^2)\geqq 0$.

Editorial note. Inequality (1) is a special case ($n=1$) of an inequality of Schur, i.e.,

$$a^n(a-b)(a-c)+b^n(b-c)(b-a)+c^n(c-a)(c-b)\geqq 0. \qquad \text{[M.S.K.]}$$

A Minimum Value

Problem 84-13, *by* M. S. KLAMKIN (University of Alberta).
     Determine the minimum value of

$$I=\frac{(y-z)^2+(z-x)^2+(x-y)^2}{(x+y+z)^2}\cdot\frac{(v-w)^2+(w-u)^2+(u-v)^2}{(u+v+w)^2}$$

subject to $ux+vy+wz=0$ and all the variables are real.

Solution by MARK KANTROWITZ (Secondary school student, Maimonides School, Brookline, MA).
     By Cauchy's inequality,

$$\sum(y-z)^2\cdot\sum(v-w)^2\geqq\left\{\sum(y-z)(v-w)\right\}^2=\left\{\sum x\cdot\sum u\right\}^2$$

where the sums are cyclic over x,y,z and u,v,w. Thus, $I\geqq 1$ with equality iff

$$(y-z,z-x,x-y)=k(v-w,w-u,u-v).$$

Solution by T.M. Hagstrom (University of Wisconsin) *and by* W.B. Jordan (Scotia, NY).

     Let $\mathbf{P}=x\mathbf{i}+y\mathbf{i}+z\mathbf{k}$, $\mathbf{Q}=u\mathbf{i}+v\mathbf{i}+w\mathbf{k}$, and $\mathbf{R}=\mathbf{i}+\mathbf{j}+\mathbf{k}$. Then

(1)
$$I = \frac{(\mathbf{P} \times \mathbf{R})^2}{(\mathbf{P} \cdot \mathbf{R})^2} \cdot \frac{(\mathbf{Q} \times \mathbf{R})^2}{(\mathbf{Q} \cdot \mathbf{R})^2} \quad \text{and} \quad \mathbf{P} \cdot \mathbf{Q} = 0.$$

If a is the angle between $\mathbf{P}$ and $\mathbf{R}$, and b is the angle between $\mathbf{Q}$ and $\mathbf{R}$, then $I = \tan^2 a \tan^2 b$. In terms of a spherical triangle whose sides are a, b, c with $c = \pi/2$, $I = \sec^2 C$. Clearly the minimum value is 1, occurring for $C = \pi$.

B. D. DORE (University of Reading, Reading, UK) shows that the minimum value 1 follows immediately from (1) by replacing

$$(\mathbf{P} \cdot \mathbf{R})^2 (\mathbf{Q} \cdot \mathbf{R})^2 \text{ by its equivalent } \{(\mathbf{P} \times \mathbf{R}) \cdot (\mathbf{Q} \times \mathbf{R})\}^2.$$

Editorial note. In the second solution, we can just as well use three vectors in $\mathbb{R}^n$, i.e.,

$$\mathbf{P} = (x_1, x_2, \cdots, x_n), \quad \mathbf{Q} = (u_1, u_2, \cdots, u_n), \quad \mathbf{R} = (1, 1, \cdots, 1).$$

Also, $(\mathbf{P} \times \mathbf{R})^2$ is to be replaced by its equivalent $[\mathbf{P}^2 \mathbf{R}^2 - (\mathbf{P} \cdot \mathbf{R})^2]$, etc. J. A. WILSON (Iowa State University) obtained this extension analytically. [M. S. K.]

Problem 71-11, An Inequality, by JOEL BRENNER (University of Arizona).

Show that

$$\left| \frac{(1 + 2\lambda + \sqrt{1 + 4\lambda})^{n+1} - (1 + 2\lambda - \sqrt{1 + 4\lambda})^{n+1}}{2^{n+1}\sqrt{1 + 4\lambda}} \right| \geq |1 + 2\lambda|\{|1 + 2\lambda| - |\lambda|\}^{n-1},$$

where $\text{Re } \lambda > -\frac{1}{4}$ and n (integer) ≥ 1. The problem arose in a numerical solution of the heat flow equation.

Solution by A. A. JAGERS (Twente University of Technology, Enschede, the Netherlands).

Consider the recurrence relation $F_{n+1} = (1 + 2\lambda)F_n - \lambda^2 F_{n-1}$ with initial conditions $F_0 = 0$ and $F_1 = 1$. Its (unique) solution is $F_n = A(x_1^n - x_2^n)$, where $A = (x_1 - x_2)^{-1} = (\sqrt{1 + 4\lambda})^{-1}$ and x_1 and x_2 are the two distinct roots of the characteristic polynomial $x^2 - (1 + 2\lambda)x + \lambda^2$, i.e., $x_{1,2} = (1 + 2\lambda \pm \sqrt{1 + 4\lambda})/2$.

In terms of F_n the given inequality takes the simple form

$$|F_{n+1}| \geq |F_2|\{|1 + 2\lambda| - |\lambda|\}^{n-1}, \qquad n \geq 1, \quad \text{Re } \lambda > -1/4,$$

and is then easily proved by induction using the evident inequalities

$$|F_{n+1}| \geq |1 + 2\lambda||F_n| - |\lambda|^2|F_{n-1}| \quad \text{and} \quad |1 + 2\lambda| - |\lambda| \geq |\lambda|, \qquad \text{Re } \lambda > -1/4.$$

Also solved by O. G. RUEHR (Michigan Technological University) who notes that the left side of the inequality can be written in terms of Chebyshev polynomials of the second kind, $U_n^*(1 + 2\lambda^{-1})$.

Editorial note. The proposer notes that the left-hand side of the inequality is the modulus of

$$\prod_{j=1}^{n} \{1 + 2\lambda - 2\lambda \cos j\pi/(n + 1)\}$$

and also corresponds to the absolute value of the determinant of a matrix used by Crank-Nicolson for solving the heat equation. For Re $(\lambda) > -1/4$, the matrix has dominant diagonal and the right-hand side of the inequality is a known bound due to G. B. Price (Proc. Amer. Math. Soc., 2 (1951), pp. 497–502).

Problem 70–10, An Inequality of Mixed Arithmetic and Geometric Means, by B. C. CARLSON (Iowa State University).

Let the arithmetic and geometric means of the real nonnegative numbers $x_1, \cdots, x_n$ taken $n - 1$ at a time be denoted by

$$A_i = \frac{x_1 + \cdots + x_n - x_i}{n - 1}, \qquad G_i = \left(\frac{x_1 \cdots x_n}{x_i}\right)^{1/(n-1)}, \qquad i = 1, \cdots, n.$$

Prove that

$$\frac{G_1 + \cdots + G_n}{n} \leq (A_1 \cdots A_n)^{1/n}, \qquad\qquad n \geq 3,$$

with equality if and only if all products $x_i x_j$, $i \neq j$, have the same value.

The problem arose in connection with an elliptic integral representing the reciprocal of the electric capacity of a conducting ellipsoid with positive semi-axes x, y, z. If the case $x = y = z$ is excluded, the inequality of the first and third members of

$$\frac{1}{2} \int_0^\infty \frac{dt}{[(t + x^2)(t + y^2)(t + z^2)]^{1/2}} < \frac{2}{[(x + y)(y + z)(z + x)]^{1/3}}$$

$$< \frac{3}{(xy)^{1/2} + (yz)^{1/2} + (zx)^{1/2}}$$

was proved by Pólya and Szegö in 1945 and the inequality of the first and second members by W. H. Greiman recently. The inequality of the second and third members was verified numerically by Miss Susan Comstock before being proved as the case $n = 3$ of the theorem.

Solution by R. K. MEANY, S. A. NELSON and the proposer (Iowa State University).

For $i = 1, \cdots, n$ and $j = 1, \cdots, n$, where $n \geq 3$, we define $A_{ii} = A_i$, $G_{ii} = G_i$, and

$$A_{ij} = \frac{x_1 + \cdots + x_n - x_i - x_j}{n - 2}, \qquad G_{ij} = \left(\frac{x_1 \cdots x_n}{x_i x_j}\right)^{1/(n-2)}, \qquad i \neq j.$$

It is easy to verify that

$$\sum_{i=1}^n A_{ij} = nA_j, \qquad \prod_{j=1}^n G_{ij} = G_i^n.$$

One form of Hölder's inequality states that if $(a), (b), \cdots, (s)$ are n-tuples of real nonnegative numbers and $\alpha, \beta, \cdots, \sigma$ are positive weights with $\alpha + \beta + \cdots + \sigma = 1$, then

$$\sum a_i^\alpha b_i^\beta \cdots s_i^\sigma \leq (\sum a_i)^\alpha (\sum b_i)^\beta \cdots (\sum s_i)^\sigma.$$

(The proof of this inequality consists in dividing each term on the left side by the right side and using the inequality of the arithmetic and geometric means.) We choose

$$\begin{bmatrix} a_1 & \cdots & s_1 \\ \vdots & & \vdots \\ a_n & \cdots & s_n \end{bmatrix} = \begin{bmatrix} G_{11} & \cdots & G_{1n} \\ \vdots & & \vdots \\ G_{n1} & \cdots & G_{nn} \end{bmatrix}$$

and put $\alpha = \beta = \cdots = \sigma = 1/n$. Hölder's inequality becomes

$$\sum_{i=1}^{n} G_i = \sum_{i=1}^{n} \prod_{j=1}^{n} G_{ij}^{1/n} \leq \prod_{j=1}^{n} \left(\sum_{i=1}^{n} G_{ij} \right)^{1/n}.$$

The inequality of the arithmetic and geometric means implies

$$\sum_{i=1}^{n} G_{ij} \leq \sum_{i=1}^{n} A_{ij} = n A_j, \qquad\qquad j = 1, \cdots, n,$$

with strict inequality for some value of j unless $x_1 = x_2 = \cdots = x_n$. Hence

$$\prod_{j=1}^{n} \left(\sum_{i=1}^{n} G_{ij} \right)^{1/n} \leq n \prod_{j=1}^{n} A_j^{1/n},$$

with strict inequality unless either (i) $x_1 = x_2 = \cdots = x_n$ or (ii) $n - 1$ of the x_i are zero (with the result that $A_j = 0$ for some value of j). We have finally

$$\sum_{i=1}^{n} G_i \leq n \prod_{j=1}^{n} A_j^{1/n},$$

with obvious equality if (i) or (ii) holds and strict inequality otherwise. The two-part condition of equality is equivalent to the condition that all products $x_i x_j$ with $i \neq j$ have the same value, for the latter condition implies that if one of the x_i is zero there cannot be two which are positive.

The theorem still holds if all geometric means (of both the x_i and the A_i) are replaced by means of order t, where t is any real number other than unity, except that the inequality is reversed if $t > 1$ and the condition of equality is $x_1 = x_2 = \cdots = x_n$ if $t > 0$. The proof uses Minkowski's inequality in place of Hölder's, and $a_1, \cdots, s_n$ are chosen to be means of order t. No generalization with unequal weights is known.

A stronger theorem for mixed means of the numbers $x_1, x_2, ..., x_n$ taken at k at a time, $k = 1, 2, ..., n$, will appear in the Pacific Journal of Mathematics, 38(1971) 343-349.

An Inequality

Problem 85-20, *by* J. L. BRENNER (Palo Alto, CA), W. A. NEWCOMB (Lawrence Livermore Laboratories) *and* O. G. RUEHR (Michigan Technological University). Prove that

(i) $G(a_1 + b_1, a_2 + b_2) > G(a_1, a_2) + G(b_1, b_2)$,

(ii) $G(\frac{1}{2}(a_1 + b_1), \frac{1}{2}(a_2 + b_2)) > \frac{1}{2}G(a_1, a_2) + \frac{1}{2}G(b_1, b_2)$, where

$$G(x, y) = (xy^r + yx^r)/(x^r + y^r)$$

and $0 < r < 1$, $0 < a_1 < a_2$, $0 < b_1 < b_2$, $a_1/b_1 \neq a_2/b_2$.

Solution by H.-J. SEIFFERT (Berlin, West Germany).

Direct computation yields

(1) $\quad G_{xy} = r(xy)^{r-1}(x^r + y^r)^{-3}[(1-r)(x^{r+1} + y^{r+1}) + (1+r)(xy^r + yx^r)] > 0$

$\quad\quad xG_{xx} + yG_{xy} = 0 \quad$ and $\quad xG_{xy} + yG_{yy} = 0$.

Now consider the function

$$g(t) =: G(x(t), y(t)), \quad 0 \leq t \leq 1,$$

where $x(t) = (1-t)a_1 + tb_1$ and $y(t) = (1-t)a_2 + tb_2$. Then

$$g''(t) = (b_1 - a_1)^2 G_{xx} + 2(b_1 - a_1)(b_2 - a_2)G_{xy} + (b_2 - a_2)^2 G_{yy},$$

and in view of (1),

$$g''(t) = -\left[(a_2 b_1 - a_1 b_2)^2/xy\right]G_{xy}.$$

Since $a_1/b_1 \neq a_2/b_2$ and $G_{xy} > 0$, we find that $g''(t) < 0$, which shows that g is strictly concave. In particular, we obtain

$$g(1/2) > \{g(0) + g(1)\}/2,$$

which proves (ii). Since $G(cx, cy) = cG(x, y)$, (i) is an easy consequence of (ii).

Some Inequalities Involving Statistical Expressions

Problem 81-10, by L. V. FOSTER (University of Pittsburgh at Johnstown).

For a set of nonnegative numbers $u_1, u_2, \cdots, u_n$, let $u = (u_1 + u_2 + \cdots + u_n)/n$ be the arithmetic mean, let $\rho = (u_1 u_2 \cdots u_n)^{1/n}$ be the geometric mean and let $s_m = [\sum_{j-1}^{n} (u_j - u)^2/(n \cdot (n-1))]^{1/2}$ be the sample standard deviation of the mean. It is well known that $u \geq \rho$. Prove the stronger inequalities

(1) $\quad u^n \geq [u + (n-1)s_m](u - s_m)^{n-1} \geq \rho^n \geq [u - (n-1)s_m](u + s_m)^{n-1}$.

These inequalities arose in studying an iterative method for finding zeros of polynomials.

Solution by J. M. BORWEIN (Carnegie-Mellon University), G. P. H. STYAN (McGill University), *and* H. WOLKOWICZ (University of Alberta).

We may, without loss of generality, assume that the u_i's are ordered: $u_1 \geq \cdots \geq u_n \geq 0$. Then $u_n \leq u - s_m$ from Samuelson's inequality [1, eq. (1.2)], and so $u - s_m \geq 0$. The first inequality in (1), therefore, follows at once from the arithmetic mean/geometric mean inequality:

(2) $\quad u^n = \left\{ \dfrac{1}{n}[u + (n-1)s_m + (n-1)(u - s_m)] \right\}^n \geq [u + (n-1)s_m](u - s_m)^{n-1}$,

with equality if and only if $s_m = 0$ or $u_1 = \cdots = u_n$.

To prove the other inequalities in (1) we first note that

(3) $u + (n-1)s_m \geq u_1 \geq u + s_m,$

cf. [1, eqs. (1.1) and (1.3)]. Equality on the left of (3) occurs if and only if

(4) $u_1 = u + (n-1)s_m$ and $u_2 = \cdots = u_n = u - s_m,$

and on the right if and only if

(5) $u_1 = \cdots = u_{n-1} = u + s_m$ and $u_n = u - (n-1)s_m.$

Let

(6) $q^2 = \sum u_i^2 = n(n-1)s_m^2 + nu^2 > 0$

and u be fixed, and let

(7) $\mathcal{U}_n = \{(u_i):u_1 \geq \cdots \geq u_n \geq 0; \sum u_i = nu, \sum u_i^2 = q^2\}.$

Since $\mathcal{U}_n$ is compact and $q^2 > 0$, the maximum of ρ^n is attained in $\mathcal{U}_n$, at $\hat{u}_1 \geq \cdots \geq \hat{u}_n > 0$, say.

Our proof of the second inequality in (1) will be by contradiction. Suppose that $\hat{u}_1$ is not at its maximum $u + (n-1)s_m$ and so $\hat{u}_2 > \hat{u}_n$. Then, by (3) and (4), there exists a set of n numbers: $v_1, v_2, \cdots, v_n$, with $v_1 \geq v_2 \geq v_n$ and such that $v_1 > \hat{u}_1$, $v_i = \hat{u}_i$ ($i \neq 1, 2, n$) and $v_1 + v_2 + v_n = \hat{u}_1 + \hat{u}_2 + \hat{u}_n$, and $v_1^2 + v_2^2 + v_n^2 = \hat{u}_1^2 + \hat{u}_2^2 + \hat{u}_n^2$; a permutation of $v_1, v_2, \cdots, v_n$ lies in $\mathcal{U}_n$.

Holding $u_3, \cdots, u_{n-1}$ fixed, the partial derivative with respect to u_1 of

$$\rho^n = u_1 u_2 \cdots u_n$$

(8)
$$= \tfrac{1}{2} u_1[(nu - u_1 - u_3 - \cdots - u_{n-1})^2$$
$$- (q^2 - u_1^2 - u_3^2 - \cdots - u_{n-1}^2)]u_3 \cdots u_{n-1}$$

is

(9) $\dfrac{\partial \rho^n}{\partial u_1} = (u_1 - u_2)(u_1 - u_n)u_3 \cdots u_{n-1} \geq 0,$

and so ρ^n is nondecreasing in u_1. Similarly, it follows that

(10) $\dfrac{\partial \rho^n}{\partial u_2} = (u_2 - u_1)(u_2 - u_n)u_3 \cdots u_{n-1} \leq 0,$

and ρ^n is nonincreasing in u_2. Thus u_2 is nonincreasing in u_1 and so

(11) $v_1 > \hat{u}_1 \geq \hat{u}_2 \geq v_2.$

Let $v_1 = \hat{u}_1 + \varepsilon$, with $\varepsilon > 0$. Then

$$\hat{u}_1 \hat{u}_2 \hat{u}_n = \tfrac{1}{2}(v_1 - \varepsilon)[(v_2 + v_n + \varepsilon)^2 - \{v_1^2 + v_2^2 + v_n^2 - (v_1 - \varepsilon)^2\}]$$

(12)
$$= -\varepsilon^3 + \varepsilon^2(2v_1 - v_2 - v_n) - \varepsilon(v_1 - v_2)(v_1 - v_n) + v_1 v_2 v_n$$

$$< v_1 v_2 v_n$$

for sufficiently small $\varepsilon > 0$ since $v_1 > v_2$ by (11). Hence the maximum of ρ^n cannot be attained at $\hat{u}_1 \geq \cdots \geq \hat{u}_n$ unless $\hat{u}_1$ equals the maximum of u_1 in $\mathcal{U}_n$. The middle inequality in (1) then follows from (4), which also characterizes equality.

The right inequality in (1) is trivial when $u \leq (n-1)s_m$. So let

(13) $u > (n-1)s_m.$

(For an interpretation of (13) see [2, p. 486].) Then by the inequality

(14) $u_n \geqq u - (n-1)s_m$

[1, (1.2)] it follows that $u_n > 0$ and hence it follows from (12), reversing the order and roles of the $\hat{u}_i$ and v_i (so that the v_i identify the minimum of ρ^n), that the right inequality in (1) holds, with equality if and only if (5) holds.

   Remarks. The nonnegative numbers u_i can be considered to be the eigenvalues of a Hermitian nonnegative definite matrix $\mathbf{A}$ (cf. [2, p. 484]), or the roots of a polynomial $p(x) = x^n - a_0 x^{n-1} + a_1 x^{n-2} + \cdots + (-1)^n a_n$. In the case of a matrix $\mathbf{A}$, we have $\rho^n = \det \mathbf{A}$, while $u = \operatorname{tr} \mathbf{A}/n$, and $s_m^2 = (\operatorname{tr} \mathbf{A}^2 - nu^2)/[n(n-1)]$, where det denotes determinant and tr denotes trace. In the case of a polynomial, we have $\rho^n = a_n$, while $u = a_0/n$ and $s_m^2 = [2a_1 - (n+1)a_0^2]/[n(n-1)]$. Foster's inequalities (1) provide information on the eigenvalues or roots. For example, if one of the inequalities among the coefficients of the polynomial fails, then the roots cannot all be nonnegative.

   Moreover, the right inequality in (1) is essentially the same as [2, eq. (2.58), p. 485], which strengthens the comparison about two upper bounds for the condition number of a Hermitian positive definite matrix.

REFERENCES

[1] H. WOLKOWICZ AND G. P. H. STYAN, *Reply: On extensions of Samuelson's inequality,* Amer. Statist., 34
      (1980), pp. 250–251.
[2] ——, *Bounds for eigenvalues using traces,* Linear Algebra Appl., 29 (1980), pp. 471–506.

Problem 70-18, A Stock Market Investment Inequality, by M. J. LEMPEL (Gulf
   Research & Development Company).

   For a_i positive and $n \geqq 1$, prove that

(1) $$\left(\sum_{i=1}^{n} a_i^p \right) \div \left(\sum_{i=1}^{n} a_i^{p-1} \right)$$

is an increasing function of p for $-\infty < p < \infty$.

   This problem arose in connection with examining a class of stock market investment strategies. Let b_i be the price per share of a given security at time $t_i, i = 1, 2, \cdots, n$. For each p, consider an investment policy S_p defined as: "Purchase $s_0(b_0/b_i)^p$ shares of stock at time t_i," where s_0, b_0 are appropriately chosen constants. Return on investment is taken as the performance measure to evaluate a strategy S_p at time t_f, $t_f \geqq t_i$, when the stock selling price is b_f. We have

$$\text{total dollar value of investment at time } t_f = b_f s_0 b_0^p \sum_{i=1}^{n} \frac{1}{b_i^p},$$

$$\text{total dollars invested} = s_0 b_0^p \sum_{i=1}^{n} \frac{1}{b_i^{p-1}},$$

$$\text{return on investment} = b_f \left(\sum_{i=1}^{n} \frac{1}{b_i^p} \right) \div \left(\sum_{i=1}^{n} \frac{1}{b_i^{p-1}} \right).$$

Letting $a_i = 1/b_i$, the condition stated in the problem implies that, if $p < q$, S_q is a better strategy than S_p according to the return on investment criterion. This result is independent of the stock price history. Finally, note that S_0 corresponds to: "Purchase a fixed number of shares at each time t_i"; and S_1 is the "dollar averaging" policy whereby a fixed number of dollars is invested at each time t_i.

Editorial note. Many solvers pointed out that the wording above should be changed from "an increasing function" to "a nondecreasing function."

One of the simplest solutions which is a generalization of the problem was given by L. CARLITZ (Duke University).

If $a_1 \geq a_2 \geq \cdots \geq a_n$, $p < q < s$, and $p + s = q + r$, then

$$(2) \qquad \frac{\sum a_i^q}{\sum a_i^p} \leq \frac{\sum a_i^s}{\sum a_i^r},$$

which follows immediately from the identity

$$\sum a_i^p \sum a_i^s - \sum a_i^q \sum a_i^r = \sum_{i<j} a_i^p a_j^p (a_i^{q-p} - a_j^{q-p})(a_i^{s-p} - a_j^{s-q}).$$

F. W. STEUTEL (T. H. Twente, Enschede, the Netherlands) employed the following more general result than (2) which is given in M. Loève, *Probability Theory*, 2nd ed., Van Nostrand, New York, 1960, p. 156:

If X is a positive random variable such that the expectation function $m(p) = EX^p$ is finite for all real p, then $m(p)/m(p - h)$ is nondecreasing in p for every positive h.

D. LIND (Stanford University) derived the latter result in his solution.

A. C. WHEELER (Southern Methodist University), in his solution, used the special case of the latter inequality for $h = 1$ which he quotes as a theorem of Sclove, Simons and Van Ryzin, Amer. Statist., 21 (1967), p. 33. The latter theorem (for $h = 1$) has also appeared in E. Beckenbach, *A class of mean value functions*, Amer. Math. Monthly, 57 (1950), p. 5. The discrete version (for all $h > 0$) was established by J. B. BRONDER (Westinghouse Electric Corporation) in his solution. However, it also has appeared previously in D. S. Mitrinovič, P. M. Vasič, *Monotonost količnika dve sredine*, Univ. Beograd. Publ. Elektrotehn. Fak. Ser. Mat. Fiz., 185 (1967), pp. 35–38.

H. J. LANDAU (Bell Telephone Laboratories) obtained (1) as a special case of the more general inequality:

Suppose that $\rho_1(x)$ and $\rho_2(x)$ are probability density functions in $[0, \infty]$, and suppose that an x_0 exists such that

$$\rho_1(x) - \rho_2(x) \geq 0 \quad \text{for } 0 \leq x \leq x_0,$$
$$\rho_1(x) - \rho_2(x) \leq 0 \quad \text{for } x_0 \leq x < \infty$$

(i.e., $\rho_1(x)$ and $\rho_2(x)$ intersect only once). Then if $h(x)$ is nondecreasing,

$$\int_0^\infty h(x)\rho_1(x)\,dx \leq \int_0^\infty h(x)\rho_2(x)\,dx$$

and also the discrete analogue holds.

He did not recall where he had seen this result before. However, the same result for $h(x) = x$ (but which can be easily extended) is given as a problem by

R. F. Wheeling and R. A. G. Mitchael, Pi Mu Epsilon J., 3 (1962), p. 339. No doubt this inequality was known much earlier.

A. H. VAN TUYL (U.S. Naval Ordnance Laboratory) in one of his two solutions noted that the proposed inequality is a special case of the following result in a paper of his, *Monotonicity of some ratios of series and integrals*, which had been submitted for publication:

Let

$$F(x) = \sum_{i=1}^{n} f_i(x) \div \sum_{i=1}^{n} g_i(x),$$

where $f_i(x) > 0$ and $g_i(x) > 0$ for $1 \leq i \leq n$. Then sufficient conditions for $F(x)$ to be nondecreasing are (1) $f_i(x)/g_i(x)$ is nondecreasing, $1 \leq i \leq n$, and (2) $f_i(x)/f_j(x) - g_i(x)/g_j(x)$ is nondecreasing either for $i > j$ or for $i < j$. At least one of these conditions must hold if $F(x)$ is nondecreasing, but neither condition alone is either necessary or sufficient. A similar result holds for the ratio of two integrals.

The special case $f_i(x) = (i + \alpha)^{-x}$, $g_i(x) = (i + \beta)^{-x}$, $n = \infty$, where $\alpha > 0$, $\beta > 0$, $x > 1$, arose in an investigation of the charge density on two charged conducting spheres. Both conditions are satisfied if $\beta > \alpha$.

For unequal a_i's, the best strategy as defined above is S_∞, i.e., one buys no shares if $b_i > b_0$, one buys S_0 if $b_i = b_0$ and finally one buys ∞ shares if $b_i < b_0$. Of course the latter is impossible and is reminiscent of the St. Petersburg paradox.

J. C. AGRAWAL (California State College, California, Pennsylvania) in his solution also gives the following more picturesque interpretation of the above best strategy:

"Don't buy when the market is bullish on this stock so that it's moving higher than b_0, and stake all your assets on the stock when everyone is bear on it and dumping it so that it has fallen below b_0."

M. H. ROTHKOPF (Shell Development Company) in his solution noted that if $a_i = $ const., the stated result is incorrect (as noted above) and this implies that dollar cost averaging does not give a better return on investment than constant share purchasing unless the market price fluctuates. He also noted that the proposal, incidentally, provides an interesting reductio ad absurdum for arguments often used in favor of dollar cost averaging. For if $p = 1$ is better than $p = 0$, why not increase p further? One answer to this is that the very factors which depress the prices of securities also tend to limit the amount available for investment in them. It's worth noting that President Nixon's recent statement about investing in the stock market began "*If* I had any money $\cdots$ ". [M.S.K.] & [W.F.T.]

Sign Pattern of Terms of a Maclaurin Series

Problem 86-17, *by* C. L. FRENZEN (Southern Methodist University).

The function

$$f(t) = \left(\frac{1 - e^{-t}}{t}\right)^{-1/2}$$

is analytic $|t| < 2\pi$ and has a Taylor series of the form

$$f(t) = \sum_{n=0}^{\infty} b_n t^n, \qquad |t| < 2\pi.$$

Prove that $b_0 = 1$, $(-1)^{n+1}b_{2n} > 0$, $(-1)^n b_{2n+1} > 0$ for all $n > 0$. The sign pattern of the terms in the Maclaurin series is thus $1 + + - - + + - - + + \cdots$.

This problem arose out of some work on error bounds for the asymptotic expansion of the ratio of two gamma functions and for least squares approximation.

Solution by W. B. JORDON (Scotia, New York).

Put $t = 4ix$ and get

$$f = (x \cot x)^{1/2} + ix(x \cot x)^{-1/2}.$$

The power series expansion of $x \cot x$ is $1 - Q$, where Q is a well-known power series in x^2 with all coefficients positive. Extracting the square roots, we have that

$$f = 1 - (\tfrac{1}{2}Q + \tfrac{1}{8}Q^2 + \cdots) + ix(1 + \tfrac{1}{2}Q + \tfrac{3}{8}Q^2 + \cdots).$$

In each ( ) all coefficients are positive, so if each is expanded in powers of x^2, all will be positive. Then,

$$f = 1 - \sum_{1}^{\infty} a_n x^{2n} + ix \sum_{0}^{\infty} b_n x^{2n} \qquad a_n, b_n > 0,$$

$$= 1 - \sum_{1}^{\infty} A_n(-t^2)^n + t \sum_{0}^{\infty} B_n(-t^2)^n \qquad A_n, B_n > 0,$$

and the proposer's pattern of signs is verified.

Problem 62–13*, On the Nonnegativity of the Coefficients of a Certain Series Expansion, by DENNIS C. GILLILAND (Goodyear Aircraft Corporation).

If U and V are two random variables with joint density function

$$F(u, v) = (2\pi\sigma_u \sigma_v)^{-1} \exp\left\{ -\frac{1}{2}\left[\left(\frac{u - a}{\sigma_u}\right)^2 + \left(\frac{v - b}{\sigma_v}\right)^2\right]\right\},$$

an offset circle probability is given by

$$\Pr\left\{U^2 + V^2 \leq r^2\right\} = \int_{-r}^{r} \int_{-\sqrt{r^2 - u^2}}^{\sqrt{r^2 - u^2}} F(u, v) \, dv \, du,$$

which is of importance in certain bombing problems.

It can be shown that

$$(1) \qquad \Pr\left\{U^2 + V^2 \leq r^2\right\} = \sum_{m=0}^{\infty} A_m g_m(r),$$

where the $g_m(r)$ are tabulated functions of r and A_m is given by

$$A_m = \int_0^{2\pi} \sum_{j=0}^{m} \frac{[x \cos 2\theta]^j [2(y \cos \theta + z \sin \theta)]^{2m-2j}}{j! \, (2m - 2j)!} \, d\theta.$$

It is conjectured that $A_m \geq 0$ for all m and $1 \geq x > 0$, $y \geq 0$, $z \geq 0$. If this conjecture is valid then a good approximation of (1) can be gotten rather easily.

Solution by ELDON HANSEN (Lockheed Missiles and Space Co.).

If we expand $(y \cos \theta + z \sin \theta)^{2m-2j}$ and $\cos^j 2\theta = (\cos^2 \theta - \sin^2 \theta)^j$ by the binomial theorem, we obtain

(2)
$$A_m = \sum_{j=0}^{m} \sum_{r=0}^{j} \sum_{k=0}^{2m-2j} \frac{(-1)^{j-r} 2^{2m-2j} x^j y^k z^{2m-2j-k}}{r!\,(j-r)!\,k!\,(2m-2j-k)!}\, I,$$

where

$$I = \int_0^{2\pi} \cos^{r+k}\theta \, \sin^{2m-2r-k}\theta \, d\theta$$

$$= \frac{[1+(-1)^k]}{m!} \left\{\frac{2m-2r-k+1}{2}\right\}!\left\{\frac{2r+k-1}{2}\right\}!.$$

Since $I = 0$ for k odd, (2) becomes, after an interchange of the order of summation,

(3)
$$A_m = \sum_{r=0}^{m} \sum_{k=0}^{m-r} \sum_{j=r}^{m-k} \frac{(-1)^{j-r} 2^{2m-2j+1} x^j y^{2k} z^{2m-2j-2k}}{r!\,(j-r)!\,(2k)!\,(2m-2j-2k)!\,m!}$$

$$\cdot \left\{\frac{2m-2r-k+1}{2}\right\}!\left\{\frac{2r+k-1}{2}\right\}!.$$

Assume, temporarily, that $x \neq 0$. Introducing new dummy indices $s = j - r$ and $n = r + k$ and using the relation $\sqrt{\pi}(2a-1)! = 2^{2a-1}(a-1)!\,(a-\frac{1}{2})!$, we rewrite A_m as

(4)
$$A_m = \frac{\pi x^m}{2^{m-1} m!} \sum_{n=0}^{m} \frac{(-1)^n}{n!\,(m-n)!} H_{2n}\left(iy\sqrt{\frac{2}{x}}\right) H_{2m-2n}\left(z\sqrt{\frac{2}{x}}\right),$$

where $H_k(t)$ denotes the Hermite polynomial defined as

$$H_k(t) = \frac{k!}{2^{k/2}} \sum_{j=0}^{[k/2]} \frac{(-1)^j}{j!\,(k-2j)!}(t\sqrt{2})^{k-2j}.$$

It is relatively easy to verify the (new?) result that

(5)
$$\sum_{n=0}^{m} \frac{(-1)^n}{n!\,(m-n)!} H_{2n}(iu) H_{2m-2n}(v) = \frac{2^m}{m!} H_m\left(\frac{v+iu}{\sqrt{2}}\right) H_m\left(\frac{v-iu}{\sqrt{2}}\right).$$

Letting $u = y(2/x)^{1/2}$ and $v = z(2/x)^{1/2}$ in (5) and substituting into (4), we get

$$A_m = \frac{2\pi x^m}{(m!)^2} H_m\left(\frac{z+iy}{\sqrt{x}}\right) H_m\left(\frac{z-iy}{\sqrt{x}}\right).$$

For $x > 0$,

$$A_m = \frac{2\pi x^m}{(m!)^2} \left| H_m\left(\frac{z+iy}{\sqrt{x}}\right) \right|^2 \geqq 0.$$

For $x < 0$, we find that

$$A_m = \frac{2\pi |x|^m}{(m!)^2} \left| H_m\left(\frac{y+iz}{\sqrt{|x|}}\right) \right|^2 \geqq 0.$$

Hence, $A_m \geqq 0$ for all y and z when $x \neq 0$. By continuity, $A_m \geqq 0$ for $x = 0$ and hence $A_m \geqq 0$ for *all* x, y and z.

A Conjectured Inequality for Hermite Interpolation at the Zeros of Jacobi Polynomials

Problem 87-7* by WALTER GAUTSCHI (Purdue University).

Let $\tau_\mu = \tau_{\mu,n}^{(\alpha,\beta)}$, $\mu = 1, 2, \cdots, n$, be the zeros of the Jacobi polynomial $P_n^{(\alpha,\beta)}$ and $h_\nu, k_\nu, \nu = 1, 2, \cdots, n$, the associated fundamental Hermite interpolation polynomials of degree $2n - 1$, defined by

$$h_\nu(\tau_\mu) = \delta_{\nu\mu}, \quad h_\nu'(\tau_\mu) = 0,$$
$$k_\nu(\tau_\mu) = 0, \quad k_\nu'(\tau_\mu) = \delta_{\nu,\mu}, \qquad \nu, \mu = 1, 2, \cdots, n$$

where $\delta_{\nu\mu}$ is the Kronecker delta. Furthermore, let $\sigma_\nu = \sigma_{\nu,n}^{(\alpha,\beta)}$ be the Christoffel numbers belonging to the Jacobi measure $d\sigma^{(\alpha,\beta)}(t) = (1 - t)^\alpha(1 + t)^\beta dt$ on $[-1, 1]$. Define

(1)
$$g_n(t) = g_n(t; d\sigma^{(\alpha,\beta)}) = \sum_{\nu=1}^{n} [h_\nu^2(t) + \sigma_\nu^{-2}k_\nu^2(t)].$$

Prove or disprove the following conjecture: There holds

(2)
$$g_n(t; d\sigma^{(\alpha,\beta)}) \leq 1 \quad \text{on} \quad [-1, 1] \quad \text{for all } n \geq 2,$$

whenever α, β satisfy $-1 < \beta \leq \alpha \leq \alpha_0$ and hence, by "symmetry," for all α, β in the rectangle $-1 < \alpha \leq \alpha_0$, $-1 < \beta \leq \alpha_0$, where $\alpha_0 = -.336945\ldots$ is the unique root on $(-1, 0)$ of the equation $g_2(1; d\sigma^{(\alpha,-1)}) = 1$, or, more explicitly,

(3)
$$\frac{1}{16}\alpha^2(3+\alpha)^4 + \frac{1}{16}(1+\alpha)^4(4+\alpha)^2 + 4^{1-\alpha}\frac{(1+\alpha)^4(2+\alpha)^2}{(3+\alpha)^2} = 1.$$

(The precise region in the α, β-plane in which (2) holds, if at all, is probably somewhat larger, but may have curved boundaries.) In particular, (2) is conjectured to hold for Chebyshev polynomials of the first kind, $\alpha = \beta = -\frac{1}{2}$.

Editorial note. The proposer has provided considerable theoretical and computational evidence to support the conjecture.

Problem 77-5*, A Conjectured Increasing Infinite Series, by M. L. GLASSER (University of Waterloo).

In a study of the heat distributions due to two spheres being maintained at a constant temperature in a conducting medium, it was found that the heat that must be supplied, as a function of the distance r between the spheres, is given in terms of the function

$$S(r) = \sum_{k=1}^{\infty} (-1)^{k+1}\sinh y \, \text{csch } ky \quad (y = \cosh^{-1}r)$$

Physical intuition and numerical evidence suggest that $S(r)$ increases steadily between the values $S(1) = \log 2$ and $S(\infty) = 1$. Prove whether or not this is the case.

The problem is also equivalent to showing that

$$\sum_{k=1}^{\infty} (-1)^{k+1} U_k'(x)[U_k(x)]^{-2} > 0$$

for all $x > 1$, where $U_k(x)$ denotes a Chebyshev polynomial of the second kind and the prime denotes differentiation with respect to x. Another equivalent formulation is to show that

$$S(r) = (q - q^{-1})[A(q) - A(q^2)]$$

is increasing for $r > 1$, where

$$q = r + (r^2 - 1)^{1/2}, \qquad A(q) = \sum_{n=0}^{\infty} (1 + q^n)^{-1}.$$

Problem 77-19*, *Two Inequalities*, by P. BARRUCAND (Universite P. et M. Curie, Paris, France).
Let

$$F_1(\theta) = \sum_{n=1}^{\infty} \frac{\cos^n \theta \cos n\theta - \cos^{2n} \theta}{n(1-2) \cos^n \theta \cos n\theta + \cos^{2n} \theta},$$

$$F_2(\theta) = \sum_{\substack{n=1 \\ n \equiv 1(2)}}^{\infty} \frac{\cos^n \theta \cos n\theta - \cos^{2n} \theta}{n(1 - 2 \cos^n \theta \cos n\theta + \cos^{2n} \theta)}.$$

It is conjectured that $F_1(\theta)$ and $F_2(\theta)$ are negative for $0 < \theta < \pi/2$. The conjecture was found from a computer computation.

Editorial note. W. Al Salam has shown that the conjectures are equivalent to showing that

$$\prod_{k=1}^{\infty} |1 - x^k e^{ik\theta}| < 1,$$

and

$$\prod_{k=1}^{\infty} \left| \frac{1 + x^k e^{ik\theta}}{1 - x^k e^{ik\theta}} \right| < 1$$

where $x = \cos \theta$ and $0 < x < 1$.

Solution by W. B. JORDAN (Scotia, NY).

To avoid confusion with the Theta functions, we use ϕ for the independent variable. Page numbers below pertain to the reference.

We first sum the series; it will turn out they are known functions. Split the general term of F_1 into partial fractions, expand each fraction in a series, and sum the resulting double series on n. Let

$$q_1^2 = q_2 = e^{i\phi} \cos \phi.$$

Then

$$F_1 = \sum_{n=1}^{\infty} \frac{1}{2n(q_2^{-n} - 1)} + \text{complex conjugate}$$

$$= \sum_{n=1}^{\infty} \sum_{j=1}^{\infty} \frac{1}{2n} q_2^{jn} + \text{conj}$$

$$= -\frac{1}{2} \sum_{j=1}^{\infty} \log(1 - q_2^j) + \text{conj}$$

$$= -\frac{1}{2} \log G(q_1) G(\bar{q}_1)$$

in which (p. 473) $G(q) = \prod_{n=1}^{\infty} (1 - q^{2n}) = (\theta_1'/2q^{1/4})^{1/3}$.
  In like manner (p. 472),

$$F_2 = -\frac{1}{4} \log \frac{\theta_1' \bar{\theta}_1'}{\theta_2 \bar{\theta}_2} \quad (\text{nome} = q_1).$$

But (p. 470)

$$[\theta_1'/\theta_2]_{q_1} = [\theta_3 \theta_4]_{q_1} = \theta_4^2(0, q_2).$$

This last relation may be verified by substituting the infinite products (p. 469) for θ_3 and θ_4.
  The conjectures are therefore true if $|G(q_1)| > 1$ and $|\theta_4(0, q_2)| > 1$ as ϕ ranges over the quadrant. Now q is small, and the series for G and θ_4 converge rapidly, for ϕ near 90°; but convergence is poor for ϕ near 0. Jacobi's imaginary transformation of G and θ_4 reverses this situation by producing a q which is small for ϕ near 0; we prove the inequalities are true as they stand for $75° < \phi < 90°$, and by a Jacobi's transformation for the remainder of the quadrant.
  Euler's identity is

$$G(q) = \sum_{n=0}^{\infty} (-1)^n q^{n(3n+1)} (1 - q^{4n+2})$$

so

$$G(q_1) = 1 - q_2 - q_2^2 + q_2^5 + q_2^7 - q_2^{12} - \cdots .$$

By the triangle inequality:

$$|G(q_1)| > |1 - q_2 - q_2^2| - (|q_2|^5 + |q_2|^7 + |q_2|^{12} + \cdots)$$

$$> |1 - q_2 - q_2^2| - \frac{|q_2|^5}{1 - |q_2|^2} = \left(1 + \frac{1}{4} \sin^2 2\phi\right)^{1/2} - \cos^3 \phi \cot^2 \phi.$$

As ϕ decreases from 90° to 75°, this expression increases from 1 to 1.030. Next,

$$\theta_4(0, q_2) = 1 - 2q_2 + 2q_2^4 - 2q_2^9 + 2q_2^{16} - \cdots ,$$

$$|\theta_4(0, q_2)| > |1 - 2q_2 + 2q_2^4| - 2(|q_2|^9 + |q_2|^{16} + \cdots)$$

$$> \left(1 + \frac{1}{4} \sin^4 2\phi\right)^{1/2} - \frac{2\cos^9 \phi}{1 - \cos^7 \phi}.$$

As ϕ decreases from 90° to 75°, this expression increases from 1 to 1.0078.

Consequently, both conjectures are true for $\phi > 75°$.

Jacobi's imaginary transformation of θ_1 is (p. 475)

$$\theta_1(z, q_1) = -i\left(-\frac{1}{\pi} \log q_1\right)^{-1/2} e^{i\tau' z^2/\pi} \theta_1(z\tau', q_3)$$

where

$$\log q_3 = \pi^2/\log q_1 \quad \text{and} \quad \tau' = -i\pi/\log q_1.$$

Differentiate with respect to z and then put $z = 0$:

$$\theta_1'(0, q_1) = -i\tau'\left(-\frac{1}{\pi} \log q_1\right)^{-1/2} \theta_1'(0, q_3)$$

so the Jacobi transformation of G is

$$G(q_1) = [\theta_1'(0, q_1)/2q_1^{1/4}]^{1/3}$$

$$= \left(-\frac{1}{\pi} \log q_1\right)^{-1/2} [\theta_1'(0, q_3)/2q_1^{1/4}]^{1/3} = \left(-\frac{1}{\pi} \log q_1\right)^{-1/2} (q_3/q_1)^{1/12} G(q_3).$$

For brevity, put

$$D = (\log \sec \phi)^2 + \phi^2.$$

Then

$$\log q_3 = -\frac{2\pi^2}{\log \sec \phi - i\phi} = -\frac{2\pi^2}{D}(\log \sec \phi + i\phi)$$

and

$$|G(q_1)| = \left(\frac{D^{1/2}}{2\pi}\right)^{-1/2} \sec^{1/24} \phi \exp\left(-\frac{\pi^2}{6D} \log \sec \phi\right)|G(q_3)|.$$

As ϕ increases from 0 to 75° the coefficient of $|G(q_3)|$ falls from ∞ to 1.032. Now

$$|G(q_3)| > 1 - (|q_3|^2 + |q_3|^4 + \cdots) = 1 - \frac{|q_3|^2}{1 - |q_3|^2}.$$

At $\phi = 75°$, $|q_3|^2 = 2.8 \times 10^{-7}$, so $|G(q_3)| = 1$ to 7 decimal places, and $|G(q_1)| > 1$ for all ϕ.

For θ_4 (p. 475, with $z = 0$)

$$\theta_4(0, q_2) = \left(\frac{-\pi}{\log q_2}\right)^{1/2} \theta_2(0, q_4) = 2q_4^{1/4}\left(\frac{-\pi}{\log q_2}\right)^{1/2} (1 + q_4^2 + q_4^6 + q_4^{12} + \cdots)$$

where

$$\log q_4 = \pi^2/\log q_2, \qquad q_4^2 = q_3,$$

so

$$|\theta_4(0, q_2)| = 2\pi^{1/2}D^{-1/4} \exp\left(-\frac{\pi^2}{4D} \log \sec \phi\right)|1 + q_3 + q_3^3 + q_3^6 + \cdots|.$$

As ϕ increases from 0 to 75°, the coefficient of the q_3-series falls from ∞ to 1.0075, while

$$|\text{series}| > 1 - \frac{|q_3|}{1 - |q_3|^2} = 1 - .00053 \quad \text{at } \phi = 75°.$$

Consequently, $|\theta_4(0, q_2)| > 1.0070$ for ϕ up to $75°$, and the proof is complete.

REFERENCE

E. T. WHITTAKER AND G. N. WATSON, *A Course of Modern Analysis*, Cambridge University Press, London, 1952.

An OC Curve Inequality

Problem 86-20*, by P. A. ROEDIGER *and* J. G. MARDO (U.S. Army Armament, Munitions and Chemical Command, Dover, NJ).

Let

$$OC(n, c, q) = \sum_{i=0}^{c} \binom{n}{i} q^{n-i} (1-q)^i$$

where $n > c > 0$ and $0 < q < 1$. Prove or disprove that

$$OC(n, c, q) < [OC(n, c, q^{1/m})]^m \quad \text{for all } m > 1.$$

In the terminology of lot-by-lot sampling inspection by attributes, e.g., per MIL-STD-105D, the Operating Characteristic (OC) curve defines the probability of accepting a lot whose true fraction effective is q, when the criterion is to accept if and only if $(n - c)$ or more effectives are found in a random n-sample. When $m > 1$ quality characteristics are distinguished, having effect rates q_i, lot quality is described by the profile $\tilde{Q} = (q_1, q_2, \cdots, q_m)$ and, generally, the accept/reject criteria are such that probability of acceptance has the form

$$PA(\tilde{Q}) = \prod_{i=1}^{m} OC(n, c, q_i).$$

Since total lot quality q is the product of the q_i's, one is naturally interested in $PA(\tilde{Q}|q)$, for a given q. The two sides of the proposed inequality can be shown to be optimal PA values, under this constraint. The difficulty is deciding which is the max and which is the min.

Solution by A. A. JAGERS (Universiteit Twente, Enschede, the Netherlands).

For fixed n and c let $h(u) = \log\{OC(n, c, e^{-u})\}$ with $u \geqq 0$. Then $h(0) = 0$, $h(u) < 0$ for $u > 0$, and the inequality can be written as $m^{-1}h(u) < h(m^{-1}u)$. It suffices to prove that h is concave, or that $h''(u) < 0$ for $u > 0$. To do so, we note that

$$\frac{c!(n-c-1)!}{n!} OC(n, c, e^{-u}) = \int_0^{\exp(-u)} t^{n-c-1}(1-t)^c \, dt$$

$$= \int_u^{\infty} e^{-(n-c)s}(1 - e^{-s})^c \, ds$$

$$:= f(u)$$

(a result which expresses a familiar relation between the binomial and beta distributions and the order statistics for a sample from an exponential distribution). It follows by a simple computation that $h''(u)$ has the same sign as

$$g(u):=(np-c)f(u)-q^{n-c}p^{c+1},$$

with $q = e^{-u}$ and $p = 1 - q$. Now

$$g'(u)=nqf(u)-q^{n-c+1}p^c$$

and

$$f(u)>p^c \int_u^\infty e^{-(n-c)s}\,ds$$

$$=\frac{q^{n-c}p^c}{n-c}.$$

Thus $g'(u)>0$. Since $g(u) \to 0$ as $u \to \infty$, we see that $g(u)<0$ for all $u>0$. The same may be said for $h''(u)$, so h is concave.

Problem 63-16, *On The Coefficients Of a Trigonometric Polynomial*, by D. J. NEWMAN (Yeshiva University) and L. A. SHEPP (Bell Telephone Laboratories).

Let

$$1 + \alpha_1 \cos x + \alpha_2 \cos 2x + \cdots + \alpha_n \cos nx$$

denote a nonnegative trigonometric polynomial. If r and s are any positive integers such that $r + s > n$, prove that

$$|\alpha_r| + |\alpha_s| \le 2.$$

Solution by LOWELL SCHOENFELD (University of Wisconsin).

We generalize the problem by considering the trigonometric polynomial

$$g(x) = a_0 + (a_1 \cos x + b_1 \sin x) + \cdots + (a_n \cos nx + b_n \sin nx)$$

which is assumed to be nonnegative for all real x. We will show that if r and s are any positive integers such that $r + s \ge n + 1$, then

$$(1) \qquad\qquad \sqrt{a_r^2 + b_r^2} + \sqrt{a_s^2 + b_s^2} \le 2a_0.$$

The case for $r + s = n + 1$ was proved by Szász [1], [2] and the proof here is merely a slight modification.

We use a result of Fejér [3], [4], that there exists a polynomial P with complex coefficients A_k such that

$$|P(e^{ix})|^2 = |A_0 + A_1 e^{ix} + \cdots + A_n e^{inx}|^2 = g(x).$$

From this, Fejér readily obtains

$$a_r + ib_r = 2\sum_{k=0}^{n-r} A_k \bar{A}_{k+r} \quad \text{if} \quad 1 \le r \le n; \qquad a_0 = \sum_{k=0}^{n} |A_k|^2.$$

Consequently, if $1 \le r, s \le n$, we have

$$\sqrt{a_r^2 + b_r^2} + \sqrt{a_s^2 + b_s^2} = 2\left|\sum_{k=0}^{n-r} A_k \bar{A}_{k+r}\right| + 2\left|\sum_{k=0}^{n-s} A_k \bar{A}_{k+s}\right|$$

$$\le \sum_{k=0}^{n-r} (|A_k|^2 + |A_{k+r}|^2) + \sum_{k=0}^{n-s} (|A_k|^2 + |A_{k+s}|^2).$$

Hence, if $r + s \geq n + 1$, then $s \geq n - r + 1$ and $r \geq n - s + 1$ so that the last expression does not exceed $2a_0$.

This result can be extended further by using a device employed by Szász [1], who shows that if m is a positive integer and if we put

$$G(x) = \frac{1}{m} \sum_{k=1}^{m} g\left(\frac{x}{m} + \frac{2\pi k}{m}\right),$$

then $G(x)$ is nonnegative and for $N = [n/m]$, $B_k = a_{km}$, $C_k = -b_{km}$, we have

$$G(x) = a_0 + \mathrm{Re}\left\{\sum_{k=1}^{N}(a_{km} + ib_{km})e^{ikx}\right\} = a_0 + \sum_{k=1}^{N}(B_k \cos kx + C_k \sin kx).$$

Hence, if $r + s \geq N + 1$, then (1) gives

$$\sqrt{B_r^2 + C_r^2} + \sqrt{B_s^2 + C_s^2} \leq 2a_0.$$

That is, if $m, r, s \geq 1$ and $r + s \geq [n/m] + 1$, then

(2) $$\sqrt{a_{rm}^2 + b_{rm}^2} + \sqrt{a_{sm}^2 + b_{sm}^2} \leq 2a_0,$$

which reduces to (1) for $m = 1$.

Many other similar results are known. For example, Szász [1] generalizes an earlier result of Fejér [3] to show that

$$\sqrt{a_r^2 + b_r^2} \leq a_0 \quad \text{if} \quad n < 2r \leq 2n.$$

REFERENCES

[1] O. Szász, *Collected Mathematical Papers*, University of Cincinnati, 1955, pp. 734–745, 656–669.
[2] ———, *Elementare Extremalprobleme über nicht-negativ trigonometrische Polynome*, S.-B. Bayer Akad. Wiss., Math.-Phys. Kl., (1927), pp. 185–196.
[3] L. Fejér, *Über trigonometrische Polynome*, J. Reine Angew. Math., 146 (1916), pp. 53–82.
[4] G. Pólya and G. Szegö, *Aufgaben und Lehrsätze aus der Analysis*, Dover, New York, 1945, II, problem 40, pp. 81, 274.

Problem 67-6, A Trigonometric Inequality, by J. N. Lyness (Argonne National Laboratory) and C. B. Moler (University of Michigan).

For all real x, show that

$$\sum_{n=1}^{\infty}(-1)^{n+1}\left\{\frac{\sin \pi n x}{n}\right\}^{2r} \geq 0, \qquad r = 1, 2, 3, \cdots.$$

Editorial note. W. A. J. Luxemburg (California Institute of Technology) has pointed out that the published solution (April, 1968) of this problem is incorrect. The error is due to some missing factors of 2.

Solution by Richard Askey and James Fitch (University of Wisconsin at Madison and Milwaukee, respectively).

We give two essentially different proofs of this inequality. The first uses a theorem of Fejér [7] which is not as well known as it should be. For this reason we shall include a proof of his result:

$$g(x, y) = \sum_{n=1}^{\infty} a_n \sin nx \sin ny \geq 0, \qquad 0 \leq x, y \leq \pi,$$

if and only if

$$f(x) = \sum_{n=1}^{\infty} na_n \sin nx \geq 0, \qquad 0 \leq x \leq \pi.$$

If $g(x, y) \geq 0$, then we have

$$f(x) = \lim_{y \to 0^+} \frac{g(x, y)}{y} \geq 0.$$

In the other direction we have

$$\frac{f(x + t) + f(x - t)}{2} = \sum_{n=1}^{\infty} na_n \sin nx \cos nt.$$

Integrating we have

$$g(x, y) = \frac{1}{2} \int_0^y [f(x + t) + f(x - t)] \, dt \geq 0 \quad \text{if} \quad 0 \leq y \leq x, \quad x + y \leq \pi.$$

By symmetry we have $g(x, y) \geq 0$ if $0 \leq x + y \leq \pi$. Letting $u = \pi - x, v = \pi - y$ we have $g(x, y) \geq 0$ for $0 \leq x, y \leq \pi$.

Next recall that

$$\frac{\pi - x}{2} = \sum_{n=1}^{\infty} \frac{\sin nx}{n}, \qquad 0 < x < \pi.$$

Using Fejér's result we have

$$\sum_{n=1}^{\infty} \prod_{j=1}^{l} \frac{\sin nx_j}{n} \geq 0, \qquad 0 \leq x_j \leq \pi.$$

If we now let $l = 2k$,

$$x_1 = \cdots = x_{l-1} = x, \qquad x_l = \pi - x,$$

we have

$$\sum_{n=1}^{\infty} (-1)^{n+1} \left(\frac{\sin nx}{n} \right)^{2k} \geq 0, \qquad 0 \leq x \leq \pi.$$

But this is an even function which is periodic of period 2π so it holds for all x. If we use stronger results than

$$\sum_{n=1}^{\infty} \frac{\sin nx}{n} \geq 0,$$

we can obtain stronger results.

Two examples follow. First

$$\sum_{n=1}^{N} \frac{\sin nx}{n} > 0, \qquad 0 < x < \pi, \quad N = 1, 2, \cdots,$$

so the partial sums of the above series are nonnegative. See [5] or [10] for proofs of $\sum_{n=1}^{N} (\sin nx)/n > 0$. Second we can use

$$\sum_{n=1}^{\infty} nr^{n-1} \sin n\theta = \frac{(1-r^2)\sin\theta}{(1-2r\cos\theta+r^2)^2}$$

to show that

$$\sum_{n=1}^{\infty} n^2 r^{n-1} \prod_{j=1}^{l} \frac{\sin nx_j}{n} > 0, \qquad -1 < r < 1, \quad 0 < x_j < \pi.$$

If $l \geq 4$ we can let $r \to -1$ to obtain

$$\sum_{n=1}^{\infty} (-1)^{n+1} n^2 \prod_{j=1}^{l} \frac{\sin nx_j}{n} \geq 0.$$

The other proof uses the Poisson integral for the circle, some elementary trigonometric identities, and an old result of Marx which we shall prove. Recall

$$P(r,x) = \frac{1}{2} + \sum_{n=1}^{\infty} r^n \cos nx = \frac{\frac{1}{2}(1-r^2)}{1-2r\cos x+r^2} > 0, \qquad -1 < r < 1.$$

Also if

$$f(x) = \frac{1}{2} + \sum_{n=1}^{\infty} a_n \cos nx \geq 0,$$

then

$$\frac{1}{2} + \sum_{n=1}^{\infty} a_n \prod_{j=1}^{l} \cos nx_j \geq 0.$$

This follows since

$$\cos nx \cos ny = \frac{1}{2}[\cos n(x+y) + \cos n(x-y)].$$

Finally Marx's result [11] is that

$$\frac{\sin(n+1)x}{(n+1)\sin x} = \int_0^{\pi} \cos ny \, d\mu(y),$$

where $d\mu(y)$ is a nonnegative measure with mass one which is independent of n. Combining these three results we see that

$$\sum_{n=1}^{\infty} r^{n-1} \prod_{j=1}^{l} \frac{\sin nx_j}{n} \geq \frac{1}{2} \prod_{j=1}^{l} \sin x_j \geq 0, \qquad |r| < 1, \quad 0 \leq x_j \leq \pi.$$

Letting $r \to -1$ we obtain a stronger inequality for $l = 2, 3, \cdots$. Marx's result is also a result of the early 1930's but no simple proof has been given of it yet. Two proofs are in [9] and a third proof of a generalization of it is due to Seidel and Szász [12]. It is sufficient to show that

$$\frac{1}{2} + \sum_{n=1}^{\infty} \frac{\sin(n+1)\theta}{(n+1)\sin\theta} \cos n\varphi \geq 0$$

and this follows if we show

$$\frac{1}{2} + \sum_{n=1}^{\infty} r^n \frac{\sin(n+1)\theta}{(n+1)\sin\theta} \cos n\varphi \geq 0, \qquad -1 < r < 1, \quad 0 \leq \theta, \varphi \leq \pi.$$

Since

$$r^n \frac{\sin(n+1)(\pi-\theta)}{(n+1)\sin(\pi-\theta)} = (-r)^n \frac{\sin(n+1)\theta}{(n+1)\sin\theta},$$

we may assume $0 \leq \theta \leq \pi/2$. By continuity or a separate calculation, we may handle $\theta = \pi/2$; thus we may assume $0 \leq \theta < \pi/2$.

Observe that

$$\frac{d}{d\theta} \frac{\sin(n+1)\theta}{(n+1)(\cos\theta)^{n+1}} = \frac{\cos n\theta}{(\cos\theta)^{n+2}}.$$

Integrating, we see that

$$\frac{\sin(n+1)\theta}{(n+1)\sin\theta} = \frac{1}{\sin\theta} \int_0^\theta \left(\frac{\cos\theta}{\cos\varphi}\right)^{n+1} \frac{\cos n\varphi}{\cos\varphi} d\varphi.$$

Thus

$$\frac{1}{2} + \sum_{n=1}^\infty r^n \frac{\sin(n+1)\theta}{(n+1)\sin\theta} \cos n\psi$$

$$= \frac{\cos\theta}{\sin\theta} \int_0^\theta \left\{ \frac{1}{2} + \sum_{n=1}^\infty \left[r\left(\frac{\cos\theta}{\cos\varphi}\right) \right]^n \cos n\varphi \cos n\psi \right\} \frac{d\varphi}{\cos^2\varphi}$$

and

$$\left| \frac{r\cos\theta}{\cos\varphi} \right| < 1 \quad \text{if} \quad |r| < 1 \quad \text{and} \quad 0 \leq \varphi \leq \theta < \frac{\pi}{2}.$$

Thus the kernel inside the integral is positive and we are finished.

All of these results have been generalized to ultraspherical polynomials and some of them to Jacobi polynomials. Fejér's positivity result has been generalized by Bochner to ultraspherical polynomials [6]. Slightly earlier than Fejér's paper, Koschmieder [8] proved that

$$\sum_{n=1}^N \frac{\sin nx \sin ny}{n^2} > 0, \qquad\qquad 0 < x, y < \pi,$$

and it was this theorem which gave Fejér his idea. Koschmieder also had the positivity of the corresponding result for Legendre polynomial series. Marx's result was generalized to ultraspherical polynomials by Seidel and Szász [12] and further generalized for ultraspherical polynomials and Jacobi polynomials by the authors [4]. The proofs, however, are no longer elementary.

Acknowledgment. The authors are grateful to the referee for pointing out several minor slips.

REFERENCES

[4] R. ASKEY AND J. FITCH, *Integral representations for Jacobi polynomials and some applications*, to appear; also M.R.C. Tech. Summary Rep. 889, Mathematics Research Center, Madison, Wisconsin, 1968.

[5] R. ASKEY, J. FITCH AND G. GASPER, *On positive trigonometric sums*, Proc. Amer. Math. Soc., 19 (1968), p. 1507.

[6] S. BOCHNER, *Positive zonal functions on spheres*, Proc. Nat. Acad. Sci. U.S.A., 40 (1954), pp. 1141–1147.

[7] L. Fejér, *Neue Eigenschaften der Mittelwerte bei den Fourierreihen*, J. London Math. Soc., 8 (1933), pp. 53–62.

[8] L. Koschmieder, *Vorzeicheneigenschafter der Abschnitte einiger physikalisch bedeutsamer Reihen*, Monatsh. Math. u. Physik, 39 (1932), pp. 321–344.

[9] L. Koschmieder and R. Stroman, *Zwei Lösungen der Aufgabe 77*, Jber. Deutsch. Math.-Verein., 43 (1933), pp. 64–66.

[10] E. Landau, *Über eine trigonometrische Ungleichung*, Math. Z., 37 (1933), p. 36.

[11] A. Marx, *Aufgaben 77*, Jber. Deutsch. Math. Verein., 39 (1930), p. 1.

[12] W. Seidel and O. Szász, *On positive harmonic functions and ultraspherical polynomials*, J. London Math. Soc., 26 (1951), pp. 36–41.

Problem 73-21, A Sine Inequality, by R. A. Askey (University of Wisconsin).

It is well known that $|(\sin n\theta)/(n \sin \theta)| \leq 1$ for $n = 1, 2, \cdots$. Show that this inequality can be strengthened to

(i) $\qquad -\dfrac{1}{3} \leq \dfrac{\sin n\theta}{n \sin \theta} \leq \dfrac{\sqrt{6}}{9}, \qquad \dfrac{\pi}{n} \leq \theta \leq \pi - \dfrac{\pi}{n}, \qquad n = 2, 3, \cdots,$

and

(ii) $\qquad -\dfrac{1}{3} \leq \dfrac{\sin n\theta}{n \sin \theta} \leq \dfrac{1}{5}, \qquad \dfrac{\pi}{n} \leq \theta \leq \dfrac{\pi}{2}, \qquad n = 2, 3, \cdots,$

and that the constants $-1/3$, $\sqrt{6}/9$, and $1/5$ are the best possible.

Solution by A. A. Jagers (Technische Hogeschool Twente, Enschede, the Netherlands).

Put $I = (\pi/n, \pi - \pi/n)$, $I_k = (k\pi/n, (k+1)\pi/n)$, and $f_n(\theta) = \sin(n\theta)/(n \sin \theta)$. Note that $I = \{\pi/2\}$ and $f_n(\pi/2) = 0$, if $n = 2$; so suppose $n > 2$.

We will establish some monotonicity properties of the relative extrema of f_n. The given problem is obtained as a corollary.

Observe that $f_n(k\pi/n) = 0$ for $k = 1, 2, \cdots, n-1$ and that, since $\sin \theta$ is positive on I, sign $f_n(\theta) = (-1)^k$ for $\theta \in I_k$ and $k = 1, 2, \cdots, n-2$. Being continuous, f_n attains a minimum or a maximum on I_k depending on whether k is odd or even, respectively. Denote this local extremum of f_n by $f_{n,k}$.

LEMMA. *The function f_n has exactly $n-2$ stationary points, say $\theta_1 < \theta_2 < \cdots < \theta_{n-2}$. Furthermore, $n \tan \theta_k = \tan(n\theta_k)$, $\theta_k \doteq \pi - \theta_{n-1-k}$, and $f_{n,k} = f_n(\theta_k) = (-1)^k \{(1 + \tan^2 \theta_k)/(1 + n^2 \tan^2 \theta_k)\}^{1/2}$.*

Proof. Let $\theta \in I$. Since $f_n'(\theta) = 0$ if and only if $n \tan \theta = \tan(n\theta)$ and since, for $\theta \neq \pi/2$,

$$\tan(n\theta) = \frac{\operatorname{Im}(1 + i \tan \theta)^n}{\operatorname{Re}(1 + i \tan \theta)^n} = \frac{\displaystyle\sum_{k=0}^{[n/2]} (-1)^k \binom{n}{2k+1} (\tan \theta)^{2k+1}}{\displaystyle\sum_{k=0}^{[n/2]} (-1)^k \binom{n}{2k} (\tan \theta)^{2k}},$$

it follows that $f_n'(\theta) = 0$ if and only if, either $\theta = \pi/2$ and n is odd, or $x = \tan^2 \theta$ is a root of (1) below, which is a polynomial of degree $[n/2] - 1$:

(1) $\qquad \displaystyle\sum_{m=0}^{[n/2]-1} (-1)^m (2m+2) \binom{n+1}{2m+3} x^m.$

On the other hand, each $f_{n,k}$ yields at least one stationary point. Hence f_n has exactly $n - 2$ stationary points. The remaining statements of the lemma follow by inspection.

PROPOSITION. (a) $f_{n,k} = (-1)^{n+1} f_{n,n-1-k}$, $\quad k = 1, 2, \cdots, n - 2$,

$\qquad$ (b) $|f_{n,k}| > |f_{n,k+1}|$, $\qquad\qquad k = 1, 2, \cdots, [(n - 1)/2]$,

$\qquad$ (c) $|f_{n,k}| > |f_{n+1,k}|$, $\qquad\qquad n = 2k + 1, 2k + 2, \cdots$.

Proof. (a) Use the lemma and $f_n(\theta) = (-1)^{n+1} f_n(\pi - \theta)$. (b) Note that $\sin \theta$ is monotone increasing for $\theta \in I$, $\theta \leq \pi/2$. (c) By the preceding argument, also $\theta_k \leq (k + \frac{1}{2})\pi/n$. Now substitute $\psi = n\theta$ in $f_n(\theta)$ and consider the denominator $g(x, \psi) = x \sin(\psi/x)$ for $k\pi \leq \psi \leq (k + \frac{1}{2})\pi$. Then for all $x > 2k + 1$,

$$\partial\psi/\partial x = \{\tan(\psi/x) - (\psi/x)\}\cos(\psi/x) > 0,$$

and a fortiori (c) holds.

Now write $A_n = \min\{f_n(\theta)|\theta \in I\}$, $B_n = \max\{f_n(\theta)|\theta \in I\}$, and $C_n = \max\{f_n(\theta)|\theta \in I, \theta \leq \pi/2\}$. Then we have the following corollary.

COROLLARY 1. $A_n = f_{n,1}$ $(n \geq 3)$, $C_n = f_{n,2}$ $(n \geq 5)$, $B_n = f_{n,n-2} = -f_{n,1}$ $(n \geq 4, n \text{ even})$, and $B_n = f_{n,2}$ $(n \geq 5, n \text{ odd})$.

In particular, $A_n = \min\{f_n(\theta)|\theta \in I, \theta \leq \pi/2\}$. For small n, $f_{n,k}$ may be evaluated using (1) and the lemma.

COROLLARY 2. $\min_n A_n = A_3 = -1/3$, $\max_n B_n = B_4 = -A_4 = \sqrt{6}/9$, and $\max_n C_n = C_5 = 1/5$.

Comments by the proposer. If $x = \cos\theta$, then

$$\frac{\sin n\theta}{n \sin\theta}$$

is a polynomial of degree $n - 1$ in x. For convenience, we will consider

$$\frac{\sin(n + 1)\theta}{(n + 1)\sin\theta} = \frac{P_n^{(1/2,1/2)}(x)}{P_n^{(1/2,1/2)}(1)}.$$

The Jacobi polynomials $P_n^{(\alpha,\beta)}(x)$ can be defined by

$$(1 - x)^\alpha(1 + x)^\beta P_n^{(\alpha,\beta)}(x) = \frac{(-1)^n}{2^n n!} \frac{d^n}{dx^n}[(1 - x)^{n+\alpha}(1 + x)^{n+\beta}], \quad \alpha, \beta > -1,$$

$$P_n^{(\alpha,\beta)}(1) = \frac{(\alpha + 1)_n}{n!} = \frac{\Gamma(n + \alpha + 1)}{\Gamma(\alpha + 1)\Gamma(n + 1)} = \binom{n + \alpha}{n}.$$

The monotonicity properties Jagers proved can be extended to many of these polynomials. Let $\alpha, \beta \geq -\frac{1}{2}$, $\alpha + \beta \neq -1$. If $-1 = x_{n,n} < x_{n,n-1} < \cdots < x_{n,1} < x_{n,0} = 1$ denote the places where $|P_n^{(\alpha,\beta)}(x)|$ takes on its relative maxima and $c_{n,k} = |P_n^{(\alpha,\beta)}(x_{n,k})|$; then $c_{n,n} > c_{n,n-1} > \cdots > c_{n,j+1}$ and $c_{n,j} < c_{n,j-1} < \cdots < c_{n,1} < c_{n,0}$, where

$$x_{n,j+1} \leqq \frac{\beta - \alpha}{\alpha + \beta + 1} \leqq x_{n,j}.$$

See the proof of Theorem 7.32.1 in Szegö [7]. This includes the monotonicity given in (b).

There is a second monotonicity result which holds in the case of the symmetric

Jacobi polynomials, i.e., $\alpha = \beta$. J. Todd conjectured, on the basis of the graphs, and G. Szegö [6] proved that $c_{n,k} > c_{n+1,k}, n = k, k+1, \cdots$, for Legendre polynomials, $P_n(x) = P_n^{(0,0)}(x)$, and O. Szász [5] proved that

$$\frac{c_{n,k}}{P_n^{(\alpha,\alpha)}(1)} > \frac{c_{n+1,k}}{P_{n+1}^{(\alpha,\alpha)}(1)}, \qquad n = k, k+1, \cdots, \alpha > -\tfrac{1}{2}.$$

The case $\alpha = \tfrac{1}{2}$ extends (c) to $n = k, k+1, \cdots$. A quadratic transformation

$$\frac{P_n^{(\alpha,-1/2)}(2x^2 - 1)}{P_n^{(\alpha,-1/2)}(1)} = \frac{P_{2n}^{(\alpha,\alpha)}(x)}{P_{2n}^{(\alpha,\alpha)}(1)}$$

applied to Szász's result gives

$$\frac{c_{n,k}}{P_n^{(\alpha,-1/2)}(1)} > \frac{c_{n+1,k}}{P_{n+1}^{(\alpha,-1/2)}(1)}, \qquad n = k, k+1, \cdots, \alpha > -\tfrac{1}{2}.$$

It seems very likely that a similar inequality is also true for $-\tfrac{1}{2} < \beta < \alpha$. But for

$$\frac{P_n(x) + P_{n-1}(x)}{2} = P_n^{(0,-1)}(x),$$

it seems likely that $c_{n,k} < c_{n+1,k}, n = k, k+1, \cdots$. If so, we could prove that $P_n^{(0,-1)}(x) > J_0(x_{1,1}) = -.402$, where $J_0(x)$ is the usual Bessel function of the first kind and $x_{1,1}$ is the first positive zero of $J_1(x)$. See [3] for a problem which would be solved by this inequality.

Applications of the inequalities in this problem are given in [1], [2], and [3]. Fejér [4] had used the weaker inequality

$$\frac{\sin n\theta}{n \sin \theta} > -\frac{2}{3}, \qquad \frac{\pi}{n} \leqq \theta \leqq \pi - \frac{\pi}{n},$$

to prove

$$\sum_{k=1}^{n} \frac{\sin k\theta}{k} \geqq \frac{\sin \theta}{3} + \frac{\sin n\theta}{2n} > 0, \qquad \frac{\pi}{n} < \theta < \pi - \frac{\pi}{n}.$$

When the inequalities in Problem 73-21 are used, we obtain

$$\sum_{k=1}^{n} \frac{\sin k\theta}{k} \geqq \frac{\sin \theta}{3} + \frac{\sin n\theta}{2n} \geqq \frac{\sin \theta}{6}, \qquad \frac{\pi}{n} \leqq \theta \leqq \pi - \frac{\pi}{n},$$

and this is an improvement on Turán's inequality [8]

$$\sum_{k=1}^{n} \frac{\sin k\theta}{k} > 4 \sin \frac{2\theta}{2} \left(\cot \frac{\theta}{2} - \frac{\pi - \theta}{2} \right), \qquad 0 < \theta < \pi,$$

for $2.165 \leqq \theta \leqq \pi - \pi/n$.

REFERENCES

[1] R. Askey and J. Fitch, *A positive Cesàro mean*, Publ. Elek. Fak. Univ. Beogradu Ser. Mat. i Fiz., 406 (1972), pp. 119–122.

[2] R. Askey, *Some absolutely monotonic functions*, to appear.

[3] R. Askey and G. Gasper, *Positive Jacobi polynomials. II*, to appear.

[4] L. Fejér, *Einige Sätze, die sich auf das Vorzeichen, u.s.w.*, Monatsh. Math. Phys., 35 (1928), pp. 305–344; reproduced in *Gesammelte Arbeiten II*, Birkhäuser-Verlag, Basel, 1970, pp. 202–237.

[5] O. Szász, *On the relative extrema of ultraspherical polynomials*, Boll. Un. Mat. Ital., 3 (1950), pp. 125–127.

[6] G. SZEGÖ, *On the relative extrema of Legendre polynomials*, Ibid., 3 (1950), pp. 120–121.

[7] ———, *Orthogonal polynomials*, Colloquium Publications, no. 23, American Mathematical Society, New York, 1959.

[8] P. TURÁN, *Über die Partialsummen der Fourierreihe*, London Math. Soc., 13 (1938), pp. 278–282.

Problem 70-24, *A Legendre Function Inequality*, by JAMES G. SIMMONDS (University of Virginia).

If an elastic spherical shell is acted upon by a concentrated radial point load at the pole $\theta = \pi$ and an equilibrating surface pressure, the solutions of the governing linear equations involve the Legendre function $P_\nu(\cos \theta)$, where $\nu = \lambda + i\mu$ and λ and μ are positive constants of the order of magnitude of the square root of the radius to thickness ratio [1]. It is useful to bound these solutions in terms of simpler functions in order to estimate, for example, the distribution of elastic energy. To this end prove that

$$|P_\nu(\cos \theta)| \leq I_0(\mu\theta) \cos (\theta/2) + e^{\mu\pi} K_0(\mu\phi) \cos (\phi/2),$$

where $0 \leq \theta < \pi$, $\phi = \pi - \theta$, and I_0 and K_0 are, respectively, modified Bessel functions of the first and second kind of order zero. Note that this inequality has the correct behavior at both ends of the domain: it is 1 at $\theta = 0$ and has a logarithmic singularity as $\theta \to \pi$.

Solution by the proposer:

Starting from the Mehler–Dirichlet formula [2, Eq. 3.7(27)],

$$(1) \qquad P_\nu(\cos \theta) = \frac{2}{\pi} \int_0^\theta \frac{\cos (\nu + \tfrac{1}{2})\alpha \, d\alpha}{\sqrt{2(\cos \alpha - \cos \theta)}},$$

valid for $0 < \theta < \pi$, we derive two different upper bounds A and B for $|P_\nu(\cos \theta)|$. The desired inequality then follows upon noting that

$$(2) \qquad \begin{aligned} |P_\nu(\cos \theta)| &= |P_\nu(\cos \theta)| \cos^2(\theta/2) + |P_\nu(\cos \theta)| \sin^2(\theta/2) \\ &\leq A(\theta) \cos^2(\theta/2) + B(\phi) \cos^2(\phi/2). \end{aligned}$$

The purpose of introducing the trigonometric functions is to suppress unwanted singularities in A and B at $\theta = \pi$ and $\phi = \pi$ respectively.

First, with the inequalities

$$(3) \qquad |\cos (\nu + \tfrac{1}{2})\alpha| \leq e^{\mu\alpha}, \qquad \nu = \lambda + i\mu,$$

and

$$(4) \qquad \begin{aligned} 2(\cos \alpha - \cos \theta) = 2 \int_\alpha^\theta \sin t \, dt &\geq 2\theta^{-1} \sin \theta \int_\alpha^\theta t \, dt \\ &\geq \cos^2(\theta/2)(\theta^2 - \alpha^2), \end{aligned}$$

and with the change of variable $\alpha = \theta t$, we obtain from (1) the inequality

$$|P_\nu(\cos\theta)| < \frac{2}{\pi}\sec(\theta/2)\int_0^1 \frac{e^{\mu\theta t}\,dt}{\sqrt{1-t^2}}$$

(5)

$$= I_0(\mu\theta)\sec(\theta/2) \equiv A(\theta).$$

Next set $\theta = \pi - \phi$, $\beta = \pi - \alpha$ in (1) to get

(6)
$$P_\nu(-\cos\phi) = \frac{2}{\pi}\int_\phi^\pi \frac{\cos(\nu+\frac{1}{2})(\pi-\beta)\,d\beta}{\sqrt{2(\cos\phi)-\cos\beta}}.$$

Use (3) again, but instead of (4), use the relation

$$2(\cos\phi - \cos\beta) \geq 2\left(\frac{\beta^2 - \phi^2}{\pi^2 - \phi^2}\right)(1 + \cos\phi)$$

(7)

$$\geq (2/\pi)^2(\beta^2 - \phi^2)\cos^2(\phi/2)$$

which is valid for $\phi \leq \beta \leq \pi$. The first line of this inequality becomes obvious upon comparing the graphs of $\cos\phi - \cos\beta$ and $a(\beta^2 - \phi^2)$ on the interval $\phi \leq \beta \leq \pi$ and then choosing the constant a so that these graphs intersect at $\beta = \pi$. It now follows that

$$|P_\nu(-\cos\phi)| < e^{\mu\pi}\sec(\phi/2)\int_\phi^\pi \frac{e^{-\pi\beta}\,d\beta}{\sqrt{\beta^2 - \phi^2}}$$

(8)

$$< e^{\mu\pi}\sec(\phi/2)\int_1^\infty \frac{e^{-\mu\phi t}\,dt}{\sqrt{t^2 - 1}}$$

$$= e^{\mu\pi}K_0(\mu\phi)\sec(\phi/2) \equiv B(\phi).$$

When (5) and (8) are inserted into (2), and it is observed that both sides are equal when $\theta = 0$, the desired inequality is obtained.

REFERENCES

[1] J. G. SIMMONDS, *Green's functions for closed elastic spherical shells: Exact and accurate approximate solutions*, Nederl. Akad. Wetensch. Proc. Ser. B, 71 (1968), pp. 236–249.
[2] A. ERDÉLYI, W. MAGNUS, F. OBERHETTINGER AND F. TRICOMI, *Higher Transcendental Functions*, vol. I, McGraw-Hill, New York, 1953.

Problem 74-19, One-sided Approximation to Special Functions, by R. ASKEY (University of Wisconsin).

It is well known that the partial sum expansions of $\sin x$ and $\cos x$ approximate these functions on one side, i.e., if

$$s_n(x) = \sum_{k=0}^n \frac{(-1)^k x^{2k+1}}{(2k+1)!}, \qquad c_n(x) = \sum_{k=0}^n \frac{(-1)^k x^{2k}}{(2k)!},$$

then

$$(-1)^{n+1}[\sin x - s_n(x)] \geq 0, \qquad x \geq 0,$$

$$(-1)^{n+1}[\cos x - c_n(x)] \geq 0, \qquad x \text{ real}.$$

According to P. Turán, L. Féjer lectured on a similar theorem for Jacobi polynomials in 1944. State and prove such a theorem, and a similar theorem for Bessel functions.

Solution by the proposer.

The Jacobi polynomials, $P_n^{(\alpha,\beta)}(x)$, can be defined by

$$P_n^{(\alpha,\beta)}(x) = \frac{(\alpha + 1)_n}{n!} {}_2F_1(-n, n + \alpha + \beta + 1; \alpha + 1; (1 - x)/2).$$

Let $\alpha \geq \beta > -1$, $\alpha \geq -\frac{1}{2}$. We will show that

(1) $(-1)^{k+1}\left[P_n^{(\alpha,\beta)}(x) - \frac{(\alpha + 1)_n}{n!} \sum_{j=0}^{k} \frac{(-n)_j(n + \alpha + \beta + 1)_j}{(\alpha + 1)_j j!}\left(\frac{1 - x}{2}\right)^j\right] \geq 0,$

$$-1 \leq x \leq 1, \quad k = 0, 1, \cdots, n.$$

(1) is equivalent to

(2) $(-1)^{k+1}[g_{n,n}^{\alpha,\beta}(y) - g_{n,k}^{\alpha,\beta}(y)] \geq 0,$

$$0 \leq y \leq 1, \quad \alpha \geq \beta > -1, \quad \alpha \geq -\frac{1}{2}, . \qquad k = 0, 1, \cdots, n,$$

where $g_{n,k}^{\alpha,\beta}(y)$ is defined by

$$g_{n,k}^{\alpha,\beta}(y) = \sum_{j=0}^{k} \frac{(-n)_j(n + \alpha + \beta + 1)_j}{(\alpha + 1)_j j!} y^j.$$

Szegö [1, Theorem 7.32.1] has proven that

(3) $|P_n^{(\alpha,\beta)}(x)| \leq P_n^{(\alpha,\beta)}(1),$

$$-1 \leq x \leq 1, \quad \alpha \geq \beta > -1, \quad \alpha \geq -\frac{1}{2},$$

and this inequality is equivalent to

$$|g_{n,n}^{\alpha,\beta}(y)| \leq g_{n,n}^{\alpha,\beta}(0) = 1 = g_{n,k}^{\alpha,\beta}(0) = g_{n,0}^{\alpha,\beta}(y),$$

$$\alpha \geq \beta > -1, \quad \alpha \geq -\frac{1}{2}, \quad 0 \leq y \leq 1,$$

so (2) holds when $k = 0$. We complete the proof by induction. A simple calculation gives

(4) $\frac{d}{dy} g_{n,k}^{\alpha,\beta}(y) \quad \frac{-n(n + \alpha + \beta + 1)}{(\alpha + 1)} g_{n-1,k-1}^{\alpha+1,\beta+1}(y),$

so

$$g_{n,k}^{\alpha,\beta}(y) - g_{n,k}^{\alpha,\beta}(0) = \frac{-n(n + \alpha + \beta + 1)}{(\alpha + 1)} \int_0^y g_{n-1,k-1}^{\alpha+1,\beta+1}(t)\, dt.$$

Then

$$g_{n,n}^{\alpha,\beta}(y) - g_{n,k}^{\alpha,\beta}(y) = \frac{-n(n + \alpha + \beta + 1)}{(\alpha + 1)} \int_0^y [g_{n-1,n-1}^{\alpha+1,\beta+1}(t) - g_{n-1,k-1}^{\alpha+1,\beta+1}(t)]\, dt,$$

and this proves (2) by induction. Observe that the condition $\alpha \geq -\frac{1}{2}$ can be replaced by $\alpha > -1$ when $k \geq 1$.

When $\beta > \alpha$, a similar theorem is true when the Jacobi polynomial $P_n^{(\alpha,\beta)}(x)$ is expanded about $x = -1$.

For Bessel functions, $J_\alpha(x)$, the proof is similar. $J_\alpha(x)$ is defined by

$$J_\alpha(x) = \sum_{n=0}^{\infty} \frac{(-1)^n (x/2)^{2n+\alpha}}{\Gamma(n + \alpha + 1)n!}.$$

The inequality which replaces (3) is

$$\left| \left(\frac{2}{x}\right)^\alpha J_\alpha(x) \right| \leq \lim_{x \to 0} \left(\frac{2}{x}\right)^\alpha J_\alpha(x) = \frac{1}{\Gamma(\alpha + 1)}, \quad \alpha \geq -\tfrac{1}{2}, \quad x \geq 0.$$

The analogue of $g_{n,k}^{\alpha,\beta}(y)$ is

$$g_n^\alpha(y) = \sum_{k=0}^{n} \frac{(-1)^k y^k}{(\alpha + 1)_k k!},$$

and the analogue of (4) is

$$\frac{d}{dy} g_n^\alpha(y) = \frac{-1}{\alpha + 1} g_{n-1}^{\alpha+1}(y).$$

The inequality is

(5) $$(-1)^{n+1} \left[J_\alpha(x) - \sum_{k=0}^{n} \frac{(-1)^k (x/2)^{2k+\alpha}}{\Gamma(k + \alpha + 1)k!} \right] \geq 0, \quad x \geq 0, \quad \alpha \geq -\tfrac{1}{2},$$

and (5) holds for $\alpha > -1$ when $n \geq 1$.

When $\alpha = -\tfrac{1}{2}$ this gives the inequality for cosines, and when $\alpha = \tfrac{1}{2}$ this gives the inequality for sines. What is involved in each of these inequalities is an inequality that certain hypergeometric functions are bounded by their values at one end of an interval. This suggests that similar inequalities should be obtained for other hypergeometric functions.

Problem 75-11, *Inequalities for the Gamma Function*, by D. K. Ross (La Trobe University, Victoria, Australia).

It is known that the Gamma function $\Gamma(x)$ satisfies the inequality

$$\begin{vmatrix} \Gamma(\alpha) & \Gamma(\alpha+\beta) \\ \Gamma(\alpha+\beta) & \Gamma(\alpha+2\beta) \end{vmatrix} > 0,$$

provided that $\alpha, \beta > 0$.

Prove that

$$\begin{vmatrix} \Gamma(\alpha) & \Gamma(\alpha+\beta) & \Gamma(\alpha+2\beta) \\ \Gamma(\alpha+\beta) & \Gamma(\alpha+2\beta) & \Gamma(\alpha+3\beta) \\ \Gamma(\alpha+2\beta) & \Gamma(\alpha+3\beta) & \Gamma(\alpha+4\beta) \end{vmatrix} > 0,$$

provided that $\alpha, \beta > 0$, and generalize the result to include higher order determinants and other classes of special functions.

Solution by A. A. JAGERS (Technische Hogeschool Twente, Enschede, the Netherlands).

The following generalization is well known: let f be a continuous function defined on the interval (a, b). For all $k \in \mathbb{N}$ and all x, δ with $a < x < x + 2(k-1)\delta < b$, let

$$F(k, x, \delta) = [f(x + (i+j)\delta)]_{i,j=0}^{k-1}.$$

Then the following four conditions on f are equivalent:

(i) $f(x + y)$ is strictly totally positive with respect to the variables x and y satisfying $a < x + y < b$.

(ii) $\det(F(k, x, \delta)) > 0$ for all $k \in \mathbb{N}$ and all x, δ with $a < x < x + 2(k-1)\delta < b$.

(iii) $F(k, x, \delta)$ is positive-definite for all $k \in \mathbb{N}$ and all x, δ with $a < x < x + 2(k-1)\delta < b$.

(iv) f can be represented as a Laplace–Stieltjes transform:

$$f(x) = \int_{-\infty}^{\infty} e^{\alpha x} \, d\mu(\alpha)$$

where μ is a nonnegative measure with an infinite number of points of increase.

Cf. Chap. 2, § 7 of: S. Karlin, *Total Positivity*, vol. I, Stanford University Press, Stanford, 1968.

Example. The Γ-function; $\Gamma(x) = \int_{-\infty}^{\infty} e^{\alpha x} e^{-e^{\alpha}} \, d\alpha, \ x > 0$.

Problem 72-15, *Inequalities for Special Functions*, by D. K. ROSS (La Trobe University, Victoria, Australia).

Prove that

(1) $$1 > \frac{{}_1F_1(a; c; x)}{{}_1F_1(a; c; y)} > e^{x-y}$$

if $y > x > 0$, $c > a > 0$ and that

(2) $$\frac{{}_1F_1(a; c; x)}{{}_1F_1(a; d; x)} > \frac{\Gamma(c)\Gamma(d-a)}{\Gamma(d)\Gamma(c-a)}$$

if $d > c > a > 0$. As usual, ${}_1F_1(a; c; x)$ denotes the confluent hypergeometric function. Also, show that

(3) $$K_\nu(x)/K_\nu(y) > e^{y-x}(x/y)^\nu$$

if $y > x > 0$, $y > 0$, where $K_\nu(x)$ denotes the modified Bessel function of the second kind. Finally, state and prove a general result from which the above and numerous other inequalities for a general class of functions can be deduced.

Solution by D. J. BORDELON (Naval Underground Systems Center, Newport, Rhode Island).

An integral representation [1, relation 4.2(1), p. 115] of ${}_1F_1(a; c; x)$ is

$$(1) \qquad {}_1F_1(a;c;x) = \frac{\Gamma(c)}{\Gamma(a)\Gamma(c-a)} \int_0^1 e^{xt} t^{a-1}(1-t)^{c-a-1}\, dt,$$

$$c > a > 0.$$

Therefore,

$$(2) \qquad \frac{{}_1F_1(a;c;x)}{{}_1F_1(a;c;y)} = \frac{\int_0^1 e^{xt} t^{a-1}(1-t)^{c-a-1}\, dt}{\int_0^1 e^{yt} t^{a-1}(1-t)^{c-a-1}\, dt} < 1$$

if $c > a > 0$, $e^{xt} < e^{yt}$ on $(0 < t < 1)$, or if $c > a > 0$, $y > x$.

Another integral representation of ${}_1F_1(a;c;x)$, obtained from representation (1) by the transformation $t = 1 - u$, is

$$(3) \qquad {}_1F_1(a;c;x) = \frac{\Gamma(c)\,e^x}{\Gamma(a)\Gamma(c-a)} \int_0^1 e^{-xu} u^{c-a-1}(1-u)^{a-1}\, du,$$

$$c > a > 0.$$

Therefore,

$$(4) \qquad \frac{{}_1F_1(a;c;x)}{{}_1F_1(a;c;y)} > e^{x-y}$$

if $c > a > 0$, $e^{-xu} > e^{-yu}$ on $(0 < u < 1)$, or if $c > a > 0$, $y > x$.

$$(5) \qquad \frac{{}_1F_1(a;c;x)}{{}_1F_1(a;d;x)} > \frac{\Gamma(c)\Gamma(d-a)}{\Gamma(d)\Gamma(c-a)},$$

if $c > a > 0$, $d > a > 0$, $(1-t)^{c-a-1} > (1-t)^{d-a-1}$ on $(0 < t <)$, or if $d > c > a > 0$.

An integral representation [1, relation 6.2.7(8), p. 213; 4.2(7), p. 116] of $K_\nu(x)$ is

$$(6) \qquad K_\nu(x) = \pi^{1/2}\, e^{-x}(2x)^\nu \psi(\nu + 1/2; 2\nu + 1; 2x),$$

$$(7) \qquad \psi(a;c;x) = [\Gamma(a)]^{-1} \int_0^\infty e^{-xt} t^{a-1}(1+t)^{c-a-1}\, dt,$$

$$a > 0, \qquad x > 0.$$

Therefore,

$$(8) \qquad K_\nu(x)/K_\nu(y) > e^{y-x}(x/y)^\nu$$

if $\nu = 1/2$, $e^{-2xt} > e^{-2yt}$ on $(0 < t < \infty)$, or if $\nu > -1/2$, $y > x > 0$.

An integral representation [1, relation 6.2.7(8), p. 213; p. 116] of the hypergeometric function, $F(a, b; c; x)$, is

$$(9) \qquad F(a, b; c; x) = \frac{\Gamma(c)}{\Gamma(c-a)\Gamma(a)} \int_0^1 t^{a-1}(1-t)^{c-a-1}(1-xt)^{-b}\, dt,$$

$$c > a > 0, \qquad 1 > x.$$

Therefore,

(10) $F(a, b; c; x)/F(a, b; c; y) < 1$

if $c > a > 0$, $(1 - xt)^{-b} < (1 - yt)^{-b}$ on $(0 < t < 1)$ or if $c > a > 0$, $1 > y > x$. Clearly,

(11) $$\lim_{b \to \infty} \frac{F(a, b; c; x/b)}{F(a, b; c; y/b)} = \frac{{}_1F_1(a, c; x)}{{}_1F_1(a, c; y)} < 1,$$

if $c > a > 0$, $y > x$. Another integral representation [1, relation 3.6(2), p. 57] of $F(a, b; c; x)$ is

(12) $$F(a, b; c; x) = \frac{\Gamma(c)}{\Gamma(a)\Gamma(c - a)} \int_0^\infty t^{a-1}(1 + t)^{b-c}(1 + t - xt)^{-b} \, dt,$$

$$c > a > 0, \qquad 1 > x.$$

Therefore,

(13) $F(a, b; c; x)/F(a, b; c; y) > (1 - x)^{-b}/(1 - y)^{-b}$

if $c > a > 0$, $(\{1/(1 - x)\} + t)^{-b} > (\{1/(1 - y)\} + t)^{-b}$ on $(0 < t < \infty)$ or if $c > a > 0$, $1 > y > x$.

Clearly,

(14) $$\lim_{b \to \infty} \frac{F(a, b; c; x/b)}{F(a, b; c; y/b)} = \frac{{}_1F_1(a, c; x)}{{}_1F_1(a, c; y)} > e^{x-y}$$

if $c > a > 0$, $y > x$. Further, from (9),

(15) $$\frac{F(a, b; c; x)}{F(a, b; d; x)} > \frac{\Gamma(c)\Gamma(d - a)}{\Gamma(d)\Gamma(c - a)}$$

if $c > a > 0$, $d > a > 0$, $(1 - t)^{c-a-1} > (1 - t)^{d-a-1}$ on $(0 < t < 1)$ or if $d > c > a > 0$, $1 > x$.

Again,

(16) $$\lim_{b \to \infty} \frac{F(a, b; c; x/b)}{F(a, b; d; x/b)} = \frac{F(a, c; x)}{F(a, d; x)} > \frac{\Gamma(c)\Gamma(d - a)}{\Gamma(d)\Gamma(c - a)}$$

if $d > c > a > 0$. From (12) with $u = (c/x)t$,

(17) $$F(a, b; c; (1 - (c/x))) = \frac{x^a \Gamma(c)}{c^a \Gamma(a)\Gamma(c - a)} \int_0^\infty u^{a-1}(1 + (x/c)u)^{b-c}(1 + u)^{-b} \, du,$$

$$c > a > 0, \qquad x > 0.$$

(18) $$\frac{F(a, b; c; 1 - (c/x))}{F(a, b; c; 1 - (c/y))} > (x/y)^a$$

if $c > a > 0$, $(1 + (x/c)u)^{b-c} > (1 + (y/c)u)^{b-c}$ on $(0 < u < \infty)$ or if $c > a > 0$, $c > b$, $y > x > 0$. Thus

(19)
$$\lim_{c \to \infty} \frac{F(a, b; c; 1 - (c/2x))}{F(a, b; c; 1 - (c/2y))} = \frac{x^a \psi(a; a - b + 1; 2x)}{y^a \psi(a; a - b + 1; 2y)} = \frac{K_\nu(x)x^{1/2} e^{-y}}{K_\nu(y)y^{1/2} e^{-x}}$$

$$> \left(\frac{x}{y}\right)^{\nu + 1/2}, \qquad \nu > -1/2, \qquad y > x > 0,$$

or

(20)
$$\frac{K_\nu(x)}{K_\nu(y)} > e^{y-x}\left(\frac{x}{y}\right)^\nu$$

if $\nu > -1/2$, $y > x > 0$.

REFERENCE

[1] YUDELL L. LUKE, *The Special Functions and Their Approximations*, vol. 1, Academic Press, New York, 1969.

Also solved by the proposer who notes the following monotonicity theorem.

Let $f(t)$ be a positive integrable function defined for $a \leq t \leq b$ and let $g_\alpha(t)$ be a sequence of integrable functions which increase (or decrease) with increasing α for each fixed t in $a \leq t \leq b$. Then, the integral

$$U_\alpha = \int_a^b g_\alpha(t) f(t)\, dt$$

exists and is a monotonic sequence of functions which increase (or decrease) with increasing α.

The proof of this result depends on the simple fact that the integrand of $U_\alpha - U_\beta$ is integrable and positive (or negative) whenever $\alpha \geq \beta$.

On the Sign of a Set of Hypergeometric Functions

Problem 86-3, by RICHARD ASKEY (University of Wisconsin).

Watson showed that

$${}_3F_2\left(\begin{matrix} -n, n+2a+2b-1, a \\ a+b, 2a \end{matrix}; 1\right)$$

$$= \frac{\Gamma(\tfrac{1}{2})\Gamma(a+\tfrac{1}{2})\Gamma(a+b)\Gamma(1-b)}{\Gamma((1-n)/2)\Gamma(a+\tfrac{1}{2}+n/2)\Gamma(a+b+n/2)\Gamma(1-b-n/2)},$$

so this series vanishes when n is an odd positive integer and $a, b > 0$. Show that if $a, b > 0$ then

$$f_n(t) = {}_3F_2\left(\begin{matrix} -n, n+2a+2b-1, a+t \\ a+b, 2a \end{matrix}; 1\right)$$

satisfies

    (i) $f(t) > 0$, $t < 0$, $n = 0, 1, \cdots$,
    (ii) $f(t) > 0$, $t \geq 0$, $n = 0, 2, \cdots$,
    (iii) $f(t) < 0$, $t > 0$, $n = 1, 3, \cdots$.

Solution by MIZAN RAHMAN (Carleton University, Ottawa, Ontario, Canada).

One of Bailey's quadratic transformation formulas [1] gives

(1)
$$f_n(t) = \frac{(-2t)_n}{(2a)_n} {}_4F_3\left[\begin{matrix} -n/2, (1-n)/2, a+t, b+t \\ a+b, t+(1-n)/2, t+(2-n)/2 \end{matrix}; 1\right].$$

Use of Whipple's transformation formula for a balanced $_4F_3$ series [1] then gives

(2)
$$f_{2m}(t) = \frac{(b)_m(\frac{1}{2})_m}{(a+\frac{1}{2})_m(a+b)_m} p_m(t^2),$$

$$f_{2m+1}(t) = -\frac{t}{a} \frac{(b+1)_m(\frac{3}{2})_m}{(a+\frac{1}{2})_m(a+b)_m} q_m(t^2),$$

$m = 0, 1, 2, \cdots$, where

(3)
$$p_m(t^2) = {}_4F_3\left[\begin{matrix} -m, m+a+b-\frac{1}{2}, t, -t \\ a, b, \frac{1}{2} \end{matrix}; 1\right],$$

$$q_m(t^2) = {}_4F_3\left[\begin{matrix} -m, m+a+b+\frac{1}{2}, 1+t, 1-t \\ a+1, b+1, \frac{3}{2} \end{matrix}; 1\right].$$

It can be easily verified that

(4)
$$p_{m+1}(t^2) - p_m(t^2) = \frac{2(2m+a+b+\frac{1}{2})}{ab} t^2 q_m(t^2).$$

It is known from Wilson's Ph.D. thesis [2] that $p_m(t^2)$ satisfies the recurrence relation

(5)
$$A_m(p_{m+1}(t^2) - p_m(t^2)) = t^2 p_m(t^2) + B_m(p_m(t^2) - p_{m-1}(t^2))$$

where

(6)
$$A_m = \frac{(m+a+b-\frac{1}{2})(m+a)(m+b)(m+\frac{1}{2})}{(2m+a+b-\frac{1}{2})(2m+a+a+b+\frac{1}{2})},$$

$$B_m = \frac{m(m+a+b-1)(m+a-\frac{1}{2})(m+b-\frac{1}{2})}{(2m+a+b-\frac{3}{2})(2m+a+b-\frac{1}{2})}.$$

Since $a, b > 0$ it is clear that

(7)
$$A_m > 0, \qquad B_m > 0 \qquad \text{for } m = 1, 2, \cdots.$$

Also $p_0(t^2) = 1$ and $p_1(t^2) = 1 + t^2((2a+2b+1)/ab) > 1$ so that $p_1(t^2) > p_0(t^2)$. It is then easily proved by induction that

(8)
$$p_m(t^2) \geqq 1,$$

(9)
$$p_{m+1}(t^2) - p_m(t^2) > 0, \qquad m = 0, 1, 2, \cdots.$$

Equation (4) then implies

(10)
$$q_m(t^2) > 0, \qquad m = 0, 1, 2, \cdots.$$

Equations (2) now lead to the solution of the problem.

REFERENCES

[1] W. W. BAILEY, *Generalized Hypergeometric Series*, Stechert-Hafner, New York and London, 1964.
[2] JAMES WILSON, Ph.D. thesis, University of Wisconsin, Madison, WI, 1978.

Solution by the proposer.
    Askey and Wilson [1] showed that when $a, b > 0$,

$$p_n(x) = i^n \, {}_3F_2\left(\begin{matrix} -n, \, n+2a+2b-1, \, a-ix \\ a+b, \, 2a \end{matrix}; \, 1\right)$$

are polynomials orthogonal with respect to $|\Gamma(a + ix)\Gamma(b + ix)|^2$ on $(-\infty, \infty)$. Thus $p_n(x)$ has n real zeros when a, $b > 0$ and the zeros are simple. Thus the only real zero of $f_n(t)$ occurs when $t = 0$ and n is odd, and it is a simple zero. When $t = -a$, $f_n(-a) = 1$. Together these imply (i), (ii) and (iii).

REFERENCE

[1] R. ASKEY AND J. WILSON, *A set of hypergeometric orthogonal polynomials*, SIAM J. Math. Anal., 13 (1982), pp. 651–655.

Problem 70-16, A Positive Functional, by RICHARD KELLNER (Los Alamos Scientific Laboratory).

If

$$\Phi[f] = 20 \int_0^\pi f^2(\theta) \sin \theta \, d\theta + 2 \int_0^\pi f'^2(\theta) \sin \theta \, d\theta$$

$$-10\left\{\int_0^\pi f(\theta) \sin \theta \, d\theta\right\}^2 - 9\left\{\int_0^\pi f(\theta) \sin 2\theta \, d\theta\right\}^2,$$

show that

$$\Phi[f] \geqq 0,$$

and determine when the equality sign holds.

This problem has arisen in some work on the Steklov eigenvalue problem.

Solution by J. E. JAMISON and C. C. ROUSSEAU (Memphis State University).

Note that if A is any constant,

$$\Phi[f(\theta) + A] = \Phi[f(\theta)].$$

Further, observe that

$$\Phi[B \cos \theta] = 0.$$

Therefore, for arbitrary constants A and B,

$$\Phi[A + B \cos \theta] = 0.$$

If $f(\theta) = A + B \cos \theta$, then by orthogonality,

(1) $$A = \frac{1}{2} \int_0^\pi f(\theta) \sin \theta \, d\theta$$

and

(2) $$B = \frac{3}{4} \int_0^\pi f(\theta) \sin 2\theta \, d\theta.$$

For an arbitrary function $f(\theta)$, we agree to *define* A and B by (1) and (2) respectively.

If C_1 and C_2 are arbitrary positive constants, the functional

$$\Phi_G[f; C_1, C_2] = \int_0^\pi \{C_1[f(\theta) - A - B\cos\theta]^2 + C_2[f'(\theta) + B\sin\theta]^2\}\sin\theta\, d\theta$$

is nonnegative and vanishes if and only if $f(\theta) = A + B\cos\theta$. Substituting A and B from (1) and (2) and integrating term by term yields

$$\Phi_G[f; C_1, C_2] = C_1 \int_0^\pi f^2(\theta)\sin\theta\, d\theta + C_2 \int_0^\pi f'^2(\theta)\sin\theta\, d\theta$$

$$-\frac{C_1}{2}\left(\int_0^\pi f(\theta)\sin\theta\, d\theta\right)^2 - \frac{3}{8}(C_1 + 2C_2)\left(\int_0^\pi f(\theta)\sin 2\theta\, d\theta\right)^2.$$

The original problem is solved by noting that $\Phi[f] = \Phi_G[f; 20, 2]$.

Solution by J. W. Riese (Kimberly–Clark Corporation).

The problem "falls apart" if the arbitrary function $f(\theta)$ is represented by the appropriate series of Legendre polynomials

$$f(\theta) = \sum_{n=0}^\infty a_n P_n(\cos\theta).$$

Direct substitution yields for the functional

$$\Phi = 4 \sum_{n=2}^\infty a_n^2 \cdot \left(\frac{n^2 + n + 10}{2n + 1}\right).$$

A sum of squares with positive coefficients obviously takes its minimum value, 0, only when each squared term equals 0:

$$a_n = 0 \quad \text{for } n \geq 2,$$

i.e., when $f(\theta) = a_0 + a_1 \cos\theta$.

Problem 77-6*, *An Integral Inequality*, by J. E. Wilkins, Jr. (EG & G Idaho, Inc.).
    To complete the solution of a certain variational problem arising in physical optics [4], it is necessary to verify that

(1) $$\left[\int_0^1 J_0(vx)x\, dx\right]^2 \int_0^1 J_0''^2(vx)x^5\, dx > \left[\int_0^1 J_0''(vx)x^3\, dx\right]^2 \int_0^1 J_0^2(vx)x\, dx,$$

at least if $0 \leq v \leq v_0$, in which $v_0 = 2.29991$ is the smallest positive zero of

$$\int_0^1 J_0''(vx)x^3\, dx.$$

Numerical calculations indicate that (1) is true when $v = 0.(0.1)2.9$, but not when $v = 3.0$. Establish the truth of (1) when $0 \leq v \leq v_0$.
    Partial solution by the proposer.
    Introduce the function

$$G(v) = \left[\int_0^v J_0(u)u\,du\right]^2 \int_0^v J_0''^2(u)u^5\,du - \left[\int_0^v J_0''(u)u^3\,du\right]^2 \int_0^v J_0^2(u)u\,du.$$

Then,

$$G'(v) = 2vJ_0(v)\int_0^v J_0(u)u\,du \int_0^v J_0''^2(u)u^5\,du + v^5 J_0''^2(v)\left[\int_0^v J_0(u)u\,du\right]^2$$

$$-2v^3 J_0''(v)\int_0^v J_0''(u)u^3\,du \int_0^v J_0^2(u)u\,du - vJ_0^2(v)\left[\int_0^v J_0''(u)u^3\,du\right]^2.$$

From the Schwarz–Buniakowski inequality,

$$\left[\int_0^v J_0''(u)u^3\,du\right]^2 \le \int_0^v J_0''^2(u)u^5\,du \int_0^v u\,du.$$

Hence,

$$G'(v) \ge vJ_0(v)\int_0^v J_0''^2(u)u^5\,du\left[2\int_0^v J_0(u)u\,du - \frac{v^2 J_0(v)}{2}\right]$$

$$+ v^5 J_0''^2(v)\left[\int_0^v J_0(u)u\,du\right]^2 - 2v^3 J_0''(v)\int_0^v J_0''(u)\,u^3\,du \int_0^v J_0^2(u)u\,du.$$

Moreover, $J_0(u) \ge J_0(v)$ if $0 \le u \le v \le j_1 = 3.8317$, and, consequently, if $0 \le v \le j_0 = 2.4048$ (so that $J_0(v) \ge 0$),

$$2G'(v) \ge v^3 J_0^2(v)\int_0^v J_0''^2(u)u^5\,du + v^5 J_0''^2(v)\left[\int_0^v J_0(u)u\,du\right]^2$$

$$-2v^3 J_0''(v)\int_0^v J_0''(u)u^3\,du \int_0^v J_0^2(u)u\,du.$$

It now follows that $G'(v) > 0$ if $1.8412 = j_1' \le v \le v_0 = 2.2999$ since $J_0''(v) \ge 0$ and

$$\int_0^v J_0''(u)u^3\,du \le 0$$

on this interval.

It is known that [1]

$$\int_0^v J_0(u)u\,du = vJ_1(v), \qquad \int_0^v J_0^2(u)u\,du = \frac{v^2}{2}\{J_0^2(v) + J_1^2(v)\},$$

$$\int_0^v J_0''(u)u^3\,du = (v^3 - 6v)J_1(v) + 3v^2 J_0(v),$$

$$\int_0^v J_0''^2(u)u^5\,du = \frac{v^2}{30}\{(3v^4 + 23v^2)J_0^2(v) + 4v(3v^2 - 23)J_0(v)J_1(v)$$

$$+ (3v^4 - 31v^2 + 92)J_1^2(v)\},$$

and that $J_0''(v) = 0$ if, and only if, $J_1(v) = vJ_0(v)$. It then follows that

$$G(v) = \frac{v^4}{30}\{-(12v^4 - 149v^2 + 448)J_1^4(v) - (78v^3 - 448v)J_1^3(v)J_0(v)$$

$$-(12v^4 - 68v^2 + 540)J_1^2(v)J_0^2(v)$$

$$-(90v^3 - 540v)J_1(v)J_0^3(v) - 135v^2J_0^4(v)\},$$

and that $G(j_1') = J_0^4(j_1')\{-12j_1'^6 + 59j_1'^4 - 22j_1'^2 - 135\}/30j_1'^4 = 0.0000277 > 0$. We conclude that $G(v) \geq G(j_1') > 0$ if $j_1' \leq v \leq v_0$, and this establishes the truth of (1) for these values of v.

If we use the identity

$$\int_a^b f^2(x)\, dx \int_a^b g^2(x)\, dx - \left[\int_0^1 f(x)g(x)\, dx\right]^2 = \frac{1}{2}\int_0^1\int_0^1 [f(x)g(y) - f(y)g(x)]^2\, dx\, dy,$$

we find that

$$2G(v) = \int_0^v J_0^2(u)u\, du \int_0^v\int_0^v [J_0''(x)x^2 - J_0''(y)y^2]^2 xy\, dx\, dy$$

$$- \int_0^v J_0''^2(u)u^5\, du \int_0^v\int_0^v [J_0(x) - J_0(y)]^2 xy\, dx\, dy$$

$$= \int_0^v\int_0^v\int_0^v J_0^2(u)[J_0(x) - J_0(y)]^2\left[\left(\frac{J_0''(x)x^2 - J_0''(y)y^2}{J_0(x) - J_0(y)}\right)^2 - \left(\frac{J_0''(u)u^2}{J_0(u)}\right)^2\right] uxy\, du\, dx\, dy.$$

For each x and y on the interval $[0, v]$, there exists a value w on the interval $(0, v)$ for which

$$\frac{J_0''(x)x^2 - J_0''(y)y^2}{J_0(x) - J_0(y)} = \frac{\{J_0''(w)w^2\}'}{J_0'(w)} = \frac{wJ_0(w)}{J_1(w)} - w^2 \equiv f_1(w).$$

It is easy to see that $f_1(0) = 2$ and that

$$f_1'(w) = w\left[\frac{J_0(w)J_2(w)}{J_1^2(w)} - 3\right].$$

Szasz [2] has shown that $2J_0(w)J_2(w) < J_1^2(w)$ for all positive w and so $f_1'(w) < 0$, $f_1(w) > f_1(v)$ if v does not exceed the smallest positive zero $j_1 = 3.8317$ of J_1. Moreover, $f_1(v) \geq 0$ if $v \leq u_0$, in which $u_0 = 1.2558$ is the smallest positive solution of the equation $J_0(u_0) = u_0 J_1(u_0)$. (Having located this value u_0 as a solution of this equation, we deduce from Dixon's theorem [3] that the positive zeros of $f_1 = w\{J_0(w) + wJ_0'(w)\}/J_1(w)$ are interlaced with the positive zeros of J_0, and so u_0 must be the smallest positive solution.) We conclude that $f_1^2(w) > f_1^2(v)$ if $0 < v \leq u_0$.

If $f_2(u) = -J_0''(u)u^2/J_0(u)$, then $f_2(u) \geq 0$ if $0 < u \leq j_1'$ and $j_1' = 1.8412$ is the smallest positive zero of $J_1'(u) = -J_0''(u)$. Also

$$f_2'(u) = u\left[1 - \frac{J_1^2(u)}{J_0^2(u)}\right],$$

and so $f_2'(u) > 0$ if $0 < u < u_1$ in which $u_1 = 1.4347$ is the smallest positive zero of the function $J_0(u) - J_1(u)$. Hence, $f_2^2(u) < f_2^2(v)$ if $0 < u < v \leq u_1$.

We conclude that if $0 < v \leq u_0$,

$$2G(v) > \int_0^v\int_0^v\int_0^v J_0^2(u)[J_0(x) - J_0(y)]^2 uxy\, du\, dx\, dy[f_1^2(v) - f_2^2(v)],$$

and so $G(v)>0$ if $0\leq v\leq v_1$, in which v_1 is the smallest positive zero of $f_1(v)-f_2(v)$. (Note that $v_1<u_0$ because $f_1(u_0)=0$, $f_2(u_0)>0$, $f_1(0)=2$, $f_2(0)=0$.) The equation satisfied by v_1 is

$$\phi(v_1)\equiv\frac{J_0(v_1)}{J_1(v_1)}+\frac{J_1(v_1)}{J_0(v_1)}-2v_1=0.$$

It follows from the identity

$$\frac{d\phi}{dv_1}=-3+\frac{J_1^2(v_1)}{J_0^2(v_1)}-\frac{\{J_0^2(v_1)-J_1^2(v_1)\}J_1'(v_1)}{J_0(v_1)J_1^2(v_1)}$$

that $\phi(v_1)$ is strictly monotone decreasing when $0<v_1<u_1$. Since $\phi(0+)=+\infty$, $\phi(u_1)=2(1-u_1)<0$, there exists a unique v_1 on the interval $(0, u_1)$ for which $\phi(v_1)=0$. We find that $v_1=1.0944$, and hence infer that $G(v)>0$ if $0\leq v\leq v_1=1.0944$.

The desired inequality is, therefore, true when $0\leq v\leq 1.0944$ and when $1.8412\leq v\leq 2.2999$, but we have been unable to establish, or deny, its truth when $1.0944<v<1.8412$.

REFERENCES

[1] R. Barakat and E. Levin, *Application of apodization to increase two-point resolution by the Sparrow criterion. II. Incoherent illumination*, J. Optical Soc. Amer., 53 (1963), pp. 274–282.

[2] O. Szasz, *Inequalities concerning ultraspherical polynomials and Bessel functions*, Proc. Amer. Math. Soc., 1 (1950), pp. 256–267.

[3] G. N. Watson, *A Treatise on the Theory of Bessel Functions*, Cambridge University Press, Cambridge, England, 1952, pp. 480–481.

[4] J. E. Wilkins, Jr., *Apodization for maximum Strehl criterion and specified Sparrow limit for coherent illumination*, J. Optical Soc. Amer., 67 (1977), pp. 553–557.

Problem 69-9*, *A Positive Integral*, by Richard Askey (University of Wisconsin).

It is conjectured that

$$(-1)^{r+s+t}\int_0^1 P_{2r+1}(x)P_{2s+1}(x)P_{2t+1}(x)\frac{dx}{x}\geq 0,$$

$r, s, t, = 0, 1, 2, \cdots$, where $P_n(x)$ is the Legendre polynomial.

Solution[2] by George Gasper (Northwestern University).

We shall prove the following generalization of Askey's conjecture. Let $c_n^\lambda(x)=C_n^\lambda(x)/C_n^\lambda(1)$, where $C_n^\lambda(x)$ is the ultraspherical polynomial [1, p. 174] of order λ, $\lambda>-\frac{1}{2}$, and let

$$a(r, s, t; \lambda)=\int_0^1 c_{2r+1}^\lambda(x)c_{2s+1}^\lambda(x)c_{2t+1}^\lambda(x)\frac{(1-x^2)^{\lambda-1/2}}{x}\,dx.$$

Then we have

(1) $a(r, s, t; \lambda)\geq 0,\qquad \lambda\geq 1,$

and

(2) $(-1)^{r+s+t}a(r, s, t; \lambda)\geq 0,\qquad -\frac{1}{2}<\lambda\leq 1,$

for $r, s, t=0, 1, 2, \cdots$. When $\lambda=0$, $c_n^\lambda(\cos\theta)$ is to be replaced by $\lim_{\lambda\to 0}c_n^\lambda(\cos\theta)$

---

[2] A previous solution appeared in the Oct. 1970 issue.

$= \cos n\theta$. Since $P_n(x) = c_n^{1/2}(x)$, Askey's conjecture is the special case $\lambda = \frac{1}{2}$ of (2).

Our proof of (1) and (2) depends on some nonnegativity results recently obtained [2] for the coefficients in the expansion

$$(3) \qquad R_r^{(\alpha,\beta)}(x)R_s^{(\alpha,\beta)}(x) = \sum_{t=|r-s|}^{r+s} g(r, s, t; \alpha, \beta)R_t^{(\alpha,\beta)}(x),$$

where $R_n^{(\alpha,\beta)}(x) = P_n^{(\alpha,\beta)}(x)/P_n^{(\alpha,\beta)}(1)$ and $P_n^{(\alpha,\beta)}(x)$ is the Jacobi polynomial of order (α, β), $\alpha, \beta > -1$. It was shown in [2] that if $\alpha \geq \beta > -1$ and $\alpha + \beta \geq -1$, then each $g(r, s, t; \alpha, \beta) \geq 0$. Since the integral

$$b(r, s, t; \alpha, \beta) = \int_{-1}^{1} R_r^{(\alpha,\beta)}(x)R_s^{(\alpha,\beta)}(x)R_t^{(\alpha,\beta)}(x)(1 - x)^{\alpha}(1 + x)^{\beta}\, dx$$

is a positive multiple of $g(r, s, t; \alpha, \beta)$, it follows that

$$(4) \qquad b(r, s, t; \alpha, \beta) \geq 0, \qquad \alpha \geq \beta > -1, \quad \alpha + \beta \geq -1.$$

This implies

$$(5) \qquad (-1)^{r+s+t}b(r, s, t; \alpha, \beta) \geq 0, \qquad \beta \geq \alpha > -1, \quad \alpha + \beta \geq -1,$$

since $P_n^{(\alpha,\beta)}(-x) = (-1)^n P_n^{(\beta,\alpha)}(x)$ (see [1, p. 169, (13)]). From

$$c_{2n+1}^{\lambda}(x) = xR_n^{(\lambda-1/2,1/2)}(2x^2 - 1), \qquad \lambda > -\tfrac{1}{2}$$

[1, p. 176, (22)], we see that $a(r, s, t; \lambda) = 2^{-\lambda-2}b(r, s, t; \lambda - \frac{1}{2}, \frac{1}{2})$. Then (1) and (2) follow immediately from (4) and (5), respectively.

The determination of all (α, β) for which $g(r, s, t; \alpha, \beta) \geq 0, r, s, t = 0, 1, 2, \cdots$, and some important applications are given in [3]. In [4] the dual of (3) is considered and an explicit formula is used to show positivity of the kernel for certain values of α and β. It would be of interest if an analogous formula could be found for $g(r, s, t; \alpha, \beta)$ from which one could easily see that it is nonnegative when $\alpha \geq \beta > -1, \alpha + \beta \geq -1$.

REFERENCES

[1] A. ERDÉLYI ET AL., *Higher Transcendental Functions*, vol. 2, McGraw-Hill, New York, 1953.
[2] G. GASPER, *Linearization of the product of Jacobi polynomials. I*, Canad. J. Math., 22 (1970), pp. 171–175.
[3] ———, *Linearization of the product of Jacobi polynomials. II*, Ibid., 22 (1970), pp. 582–593.
[4] ———, *Positivity and the convolution structure for Jacobi series*, Ann. Math., 93 (1971), pp. 112–118.

Problem 71–13, A Positive Integral, by L. CARLITZ (Duke University).

Show that

$$\int_{0}^{\infty} e^{-x^2} H_{2k+1}(x)H_{2m+1}(x)H_{2n+1}(x)\frac{dx}{x} \geq 0,$$

where $H_n(x)$ is the Hermite polynomial defined by

$$\exp\{2xt - t^2\} = \sum_{n=0}^{\infty} H_n(x)\frac{t^n}{n!}.$$

Solution by R. ASKEY (University of Wisconsin) and G. GASPER (Northwestern University).

Using $H_{2n+1}(x) = (-1)^n 2^{2n+1} n! x L_n^{1/2}(x^2)$ (see [6, (5.6.1)]), where $L_n^{\alpha}(y)$ is defined by

(1) $$\frac{\exp[-yr/(1-r)]}{(1-r)^{\alpha+1}} = \sum_{n=0}^{\infty} L_n^{\alpha}(y) r^n$$

(see [6, (5.1.9)]), and letting $x^2 = y$ gives

$$A(k, m, n) = \int_0^{\infty} e^{-x^2} H_{2k+1}(x) H_{2m+1}(x) H_{2n+1}(x) \frac{dx}{x}$$

(2) $$= 4^{k+m+n+1} (-1)^{k+m+n} k! m! n!$$

$$\cdot \int_0^{\infty} L_k^{1/2}(y) L_m^{1/2}(y) L_n^{1/2}(y) y^{1/2} e^{-y} \, dy.$$

There are many ways to show that

$$(-1)^{k+m+n} \int_0^{\infty} L_k^{\alpha}(x) L_m^{\alpha}(x) L_n^{\alpha}(x) x^{\alpha} e^{-x} \, dx \geqq 0, \qquad \alpha > -1.$$

The most natural one is to use the generating function (1) to obtain

$$\sum_{k,m,n=0}^{\infty} \int_0^{\infty} L_k^{\alpha}(x) L_m^{\alpha}(x) L_n^{\alpha}(x) x^{\alpha} e^{-x} \, dx \, r^k s^m t^n$$

$$= \int_0^{\infty} \frac{\exp[-xr/(1-r)]}{(1-r)^{\alpha+1}} \frac{\exp[-xs/(1-s)]}{(1-s)^{\alpha+1}} \frac{\exp[-xt/(1-t)]}{(1-t)^{\alpha+1}} x^{\alpha} e^{-x} \, dx$$

$$= \frac{1}{[(1-r)(1-s)(1-t)]^{\alpha+1}} \int_0^{\infty} \exp\left[-\left(\frac{r}{1-r} + \frac{s}{1-s} + \frac{t}{1-t} + 1\right)x\right] x^{\alpha} \, dx$$

$$= \frac{\Gamma(\alpha+1)}{[r(1-s)(1-t) + (1-r)s(1-t) + (1-r)(1-s)t + (1-r)(1-s)(1-t)]^{\alpha+1}}$$

$$= \frac{\Gamma(\alpha+1)}{[1 - (rs + rt + st) + 2rst]^{\alpha+1}}.$$

Replacing r by $-r$, s by $-s$, t by $-t$ yields

$$\sum_{k,m,n=0}^{\infty} (-1)^{k+m+n} \int_0^{\infty} L_k^{\alpha}(x) L_m^{\alpha}(x) L_n^{\alpha}(x) x^{\alpha} e^{-x} \, dx \, r^k s^m t^n$$

$$= \frac{\Gamma(\alpha+1)}{[1 - (rs + rt + st) - 2rst]^{\alpha+1}}$$

$$= \Gamma(\alpha+1) \sum_{a,b,c,d \geqq 0} \frac{(\alpha+1)_{a+b+c+d}}{a! b! c! d!} (rs)^a (rt)^b (st)^c (2rst)^d,$$

and so,

$$B(k, m, n; \alpha) = (-1)^{k+m+n} \int_0^{\infty} L_k^{\alpha}(x) L_m^{\alpha}(x) L_n^{\alpha}(x) x^{\alpha} e^{-x} \, dx \geqq 0, \qquad \alpha > -1.$$

A change of variables in the above sum gives

$$B(k, m, n; \alpha) = \Gamma(\alpha + 1) \sum_{j \geq 0} \frac{(\alpha + 1)_j 2^{k+m+n-2j}}{(j-k)!(j-m)!(j-n)!(k+m+n-2j)!},$$

where $1/(-r)! = 0, r = 1, 2, \cdots$. When $\alpha = \frac{1}{2}$ this gives

$$A(k, m, n) = 2^{3(k+m+n)+2} k! m! n! \sum_{j \geq 0} \frac{\Gamma(j + 3/2)2^{-2j}}{(j-k)!(j-m)!(j-n)!(k+m+n-2j)!},$$

and so,

$$A(k, m, n) > 0, \qquad |n - m| \leq k \leq m + n,$$

$$A(k, m, n) = 0, \qquad k < |n - m| \quad \text{or} \quad k > m + n.$$

Another proof which gives an explicit formula was found by Erdélyi [4]. There is a completely different proof using recurrence formulas which extends to many other orthogonal polynomials.

Any set of monic orthogonal polynomials $p_n(x) = x^n + \cdots$ satisfies

$$(3) \qquad xp_n(x) = p_{n+1}(x) + \alpha_n p_n(x) + \beta_n p_{n-1}(x), \qquad \beta_n > 0, \quad n = 1, 2, \cdots, \beta_0 = 0,$$

α_n real. Conversely any set of polynomials which satisfies (3) is orthogonal with respect to a positive measure $d\psi(x)$. The polynomials $p_n(x)$ satisfy $p_n(-x) = (-1)^n p_n(x)$ if and only if $\alpha_n = 0, n = 0, 1, \cdots$, and then the measure $d\psi(x)$ is even. Thus this problem is a special case of

$$\int_0^\infty p_{2k+1}(x) p_{2m+1}(x) p_{2n+1}(x) \frac{d\psi(x)}{x} \geqq 0,$$

where

$$xp_n(x) = p_{n+1}(x) + \beta_n p_{n-1}(x), \qquad \beta_n > 0, \quad n = 1, 2, \cdots, \beta_0 = 0,$$

and

$$\int_{-\infty}^\infty p_m(x) p_n(x) \, d\psi(x) = 0, \qquad\qquad m \neq n.$$

For Hermite polynomials $\beta_n = n/2$.

Letting $xq_n(x^2) = p_{2n+1}(x), n = 0, 1, \cdots$, we find that the $q_n(x)$ satisfy

$$\int_0^\infty q_m(x) q_n(x) x \, d\psi(x^{1/2}) = \int_0^\infty q_m(x) q_n(x) \, d\mu(x) = 0, \qquad m \neq n,$$

and

$$(4) \qquad\qquad q_1(x) q_n(x) = q_{n+1}(x) + \gamma_n q_n(x) + \delta_n q_{n-1}(x),$$

where

$$\gamma_n = \beta_{2n+1} + \beta_{2n+2} - \beta_1 - \beta_2, \qquad \delta_n = \beta_{2n} \beta_{2n+1}.$$

The following result is known [1], [2]:

THEOREM 1. *If $q_n(x)$ is defined by* (4), $q_0(x) = 1, q_1(x) = x + a$, *and if*

$$(5) \qquad\qquad 0 \leqq \gamma_n \leqq \gamma_{n+1}, \quad 0 < \delta_n \leqq \delta_{n+1}, \quad \delta_0 = 0,$$

then

(6) $$q_m(x)q_n(x) = \sum_{k=|m-n|}^{m+n} a(k, m, n)q_k(x)$$

with $a(k, m, n) \geqq 0$.

By orthogonality,

$$a(k, m, n) = \int_0^\infty q_k(x)q_m(x)q_n(x)\, d\mu(x) \bigg/ \int_0^\infty q_k^2(x)\, d\mu(x),$$

and so

(7) $$\int_0^\infty q_k(x)q_m(x)q_n(x)\, d\mu(x) \geqq 0.$$

For $p_{2k+1}(x)$, (7) becomes

(8) $$\int_0^\infty p_{2k+1}(x)p_{2m+1}(x)p_{2n+1}(x) \frac{d\psi(x)}{x} \geqq 0.$$

A sufficient condition for (5) to hold (and thus for (8) to hold) is $\beta_n \leqq \beta_{n+1}$, or even $\beta_{2n} \leqq \beta_{2n+2}$, $\beta_{2n+1} \leqq \beta_{2n+3}$. These conditions are satisfied for Hermite polynomials, since $\beta_n = n/2$, and so

$$\int_0^\infty H_{2k+1}(x)H_{2m+1}(x)H_{2n+1}(x)\, e^{-x^2}\frac{dx}{x} \geqq 0.$$

One other classical case is quite interesting. The ultraspherical polynomials $P_n^{(\alpha,\alpha)}(x)$ satisfy

(9) $$\int_0^1 P_{2k+1}^{(\alpha,\alpha)}(x)P_{2m+1}^{(\alpha,\alpha)}(x)P_{2n+1}^{(\alpha,\alpha)}(x)(1-x^2)^\alpha\frac{dx}{x} \geqq 0, \qquad \alpha \geqq \tfrac{1}{2},$$

but

$$(-1)^{k+m+n}\int_0^1 P_{2k+1}^{(\alpha,\alpha)}(x)P_{2m+1}^{(\alpha,\alpha)}(x)P_{2n+1}^{(\alpha,\alpha)}(x)(1-x^2)^\alpha\frac{dx}{x} \geqq 0,$$
$$-1 < \alpha \leqq \tfrac{1}{2}.$$

See Carlitz [3] for the case $\alpha = 0$ and Gasper [5] for the remaining cases. Since

$$\lim_{\alpha \to \infty} \alpha^{-n/2} P_n^{(\alpha,\alpha)}(\alpha^{-1/2}x) = H_n(x)/(2^n n!),$$

the nonnegativity of (2) also follows from (9).

We take this opportunity to point out that in order to prove inequality (2.3) in [2] one also needs to assume that $\sup_n |p_n(x)| < \infty$.

REFERENCES

[1] R. Askey, *Linearization of the product of orthogonal polynomials*, Problems in Analysis, A Symposium in Honor of Salomon Bochner, R. C. Gunning, ed., Princeton Univ. Press, Princeton, N.J., 1970, pp. 131–138.

[2] ———, *Orthogonal polynomials and positivity*, Studies in Applied Mathematics 6, Special Functions and Wave Propagation, D. Ludwig and F. W. J. Olver, eds., SIAM, Philadelphia, 1970, pp. 64–85.

[3] L. Carlitz, *Problem 69-9*, this Review, 12 (1970), pp. 588–590.

[4] A. Erdélyi, *On some expansions in Laguerre polynomials*, J. London Math. Soc., 13 (1938), pp. 154–156.

[5] G. Gasper, *Problem 69-9*, this Review, 13 (1971), pp. 396–397.

[6] G. Szegö, *Orthogonal Polynomials*, Colloquium Publications, vol. 23, Amer. Math. Soc., Providence, R.I. 1959.

An Integral Inequality

Problem 78-18*, *by* A. MEIR (University of Alberta).
  $F(x)$ is nonnegative and integrable on $[0, a]$ and such that

$$\left\{ \int_0^t F(x)\, dx \right\}^2 \geqq \int_0^t F(x)^3\, dx$$

for every t in $[0, a]$. Prove or disprove the conjecture:

$$\frac{a^3}{3} \geqq \int_0^a \{F(x) - x\}^2\, dx.$$

(Reference: Amer. Math. Monthly, 83 (1976), pp. 26–30.)

  Solution by J. VAN KAN (University of Technology, Delft, The Netherlands).
  There exists a constant $c > 0$, independent of F and a such that

$$\int_0^a \{F(x) - x\}^2\, dx \leqq \tfrac{1}{3} a^3 - c \left\{ \int_0^a F(x)\, dx \right\}^{3/2}.$$

  Proof. Applying Hölder's inequality we find that

$$\int_0^t F(x)\, dx \leqq \left(\int_0^t F^3(x)\, dx \right)^{1/3} t^{2/3} \leqq \left(\int_0^t F(x)\, dx \right)^{2/3} t^{2/3}.$$

Whence,

$$\int_0^t F(x)\, dx \leqq t^2 \quad \forall t \in [0, a].$$

Define $\phi(t) = \int_0^t F(x)\, dx$. Now

(1) $$\int_0^t xF(x)\, dx \geqq \int_0^t \phi^{1/2}(x) F(x)\, dx = \tfrac{2}{3} \phi^{3/2}(t).$$

Also,

(2) $$\int_0^t F^2(x)\, dx \leqq \left(\int_0^t F^3(x)\, dx \right)^{1/2} \left(\int_0^t F(x)\, dx \right)^{1/2} \leqq \phi^{3/2}(t).$$

Whence,

$$\int_0^a (F(x) - x)^2\, dx \leqq \tfrac{1}{3} a^3 - \tfrac{4}{3} \phi^{3/2}(a) + \phi^{3/2}(a) = \tfrac{1}{3} a^3 - \tfrac{1}{3} \phi^{3/2}(a),$$

establishing the result.
  Remark. The value $\tfrac{1}{3}$ for the constant c is not optimal. It is possible to obtain the following estimates:

  A. $\phi(t) \leqq \tfrac{1}{2} t^2 \quad \forall t \in [0, a];$

  B. $\displaystyle \int_0^t F^\gamma(x)\, dx \leqq \frac{1}{\gamma + 1} \{2\phi(t)\}^{(\gamma+1)/2} \quad \forall \gamma \in [1, 3], \forall t \in [0, a].$

These estimates give

$$\int_0^t xF(x)\, dx \geqq \frac{2\sqrt{2}}{3} \phi^{3/2}(t), \qquad \int_0^t F^2(x)\, dx \leqq \frac{2\sqrt{2}}{3} \phi^{3/2}(t).$$

Using these instead of (1) and (2), respectively, yields the result:

$$\int_0^a (F(x) - x)^2 \, dx \leq \tfrac{1}{3}a^3 - \frac{2\sqrt{2}}{3} \phi^{3/2}(a).$$

This value of c is optimal as the example $F(x) = x$ shows.

J. STRIKWERDA (NASA Langley Research Center) obtained the slightly different and better inequality

$$a^3 \geq 12 \int_0^a \{F(x) - \tfrac{1}{2}x\}^2 \, dx.$$

Problem 69-10, An Integral Inequality, by D. J. NEWMAN (Yeshiva University).

Show that

$$\int_0^\infty \{|F(x)| + |F''(x)|\} \, dx \geq \sqrt{2}|F(0)|, \qquad F \in C^2,$$

and that $\sqrt{2}$ is the best possible constant.

This problem arose in some work on numerical differentiation.

Solution by DAVID W. BOYD (California Institute of Technology).

We assume that $|F|$ and $|F''|$ are integrable since otherwise the result is trivial.

In this case, $|F'(x) - F'(y)| \leq \int_x^y |F''| \, dx$ shows that $\lim F'(x)$ exists as $x \to \infty$, and since $|F|$ is integrable this limit is zero.

We may now establish the following identity:

$$(1) \qquad \sqrt{2}F(0) = \int_0^{\sqrt{2}} F(x) \, dx + \int_0^{\sqrt{2}} \left(\sqrt{2}x - \frac{x^2}{2}\right) F''(x) \, dx + \int_{\sqrt{2}}^\infty F''(x) \, dx.$$

This may be verified by integration by parts twice in the second integral and observing that the third integral is $-F'(\sqrt{2})$. Since $|\sqrt{2}x - x^2/2| \leq 1$ for $0 \leq x \leq \sqrt{2}$, we obtain from (1) that

$$(2) \qquad \sqrt{2}|F(0)| \leq \int_0^{\sqrt{2}} |F(x)| \, dx + \int_0^\infty |F''(x)| \, dx,$$

which is slightly stronger than the required result.

That $\sqrt{2}$ is the best constant may be seen by considering the functions $F_\alpha(x) = (\sqrt{2} - x)^\alpha$ for $x \leq \sqrt{2}$, $F_\alpha(x) = 0$ for $x \geq \sqrt{2}$, with $\alpha > 1$. Then $F_\alpha(0) \to \sqrt{2}$ as $\alpha \to 1$, and $\int_0^\infty \{|F_\alpha| + |F_\alpha''|\} \, dx \to 2$ as $\alpha \to 1$.

Problem 74–10, An Integral Inequality, by R. ASKEY (University of Wisconsin).

It is clear that

$$R(x) = \int_0^x t^\alpha J_\alpha(t) \sin t \, dt \geq 0 \quad \text{for } x \geq 0$$

when $\alpha = \pm\frac{1}{2}$. Show that the inequality is also valid for $-\frac{1}{2} < \alpha < \frac{1}{2}$.

Solution by A. R. HOLT (University of Essex, Colchester, England).

(i) It is easily shown that

(1) $$R(x) = \frac{x^{\alpha+1}}{\alpha+1}[\sin x \, J_\alpha(x) - \cos x \, J_{\alpha+1}(x)], \quad \alpha > -\frac{1}{2}.$$

(ii) Since

(2) $$J_\alpha(x) = \frac{x^\alpha}{2^{\alpha-1}\Gamma(\alpha+\frac{1}{2})\Gamma(\frac{1}{2})} \int_0^{\pi/2} \cos(x \cos \theta) \sin^{2\alpha} \theta \, d\theta,$$

integrating the expression for $J_{\alpha+1}(x)$ by parts and using (2) gives

(3) $$R(x) = \frac{x^{2\alpha+1}}{2^{\alpha+1}\Gamma(\alpha+\frac{3}{2})\Gamma(\frac{1}{2})} \int_0^{\pi/2} \sin^{2\alpha} \theta \left[2 \sin^2 \frac{\theta}{2} \sin\left(2x \cos^2 \frac{\theta}{2}\right) \right.$$
$$\left. + 2 \cos^2 \frac{\theta}{2} \sin\left(2x \sin^2 \frac{\theta}{2}\right) \right] d\theta.$$

If $0 \le x \le \pi/2$, the integrand is positive for $0 \le \theta \le \pi/2$ and hence $R(x) \ge 0$ for all $\alpha > -\frac{1}{2}$.

(iii) Using the theory of zeros of Bessel functions given in [1, § 15.33–34] and writing

$$a(m) = m\pi + 3\frac{\pi}{4} + \alpha\frac{\pi}{2},$$

$$b(m) = m\pi + 7\frac{\pi}{8} + \alpha\frac{\pi}{4},$$

$$c(m) = m\pi + \frac{\pi}{8} + \alpha\frac{\pi}{4},$$

$$d(m) = m\pi + \frac{\pi}{4} + \alpha\frac{\pi}{2},$$

we see that when $-\frac{1}{2} < \alpha < \frac{1}{2}$, the only zeros of $J_\alpha(x)$ lie in the intervals $(a(m), b(m))$, $m = 0, 1, 2, \cdots$, and the only zeros of $J_{\alpha+1}(x)$ lie in $(c(m), d(m))$, $m = 1, 2, \cdots$. Hence

(4)
$$J_\alpha(x) \sin x \ge 0, \qquad m\pi \le x \le a(m),$$
$$J_{\alpha+1}(x) \cos x \le 0 \qquad (m+\tfrac{1}{2})\pi \le x \le c(m+1),$$

$m = 0, 1, 2, \cdots$.

From (1) and (4), if $x > \pi/2$, $R(x) > 0$ except possibly for the intervals

I: $c(m+1) < x < (m+\frac{3}{2})\pi$,
II: $a(m) < x < (m+1)\pi$.

Since $R'(x) = x^\alpha J_\alpha(x) \sin x$, in interval I $R'(x) > 0$ and, moreover, $R(x) > 0$ at both endpoints. Thus $R(x) > 0$ throughout I.

Since the zeros of $J_\alpha(x)$ and $J_{\alpha+1}(x)$ are interlaced (1, § 15.22), and since intervals I and II are nonoverlapping, $R'(x)$ has only one zero, β, in $a(m) < x < b(m)$ and one zero at $(m+1)\pi$. Moreover, the sign of $R'(x)$ indicates that $R(x)$ has a maximum at β and a minimum at $(m+1)\pi$. Moreover, $R(x)$ is positive at both endpoints, and hence $R(x) > 0$ throughout II. Thus $R(x) > 0 \forall x > \pi/2$, which completes the proof.

A slight modification of this proof holds for $\frac{1}{2} < \alpha < \frac{3}{2}$, and the case $\alpha = \frac{1}{2}$ is fairly obvious. However, investigation of the asymptotic form for large x shows that $R(x)$ becomes negative if $\alpha > \frac{3}{2}$.

REFERENCES

[1] G. N. WATSON, *A Treatise on the Theory of Bessel Functions*, 2nd ed., Cambridge University Press, London, 1966.

Solutions were also obtained by R. A. Ross (University of Toronto) and M. L. GLASSER (University of Waterloo), who each reduced the integral to

$$\frac{2^\alpha x^{2\alpha+1}}{\Gamma(\frac{1}{2}-\alpha)\pi^{1/2}} \int_0^\infty (s^2 + 2sx)^{-\alpha-1/2} \left[\frac{\sin s}{s} - \frac{\sin(s+2x)}{s+2x} \right] ds,$$

and by O. G. RUEHR (Michigan Technological University), who reduced the integral to

$$\frac{2^\alpha}{\sqrt{\pi}\Gamma(\alpha+\frac{3}{2})} \int_0^x (\sin^2 t)(x-t)^{\alpha-1/2} t^{\alpha-3/2} [(\alpha+\tfrac{1}{2})t + (\tfrac{1}{2}-\alpha)(x-t)] \, dt.$$

From here the argument is easy. A solution was also given by the proposer.

Comment by the proposer. The positivity of

$$\int_0^x (x-t)^{\alpha+1/2} t^{\alpha-1/2} \sin 2t \, dt \quad \text{for } -\tfrac{1}{2} < \alpha \le \tfrac{3}{2}$$

is a corollary of a more general result of Gasper, *Positive integrals of Bessel functions*, SIAM J. Math. Anal., to appear. The more general problem of the positivity of $\int_0^x t^\gamma J_\alpha(t) J_\beta(t) \, dt$ also reduces to the positivity of a $_3F_4$ on the negative real axis. It would be interesting to see how close one can come to best possible results on this problem.

Problem 73-13, *An Integral Inequality*, by RICHARD ASKEY (University of Wisconsin).

Heisenberg's inequality can be stated as

$$M_0 \left\{ \int_{-\infty}^\infty x^2 |F(x)|^2 \, dx, \int_{-\infty}^\infty t^2 |\bar{F}(t)|^2 \, dt \right\} \ge \frac{1}{4\pi} \int_{-\infty}^\infty |F(x)|^2 \, dx,$$

where

$$M_0\{a, b\} = \sqrt{ab}$$

and

$$\bar{F}(t) = \int_{-\infty}^\infty F(x) e^{2\pi ixt} \, dx.$$

By Schlömilch's inequality,

(1) $$M_p\left\{\int_{-\infty}^{\infty} x^2|F(x)|^2 \, dx, \int_{-\infty}^{\infty} t^2|\bar{F}(t)|^2 \, dt\right\} \geq \frac{1}{4\pi} \int_{-\infty}^{\infty} |F(x)|^2 \, dx$$

holds for $p > 0$, where $M_p\{a, b\} = \{(a^p + b^p)/2\}^{1/p}$. Show that even if we replace the $1/4\pi$ by any positive constant A_p in (1), the inequality will be invalid for any $p < 0$.

Solution by A. A. JAGERS (Technische Hogeschool Twente, Enschede, the Netherlands).

For $p < 0$, even the following holds:
If F_α is given by $F_\alpha(x) = F(\alpha x)$, $\alpha > 0$, then

$$\int_{-\infty}^{\infty} |F_\alpha(x)|^2 \, dx = \alpha^{-1} \int_{-\infty}^{\infty} |F(x)|^2 \, dx,$$

whereas

$$2^{1/p} M_p\left\{\int_{-\infty}^{\infty} x^2|F_\alpha(x)|^2 \, dx, \int_{-\infty}^{\infty} t^2|\bar{F}_\alpha(t)|^2 \, dt\right\}$$

$$\leq \int_{-\infty}^{\infty} x^2|F_\alpha(x)|^2 \, dx = O(\alpha^{-3}), \qquad \alpha \to \infty.$$

Comment by the proposer. The most attractive proof I know of Heisenberg's inequality is due to de Bruijn [*Uncertainty principles in Fourier analysis*, Inequalities, O. Shisha, ed., Academic Press, New York, 1967, pp. 57–71]. He first proves (1) for $p = 1$ by means of Hermite series, then scales the function and its Fourier transform, and proves (1) for $p = 0$ by minimizing with respect to the scale factor. This naturally suggested the question of trying to extend the inequality further in the same direction, and Jagers' solution shows how hopeless this task is.

Problem 75–16, An Integral Inequality, by J. WALTER (Technical University, Aachen, Federal Republic of Germany).

Let G denote a continuously differentiable positive function defined in some interval $[t_0, \infty)$ and a, b, c, x, y, z, w be real numbers such that $0 < a \leq b$, $t_0 \leq x \leq y \leq z \leq w$. Prove the existence of a continuous function $H(a, b, c)$ of three variables such that

$$\int_y^z \frac{dt}{G(t)} = a, \qquad \int_x^w \frac{dt}{G(t)} \leq b, \ |G'(t)| \leq c \qquad \text{for } t \in [t_0, \infty)$$

imply that

$$\int_x^w G(t)dt \leq H(a, b, c) \int_y^z G(t) \, dt.$$

The problem arose in some work on spectral theory of Sturm–Liouville differential operators. The last inequality can be considered as an inversion of trivial inequalities like $(c, d) \subset (a, b) \Rightarrow \int_c^d |f| \, dt \leq \int_a^b |f| \, dt.$

Solution by I. I. KOLODNER (Carnegie-Mellon University).

It will be shown that $H(a, b, c) = (e^{2bc} - 1)/2ac$ will do; also, it suffices to assume that G is merely Lipschitz continuous, with constant c.

On $[x, w]$, G assumes a positive minimum m, at some s. We have then:

(i) $m \leq G(t) \leq m + c|t - s|$ for all $t \in [x, w]$;

(ii)
$$a = \int_y^z \frac{dt}{G(t)} \leq \frac{z - y}{m},$$

whence $(z - y) \geq ma$;

(iii) by C.S.,

$$\int_y^z G(t) \, dt \geq \left(\int_y^z dt \right)^2 \bigg/ \int_y^z \frac{dt}{G(t)} = \frac{(z - y)^2}{a} \geq m^2 a,$$

using (ii);

(iv)
$$b \geq \int_x^w \frac{dt}{G(t)} \geq \int_x^w \frac{dt}{m + c|t - s|} = \frac{1}{c} \log \left(1 + \frac{c}{m}(w - s) \right)\left(1 + \frac{c}{m}(s - x) \right)$$

$$\geq \frac{1}{c} \log \left(1 + \frac{c}{m}(w - x) \right),$$

whence $(w - x) \leq (m/c)(e^{bc} - 1)$.

Finally, using (i), (iv) and (iii), we get:

$$\int_x^w G(t) \, dt \leq \int_x^w (m + c|t - s|) \, dt = m(w - x) \frac{c}{2}[(w - s)^2 + (x - s)^2]$$

$$\leq m(w - x)\left[1 + \frac{c}{2m}(w - x) \right] \leq \frac{m^2}{2c}(e^{2bc} - 1)$$

$$\leq \frac{1}{2ac}(e^{2bc} - 1) \int_y^z G(t) \, dt$$

Problem 74-1*, *Monotonicity of a Function*, by M. L. MEHTA (Centre d'Etudes Nucleaires de Saclay, Gif-sur-Yvette, France).

Let n be a positive integer and let $F(t)$ and $G(t)$ be the multiple integrals

$$F(t) = \int_0^\infty \cdots \int_0^\infty P(x_1, \cdots, x_n) \prod_{j=1}^n (x_j + t)^{-1/2} \, dx_j,$$

$$G(t) = \int_0^\infty \cdots \int_0^\infty P(x_1, \cdots, x_n) \prod_{j=1}^n (x_j + t)^{1/2} \, dx_j$$

where $P(x_1) = \exp(-x_1)$ for $n = 1$, and

$$P(x_1, \cdots, x_n) = \exp\left(-\sum_1^n x_j \right) \prod_{n \geq j > k \geq 1}^n (x_j - x_k)^2$$

for $n > 1$. Then is it true that the product $F(t) \cdot G(t)$ is a monotonically decreasing function of t for $0 \leq t < \infty$?

The problem can be rephrased as follows. Let

$$H(t) \equiv 2 \frac{d}{dt}\{F(t) \cdot G(t)\},$$

and note that by differentiation of the product of integrals,

$$H(t) = \int_0^\infty \cdots \int_0^\infty P(x_1, \cdots, x_n)P(y_1, \cdots, y_n) \sum_{j=1}^n \left(\frac{1}{x_j + t} - \frac{1}{y_j + t}\right)$$
$$\cdot \prod_{j=1}^n \left(\frac{x_j + t}{y_j + t}\right)^{1/2} dx_j \, dy_j.$$

Is $H(t)$ negative for all $t > 0$?

The proposition is true for $t \ll 1$ and for $t \gg 1$, as one may demonstrate by expanding in powers of t and $t^{-1/2}$, respectively. Also, it is true for all $t > 0$ when $n = 1$. To see this, write $H(t)$ in the form

$$H(t) = \int_0^\infty \int_0^\infty P(x_1)P(y_1)\{(x_1 + t)(y_1 + t)\}^{1/2} A(x_1, y_1, t) \, dx_1 \, dy_1$$

and note that

$$A(x_1, y_1, t) = \frac{1}{y_1 + t}\left\{\frac{1}{x_1 + t} - \frac{1}{y_1 + t}\right\}$$

can be replaced by

$$\frac{1}{2}\{A(x_1, y_1, t) + A(y_1, x_1, t)\} = -\frac{1}{2}\left\{\frac{1}{x_1 + t} - \frac{1}{y_1 + t}\right\}^2$$

The integrand is now ≤ 0 for all $t > 0$, and hence the integral itself is negative.

Editorial note. This problem and the one to follow have their origin in the statistical theory of energy levels in complex systems. Since the early work on the subject, the most popular mathematical model for the study of such systems has been the Gaussian ensemble introduced by Wigner and extensively studied by Dyson, Gaudin, Mehta, Porter and others [1], [2]. The proposer describes the background of the present problem in the following way. "Consider an ensemble of $n \times n$ Hermitian matrices with the probability $P(H) \, dH$ of finding a matrix between H and $H + dH$, where $P(H) = \exp(-\operatorname{tr} H^2)$. The probability density is Gaussian and has certain invariance properties. For example, if H is restricted to be real, then $P(H)$ is invariant under all real orthogonal transformations of H; if H is a general complex matrix, then $P(H)$ is invariant under all unitary transformations of H. The eigenvalues of H have the joint probability density

$$P(x_1, \cdots, x_n) = C_{n\beta} \exp\left(-\frac{\beta}{2}\sum_{i=1}^n x_i^2\right) \prod_{n \geq i > j \geq 1} |x_i - x_j|^\beta,$$

where $\beta = 1$ if we are dealing with real Hermitian (symmetric) matrices, $\beta = 2$ if we are dealing with general complex Hermitian matrices, and $\beta = 4$ for still another particular class of Hermitian matrices. The spacings between successive eigenvalues of H have a distribution related to the multiple integral F if $\beta = 1$, to the product of F and G if $\beta = 2$, and to the sum of F and G if $\beta = 4$." [C.C.R.]

REFERENCES

[1] C. E. PORTER, *Statistical Theories of Spectra: Fluctuations,* Academic Press, New York, 1965.
[2] M. L. MEHTA, *Random Matrices and the Statistical Theory of Energy Levels,* Academic Press, New York, 1967.

Problem 74-7*, Positivity Conditions for Orthogonal Polynomials, by M. L. MEHTA (Centre d'Etudes Nucleaires de Saclay, Gif-sur-Yvette, France).

Let $w(x)$ be a function on $[a, b]$ which fulfills the following requirements:

(i) $w(x) \geqq 0$ on $[a, b]$.

(ii) $\mu_n = \int_a^b x^n w(x)\, dx$ exists for each $n = 0, 1, \cdots, .$

For $n = 0, 1, \cdots$, let $p_n(x) = a_n x^n + \cdots$ be a family of polynomials which satisfy the following conditions:

(i) $\int_a^b p_m(x) p_n(x) w(x)\, dx = \delta_{mn}$, $m, n = 0, 1, \cdots,$

(ii) $a_{2n} > 0$ and $a_1 \cdot a_{2n+1} > 0$, $n = 0, 1, \cdots.$

What requirements on $w(x)$ will insure that, with an appropriate choice of sign for a_1, the condition

(A) $I(j, k, l) \equiv \int_a^b p_j(x) p_k(x) p_l(x) w(x)\, dx \geqq 0$

holds for every triple of nonnegative integers j, k, l? If condition (A) holds, then is it necessarily true that the inequality

(B) $\int_a^b \int_a^b \prod_{j=1}^k \{p_{n_j}(x) + \varepsilon_j p_{n_j}(y)\} w(x) w(y)\, dx\, dy \geqq 0$

holds for every set of nonnegative integers $n_1, n_2, \cdots, n_k$ and every set of signs $\varepsilon_j = \pm 1, j = 1, 2, \cdots, k$?

Editorial note. Questions arising in harmonic analysis and in other areas of mathematics provide motivation for the study of condition (A). The following results are known. In [2], Gasper has determined all α and β such that condition (A) holds for the Jacobi polynomials $P_n^{(\alpha, \beta)}(x)$. Askey [1] has obtained the following general result: if $p_n(x) = x^n + \cdots$ satisfies $p_1(x) p_n(x) = p_{n+1}(x) + \alpha_n p_n(x) + \beta_n p_{n-1}(x)$ with $\alpha_{n+1} \geqq \alpha_n \geqq 0$ and $\beta_{n+1} \geqq \beta_n > 0$, then condition (A) holds. This gives the result for Laguerre, Hermite, Charlier and Meixner polynomials. Interest in condition (B) derives from the work of Ginibre concerning Griffiths' inequalities [3]. It is known that condition (B) is satisfied by the Hermite polynomials and by the Chebyshev polynomials of the first kind. [C.C.R.]

REFERENCES

[1] R. ASKEY, *Linearization of the product of orthogonal polynomials*, Problems in Analysis, A Symposium in Honor of Solomon Bochner, R. C. Gunning, ed., Princeton Univ. Press, Princeton, N.J., 1970, pp. 131–138.

[2] G. GASPER, *Linearization of the products of Jacobi polynomials. II*, Canad. J. Math., 22 (1970), pp. 582–593.

[3] J. GINIBRE, *General formulation of Griffiths' inequalities*, Comm. Math. Phys., 16 (1960), pp. 310–328.

A Lagrangian Inequality

Problem 89-7*, by ANDREW N. NORRIS (Rutgers University).

The exterior problem in acoustics involves the solution of the Helmholtz equation

$$\Delta u + u = 0$$

for the complex-valued pressure $u(\underline{x})$ in the infinite region outside some compact source region V. Let the surface S of V be smooth, and either u or $\partial u/\partial n$ be given at every point on S. In addition, u must satisfy the radiation condition

$$u(\underline{x}) \sim \frac{e^{ir}}{r^{(d-1)/2}} f(\underline{x}/r)$$

as $r \equiv |\underline{x}| \to \infty$, where $d = 2$ or 3 is the spatial dimension, and f some smooth function. Define the integrated Lagrangian of the exterior field as

$$L = \frac{1}{2} \int_{R^d/V} [|u|^2 - |\nabla u|^2] \, d\underline{x}$$

where $|u|^2 = uu^*$, $|\nabla u|^2 = (\nabla u) \cdot (\nabla u)^*$, and $*$ denotes complex conjugate. Prove or disprove that

$$L < 0$$

for nonzero conditions on S.

The proposer notes that he has found $L = 0$ for $d = 1$, and $L < 0$ for a circle ($d = 2$) or a sphere ($d = 3$) but has not been able to extend the results to arbitrary compact shapes.

Editorial note. It is likely that S will also have to be convex.

Problem 79-1*, An Inequality, by I. LUX (Central Research Institute for Physics, Budapest, Hungary).

Let V be an arbitrary three-dimensional spatial region. Let $P = (\mathbf{r}, \boldsymbol{\omega})$, a six-dimensional phase space point, where $\mathbf{r} \in V$ and $\boldsymbol{\omega}$ is a directional unit vector. Define a function $M_\lambda(P)$ through the following integral equation

$$M_\lambda(P) = 1 - e^{-D} + \frac{\lambda}{4\pi} \int_0^D e^{-\lambda x} \, dx \int M_\lambda(P') \, d\boldsymbol{\omega}'$$

where $P' = (\mathbf{r} + x\boldsymbol{\omega}, \boldsymbol{\omega}')$, λ is an arbitrary but positive parameter, D is the distance between the point $\mathbf{r}$ and the boundary of V along the direction $\boldsymbol{\omega}$ and the integral over $d\boldsymbol{\omega}'$ is a double integral over the surface of a unit sphere. Prove or disprove that

$$\frac{d}{d\lambda} M_\lambda(P) \bigg]_{\lambda=1} \geqq 0.$$

The problem arose in connection with a variance study of certain Monte Carlo estimators of reaction rates in nuclear reactors.

Bounds for Nonlinear Heat Conduction

Problem 87-16, *by* A. A. JAGERS (Universiteit Twente, Enschede, the Netherlands).
Let $p > 1$ and let g be a nondecreasing, nonnegative function on $[0, 1)$ with

$$a = \liminf_{t\uparrow 1} g(t)(1-t)^{1/(p-1)} > 0$$

and

$$A = \limsup_{t\uparrow 1} g(t)(1-t)^{1/(p-1)} < \infty.$$

Determine sharp upper and lower bounds, in terms of a and A, for

$$\underline{\sigma} = \liminf_{t\uparrow 1} \frac{\int_0^t g(s)^p\, ds}{g(t)}$$

and

$$\bar{\sigma} = \limsup_{t\uparrow 1} \frac{\int_0^t g(s)^p\, ds}{g(t)}.$$

This problem arose in the analysis of a model of nonlinear heat conduction.

In the reference below a semi-infinite bar is considered in which the heat conductivity is proportional to T^{p-1} where T denotes the temperature. The function g represents an imposed temperature at the boundary point of the bar. The property that $g \uparrow \infty$ as time $t \uparrow 1$ is compatible with a combustion process. The problem was to relate some independent estimates involving a, A, and $\bar{\sigma}$ for the size of the part of the bar where the temperature rise is infinite as $t \uparrow 1$.

REFERENCE

B. H. GILDING AND M. A. HERRERO, *Localization and blow-up of thermal waves in nonlinear heat conduction with peaking*, Math. Ann., to appear.

Solution by the proposer.
We shall establish the sharp bounds $\underline{\sigma} \geqq (p-1)a^p/A$ and $\bar{\sigma} \leqq pA^{p-1} - a^{p-1}$. Let $f(t) = (1-t)^{-1/(p-1)}$ and note for future reference that

$$(1) \qquad\qquad f(t)^p = \frac{f(t)}{1-t},$$

$$(2) \qquad\qquad \int_0^t f(s)^p\, ds = (p-1)\{f(t) - 1\}$$

for $0 \leqq t < 1$. Let g be a nondecreasing, nonnegative function satisfying the given lim inf and lim sup conditions and let $0 < \varepsilon < a$ be fixed. By the definitions of lim inf and lim sup, there exists a real number $\delta > 0$ such that

(3) $$(a-\varepsilon)f(t)\leqq g(t)\leqq(A+\varepsilon)f(t)$$

for all $t \in (1 - \delta, 1)$. Using (2) and (3) together with the fact that g is nonnegative, we see that

$$\frac{\int_0^t g(s)^p\, ds}{g(t)} \geqq \frac{(p-1)(a-\varepsilon)^p\{f(t)-f(1-\delta)\}}{(A+\varepsilon)f(t)}$$

for all $t \in (1 - \delta, 1)$. Since $f(t) \to \infty$ as $t \uparrow 1$ and ε is arbitrary, we conclude that $\underline{\sigma} \geqq (p-1)a^p/A$.

We now prove that $\bar{\sigma} \leqq pA^{p-1} - a^{p-1}$. To avoid another series of inessential epsilon arguments, we shall assume that $af(t) \leqq g(t) \leqq Af(t)$ for all t. For $0 < t < 1$, set $b = b(t) = g(t)/f(t)$ and define x through $g(t) = bf(t) = Af(x)$. Then $a \leqq b(t) \leqq A$, and from $bf(t) = Af(x)$ we obtain $(A/b)^{p-1} = (1-x)/(1-t)$. In view of (1), we have

$$(t-x)g(t)^p = \frac{t-x}{1-t}b^{p-1}g(t) = (A^{p-1}-b^{p-1})g(t).$$

Using this relation, we obtain

$$\int_0^t g(s)^p\, ds \leqq A^p \int_0^x f(s)^p\, ds + (t-x)g(t)^p$$

$$< A^p(p-1)f(x) + (A^{p-1}-b^{p-1})g(t)$$

$$= \{pA^{p-1}-b^{p-1}\}g(t)$$

$$\leqq \{pA^{p-1}-a^{p-1}\}g(t),$$

from which $\bar{\sigma} \leqq pA^{p-1} - a^{p-1}$ follows.

In case $A = a$, the two bounds just proved give $\underline{\sigma} = \bar{\sigma} = (p-1)a^{p-1}$. This condition is realized by the example $g(t) = af(t)$. Henceforth we assume that $a < A$. We now describe an example that yields $\underline{\sigma} = (p-1)a^p/A$. Let (t_k) be an increasing sequence such that $t_k \to 1$ and $f(t_{k+1})/f(t_k) \to \infty$ as $k \to \infty$. (For example, let $t_k = 1 - r^{k^2}$, where $r = (a/A)^{p-1}$. Then $f(t_{k+1})/f(t_k) = (A/a)^{2k+1}$.) Define a second sequence (u_k) by the condition $Af(t_k) = af(u_k)$. (With the specific choice made above, $u_k = (1 - r^{k^2+1})$.) For each $k \geqq 0$, define $g(t)$ for $t_k \leqq t < t_{k+1}$ by

$$g(t) = \begin{cases} Af(t_k) & \text{for } t_k \leqq t \leqq u_k, \\ af(t) & \text{for } u_k < t < t_{k+1}. \end{cases}$$

Using (1) and the fact that $1 - u_k = (a/A)^{p-1}(1 - t_k)$, we get

$$\int_{t_k}^{u_k} g(s)^p\, ds = A^p f(t_k)[1 - (a/A)^{p-1}].$$

On the other hand,

$$\int_{u_k}^{t_{k+1}} g(s)^p\, ds = (p-1)a^p[f(t_{k+1}) - (A/a)f(t_k)].$$

Since $f(t_{k+1})/f(t_k) \to \infty$ as $k \to \infty$ and $g(t_{k+1}) = Af(t_{k+1})$, we find that

$$\int_0^{t_{k+1}} g(s)^p\, ds \sim (p-1)a^p g(t_{k+1})/A \qquad (k \to \infty).$$

It follows that this example yields the value $\underline{\sigma} = (p-1)a^p/A$.

A similar construction yields an example in which $\bar{\sigma} = pA^{p-1} - a^{p-1}$. Let (t_k) be as before and define (s_k) through $Af(s_k) = af(t_k)$. For each $k \geq 1$, define $g(t)$ for $t_{k-1} < t \leq t_k$ by

$$g(t) = \begin{cases} Af(t) & \text{for } t_{k-1} < t < s_k, \\ af(t_k) & \text{for } s_k \leq t \leq t_k. \end{cases}$$

Then

$$\int_{t_{k-1}}^{s_k} g(s)^p \, ds = (p-1)A^p[(a/A)f(t_k) - f(t_{k-1})]$$

and

$$\int_{s_k}^{t_k} g(s)^p \, ds = a^p f(t_k)[(A/a)^{p-1} - 1].$$

Since $f(t_k)/f(t_{k-1}) \to \infty$ as $k \to \infty$ and $g(t_k) = af(t_k)$, we find that

$$\int_0^{t_k} g(s)^p \, ds \sim (pA^{p-1} - a^{p-1})g(t_k) \qquad (k \to \infty).$$

Thus $\bar{\sigma} = pA^{p-1} - a^{p-1}$ is attained in this case. I thank the referee for improving the presentation considerably.

A Nonnegative Integral

Problem 87-13, *by* J. GEVIRTZ *and* O. G. RUEHR (Michigan Technological University), Houghton, Michigan).

For $t > 1$, show that $F(t) \geq 0$, where

$$F(t) = \int_0^1 \left\{ z \ln\left(\frac{1+z}{1-z}\right) - \frac{t(z^2+1)}{1+t^2} \ln\left(\frac{2}{1-z^2}\right) \right\} \frac{dz}{t^2 - z^2}.$$

The problem arose in connection with the development of sharp mean value theorems in the unit disk.

Solution by M. L. GLASSER (Clarkson University, Potsdam, New York).

To simplify matters we map the region $t > 1$ onto the (open) unit interval by the substitution $u = (t-1)/(t+1)$. Then straightforward integration [1] gives

$$F(t) = (2 - \ln 2)(u^2 + 1)^{-1} - \tfrac{1}{2} \ln 2 \ln u + \ln u \ln(1-u) + \mathrm{Li}_2(u)$$

$$- 1 + \frac{\pi^2}{6} - \frac{1}{2} \ln 2 \equiv \phi(u)$$

where Li_2 denotes the dilogarithm. We have

$$\phi(u) \cong \tfrac{1}{2} \ln 2 \ln u \quad \text{as } u \to 0^+ \quad (t \to 1^+),$$

$$\phi(1^-) = 0 \quad (t \to \infty)$$

since

$$-\phi'(u) = (4 - \ln 4)u(u^2 + 1)^{-2} + \tfrac{1}{2}u^{-1} \ln 2 + (1-u)^{-1} \ln u$$

does not change sign on $(0, 1)$, $\phi(u) > 0$ on this interval. $(-\phi'(u))$ drops from infinity at $u = 0$ to a local minimum value $0.0950 \cdots$ at $u = 0.30756 \cdots$, rises to the local maximum value $0.1495 \cdots$ at $u = 0.57062$, and then drops monotonically to zero as $u \to 1^-$.)

REFERENCE

[1] L. LEWIN, *Dilogarithms and Associated Functions*, MacDonald, London, 1958, Chap. 8.

Problem 67-11, An Integral Inequality, by D. J. NEWMAN (Yeshiva University).

If $\mathbf{X}$ and $\mathbf{Y}$ are n-vectors and the appropropriate integrals exist, show that

$$\iint_R |\mathbf{X} - \mathbf{Y}| \Phi(\mathbf{X})\Phi(\mathbf{Y}) \, d\mathbf{X} \, d\mathbf{Y} \leqq \iint_R \{|\mathbf{X}| + |\mathbf{Y}|\}\Phi(\mathbf{X})\Phi(\mathbf{Y}) \, d\mathbf{X} \, d\mathbf{Y}.$$

Solution by the proposer.
We first show that

(1) $$\int e^{i\mathbf{Z}\cdot\mathbf{T}}(1 + |\mathbf{T}|^2)^{-(n+1)/2} \, d\mathbf{T} = Ce^{-|\mathbf{Z}|},$$

where

$$C = \int (1 + |\mathbf{T}|^2)^{-(n+1)/2} \, d\mathbf{T} = \pi^{(n+1)/2}\Gamma^{-1}\left(\frac{n+1}{2}\right).$$

Choose a t-coordinate system such that the first axis is in the $\mathbf{Z}$ direction. Then the given integral (1) is equal to

$$\int e^{i|\mathbf{Z}|t_1} \, dt_1 \int (1 + t_1^2 + t_2^2 + \cdots + t_n^2) \, dt_2 \, dt_3 \cdots dt_n.$$

Changing to spherical coordinates, we can write the inner integral as

$$C_1 \int (1 + t_1^2 + r^2)^{-(n+1)/2} r^{n-1} \, dr,$$

where C_1 is a constant. Letting $r = s\sqrt{1 + t_1^2}$ yields

$$\frac{C_1}{1 + t_1^2} \int (1 + s^2)^{-(n+1)/2} s^{n-1} \, ds = \frac{C_2}{1 + t_1^2}.$$

Equation (1) now follows by using the known result

$$\int \frac{e^{i|\mathbf{Z}|t_1}}{1 + t_1^2} \, dt_1 = \pi e^{-|\mathbf{Z}|}.$$

We therefore have

$$C(1 - e^{-\epsilon|\mathbf{X}|} - e^{-\epsilon|\mathbf{Y}|} + e^{-\epsilon|\mathbf{X}-\mathbf{Y}|}) = \int \frac{(1 - e^{i\epsilon\mathbf{X}\cdot\mathbf{T}})(1 - e^{-i\epsilon\mathbf{Y}\cdot\mathbf{T}}) \, d\mathbf{T}}{(1 + |\mathbf{T}|^2)^{(n+1)/2}},$$

so that

$$\frac{1}{\epsilon}\iint \{1 - e^{-\epsilon|\mathbf{X}|} - e^{-\epsilon|\mathbf{Y}|} + e^{-\epsilon|\mathbf{X}-\mathbf{Y}|}\}\Phi(\mathbf{X})\Phi(\mathbf{Y}) \, d\mathbf{X} \, d\mathbf{Y}$$

$$= \frac{1}{\epsilon C} \iiint (1 - e^{-i\epsilon \mathbf{X} \cdot \mathbf{T}})\Phi(X)(1 - e^{-i\epsilon \mathbf{Y} \cdot \mathbf{T}})\Phi(Y) \, \frac{d\mathbf{X} \, d\mathbf{Y} \, d\mathbf{T}}{(1 + |\mathbf{T}|^2)^{(n+1)/2}}$$

$$= \frac{1}{\epsilon C} \int \left| \int (1 - e^{i\epsilon \mathbf{X} \cdot \mathbf{T}})\Phi(\mathbf{X}) \, d\mathbf{X} \right|^2 \frac{d\mathbf{T}}{(1 + |\mathbf{T}|^2)^{(n+1)/2}} \geqq 0.$$

Finally, letting $\epsilon \to 0$, we have

$$\iint \{|\mathbf{X}| + |\mathbf{Y}| - |\mathbf{X} - \mathbf{Y}|\}\Phi(\mathbf{X})\Phi(\mathbf{Y}) \, d\mathbf{X} \, d\mathbf{Y} \geqq 0.$$

(The interchange of limits and orders of integration can be justified by summability.)

Supplementary References
Inequalities

[1]   C.Bandle, *Isoperimetric Inequalities*, Pitman, London, 1980.
[2]   Y.D.Burago, and V.A.Zalgaller, *Geometric Inequalities*, Springer-Verlag, Heidelberg, 1988.
[3]   E.F.Beckenbach, ed., *General Inequalities I*, Birkhauser Verlag, Basel, 1978.
[4]   E.F.Beckenbach, ed., *General Inequalities II*, Birkhauser Verlag, Basel, 1980.
[5]   E.F.Beckenbach and W. Walter, ed., *General Inequalities III*, Birkhauser Verlag, Basel, 1983.
[6]   W.Walter, ed., *General Inequalities IV*, Birkhauser Verlag, Basel, 1984.
[7]   E.F.Beckenbach and R.Bellman, *Inequalities*, Springer-Verlag, Heidelberg, 1965.
[8]   O.Bottema et al, *Geometric Inequalities*, Wolters-Noordhoff, Groningen, 1969.
[9]   P.S.Bullen, D.S.Mitrinovic and P.M.Vasic, *Means and their Inequalities*, Reidel, Dortrecht, 1988.
[10] G.H.Hardy, J.E.Littlewood, and G.Polya, *Inequalities*, Cambridge University Press, Cambridge, 1934.
[11] D.S.Mitrinovic, *Elementary Inequalities*, Noordhoff, Groningen, 1964.
[12] D.S.Mitrinovic, *Analytic Inequalities*, Springer-Verlag, Heidelberg, 1970.
[13] D.S.Mitrinovic, J.E.Pecaric, V.Volenec, *Recent Advances in Geometric Inequalities*, Kluwer Academic, Dordrecht, 1989.
[14] G.Polya and G.Szego, *Isoperimetric inequalities in Mathematical Physics*, Princeton University Press, Princeton, 1951.
[15] M.H.Protter and H.F.Weinberger, *Maximum Principles in Differential Equations*, Prentice-Hall, N.J., 1967.

14. OPTIMIZATION

An Optimal Control Problem

Problem 86-1*, by J. C. MOORHOUSE *and* J. V. BAXLEY (Wake Forest University).

Consider the optimal control problem of choosing the terminal time T and the function $q(t)$, $t_0 \leq t_1 \leq T$, which maximizes

$$V = \int_{t_0}^{t_1} [(1-\lambda)Pq(t) - c(q(t), Q(t))]e^{-rt}\,dt + \int_{t_1}^{T} [Pq(t) - c(q(t), Q(t))]e^{-rt}\,dt$$

where t_1 is given and known at t_0, T is free, $q(t) = \dot{Q}(t)$, $Q(0) = 0$, $Q(T) \leq R$ (R given), $q(t) \geq 0$, $c_q > 0$, $c_{qq} > 0$, $c_Q > 0$, $c_{QQ} > 0$, $c(0, Q) = 0$; P, λ, r, R are positive parameters with $\lambda < 1$.

(1) Does this optimal control problem have a (unique) solution?

(2) What qualitative conclusions can be drawn from the model? In particular, how does the optimal choice of $q(T)$ and T differ from the case where $\lambda = 0$ and t_1 becomes irrelevant?

Progress on the second question is of interest even in the absence of a clear answer to the first.

The problem arises in an economic study of the impact of the "Windfall Profits Tax" on the rate of production from an oil well. The unique feature of this optimal extraction problem is that oil producers know at t_0 the time t_1 when the tax λ will terminate. Thus the central issue is, how do the presence of the tax and prior knowledge of when the tax will end influence the optimal extraction rate $q(t)$ and the date T for depleting the reserves of a given oil pool? Partial results in a closely related problem have been given by Kemp and Long (*Economic Record*, September 1977).

Editorial note. Although the following solution is a heuristic one it is being published for possible interest for applications.

Solution by JAMES H. CASE (Baltimore, Maryland).

The optimal extraction schedule $q(t) = \dot{Q}(t)$ for an oil well maximizes the functional

(1) $$J = J(t_0, Q_0) = \int_{t_0}^{T} [p(t)Q(t) - C(Q(t), \dot{Q}(t))]e^{-rt}\,dt,$$

$p = p(t)$ being the (known) price of crude, $Q(t_0) = Q_0$, and $C = C(Q, q)$ the cost of extracting q barrels per day after Q barrels have been recovered. Presumably C_q, C_Q, C_{qq}, and C_{QQ} are everywhere positive. T is the time of exhaustion of the resource, and is to be chosen by the extractor.

We consider here the case wherein $p(t) = P$ before t_1, and again after t_2, but $p(t) = (1 - \lambda)P$ in the intervening interval. λ represents a windfall profits tax (WPT), and we seek to discover the extent to which its imposition distorts extraction schedules. Obviously there can be no distortion prior to the time $t = 0$ at which the tax is first announced.

The "value function" for the problem at hand is just

(2) $$V(t_0, Q_0) = \sup_{Q(\cdot)} J(t_0, Q_0),$$

the supremum being taken over all absolutely continuous and nondecreasing $Q(t)$, $t_0 < t < \infty$ which are constant save on a finite t-interval. $V(t, Q)$ satisfies the following partial differential equation:

(3) $$\max_q [p(t)q - C(Q, q)]e^{-rt} + qV_Q + V_t = \max_q [Rq - C(Q, q)]e^{-rt} + V_t = 0,$$

$R = R(t, Q) = p(t) + e^{rt}V_Q(t, Q)$ being a sort of "corrected price" designed to account for the depletion effect, as well as the revenue effect, of extraction.

To bring (3) into standard form, one must perform the indicated maximization. Consider accordingly

(4) $$L(Q, q, R) = qR - C(Q, q).$$

Obviously $L^*(Q, R) = \max_q L(Q, q, R) > L(Q, 0, R) = 0$. Indeed for each fixed Q there is a greatest $R = f(Q)$ for which $L^*(Q, R) = 0$. For each $R > f(Q)$ there is an interval I: $a(Q, R) < q < b(Q, R)$ whereon $L(Q, q, R) > 0$. I shrinks to a point $q^* = q^*(Q) = a(Q, f(Q)) = b(Q, f(Q))$, as R shrinks back to $f(Q)$; $q^*(Q) = 0$ only in the unlikely event $C(Q, 0+) = C(Q, 0) = 0$. $\phi(Q, R) = \text{argmax}_q L(Q, q, R)$ belongs to I when $R > f(Q)$, equals $q^*(Q)$ when $R = f(Q)$, and is zero otherwise.

Because $C = qR$ and $C_q = R$ must both hold wherever $R = f(Q), f'(Q) = C_Q/q > 0$. Likewise, becuase $C_q = R$ whenever $q = \phi(Q, R)$, it follows that $\phi_R = 1/C_{qq} > 0$, and $\phi_Q = -C_{qQ}/C_{qq}$, which is negative in the usual case $C_{qQ} > 0$.

Next, writing $S(t) = V_Q(t, Q(t))$, and introducing the "Hamiltonian" $H(t, Q, q, S) = [Pq - C(Q, q)]e^{-rt} + Sq$, we may write Hamilton's canonical equations

(5) $$\dot{Q} = H_s = q,$$
$$\dot{S} = -H_Q = e^{-rt}C_Q(Q, q).$$

These may be rewritten in the form

(6) $$\dot{Q} = \phi(Q, R),$$
$$\dot{R} = r(R - p(t)) + C_Q(Q, \phi(Q, R)).$$

During periods $A < t < B$ wherein $p(t)$ remains constant, the latter constitute a pair of autonomous ODE's with right-hand sides that are discontinuous along $R = f(Q)$. We distinguish two particularly relevant special cases, namely (6') in which $p(t) \equiv P$ and (6'') in which $p(t) \equiv (1 - \lambda)P$. Their (entirely unique) solutions in the semi-infinite strip $0 < R < p; Q > 0$ are sketched in Fig. 1.

In region I where $L^*(Q, R) > 0$, the solutions move from left to right, exiting either through the top, whereon $R = p$, or through the right-hand boundary $R = f(Q)$. In region II where $L^*(Q, R) = 0$, the solutions are just vertical line segments. The curve $R = f(Q)$, which separates I and II, is an additional solution, to which the ones in I and II are tributary.

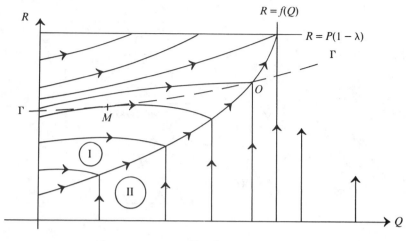

FIG. 1

It is now rather easy to solve the case $\lambda = 0$, in which there is no WPT, for then $p(t) = P$ always, and $Q(t)$, $R(t)$ must satisfy (6′). Also extraction continues until the interval $a(Q, P) < q < b(Q, P)$ of "feasible extraction rates" shrinks to a point, then ceases permanently. This occurs when $Q(t) = f^{-1}(P)$ and $\dot{Q}(t) = q^*(Q) > 0$. Thus $Q(T) = f^{-1}(P)$, $R(T) = P$, and the relevant solution of (6′) is the one simultaneously crossing the graph $R = f(Q)$ and the line $R = P$. That solution is itself the graph of a function $R = F(Q)$ in terms of which the optimal extraction policy is $q(t) = \dot{Q}(t)$ if $Q(t)$ solves $q = \dot{Q} = \phi(Q, F(Q))$: $Q(t_0) = Q_0$.

The case $\lambda > 0$ is marginally more complex. Observe first that, because the optimal trajectories can never meet tangentially the lines $t = t_1$ and $t = t_2$ on which the integrand in (1) is discontinuous, $V_Q(t, Q)$ is continuous across those lines. Thus $R = p(t) + V_Q(t, Q)e^{rt}$ is discontinuous across them, as is $p(t)$. Specifically, $R(t_2-) = R(t_2+) - P\lambda$, and $R(t_1-) = R(t_1+) + P\lambda$. But $R = F(Q)$ at t_2, and all subsequent times, because the WPT has no effect after expiration. So $R = F(Q) - P\lambda$ just before $t = t_2$.

One possible result of all this is indicated in Fig. 1, where the graph Γ of $R = F(Q) - P\lambda$ is shown crossing $R = f(Q)$ at 0 and meeting tangentially the solutions of (6″) at M. The relevant solutions of (6″) are just those which end on Γ at $t = t_2$. Most of them remain only briefly in I or II, before striking the graph of $R = f(Q)$ and remaining on it until $t = t_2$.

The corresponding trajectories in tQ-space are shown in Fig. 2. Most of the region between $t = t_1$ and $t = t_2$ is filled with tributaries to a single "turnpike," or "universal curve," which corresponds to the graph of $R = f(Q)$ in Fig. 1. The rest is covered with curves either purely of type I or purely of type II.

The positions occupied at time $t = t_1$ by the pairs $Q(t)$, $R(t)$ which reach the graph $R = F(Q) - P\lambda$ in Fig. 1 precisely at time $t = t_2$, typically form the graph of another function $R = G(Q)$. Translating that upwards to $R = G(Q) + P\lambda$ yields additional boundary data, from which first the phase trajectories $Q(t)$, $R(t)$, and later the ordinary ones t, $Q(t)$ can be continued backwards into $0 < t < t_1$. A few such trajectories are sketched in Fig. 2.

Again there is a turnpike which, together with its tributaries, fills most of the area to the left of $t = t_1$. Above and below these, additional type I and type II solutions

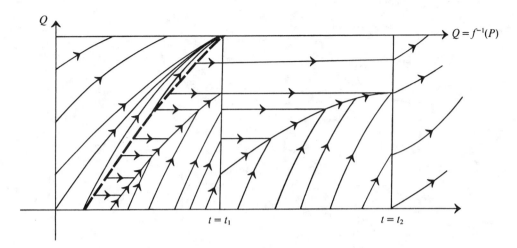

FIG. 2

complete the covering of the strip $0 < Q < f^{-1}(P)$. But not all of these are optimal! Obviously, solutions of the problem $\lambda = 0$ which complete extraction prior to $t = t_1$ are preferable to those which delay completion unnecessarily. And solutions of the $\lambda = 0$ problem to which is adjoined the additional constraint $T \leqq t_1$ may also be optimal for the given problem.

In effect there are two candidates for optimality starting at points (t, Q) to the left of $t = t_1$; one of them terminates before t_1, and one does not. A "dispersal curve" separates the region from which it is better to "hurry" and complete extraction before the tax goes into effect from that wherein it is better to wait.

It is not asserted that no additional phenomena can arise in the present problem, because the given hypotheses reveal very little about the mode of crossing between $R = f(Q)$ and $R = F(Q) - P\lambda$ or $R = G(Q) + P\lambda$. But the qualitative behaviour indicated in Fig. 2 does seem moderately typical.

Observe that no wells close down permanently during $t_1 < t < t_2$. This is due to the lack of a reopening cost in the problem specification; in reality, some wells do shut down while the tax is in force, causing some reserves to remain permanently unrecovered. Moreover, the plethora of horizontal trajectories in Fig. 2 along which no extraction takes place, as well as the "hurry up" ones, serve to emphasize that the WPT is only defensible (and historically was only proposed) in times of generally rising prices, rather than the stable ones foreseen by the posers of the present problem.

Extreme Value of an Integral

Problem 80-11, *by* W. W. MEYER (General Motors Research Laboratories).

Find the infimum of

$$I(f) = \int_0^a \sqrt{(f(t) - t)^2 + (f'(t))^2} \, dt$$

over all real-valued and differentiable functions f on a given interval $(0, a)$.

Solution by O. P. LOSSERS (Eindhoven University of Technology, Eindhoven, the Netherlands).

We may assume that $f(t) - t$ has a zero in the interval $[0, a]$. Define

$$x := (f(t) - t) \cos t + \sin t,$$

$$y := (f(t) - t) \sin t - \cos t.$$

We then have to find the infimum of

$$\int_0^a \sqrt{\dot{x}^2 + \dot{y}^2}\, dt.$$

This is the length of a curve and suggests a straight line for the infimum. However, the curve is subjected to the conditions that, first, $x \sin t - y \cos t = 1$, i.e., the point $(x(t), y(t))$ lies on the line that touches the circle $x^2 + y^2 = 1$ in the point $(\sin t, -\cos t)$; and second, $x^2 + y^2 = (f(t) - t)^2 + 1$, i.e., the curve lies outside the unit circle. In view of the opening remark, the curve touches the unit circle at least once.

Now if $a < \pi$, the shortest curve joining the tangents corresponding to $t = 0$ and $t = a$ via a point of the circle consists of two straight pieces of equal length meeting on the circle. This gives the minimum length $2(1 - \cos \tfrac{1}{2} a)$. If $a > \pi$, then the first condition forces the minimal curve to wind around the circle up to the point from where the final tangent can be reached by a straight line. In this case it is easy to show that the length of the shortest curve is $a - \pi + 2$.

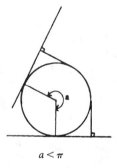

$a < \pi$

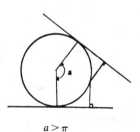

$a > \pi$

Also solved by H. G. MOYER (Grumman Aerospace, NY) *and the proposer,* who also finds the infimum of

$$\int_0^a \sqrt{|f(t) - g(t)|^2 + |f'(t)|^2}\, dt,$$

where g is a given continuous nonnegative, monotone increasing function on $[0, a]$.

Problem 74-18, Constrained Minimization of an Integral Functional, by MICHAEL H. MOORE (Vector Research, Inc.).

Let $N(a, b)$ denote the set of all nonnegative functions n on R^1 with support $[-a, a]$ for which

$$\int_{-a}^{a} n(t)\, dt = b.$$

Show that

$$\inf_{n\in N(a,b)} \sup_x \int_{x-c}^{x+c} n(t)\,dt = \frac{b}{\{a/c\}},$$

where $\{a/c\}$ denotes the smallest integer $\geq a/c$.

The problem has an interpretation in terms of minimizing the vulnerability of a military force on a one-dimensional nuclear battlefield subject to a constraint on the total force size.

Solution by C. F. SCHUBERT (Queen's University, Ontario, Canada).

If $k = \{a/c\}$, then $\delta = 2(a - (k-1)c)$ satisfies the inequality $0 < \delta \leq 2c$. Starting with the left endpoint, $-a$, partition the interval $[-a, a]$ into k sub-intervals $I_1, I_2, \cdots, I_k$, of which the first $k - 1$ have length $2c$ and the last, with a as its right endpoint, has length δ. If $n \in N(a, b)$ and

$$\sup_x \int_{x-c}^{x+c} n(t)\,dt < \frac{b}{k},$$

then

$$\int_{-a}^{a} n(t)\,dt = \sum_{j=1}^{k} \int_{I_j} n(t)\,dt < (k-1)\frac{b}{k} + \int_{a-\delta}^{a} n(t)\,dt \leq b,$$

which is a contradiction. Thus

$$\sup_x \int_{x-c}^{x+c} n(t)\,dt \geq \frac{b}{k}$$

and

$$\inf_{n\in N(a,b)} \sup_x \int_{x-c}^{x+c} n(t)\,dt \geq \frac{b}{k}.$$

The converse inequality is established as follows. Choose λ such that $0 < \lambda < \delta$, and then with $m(t) = b/(\lambda k)$ for $t \in [-a, -a + \lambda]$ and zero otherwise, put

$$\bar{n}(t) = \sum_{j=1}^{k} m(t - 2(j-1)c).$$

Then $\bar{n} \in N(a, b)$ and

$$\int_{x-c}^{x+c} \bar{n}(t)\,dt \leq \frac{b}{k}$$

for all x. Thus

$$\inf_{n\in N(a,b)} \sup_x \int_{x-c}^{x+c} n(t)\,dt \leq \frac{b}{k},$$

which proves the desired inequality.

Problem 72-9, An Extremum Problem, by R. Tapia (Rice University).

Determine the extrema in a nonvariational manner for the functional

(1) $$J(y) = \int_0^a (y'^2 - y^2)\,dx,$$

where y is continuously differentiable and $y(0) = y(a) = 0, 0 < a$.

It is to be noted that the problem has been used in a number of texts in the calculus of variations to illustrate the Jacobi condition, i.e., the application of conjugate points. In particular, since the points 0 and π are conjugate, a quick look at the well-known necessary and sufficient conditions show that (1) has a unique minimum if $a < \pi$ and no minimum if $a > \pi$. However, no conclusion can be made for $a = \pi$.

I. Solution by A. G. Konheim (IBM Research Center).

If we expand $y(x)$ in a Fourier-sine series

$$y(x) \sim \sum_{k=1}^{\infty} a_k \sin \frac{\pi k}{a} x,$$

then $J(y) = \int_0^a ((y'(x))^2 - y^2(x))\,dx$ is given by

$$J(y) = \frac{a}{2} \sum_{k=1}^{\infty} a_k^2 \left\{ \left(\frac{k\pi}{a} \right)^2 - 1 \right\}.$$

Thus for

$a < \pi$ there is a unique minimum, $y(x) = 0$;

$a > \pi$ there is no minimum, i.e. $\inf J(y) = -\infty$;

while for

$a = \pi$ the functions $y(x) = b \sin (\pi x/a)$ satisfy $J(y) = 0$ (for all $b, -\infty < b < \infty$) and $0 = \inf J(y)$.

Also solved in essentially the same way by K. Park (NASA, Langley Research Center), O. Ruehr (Michigan Technological University) and the proposer. However, only the proposer showed that J has no maxima, by observing that if $\Phi_n(x) = \sin (n\pi x/a)$, then $J(\Phi_n) = (a/2)((n\pi/a)^2 - 1) \to \infty$ as $n \to \infty$, for all $a > 0$.

II. Solution by R. T. Shield (University of Illinois).

The substitution $y = u \sin (\pi x/a)$ (cf. Jacobi's method of multiplicative variation [1]) is permissible in view of the properties of $y(x)$. The integral becomes

$$J = \int_0^a \left[u'^2 + \left(\frac{\pi^2}{a^2} - 1 \right) u^2 \right] \sin^2 \frac{\pi}{a} x\,dx,$$

and the extrema for $a < \pi, a = \pi$ and $a > \pi$ can be determined by inspection.

REFERENCE

[1] R. Courant and D. Hilbert, *Methods of Mathematical Physics*, vol. I, Interscience, New York, 1953, p. 458.

An Optimal Raster

Problem 81-6, *by* J. S. LEW (IBM T. J. Watson Research Center).

An electromagnetic device accepts a time-dependent locally integrable input $f(t)$ with period 1. It produces the corresponding periodic output

$$(1) \qquad\qquad g(t) = \int_0^{p/q} f(t - u) \, du$$

for specified relatively prime positive integers p, q. Limitations of the device require

$$(2) \qquad\qquad 0 \le f(t) \le 1$$

for all times t. Let the *raster interval,* for any resulting $g(t)$, be its longest interval of constant positive slope. Many such $f(t)$ yield functions $g(t)$ having the same longest raster interval. Which $f(t)$, among these, makes $g(t)$, on its raster interval, have the largest possible increase as well?

A raster design for an ink-jet printer [1] was the motivation for this problem.

REFERENCE

[1] LARRY KUHN AND ROBERT A. MYERS, *Ink-jet printing,* Scientific American 240, 4 (April 1979), pp. 162–178.

Solution by the proposer.

If $q = 1$, then $g =$ constant. Hence let $q > 1$ and let $\Delta = q^{-1}$. If $f(t)$ is any locally integrable function of t, and

$$(3) \qquad\qquad F(t) = \int_c^t f(u) \, du$$

for any fixed c, then $F(t)$ is an absolutely continuous function of t, and

$$(4) \qquad\qquad g(t) = F(t) - F(t - p\Delta)$$

by (1). If $f(t)$ is a function with period 1, then

$$(5) \qquad\qquad F(t) - F(t - q\Delta) = A = \text{constant}$$

independent of t. Moreover $f(t) = F'(t)$ except on a null set [3, Chap. V], and any such set is irrelevant in this discussion. We shall deduce properties of an optimal solution and thereby construct a unique function with these properties. Hence we let $g(t)$, in the following arguments, have linear increase on a maximum interval, and we let $g(t)$, by translation-invariance, commence its linear increase at the origin.

The linear interval associated with $g(t)$ cannot have length greater than $(q - 1)\Delta$. Indeed, if $g(t) = at + b$ with $a > 0$ when $0 \le t \le v + (q - 1)\Delta$ with $v > 0$, then clearly

$$(6) \qquad\qquad g(v + t) - g(t) = av > 0$$

when $0 \le t \le (q - 1)\Delta$. However, relations (4), (5), (6) yield a contradiction:

$$
\begin{aligned}
(7) \qquad 0 < qav &= \sum_{k=1}^{q} [g(v + kp\Delta) - g(kp\Delta)] \\
&= \sum_{k=1}^{q} [F(v + kp\Delta) - F(v + (k - 1)p\Delta) - F(kp\Delta) + F((k - 1)p\Delta)] \\
&= F(v + pq\Delta) - F(v) - F(pq\Delta) + F(0) = pA - pA = 0.
\end{aligned}
$$

If the linear interval has length $(q-1)\Delta$, then the total increment cannot exceed this same Δ, because (2), (3), (4) yield the estimate

$$
\begin{aligned}
g((q-1)\Delta) - g(0) &= g((q-1)\Delta) - g(q\Delta) \\
&= F((q-1)\Delta) - F(q\Delta) - F((q-p-1)\Delta) + F((q-p)\Delta) \\
&= \int_{(q-p-1)\Delta}^{(q-p)\Delta} f(u)\,du - \int_{(q-1)\Delta}^{q\Delta} f(u)\,du \le \Delta - 0.
\end{aligned}
$$
(8)

Indeed this increment cannot equal Δ, by (2) and (8), unless

$$
\begin{aligned}
f(t) &= 0 \quad \text{(a.e.) on } (q-1)\Delta < t < q\Delta, \\
f(t) &= 1 \quad \text{(a.e.) on } (q-p-1)\Delta < t < (q-p)\Delta.
\end{aligned}
$$
(9)

Moreover, if $0 < t < \Delta$, then similarly

$$
\begin{aligned}
g(t+k\Delta) - g(k\Delta) &= F(t+k\Delta) - F(k\Delta) - F(t+k\Delta-p\Delta) + F(k\Delta-p\Delta) \\
&= \int_{k\Delta}^{t+k\Delta} f(u)\,du - \int_{k\Delta-p\Delta}^{t+k\Delta-p\Delta} f(u)\,du.
\end{aligned}
$$
(10)

Relations (9) and (10) with $k = q-1$ imply linear *decrease* on $(q-1)\Delta < t < q\Delta$, and hence yield

$$
\begin{aligned}
g(t) &= g(0) - t \quad \text{on } -\Delta \le t \le 0, \\
g(t) &= g(0) + t(q-1)^{-1} \quad \text{on } 0 \le t \le (q-1)\Delta.
\end{aligned}
$$
(11)

Therefore $f(t)$ is essentially constant on $(k\Delta, (k+1)\Delta)$, by (10), whenever $f(t)$ is essentially constant on $((k \pm p)\Delta, (k+1 \pm p)\Delta)$ with either sign. However, p and q, by hypothesis, are relatively prime, whence $f(t)$, by (9), is essentially piecewise constant. Thus we may assume (9) for all t, and seek a piecewise constant solution for $f(t)$.

Next we differentiate (10) with $0 < t < \Delta$, and invoke (11) for $k = 0, \cdots, q-2$. This yields

$$
f(t+k\Delta) = (q-1)^{-1} + f(t+k\Delta - p\Delta),
$$
(12)

which determines $f(t)$. Indeed, $f(t) = 0$ on $(-\Delta, 0)$, by (9), and if t increases by $p\Delta$ then $f(t)$ increases by $(q-1)^{-1}$, until $f(t) = 1$ on $((q-p-1)\Delta, (q-p)\Delta)$, whence, by (9), if t increases by $p\Delta$ then $f(t)$ decreases by unity. To give an explicit formula for $f(t)$ we find the unique integer m among $1, \cdots, q-1$ such that $mp + nq = 1$ for some integer n, or equivalently

$$
mp \equiv 1 \pmod{q}.
$$
(13)

Now, if $(k-1)\Delta < t < k\Delta$, then clearly

$$
f(t) = (q-1)^{-1} [\text{residue of } km \text{ modulo } q];
$$
(14)

since (14) agrees with (9) when $k = 0$, and (14) agrees with (12) when $k \to k + p$. This $f(t)$, by (10), produces a $g(t)$ with the properties (11).

REFERENCES

[1] LARRY KUHN AND ROBERT A. MYERS, *Ink-jet printing*, Scientific American, 240, 4 (April 1979), pp. 162–178.

[2] J. S. LEW AND D. A. QUARLES, JR., *An ink-jet raster design for near-optimal efficiency*, IBM Technical Disclosure Bulletin, 18 (Jan. 1976), pp. 2718–2721.

[3] E. J. MCSHANE, *Integration*, Princeton Univ. Press, Princeton, NJ, 1944.

Problem 74-3, *Davidon's Cubic Interpolation*, by S. K. PARK and T. A. STRAETER (NASA/Langley Research Center).

The classic paper by W. C. Davidon (*Variable metric method for minimization*, ANL-5990, Rev. 1959, AEC Research Development) contained several important ideas, one of which was an algorithm for solving the one-dimensional search problem. That is, suppose that one has a real valued function with a minimum in the interval $0 \leq \alpha \leq \lambda$. Davidon uses f, g_s and f', g'_s to denote the function value and slope at the two points $\alpha = 0$ and $\alpha = \lambda$, respectively. With this notation, Davidon says "$\cdots$ we interpolate for the location of the minimum by choosing the 'smoothest' curve satisfying the boundary conditions $\cdots$, namely the curve defined as the one which minimizes

$$\int_0^\lambda \left(\frac{d^2 f}{d\alpha^2}\right)^2 d\alpha$$

over the curve. This is the curve formed by a flat spring fitted to the known ordinates and slope at the end points provided the slope is small. The resulting curve is a cubic, and its slope at any point α $(0 \leq \alpha \leq \lambda)$ is given by

(i) $$g_s(\alpha) = g_s - \frac{2\alpha}{\lambda}(g_s + z) + \frac{\alpha^2}{\lambda^2}(g_s + g'_s + 2z)$$

where

$$z = \frac{3(f - f')}{\lambda} + g_s + g'_s.$$

The root of (i) that corresponds to a minimum lies between 0 and λ by virtue of the fact that $g_s < 0$ and either $g'_s > 0$ or $z < g_s + g'_s$. It can be expressed as

$$\alpha_{\min} = \lambda(1 - a)$$

where

(ii) $$a = \frac{g'_s + Q - z}{g'_s - g_s + 2Q}$$

and

$$Q = (z^2 - g_s g'_s)^{1/2}.$$

The particular form of (ii) is chosen to obtain maximum accuracy which might otherwise be lost in taking the difference of nearly equal quantities."

Where does (ii) come from?

Solution by WILLIAM C. DAVIDON (Haverford College).

The solution to this problem can be obtained as a special case of the following simple theorem.

THEOREM. *Let* $\phi(x) = Ax^2 + 2Bx + C$ *be any polynomial of degree at most two and let* Δ *be a solution to* $\Delta^2 = B^2 - AC$. *If* ω *is any number with* $\omega A + B + \Delta \neq 0$, *then*

$$x = -\frac{\omega(B - \Delta) + C}{\omega A + B + \Delta}$$

is the root of $\phi(x) = 0$ *for which* $\phi'(x) = 2\Delta$.

Proof. If $A \neq 0$, then this root is given by the usual expression, $(-B + \Delta)/A$. This is not defined for $A = 0$, and gives excessive rounding error when B is near Δ. Similarly, if $B + \Delta \neq 0$, then $-C/(B + \Delta)$ also gives this root. Now, for any four numbers a, b, c and d, if $a/b = c/d$, then $a/b = c/d = (\omega a + c)/(\omega b + d)$ for any ω with $\omega b + d \neq 0$. Since $x = -(B - \Delta)/A = -C/(B + \Delta)$, the theorem follows.

The point of the theorem is that in all cases other than $A = 0$ and $B = -\Delta$, we can choose a number ω which not only satisfies $\omega A + B + \Delta \neq 0$, but which also gives an expression for the desired root which can be evaluated with less rounding. The solution to the proposed problem is obtained from $a = 1 - x$, where x is computed using the result of the theorem by setting $A = g_s + g_s' + 2z$, $B = -(g_s + z)$, $C = g_s$, $\Delta = Q = (z^2 - g_s g_s')^{1/2}$ and by making the special choice $\omega = \frac{1}{2}$.

Problem 78-4*, *Minimizing an Integral*, by C. L. MALLOWS (Bell Telephone Laboratories).

Find the symmetric cumulative distribution function $G(x)$ satisfying $dG(0) = \alpha$, $0 < \alpha < 1$ that minimizes the integral

$$I_f = \int_{-\infty}^{\infty} \frac{(f'(x))^2}{f(x)} \, dx$$

where $f(x)$ is the convolution $f(x) = \int_{-\infty}^{\infty} \phi(x - u) dG(u)$, with $\phi(u)$ the standard Gaussian density $\phi(u) = (2\pi)^{-1/2} \exp[-\frac{1}{2}u^2]$. It is believed that G is a step function, so that $f(x) = \sum p_j \phi(x - g_j)$ with $g_{-j} = -g_j$, $p_{-j} = p_j > 0$, $p_0 = \alpha$. The problem arose in a study of the robustness of certain statistical procedures; I_f is the Fisher Information associated with the distribution f of a Gaussian random variable that is occasionally contaminated by additive noise.

Comment by the author. A plausible guess is that $p_j = cp^j$ and $g_j = jg$ for $j > 0$, but I don't know whether this is correct. It is especially frustrating that if we consider *replacing* noise rather than *additive* noise, so that now $f(x) = \alpha\phi(x) + (1 - \alpha)g(x)$, g being an arbitrary symmetric density, the problem has a very simple solution—see P. J. Huber, *Robust estimation of a location parameter*, Ann. Math. Statist., 35 (1964), pp. 73–101. This analogy suggests that G must be asymptotically geometric in its far tails.

Problem 59-4*, Minimum-Loss Two Conductor Transmission Lines, by Gordon Raisbeck (Bell Telephone Laboratories).

Let C_1 and C_2 be two closed curves in a plane, one totally surrounding the other, bounding an annular region. Let ψ be a harmonic function within the annular region, having a constant value on C_1 and a constant value on C_2. If this configuration is regarded as the cross-section of a transmission line carrying a TEM wave, with cylindrical conductors of section C_1 and C_2, with a lossless dielectric between them, then the attenuation[1] is proportional to

---

[1] Gordon Raisbeck, *Minimum-Loss Two-Conductor Transmission Lines*, Trans. IRE PGCT, Sept. 1958.

$$\alpha = \frac{\int |\nabla \psi|^2 \, ds}{\iint |\nabla \psi|^2 \, da},$$

where the single integral is taken over the boundary $C_1 + C_2$ and the double integral over the area between them.

It has been shown[1] that under the constraint

$$\left\{ \int_{C_1} ds \right\}^{-1} + \left\{ \int_{C_2} ds \right\}^{-1} = \text{constant},$$

i.e., that the harmonic mean of the perimeters of the conductors is fixed, the minimum of the attenuation α is attained when the boundaries are concentric circles. Prove (or disprove) that the same conclusion holds under the alternative constraint that the area bounded between the curves is fixed.

Problem 66-6, The Minimum of a Double Integral, by A. W. OVERHAUSER (Ford Scientific Laboratory).

If $D(x)$ is an arbitrary even positive function, determine a monotonic function $\theta(x)$ whose range is equal to or less than $\pi/2$ such that the double integral

$$\int_{-1}^{1} \int_{-1}^{1} D(x - y) \cos^2 \{\theta(x) - \theta(y)\} \, dx \, dy$$

takes on its least value. This problem has arisen in a study of an electron gas.

Solution by PAUL I. RICHARDS (Technical Operations, Inc.).

The integral to be minimized can be written as

$$I = 2 \int_{-1}^{1} dx \int_{-1}^{x} D(x - y) \cos^2 [\theta(x) - \theta(y)] \, dy,$$

since $D(x)$ is even. Also, without loss of generality, we can assume $\theta(x)$ is monotone increasing with $0 \le \theta(x) \le \pi/2$. Let $y = x - v$ in the inner integral and interchange the order of integration to give

$$I = 2 \int_{0}^{2} D(v) J(v) \, dv,$$

where

$$J(v) = \int_{-1+v}^{1} \cos^2 [\theta(x) - \theta(x - v)] \, dx$$

$$= 2 - v - \int_{-1+v}^{1} \sin^2 [\theta(x) - \theta(x - v)] \, dx.$$

Now let

$$\theta_0(x) = \begin{cases} 0 & \text{for} \quad x < 0, \\ \pi/2 & \text{for} \quad x > 0, \end{cases}$$

so that the corresponding J function is

$$J_0(v) = \begin{cases} 2(1 - v) & \text{for} \quad v < 1, \\ 0 & \text{for} \quad v > 1. \end{cases}$$

For any other admissible $\theta(x)$, it is asserted that $J(v) \geq J_0(v)$ for all v in $[0, 2]$, so that $\theta_0(x)$ is a solution to the problem. To prove this, we use the trigonometric identity

$$\sin^2(a - b) = \frac{\sin (a - b)}{\sin (a + b)} [\sin^2 a - \sin^2 b].$$

Let $a = \theta(x)$ and $b = \theta(x - v)$. If $c = a - b$, then $c \geq 0$ because $\theta(x)$ is monotone and

$$c \leq 2b + c = a + b = 2a - c \leq \pi - c.$$

Thus,

$$\frac{\sin (a - b)}{\sin (a + b)} = \frac{\sin c}{\sin (a + b)} \leq 1$$

and

$$\sin^2 [\theta(x) - \theta(x - v)] \leq \sin^2 \theta(x) - \sin^2 \theta(x - v).$$

Then, if $z = x - v$,

$$J(v) \geq 2 - v - \int_{-1+v}^{1} \sin^2 \theta(x)\, dx + \int_{-1}^{1-v} \sin^2 \theta(z)\, dz$$

$$= 2 - v - \int_{1-v}^{1} \sin^2 \theta(x)\, dx + \int_{-1}^{-1+v} \sin^2 \theta(x)\, dx$$

$$\geq 2 - v - \int_{1-v}^{1} dx = 2(1 - v).$$

Hence, $J(v) \geq J_0(v)$ for $0 \leq v \leq 1$, while $J(v) \geq 0 = J_0(v)$ for $1 \leq v \leq 2$; and it follows that $\theta_0(x)$ yields the minimum of I with an arbitrary positive, even $D(x)$.

Moreover, the solution is unique, for when v is near 1, equality cannot hold in the last two manipulations above unless $\sin^2 \theta(x) = 0$ for $x < 0$ as well as $\sin^2 \theta(x) = 1$ for $x > 0$.

Problem 60–10*, A Parking Lot Design, by D. J. NEWMAN (Yeshiva University).

A factory and a parking lot are to be designed in such a way (unlike current practise) as to minimize the walking distance from car to desk. Variants of this problem can also be investigated; e.g., the factory is already designed and only the lot is to be mapped out, etc.

For the problem here, we will assume that the parking is done at random (with a uniform distribution) in the lot, that the desks are distributed at random (with a uniform distribution) throughout the factory, and we wish to minimize the average walking distance.

Again, variants appear if we change average to maximum, etc. But sticking to the original formulation and translating into mathematical symbolism, we obtain the problem:

To find two disjoint regions S_1, S_2 (factory, lot) such that

$$\int_{S_1} dx = F \qquad\qquad \text{size of factory specified,}$$

$$\int_{S_2} dy = L \qquad\qquad \text{size of lot specified,}$$

which minimize

(1) $$\iint_{S_1 \times S_2} |x - y| \, dx \, dy.$$

In this formulation, S_1 and S_2 are understood to be plane sets, x and y vectors in the plane, and dx and dy are infinitesimals of area.

In particular, show that for the one-dimensional analogue (where S_1, S_2 are linear sets, x, y real numbers, and dx, dy are ordinary infinitesimals) that

$$\bar{D} = \frac{\displaystyle\iint_{S_1 \times S_2} |x - y| \, dx \, dy}{\displaystyle\iint_{S_1 \times S_2} dx \, dy} \geq \frac{L}{2} + \frac{F}{4} - \frac{L^2}{12F}$$

(it is assumed without loss of generality that $F \geq L$). Also, give a construction for the sets $S_1(n)$, $S_2(n)$ such that

$$\bar{D} \to \frac{L}{2} + \frac{F}{4} - \frac{L^2}{12F}.$$

Partial solution by the proposer.

In the given problem, there is generally no minimum but there is an infimum, such that

$$\iint_{S_1 \times S_2} |x - y| \, dx \, dy > \alpha,$$

and a sequence of sets $S_1(n)$, $S_2(n)$, such that

$$\iint_{S_1(n) \times S_2(n)} |x - y| \, dx \, dy \to \alpha.$$

We now find α and $S_1(n)$, $S_2(n)$.

The crux of the derivation comes from the following

LEMMA 1. *Suppose* $f(0) = 0, f(1) = 1, |f'(x)| \leq M$, *then*

$$\int_0^1 f(x)(1 - f(x)) \, dx \leq \tfrac{1}{4} - (1/12M).$$

Proof: Call $g(x) = 2f(x) - 1$, then $g(0) = -1$, $g(1) = 1$, $|g'(x)| \leq 2M$ and

$$\int_0^1 g^2 \, dx = \int_0^1 \{1 - 4f(1 - f)\} \, dx = 1 - 4 \int_0^1 f(1 - f) \, dx.$$

So that we need only prove, $\int_0^1 g^2 \, dx \geq 1/3M$. Now for x in $(0, 1/2M)$, we have $g(x) = g(0) + xg'(\xi)$ or $|g(x)| \geq 1 - 2Mx$. Similarly for x in $[1 - (1/2M), 1]$, we find $|g(x)| \geq 1 - 2M(1 - x)$.

Hence, $\displaystyle\int_0^1 g^2 \, dx \geqq \int_0^{1/2M} (1 - 2Mx)^2 \, dx$

$$+ \int_{1-1/2M}^1 (1 - 2M(1-x))^2 \, dx = \frac{1}{3M}.$$

(M is automatically $\geqq 1$) which proves the lemma.

An immediate extension of this lemma is:

LEMMA 2. *If* $U'(0) = 0$, $|U''(x)| \leqq M$, $U'(\lambda) > 0$ *then*

$$\int_0^\lambda U(x) U''(x) \, dx \leqq \frac{\lambda U'(\lambda)^2}{4} - \frac{U'(\lambda)^3}{12M}.$$

Proof. By integration by parts, the given integral becomes

$$U(\lambda) U'(\lambda) - U(0) U'(0) - \int_0^\lambda U'^2(x) \, dx$$

$$= \int_0^\lambda U'(x)(U'(\lambda) - U'(x)) \, dx$$

$$= \lambda U'^2(\lambda) \int_0^1 \frac{U'(\lambda t)}{U'(\lambda)} \left\{ 1 - \frac{U'(\lambda t)}{U'(\lambda)} \right\} dt$$

and this is $\leqq U'^2(\lambda)\{(\lambda/4) - (U'(\lambda)/12M)\}$ by the previous lemma.

If we denote the characteristic functions of S_1 and S_2 by $\phi(x)$ and $\psi(x)$, respectively, then (1) becomes

$$\int_{-\infty}^\infty \int_{-\infty}^\infty |x - y| \, \phi(x)\psi(y) \, dx \, dy.$$

Now form

$$\int_{-\infty}^t \{\phi(x) + \psi(x)\} \, dx = H(t)$$

and define

$$\Phi(x) = 1, \quad \text{if} \quad x = H(t), \quad t \in S_1 \, ; \quad \text{otherwise} \quad \Phi(x) = 0,$$

$$\Psi(y) = 1, \quad \text{if} \quad y = H(t), \quad t \in S_2 \, ; \quad \text{otherwise} \quad \Psi(y) = 0.$$

Some elementary properties of Φ and Ψ are:

a) $\Phi(x)$ and $\Psi(x)$ are both zero except in the interval $(0, L + F)$.

b) $\Phi(x) + \Psi(x) = 1$, a.e., in the interval $(0, L + F)$.

c) $\displaystyle\int \Phi(x) \, dx = F, \int \Psi(x) \, dx = L.$

For a proof of (c), consider

$$\int \Phi \, dx = \int_{-\infty}^\infty \Phi(H(t)) H'(t) \, dt = \int_{-\infty}^\infty \Phi(H(t))\{\phi(t) + \psi(t)\} \, dt$$

$$= \int_{S_1} \{\phi(t) + \psi(t)\} \, dt = \int_{S_1} dt = F,$$

and similarly for Ψ.

A less obvious property is that

d) $\displaystyle\int_{-\infty}^{\infty}\int_{-\infty}^{\infty} |x - y| \, \phi(x)\psi(y) \, dx \, dy \geqq \int_{0}^{L+F}\int_{0}^{L+F} |x - y| \, \Phi(x)\Psi(y) \, dx \, dy,$

whose proof is as follows:

$$\int_{0}^{L+F}\int_{0}^{L+F} |x - y| \, \Phi(x)\Psi(y) \, dx \, dy = \int_{-\infty}^{\infty} |H(t)$$
$$- H(s)| \, \Phi(H(t))\Psi(H(s))H'(t)H'(s) \, dt \, ds.$$

Now $\qquad\qquad\qquad \Phi(H(t))\Psi(H(s))H'(t)H'(s) = 0,$

unless $t \in S_1$, $s \in S_2$, in which case it equals

$$[\phi(t) + \psi(t)][\phi(s) + \psi(s)] = 1.$$

Thus, $\qquad\qquad \Phi(H(t))\Psi(H(s))H'(t)H'(s) = \phi(t)\psi(s),$

and the last double integral becomes

$$\int_{-\infty}^{\infty}\int_{-\infty}^{\infty} |H(x) - H(y)| \, \phi(x)\psi(y) \, dx \, dy.$$

(d) will now follow once we show that $|H(x) - H(y)| \leq |x - y|$ which is equivalent to showing $|H'(t)| \leq 1$ a.e. $H'(t)$, however, is a.e. equal to $\phi(t) + \psi(t)$ which is either 0 or 1. Now,

$$\int_{-\infty}^{\infty}\int_{-\infty}^{\infty} |x - y| \, \Phi(x)\Psi(y) \, dx \, dy$$
$$= \int_{0}^{L+F}\int_{0}^{x} (x - y)\Phi(x)\Psi(y) \, dy \, dx + \int_{0}^{L+F}\int_{0}^{y} (y - x)\Phi(x)\Psi(y) \, dx \, dy$$
$$= \int_{0}^{L+F}\int_{0}^{x} (x - y)\{\Phi(x)\Psi(y) + \Phi(y)\Psi(x)\} \, dy \, dx.$$

Now let $\Phi(x) = [1 + u(x)]/2$, $\Psi(y) = [1 - u(y)]/2$ (the same u) and the last integral becomes

(2) $\qquad\qquad \dfrac{(L + F)^3}{12} - \dfrac{1}{2}\displaystyle\int_{0}^{L+F}\int_{0}^{x} (x - y)u(x)u(y) \, dy \, dx.$

Letting

$$\int_{0}^{x} (x - y)u(y) \, dy = U(x),$$

it follows that $U''(x) = u(x)$ and then (2) becomes

(3) $\qquad\qquad \dfrac{(L + F)^3}{12} - \dfrac{1}{2}\displaystyle\int_{0}^{L+F} U(x)U''(x) \, dx.$

Also, $U'(0) = 0$, $|U''(x)| = |2\Phi(x) - 1| \leq 1$, and

$$U'(L + F) = \int_{0}^{L+F} U''(x) \, dx = \int_{0}^{L+F} (2\Phi(x) - 1) \, dx = 2F - (L + F) > 0.$$

Consequently Lemma 2 applies which gives

$$\int_0^{L+F} U(x)U''(x)\, dx \leqq \frac{(L+F)(F-L)^2}{4} - \frac{(F-L)^3}{12},$$

and finally that

$$\int_{S_1 \times S_2} |x-y|\, dx\, dy \geqq \frac{(L+F)^3}{12} - \frac{(L+F)(F-L)^2}{8} + \frac{(F-L)^3}{24}$$

or

$$\int_{S_1 \times S_2} |x-y|\, dx\, dy \geqq \frac{FL^2}{2} + \frac{F^2 L}{4} - \frac{L^3}{12}.$$

The average distance, $\bar{D}$, is given by

$$\frac{\displaystyle\iint_{S_1 \times S_2} |x-y|\, dx\, dy}{\displaystyle\iint_{S_1 \times S_2} dx\, dy}$$

and thus

$$\bar{D} \geqq \frac{L}{2} + \frac{F}{4} - \frac{L^2}{12F}.$$

We now construct $S_1(n)$ and $S_2(n)$ such that

$$\bar{D} \to \frac{L}{2} + \frac{F}{4} - \frac{L^2}{12F}.$$

Simply choose S_2 in the interval $(0, L+F)$ to constitute half the inner interval $((F-L)/2, (F+3L)/2)$ homogeneously and S_1 to be the other half of this inner interval plus the outer intervals $(0, (F-L)/2)$ and

$$((F+3L)/2, L+F).$$

A simple computation shows that, in the limit,

$$\bar{D} = \frac{L}{2} + \frac{F}{4} - \frac{L^2}{12F}.$$

The two-dimensional problem is still open.

Problem 67-17, *An Extremal Problem*, by J. NEURINGER (AVCO Corporation) AND D. J. NEWMAN (Yeshiva University).

Consider the differential equation

$$\{D^2 - F(x)\}y = 0, \qquad y(0) = 1, \qquad y'(0) = 0.$$

Choose $F(x)$ subject to the conditions

$$F(x) \geqq 0, \qquad \int_0^1 F(x)\, dx = M,$$

so as to maximize $y(1)$.

This problem has arisen in connection with the construction of an optimal refracting medium.

Solution by J. ERNEST WILKINS, JR. (Gulf General Atomic Incorporated).

Let $\mathfrak{M}$ be the class of integrable, almost everywhere nonnegative, functions $F(x)$ defined on the interval $(0, 1)$ whose integral over $(0, 1)$ is M. For each such $F(x)$ there exists a unique "associated" function $y(x)$ with an absolutely continuous first derivative $y'(x)$ which satisfies the differential equation $y'' = Fy$ almost everywhere on $(0, 1)$ and for which $y(0) = 1$, $y'(0) = 0$. Let A be the least upper bound of the values $y(1)$ as $F(x)$ ranges over $\mathfrak{M}$. I shall prove that $A = M + 1$ if $0 \leq M \leq 1$, $A = 2 \exp(M^{1/2} - 1)$ if $M \geq 1$.

The proof of this assertion will depend on the following lemma, which is suggested by the usual calculus of variations devices.

LEMMA 1. *If a constant v and two absolutely continuous functions $u(x)$ and $\lambda(x)$ can be found such that* (i) $\lambda \leq 0$, (ii) $\lambda(u' + u^2) \geq 0$, (iii) $\lambda' = 2u(\lambda + v) - 1$,

(iv) $\lambda(1) + v = 0$, (v) $v\left\{ \int_0^1 (u' + u^2)\, dx - M + u(0) \right\} + \lambda(0)u(0) = 0$, *then*

$$A \leq \exp \int_0^1 u(x)\, dx.$$

Suppose that $F(x)$ is an element of $\mathfrak{M}$ and that $y(x)$ is associated with $F(x)$. It is easy to show that $y(x) \geq 1$ and hence the function $V(x) = y'(x)/y(x)$ is absolutely continuous. Moreover, $V(0) = 0$ and $F = V' + V^2$. Therefore, by virtue of the hypotheses (i) through (iv) and the definition of $\mathfrak{M}$,

$$\log y(1) = \int_0^1 V(x)\, dx$$

$$\leq \int_0^1 \{V - \lambda(V' + V^2) - v(V' + V^2 - M)\}\, dx$$

$$= \int_0^1 \{u - \lambda(u' + u^2) - v(u' + u^2 - M)\}\, dx$$

$$\quad - \int_0^1 \{(V - u)[2u(\lambda + v) - 1] + (V' - u')(\lambda + v)\}\, dx$$

$$\quad - \int_0^1 (V - u)^2(\lambda + v)\, dx$$

$$= \int_0^1 u\, dx - v\int_0^1 (u' + u^2 - M)\, dx - (V - u)(\lambda + v)\Big|_0^1$$

$$\quad - \int_0^1 (V - u)^2(\lambda + v)\, dx$$

$$= \int_0^1 u\, dx - \int_0^1 (V - u)^2(\lambda + v)\, dx.$$

The conclusion of the lemma now follows from the inference from (iii) and (iv) that

$$\lambda + v = \int_x^1 \exp\left\{ -2\int_x^t u(s)\, ds \right\} dt \geq 0.$$

When $0 \leq M \leq 1$, the quantities

$$v = \frac{1}{M+1}, \qquad u(x) = \frac{M}{Mx+1}, \qquad \lambda(x) = -\frac{x(Mx+1-M)}{M+1}$$

are easily seen to satisfy the hypotheses of the lemma, and hence

$$A \leq \exp \int_0^1 u(x)\,dx = M + 1.$$

When $M \geq 1$, the quantities $v = 1/(2M^{1/2})$,

$$u(x) = M^{1/2}, \qquad \lambda(x) = 0 \quad \text{if} \quad 0 \leq x \leq 1 - M^{-1/2},$$

$$u(x) = \frac{M^{1/2}}{2 - M^{1/2}(1-x)},$$

$$\lambda(x) = -\frac{\{M^{1/2}(1-x) - 1\}^2}{2M^{1/2}} \quad \text{if} \quad 1 - M^{-1/2} \leq x \leq 1,$$

also satisfy the hypotheses of the lemma, and hence

$$A \leq \exp \int_0^1 u(x)\,dx = 2 \exp(M^{1/2} - 1).$$

To see that the upper bounds just established are in fact least upper bounds, it is sufficient to exhibit a sequence of absolutely continuous functions $V_n(x)$ defined for sufficiently large n when $0 \leq x \leq 1$, such that $V_n(0) = 0$, $F_n = V_n' + V_n^2$ is in $\mathfrak{M}$ and

$$\exp\left\{\lim_{n\to\infty} \int_0^1 V_n(x)\,dx\right\} = \exp \int_0^1 u(x)\,dx.$$

When $M \leq 1$, define

$$V_n(x) = \begin{cases} n \tanh nx & \text{if} \quad 0 \leq x \leq M/n^2, \\ (x + \alpha_n)^{-1} & \text{if} \quad M/n^2 \leq x \leq 1, \end{cases}$$

in which

$$\alpha_n = -\frac{M}{n^2} + \frac{1}{n}\coth\frac{M}{n},$$

so that V_n is absolutely continuous on $(0, 1)$. Since $F_n = n^2$ on $(0, M/n^2)$ and $F_n = 0$ on $(M/n^2, 1)$, it is clear that F_n is in $\mathfrak{M}$. Moreover $V_n(0) = 0$ and

$$\exp \int_0^1 V_n(x)\,dx = \left\{n\left(1 - \frac{M}{n^2}\right)\sinh\frac{M}{n} + \cosh\frac{M}{n}\right\} \to M + 1.$$

When $M > 1$, we define

$$V_n(x) = \begin{cases} M^{1/2}nx & \text{if} \quad 0 \leq x \leq 1/n, \\ M^{1/2} & \text{if} \quad 1/n \leq x \leq 1 - M^{-1/2} + 2/(3n), \\ (x + \alpha_n)^{-1} & \text{if} \quad 1 - M^{-1/2} + 2/(3n) \leq x \leq 1, \end{cases}$$

for values of n such that $n \geq 2M^{1/2}/3$, $n \geq M^{1/2}/(3(M^{1/2} - 1))$. The function $V_n(x)$ is absolutely continuous if

$$\alpha_n = 2M^{-1/2} - 1 - 2/(3n)$$

and $V_n(0) = 0$. Moreover, $F_n = M^{1/2}n + Mn^2x^2$, M, and 0 on the indicated subintervals of $(0, 1)$, and hence F_n is in $\mathfrak{M}$. Finally,

$$\exp \int_0^1 V_n(x)\,dx = 2\left(1 - \frac{M^{1/2}}{3n}\right)\exp\left\{M^{1/2}\left(1 + \frac{1}{6n}\right) - 1\right\}$$

$$\to 2\exp(M^{1/2} - 1).$$

An Optimum Multistage Rocket Design

Problem 87-15, *by* M. S. KLAMKIN (University of Alberta).

In a recent paper [1] on optimization in multistage rocket design, Peressini reduced his problem to minimizing the function

$$\ln\left\{\frac{M_0 + P}{P}\right\} = \sum_{i=1}^{n}\{\ln N_i + \ln(1 - S_i) - \ln(1 - S_iN_i)$$

subject to the constraint condition

$$\sum_{i=1}^{n} c_i \ln N_i = v_f \text{ (constant)}.$$

Here the S_i's and the c_i's are given structural factors and engine exhaust speeds, respectively. He then obtains the necessary optimization equations using Lagrange multipliers. These equations are then solved explicitly for the special case when $S_i = S$, $c_i = c$ for all i.

For the latter special case, show how to determine the minimum in a simpler fashion without using calculus.

REFERENCES

[1] A. L. PERESSINI, *Lagrange multipliers and the design of multistage rockets*, UMAP J., 7 (1986), pp. 249–262.

Solution by the proposer.

More generally, the S_i's need not be the same. Thus we wish to maximize $P \equiv \Pi(1 - x_i)$, where $x_i = S_iN_i$ and where the product and sums here and subsequently are all from $i = 1$ to n. Since $\sum \ln N_i = v_f/c = $ constant, the S_i's are specified, tacitly $0 < x_i < 1$, and our constraint condition is $\Pi x_i = $ constant.

Using the concavity of $\ln(1 - x)$ for $0 < x < 1$ and the A.M.–G.M. inequality, we have that

$$\sum\ln(1 - x_i) \leqq n\ln(1 - \Sigma x_i/n) \leqq n\ln(1 - \Pi x_i^{1/n}).$$

Thus the maximum is taken on for $x_i = S_iN_i = $ constant or $N_i = \lambda/S_i$, where $\lambda^n = e^{v_f/c}\Pi S_i$. Finally,

$$\text{Min}\,(M_0 + P)/P = e^{v_f/c}(1 - \lambda)^{-n}\Pi(1 - S_i).$$

Constructing Nodes for Optimal Quadrature

Problem 85-6*, *by* J. P. LAMBERT (University of Alaska).

(a) Let

$$f(s,t) = \frac{1}{2}\left[\frac{1}{3-s^2} + \frac{1}{3-t^2} + \frac{2}{3-st}\right] - \left[\frac{1}{s}\ln\left(\frac{3+s}{3-s}\right) + \frac{1}{t}\ln\left(\frac{3+t}{3-t}\right)\right]$$

for $s,t \in [-1,1]$. Prove (or disprove) that $f(x,t)$ is minimal when $s = -t$ ($\approx \pm .58324119$).

(b) More generally, the following problem arises in the construction of a certain n-point optimal quadrature scheme. Let

$$f(t_1,\cdots,t_n) = \frac{1}{n}\sum_{i,j=1}^{n}\frac{1}{3-t_it_j} - \sum_{i=1}^{n}\frac{1}{t_i}\ln\left(\frac{3+t_i}{3-t_i}\right)$$

for $t_1,\cdots,t_n \in [-1,1]$. Prove (or disprove) that f achieves a minimum at a point $(t_1,\cdots,t_{n/2}, -t_1,\cdots,-t_{n/2})$ if n is even, and at a point $(t_1,\cdots,t_{(n-1)/2}, 0, -t_1,\cdots,-t_{(n-1)/2})$ if n is odd. That is, prove (or disprove) that the $t_1,\cdots,t_n$ which minimize the value of f are symmetrically disposed about the origin. There t's are the nodes for an optimal quadrature formula.

Note. If $t=0$, it is understood that

$$\frac{1}{t}\ln\frac{3+t}{3-t}$$

is to be replaced by

$$\lim_{t\to 0}\frac{1}{t}\ln\frac{3+t}{3-t} = \frac{2}{3}.$$

Editorial comment. The proposer notes that there is compelling numerical evidence to indicate the validity of the above two assertions.

Solution by T. L. MCCOY (Michigan State University).

We shall prove that, as conjectured, $f(s,t)$ takes its minimum over the rectangle $R = [-1,1] \times [-1,1]$ at an interior point of R for which $s = -t$.

We observe that $f(s,t) = f(-s,-t)$. Also, let us notice that replacing the point (s,t) by $(s,-t)$ only changes the term $(3-st)^{-1}$ in the definition of f; that is

$$f(s,-t) = f(s,t) + \frac{1}{3+st} - \frac{1}{3-st}.$$

Therefore, s and t have opposite sign at a point where the minimum occurs. In the rest of this note, we assume $0 \le s \le 1$ and $-1 \le t \le 0$.

The equations $\partial f/\partial s = \partial f/\partial t = 0$ can be written

$$t(3-st)^{-2} = \phi(s),$$
$$s(3-st)^{-2} = \phi(t),$$

where

$$\phi(u)=\frac{6}{u(9-u^2)}-\frac{u}{(3-u^2)^2}+\frac{1}{u^2}\ln\left(\frac{3-u}{3+u}\right).$$

Thus, for a minimum to occur at an interior point, it is necessary that

(1) $$s\phi(s)=t\phi(t).$$

It is easy to see that $\phi(-u)=-\phi(u)$. Thus, if we set $t=-r$, we get

(2) $$s\phi(s)=r\phi(r),$$

where we may assume $s,r\geq 0$.

The function $u\phi(u)$ has the Maclaurin expansion

$$u\phi(u)=\sum_{n=1}^{\infty}\left\{\frac{4}{(2n+1)3^n}-1\right\}n3^{-(n+1)}u^{2n},$$

valid for $|u|<\sqrt{3}$. Since all the coefficients of this series are negative, it follows that $u\phi(u)$ is monotone decreasing on $[0,1]$. Thus, (2) only holds for $r=s$. Consequently, for $f(r,s)$ to have a minimum at an interior point of R, it is necessary that $s=-t$.

Next, we rule out the possibility that f assumes its minimum at a boundary point of R. Since $f(s,t)=f(t,s)$, we need only examine the function $f(1,t)$ with $-1\leq t\leq 0$. We have

$$f(1,-r)=\left(\frac{1}{4}-\ln 2\right)+\frac{1}{2(3-r^2)}+\frac{1}{3+r}+\frac{1}{r}\ln\left(\frac{3-r}{3+r}\right).$$

Thus, for $0<r<\sqrt{3}$,

$$\frac{d}{dr}f(1,-r)=\frac{r}{(3-r^2)^2}-\frac{1}{(3+r)^2}-\frac{1}{r^2}\ln\left(\frac{3-r}{3+r}\right)+\frac{6}{r(9-r^2)}$$

$$>\frac{6}{r(9-r^2)}-\frac{1}{(3+r)^2}=\frac{r^2+3r+18}{r(3-r)(3+r)^2}>0.$$

Therefore

$$\min_{|t|\leq 1}f(1,t)=f(1,0)=\frac{1}{12}-\ln 2,$$

while a computation shows that $f(.6,-.6)$ has a smaller value. Consequently, the minimum does indeed occur in the interior.

Editorial note: No complete solutions of part (b) were received. [C. C. R.]

Note added in proof. There is a sign error in computing the derivative of $f(1,-r)$, therefore, it is still possible that f assumes its minimum on the boundary of the rectangle.

Problem 70-20, A Minimization for Markov Chains, by PAUL J. SCHWEITZER (Technion, Israel).

Consider

$$\min_{P} \sum_{i=1}^{N} \sum_{j=i}^{N} \pi_i P_{ij},$$

where the minimum is taken over all $N \times N$ single-chained stochastic matrices P and where π denotes the unique stationary distribution for P (i.e., $P \geq 0$, $\sum_{j=1}^{N} P_{ij} = 1$, $\pi = \pi P$, $\sum_{i=1}^{N} \pi_i = 1$). Show that the minimum is $1/N$ and is achieved uniquely for

$$P^* = \begin{Vmatrix} 0 & 0 & 0 & \cdots & 0 & 1 \\ 1 & 0 & 0 & \cdots & 0 & 0 \\ 0 & 1 & 0 & \cdots & 0 & 0 \\ \vdots & & & & & \vdots \\ 0 & 0 & 0 & \cdots & 1 & 0 \end{Vmatrix}.$$

Solution by A. A. JAGERS (Technische Hogeschool Twente, Enschede, Netherlands.

Since $\pi P = \pi$ and $\sum_{j=1}^{N} P_{ij} = 1$, the matrix R defined by $R_{ij} = \pi_i P_{ij}$ satisfies

$$\sum_{j=1}^{N} R_{ij} = \sum_{j=1}^{N} R_{ji} (= \pi_i)$$

for all i. Thus

$$\sum_{i=1}^{N} \sum_{j=i}^{N} N R_{ij} = \sum_{i=1}^{N} \sum_{j=1}^{N} t_{ij} R_{ij},$$

where the t_{ij} are defined by $1 \leq t_{ij} \leq N$ and $t_{ij} \equiv N + i - j \pmod{N}$. Since $R \geq 0$, we have

$$\sum_{i=1}^{N} \sum_{j=i}^{N} R_{ij} \geq \frac{1}{N} \sum_{i=1}^{N} \sum_{j=1}^{N} R_{ij} = \frac{1}{N} \sum_{i=1}^{N} \pi_i = \frac{1}{N},$$

where equality holds if and only if $R_{ij} = 0$ for all i, j with $t_{ij} > 1$.

Finally, the desired result now follows by noting that $\pi_i \neq 0$ and $P_{ij} \neq 0$ at least once in every row.

Also solved by the proposer by means of Markov renewal programming (see the author's paper, *Perturbation theory and undiscounted Markov renewal programming*, Operations Res., 17 (1969), pp. 716–727, Eq. (26)).

A Multi-Variable Extremum Problem

Problem 85-11*, *by* V. BALAKOTAIAH (University of Houston).
   Conjecture. Let

$$F(x_1, x_2, \cdots, x_{n+1}) = \sum_{j=1}^{n+1} (1 + x_j) \prod_{\substack{l=1 \\ l \neq j}}^{n+1} \frac{(1 + x_l)}{(x_l - x_j)} e^{-x_j}$$

where $x_j > 0$ for $j = 1, 2, \cdots, n+1$. The function is not defined for $x_l = x_j$ but the limit

$x_l \to x_j$ exists. F attains its supremum value when $x_j = n$ for all j and

$$F_{max} = \frac{(n+1)^{n+1} e^{-n}}{n!}.$$

Editorial note. There is a misprint in the defining statement for F which affects the value of F_{max} but not the technique by which it may be obtained. As corrected, F is defined by

$$F(x_1, \cdots, x_{n+1}) = \sum_{j=1}^{n+1} (1 + x_j) \exp(-x_j) \prod_{\substack{l=1 \\ l \neq j}}^{n+1} \frac{(1 + x_l)}{(x_l - x_j)}.$$

The following argument shows that for this function the conjectured maximum is correct. [C.C.R.]

Solution by DAVID G. CANTOR (Malibu, CA).

The conjecture is true. We first recall some elementary facts about divided differences. Given a function g, define its *divided-difference* of order n inductively by

(1) $$g^{(1)}[x_1] = g(x_1),$$

(2) $$g^{(n+1)}[x_1, \cdots, x_{n+1}] = \frac{g^{(n)}[x_1, \cdots, x_{n-1}, x_n] - g^{(n)}[x_1, \cdots, x_{n-1}, x_{n+1}]}{(x_n - x_{n+1})}.$$

Then it is easy to verify that

(3) $$g^{(n)}[x_1, \cdots, x_n] = \sum_{j=1}^{n} \frac{g(x_j)}{\prod_{i \neq j}(x_j - x_i)}$$

is a symmetric function of the x_i. While initially defined only for distinct x_i, if $g \in C^n$, it has a unique continuous extension to vectors $(x_1, \cdots, x_n)$ with some or all of the x_i equal. Moreover if $x_1 \leq x_2 \leq \cdots \leq x_n$, there exists θ satisfying $x_1 \leq \theta \leq x_n$, such that

(4) $$g^{(n)}[x_1, \cdots, x_n] = g^{(n)}[\theta, \cdots, \theta] = g^{(n-1)}(\theta)/(n-1)!.$$

Observe that

(5) $$F(x_1, \cdots, x_{n+1}) = g^{(n+1)}[u_1, \cdots, u_{n+1}],$$

where $g(u) = \exp((u-1)/u) u^{n-1}$ and $u_j = 1/(1 + x_j)$, $j = 1, \cdots, n+1$. Also

(6) $$F(x_1, \cdots, x_{n+1}) = (-1)^n \prod_{j=1}^{n+1} (1 + x_j) h^{(n+1)}[x_1, \cdots, x_{n+1}],$$

where $h(x) = e^{-x}$. According to (4) and (5) and (6), for each $x_1, \cdots, x_{n+1}$ there is a θ such that

(7) $$F(x_1, \cdots, x_{n+1}) = F(\theta, \cdots, \theta) = (1 + \theta)^{n+1} e^{-\theta}/n!.$$

Hence

$$\sup F(x_1, \cdots, x_{n+1}) = \sup (1+\theta)^{n+1} e^{-\theta}/n! = (n+1)^{n+1} e^{-n}/n!.$$

Problem 60-3,* *A Center of Gravity Perturbation,* by M.S. Klamkin (AVCO Research and Advanced Development Division).

Determine a vector $\mathbf{Z} = (z_1, z_2, \cdots, z_n)$ which maximizes $(\mathbf{A} \cdot \mathbf{Z})^2 + (\mathbf{B} \cdot \mathbf{Z})^2$, where $\mathbf{A}$ and $\mathbf{B}$ are given vectors and $|z_r| \leq 1, r = 1, 2, \cdots, n$. This problem arises from the following physical situation:

A composite body consists of n component masses $\{m_r\}$ with individual $C.G.$'s at (x_r, y_r). The masses will not be known exactly but can vary within a tolerance of $\pm \epsilon_r m_r (\epsilon_r \ll 1)$. What is the greatest distance the $C.G.$ can be from the $C.G.$ which is calculated by using the nominal masses?

If the origin of our coordinate system is taken at the nominal $C.G.$, then to first order terms the perturbation in the position of the $C.G.$ due to perturbations in the masses will be given by

$$\Delta x = \frac{\sum x_r \, \Delta m_r}{\sum m_r},$$

$$\Delta y = \frac{\sum y_r \, \Delta m_r}{\sum m_r}.$$

Then,

$$\Delta x^2 + \Delta y^2 = (\mathbf{A} \cdot \mathbf{Z})^2 + (\mathbf{B} \cdot \mathbf{Z})^2,$$

where

$$\mathbf{A} = \left\{ \frac{\epsilon_r \, m_r \, x_r}{\Sigma m_r} \right\}, \qquad\qquad r = 1, 2, \cdots, n$$

$$\mathbf{B} = \left\{ \frac{\epsilon_r \, m_r \, y_r}{\Sigma m_r} \right\}, \qquad\qquad r = 1, 2, \cdots, n$$

$$\mathbf{Z} = \{z_r\}, \qquad\qquad |z_r| \leq 1, r = 1, 2, \cdots, n.$$

It follows that the maximizing vector $\mathbf{Z}$ (emanating from the origin) will terminate on one of the vertices of the hypercube $(\pm 1, \pm 1, \cdots, \pm 1)$. The difficulty in the problem resides in the fact that in the actual problem involved, $n = 43$ and thus the number of vertices is 2^{43} which is much too large to check each one. Crude upper and lower bounds can be immediately obtained by considering $\mathbf{Z}$ to terminate on the circumscribed and inscribed hyperspheres, respectively. In these cases the maximizing vector will lie in the plane of $\mathbf{A}$ and $\mathbf{B}$ and is easily determined. The ratio of these bounds will be $\sqrt{n}$. The lower bound can be improved by extending the maximizing vector until it hits the cube. This in turn can probably be improved by choosing the "closest" vertex vector to the latter maximizing vector.

Problem 66-3, *A Least Absolute Fit,* by DERALD WALLING (University of Arizona).

Let (x_1, y_1), (x_2, y_2) and (x_3, y_3) be three points in the x, y-plane such that $x_1 < x_2 < x_3$. Determine the straight line $y = a + bx$ which is the best fit to these points in the sense of least absolute values, i.e., find

$$\min_{a,b} \sum_{i=1}^{3} |y_i - a - bx_i|.$$

Editorial note. Obvious extensions would be to minimize the sum of the absolute values of the distances to the line and also to increase the number of points.

Solution by IAN BARRODALE (University of Liverpool, England).

It is known [1] that a best approximation occurs when the straight line $y = a + bx$ passes through some pair of the three points (x_i, y_i).

It follows that the best approximation interpolates (x_1, y_1) and (x_3, y_3), and that it is unique and independent of the ordinates.

The generalization to more than three points and to any linear approximating function has been solved as a linear program [2].

REFERENCES

[1] J. R. RICE, *The Approximation of Functions*, vol. 1, Addison-Wesley, Reading, Massachusetts, 1964, p. 114, Th. 4–7.

[2] I. BARRODALE AND A. YOUNG, *Algorithms for best L_1 and L_∞ linear approximations on a discrete set*, Numer. Math., 8 (1966), pp. 295–306.

In the solution by R. R. SINGLETON (Wesleyan University), he refers to his paper [3] of which we give part of the introduction.

"In the Philosophical Magazine, 7th series, May 1930, E. C. Rhodes described a method of computation for the estimation of parameters by minimizing the sum of absolute values of deviations. His is an iterative and recursive method, in the following sense. There is a direct method for minimization with one parameter. Assuming a method for minimization with $n - 1$ parameters, Rhodes imposes a relation between the n parameters (in an n-parameter problem) and finds a restricted minimum by the method for $n - 1$ parameters. In this sense his method is recursive. He then repeats the process by imposing on the n parameters a new relation determined by the restricted minimum. In this sense his method is iterative. The process is finite, ending when a restricted minimum immediately succeeds itself, indicating a true minimum.

Rhodes' paper presents the method without proof. The purpose of the present paper is to analyze the situation in detail sufficient to indicate proofs for various methods, and to present a new method which reduces the labor of solution by eliminating the recursive feature. The iterative approach is retained. The solution of Rhodes' illustrative probem will be given for comparison between the two methods.

The paper uses geometric terminology and develops to quite an extent the geometry of a surface representing the summed absolute deviations. This seems the clearest means of presenting the relationships. Further analysis of the properties of this surface should lead to an even more direct method for attaining the minimum than the one here presented."

REFERENCE

[3] R. R. SINGLETON, *A method for minimizing the sum of the absolute values of deviations*, Ann. Math. Statis., 11 (1940), pp. 301–310.

Problem 61–12, On a Least Square Approximation, by D. J. Newman (Yeshiva University).

$F(x)$ is given in the interval $[0, 1]$ such that $| F^{(n)}(x) | < M$. $P(x)$ is the $(n - 1)$st degree polynomial passing through the n points $(a_r, F(a_r))$, $(r = 1, 2, \cdots, n)$. If $\lambda(a_1, a_2, \cdots, a_n)$ is defined by the inequality

$$| F(x) - P(x) | \leqq \lambda(a_1, a_2, \cdots, a_n)M,$$

show that over all selection of the a_r's

$$\min \lambda(a_1, a_2, \cdots, a_n) = \frac{2}{4^n n!}.$$

The solutions by W. Fraser (University of Toronto) and Pierre Robert (Universite de Montreal) were virtually identical and is given below.

It is a well known result that the error term in the approximation of $F(x)$ by $P(x)$ is

$$(x - a_1)(x - a_2) \cdots (x - a_n) \frac{F^{(n)}(\xi)}{n!}$$

for $0 \leqq \xi \leqq 1$, if $0 \leqq x \leqq 1$ and $0 \leqq a_i \leqq 1$ for all i. (See Hildebrand: Introduction to Numerical Analysis, pp. 60–63.) Hence

$$| F(x) - P(x) | \leqq \frac{M}{n!} | (x - a_1)(x - a_2) \cdots (x - a_n) |.$$

There remains to show that $| (x - a_1)(x - a_2) \cdots (x - a_n) |$ has minimum $2/4^n$. It is also a classical result that the polynomial $p_n(x)$ of degree n with coefficient of x^n equal to 1 and such that $\max_{0 \leqq x \leqq 1} | p(x) |$ is minimum is the polynomial

$$\frac{T_n(2x - 1)}{2^n 2^{n-1}} \quad \text{where} \quad T_n(y) = \cos(n \cos^{-1} y)$$

is the nth Tchebychev polynomial, whose maximum absolute value is 1. (See Householder: Principles of Numerical Analysis, pp. 197–200.)

Therefore

$$\min \lambda(a_1, a_2, \cdots, a_n) = \max_{0 \leqq x \leqq 1} \left| \frac{T_n(2x - 1)}{2^n 2^{n-1}} \right| = \frac{2}{4^n}$$

which completes the proof.

The stated relation becomes an equality if $F(x)$ is a polynomial of degree n, and hence can be considered a best possible result.

Problem 61-11, On a Least Square Approximation, by D. J. Newman (Yeshiva University).

In certain experiments relating a variable x to the time t, it is known a priori (from physical considerations) that x is a nondecreasing function of t. However, the experimental results give x as an oscillatory function of t due to random errors, etc. This leads to the following problem.

Given an arbitrary function $F(t)$ in $[0, 1]$, determine the best least square approximation among the class of nondecreasing functions. (As usual this least square problem leads to the maximum likelihood solution.)

Comment by GEORGE M. EWING (University of Oklahoma).

This type of problem was encountered by a group of us in the Systems Evaluation Department of Sandia Corporation in 1953 in connection with the estimation of casualty functions from empirical data. Since at that time we could not find any literature on the problem, we published several papers on the subject; in particular, MIRIAM AYER, H. D. BRUNK, G. M. EWING, W. T. REID and EDWARD SILVERMAN, *An empirical distribution function for sampling with incomplete information*, Ann. Math. Statist., 26 (1955), pp. 641–647, and H. D. BRUNK, G. M. EWING and W. R. UTZ, *Minimizing integrals in certain classes of monotone functions*, Pacific J. Math., 7 (1957), pp. 833–847. Theorem 4.3 of the latter paper can be developed into a computational procedure. It covers Problem 61-11 here as well as a wider class of similar problems. H. D. BRUNK has extended the above in several directions in his papers, *Maximum likelihood estimates of monotone parameters*, Ann. Math. Statist., 26 (1955), pp. 607–616, and *Best fit by a random variable measurable with respect to a σ-lattice*, Pacific J. Math., 11 (1961), pp. 785–802.

A Min-Max Problem

Problem 79-17*, *by* W. R. UTZ (University of Missouri).

Determine an algorithm, better than complete enumeration, for the following problem: given a nonnegative integer matrix, permute the entries in each column independently so as to minimize the largest row sum. This problem had arisen in determining an optimal scheduling for a factory work force.

Problem 59-1*, *The Ballot Problem*, by Mary Johnson (American Institute of Physics) and M. S. Klamkin.

A society is preparing 1560 ballots for an election for three offices for which there are 3, 4, and 5 candidates, respectively. In order to eliminate the effect of the ordering of the candidates on the ballot, there is a rule that each candidate must occur an equal number of times in each position as any other candidate for the same office. What is the least number of different ballots necessary?

It is immediately obvious that 60 different ballots would suffice. However, the following table gives a solution for 9 different ballots:

No. of Ballots	312	78	130	234	182	104	208	286	26
Office									
1........	A	A	A	B	B	B	C	C	C
2........	D	D	E	E	F	G	F	G	E
3........	H	I	K	I	K	J	J	L	L

Another solution (by C. Berndtson) is given by

No. of Ballots	260	182	78	234	52	130	104	312	208
Office									
1........	A	A	A	B	B	B	B	C	C
2........	D	F	E	G	G	D	G	E	F
3........	H	I	J	J	H	I	K	L	K

The above tables just give the distribution for the first position on the ballot for each office. The distributions for the other positions are obtained by cyclic permutations.

We now show that 9 is the least possible number of ballots. Let us consider the distribution for office 3 using only 8 different ballots. We must have the following (for simplicity we consider a total of 60 ballots):

No. of Ballots	x	$12 - x$	y	$12 - y$	z	$12 - z$	12	12
Office								
3........	H	H	I	I	J	J	K	L

Now to get a total of 15 representations for each position for office 2, we must have $x = y = 3$, $z = 6$. But this does not satisfy the requirements for office 1. Similarly no number of ballots fewer than 8 will suffice.

It would be of interest to solve this problem in general. The problem is to determine a distribution of the candidates such that the system of linear equations for the number of each type of ballot, which contains more equations than unknowns, is solvable in positive integers.

A trick solution to the problem can be obtained using 5 different ballots: add two fictitious names to the group of 3 and one to the group of 4. We then have 3 offices for which there are 5 "candidates" for each. This would also provide a survey on the effect of ordering of the candidates on the ballot.

Editorial note. A still smaller solution with 3 ballots was given by Fred Kohler (private communication). Just use an equilateral triangle, a square and a regular pentagon and place the names along the edges of the appropriate figure. The three different ballots are then placed into each mailing envelope in a random fashion. A general treatment of the ballot problem, without the above two dodges, for the special case of two offices with α and β candidates, respectively, was solved subsequently [1]. Here the minimum number is $\alpha + \beta - (\alpha, \beta)$. For the special case of 3 offices with $2n - 1, 2n$, and $2n + 1$, candidates, respectively, O. Jacobson (private communication) has shown that $6n - 3$ ballots suffice for $n = 2, 3$, and 4. Whether or not $6n - 3$ ballots are necessary here is still an open problem. Unfortunately, Jacobson's method failed for $n = 5$. For this case a computer solution showed that his equations were inconsistent. Subsequently, B. Cipra [2] gave a simpler proof for the case of two offices.

REFERENCES

[1] M.S. Klamkin, A. Rhemtulla, *The ballot problem*, Math. Modelling 5(1984) 1-6.
[2] Barry A. Cipra, *The ballot problem*, Math. Modeling 7(1986) 197-201.

Problem 76-7*, A Facility Location Problem, by R. D. Spinetto (University of Colorado).

Suppose a company wants to locate k service centers that will service n communities and suppose that the company wants to locate these k centers in k of the communities so that the total population distance traveled by the people in the $n - k$ communities without service centers to those communities with service centers is minimized [1]. This problem can be set up as a 0-1 integer programming problem as follows: Let

$$x_{jj} = \begin{cases} 1 & \text{if community } j \text{ gets a service center,} \\ 0 & \text{otherwise,} \end{cases}$$

and let

$$x_{ij} = \begin{cases} 1 & \text{if community } i \text{ is to be serviced by a center in community } j, \\ 0 & \text{otherwise.} \end{cases}$$

Let p_i be the population of community i and let d_{ij} be the distance from community i to community j. The problem then is to minimize

$$\sum_{i=1}^{n} \sum_{j=1}^{n} p_i d_{ij} x_{ij},$$

subject to constraints

$$\sum_{j=1}^{n} x_{ij} = 1 \quad \text{for } i = 1, 2, 3, \ldots, n;$$

(1) $$x_{ij} - x_{jj} \leqq 0 \quad \text{for } i = 1, 2, 3, \cdots, n \text{ and} \\ \text{for } j = 1, 2, 3, \cdots, n;$$

$$\sum_{j=1}^{n} x_{jj} = k,$$

and with the added condition that each of the variables x_{ii} and x_{ij} takes on only the values of 0 or 1.

If one ignores this last 0-1 condition and solves the problem as though it were a linear programming problem, then one will find that very often (but not always) an optimal extreme point solution to this linear programming problem will in fact be a 0-1 extreme point [2]. I suspect that this is due to the fact that most of the extreme points of the polyhedron determined by constraints (1) are in fact 0-1 extreme points, but I cannot prove this. This in turn suggests the following problems:

1. What are the smallest n and k for which there exists a linear programming problem of the above form which will have only non-0-1 optimal extreme point solutions?
2. Can the non-0-1 extreme points of polyhedrons determined by the constraints in (1) be characterized in any set theoretic way that would be useful in developing more efficient algorithms for solving this facility location problem?

REFERENCES

[1] C. Revelle and R. Swain, *Central facilities location*, Geographical Analysis, 2(1970), Jan.
[2] C. Revelle, D. Marks and J. C. Liebman, *An analysis of private and public sector location models*, Management Sci., 16 (1970), pp. 6292–6307.

Problem 62-11*, *The Optimum Location of a Center*, by KURT EISEMANN (Remington Rand Univac).

Optimizing the location of a warehouse leads to the following problem: For known x_r, y_r, w_r, n, devise an effective computational method for determining x, y so as to minimize

$$f = \sum_{r=1}^{n} w_r \sqrt{(x - x_r)^2 + (y - y_r)^2},$$

which may be interpreted as costs of shipment or moments of masses. Straightforward elementary calculus or stepwise trial and error is not adequate.

While this is a somewhat celebrated problem and appears simple, there are not, apparently, very good methods for its solution, especially for large n. Since this problem becomes increasingly important in an economic context, better methods are needed.

Editorial note: While the author has submitted two algorithms for the solution, one due to himself and the other to E. V. Kogbetliantz (IBM Corp.), he has not shown their convergence or effectiveness.

The solution to the problem is simple for the special cases:

1. (x_r, y_r) lie on a straight line,
2. (x_r, y_r) and w_r are n-fold symmetric,
3. $n = 3$,
4. $n = 4$ and $w_1 = w_3$, $w_2 = w_4$.

A physical solution to the problem can be gotten by using a force system (G, POLYA, *Induction and Analogy in Mathematics*, Princeton University Press, 1954, pp. 147–148). Consider the case $n = 3$.

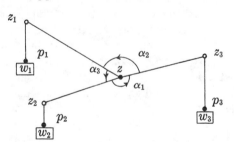

The strings are perfectly flexible and inextensible, the friction, the weight of the strings and dimensions of the pulleys are negligible.

The center z will come to rest at a point such that the system is in equilibrium. Since the potential energy will be a minimum, $\sum w_i \overline{p_i z_i}$ will be a maximum. Also, since the length of each string is invariable, $\sum w_i \overline{z_i z}$ will be a minimum. From Lame's theorem, it follows that z will be given by the angles α_1, α_2, α_3 where

$$\frac{w_1}{\sin \alpha_1} = \frac{w_2}{\sin \alpha_2} = \frac{w_3}{\sin \alpha_3}.$$

For the special case $w_1 = w_2 = w_3$, $\alpha_1 = \alpha_2 = \alpha_3 = 120°$ which is well known.

It should be noted that there are extreme cases when z will coincide with one of the vertices z_i (i.e., if $w_1 = w_2 = w_3$, and $\angle z_1 z_2 z_3 \geq 120°$, then z will be the vertex of the obtuse angle).

For $n = 4$, $w_1 = w_3$, $w_2 = w_4$, it now follows that z will be the intersection of the diagonals of the quadrilateral formed (if convex).

How well one can approximate to the point z by this physical method, when n is large, will depend on how closely one can approximate the idealization one made on the physical system. It would be of interest to carry this experiment out.

The case $w_r =$ constant leads to the determination of the minimum sum of distances from a set of given points. Another physical solution (for $n = 3$) is given by a soap film experiment (R. Courant, and H. Robbins, *What is Mathematics*, Oxford University Press, N. Y., 1941, pp. 354–361, 392). For $n > 3$, the soap film experiment leads to the solution of another allied problem, i.e., "Given n points z_1, z_2, $\cdots$, z_n, to find a connected system of straight line segments of shortest total length such that any two of the given points can be joined by a polygonal path consisting of segments of the system". The continuous version of the proposed problem would be to find

$$\min_{h,k} \int_R \int w(x, y) \sqrt{(x - h)^2 + (y - k)^2} \, dA.$$

Immediate extensions of the discrete and continuous problems, which are also well known, is the case when there is more than one warehouse.

Problem 74-23*, An Optimal Strategy, by M. S. Klamkin (University of Waterloo, Ontario, Canada).

A father, mother and son decide to hold a certain type of board game family tournament. The game is a two-person one with no ties. Since the father is the weakest player, he is given the choice of deciding the two players of the first game. The winner of any game is to then play the person who did not play in that game, and so on. The first player to win N games wins the tournament. If the son is the strongest player, it is intuitive that the father will maximize his probability of winning the tournament if he chooses to play in the first game and with his wife. Prove either that this strategy is, indeed, optimal or that it is not. It is assumed that any player's probability of winning an individual game from another player does not change throughout the tournament.

The special case corresponding to $N = 2$ was set as a problem in the Third U.S.A. Mathematical Olympiad, May 1974.

For another related problem, prove either that the previous strategy is still optimal or it is not, if now the tournament is won by the first player who wins N consecutive games. This latter problem is a generalization of a variant of a chess problem due to the late Leo Moser (see M. Gardner, *The Unexpected Hanging*, Simon and Schuster, New York, 1969, pp. 171–172).

A Minimization Problem

Problem 81-18, *by* M. Goldberg (Washington, DC).

Consider a square prism of dimensions $1 \times 1 \times 2$. The diagonal $A_1 B_1$ of length $\sqrt{2}$, at one end of the prism, is at right angles to the diagonal $A_2 B_2$ at the other end of the prism. The line segment $A_1 B_1$ can be rotated through a right angle and translated so that its ends trace helical paths until it coincides with $A_2 B_2$. Each helical path has length L, where

$$L^2 = \left(\frac{\sqrt{2}\pi}{4}\right)^2 + 2^2 = \frac{\pi^2}{8} + 4 = 5.2337,$$

$$L = 2.28773 \quad \text{and} \quad 2L = 4.5754565.$$

Find a method of moving the line $A_1 B_1$ to $A_2 B_2$ so that the sum of the motions of the ends A and B is less than the sum of the two helical paths.

The above proposal is a special case of a class of problems in the calculus of variations suggested by S. M. Ulam, *Problems in Modern Mathematics*, John Wiley, New York, 1960, p. 79.

Solution by G. N. LEWIS (Michigan Technological University).

Suppose A_1 is at $(0, 0, 0)$, B_1 is at $(1, 1, 0)$, A_2 is at $(0, 1, 2)$ and B_2 is at $(1, 0, 2)$. Move B_1 along the shortest arc of the sphere $x^2 + y^2 + z^2 = 2$ to the point C_1 at $(1, \frac{3}{5}, \frac{4}{5})$, a distance $\sqrt{2}\cos^{-1}(\frac{4}{5})$. Then move B_1 along the line segment from C_1 to B_2, a distance $3\sqrt{5}/5$, while at the same time moving A_1 directly along the line segment from $(0, 0, 0)$ to the point C_2 at $(0, \frac{3}{5}, \frac{6}{5})$. This can be done as the following analysis shows. Let $\mathbf{r}$ and $\mathbf{r}_1$ represent position vectors. Then $\mathbf{r}(t) = (1, \frac{3}{5} - \frac{3}{5}t, \frac{4}{5} + \frac{6}{5}t), 0 \le t \le 1$, is a parameterization of the line segment from C_1 to B_2. We need to find a parameterization of the line segment from A_1 to C_2 in terms of t, $0 \le t \le 1$, which is consistent with that for $\mathbf{r}$. Thus, $\mathbf{r}_1 = (0, y_1, z_1) = (0, y_1, 2y_1)$, and we must have $\|\mathbf{r} - \mathbf{r}_1\| = \sqrt{2}$. This is a quadratic equation for y_1, whose solution is $y_1 = (22 + 18t - \sqrt{484 + 192t - 576t^2})/50$. Since y_1 is increasing for $0 \le t \le 1$, then the point $(0, y_1(t), 2y_1(t))$ moves in one direction from A_1 to C_2. Finally, move A_1 along the sphere $(x - 1)^2 + y^2 + (z - 2)^2 = 2$ to A_2. A_1 has moved the same distance as B_1. The total distance travelled by A_1 and B_1 is $2(\sqrt{2}\cos^{-1}(\frac{4}{5}) + 3\sqrt{5}/5) \cong 4.503$.

Editorial note. Deleted from the above solution were detailed calculations showing that the points could actually be moved without back and forth motion. Also, one could probably do better by first moving to the point $(1, \cos\theta, \sin\theta)$ instead of $(1, \frac{3}{5}, \frac{4}{5})$ and then minimizing over θ. It would then be of interest to know whether or not this solution would give the absolute minimum over all motions. [M.S.K.]

Problem 63-12*, A Sharp Autocorrelation Function, by HENRY D. FRIEDMAN (Sylvania Electronics Systems).

How does one design a finite sequence $x_1, x_2, \cdots, x_n$ of 0's and 1's whose autocorrelation function is sharp? More specifically, if

$$\phi_k = \sum_{i=1}^{n-k} x_i \, x_{i+k}, \qquad\qquad 0 \le k \le n - 1,$$

we wish to make ϕ_0 (the number of 1's used) as large as possible while keeping $\phi_k \le 1$ for $k \ne 0$. ($n < 50$ would seem to be a practical limit.) It can be shown that the best possible result is obtained when $\phi_k = 1$ for all $k \ne 0$, in which case we must have $n = 1 + \phi_0(\phi_0 - 1)/2$. Simple optimum sequences are:

$$n = 1 \qquad 1,$$
$$n = 2 \qquad 1, 1,$$
$$n = 4 \qquad 1, 1, 0, 1,$$
$$n = 7 \qquad 1, 1, 0, 0, 1, 0, 1.$$

and the reversed sequences are also optimum. Are there any others, and what are good but non-optimum sequences?

The problem arose in the design of an optical device used to measure the inch-marks in the movement of an automatic drilling machine. A periodic system of thousandth-inch apertures was used on two identical strips of black glass. One strip was fixed while the other moved with the drill-head, and at each inch the perfect correlation of apertures would allow a sharp pulse of light to strike a photo-cell. Otherwise, little or no light was transmitted. In practice, an optimum solution was not necessary.

Solution by ARNOLD DUMEY (Roslyn Heights, N.Y.)

There are no optimum sequences for $\phi_0 > 4$. The sequences exhibited are the only ones possible.

Let $r = \phi_0$ and let $a_1 < a_2 < \cdots < a_r$ be the positions, or subscripts, at which $x_i = 1$. In an optimum sequence the $r(r-1)/2$ differences $a_v - a_u$, $u < v$, must all be different. Because of the length of the optimum sequence, the largest difference is 1 and every one must be used.

For an optimum sequence, we must have $\phi_k = 1$ for all $0 < k < n$. Thus, $\phi_{n-1} = x_1 x_n = 1$, so that $x_1 = x_n = 1$. Also, $\phi_{n-2} = x_1 x_{n-1} + x_2 x_n = x_{n-1} + x_2 = 1$. Because of symmetry, the choice here is arbitrary, and let us take $x_2 = 1$, $x_{n-1} = 0$. Further, $\phi_{n-3} = x_1 x_{n-2} + x_2 x_{n-1} + x_3 x_n = x_{n-2} + x_3 = 1$. But $x_3 \neq 1$, or else the sequence begins with 111, which is inadmissible since the difference 1 is used twice. Thus, $x_3 = 0$, $x_{n-1} = 1$. Continuing on with the equations $\phi_{n-4} = 1$, $\phi_{n-5} = 1$, it is easy to show that $x_4 = x_{n-3} = 0$ and $x_5 = 1$, $x_{n-4} = 0$. At the next step, we have $\phi_{n-6} = x_6 + x_{n-5} = 1$. If $x_6 = 1$, we have pairs (x_1, x_2) and (x_5, x_6). If $x_{n-5} = 1$, we have pairs (x_2, x_5) and (x_{n-5}, x_{n-2}). Both possibilities are inadmissible by the difference argument and we are stopped. Note that the sequences for $\phi_0 = 1, 2, 3, 4$ are obtained by ending the derivations at different steps, and that there are no optimal sequences for $\phi_0 > 4$.

If optimality is not required, ϕ_0 can be as large as desired. For example, if the 1's are positioned at $1, 2, 4, 8, \cdots, 2^r$, then $\phi_0 = r$, $\phi_k \leq 1$ for $0 < k < 2^r$.

LORENZO LUNELLI (Istituto di Elettrotecnica Generale del Politecnico, Milano, Italy) used a digital computer to obtain some best possible (non-optimum) sequences for $\phi_0 > 4$. For $\phi_0 = 5, 6, 7$, he found that the shortest possible sequences have lengths 12, 18, and 26, respectively. Examples are:

$$1, 2, 5, 10, 12.$$

$$1, 2, 5, 11, 16, 18.$$

$$1, 2, 5, 11, 19, 24, 26.$$

For $\phi_0 = 8$, he also obtained some sequences of length 36, e.g., 1, 2, 9, 21, 25, 31, 34, 36. However, a full study of this case has not been made.

In addition he programmed for the condition $\phi_k \leq 2$, $0 < k < n$, and obtained a number of best possible results. Examples are:

$$1, 2, 4, 5.$$

$$1, 2, 4, 6, 7.$$

$$1, 2, 4, 6, 9, 10.$$

$$1, 2, 4, 8, 9, 12, 14.$$

P. M. WILL (AMF British Research Laboratory, Berks, England) noted that the identical problem came up in England about 1935, in the photoelectric measurement of the period of a pendulum. He also pointed out that this problem is similar to two well-known problems in communication theory: the design of sequences of 1's and -1's with sharp autocorrelation and the construction of Barker sequences.

Subsequent comment by Solomon A. Zadoff (Sperry Gyroscope Company)

This problem has also occurred in certain radar problems.

The synthesis of optimum sequences undoubtedly becomes a computer search problem for $n > 7$ since there are no sequences that achieve the lower bound $n \geq 1 + \phi_0(\phi_0 - 1)/2$ for $\phi_0 > 4$. However, it can be shown that there is an upper bound on the order of twice the lower bound, i.e., $n \leq 1 + (\phi_0 - 1)^2$. Sequences that meet this bound exist for all $\phi_0 = p^s + 1$, where p is a prime and s an integer. Whether or not these sequences are optimum in general is not known.

These sequences and the bound are derived from the properties of difference sets and periodic sequences.

Following Chowla's [1] definition, a perfect difference set of order $m + 1$ is a set of $m + 1$ integers $D = \{d_1, d_2, \cdots, d_{m+1}\}$ such that the congruence $d_i - d_j \equiv k(\bmod\ m^2 + m + 1)$ has exactly one solution for every $k \not\equiv 0\ (\bmod\ m^2 + m + 1)$.

Now if D is a perfect difference set and x_i is a periodic sequence with

$$x_i = \begin{cases} 1 & \text{if } i \in D, \\ 0 & \text{if } i \notin D, \end{cases}$$

then it follows that the periodic autocorrelation function ψ_k has the property

$$\psi_k = \sum_{i=1}^{m^2+m+1} x_i x_{i+k} = \begin{cases} m + 1 & \text{for } k \equiv 0\ (\bmod\ m^2 + m + 1), \\ 1 & \text{otherwise.} \end{cases}$$

Note now that

$$\psi_k = \phi_k + \sum_{i=m^2+m-k}^{m^2+m+1} x_i x_{i+k} \geq \phi_k$$

for $k \neq 0$.

Hence, a finite sequence, derived from a perfect difference set, will have the desired properties. Furthermore, if the periodic sequence is indexed so that it has a maximum run of zeros at the end of its cycle, the finite sequence can be shortened by at least m elements (which are all zero) and still have the required property. This gives

$$n \leq m^2 + m + 1 - m = m^2 + 1 = (\phi_0 - 1)^2 + 1.$$

It has been shown [2] that a perfect difference set of order $m + 1$ exists for every $m = p^s$, where p is a prime and s an integer.

A perfect difference set of order 18 is given in a paper by Hall and Ryser [3] as

$D = \{1, 2, 4, 8, 16, 32, 64, 91, 117, 128,$

$$137, 182, 195, 205, 234, 239, 256\}\ (\bmod\ 273).$$

Now $D^1 = D - 181\ (\bmod\ 273)$ will also be a perfect difference set. Here

$D^1 = \{1, 14, 24, 53, 58, 75, 93, 94, 95, 100,$

$$108, 124, 156, 183, 209, 220, 229\} \pmod{273}.$$

This set of integers can be used as the indices of a finite sequence with $n = 229$ which is less than the upper bound of

$$(\phi_0 - 1)^2 + 1 = 257.$$

REFERENCES

[1] S. CHOWLA, *On difference sets*, Proc. Nat. Acad. Sci. U.S.A., 35 (1949), pp. 92–94.
[2] J. SINGER, *A theorem in finite projective geometry and some applications to number theory*, Trans. Amer. Math. Soc., 43 (1938), pp. 377–385.
[3] M. HALL AND H. V. RYSER, *Cyclic incidence matrices*, Canad. J. Math., 3 (1951), pp. 495–502.

RON GRAHAM (Bell Telephone Laboratories) noted that relevant references to the question are given in "The Generation of Impulse-Equivalent Pulse Trains", by D. A. Huffman, IRE Trans. Information Theory, September, 1962.

Problem 59-5, *A Minimum Switching Network*, by RAPHAEL MILLER (Hermes Electronic Corp.).

In a recent issue of an automation periodical[1], the problem is posed of obtaining a switching network which will actuate a device for a selected range of binary inputs. In particular, it is asked that a valve be opened for binary inputs cor-

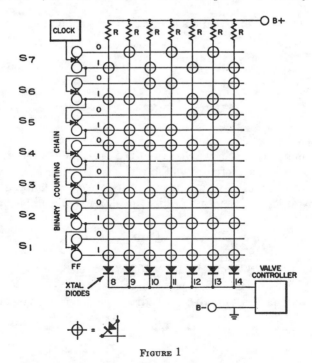

FIGURE 1

---

[1] Klein, Williams and Morgan, *Digital process control*. Instruments and Automation, October, 1956, pp. 1979–1984.

responding to the integers 8, 9, 10, 11, 12, 13, and 14, where possible inputs go up to 7 binary digits. The proposed network is shown in the figure.

The binary significance of a vertical line is read bottom to top from the unencircled crossovers of the vertical line with horizontal lines. Note that such crossovers are *not* electrical connections. If a setting of the switches includes a circled crossover in a vertical line, then the line is shorted directly to ground through the switch. (The short is not indicated on the diagram.) Otherwise, current flows through the vertical line from the B+ terminal and a potential drop is produced across the valve controller. A circled crossover represents a diode inserted as shown in Fig. 1. Accordingly, current can flow from a vertical line to a horizontal one, but not vice versa. This feature prevents the deliberate shorting of one vertical line from inadvertently shorting another vertical line. For example, the reader should verify that if the diodes were replaced by tie points, then shorting of lines 5, 6, and 7 by switch S_5 would also short lines 1, 2, 3, and 4.

Show that the proposed network can be considerably simplified so as to involve the minimum number of diodes.

Editorial Note: In Ralph Miller's solution (July 1960), the number of diodes needed are 15. In A. H. McMorris' solution (July 1961), 9 diodes are required. In the solution below, only 8 diodes are required and this is the minimum number.

Solution by Layton E. Butts (Systems Laboratories).

McMorris' simplification of the logical function F is certainly correct, but it seems somewhat simpler to me to start with the logical function $\bar{F}$ on p. 221 (July 1960), and use DeMorgan's theorem to find its negative F. Thus (in the notation of Miller)

$$\bar{F} = p_1 \lor p_2 \lor p_3 \lor \overline{p_4} \lor (p_5 p_6 p_7),$$

and hence

$$F = \overline{p_1 p_2 p_3 p_4 (\overline{p_5 p_6 p_7})} = \overline{p_1 p_2 p_3 p_4} (\overline{p_5} \lor \overline{p_6} \lor \overline{p_7}).$$

Furthermore, the logical diagram given by McMorris involves a redundancy; it should be (in his notation)

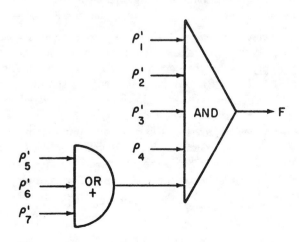

for which the circuit diagram is

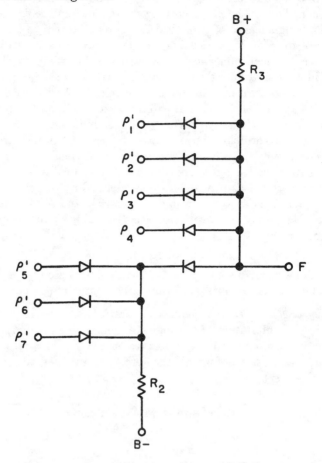

with only 8 diodes.

This is believed to be the minimum number of diodes inasmuch as one diode must be employed for each stage of the counter in order to inhibit the valve response to the integer 15, while an additional diode is required to isolate the And gate from the Or gate. The imposing of the condition $R_2 \leqq R_3$ assumes that $|B-| = |B+|$ and that the output voltage does not exceed zero when F is low.

Comment on the above solution by A. H. McMorris (University of Houston):

Butts' belief that 8 is the minimum number of diodes required can be substantiated by applying the methods of Caldwell to obtain the *minimum sum expression* (MSE). The MSE obtained here will be the unfactored form for F. The MSE can then be factored to obtain the *simplest factored form* (SFF) where the SFF is defined as that form which does not contain any repetition of literals. The number of diodes required will be the total number of literals in the SFF plus (L-1) where L is the number of logic levels required by the SFF. In this case, the number of literals is 7 and the number of levels is 2. It should be noted that it is not always possible to obtain the SFF. For example, the function

$$F = WX + WY + XY$$

cannot be factored so that repeated literals do not occur. The three factored forms are:

$$F = W \ (X+Y) \ + \ XY,$$
$$= Y \ (W+X) \ + \ WX,$$
$$= X \ (W+Y) \ + \ WY.$$

In each case, two literals are duplicated.

Supplementary References
Optimization

I. Linear, Nonlinear, Dynamic, and Geometric Programming

[1] J. ABADIE, ed., *Nonlinear Programming*, Interscience, N.Y., 1967.
[2] R. ARIS, *Discrete Dynamic Programming*, Blaisdell, Waltham, 1964.
[3] E. M. L. BEALE, ed., *Applications of Mathematical Programming Techniques*, English Universities Press, London, 1970.
[4] C. S. BEIGHTLER AND D. T. PHILLIPS, *Applied Geometric Programming*, Wiley, N.Y., 1976.
[5] R. BELLMAN, *Dynamic Programming*, Princeton University Press, Princeton, 1957.
[6] R. E. BELLMAN AND S. E. DREYFUS, *Applied Dynamic Programming*, Princeton University Press, Princeton, 1962.
[7] R. G. BLAND, "The allocation of resources by linear programming," Sci. Amer. (19181) pp. 126–144.
[8] J. BRACKEN AND G. P. McCORMICK, *Selected Applications of Nonlinear Programming*, Wiley, N.Y., 1968.
[9] A. CHARNES AND W. W. COOPER, *Management Models and Industrial Applications of Linear Programming I, II*, Wiley, N.Y., 1961.
[10] L. COOPER AND M. W. COOPER, *Introduction to Dynamic Programming*, Pergamon, Oxford, 1981.
[11] G. B. DANTZIG, *Linear Programming and Extensions*, Princeton University Press, Princeton, 1963.
[12] G. B. DANTZIG AND B. C. EAVES, eds., *Studies in Optimization*, MAA, 1974.
[13] J. B. DENNIS, *Mathematical Programming and Electrical Networks*, Technology Press, N.Y., 1959.
[14] S. E. DREYFUS, *Dynamic Programming and the Calculus of Variations*, Academic Press, N.Y., 1965.
[15] S. DREYFUS AND A. M. LAW, *The Art and Theory of Dynamic Programming*, Academic Press, N.Y., 1977.
[16] R. J. DUFFIN, E. L. PETERSON AND C. M. ZENER, *Geometric Programming*, Wiley, N.Y., 1967.
[17] J. C. ECKER, "Geometric programming: Methods, computers and applications," SIAM Rev. (1980) pp. 338–362.
[18] A. V. FIACCO AND G. P. McCORMICK, *Nonlinear Programming, Sequential Unconstrained Minimization Techniques*, Wiley, N.Y., 1968.
[19] G. HADLEY, *Nonlinear and Dynamic Programming*, Addison-Wesley, Reading, 1964.
[20] D. H. JACOBSON AND D. Q. MAYNE, *Differential Dynamic Programming*, Elsevier, N.Y., 1970.
[21] S. KARLIN, *Mathematical Methods and Theory in Games, Programming and Economics*, Addison-Wesley, Reading, 1959.
[22] A. KAUFMANN, *Graphs, Dynamic Programming, and Finite Games*, Academic, N.Y., 1967.
[23] E. L. PETERSON, "Geometric programming – a survey," SIAM Rev. (1976) pp. 1–51.
[24] S. M. ROBERTS, *Dynamic Programming in Chemical Engineering and Process Control*, Academic, N.Y., 1964.
[25] H. THEIL AND C. VAN DE PANNE, "Quadratic programming as an extension of classical quadratic maximization," Management Sci. (1960) pp. 1–20.
[26] L. E. WARD, JR., "Linear programming and approximation problems," AMM (1961) pp. 46–52.
[27] G. ZOUTENDIJK, "Nonlinear programming: A numerical survey," SIAM J. Control (1966) pp. 194–210.

II. Control Theory

[28] D. J. BELL, ed., *Recent Mathematical Developments in Control*, Academic, London, 1973.
[29] R. BELLMAN, *Adaptive Control Processes: A Guided Tour*, Princeton University Press, Princeton, 1961.
[30] ———, *Introduction to the Mathematical Theory of Control Processes I, II*, Academic Press, N.Y., 1967 and 1971.
[31] L. D. BERKOVITZ, "Optimal control theory," AMM (1976) pp. 225–239.
[32] D. N. BURGHES AND A. GRAHAM, *Introduction to Control Theory, Including Optimal Control*, Horwood, Chichester, 1980.
[33] F. CSAKI, *Modern Control Theories*, Akademiai Kiado, Budapest, 1972.

[34] I. FLUGGE-LOTZ, *Discontinuous and Optimal Control*, McGraw-Hill, N.Y., 1968.

[35] C. M. HAALAND AND E. P. WIGNER, *"Defense of cities by antiballistic missiles,"* SIAM Rev. (1977) pp. 279–296.

[36] C. S. JONES AND A. STRAUSS, *"An example of optimal control,"* SIAM Rev. (1968) pp. 25–55.

[37] M. S. KLAMKIN, *"A student optimum control problem and extensions,"* Math. Modelling (1985) pp. 49–64.

[38] G. KNOWLES, *An Introduction to Applied Optimal Control Theory*, Academic, N.Y., 1981.

[39] P. KOKOTOVIC, *"Application of singular perturbation techniques to control problems,"* SIAM Rev. (1984) pp. 501–550.

[40] J. MACKI AND J. STRAUSS, *An Introduction to Optimal Control Theory*, Springer-Verlag, N.Y., 1982.

[41] P. C. PARKS, *"Mathematicians in control,"* BIMA (1977) pp. 180–188.

[42] L. S. PONTRYAGIN, V. G. BOL'TANSKII, R. S. GAMKRELIDZE AND E. F. MISCHENKO, *The Mathematical Theory of Optimal Processes*, Pergamon, Oxford, 1964.

[43] L. C. YOUNG, *Lectures on the Calculus of Variations and Optimal Control Theory*, Saunders, Philadelphia, 1969.

III. Games

[44] R. BARTON, *A Primer on Simulation and Gaming*, Prentice-Hall, N.J., 1970.

[45] E. R. BERLEKAMP, J. H. CONWAY AND R. K. GUY, *Winning Ways I, II*, Freeman, San Francisco, 1977.

[46] I. BUCHLER AND H. NOTINI, *Game Theory in the Behavioral Sciences*, University of Pittsburgh Press, Pittsburgh, 1969.

[47] J. H. CONWAY, *"All games bright and beautiful,"* AMM (1977) pp. 417–434.

[48] ———, *On Numbers and Games*, Academic, London, 1976.

[49] H. S. M. COXETER, *"The golden section, phyllotaxis, and Wythoff's game,"* Scripta Math. (1953) pp. 135–143.

[50] W. H. CUTLER, *"An optimal strategy for pot-limit poker,"* AMM (1975) pp. 368–376.

[51] M. DRESHER, A. W. TUCKER AND P. WOLFE, eds., *Contributions to the Theory of Games III*, Princeton University Press, Princeton, 1957.

[52] N. V. FINDLER, *"Computer poker,"* Sci. Amer. (1978) pp. 112–119.

[53] O. HAJEK, *Pursuit Games*, Academic, N.Y., 1975.

[54] J. C. HOLLADAY, *"Some generalizations of Wythoff's game and other related games,"* MM (1968) pp. 7–13.

[55] H. W. KUHN AND A. W. TUCKER, eds., *Contributions to the Theory of Games I, II*, Princeton University Press, Princeton, 1950, 1953.

[56] R. D. LUCE AND H. RAIFFA, *Games and Decisions*, Wiley, N.Y., 1958.

[57] R. D. LUCE AND A. W. TUCKER, eds., *Contributions to the Theory of Games IV*, Princeton University Press, Princeton, 1959.

[58] J. C. C. McKINSEY, *Introduction to the Theory of Games*, McGraw-Hill, N.Y., 1952.

[59] W. H. RUCKLE, *Geometric Games and their Applications*, Pitman, London, 1983.

[60] ———, *"Geometric games of search and ambush,"* MM (1979) pp. 195–206.

[61] C. A. B. SMITH, *"Graphs and composite games,"* J. Comb. Th. (1966) pp. 51–81.

[62] J. VON NEUMANN AND O. MORGENSTERN, *Theory of Games and Economic Behaviour*, Princeton University Press, Princeton, 1947.

[63] J. D. WILLIAMS, *The Compleat Strategyst*, McGraw-Hill, N.Y., 1954.

IV. Miscellaneous

[64] M. AURIEL AND D. J. WILDE, *"Optimality proof for the symmetric Fibonacci search technique,"* Fibonacci Quart. (1966) pp. 265–269.

[65] B. AVI-ITZHAK, ed., *Developments in Operations Research I, II*, Gordon and Breach, N.Y., 1971.

[66] A. V. BALAKRISHMAN, *Kalman Filtering Theory*, Optimization Software Inc., N.Y., 1984.

[67] E. BALAS, *"An additive algorithm for solving linear programs with zero-one variables,"* Oper. Res. (1965) pp. 517–546.

[68] M. L. BALINSKY, *"Integer programming: Methods, uses, computation,"* Management Sci. (1965) pp. 253–313.

[69] C. BANDLE, *"A geometric isoperimetric inequality and applications to problems of mathematical physics,"* Comment. Math. Helv. (1974) pp. 253–261.

[70] R. BELLMAN, A. O. ESOGBUE AND I. NABESHIMA, *Mathematical Aspects of Scheduling & Applications*, Pergamon, Oxford, 1982.

[71] M. BELLMORE AND G. NEMHAUSER, *"The travelling salesman problem,"* Oper. Res. (1968) pp. 538–558.

[72] H. J. BRASCAMP AND E. H. LIEB, *"On extentions of the Brunn-Minkowski and Prekopa-Leindler theorems, including inequalities for log concave functions, and with an application to the diffusion equation,"* J. Funct. Anal. (1976) pp. 366–389.

[73] S. T. BROOKS, *"A comparison of maximum-seeking methods,"* Oper. Res. (1959) pp. 430–457.

[74] S. S. BROWN, *"Optimal search for a moving target in discrete space and time,"* Oper. Res. (1980) pp. 1275–1289.

[75] E. G. COFFMAN, ed., *Computer and Job-Shop Scheduling Theory*, Wiley, N.Y., 1976.

[76] L. COLLATZ AND W. WETTERLING, *Optimization Problems*, Springer-Verlag, N.Y., 1975.

[77] L. COOPER AND D. I. STEINBERG, *Introduction to Methods of Optimization,* Saunders, Philadelphia, 1970.

[78] L. COOPER AND M. W. COOPER, *"Nonlinear integer programming,"* Comp. Math. Appl. (1975) pp. 215-222.

[79] COURANT, R., *"Soap film experiments with minimal surfaces,"* AMM (1940) pp. 167-174.

[80] H. G. DAELLENBACH AND J. A. GEORGE, *Introduction to Operations Research Techniques,* Allyn and Bacon, Boston, 1978.

[81] N. J. FINE, *"The jeep problem,"* AMM (1946) pp. 24-31.

[82] L. R. FOULDS, *Combinatorial Optimization for Undergraduates,* Springer-Verlag, N.Y., 1984.

[83] ——, *"Graph theory: A survey of its use in operations research,"* New Zeal. Oper. Res. (1982) pp. 35-65.

[84] J. N. FRANKLIN, *"The range of a fleet of aircraft,"* SIAM J. Appl. Math. (1960) pp. 541-548.

[85] D. R. FULKERSON, *"Flow networks and combinatorial operations research,"* AMM (1966) pp. 115-137.

[86] D. GALE, *"The jeep once more or jeeper by the dozen,"* AMM (1970) pp. 493-500.

[87] S. I. GASS, *"Evaluation of complex models,"* Comp. Oper. Res. (1977) pp. 27-35.

[88] D. P. GAVER AND G. L. THOMPSON, *Programming and Probability Models in Operations Research,* Brooks/Cole, Monterey, 1973.

[89] R. L. GRAHAM, *"The combinatorial mathematics of scheduling,"* Sci. Amer. (1978) pp. 124-132, 154.

[90] G. Y. HANDLER AND P. B. MIRCHANDANI, *Location on Networks: Theory and Algorithms,* M.I.T. Press, Cambridge, 1979.

[91] M. R. HESTENES, *Optimization Theory,* Interscience, N.Y., 1975.

[92] F. HILLIER AND G. J. LIEBERMAN, *Introduction to Operations Research,* Holden-Day, San Francisco, 1967.

[93] J. M. HOLTZMAN AND H. HALKIN, *"Directional convexity and the maximum principle for discrete systems,"* SIAM J. Control (1966) pp. 263-275.

[94] R. JACKSON, *"Some algebraic properties of optimization problems in complex chemical plants,* Chem. Eng. Sci. (1964) pp. 19-31.

[95] P. B. JOHNSON, *"The washing of socks,"* MM (1966) pp. 77-83.

[96] A. KAUFMANN AND R. FAURE, *Introduction to Operations Research,* Academic, N.Y., 1968.

[97] H. J. KELLEY, *"Gradient theory of optimal flight paths,"* Amer. Rocket Soc. J. (1960) pp. 947-954.

[98] Z. J. KIEFER, *"Sequential minimax search for a max,"* Proc. AMS (1953) pp. 502-506.

[99] S. KIRKPATRICK, et al., *"Optimization by simulated annealing,"* Science (1983) pp. 671-680.

[100] M. S. KLAMKIN, *"A physical application of a rearrangement inequality,"* AMM (1970) pp. 68-69.

[101] ——, *"On Chaplygin's problem,"* SIAM J. Math. Anal. (1977) pp. 288-289.

[102] ——, *"Optimization problems in gravitational attraction,"* J. Math. Anal. Appl. (1972) pp. 239-254.

[103] M. S. KLAMKIN AND A. RHEMTULLA, *"The ballot problem,"* Math. Modelling (1984) pp. 1-6.

[104] R. C. LARSON AND A. R. ODON, *Urban Operations Research,* Prentice-Hall, N.J., 1981.

[105] E. LAWLER, *Combinatorial Optimization: Networks and Matroids,* Holt, Rinehart & Winston, N.Y., 1976.

[106] G. LEITMANN, ed., *Topics in Optimization,* Academic Press, N.Y., 1967.

[107] A. G. J. MACFARLANE, *"An eigenvector solution of an optimal linear regulator problem,"* J. Elec. Control (1963) pp. 643-653.

[108] A. MIELE, ed., *Theory of Optimum Aerodynamic Shapes,* Academic Press, N.Y., 1965.

[109] D. J. NEWMAN, *"Location of the maximum on a unimodal surface,"* J. Assn. Comp. Mach. (1965) pp. 395-398.

[110] J. C. C. NITSCHE, *"Plateau's problem and their modern ramifications,"* AMM (1974) pp. 945-968.

[111] I. NIVEN, *Maxima and Minima Without Calculus,* MAA, 1981.

[112] L. T. OLIVER AND D. J. WILDE, *"Symmetric sequence minimax search for a maximum,"* Fibonacci Quart. (1964) pp. 169-175.

[113] R. OSSERMAN, *"The isoperimetric inequality,"* Bull. AMS (1978) pp. 1182-1238.

[114] C. H. PAPADIMITRIOU AND K. STEIGLITZ, *Combinatorial Optimization: Algorithms and Complexity,* Prentice-Hall, N.J., 1982.

[115] L. E. PAYNE, *"Isoperimetric inequalities and their applications,"* SIAM Rev. (1967) pp. 453-488.

[116] C. G. PHIPPS, *"The jeep problem: A more general solution,"* AMM (1946) pp. 458-462.

[117] G. POLYA AND G. SZEGO, *Isoperimetric Inequalities in Mathematical Physics,* Princeton University Press, Princeton, 1951.

[118] G. RAISBECK, *"An optimum shape for fairing the edge of an electrode,"* AMM (1961) pp. 217-225.

[119] V. J. RAYWARD-SMITH AND M. SHING, *"Bin packing,"* BIMA (1983) pp. 142-148.

[120] T. L. SAATY, *Mathematical Methods of Operational Research,* McGraw-Hill, N.Y., 1959.

[121] ——, *Optimization in Integers and Related Extremal Problems,* McGraw-Hill, N.Y., 1970.

[122] B. V. SHAH, R. J. BUEHLER AND O. KEMPTHORNE, *"Some algorithms for minimizing a function of several variables,"* SIAM J. Appl. Math. (1964) pp. 74-92.

[123] D. K. SMITH, *Network Optimization Practice: A Computational Guide,* Horwood, Chichester, 1982.

[124] D. R. SMITH, *Variational Methods in Optimization,* Prentice-Hall, N.J., 1974.

[125] L. J. STOCKMEYER AND A. K. CHANDRA, *"Intrinsically difficult problems,"* Sci. Amer. (1979) pp. 140-159.

[126] L. D. Stone, *Theory of Optimal Search*, Academic, N.Y., 1975.

[127] H. A. Taha, *Integer Programming*, Macmillan, N.Y., 1978.

[128] ———, *Operations Research, An Introduction*, Macmillan, N.Y., 1976.

[129] H. Wacker, ed., *Applied Optimization Techniques in Energy Problems*," Teubner, Stuttgart, 1985.

[130] H. M. Wagner, *Principles of Operations Research with Applications to Managerial Decisions*, Prentice-Hall, N.J., 1969.

[131] D. J. Wilde, *Optimum-Seeking Methods*, Prentice-Hall, N.J., 1964.

[132] D. J. Wilde and C. S. Beightler, *Foundations of Optimization*, Prentice-Hall, N.J., 1967.

[133] D. Wildfogel, *"The maximum brightness of Venus,"* MM (1984) pp. 158–164.

15. GRAPH THEORY

"Bad" Triangles in Hilbert Space

Problem 82-18*, *by* M. BURNASHEV (Institute for Problems of Information Transmission of U.S.S.R. Academy of Sciences, Moscow).

Let H be a Hilbert space and let any $\delta > 0$ be given. We say that the triangle with vertices at x_i, x_j, $x_k \in H$ is "bad," if two of its sides have lengths ≤ 1 and the third side $\geq 1 + \delta$.

Prove or disprove the conjecture that there exists $\lambda(\delta) > 0$ such that from any set of elements $x_1, x_2, \cdots, x_n \in H$ it is possible to choose a subset of no less than λn elements which does not contain all the vertices of any "bad" triangle.

Remark. Of course, it is possible to replace a Hilbert space H by Euclidean space E^{n-1}, but it is important that $\lambda(\delta)$ does not depend on n.

The problem arose in information theory (optimal choice of signals, strategy, etc.).

Solution by P. ERDÖS (Hungarian Academy of Sciences, Budapest).

The conjecture is true for $\delta > 1$ with $\lambda = 1$ (triangle inequality), but it is false for $\delta \leq \sqrt{2} - 1$. Let $H = l^2$. Let $\delta \leq \sqrt{2} - 1$ be given and suppose that $\lambda(\delta) > 0$ exists. Choose n so that $\lambda(\delta) \binom{n}{2} \geq n + 1$ and consider the $\binom{n}{2}$ elements of H where exactly two of the first n coordinates are $\sqrt{2}/2$ and all others are 0. Each of these elements of H corresponds to an edge joining two vertices in $V = \{1, 2, \cdots, n\}$. A subset of at least $\lambda(\delta) \binom{n}{2} \geq n + 1$ elements corresponds to a graph with n vertices and at least $n + 1$ edges. By a well-known theorem of Erdös and Gallai [*On maximal paths and circuits of graphs,* Acta Math., Acad. Sci. Hungar., 10 (1959), pp. 337–357] this graph contains a path of length three. Let x_i, x_j, x_k be the elements of H corresponding to the edges of this path. Note that if two edges are adjacent in the graph then the distance between the corresponding Hilbert space elements is 1; otherwise, the distance is $\sqrt{2}$. Thus x_i, x_j, x_k yield a "bad" triangle with two sides of length 1 and one side of length $\sqrt{2}$. Thus we have a contradiction and this proves that no $\lambda(\delta)$ exists for $\delta \leq \sqrt{2} - 1$. An interesting problem is that of finding the largest δ for which the conjecture is false.

Problem 72-6*, *A Solved and Unsolved Graph Coloring Problem,* by P. ERDÖS (University of Waterloo, Waterloo, Ontario, Canada).

Show that if one two-colors the edges of a K_{2n-1} (complete graph of $2n - 1$ vertices), then there is a C_n (cycle of order n) of one color.

Editorial note. The proposer notes that his problem was recently solved by J. A. BONDY (University of Waterloo) who also conjectures the related result:

If $k \leq n$ and one two-colors the edges of a K_{2n-1}, then either there is a C_k of the first color or else there is a C_n of the second color.

The proposer has proved the latter result for fixed k if $n > n_0(k^k)$ and notes that this weaker result was first conjectured by W. G. BROWN.

Solution by R. J. FAUDREE and R. H. SCHELP (Memphis State University).

It is convenient to rephrase this problem and its solution in terms of Ramsey numbers for cycles in graphs. Let G be a graph of order m with edge set E, and let $\bar{G}$ and $\bar{E}$ denote the complementary graph with its edge set. ($G, \bar{G}, E, \bar{E}$ and m will retain these meanings throughout the solution presented.) The Ramsey number, $R(C_k, C_n)$, for the cycles C_k, C_n is the minimal number m such that either G contains a C_k or $\bar{G}$ contains a C_n. In terms of Ramsey numbers the problem is: *show* $R(C_k, C_n) \leq 2n - 1$ *for* $k \leq n$ *and* $n > 3$.

A detailed solution of the problem is quite lengthy so that we shall simply present a sketch of the proof, giving only the essentials. The detailed solution of this result together with Ramsey numbers for all pairs of cycles will be published elsewhere by the authors.

Several known results are used in the proof presented and are now stated.

LEMMA 1 (Chartrand, Schuster [3]).

$$R(C_3, C_n) = \begin{cases} 6, & n = 3, \\ 2n - 1, & n > 3, \end{cases} \qquad R(C_4, C_n) = \begin{cases} 6, & n = 4, \\ 7, & n = 5, \\ n + 1, & n > 5, \end{cases}$$

$$R(C_5, C_n) = 2n - 1, \quad n > 2, \qquad R(C_6, C_6) = 8.$$

LEMMA 2 (Erdös, Bondy [2]). $R(C_n, C_n) \leq 2n - 1, n > 3$.

LEMMA 3 (Erdös, Gallai [4]). *If G_1 is a graph of order n and size at least* $[(c - 1)(n - 1) + 1]/2$, *then G_1 contains a cycle of length at least c.*

LEMMA 4 (Bondy [1]). *If G_1 is a graph of order n and size at least $(n^2 + 1)/4$, then G_1 contains cycles of all lengths j, $3 \leq j \leq (n + 3)/2$.*

In light of Lemmas 1 and 2, we know that if the given bound on $R(C_k, C_n)$ fails to hold, then $5 < k < n$. Thus we assume throughout the remainder of this solution that $m = 2n - 1$ is such that G contains no C_k, $5 < k < n$, and $\bar{G}$ contains no C_n, i.e., $R(C_k, C_n) \nleq 2n - 1 = m$. We shall of course eventually contradict this assumption.

As an initial step, we observe by Lemmas 3 and 4 that $|E| < (m^2 + 1)/4$ $= n(n - 1) + 1/2$ and unless $\bar{G}$ contains a C_j, $j > n$, also $|\bar{E}| < [(n - 1)(m - 1) + 1]/2 = (n - 1)^2 + 1/2$. But $|E| + |\bar{E}| = (2n - 1)(2n - 2)/2 = (n - 1)^2 + n(n - 1)$, so that for $|E| < (m^2 + 1)/4$ and $|\bar{E}| < [(n - 1)(m - 1) + 1]/2$ we must have $|E| = n(n - 1)$ and $|\bar{E}| = (n - 1)^2$. Thus, consider the case where $|E| = n(n - 1)$ and $|\bar{E}| = (n - 1)^2$. By applying Lemma 4 to the graph $G \setminus \{u\}$ for $u \in G$, it can be shown that the degree of v in G, denoted $d_G(v)$, satisfies $d_G(v) \geq n - 1$ for all points in G. Furthermore, applying the same lemma to $G \setminus \{u, v\}$ for $u, v \in G, u \neq v$, there exist n vertices of degree $n - 1$ in G no two of which are adjacent. From these facts, it can be shown that $\bar{G} = K_n \cup K_{n-1}$, a contradiction to $\bar{G}$ containing no C_n. Therefore, $|E| = n(n - 1)$ and $|\bar{E}| = (n - 1)^2$ never occurs so that $|\bar{E}| \geq [(n - 1) \cdot (m - 1) + 1]/2$. This gives the following lemma.

LEMMA 5. *$\bar{G}$ contains a $C_j, j > n$.*
This lemma can be strengthened to the following statement.

LEMMA 6. $\bar{G}$ contains a C_{n+1}.

Proof. Let $C = (x_1, x_2, \cdots, x_j, x_1)$ be a cycle of length $j > n + 1$ in $\bar{G}$. The lemma follows when we show that $\bar{G}$ contains a C_{j-1} or a C_{j-2}. Suppose $\bar{G}$ contains neither a C_{j-1} or C_{j-2}. Then for every i, $1 \leq i \leq j$, $(x_1, x_{i+2}), (x_i, x_{i+3}) \in E$, where subscript addition is taken, as usual, modulo the cycle length j. Letting $k = 2p + 1, 2p$, when k is odd, even, we then have

$$(x_1, x_4, x_6, \cdots, x_{2p}, x_{2p+3}, x_{2p+1}, x_{2p-1}, \cdots, x_3, x_1),$$

$$(x_1, x_4, x_6, \cdots, x_{2p+2}, x_{2p-1}, x_{2p-3}, \cdots, x_3, x_1),$$

is a cycle of length k in G, a contradiction.

The remainder of the proof is specifically aimed at showing that $\bar{G}$ contains a C_{n+1} implies $\bar{G}$ must also contain a C_n or G, a C_k. Hence, we pinpoint the general idea of showing that the existence of cycles $j > n$ in $\bar{G}$ implies (when G contains no C_k) the existence of a C_n in $\bar{G}$.

We next make the following observation. If $\bar{G}$ contains the $n + 1$ cycle $(x_1, x_2, \cdots, x_{n+1}, x_1)$, $n \geq 7$, then $|\{i | 1 \leq i \leq n + 1, (x_i, x_{i+3}) \in \bar{E}\}| \geq (n + 1)/2$. To see this let k be odd, $k = 2p + 1$. Then for $1 \leq p \leq n + 1$ and $(x_i, x_{i+3}) \in E$ we must have $(x_{i+2p-1}, x_{i+2p+2}) \in \bar{E}$, otherwise (since $\bar{G}$ contains no C_n implies $(x_j, x_{j+2}) \in E$ for all j) we get the k cycle

$$(x_i, x_{i+3}, x_{i+5}, \cdots, x_{i+2p-1}, x_{i+2p+2}, x_{i+2p}, \cdots, x_{i+2}, x_i)$$

in G. When k is even the argument is similar.

One more idea is needed to prove the result. Let $C = (x_1, x_2, \cdots, x_j, x_1)$ be a cycle in G and $x \in G \backslash C$. Vertex x is said to be *dominant* (*strongly dominant*) to C in G, when $|\{i | 1 \leq i \leq n, (x, x_i) \in E\}|$ dominates (strictly dominates) $j/2$.

LEMMA 7. *Let $\bar{G}$ contain the $n + 1$ cycle $C = (x_1, x_2, \cdots, x_{n+1}, x_1), n \geq 7$.*
Then:
1. *If $x \in \bar{G} \backslash C$ we have*
   (a) *x is strongly dominant to $(x_1, x_3, \cdots, x_{n+1}, x_2, x_4, \cdots, x_n, x_1)$ in G for n even, and*
   (b) *x is strongly dominant to $C^{(1)} = (x_1, x_3, \cdots, x_n, x_1)$ or $C^{(2)} = (x_2, x_4, \cdots, x_{n+1}, x_2)$ in G for n odd, and dominant to their union.*
2. *If $x, y \in \bar{G} \backslash C, (x, y) \in \bar{E}$, and x is dominant to $C^{(1)}$ but not to $C^{(2)}$ in G, then y is dominant to $C^{(2)}$ in G, where $C^{(1)}$ and $C^{(2)}$ are as in part 1.*

Proof. The flavor of the proofs of parts 1 and 2 are similar so that we only prove part 1. Since $\bar{G}$ contains no $C_n, (x_i, x_{i+2}) \in E$ for $1 \leq i \leq n + 1$. Thus in G we obtain the cycles shown in part 1. Let $x_i \in C$. By the observation noted in a previous paragraph and since $n \geq 7$, there exists a $j, 1 \leq j \leq n + 1$, such that $(x_j, x_{j+3}) \in \bar{E}$ and $x_{j+1}, x_{j+2} \notin \{x_i, x_{i+1}\}$. Therefore $(x, x_i) \in \bar{E}$ implies $(x, x_{i+1}) \in E$; otherwise $(x, x_{i+1}, x_{i+2}, \cdots, x_j, x_{j+3}, \cdots, x_i, x)$ is an n cycle in $\bar{G}$. Hence $|\{i | (x, x_i) \in \bar{E}, 1 \leq i \leq n + 1\}| \leq |\{i | (x, x_i) \in E, 1 \leq i \leq n + 1\}|$.

THEOREM 1. $R(C_k, C_p) \leq 2p - 1, p > 3, p \geq k$.

Proof (Sketch). Suppose the theorem is false so that as before we set $p = n$, $m = 2n - 1$. Thus by Lemma 6, $\bar{G}$ contains a C_{n+1}. Furthermore, by Lemma 1 we take $n \geq 7$. Let $C = (x_1, x_2, \cdots, x_{n+1}, x_1)$ be a cycle in $\bar{G}$.

Case I. *n is even.*

Since $\bar{G}$ contains no $C_n, (x_i, x_{i+2}) \in E$ and $(x_1, x_3, \cdots, x_{n+1}, x_2, \cdots, x_n, x_1)$ is an $n + 1$ cycle in G. Let $x \in \bar{G} \backslash C$. Then by Lemma 7, part 1, x is strongly dominant to C in G, so that there exists a subset $\{x_{i_1}, x_{i_2}, \cdots, x_{i_q}\} \subseteq C$,

$q \geq (n + 1)/2$, such that $(x, x_{i_j}) \in E$ for $j = 1, 2, \cdots, q$. But then, since $q \geq (n + 1)/2$, there exists an i such that $(x, x_i), (x, x_{i-2k+2}) \in E$. Hence $(x, x_i, x_{i-2}, \cdots, x_{i-2k+2}, x)$ is a cycle of length k in G, a contradiction.

   Case II. n is odd.

   As in Case I, $\bar{G}$ contains no C_n implies that $C^{(1)} = (x_1, x_3, \cdots, x_n, x_1)$ and $C^{(2)} = (x_2, x_4, \cdots, x_{n+1}, x_2)$ are cycles in G. Let

$$L = \{x \in \bar{G} \setminus \dot{C} | x \text{ is strongly dominant to } C^{(1)} \text{ in } G \text{ and not dominant to } C^{(2)} \text{ in } G\},$$

$$M = \{x \in \bar{G} \setminus C | x \text{ is dominant to both } C^{(1)} \text{ and } C^{(2)} \text{ in } G\}$$

and

$$R = \{x \in \bar{G} \setminus C | x \text{ is strongly dominant to } C^{(2)} \text{ in } G \text{ and not dominant to } C^{(1)} \text{ in } G\}.$$

By Lemma 7, part 2, L and R are complete graphs in G and by Lemma 7, part 1b, $\bar{G} \setminus C = L \cup M \cup R$. If $|L|$ or $|R|$ is greater than or equal to $(n - 3)/2$, then either $L \cup C^{(1)}$ or $L \cup C^{(2)}$ contains all C_i, $3 \leq i \leq n - 1$, in G and in particular a C_k. Thus we assume $|L \cup R| \leq n - 5$ so that $|M| \geq (n - 2) - (n - 5) = 3$. Pick $x, y \in M, x \neq y$. Using Lemma 7, part 1b, it can be shown by a proper choice of i, j, s, r that $(x, x_j, x_{j+2}, \cdots, x_{j+2r}, y, x_{i+2s}, x_{i+2s-2}, \cdots, x_i, x)$ is a cycle of length $k = 4 + r + s$ in G, a contradiction.

   Thus the supposition is false and the proof is complete.

   For k odd the graph $G = K_{p-1,p-1}$ shows that $R(C_k, C_p) = 2p - 1, p > 3$, $p \geq k$. The remaining Ramsey numbers, where k is even, are as follows:

$$R(C_{2k}, C_{2p}) = 2p + k - 1 \quad \text{for } p \geq k \geq 3,$$

$$R(C_{2k}, C_{2p+1}) = 2p + k \quad \text{for } 2p + 1 \geq 3k,$$

$$R(C_{2k}, C_{2p+1}) = 4k - 1 \quad \text{for } 2k \leq 2p + 1 \leq 3k - 1.$$

A detailed proof that these are the remaining Ramsey numbers will be included in the paper *All Ramsey numbers of cycles in graphs*, to be published elsewhere.

   Editorial note. Another such paper by V. ROSTA (Mathematical Institute of the Hungarian Academy of Sciences) is to appear in J. Combinatorial Theory.

REFERENCES

[1] J. A. BONDY, *Large cycles in graphs*, Discrete Math., 1 (1971), pp. 121–132.
[2] J. A. BONDY AND P. ERDÖS, *Ramsey numbers for cycles in graphs*, submitted for publication.
[3] G. CHARTRAND AND S. SCHUSTER, *On the existence of specified cycles in complementary graphs*, Bull. Amer. Math. Soc., 77 (1971), pp. 995–998.
[4] P. ERDÖS AND T. GALLAI, *On maximal paths and circuits of graphs*, Acta. Math. Acad. Sci. Hungar., 10 (1959), pp. 337–356.

An Inequality for Walks in a Graph

Problem 83-15, *by* J. C. LAGARIAS, J. E. MAZO, L. A. SHEPP (AT&T Bell Laboratories, Murray Hill, NJ) *and* B. McKAY (Vanderbilt University).

   Let G be a finite undirected graph with N vertices, which is permitted to have multiple edges and multiple loops. A *directed walk* of length k is a sequence of edges $e_1, \cdots, e_k$ of G together with vertices $v_1, \cdots, v_{k+1}$ such that edge e_i connects vertex v_i and v_{i+1}. Let $w_k = w_k(G)$ denote the number of distinct directed walks of length k in G. For which pairs (r, s) is the inequality

$$w_r w_s \leq N w_{r+s}$$

true for all graphs G?

Solution by the proposers.

We show that the inequality holds when $r + s$ is even, and exhibit counterexamples whenever $r + s$ is odd.

Let A denote the adjacency matrix of the graph G (for a fixed ordering of its vertices), i.e., $a_{ij} = 1$ if there is an edge between i and j in G and $a_{ij} = 0$ otherwise. Then A is a nonnegative symmetric matrix with integer entries. It turns out that only the symmetry of A is important in proving the result for $r + s$ even.

THEOREM 1. *For any graph G with N vertices, the inequality*

$$w_r w_s \leq N w_{r+s}$$

holds for all positive integers r, s for which $r + s$ is even. If the adjacency matrix A of G is positive definite, this inequality holds for all positive integers r and s.

Proof. It is well known that the (i, j)th entry of A^k counts the number of directed walks of length k in A which start at vertex i and finish at vertex j. Consequently

$$w_k = 1^t A^k 1,$$

where 1 is the $N \times 1$ column vector with all entries 1. The theorem is an immediate consequence of the following result, after noting that $1^t 1 = N$. $\square$

THEOREM 2. *Let A be a real $N \times N$ symmetric matrix and ψ a real $N \times 1$ column vector. If $z_k = \psi^t A^k \psi$ then*

$$z_r z_s \leq (\psi^t \psi) z_{r+s}$$

if $r + s$ is even. It holds for all r, s if A is positive definite.

Proof. Since A is symmetric, A is diagonalizable by an orthogonal matrix, and all its eigenvalues are real. Let $\phi_1, \cdots, \phi_N$ be an orthonormal set of (real) eigenvectors with corresponding eigenvalues $\lambda_1, \cdots, \lambda_N$, i.e., $\phi_i^t \phi_j = 1$ if $i = j$, 0 otherwise. Then we have

(1)
$$\psi = \sum_{i=1}^{N} c_i \phi_i$$

where

$$c_i = \langle \psi, \phi_i \rangle = \psi^t \phi_i.$$

Then

$$A^r \psi = A^r \left(\sum_{i=1}^{N} c_i \phi_i \right) = \sum_{i=1}^{n} c_i \lambda_i^r \phi_i.$$

Hence

$$w_r = \psi^t (A^r \psi) = \sum_{i=1}^{n} c_i^2 \lambda_i^r.$$

Now define a discrete random variable X on the set $\{\lambda_1, \cdots, \lambda_n\}$ by

$$\Pr(X = \lambda_i) = \frac{c_i^2}{(\psi, \psi)}.$$

Note that

$$\sum_{i=1}^{N} c_i^2 = (\psi, \psi)$$

using (1). Consequently,

$$E(X^r) = \frac{1}{(\psi, \psi)} \sum_{i=1}^{N} c_i^2 \lambda_i^r = w_r.$$

We also note that if A is positive semidefinite, then all the eigenvalues λ_i are nonnegative so that X is a nonnegative random variable. The proof is completed using the following well-known fact.

FACT. *The inequality*

(2) $$E(X^r)E(X^s) \leq E(X^{r+s})$$

holds for all random variables X when $r + s$ is even, and for nonnegative random variables for all positive r and s.

Proof. Let X, Y be independent random variables both having the same distribution. For any realizations X_0, Y_0 we have

(3) $$(X_0^r - Y_0^r)(X_0^s - Y_0^s) \geq 0$$

if $r + s$ is even. If Y_0, X_0 are nonnegative, then (3) holds for all positive integers r and s. Now take expected values of both sides of (3) and note that

$$E((X_0^r - Y_0^r)(X_0^s - Y_0^s)) = 2E(X^{r+s}) - 2E(X^r)E(X^s)$$

using independence. The fact follows.   □

THEOREM 3. *For any pair r, s with $r + s$ odd there is a graph G with N vertices with*

$$w_r w_s > N w_{r+s}.$$

Proof. We may suppose $r = 2k - 1$, $s = 2l$ with $k, l \geq 1$. Let K_m denote the complete graph with m vertices and S_p the p-star, the unique tree with p vertices that has a vertex of degree $p - 1$.

We consider the graph G_{m+1,m^2+t+1} which is the disjoint union of K_{m+1} and S_{m^2+t+1}. We use the fact that if G is the disjoint union of the graphs G_1 and G_2 then

(4) $$w_r(G) = w_r(G_1) + w_r(G_2).$$

It is easy to verify that

(5) $$w_r(K_{m+1}) = m^r(m + 1),$$

and that

(6)
$$w_{2r-1}(S_{p+1}) = 2p^r,$$
$$w_{2r}(S_{p+1}) = p^r(p + 1).$$

Now consider G_{m+1,m^2+t+1} where $t \geq 1$ is viewed as fixed and we let $m \to \infty$. A straightforward algebraic calculation using (4)–(6) gives

$$w_{2k-1}w_l = 3m^{2k+2l+2} + 4m^{2k+2l+1} + ((2k + 3l + 3)t + 7)m^{2k+2l} + O(m^{2k+2l-1}),$$

$$Nw_{2k+2l-1} = 3m^{2k+2l+2} + 4m^{2k+2l+1} + ((2k + 2l + 3)t + 7)m^{2k+2l} + O(m^{2k+2l-1}).$$

This shows that for sufficiently large m (depending on t) G_{m+1,m^2+t+1} has

$$v_{2k-1}v_{2l} > Nv_{2k+2l-1},$$

the desired counterexample.

These counterexamples are disconnected graphs. One may obtain connected counter-examples by considering the graph G^*_{m+1,m^2+t+1} obtained by adding an edge connecting K_{m+1} to a vertex of degree of one in S_{m^2+t+1}. More explicitly, for $r = 1, s = 2$, we may take the disjoint union of K_3 and S_6. This has 9 vertices, and $w_1 = 16$, $w_2 = 42$, $w_3 = 74$, so that

$$w_1w_2 = 672 > Nw_3 = 666.$$

We can also take $m = 8, t = 7$ to obtain a connected counterexample, i.e. take K_9 and S_{89} with an edge connecting a vertex of K_9 to a vertex of degree 1 of S_{89}.   □

Although there are counterexamples to the inequality when $r + s$ is odd, it is possible to show that it is "almost true" in the following sense.

THEOREM 4. *For any graph G there is a constant $n_0 = n_0(G)$ such that for all* $r, s \geq n_0$

(7) $w_rw_s \leq Nw_{r+s}.$

We omit the proof.

Edge Three-Coloring of Tournaments

Problem 78-11*, *by* N. MEGIDDO (University of Illinois).

We define an edge *k-coloring* of a tournament (i.e. a directed graph with a unique edge between every pair of vertices) to be a coloring of the edges in k colors such that every directed cycle of length n contains at least min (k, n) edges of distinct colors.

It can be easily seen that every tournament has a 2-coloring. Specifically, if the vertices are numbered $1, \cdots, m$ $(m \geq 2)$, then color every "ascending" edge black and every "descending" edge white.

We shall show that for each $k \geq 4$ there are tournaments which do not have k-colorings. Given $k \geq 4$, let $m = k + 1$ and consider a directed graph whose vertices are $1, \cdots, m, m+1, \cdots, 2m$ and whose edges are $(i, m+i)$ $(i = 1, \cdots, m)$ and $(m+j, i)$ $(i = 1, \cdots, m, j = 1, \cdots, m. i \neq j)$. Every pair of edges $(i, m+i)$, $(j, m+j)$ $(1 \leq i < j \leq m)$ lies on some cycle of length 4, namely, $(i, m+i)$, $(m+i, j)$, $(j, m+j)$, $(m+j, i)$. Thus, in every k-coloring the edges $(i, m+1)$, $(2, m+2), \cdots, (m, 2m)$ must have distinct colors. This implies that there is no k-coloring for such graphs.

The remaining open question is: Does every tournament have a three-coloring? I conjecture that it does.

Editorial note. G. K. KRISTIANSEN (Roskilde, Denmark) gives a counterexample, disproving the conjecture. It is the tournament T with vertices $1, 2, \cdots, 9$ that contains the directed edge (i, j) if and only if $j - 1 \equiv 1, 2, 4, 6$ (mod 9). J. MOON (University of Alberta) simplified Kristiansen's proof from one involving four cases to one of two cases. Although it is straightforward it still is a tedious exercise to show that T has no 3-coloring. L. L. KEENER (University of Waterloo) described a construction for a larger counter example. [M.S.K.]

Problem 77-15*, *A Conjectured Minimum Valuation Tree*, by I. CAHIT (Turkish Telecommunications, Nicosia, Cyprus).

Let T denote a tree on n vertices. Each vertex of the tree is labeled with distinct integers from the set $1, 2, \cdots, n$. The weight of an edge of T is defined as the absolute value of the difference between the vertex numbers at its endpoints. If S denotes the sum of all the edge weights of T with respect to a given labeling, it is conjectured that for a k-level complete binary tree, the minimum sum is given by

$$S_{min}^k = \min_{\substack{\text{all labelings}}} \sum_{(i,j)\in T} |i-j| = (k-1)2^{(k-1)}, \quad k > 1.$$

Examples of minimum valuation trees for $k = 2, 3, 4$ are given by

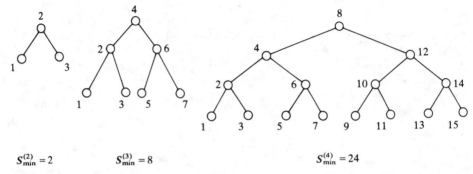

$S_{min}^{(2)} = 2 \qquad\qquad S_{min}^{(3)} = 8 \qquad\qquad\qquad S_{min}^{(4)} = 24$

Editorial note. A. Meir suggested the related problem of determining min $\sum (i-j)^2$. More generally, one can also consider max and min $\sum |i-j|^m$. [M.S.K.]

Solution by F. R. K. CHUNG (Bell Laboratories).

The conjecture is not true for $k > 4$. The following labeling for the 5-level complete binary tree shows that $S_{min}^{(5)} \le 60 < 4.2^4$:

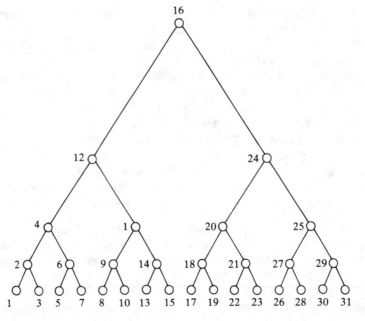

FIG. 1

We let $S_k = S_{\min}^{(k)}$. It can be shown that

$$S_k = 2^k(k/3 + 5/18) + (-1)^k(2/9) - 2.$$

Proof. The optimal labeling L_k for the k-level binary tree T_k satisfies the following properties. The proof can be found in [1], [2] or can be easily verified.

   Property 1. The vertex labeled by 1 or $n = 2^k - 1$ in L_k is a leaf (a *leaf* is a vertex of degree 1).

   Property 2. Let P denote the path connecting the two vertices labeled by 1 and n in L_k. Let P have vertices $v_0, \cdots, v_t$. Then the labeling of the vertices of P is monotone, i.e.,

$$L(v_i) < L(v_{i+1}) \quad \text{for } i = 0, \cdots, t-1$$

or

$$L(v_i) > L(v_{i+1}) \quad \text{for } i = 0, \cdots, t-1.$$

   Property 3. In T_k, we remove all edges of P. The resultant graph is a union of vertex disjoint subtrees. Let $\bar{T}_i$ denote the subtree which contains the vertex v_i, $i = 1, \cdots, t-1$. Then for a fixed i, the set of labelings of vertices in $\bar{T}_i$ consists of consecutive integers. Moreover, the labeling on each $\bar{T}_i$ are optimal.

   Property 4. Let $\bar{v}$ be the only vertex of T_k with degree 2. Then P passes $\bar{v}$.

   Property 5. Let T'_k denote the tree which contains T_k as subtree and T'_k has one more vertex than T_k, which is a leaf adjacent to $\bar{v}$. Then $S(T'_k) = S_k + 2$.

   From Properties 1 to 5, the following recurrence relation holds:

$$S_k = 2^{k-1} + 4 + S_{k-1} + 2S_{k-2} \quad \text{for } k \geqq 4$$

and

$$S_2 = 2, \qquad S_3 = 8.$$

It can be easily verified by induction that

$$S_k = 2^k(k/3 + 5/18) + (-1)^k(2/9) - 2. \quad \square$$

If we consider k-level complete p-nary trees T_k^p, some asymptotic estimates for S_k^p, minimum sum of all edge weights of T_k^p over all labelings, have been obtained in [1], [2]. We will briefly discuss the case that $p = 3$.

Let $T_p(k, i)$ denote a k-level tree which has the root connected to i copies of $(k-1)$-level p-nary tree. For example, the graph as shown in Fig. 2 is $T_3(3, 2)$. Let $S_p(k, i)$ denote the minimum value of all edge weights of $T_p(k, i)$ over all labelings of $T_p(k, i)$.

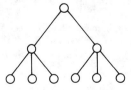

FIG. 2

It can be easily verified that

$$S_3(k, 2) = 3S_3(k-1, 2) + 2 \cdot 3^{k-2} \quad \text{for } k \geq 3$$

and

$$S_3(2, 2) = 2.$$

Therefore,

$$S_3(k, 2) = 2(k-1)3^{k-2} \quad \text{for } k \geq 2.$$

In general, it can be shown that, for p odd,

$$S_p(k, 2) = k(p+1)p^{k-2}/2 + (-2p^k + 3p^{k-1} + p^{k-2} + p - 3)/(2(p-1)).$$

and

$$S_p(k, p-1) = (k-1)(p^2-1)p^{k-2}/4.$$

The recurrence relation for S_k^3 is as follows: Let $f(k)$ be the integer l satisfying

$$(l-1)3^{l-2} + l + 1 \leq k < l3^{l-1} + l, \qquad k \geq 3.$$

Then we have

$$S_k^3 = 3^{k-2}(2k - 1/2) - 1/2 + k - f(k) + S_{k-1}^3 \quad \text{for } k \geq 3,$$

and

$$S_2^3 = 4.$$

This complicated recurrence relation reveals the possible difficulty in getting explicit expression for S_k^p for general p.

REFERENCES

[1] M. A. IORDANSKII, *Minimal numberings of the vertices of trees*, Soviet Math. Dokl., 15(1974), no. 5, pp. 1311–1315.

[2] M. A. ŠEĬDVASSER, *The optimal numbering of the vertices of a tree*, Diskret. Analiz., 17(1970) pp. 56–74.

Also solved by W. F. SMYTH (Winnipeg, Manitoba) who sent a copy of a long paper, *A labelling algorithm for minimum edge weight sums of complete binary trees*, which had been submitted to Comm. ACM, whose interests were felt to be more directly related to the subject matter. An abstract of the paper is as follows:

Given a K-level complete binary tree $T_K = (V_K, E_K)$ on $2^K - 1$ vertices and a set $\mathcal{W}_K \equiv \{N+1, N+2, \cdots, N+2^K - 1\}$ of integers, it is desired to label the vertices V_K from the set $\mathcal{W}_K$ without replacement, in such a way that the sum $S_K = \sum |n(u) - n(v)|$, taken over all edges $(u, v) \in E_K$, is a minimum, where $n(u)$ denotes the label assigned to vertex u. The labeled tree is called a valuation tree and, corresponding to a minimum labeling, a minimum valuation tree. An algorithm for this purpose is specified, with execution time $O(2^K - 1)$. An expression is derived for $S_K^{\min}$, and it is shown that in fact the algorithm achieves this minimum. Connections to the minimum bandwidth and minimum profile problems are outlined. Some open problems are stated.

Supplementary References
Graph Theory

[1] B. ANDRÁSFAI, *Introductory Graph Theory*, Pergamon, N.Y., 1977

[2] A. T. BALABAN, ed., *Chemical Applications of Graph Theory*, Academic, London, 1976.

[3] A. Battersby, *Network Analysis*, Macmillan, London, 1964.

[4] M. BEHZAD, G. CHARTRAND AND L. LESNIAK-FOSTER, *Graphs and Digraphs*, Wadsworth, Belmont, 1980.

[5] R. BELLMAN, K. L. COOKE AND J. A. LOCKETT, *Algorithms, Graphs, and Computers*, Academic Press, N.Y., 1970.

[6] C. BERGE, *Graphs and Hypergraphs*, North-Holland, Amsterdam, 1973.

[7] ———, *Theory of Graphs and its Applications*, Methuen, London, 1962.

[8] B. Bollobás, *Graph Theory, An Introductory Course*, Springer-Verlag, 1979, N.Y.

[9] ———, *Extermal Graph Theory*, Academic Press, London, 1978.

[10] J. A. BONDY AND U. S. R. MURTY, *Graph Theory with Applications*, Elsevier, Great Britain, 1976.

[11] R. G. BUSACKER AND T. L. SAATY, *Finite Graphs and Networks: An Introduction with Applications*, McGraw-Hill, N.Y., 1965.

[12] B. CARRE, *Graphs and Networks*, Oxford University Press, Oxford, 1979.

[13] K. W. CATTERMODE, "*Graph theory and the telecommunications network*," BIMA (1975) pp. 94–107.

[14] V. CHACHRA, P. M. GHARE AND J. M. MOORE, *Applications of Graph Theory Algorithms*, Elsevier, N.Y., 1979.

[15] G. CHARTRAND, *Graphs as Mathematical Models*, Prindle, Weber & Schmidt, Boston, 1977.

[16] G. CHARTRAND AND S. F. KAPOOR, *The Many Facets of Graph Theory*, Springer-Verlag, Berlin, 1969.

[17] W. CHEN, *Applied Graph Theory*, Elsevier, N.Y., 1971.

[18] N. CHRISTOFIDES, *Graph Theory and the Algorithmic Approach*, Academic, London, 1977.

[19] N. DEO, *Graph Theory with Applications to Engineering and Computer Science*, Prentice-Hall, N.J., 1974.

[20] C. FLAMENT, *Applications of Graph Theory to Group Structure*, Prentice-Hall, N.J., 1963.

[21] D. R. FULKERSON, ed., *Studies in Graph Theory I, II*, MAA, 1975.

[22] R. E. GOMORY AND T. C. HU, "*Multi-terminal focus in a network*," SIAM J. Appl. Math. (1961) pp. 551–570.

[23] P. HAGGERT AND R. J. CHORLEY, *Network Analysis in Geography*, St. Martin's Press, N.Y., 1969.

[24] F. HARARY, *Graph Theory*, Addison-Wesley, Reading, 1969.

[25] ———, *Graph Theory and Theoretical Physics*, Academic Press, London, 1967.

[26] F. HARARY, R. Z. NORMAN AND D. CARTWRIGHT, *Structural Models: An Introduction to the Theory of Directed Graphs*, Wiley, N.Y., 1965.

[27] F. HARARY AND L. MOSER, "*The theory of round robin tournaments*," AMM (1966) pp. 231–246.

[28] E. J. HENLEY AND R. A. WILLIAMS, *Graph Theory in Modern Engineering*, Academic Press, N.Y., 1973.

[29] D. E. JOHNSON AND J. R. JOHNSON, *Graph Theory with Engineering Applications*, Ronald Press, N.Y., 1972.

[30] J. MALKEVITCH AND W. MEYER, *Graphs, Models, and Finite Mathematics*, Prentice-Hall, N.J., 1974.

[31] R. B. MARIMONT, "*Applications of graphs and Boolean matrices to computer programming*," SIAM Rev. (1960) pp. 259–268.

[32] J. W. MOON, "*A problem of rankings by committees*," Econometrica (1976) pp. 241–246.

[33] ———, *Topics on Tournaments*, Holt, Rinehart and Winston, N.Y., 1968.

[34] O. ORE, *Graphs and Their Uses*, MAA, 1963.

[35] W. L. PRICE, *Graphs and Networks*, Auerbach, London, 1971.

[36] F. S. ROBERTS, *Graph Theory and its Applications to Problems of Society*, SIAM, 1978.

[37] F. S. ROBERTS AND T. A. BROWN, "*Signed digraphs and the energy crisis*," AMM (1975) pp. 577–594.

[38] D. F. ROBINSON AND L. R. FOULDS, *Digraphs: Theory and Techniques*, Gordon and Breach, N.Y., 1978.

[39] P. D. STRAFFIN, JR., "*Linear algebra in geography: Eigenvalues of networks*," MM (1960) pp. 269–276.

[40] K. G. TINKLER, "*Graph theory*," Prog. Human Geography (1979) pp. 85–116.

[41] ———, *Introduction to Graph Theoretical Methods in Geography*, Geographical Abstracts Ltd., University of East Anglia, Norwich, England, 1977.

[42] R. J. TRUDEAU, *Dots and Lines*, Kent State University Press, Ohio, 1976.

[43] J. TURNER AND W. H. KAUTZ, "*A survey of progress in graph theory in the Soviet Union*," SIAM Rev. Supplement (1970) pp. 1–68.

16. GEOMETRY

Problem 66-12, Stability of Polyhedra, by J. H. Conway (Cambridge University, England) and R. K. Guy (University of Alberta, Calgary).

It is obvious that a regular homogeneous tetrahedron will rest in stable position on top of a horizontal table when lying on any one of its faces.

(a) Show that any homogeneous tetrahedron will rest in stable position when lying on any one of at least two of its faces.

(b) Give an example of a homogeneous convex polyhedron which will rest in a stable position when lying on *only one* of its faces.

Solution of part (a) by Michael Goldberg (Washington, D.C.).

(a) A tetrahedron is always stable when resting on the face nearest to the center of gravity (C.G.) since it can have no lower potential. The orthogonal projection of the C.G. onto this base will always lie within this base. Project the apex V to V' onto this base as well as the edges. Then, the projection of the C.G. will lie within one of the projected triangles or on one of the projected edges. If it lies within a projected triangle, then a perpendicular from the C.G. to the corresponding face will meet within the face making it another stable face. If it lies on a projected edge, then both corresponding faces are stable faces.

Editorial note. Goldberg also gives a construction for a 21-sided polyhedron satisfying part (b).

Solution of part (b) by R. K. Guy (University of Alberta, Calgary).

A problem which was popular some years ago was to prove that a convex polyhedron will always stand in stable equilibrium on at least one of its faces. One answer can be made by an appeal to the impossibility of perpetual motion. One is led to ask if there are polyhedra which will stand on only one face. Manufacturers of kelly-lamps and unspillable saltcellars will reply in the affirmative, so we add the condition that the polyhedron must be made of material of uniform density.

The present solution exhibits such a solid and improves the previous best known result for the least number of faces such a solid can have. We conjecture that we are close to best possible in this respect, but this is an open question.

Our solid is a 17-sided prism, half of whose section is illustrated in Fig. 5, truncated obliquely as in the side view of Fig. 6, so it is an enneadecahedron. In order to analyze the problem, we shall consider more generally a $(2m - 1)$-sided prism, whose section is made up of $2m$ similar right triangles, each having an angle $\beta = \pi/m$ at the point O. The hypotenuse of the largest pair of triangles has length $r = r_0$, and this will be vertical when the solid stands in stable equilibrium. Other hypotenuses are $r_n = r \cos^n \beta$, $0 < n < m$, and the side of length

478

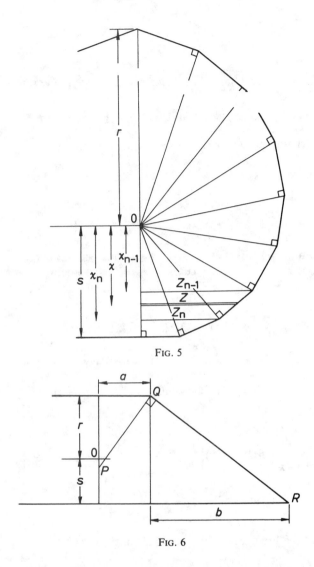

Fig. 5

Fig. 6

$r_m = s = r \cos^m \beta$ is collinear with that of length $r = r_0$. The vertices of the triangles other than O lie on two symmetrical equiangular spirals. The half-length of the cylinder is $y = fx + g$, where x is measured vertically downwards from O and $f = b/(r + s)$, $g = a + br/(r + s)$. We shall choose b sufficiently large compared with a to ensure that the center of mass is below O, but above P, where PQR in Fig. 6 is a right angle. Then the solid will stand on only one lateral face and on neither of the oblique ones.

The half-width of the section at depth x below O is $z = px + q$, where

$$p = (z_n - z_{n-1})/(x_n - x_{n-1}), \quad q = (z_{n-1}x_n - z_n x_{n-1})/(x_n - x_{n-1})$$

and $x_n = -r_n \cos n\beta$, $z_n = r_n \sin n\beta$. The first moment, below O, of the material of the solid is thus

(2)
$$4 \sum_{n=1}^{m} \int_{x_{n-1}}^{x_n} xyz \, dx,$$

and we shall choose m (and b) so that this is positive. The integral in (2) has value

$$\tfrac{1}{4} fp(x_n^4 - x_{n-1}^4) + \tfrac{1}{3}(fq + gp)(x_n^3 - x_{n-1}^3) + \tfrac{1}{2} gq(x_n^2 - x_{n-1}^2).$$

Algebraic and trigonometric manipulation reduces this to

$$\tfrac{1}{4} f(x_n^3 z_n - x_{n-1}^3 z_{n-1}) + \tfrac{1}{3} g(x_n^2 z_n - x_{n-1}^2 z_{n-1})$$
$$+ \tfrac{1}{12} f(x_n^3 z_{n-1} + x_n^2 x_{n-1} z_{n-1} + x_n x_{n-1}^2 z_{n-1} - x_n^2 x_{n-1} z_n - x_n x_{n-1}^2 z_n - x_{n-1}^3 z_n)$$
$$+ \tfrac{1}{6} g(x_n^2 z_{n-1} + x_n x_{n-1} z_{n-1} - x_n x_{n-1} z_n - x_{n-1}^2 z_n).$$

If we sum from $n = 1$ to m, the first two terms yield $\tfrac{1}{4} f(x_m^3 z_m - x_0^3 z_0)$ and $\tfrac{1}{3} g(x_m^2 z_m - x_0^2 z_0)$, which vanish, since $z_0 = z_m = 0$. It remains to sum the other two terms, which, on writing $\cos \beta = k$, may be thrown into the form

$$\frac{r^3 \sin \beta}{24 k^3} \{ fr[(k^4)^n + 2k^2(k^4)^n + k^2(k^4)^n \cos 2n\beta$$
$$+ k(k^4)^n \cos (2n - 1)\beta + (k^4)^n \cos (2n - 2)\beta]$$
$$- 4gk[k(k^3)^n \cos n\beta + (k^3)^n \cos (n - 1)\beta] \},$$

whose sum is

$$\frac{r^3 k}{24 \sin \beta} \left\{ fr(1 - k^{4m}) \left[\frac{1 + 2k^2}{1 + k^2} + \frac{1 + 3k^4 - 4k^6}{1 + k^2 + 3k^4 - k^6} \right] \right.$$
$$\left. - 4g(1 + k^{3m}) \frac{(1 - k^2)(1 + 2k^2)}{1 + k^2 - k^4} \right\}.$$

The contents of the braces, putting f and g in terms of a and b, is

$$-4a(1 + k^{3m}) \frac{(1 - k^2)(1 + 2k^2)}{1 + k^2 - k^4}$$
$$+ \frac{2br}{r + s} \left\{ (1 - k^{4m}) \frac{1 + 2k^2 + 4k^4 + 2k^6 - 3k^8}{(1 + k^2)(1 + k^2 + 3k^4 - k^6)} - 2(1 + k^{3m}) \frac{(1 - k^2)(1 + 2k^2)}{1 + k^2 - k^4} \right\},$$

and this can be made positive, by taking b large compared with a, provided the content of the braces is positive, i.e., provided

$$(1 - k^{4m})(1 + k^2 - k^4)(1 + 2k^2 + 4k^4 + 2k^6 - 3k^8)$$
$$> 2(1 + k^{3m})(1 - k^4)(1 + 3k^2 + 5k^4 + 5k^6 - 2k^8),$$

which calculation shows to be true for $m \geq 9$, but for no smaller value.

In the opposite direction J. H. Conway has shown that every uniform tetrahedron will rest on at least two of its faces. Paradoxically, A. Heppes (see Problem 66-13, July 1967) has constructed a "2-tip" tetrahedron that will stand on either of two faces, but when placed on a third face, falls first onto the fourth, before finding one of the stable faces. The next questions to ask in this direction are:

(a) Can it be proved that each uniform pentahedron will always stand on either of two faces? The present construction, using a prism whose section is an obtuse-angled triangle with a sufficiently short side, shows that there are pentahedra which will stand on only one of two faces.

(b) What is the least dimension, if any, in which a simplex can be unstable?

If the condition of uniform density of material is relaxed, then a simplex can be unstable in 3 dimensions. However, Conway has shown that even with this relaxation no regular polygon cap be unstable. In the uniform case, Conway has shown that no 2-dimensional body (convex polygon) is unstable, and that no body of revolution (in 3 dimensions) can be unstable. He asks:

(c) Can a unistable polyhedron have an n-fold axis of symmetry for $n > 2$?

(d) What is the smallest possible ratio of diameter to girth for a convex unistable polyhedron?

(e) Can the resting face of a 3-dimensional unistable convex polyhedron be the face of least diameter?

(f) What is the set of convex bodies uniformly approximable by unistable polyhedra, and does this contain the sphere?

In connection with (d), by taking m sufficiently large in our present construction, the section can be made to approximate a circle; s can be made arbitrarily close to r, and then a and b can each be taken arbitrarily close to $\sqrt{2}r$. If the oblique truncations are made (almost) parallel, instead of symmetrical, then the diameter and girth can be approximated arbitrarily closely to $6r$ and $2\pi r$, so their ratio can be made less than $(3/\pi) + \varepsilon$.

We can answer (e) by making the base as small as we please, both in diameter and in area, and both absolutely and in comparison with other faces. We may begin the construction of the section, not with two right triangles of angle $\beta = \pi/m$, but with two of angle δ, which may be chosen as small as we wish, making the width of the base arbitrarily small. To make its length arbitrarily small also, taper the solid with 2 faces making a sufficiently small angle with the base and separated slightly so that it will stand on neither of the tapering faces. Each of these processes requires two more faces, so the result may be achieved with a solid having no more than 23 faces. The tapering process alone achieves the weaker requirement that the base should be the face with the least diameter (and area), and this can be realized with a solid of 21 faces.

Editorial note. R. J. WALKER (Cornell University) notes that a geometric proof that a convex polyhedron will always stand in stable equilibrium on at least one of its faces follows by considering spheres of increasing radii which are centered at the centroid. However, it is not easy to see how this argument can be extended to show that a tetrahedron is stable on at least two faces.

In connection with (b), since a triangle is stable on at least one edge and a tetrahedron is stable on at least two faces, a "daring" conjecture would be that any n-simplex is stable on at least $n - 1$ faces and this number cannot be improved. (M.S.K.)

Postscript. That the latter conjecture is not valid and even not daring follows from some recent results of R.J.M. Dawson, Monostatic Simplexes, Amer. Math. Monthly 92 (1985) 541–546. He shows that no simplex is monostatic in $\mathbf{R}^6$ and also that there exist monostatic simplexes in $\mathbf{R}^{10}$. I am grateful to John H. Conway for bringing Dawson's results to my attention.

Problem 66-13, A Double Tipping Tetrahedron, by A. Heppes (Mathematical Institute of the Hungarian Academy of Sciences).

Design a homogeneous tetrahedron which, when placed lying on one of its faces on the top of a horizontal table, will tip over to another face and then tip over again, finally coming to rest on a third face.

Solution by the proposer.

Let A, B, C, D, denote the vertices of a tetrahedron T and let G' denote the orthogonal projection of its centroid on the "horizontal" coordinate plane $z = 0$. The size of T is specified by the following coordinates of its vertices in the initial position:

$$A(-7, -8(1 - \epsilon), 0), \qquad\qquad 0 < \epsilon < 1,$$
$$B(-1, \quad 0, \quad 0),$$
$$C(\ 1, \quad 0, \quad 0),$$
$$D(\ 7, \quad 8, \quad 8).$$

We then have $G'(0, 2\epsilon, 0)$.

G' lying outside of $\triangle ABC$ and on the y-axis, implies that it will tip over about edge BC. This actually means a rotation of $45°$ keeping vertices B and C fixed. The new coordinates will be as follows:

$$A(-7, -4\sqrt{2}(1 - \epsilon), 4\sqrt{2}(1 - \epsilon)),$$
$$B(-1, \quad 0, \quad 0\quad),$$
$$C(\ 1, \quad 0, \quad 0\quad),$$
$$D(\ 7, \quad 8\sqrt{2}, \quad 0\quad),$$
$$G'(\ 0, \quad \sqrt{2}(1 - \epsilon), \quad 0\quad).$$

Since G' is again outside the new base $\triangle BCD$, T will tip over once more.

It follows of course that there are many other tetrahedrons with the desired property.

Problem 67-8*, A Double-Cone Radar Search, by D. J. Newman (Yeshiva University).

Given any n points in space, it is conjectured that all the points can be caught simultaneously by a double-cone radar device (see Fig. 1) whose semi-angle α is $(\pi/2)(1 - 1/n)$.

Editorial Note. The proposer has established the case $n = 4$ by tedious calculations and also the case of the n points being coplanar.

Problem 63-5*, The Smallest Covering Cone or Sphere, by C. Groenewoud and L. Eusanio (Cornell Aeronautical Laboratory).

An investigation of multiple airborne target tracking with a ground based radar leads to the following problem:

Let S be a fixed set of n points in E_3 and let P be a point sufficiently remote from S so that all of S lies on the same side of a suitably chosen plane through P. Find the right circular cone of minimum vertex angle (vertex at P) containing on or in it all the points of S.

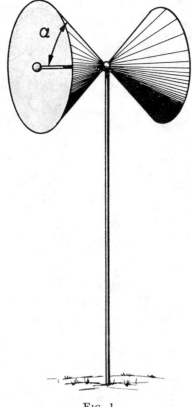

FIG. 1

A related problem is to find the smallest covering sphere.

As a generalization, one can consider the above two problems in E_n .

Editorial Note: Another related class of problems arise if we no longer consider the set S to be a set of n fixed points but a set of n random points from a uniform or a normal distribution.

Solution by C. L. Lawson (Jet Propulsion Laboratory).

The cone problem will be made to depend upon the sphere problem and an algorithm will be given for the sphere problem.

The cone problem is unchanged if each point $s \in S$ is replaced by the point t at which the half ray $\overrightarrow{P_s}$ intersects the surface of the unit sphere H_1 centered at P. Denoting this new set by T, let H_2 denote the smallest sphere containing T. From the assumption that S(and thus T) lies strictly on one side of some plane through P it follows that H_2 is smaller than H_1 , so that the intersection of H_1 and H_2 is a circle K. Clearly the cone C with vertex at P and containing the circle K on its surface contains T but no smaller right circular cone does.

The sphere problem is a very special example of a linear least maximum (also called uniform or Chebyshev) approximation problem involving vector valued functions. The theory of this type of approximation problem as well as a practical computational algorithm was treated in the writer's dissertation (*Contributions to the theory of linear least maximum approximation*, U.C.L.A., 1961) and reported upon at the 1961 National Meeting of the A.C.M.

The sphere problem may be formulated as follows: the given data consists of real numbers t_{ij}; $i = 1, \cdots, n$; $j = 1, \cdots, m$, denoting the jth coordinate of the ith data point $\mathbf{t}_i$ in m-dimensional Euclidean space E_m. It is desired to find a point $\mathbf{x} = (x_1, \cdots, x_m)$ in E_m satisfying

$$\max_i |\mathbf{x} - \mathbf{t}_i| = \min_{y \in E_m} \max_i |\mathbf{y} - \mathbf{t}_i| \equiv \tau^*,$$

where $|\cdot|$ denotes the Euclidean norm in E_m. Then $\mathbf{x}$ is the center and τ^* is the radius of the smallest sphere containing the points $\mathbf{t}_1, \cdots, \mathbf{t}_n$.

The algorithm takes the following form for the sphere problem.

1. $w_i := 1/n, \quad i = 1, \cdots, n$.
2. $x_j := \sum_i w_i t_{ij}, \quad j = 1, \cdots, m$,
3. $r_i := \mathrm{sqrt}\,(\sum_j (x_j - t_{ij})^2), \quad i = 1, \cdots, n$.
4. $\sigma := \mathrm{sqrt}\,(\sum_i w_i r_i^2)$.
5. $\tau := \max_i r_i$.
6. If $(\tau - \sigma)$ is sufficiently small, quit.
7. $u := 1/\sum_i w_i r_i$.
8. $w_i := w_i r_i u, \quad i = 1, \cdots, n$.
9. Go to 2.

It can be shown that σ and τ always satisfy $\sigma \leq \tau^* \leq \tau$. The number σ increases strictly with each iteration unless a stage is reached at which r_i is constant for all i for which $w_i > 0$, in which case all quantities clearly remain unchanged in all successive iterations.

Theoretically this algorithm can fail only if at some iteration it occurs that $r_i = 0$ for some critical index i, that is, an index such that if t_i were removed from the problem the value of τ^* would be smaller. In the general problem this is possible but probably very unlikely to occur in practice.

For the sphere problem, however, this cannot happen since the critical points must necessarily be extreme points of the convex hull of $\{\mathbf{t}_1, \cdots, \mathbf{t}_n\}$ and, due to Step 2, $\mathbf{x}$ could coincide with an extreme point only if the weights were already zero at all other points. This would then result in $\sigma = 0$ which is impossible if σ was ever previously positive since, by the general theory, σ never decreases. The number σ is always positive on the first iteration except in the trivial case in which all $\mathbf{t}_i$ are identical.

A FORTRAN program was written implementing the above algorithm and an example was run using 20 points on the unit hemisphere in E_3. Each point was defined by cartesian coordinates $(\rho \cos \theta, \rho \sin \theta, (1 - \rho^2)^{1/2})$ where ρ and θ were drawn from uniform distributions on $(0, 0.75)$ and $(0, 2\pi)$ respectively. The nature of the convergence is indicated by the abridged table in Table 1. Anyone intending to make much use of this algorithm would probably want to incorporate some procedure for accelerating the convergence.

Problem 61-6*, *Satellite Communications*, by FRANK W. SINDEN (Bell Telephone Laboratories).

This problem arose out of an effort to find the best orbit arrangement for a satellite system.

Choose n great circles so that every point on the unit sphere is within distance d (measured on the surface) of at least one one of them. For each n, what is the

TABLE 1

Iteration k	τ_k	σ_k	$\delta_k = \tau_k - \sigma_k$	δ_k/δ_{k-1}
1	.796	.435	.361	$\cdots$
2	.753	.517	.236	.65
3	.709	.566	.144	.61
4	.699	.595	.105	.73
5	.706	.613	.093	.89
6	.707	.625	.082	.88
7	.705	.634	.071	.87
8	.701	.641	.061	.86
9	.698	.647	.051	.85
10	.695	.651	.043	.84
$\vdots$				
20	.680	.673	.0068	.83
$\vdots$				
30	.678	.678	.0014	.85
$\vdots$				
40	.677	.677	.0003	.85

smallest possible d?

Dual form of problem. Choose n pairs of antipodal points so that every great circle is within distance d of at least one of them. For each n, what is the smallest possible d?

Algebraic form of problem. Let $|\,\mathbf{A},\,\mathbf{X}\,|$ denote the absolute value of the scalar product. Choose n unit vectors $\mathbf{A}_n$ so that the inequalities $|\,\mathbf{A}_n\,,\,\mathbf{X}\,| > d'$ ($n = 1, 2, \cdots, n$) have no unit vector solution $\mathbf{X}$. What is the smallest d' for which this is possible?

Conjecture.

$$\arcsin d'_{\min} = d_{\min} = \frac{\pi}{2n}.$$

This is achieved by letting $\mathbf{A}_1\,,\,\cdots\,,\,\mathbf{A}_n\,,\,-\,\mathbf{A}_1\,,\,\cdots\,,\,-\,\mathbf{A}_n$ be uniformly spaced around a great circle.

Problem 61-7*, Another Satellite Communications Problem, by ISIDORE SILBERMAN (Raytheon Mfg. Co.).

Determine the minimum number $N(r, R)$ of circles of radius r necessary to cover the entire surface of a sphere of radius R. *Editoral Note:* A lower bound can be gotten immediately from area considerations. If A_r denotes the area of a spherical cap (of radius R) whose base is a circle of radius r, then $N(r, R) > 4\pi R^2/A_r$ (since there must be overlap). Since an exact determination of $N(r, R)$ appears to be extremely difficult, reasonably close upper and lower bounds will be acceptable as a solution.

Also, closely allied to this problem, is the packing problem of determining the maximum number of circles of radius r which can be placed on the entire surface of a sphere of radius R without overlap.

Solution by LOUIS D. GREY (The Teleregister Corp.).

We shall determine an upper bound for the minimum number $N(r, R)$ of circles of radius r necessary to cover the entire surface of a sphere of radius R.

It is convenient when working with spherical caps to work with the angular radius A rather than the linear radius r. The relation between these two is given by

$$(1) \qquad\qquad \cos A = \frac{\sqrt{R^2 - r^2}}{R}$$

By suitably choosing r the problem can be restricted to the unit sphere without any loss of generality.

To determine an upper bound for the minimum number $N^*(A, 1)$ of spherical caps of angular radius A necessary to cover the entire surface of a unit sphere, we consider a closely allied problem. This latter problem is the determination of the maximum number $K(A)$ of spherical caps of angular radius A which can be placed on a unit sphere without overlapping. The inverse of this latter problem, namely, what is the largest angular radius $A(K)$ such that K spherical caps of angular radius A can be placed on the unit sphere without overlapping, has been treated by L. Fejes Toth [1] who obtained the result.

$$(2) \qquad\qquad A(K) \leq \tfrac{1}{2} \cos^{-1} \tfrac{1}{2}\left[\cot^2 \frac{K\pi}{6(K-2)} - 1 \right].$$

This is an asymptotic upper bound which is exact for $K = 3, 4, 6,$ and 12.

For $K > 2$ which implies $0 < A < \frac{\pi}{3}$, we can solve for K to obtain

$$(3) \qquad\qquad K(A) \leq \frac{12 \cot^{-1} \sqrt{4 \cos^2 A - 1}}{6 \cot^{-1} \sqrt{4 \cos^2 A - 1} - \pi}.$$

We shall show that

$$N^*(2A, 1) \leq \frac{12 \cot^{-1} [\sqrt{4 \cos^2 A - 1}]}{6 \cot^{-1} [\sqrt{4 \cos^2 A - 1}] - \pi}, \qquad 0 < A < \frac{\pi}{3}.$$

The argument is simple. Imagine that we have placed $K(A)$ nonoverlapping spherical caps of angular radius on the surface of a sphere. If we replace these caps by concentric caps of angular radius $2A$, then we claim the surface of the sphere is completely covered. If it is not, there is a point whose distances from the centers of the caps is greater than $2A$. This implies that we could center a cap of angular radius A at this point and that this cap would not overlap any of the original $K(A)$ caps which contradicts the assumption that $K(A)$ is maximum.

The problem of determining an upper bound for $K(A)$ has been generalized to n dimensions. This bound will provide a bound for $N^*(2A, 1)$ by the argument given above. It is shown [2] that

$$(4) \qquad N_n^*(2A, 1) \leq \frac{\pi^{1/2}\, \Gamma\left(\dfrac{n-1}{2}\right) \sin B \tan B}{2\, \Gamma\left(\dfrac{n}{2}\right) \displaystyle\int_0^B (\sin \theta)^{n-2}(\cos \theta - \cos B)\, d\theta},$$

$$0 < A < \frac{\pi}{4}, B = \sin^{-1}\left(\sqrt{2 \sin A}\right).$$

A summary of results concerning the determination of $K(A)$ and the inverse problem in n-dimensions appears in a paper by the author [3].

REFERENCES

1. L. FEJES TOTH, *On The Densest Packing of Spherical Caps*, American Mathematical Monthly, Vol. 56, pp. 330–331. 1949.
2. R. A. RANKIN, *The Closest Packing Of Spherical Caps in N Dimensions*, Proc. Glasgow Math. Assoc., Vol. 2, p. 139, 195.
3. FLORES AND GREY, *Reference Signals For Character Recognition Systems*, IRE Transactions on Electronic Computers, Vol. EC-9, March 1960, p. 57–60.

Problem 66-11*, *Moving Furniture Through a Hallway*, by LEO MOSER (University of Alberta, Edmonton).

What is the largest area region which can be moved through a "hallway" of width one (see Fig. 1)?

Editorial note. For the three-dimensional version, in which the hallway has a fixed height h, it seems reasonable to suppose the answer would be a cylinder of height h whose cross section is given by the solution of the above problem.

Another similar problem would be to find the longest rigid convex thin rod which can be moved through the above hallway. (M.S.K.)

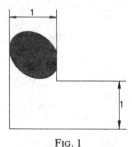

FIG. 1

Editorial Note. In a previous solution, Michael Goldberg obtains a lower bound for the largest area of ≈ 2.044. A better lower bound of $\pi/2 = 2/\pi \approx 2.207$ is given by J.M. Hammersley in his paper referred to in Problem 69-1 (1969, p. 73). This is obtained for the following figure 1A.

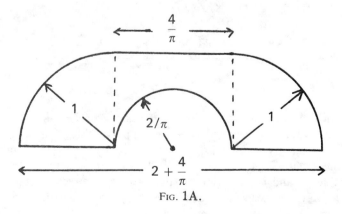

FIG. 1A.

Partial solution   by JAMES D. SEBASTIAN (University of Illinois, Urbana).

Upper bounds on the maximum area of a region R and the length of a rigid convex arc C which can be moved around the corner are easily obtained when certain mild restrictions are placed on the classes of permissible regions and arcs. For the regions these are: (i) R has an axis of symmetry S; (ii) when R is not in the corner, S may be oriented perpendicular to the direction of the corridor; and (iii) as R moves around the corner, S becomes coincident with the line through the inner and outer vertices of the corner, i.e., the y-axis in Fig. 2.

The implication of conditions (i) and (ii) is that R has a maximum width of 1 in the direction of S. By condition (iii), then, R must lie entirely in the intersection I of a unit strip with edges parallel to the x-axis and the corridor, the bottom edge of the strip being some distance d from the x-axis. In order for I to be non-empty and connected we must have $-1 \leqq d \leqq \sqrt{2}$.

The area of I is easily expressed as a function of d, and application of elementary calculus yields the intuitively obvious result that the maximum area of I, and hence an upper bound on the maximum area of R, is $2\sqrt{2}$, where $d = \sqrt{2}$. The figure due to Hammersley satisfies the stated conditions and achieves approximately 78% of the maximum area.

For the arc problem we impose conditions (i) through (iii), with R replaced by C, and a fourth condition, which is perhaps part of the problem definition, namely that C may be described as a convex function of x when S coincides with the y-axis. Clearly any convex function C satisfying these conditions must lie in the region I for some d, $-1 \leqq d \leqq \sqrt{2}$.

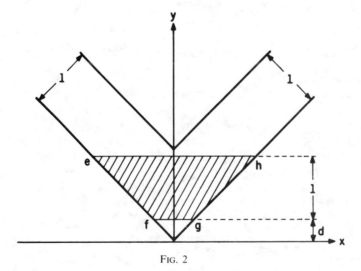

FIG. 2

Let L denote the lower boundary of I, i.e., the broken line *efgh* in Fig. 2. Then it is easy to see that the length of L is an upper bound on the length of any C lying in I. For, let $[p_0, p_1, \cdots, p_n]$ be the consecutive vertices of any polygon inscribed in C, with $p_k p_{k+1}$ denoting the line segment between p_k and p_{k+1}. In addition, let p'_k and p''_{k+1} denote the points on L where lines through p_k and p_{k+1}, respectively, and perpendicular to $p_k p_{k+1}$ intersect L, with $p'_k p''_{k+1}$ denoting that part of L which joins p'_k and p''_{k+1} (see Fig. 3).

Since C is wholly within I, such points certainly exist, and, by the convexity of C, the slopes of the line segments $p_k p_{k+1}$ are nondecreasing as k increases. Thus any two of the (possibly broken) line segments $p'_k p''_{k+1}$ have at most one

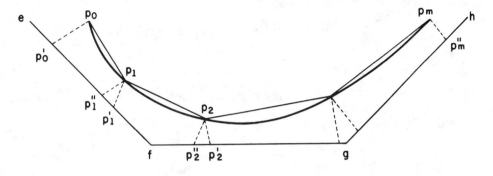

FIG. 3

point in common. Further, for each $k = 0, 1, \cdots, n - 1$, the length of $p'_k p''_{k+1}$ is $\geq$ to that of $p_k p_{k+1}$, for $p'_k p''_{k+1}$ is at worst parallel to $p_k p_{k+1}$. Summing on k, then, we have that the length of the inscribed polygon is $\leq$ the sum of the lengths of the $p'_k p''_{k+1}$, whis is $\leq$ to the length of L. Since this holds for any inscribed polygon it follows that the length of C is $\leq$ to that of L.

Now the maximum length of L occurs when $d = \sqrt{2}$, where it is $4\sqrt{2}$. Thus $4\sqrt{2}$ is an upper bound on the length of a convex arc satisfying conditions (i) through (iv).

We now proceed to construct a convex arc which satisfies conditions (i) through (iv), will move around the corner, and whose length is greater than 96% of the maximum.

The base of the arc we take to be a line segment of length $2\sqrt{2}$, AB in Fig. 4. We erect a rectangular coordinate system $\{v, H\}$ with origin at B and H-axis along AB. The aim is to determine a function $H(v)$, $0 \leq v \leq 1$, which describes the edge of the arc, and to this end we prescribe the following motion:

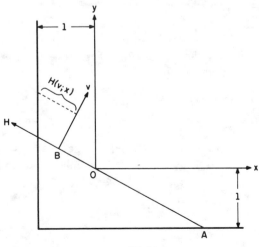

FIG. 4

Starting from the position of symmetry, A at $(1, -1)$, A moves to the right, remaining on the line $y = -1$, while AB maintains contact with the corner 0. The coordinates of A are thus $(x, -1)$, and, since when $x = \sqrt{7}$ the point B coincides with 0, only values of x such that $1 \leq x \leq \sqrt{7}$ are of interest.

Let $H(v; x)$ denote the distance parallel to the H-axis from a point on the v-axis to the line $x = -1$. Then, from Fig. 4,

$$\frac{H(v; x) - H(0; x)}{v} = \frac{1}{x}, \qquad v > 0,$$

or

(1) $$H(v; x) = H(0; x) + v/x, \qquad v \geq 0.$$

Further, by similar triangles,

$$\frac{H(0; x) + 2\sqrt{2}}{x + 1} = \frac{(1 + x^2)^{1/2}}{x},$$

from which

(2) $$H(0: x) = [(1 + x)^2 + (1 + 1/x)^2]^{1/2} - 2\sqrt{2}.$$

Thus, combining (1) and (2), we have

(3) $$H(v; x) = [(1 + x)^2 + (1 + 1/x)^2]^{1/2} - 2\sqrt{2} + v/x, \qquad v \geq 0.$$

Now for any fixed value of v, the largest value of H which will permit the prescribed motion is the minimum value of $H(v; x)$ for $1 \leq x \leq \sqrt{7}$. Where this minimum occurs, we have

$$\frac{\partial H(v; x)}{\partial x} = 0,$$

which yields, after some simplification,

(4) $$v = (x^3 - 1)/(x^2 + 1)^{1/2}.$$

There is no obvious way to eliminate x between (3) and (4). However, from (4),

$$\frac{dv}{dx} = (2x^4 + 3x^2 + x)/(x^2 + 1)^{3/2},$$

which is clearly positive for the values of x of interest. Thus v is a monotonic increasing function of x, and the equations

(5)
$$v(t) = (t^3 - 1)/(t^2 + 1)^{1/2},$$
$$H(t) = [(1 + t)^2 + (1 + 1/t)^2]^{1/2} - 2\sqrt{2} + v(t)/t$$

may be taken as parametric equations of $H(v)$. Since we must have $0 \leq v \leq 1$, the range of values of t is $1 \leq t \leq t_1$, where $t_1 \approx 1.39534$ is the positive root of the equation

$$t^4 - 2t - 1 = 0,$$

obtained by setting $v = 1$ in (4).

The convexity of $H(v)$ may be shown in several ways. Perhaps the simplest is to observe that $H(v)$ is defined by (3) and (4) as the envelope of a one-parameter family of straight lines (really the line $x = -1$ viewed from the counterclockwise rotating coordinate system $\{v, H\}$). Since v increases as x increases, and the slope of the line $x = -1$, which is the tangent to $H(v)$ at v, decreases relative to the coordinate system $\{v, H\}$ as x increases, the slopes of the tangents to $H(v)$ decrease monotonically from 1 at $v = 0$ to $1/t_1$ at $v = 1$ as v increases. Hence $H(v)$ is convex in the system $\{v, H\}$. Further, since the tangent lines are nowhere horizontal or vertical, the curve described by $H(v)$ is a convex function in the system $\{-H, v\}$, which corresponds to that of Fig. 2.

The shape of the arc is thus completely determined, the right edge following from the symmetry conditions. While we have considered motion over only a relatively short range, $1 \leq x \leq t_1$, it is precisely here that the constraints are most difficult to satisfy, and subsequent motion to the right may clearly be accomplished in an infinity of ways.

Values of $H(v)$ were calculated using equations (5) at 100 equally spaced values of t, $1 \leq t \leq t_1$, as well as the length l of the polygon connecting the points $(v(t), H(t))$. The value obtained for l was 1.30435, so that

$$100(2l + 2\sqrt{2})/4\sqrt{2} > 96\%.$$

The arc derived here, as well as the arc with maximum length, are shown in Fig. 5.

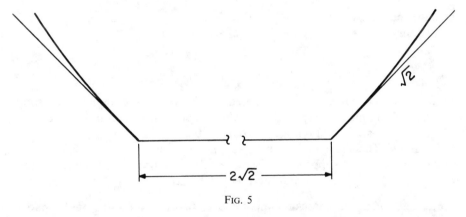

FIG. 5

By way of generalization, we note that the same method may be applied in cases where the two corridors entering the corner are of different widths, and/or the angle at which they meet is different from 90°, to obtain equations analogous to (5) which specify the shape of the edges of an arc, one part of which is a line segment.

Finally, we make the conjecture, which seems reasonable in view of the inherent symmetry of the problem, that the region of maximum area and arc of maximum length will be found among those which satisfy the stated conditions.

A Hyper-Volume

Problem 84-20, by E. E. DOBERKAT (Clarkson College of Technology).
Determine the $(n+1)$-dimensional volume of the set of points defined by

$$A_n := \left\{ (x_0, x_1, \cdots, x_n) \in [0,1]^{n+1} : x_{i-1} + x_i < 1 \text{ for } 1 \leqq i \leqq n \right\}.$$

This problem arose in the analysis of an algorithm.

Comment by RICHARD STANLEY (Massachusetts Institute of Technology).

The same problem was proposed by me as *problem* E2701, American Mathematical Monthly, 85 (1978), p. 197. A solution by I. G. Macdonald and (independently) R. B. Nelsen was published in vol. 86 (1979), p. 396. The solution may be written in the form (setting $V_{n+1} = \text{vol} A_n$)

$$1 + \sum_{n \geq 1} V_n x^n = \tan x + \sec x.$$

It is then well known (e.g., L. Comtet, *Advanced Combinatorics*, pp. 258–9) that $n! \, V_n$ is equal to the number of *alternating permutation* $a_1, a_2, \cdots, a_n$ of $1, 2, \cdots, n$, i.e., $a_1 < a_2 > a_3 < \cdots$.

The following is a sketch of another proof of the proposed problem. Define a set

$$B_n = \left\{ (x_0, x_1, \cdots, x_n) \in [0,1]^{n+1} : x_0 \leqq x_1 \geqq x_2 \leqq x_3 \geqq \cdots \right\}.$$

Define a map $\phi: A_n \to B_n$ by $\phi(x_0, x_1, \cdots, x_n) = (y_0, y_1, \cdots, y_n)$, where

$$y_{2i} = x_{2i},$$
$$y_{2i+1} = \max\{ x_{2i} + x_{2i+1}, x_{2i+1} + x_{2i+2} \}$$

(where if necessary we set $x_{n+1} = 0$). Then ϕ is easily seen to be a continuous, piecewise linear, volume-preserving bijection. Its inverse is given by

$$x_{2i} = y_{2i},$$
$$x_{2i+1} = \min\{ y_{2i+1} - y_{2i}, \, y_{2i+1} - y_{2i+2} \}.$$

Now there is a canonical way of triangulating B_n into $(n+1)$-dimensional simplices, such that each simplex corresponds to an alternating permutation $a_0 < a_1 > a_2 < \cdots$ of $0, 1, \cdots, n$ and has volume $1/(n+1)!$. From this the proof follows.

Postscript: The above proof can be simplified by using Theorem 2.3 of my paper *Two poset polytopes*, (to appear in Discrete & Computational Geometry) to provide a simpler map $A_n \to B_n$, and also this result is generalized.

Editorial note: J. B. LASSERE, in his solution, first uses a result from his paper, *An analytical expression and an algorithm for the volume of a convex polyhedron in R^n*, J. Optim. Theory Appl., 39 (1983), to obtain a recurrence relation for A_n.

Problem 59-2*, *N*-dimensional Volume, By Maurice Eisenstein and M.S. Klamkin.

Determine the volume in *N*-space bounded by the region

$$0 \leq a_1 x_1 + a_2 x_2 + \cdots + a_N x_N < 1, \qquad\qquad a_r \geq 0$$
$$b_r \geq x_r \geq c_r, \qquad\qquad r = 1, 2, \cdots, N.$$

This problem has arisen from the following physical situation: a series-parallel circuit of N resistances is given where each of the resistances R_i are not known exactly but are uniformly distributed in the range $R_i \pm \epsilon_i R_i (\epsilon_i \ll 1)$. We wish to determine the distribution function for the circuit resistance

$$R = F(R_1, R_2, \cdots, R_N).$$

To first order terms

$$\Delta R = \sum_{i=1}^{N} \frac{\partial F}{\partial R_i} dR_i, \qquad |dR_i| \leqq \epsilon_i R_i.$$

The probability that the circuit resistance lies between R and $R + \Delta R$ will be proportional to the volume bounded by the region

$$0 \leqq \sum_{i=1}^{N} \frac{\partial F}{\partial R_i} x_i \leqq \Delta R, \qquad -\epsilon_i R_i \leqq x_i \leqq \epsilon_i R_i.$$

Special cases of the problem arise in the two following examples:

(A) A sequence of independent random variables with a uniform distribution is chosen from the interval $(0, 1)$. The process is continued until the sum of the chosen numbers exceeds L. What is the expected number of such choices? The expected number E will be given by

$$E = 1 + F_1 + F_2 + F_3 + \cdots,$$

where F_i is the probability of failure up to and including the ith trial. Geometrically, F_i will be given by the volume enclosed by

$$x_1 + x_2 + \cdots + x_i \leqq L,$$

$$0 \leqq x_r \leqq 1, \qquad\qquad r = 1, 2, \cdots, i.$$

For the case $L = 1$,

$$E = \sum_{m=0}^{\infty} \frac{1}{m!} = e.$$

(D. J. Newman and M. S. Klamkin, *Expectations for Sums of Powers*, American Mathematical Monthly, January, 1959, pp. 50–51.)

(B) What is the probability that N points picked at random in a plane form a convex polygon?

If we denote the interior angles by θ_i, the probability that the polygon will be convex will be proportional to the volume of the region given by

$$\theta_1 + \theta_2 + \theta_3 + \cdots + \theta_N = (N - 2)\pi,$$

$$0 \leqq \theta_r \leqq \pi$$

The normalizing constant will be given by the volume of the region

$$\theta_1 + \theta_2 + \cdots \theta_N = (N - 2)\pi,$$

$$0 \leqq \theta_r < 2\pi,$$

(we are assuming that the angles are uniformly distributed).
Solution by I. J. Schoenberg, University of Pennsylvania.

Let B_ω denote the volume of the n-dimensional polyhedron

(1) $$0 \leqq x_i \leqq a_i \qquad\qquad i = 1, 2, \cdots, n,$$

(2) $$0 \leqq \lambda_1 x_1 + \lambda_2 x_2 + \cdots + \lambda_n x_n \leqq \omega,$$

where $\lambda_1^2 + \lambda_2^2 + \cdots + \lambda_n^2 = 1$ and $a_r, \lambda_r, \omega \geqq 0$. Also, let $b_i = a_i \lambda_i$. If $F(u)$ is a function of one variable u, we define the operator L_n by

(3) $$L_n F(u) = \sum_{(\alpha_1,\alpha_2,\cdots,\alpha_i)} (-1)^{n-i} F(b_{\alpha_1} + b_{\alpha_2} + \cdots + b_{\alpha_i})$$

where $(\alpha_1, \alpha_2, \cdots, \alpha_i)$ runs through all the 2^n combinations of the n quantities $b_1, b_2, \cdots, b_n$. For example,

$$L_1 F(u) = F(b_1) - F(0),$$

$$L_2 F(u) = F(b_1 + b_2) - F(b_1) - F(b_2) + F(0).$$

It follows that

$$L_n F(u) = L_{n-1} F(u + b_n) - L_{n-1} F(u).$$

If $F(u)$ is sufficiently smooth,

(4) $$\int \cdots \int_B F^{(n)}(\lambda_1 x_1 + \cdots + \lambda_n x_n) \, dx_1 \cdots dx_n = \prod_{r=1}^{n} \lambda_r^{-1} L_n F(u)$$

where B denotes the box defined by (1). To establish (4), we assume it holds for $n = 1, 2, \cdots, n - 1$. Then

$$\int_0^{a_n} dx_n \int \cdots \int_{x_n \text{fixed}} F^{(n)}(\lambda_1 x_1 + \cdots + \lambda_n x_n) \, dx_1 \cdots dx_{n-1}$$

$$= \frac{1}{\lambda_1 \cdots \lambda_{n-1}} \int_0^{a_n} L_{n-1} F'(u + \lambda_n x_n) \, dx_n$$

$$= \frac{1}{\lambda_1 \cdots \lambda_n} \left\{ L_{n-1} F(u + b_n) - L_{n-1} F(u) \right\}$$

$$= \prod \lambda_r^{-1} L_n F(u).$$

Since (4) is valid for $n = 1$, it is valid for all n by induction.

One consequence of (4) is that $L_n F(u) = 0$ whenever $F(u)$ is a polynomial of degree less than n. By a known theorem of Peano, we can write

(5) $$L_n F(u) = \int_{-\infty}^{\infty} \Phi_n(x) F^{(n)}(x) \, dx,$$

where the kernel Φ_n may be described as follows: If we define the truncated power function x_+^k by

(6) $$x_+^k = \begin{cases} x^k & \text{if } x \geqq 0, \\ 0 & \text{if } x < 0, \end{cases} \qquad\qquad k = 0, 1, 2, \cdots$$

then

(7) $$\Phi_n(x) = L_n \frac{(u - x)_+^{n-1}}{(n-1)!},$$

where on the right side x is treated as a parameter and L_n operates on the variable u. Since $\Phi_n(x) = 0$ if $x < 0$ or $x > \sum_1^n b_r = b$,

(8)
$$\int \cdots \int_B F^{(n)}(\lambda_1 x_1 + \cdots + \lambda_n x_n) \, dx_1 \cdots dx_n$$

$$= \prod \lambda_r^{-1} \int_0^b \Phi_n(x) F^{(n)}(x) \, dx.$$

Equation (8) shows that $\prod \lambda_r^{-1} \Phi_n(x)$ is the area of the intersection of the box B

with the hyperplane $\lambda_1 x_1 + \cdots + \lambda_n x_n = x$ (x-fixed). To see this more clearly, we choose $F(x)$ in (8) such that

$$F^{(n)}(x) = \begin{cases} 1 & \text{if} \quad x \leqq \omega, \\ 0 & \text{if} \quad x > \omega; \end{cases}$$

i.e.

$$F(x) = (-1)^n \frac{(\omega - x)_+^n}{n!}$$

Equation (8) now reduces to

(9) $$B_\omega = \prod \lambda_r^{-1} \int_0^w \Phi_n(x)\, dx.$$

Since the operator L_n commutes with the integration,

$$\prod \lambda_r B_\omega = L_n \int_0^\omega \frac{(u - x)_+^{n-1}}{(n - 1)!}\, dx = -L_n \left\{ \frac{(u - x)_+^n}{n!} \right\}_{x=0}^{x=\omega}$$

$$= \frac{1}{n!} L_n u_+^n - \frac{1}{n!} L_n (u - \omega)_+^n.$$

Writing $B = a_1 a_2 \cdots a_n$ and observing that if $\omega \geqq b$ then $L_n(u - \omega)_+^n = 0$ and $B_\omega = B$. We may now write our final result as

(10) $$B_\omega = B - \frac{\prod \lambda_r^{-1}}{n!} L_n (u - \omega)_+^n.$$

As an example, let us consider the hypercube when $a_r = 1$ and $\lambda_r = n^{-1/2}$, $r = 1, 2, \cdots, n$. Then also $b_r = n^{-1/2}$ and (10) gives

(11) $$B_\omega = 1 - \frac{n^{n/2}}{n!} \Delta^n (u - \omega)_+^n \Big|_{u=0},$$

where Δ^n is the ordinary n^{th} order advancing difference operator of step $h = n^{-1/2}$. Now, if $\omega = 0$ then $B_\omega = 0$ and (11) gives

$$\Delta^n u_+^n \Big|_{u=0} = \Delta^n u^n \Big|_{u=0} = n^{-n/2} n!$$

which is a known relation. If $\omega = n^{-1/2}$ then again for the ordinary power function

(12) $$\Delta^n (u - \omega)^n \Big|_{u=0} = n^{-n/2} n!.$$

Passing to the truncated power function only one term of the left side of (12) drops out so that

$$\Delta^n (u - \omega)_+^n \Big|_{u=0} = n^{-n/2} \{n! - 1\}.$$

Finally (11) gives for $\omega = n^{-1/2}$ the value

$$B_{n^{-1/2}} = \frac{1}{n!}$$

which is also known.

The expression (7) shows that $\Phi_n(x)$ is what has been called elsewhere[1] **a** spline curve of degree $n - 1$, i.e. a composite of different polynomials of degree

---

[1] Bull. Amer. Math. Soc., vol. 64 (1958), pp. 352–357.

$n - 1$ having $n - 2$ continuous derivatives while $\Phi_n^{(n-1)}(x)$ has jumps at the "knots" $x = b_{\alpha_i} + \cdots + b_{\alpha_i}$. The Laplace transform of $\Phi_n(x)$, however, has the simple form

$$(13) \qquad \int_{-\infty}^{\infty} e^{-sx} \Phi_n(x)\, dx = \prod_{r=1}^{n} \frac{1 - e^{-sb_r}}{s}.$$

This transform is particularly useful if we wish to discuss the limit properties of the distribution $\Phi_n(x)$ for large n.

Remark: No originality is claimed for the matters presented here. The operator L_n was studied by M. Frechet, T. Popoviciu, and others. Laplace transforms of the kind obtained here were already derived by Laplace himself. Finally, G. Polya's Hungarian doctoral dissertation is devoted to an intensive study of the transforms (13).[2] As a matter of fact, Polya starts from the problem of determining the volume B_ω and also stresses the relations with probability theory which are obtained if n is allowed to tend to infinity.

Also solved by Larry Shepp who shows that the probability that an $n + 1$ sided polygon be convex (the angles of which are assumed uniformly distributed) is

$$P_{n+1} = \frac{2^n - n - 1}{(n-1)^n - \binom{n+1}{1}(n-3)^n + \cdots} .$$
$$+ (-1)^{[n/2]+1} \binom{n+1}{[n/2]-1}(n - 2[n/2] + 1)^n$$

This generalizes the result of H. Demir for the case $n = 3$ (Pi Mu Epsilon Journal, Spring 1958).

Editorial note: E. G. Olds in "A Note on the Convolution of Uniform Distributions," Annals of Mathematical Statistics, v. 23, 1952, pp. 282–285, gives a derivation for the probability density function for a sum of independent rectangularly distributed random variables.

A Volume Problem

Problem 82-4, *by* M. K. Lewis (Memorial University of Newfoundland).

An asymmetrically positioned hole of radius b is drilled at right angles to the axis of a solid right circular cylinder of radius a ($a > b$). If the distance between the axis of the drill and the axis of the cylinder is p, determine the volume of material drilled out.

Solution by J. Boersma, P. J. de Doelder *and* J. K. M. Jansen (Department of Mathematics and Computing Science, Eindhoven University of Technology, Eindhoven, the Netherlands).

Introduce Cartesian coordinates x, y, z, then the cylinder and the drill may be described by

$$(1) \qquad\qquad C_a : x^2 + y^2 = a^2, \qquad C_b : (x - p)^2 + z^2 = b^2.$$

The volume of the intersection of C_a and C_b is denoted by $V(a, b, p)$. Without loss of generality, we may assume $0 \le b \le a, p \ge 0$.

---

[2] Mathematikai es Physikai Lapok, vol. XXII.

From a cross-section of C_a and C_b with the plane $z = 0$, it is readily seen that

(2)
$$V(a, b, p) = 2 \int\!\!\int_G [b^2 - (x - p)^2]^{1/2}\, dx\, dy,$$

where the domain of integration is

(3)
$$G = \{(x, y) \mid x^2 + y^2 \le a^2, |x - p| \le b\}.$$

We now distinguish three cases:

I. $0 \le p \le a - b$. In this case the integral (2) reduces to

(4)
$$V(a, b, p) = 4 \int_{p-b}^{p+b} [(a - x)(b + p - x)(x + b - p)(x + a)]^{1/2}\, dx.$$

The latter integral can be expressed in terms of elliptic integrals by means of Byrd and Friedman [1, form. 254.38]. Omitting the details of the tedious though straightforward calculation, we present the result

(5)
$$\begin{aligned}
V(a, b, p) = {}& 4(a + b + p)^{-1/2}(a + b - p)^{-1/2} \\
&\cdot [\tfrac{1}{6}(p^2 - 2ap - 2a^2 + 2b^2)(a + b + p)(p + a - b)K(k) \\
&+ \tfrac{1}{6}(p^2 + 2a^2 + 2b^2)(a + b + p)(a + b - p)E(k) \\
&+ (a^2 - b^2)p(p + a - b)\Pi(\alpha^2, k)],
\end{aligned}$$

where

(6)
$$k^2 = \frac{4ab}{(a + b + p)(a + b - p)}, \qquad \alpha^2 = \frac{2b}{a + b + p}.$$

In (5), $K(k)$, $E(k)$ and $\Pi(\alpha^2, k)$ denote Legendre's complete elliptic integrals of the first, second and third kinds, respectively, as defined in [1, form. 110.06–08].

II. $a - b \le p \le a + b$. In this case, the integral (2) reduces to

(7)
$$V(a, b, p) = 4 \int_{p-b}^{a} [(b + p - x)(a - x)(x + b - p)(x + a)]^{1/2}\, dx,$$

which can again be evaluated by means of [1, form. 254.38]. Thus we obtain

(8)
$$\begin{aligned}
V(a, b, p) = {}& 2a^{-1/2}b^{-1/2}[-\tfrac{1}{3}a(p + a - b)(2ab - 2b^2 + (3a + b)p)K(k) \\
&+ \tfrac{2}{3}ab(p^2 + 2a^2 + 2b^2)E(k) \\
&+ (a^2 - b^2)p(p + a - b)\Pi(\alpha^2, k)]
\end{aligned}$$

where

(9)
$$k^2 = \frac{(a + b + p)(a + b - p)}{4ab}, \qquad \alpha^2 = \frac{a + b - p}{2a}.$$

III. $p \ge a + b$. In this case the drill C_b is outside the cylinder C_a, hence

(10)
$$V(a, b, p) = 0.$$

The results in (5) and (8) may be combined into a single formula, viz.,

(11)
$$\begin{aligned}
V(a, b, p) = {}& 2^{3/2}[ab(a + b + p)(a + b - p)]^{-1/4} \\
&\cdot \{\tfrac{1}{3}a(a + b + p)(2ab + 2b^2 - (3a - b)p)k^{1/2}K(k) \\
&+ \tfrac{1}{3}(p^2 + 2a^2 + 2b^2)[ab(a + b + p)(a + b - p)]^{1/2}k^{-1/2}[E(k) - K(k)] \\
&+ (a^2 - b^2)p(p + a - b)k^{1/2}\Pi(\alpha^2, k)\}
\end{aligned}$$

where k^2, α^2 are given by (6) in case I, and by (9) in case II. It has been verified that the results (11) for cases I and II are related by the reciprocal modulus transformation (cf. [1, form. 162.02])

(12)
$$\operatorname{Re} k_1^{1/2} K(k_1) = k^{1/2} K(k), \operatorname{Re} k_1^{-1/2}[E(k_1) - K(k_1)] = k^{-1/2}[E(k) - K(k)],$$
$$\operatorname{Re} k_1^{1/2} \Pi(\alpha_1^2, k_1) = k^{1/2} \Pi(\alpha^2, k)$$

where $k_1 = 1/k, \alpha_1^2 = \alpha^2/k^2, \alpha^2 \leq k^2 \leq 1$.

The previous general results simplify in the following special cases:

1. $p = 0$. Then it is found from (5) that

(13) $$V(a, b, 0) = \tfrac{4}{3}(a + b)[(a^2 + b^2)E(k) - (a - b)^2 K(k)],$$

where $k^2 = 4ab/(a + b)^2$. By means of Gauss' transformation [1, form. 164.02] the latter result can be reduced to

(14) $$V(a, b, 0) = \tfrac{8}{3} a\left[(a^2 + b^2)E\left(\frac{b}{a}\right) - (a^2 - b^2)K\left(\frac{b}{a}\right)\right].$$

The same result can also be found directly from (4) with $p = 0$, by use of [1, form. 219.11].

2. $p = a - b$. The expression for $V(a, b, p)$ in this case should follow continuously from (5) and (8). Indeed, by taking the limit for $p \to a - b$ in (5) and (8) we find

(15) $$V(a, b, a - b) = \tfrac{4}{3}(ab)^{1/2}(3a^2 + 3b^2 - 2ab) - 4(a - b)^2(a + b) \log\left(\frac{a^{1/2} + b^{1/2}}{(a - b)^{1/2}}\right),$$

which has been checked by a direct calculation from (4) with $p = a - b$.

3. $p = a + b$. Then C_b is tangent to C_a on the outside, and it is found from (8) and (10) that

(16) $$V(a, b, a + b) = 0.$$

4. $a = b$. Then the result in (8) simplifies to

(17) $$V(a, a, p) = \tfrac{4}{3} a[(p^2 + 4a^2)E(k) - 2p^2 K(k)],$$

where $k^2 = 1 - p^2/4a^2$.

5. $a = b, p = 0$. Then it follows from (14), (15) or (17) that

(18) $$V(a, a, 0) = \frac{16}{3} a^3,$$

which result can also be found immediately from (4).

By means of (5) and (8) we have computed a table of $V(a, b, p)$ to four decimal places, for values of the argument

$$a = 1, \quad b = 0.1(0.1)1, \quad p = 0(0.1)1 + b.$$

The required elliptic integrals were calculated by standard procedures taken from Bulirsch [2].

Boersma and Kamminga [3] calculated the volume of intersection, $V(\rho, \eta)$, of a sphere of unit radius and a cylinder of radius ρ, with η denoting the distance of the center of the sphere and the axis of the cylinder. Their expressions for $V(\rho, \eta)$ in terms of elliptic integrals are quite similar in form to the present expressions (5) and (8). In addition it was shown in [3] that $V(\rho, \eta)$ can be represented by an infinite integral of a product of three

TABLE
$V(1, b, p)$

p \ b	0.1	0.2	0.3	0.4	0.5	0.6	0.7	0.8	0.9	1.0
0	0.0628	0.2501	0.5591	0.9848	1.5200	2.1550	2.8763	3.6664	4.4999	5.3333
0.1	0.0624	0.2488	0.5561	0.9794	1.5113	2.1416	2.8566	3.6373	4.4534	5.2557
0.2	0.0615	0.2449	0.5472	0.9630	1.4846	2.1006	2.7955	3.5436	4.3029	5.0786
0.3	0.0598	0.2383	0.5320	0.9350	1.4384	2.0290	2.6842	3.3665	4.0833	4.8341
0.4	0.0575	0.2287	0.5098	0.8938	1.3698	1.9180	2.5042	3.1363	3.8161	4.5399
0.5	0.0543	0.2157	0.4795	0.8368	1.2707	1.7501	2.2828	2.8716	3.5149	4.2089
0.6	0.0501	0.1985	0.4389	0.7571	1.1253	1.5509	2.0373	2.5843	3.1900	3.8515
0.7	0.0447	0.1758	0.3826	0.6411	0.9577	1.3364	1.7785	2.2836	2.8505	3.4767
0.8	0.0373	0.1436	0.2997	0.5111	0.7826	1.1168	1.5151	1.9777	2.5040	3.0922
0.9	0.0262	0.0947	0.2107	0.3809	0.6097	0.9002	1.2546	1.6736	2.1576	2.7055
1.0	0.0085	0.0472	0.1283	0.2596	0.4467	0.6937	1.0036	1.3780	1.8179	2.3230
1.1	0.0000	0.0130	0.0611	0.1546	0.3004	0.5037	0.7684	1.0969	1.4910	1.9511
1.2		0.0000	0.0163	0.0724	0.1769	0.3362	0.5548	0.8363	1.1828	1.5958
1.3			0.0000	0.0190	0.0821	0.1967	0.3685	0.6016	0.8990	1.2628
1.4				0.0000	0.0214	0.0908	0.2147	0.3982	0.6449	0.9576
1.5					0.0000	0.0235	0.0987	0.2313	0.4258	0.6855
1.6						0.0000	0.0255	0.1060	0.2468	0.4517
1.7							0.0000	0.0273	0.1129	0.2613
1.8								0.0000	0.0290	0.1193
1.9									0.0000	0.0306
2.0										0.0000

Bessel functions, viz.,

$$(19) \qquad V(\rho, \eta) = 2\pi\rho\sqrt{2\pi} \int_0^\infty J_1(\rho t) J_0(\eta t) \frac{J_{3/2}(t)}{t^{3/2}} \, dt.$$

We shall now derive a similar integral representation for the volume of intersection $V(a, b, p)$ of two cylinders. Following [3], we introduce the function $f(\rho)$ defined by

$$(20) \qquad f(\rho) = b \int_0^\infty J_0(\rho t) J_1(bt) \, dt = \begin{cases} 0 & \text{for } \rho > b, \\ \tfrac{1}{2} & \text{for } \rho = b, \\ 1 & \text{for } 0 \le \rho < b \end{cases}$$

(cf. Watson [4, form. 13.42(9)]). Let ρ denote the distance to the axis of the cylinder C_b, then $V(a, b, p)$ may be determined by integrating $f(\rho)$ over the volume of the cylinder C_a. Thus, employing cylindrical coordinates r, ϕ, z, one has

$$(21) \qquad \rho = \sqrt{(r \cos \phi - p)^2 + z^2},$$

$$(22) \qquad \begin{aligned} V(a, b, p) &= 2 \int_0^a r \, dr \int_0^{2\pi} d\phi \int_0^\infty f(\sqrt{(r \cos \phi - p)^2 + z^2}) \, dz \\ &= 2b \int_0^a r \, dr \int_0^{2\pi} d\phi \int_0^\infty dz \int_0^\infty J_0(t\sqrt{(r \cos \phi - p)^2 + z^2}) J_1(bt) \, dt. \end{aligned}$$

By interchanging the order of integration, we may successively perform the integrations with respect to z, ϕ and r by means of [4, form. 13.47(5), 2.3(1), 5.1(1)]:

$$(23) \qquad \int_0^\infty J_0(t\sqrt{\sigma^2 + z^2}) \, dt = \left(\frac{\pi\sigma}{2t}\right)^{1/2} J_{-1/2}(\sigma t) = \frac{\cos(\sigma t)}{t},$$

(24) $$\int_0^{2\pi} \cos\left(t(r\cos\phi - p)\right) d\phi = 2\pi \cos\left(pt\right) J_0(rt),$$

(25) $$\int_0^a r J_0(rt)\, dr = \frac{a}{t} J_1(at).$$

As a result we obtain the integral representation

(26) $$V(a,b,p) = 4\pi a b \int_0^\infty \frac{J_1(at)\,J_1(bt)}{t^2} \cos\left(pt\right) dt,$$

and implicitly it is found that the latter Bessel function integral is given by (5), (8) and (10). Infinite integrals of products of Bessel functions have been studied by Bailey [5], and listed by Luke [6, §13.4], Okui [7]; however none of these references contains the particular integral (26). According to [5, form. (7.1)] the integral (26) can be expressed in terms of an Appell function of the two variables a^2/p^2, b^2/p^2; furthermore, by [5, form (8.3)] the integral vanishes if $p \geq a + b$, in accordance with (10). The special case $p = 0$ of (26) is given in [7, form. I.2.7(1)] and the result agrees with (13); see also [4, form. 13.4(2)] for a representation in terms of a hypergeometric function of argument b^2/a^2, which is equivalent to (14). The special case $a = b$ of (26) can be derived from [7, form. I. 2.3(3)] and the result agrees with (17).

REFERENCES

[1] P. F. Byrd and M. D. Friedman, *Handbook of Elliptic Integrals for Engineers and Physicists*, Springer-Verlag, Berlin, 1954.

[2] R. Bulirsch, *Numerical calculation of elliptic integrals and elliptic functions* I, II, Numer. Math., 7 (1965), pp. 78–90; 353–354.

[3] J. Boersma and W. Kamminga, *Calculation of the volume of intersection of a sphere and a cylinder*, Proc. Kon. Ned. Akad. Wet., A64 (1961), pp. 496–507.

[4] G. N. Watson, *A Treatise on the Theory of Bessel Functions*, Cambridge University Press, Cambridge, 1958.

[5] W. N. Bailey, *Some infinite integrals involving Bessel functions*, Proc. London Math. Soc., (2) 40 (1936), pp. 37–48.

[6] Y. L. Luke, *Integrals of Bessel Functions*, McGraw-Hill, New York, 1962.

[7] S. Okui, *Complete elliptic integrals resulting from infinite integrals of Bessel Functions* I, II, J. Res. Nat. Bur. Stand., 78B (1974), pp. 113–135; 79B (1975), pp. 137–170.

Problem 65-9*, The Shape of Milner's Lamp, by Roland Silver (The MITRE Corporation).

Determine the shape of the lamp in the following quotation [1]: "The same gentleman vouches for Milner's lamp: but this had visible *science* in it. . . . A hollow semi-cylinder, but not with a circular curve, revolved on pivots. The curve was calculated on the law that, whatever quantity of oil might be in the lamp, the position of equilibrium just brought the oil up to the edge of the cylinder, at which a bit of wick was placed. As the wick exhausted the oil, the cylinder slowly revolved about the pivots so as to keep the oil always touching the wick." See Fig. 1.

REFERENCE

[1] Augustus De Morgan, *A Budget of Paradoxes*, vol. 1, Dover, New York, 1954, p. 252.

Solution by J. D. Lawson (University of Waterloo).

We assume for definiteness that the empty lamp has mass m and that the center of gravity is at the lip. We assume further that the distance from pivot to lip is unity, that the lamp is of unit thickness and that the fluid is of unit density.

We consider static equilibrium of the lamp as shown in Fig. 2. The angle θ is measured counterclockwise from the x axis. The equation of the liquid surface is $r = \cos\alpha/\cos\theta$ and the angle β is found from $\rho(\beta + \alpha) = \cos\alpha/\cos\beta$, where $r = \rho(\theta + \alpha)$ is the equation of the lamp. For equilibrium, we have

$$(1) \qquad m \sin\alpha = \int_{\theta=-\alpha}^{\beta} \int_{r=\cos\alpha/\cos\theta}^{\rho(\theta+\alpha)} r \sin\theta \, r dr \, d\theta.$$

Simplifying,

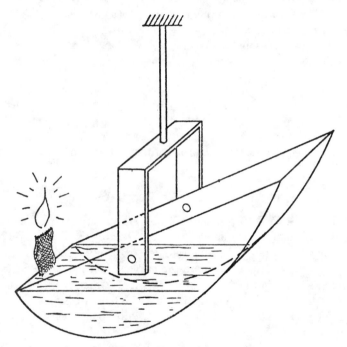

Fig. 1

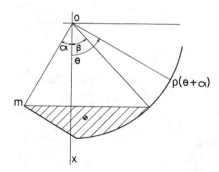

Fig. 2

$$(2) \qquad 6m \sin\alpha = 2 \int_{0}^{\alpha+\beta} \rho^3(\theta) \sin(\theta - \alpha) \, d\theta - \cos^3\alpha/\cos^2\beta + \cos\alpha.$$

This equation defines β as a function of α. Differentiating (2) twice with respect to α, noting that $\rho(\beta + \alpha) = \cos \alpha / \cos \beta$, and adding the result to (2) gives the differential equation

$$(3) \qquad\qquad \frac{d\beta}{d\alpha} = \tan \alpha / \tan \beta.$$

The solution of (3) is

$$(4) \qquad\qquad \cos \beta = c \cos \alpha, \quad c \leqq 1.$$

Thus $\rho(\beta + \alpha) = c^{-1}, \beta + \alpha \geqq \cos^{-1} c$. For $\alpha = 0, \beta = \cos^{-1} c$. The shape of the lamp for $\theta \leqq \cos^{-1} c$ is constrained only by $\rho(\theta) \geqq [\cos \theta]^{-1}$. Choosing $\rho(\theta) = [\cos\theta]^{-1}$, a straight line, for $\theta \leqq \cos^{-1} c$ ensures that the lamp will empty entirely, and is thus a reasonable choice.

Substituting $\rho(\theta) = [\cos \theta]^{-1}, \theta \leqq \cos^{-1} c$, and $\rho(\theta) = c^{-1}, \theta \geqq \cos^{-1} c$, in (2) gives

$$m = (1 - c^2)^{3/2}/3c^3.$$

The shape is thus a straight line segment plus a circular arc, the length of the line segment depending upon the mass of the empty lamp.

Designing a Three-Edged Reamer

Problem 80-17, *by* B. C. RENNIE (James Cook University, N.Q., Australia).

A simple closed plane curve and a triangle ABC (regarded as a rigid body movable in the plane) have the property that the triangle can be moved continuously so that each vertex moves monotonically once round the curve. Must the curve be a circle?

The problem arises in the design of a "reamer," which is a tool used by a fitter to finish-cut a hole to an accurate diameter; it may be fixed or adjustable. An adjustable reamer usually has six straight cutting edges equally spaced parallel to the axis of the tool. In reaming a hole it is not enough to enlarge the diameter when it is too small; we also require to ensure that the hole is circular. A two-edged reamer would be of no use since there are noncircular plane curves of constant diameter. Designing a three-edged reamer leads one to the above question in plane geometry.

Solution by M. GOLDBERG (Washington, DC).

The curve does not have to be a circle. It can be a square with rounded corners, as shown in the following example.

The Reuleaux triangle, made by three circular arcs, can be rotated within a square while keeping contact with all the sides of the square. Each of the corners of the Reuleaux triangle traces the square with rounded corners. This is the basis for a commercial drill made by Watts Brothers Tool Works of Wilmerding, PA. It is pictured and described by Martin Gardner in his column in Scientific American, February 1963, p. 150, and it is mentioned in my paper *Rotors in polygons and polyhedra*, Math. Comp., 14 (1960), pp. 229–239.

There are many other possibilities. My papers describe various methods of obtaining noncircular ovals which can rotate within regular polygons,. If a rotor is held fixed while the polygon is rotated about it, then all the vertices of the polygon trace another noncircular curve (see Figures). Therefore, if this new curve is fixed, then the

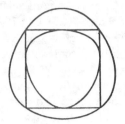

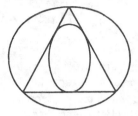

(a) *Triangle rotor in an oval.*   (b) *Square rotor in an oval.*

regular polygon can be rotated within it while all the vertices trace the curve. Any three of the vertices can be the vertices of a triangle rotating within the noncircular oval. Another application of a triangular rotor in an oval is the rotary engine designed by Felix Wankel (presently being used in the sports car, Mazda RX7).

A Covering Problem

Problem 83-4*, by J. VIJAY (University of Florida).

Let M be a set of m fixed and distinct points in the plane. Define a hoop of radius $r > 0$ to be any maximal subset of M that is simultaneously coverable by a circle of radius r. Obviously, the number of such hoops depends on both r and the locations of the m points. For example, if r is very small then each point is a hoop, and if r is very large then the only hoop is M itself.

It is conjectured that for a given m, the number of hoops cannot exceed $3m$ irrespective of the locations of the m points and the value of r. This bound is attained in the limit when $m \to \infty$ and the points are in an equilateral triangular lattice with $2r$ being the distance between any point and any of its six closest neighbors.

Solution by O. P. LOSSERS (Eindhoven University of Technology, Eindhoven, the Netherlands).

The conjecture is false!

Consider in the equilateral triangular lattice the following configuration.

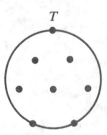

It is a matter of straightforward checking that every hoop has the above form (apart from a rotation over a multiple of $\pi/3$). Since every hoop has a unique "top" T and any point is the top of 6 hoops, it follows that $6m(1 + \sigma(1))$ is the number of hoops for $m \to \infty$.

Subsequent comment by A. J. Schwenk (Office of Naval Research, Arlington, VA).

Vijay conjectured that the number of hoops cannot exceed $3m$, but O. P. Lossers (Eindhoven) constructed an example with approximately $6m$ hoops. We shall prove that for any constant c, it is possible to have more than cm hoops.

Let M be a large square subset of the lattice of points in the plane with integer coordinates centered at the origin. Select n so that $2^{n+3} > c$ and select n distinct primes each congruent to $1 \bmod 16$. Define $R = \sqrt{5 p_1 p_2 \cdots p_n}$. The radius r used to define the hoops of M will be just slightly smaller than R.

Now for the circle C centered at the origin with radius R, let $I(C)$ denote the set of interior lattice points, and let $B(C)$ denote the set of lattice points on C itself. We wish to determine the order of $B(C)$, that is, how many integer solutions does $x^2 + y^2 = R^2$ have? It is well known that 5 and each prime p_i has a unique representation as a sum of the two squares $a^2 + b^2$ with $a > b > 0$. Then a trick that goes back to Euler allows us to construct 2^n solutions to $x^2 + y^2 = R^2$ with $x > y > 0$. But each solutions gives rise to 8 lattice points on C, namely $(\pm x, \pm y)$ and $(\pm y, \pm x)$. Thus, there are $|B(C)| = 2^{n+3}$ lattice points on C.

In order to construct hoops, we shall move the center of the circle a small distance $\varepsilon = 1/(4R)$ in some direction θ. We claim that every point of $I(C)$ remains inside the new circle, since the shortest distance from any interior lattice point to the original circle is $R - \sqrt{R^2 - 1} \approx 1/(2R) = 2\varepsilon$. As long as θ is not perpendicular to one of the 2^{n+2} diameters joining opposite lattice points in $B(C)$, the new circle contains $I(C)$ and exactly half of $B(C)$. (If we happened to choose θ perpendicular to one of these diameters the new circle contains $I(C)$ plus one less than half of $B(C)$. But perturbing θ by a trifle will pick up one of the diametric points, so such a subset is not maximal, i.e., cannot be a hoop.) Thus, there are 2^{n+3} sets formed by taking half of $B(C)$ and all of $I(C)$ that we wish to qualify as hoops. If $r = R$ were unchanged, all these candidates would be subsets of $I(C) \cup B(C)$, so none would be hoops. But we select one suitable θ to define a representative circle for each of the 2^{n+3} candidate sets. Since each candidate set lies in the interior of its representative circle, we may define $\delta > 0$ to be the minimum distance from any lattice point inside any candidate set to its own representative circle. We now set $r = R - \delta$ to reduce the radius just a bit. Each candidate set remains within its own representative circle, but $I(C) \cup B(C)$ is not contained in any circle of radius r.

It remains to show that each candidate set is maximal. We shall accomplish this by showing that any attempt to move the representative circle far enough to pick up an additional lattice point will necessarily force us to lose one of the original interior points. Since $I(C)$ is already a subset of each candidate set, we cannot pick up an extra point from $I(C)$. Since we already have half of $B(C)$ in each candidate set, adding a point from $B(C)$ would imply that we have included a diametric pair at distance $2R > 2r$, so $B(C)$ cannot provide any additional point. Thus, any point to be added must come from lattice points exterior to the original circle C. Now R was carefully defined so that $R^2 + 1$ and $R^2 + 2$ are congruent to 6 and $7 \bmod 8$, and so cannot be written as a sum of two squares. Thus, the nearest exterior lattice point lies on a circle concentric with C and having radius $\sqrt{R^2 + 3} \approx R + 3/(2R) = R + 6\varepsilon$. Thus, to add a new point to the candidate set the center must be moved at least 6ε from the original center of C. But since $R^2 \equiv 5 \bmod 16$, $R^2 - 1 = 4(4k+1)$ which implies that there are integer solutions to $x^2 + y^2 = R^2 - 1$. In fact, at the very least, each 45° sector of the form $\pi i/4 < \phi < \pi(i+1)/4$ must contain an integer solution. Consequently, the sector most nearly opposite the direction of θ must include a lattice point P. Now the distance from P to C was already found to be about 2ε. Of course, when we slide the circle in the direction of θ, the motion may not be along the radius through P, but its direction is guaranteed to form an angle $\alpha < 45°$. Elementary geometry confirms that P will be excluded from the circle after a translation of $2\varepsilon\sec\alpha<2\sqrt{2}\,\varepsilon$. Since we must translate by at least 6ε to pick up an exterior point, we will necessarily lose one of the interior points first. We conclude that each of the 2^{n+3} candidates is indeed a hoop.

We have found 2^{n+3} hoops whose centers lie a distance $\varepsilon=1/(4R)$ away from the origin. Clearly the same argument applies to hoops with centers near each lattice point, so long as the circles do not reach the boundary of our large square. Near the boundary it will be very difficult to identify hoops. Still, we have shown that the square of lattice points $M=\{(a,b)\mid |a|\leq K, |b|\leq K\}$ has at least $2^{n+3}(2K+1-2R)^2$ distinct hoops and $m=(2K+1)^2$ points. Thus as K and hence m increase without bound the number of hoops is of order $2^{n+3}m>cm$.

Problem 70-25 corrected*, *On the Diagonals of a Polygon*, by M. S. KLAMKIN (Ford Motor Company).

If D_i, D_e, D_m denote the number of diagonals which except for their end-points lie in the interior, the exterior, or neither in the interior or exterior, respectively, of a simple n-gon P, then

$$D_i + D_e + D_m = \binom{n}{2} - n.$$

It is obvious and well known that max $D_i = \binom{n}{2} - n$ occurring when P is convex.

The determination of max D_e is given by Problem E2214 (Amer. Math. Monthly, 77 (1970), p. 79). To complete this classification, determine max D_m. Also, consider the corresponding problem for higher-dimensional polytopes.

Problem 73-1*, *On the Characterization of a Curve*, by FRED G. STOCKTON (Houston, Texas).

An idealized vehicle sets out from point P to go to point Q over a path PQ. The vehicle is a line segment of unit length and it must move so that both its endpoints are always on PQ. The path is continuous between P and Q; outside P and Q, it extends indefinitely along the straight line determined by P and Q.
   (a) Is any additional characterization of the path PQ required to insure that the vehicle can reach point Q?
   (b) Give an algorithm for movements of the vehicle.
   (c) Replace the vehicle with an articulated train of n equal segments. The end points of each segment must remain on PQ. Is there a path which is feasible for the original vehicle, but infeasible for the train?
   (d) Same as (c) above, but with unequal segments.

Editorial note. The proposer notes that initial conception of the problem occurred in connection with a problem assigned in a mathematics course at Stanford University in 1957–58 which had to do with the self-intersection of a Jordan domain after repeated translations. The corresponding problem of part (a) for closed curves has just been treated by R. Fenn [*Sliding a Chord and the Width and Breadth of a Closed Curve*, J. London Math. Soc., 5 (1972), pp. 39–47].

For part (a), it is also of interest to determine the maximum of the sum of the distances moved by both ends of the vehicle in traversing a feasible curve PQ of given length L and given span $S = \overline{PQ}$. [M.S.K.]

Problem 73-18*, A Minimum Number of Normals, by S. STEIN (University of California, Davis).

Let K be a plane convex region. It is known that from its centroid there are at least four normals to the border of K. The same result holds for the centroid of the border [1]. The analogous questions in dimension three are open. N.C. Kuiper [2] has shown that if K is centrally-symmetric, then from its center there are at least six normals to the surface. The ellipsoid with unequal axes shows that "six" is best possible. However, Gregory Yob and Peter Cooper have, independently, constructed homogeneous solids from whose center of gravity there are only four normals. Cooper simply cut off a section of a right-circular broom stick by two oblique planes, one cut being the mirror image of the other across a plane perpendicular to the axis. Is there a homogeneous solid from whose centroid there are fewer than four normals to the surface?

REFERENCES

[1] G. D. CHAKERIAN AND S. K. STEIN, *On the centroid of a homogeneous wire*, Michigan Math. J., 11 (1964), pp. 189–192.
[2] N. C. KUIPER, *Double normals of a convex body*, Israel J. Math., 2 (1964), pp. 71–80.
[3] S. K. STEIN, *Higher moments of plane convex sets*, Elem. Math., 23/5 (1968), pp. 108–110.

Editorial note. A related result and conjecture are: The circle is the only oval with a unique "four-normal" point. It is conjectured that the sphere is the only ovaloid with a unique "six-normal" point. [M.S.K.]

Note added in proof. In a note, *Existence of four concurrent normals to a smooth closed hypersurface of E^n*, to appear in Amer. Math. Monthly, B. Wagner shows that every neighborhood of a focal point of a smooth closed hypersurface F of E^n contains a four-normals-point of F if F is an immersed $(n-1)$-sphere and also that the $(n-1)$-sphere is the only hypersurface of E^n with even Euler characteristic which has only one four-normals-point. The proof uses Morse theory. It would be of interest to obtain more elementary proofs, at least in E^3.

A Triangle Inequality

Problem 79-19, by M. S. KLAMKIN (University of Alberta).

If a_1, a_2, a_3 and m_1, m_2, m_3 denote the sides and corresponding medians of a triangle, respectively, prove that

$$(1) \qquad (a_1^2 + a_2^2 + a_3^2)(a_1 m_1 + a_2 m_2 + a_3 m_3) \geq 4 m_1 m_2 m_3 (a_1 + a_2 + a_3).$$

Solution by the proposer.

To prove (1) as well as to give a dual inequality, we will use the known duality theorem that

(2) $$F(a_1, a_2, a_3, m_1, m_2, m_3) \geqq 0 \Leftrightarrow F\left(m_1, m_2, m_3, \frac{3a_1}{4}, \frac{3a_2}{4}, \frac{3a_3}{4}\right) \geqq 0.$$

This follows immediately from the fact that the three medians m_1, m_2, m_3 of any triangle are themselves sides of a triangle with respective medians, $3a_1/4, 3a_2/4, 3a_3/4$. For a more general duality result, see [1].

We will now prove successively that

(3) $$\sum a_1^2 \sum a_1 m_1 \geqq \sum a_1 \sum a_1^2 m_1 \geqq 4m_1 m_2 m_3 \sum a_1,$$

where the summations are to be understood as cyclic sums over the indices 1, 2, 3. The right-hand inequality of (3) follows immediately from the known inequality

(4) $$a_1^2 m_1 + a_2^2 m_2 + a_3^2 m_3 \geqq 4m_1 m_2 m_3,$$

which was obtained by Bager [2] by first establishing its dual, i.e.,

(5) $$4\{a_1 m_1^2 + a_2 m_2^2 + a_3 m_3^2\} \geqq 9a_1 a_2 a_3.$$

However, if we make the substitutions

$$4m_1^2 = 2a_2^2 + 2a_3^2 - a_1^2, \quad \text{etc.,}$$

we obtain

(5′) $$2 \sum a_1 \sum a_1^2 \geqq 3\{3a_1 a_2 a_3 + \sum a_1^3\},$$

which was obtained by Colins in 1870 [3, p. 13].

To establish the left-hand inequality of (3), we expand out and collect terms to give

(6) $$\sum a_1 a_2 (a_1 - a_2)(m_2 - m_1) \geqq 0.$$

The latter inequality is valid, since if $a_1 \geqq a_2 \geqq a_3$, then $m_1 \leqq m_2 \leqq m_3$. There is equality in (1) if, and only if, the triangle is equilateral. However, if we allow degenerate triangles, there is another case of equality:

$$(a_1, a_2, a_3) = (2, 2, 0), \qquad (m_1, m_2, m_3) = (1, 1, 2).$$

The given inequality (1) has the following nice geometric interpretation. Let the medians of triangle $A_1 A_2 A_3$ be extended to intersect the circumcircle again in points A_1', A_2', A_3'. Then,

(7) $$\text{Perimeter } (A_1', A_2', A_3') \geqq \text{Perimeter } (A_1 A_2 A_3).$$

For a related result concerning the repetition of the above operation, see [4]. Another related result due to Janić [3, p. 90] is that

$$\text{Area } (A_1' A_2' A_3') \geqq \text{Area } (A_1 A_2 A_3).$$

The latter result was extended to simplexes by the author (this Journal, 21 (1979), pp. 569–570).

From (2), the dual of (1) is

(8) $$4(m_1^2 + m_2^2 + m_3^2)(a_1 m_1 + a_2 m_2 + a_3 m_3) \geqq 9a_1 a_2 a_3 (m_1 + m_2 + m_3).$$

There is equality if $a_1 = a_2 = a_3$ or $2a_1 = 2a_2 = a_3$.

The author has also shown [3, p. 90] that if the angle bisectors of a triangle $A_1 A_2 A_3$ are extended to intersect the circumcircle again in points A_1', A_2', A_3', then

$$\text{Area } (A_1' A_2' A_3') \geqq \text{Area } (A_1 A_2 A_3).$$

We now also establish that

(9)   Perimeter $(A_1' A_2' A_3') \geqq$ Perimeter $(A_1 A_2 A_3)$.

In what follows O, I, R, r will denote the circumcenter, incenter, circumradius and inradius, respectively, of $A_1 A_2 A_3$. Let, $R_i = A_i I$, $R_i' = A_i' I$; then from the power of a point property of circles,

$$R_i R_i' = R^2 - OI^2 = 2Rr \quad (i = 1, 2, 3).$$

Then, by way of the law of cosines,

$$a_i' = \left\{ \frac{2Rr}{R_1 R_2 R_3} \right\} a_i R_i,$$

(for more extensive properties of this transformation and related ones, see [5]). Inequality (9) is now equivalent to

(9') $\dfrac{2Rr}{R_1 R_2 R_3} \{a_1 R_1 + a_2 R_2 + a_3 R_3\} \geqq a_1 + a_2 + a_3.$

From the known relations,

$$r = R_i \sin A_i/2, \qquad a_i = 2R \sin A_i, \qquad r = 4R \textstyle\prod \sin A_i/2,$$

we can transform (9') into the trigonometric form

$$\cos A_1/2 + \cos A_2/2 + \cos A_3/2 \geqq \sin A_1 + \sin A_2 + \sin A_3,$$

or, equivalently,

(9'') $\Sigma \cos A_i/2 \geq 4\Pi \cos A_i/2.$

Using the Arithmetic–Geometric Mean Inequality, it suffices to prove that

(10) $(\sqrt{3}/2)^3 \geqq \prod \cos A_i/2.$

Although (10) is a known inequality [3, p. 26], another proof follows immediately from the concavity of $\log \cos x/2$.

Finally, it would be of interest to extend (7) and (9) to simplexes for which we consider total edge length as the perimeter.

REFERENCES

[1] M. S. KLAMKIN, *Solution to Aufgabe 677*, El. der Math., 28 (1973), p. 130.

[2] A. BAGERS, *Some inequalities for the medians of a triangle*, Univ. Beograd Publ. Elektrotehn. Fak. Ser. Mat. Fiz. No. 341 (1971), pp. 37–40.

[3] O. BOTTEMA, R. Ž. DJORDJEVIĆ, R. R. JANIĆ, D. S. MITRINOVIĆ, AND P. M. VASIĆ, *Geometric Inequalities*, Wolters-Noordhoff, Groningen, 1969.

[4] J. GARFUNKLE, A. BROUSSEAU, AND E. ITORS, *Problem 913*, Math. Mag. 48 (1975), pp. 246–247.

[5] M. S. KLAMKIN, *Triangle inequalities via transforms*, Notices of Amer. Math. Soc., Jan., 1972, pp. A-103, 104.

Postscript For an excellent and very comprehensive reference to geometric inequalities, see D. S. Mitrinović, J. E. Pečarić and V. Volenec, *Recent Advances in Geometric Inequalities*, Kluwer Academic, Dordrecht, 1989.

Problem 77-9, *A Triangle Inequality*, by I. J. SCHOENBERG (University of Wisconsin).

Let $P_i = (x_i, y_i)$, $i = 1, 2, 3$, $x_1 < x_2 < x_3$, be points in the Cartesian (x, y) plane and let R denote the radius of the circumcircle Γ of the triangle $P_1P_2P_3$ ($R = \infty$ of the triangle is degenerate). Show that

(1)
$$\frac{1}{R} < 2 \left| \frac{y_1}{(x_1 - x_2)(x_1 - x_3)} + \frac{y_2}{(x_2 - x_3)(x_2 - x_1)} + \frac{y_3}{(x_3 - x_1)(x_2 - x_2)} \right|$$

unless both sides vanish and that 2 is the best constant in (1).

Solution by W. J. BLUNDON (Memorial University of Newfoundland, Canada).
First note that

$$|P_1P_2| \geq x_2 - x_1 > 0, \quad |P_2P_3| \geq x_3 - x_2 > 0, \quad |P_3P_1| \geq x_3 - x_1 > 0,$$

and strict inequality holds in at least two of these (excluding the straightforward case $R = \infty$). The required inequality now follows immediately by means of the well known formulae for the area Δ of triangle $P_1P_2P_3$:

$$2\Delta = |y_1(x_3 - x_2) + y_2(x_1 - x_3) + y_3(x_2 - x_1)|,$$
$$4\Delta R = |P_1P_2| \cdot |P_2P_3| \cdot |P_3P_1|.$$

That 2 is the best constant in (1) follows by considering the special case $P_1(-2k, 0)$, $P_2(0, 2k^2)$, $P_3(2k, 0)$ where k is arbitrarily small. Here, $R = 1 + k^2$ and the right hand side of (1) reduces to 1.

Editorial note. The proposer notes that the problem had arisen in determining conditions that would insure that an $F(x) \in C[0, 1]$ is a linear function. He then lets $F(x_1, x_2, x_3)$ be the 2nd order divided difference of $F(x)$ where $0 \leq x_1 < x_2 < x_3 \leq 1$. If $F(x_1, x_2, x_3) \to 0$ whenever x_1, x_2, x_3 converge to a common limit l in $[0, 1]$, for all l, then inequality (1) shows that the plane arc $y = F(x)$ has Menger curvature zero at all of its points. It then follows by a theorem of Menger that the arc is straight. Also, the proposer in his proof establishes the equivalent interesting result:

If P_1, P_2, P_3 are three distinct points on a parabola, then their circumcircle is larger than the circle of curvature of the parabola at its vertex. [M.S.K.]

Problem 77-10. *A Two Point Triangle Inequality*, by M. S. KLAMKIN (University of Alberta).

Let P and P' denote two arbitrary points and let $A_1A_2A_3$ denote an arbitrary triangle of sides a_1, a_2, a_3. If $R_i = PA_i$ and $R_i' = P'A_i$, prove that

(1)
$$a_1 R_1 R_1' + a_2 R_2 R_2' + a_3 R_3 R_3' \geq a_1 a_2 a_3$$

and determine the conditions for equality. It is to be noted that when P' coincides with P, we obtain a known polar moment of inertia inequality.

Solution by the proposer.

We start with the known identity for five arbitrary complex numbers

(2)
$$\frac{z - z_1}{z_1 - z_2} \cdot \frac{z' - z_1}{z_1 - z_3} + \frac{z - z_2}{z_2 - z_3} \cdot \frac{z' - z_2}{z_2 - z_1} + \frac{z - z_3}{z_3 - z_1} \cdot \frac{z' - z_3}{z_3 - z_2} = 1.$$

It now follows by the triangle inequality that

(3)
$$|z-z_1||z'-z_1||z_2-z_3|+|z-z_2||z'-z_2||z_3-z_1|+|z-z_3||z'-z_3||z_1-z_2|$$

with equality if and only if each of the three terms on the left hand side of (2) are real. Let z_1, z_2, z_3, z, z' be the complex numbers corresponding to the points A_1, A_2, A_3, P, P', respectively, then (3) is equivalent to (1). The equality condition requires that

$$\angle A_2 A_1 P = \angle P' A_1 A_3, \quad \angle A_1 A_2 P = \angle P' A_2 A_3, \quad \angle A_2 A_3 P = \angle P' A_3 A_1.$$

Thus, the two points P and P' must be isogonal conjugates with respect to the given triangle. If P is the center of the inscribed circle of the triangle, then P' coincides with P. If P is the center of the circumcircle, then P' is the orthocenter.

It is a known result [1] that if P and P' are foci of an ellipse inscribed in the triangle $A_1 A_2 A_3$, then we have the equality condition of inequality (1). The proof given was geometric. Fujiwara [2] using identity (2) easily establishes the equality condition of inequality (1) that P and P' must be isogonal conjugates. The ellipse result of [1] also follows easily from (2) by using the known general angle property of an ellipse that two tangents to an ellipse from a given point make equal angles with the focal radii to the given point.

Many other triangle inequalities can be obtained in similar fashion. These are given in a paper *Triangle inequalities from the triangle inequality*, Elem. Math. 34 (1979) 49–55.

REFERENCES

[1] W. J. MILLER, *Mathematical Questions and Solutions from the Educational Times*, vol. 7, F. Hodgson, London, 1876, Problem 210, p. 43.
[2] M. FUJIWARA, *On the deduction of geometrical theorems from algebraic identities*, Tôhoku Math. J., 4 (1913–1914), pp. 75–77.

A Maximum, Minimum Problem

Problem 82-5, by T. SEKIGUCHI (University of Arkansas).

In Euclidean space E_3, with origin at O and coordinates (x_1, x_2, x_3), let $\overrightarrow{OX}$ be a ray in the closed first orthant, and α_i be the angles between the positive x_i axes and $\overrightarrow{OX}, i = 1, 2, 3.$

Determine the maximum and minimum values of $\alpha_1 + \alpha_2 + \alpha_3$.

Solution by O. P. LOSSERS (Eindhoven University of Technology, Eindhoven, the Netherlands).

Without loss of generality we assume (x_1, x_2, x_3) to be on the unit sphere in the closed first octant, denoted by

$$S: \begin{cases} x_1^2 + x_2^2 + x_3^2 = 1, \\ x_1 \geq 0, x_2 \geq 0, x_3 \geq 0. \end{cases}$$

The function to be considered then reads

$$F(x_1, x_2, x_3) := \arccos x_1 + \arccos x_2 + \arccos x_3.$$

(i) At the boundary points of S, we have $F(x_1, x_2, x_3) = \pi$ since

$$\arccos 0 + \arccos x + \arccos \sqrt{1 - x^2} = \pi.$$

(ii) In the interior of S, we find stationary points for F in those points where grad $F = \lambda$ grad $(x_1^2 + x_2^2 + x_3^2 - 1)$, i.e.,

$$x_1\sqrt{1 - x_1^2} = x_2\sqrt{1 - x_2^2} = x_3\sqrt{1 - x_3^2}.$$

Squaring, we obtain

$$x_1^2 - x_2^2 - x_1^4 + x_2^4 = 0, \quad \text{etc.}$$

$$(x_1 - x_2)(x_1 + x_2)(1 - x_1^2 - x_2^2) = 0, \quad \text{etc.}$$

In this product for the interior points only the first factor can be zero, so $x_1 = x_2 = x_3 = 3^{-1/2}$.

At this point, F attains the value $3 \arccos 3^{-1/2}$.

Hence

$$3 \arccos 3^{-1/2} \leq F(x_1, x_2, x_3) \leq \pi.$$

An Extremal Sphere

Problem 80-14, *by* D. SINGMASTER (The Open University, Milton Keynes, England).
Determine the unit sphere of maximum volume (with respect to dimension).

The solutions of L. N. HOWARD (Massachusetts Institute of Technology) *and* M. TORTORELLA (Bell Laboratories, Holmdel, NJ) were essentially the same, as follows:

As is well known, the volume V_n of a unit sphere in E^n is

$$V_n = \frac{\pi^{n/2}}{\Gamma\left(1 + \frac{n}{2}\right)},$$

which gives

$$V_1 = 2, \qquad\qquad V_4 = \frac{\pi^2}{2} = 4.93480,$$

$$V_2 = \pi, \qquad\qquad V_5 = \frac{8\pi^2}{15} = 5.26378901,$$

$$V_3 = \frac{4\pi}{3} = 4.18879, \qquad V_6 = \frac{\pi^3}{6} = 5.16771.$$

Since $V_{n+2} = \pi V_n/(1 + n/2)$, it follows that $V_{n+2} < V_n$ for $n \geq 5$ and $V_{n+2} > V_n$ for $n \leq 4$. Thus, $V_5 = 8\pi^2/15$ is the maximum volume.

Additionally, since the surface S_n of the unit sphere in E^n is given by $S_n = nV_n$, we obtain $S_{n+2}/S_n = 2\pi/n$. Thus, S_n is a maximum for $n = 7$ and is given by

$$S_7 = 16\pi^3/15 = 33.07336179.$$

Editorial note. R. P. BOAS (Northwestern University) notes that he recalls seeing a graph of V_n vs. n, showing the maximum, in a book when he was an undergraduate, with the comment "This is one of the most remarkable phenomena in the wonderland of higher space". Unfortunately, he does not recall the book. R. T. SHIELD (University of Illinois) notes that the problem is solved in the paper by D. R. Barr, *A packing problem revisited*, (Operations Research, 18 (1970), pp. 746–747). Also, it is of interest to note that the same problem had arisen in a recent report of M. L. Glasser and J. Boersma, *Exchange energy of an electron gas of arbitrary dimension*. The maximum volume was determined to occur for $n = 5.26$ (assuming n continuous), which was also calculated by the proposer.

A Volume Inequality for a Pair of Associated Simplexes

Problem 78-20, *by* M. S. KLAMKIN (University of Alberta).

The lines joining the vertices $\{A_i\}$, $i = 0, 1, \cdots, n$ of a simplex S to its centroid G meet the circumsphere of S again in points $\{A'_i\}$, $i = 0, 1, \cdots, n$. Prove that the volume of simplex S' with vertices A'_i is $\geq$ the volume of S.

Solution by the proposer.

Let V_i (V'_i) denote the volume of the simplex whose vertices are G and those of the face F_i (F'_i) of S (S') opposite A_i (A'_i). It follows that

$$\frac{V'_i}{V_i} = \frac{GA_i}{GA'_i} \prod \frac{GA'_j}{GA_j}$$

where in the product here and subsequently (also sums), the index runs from 0 to n. By the power of a point theorem for spheres,

$$GA_i \cdot GA'_i = R^2 - OG^2 \equiv k$$

where R, O are the circumradius and circumcenter, respectively, of S. Also, $V_i = V/(n+1)$ where $V =$ volume of S. Then,

$$\frac{V'_i}{V} = \frac{k^n GA_i^2}{\prod GA_j^2}$$

and

$$V' = \sum V'_i = \frac{k^n V}{n+1} \frac{\sum GA_j^2}{\prod GA_j^2}.$$

We now want to show that

$$\frac{k^n}{n+1} \sum GA_j^2 \geq \prod GA_j^2.$$

By the arithmetic-geometric mean inequality, it suffices to establish the stronger inequality

$$\frac{k^n}{n+1} \sum GA_j^2 \geq \left\{ \frac{\sum GA_j^2}{n+1} \right\}^{n+1}.$$

Actually, the latter is an equality since it is known that

$$k = \sum GA_i^2/(n+1)$$

and which follows from (where $\mathbf{A}_i = \overrightarrow{OA_i}$)

$$\sum_i \sum_j A_i A_j^2 = \sum_i \sum_j (\mathbf{A}_i - \mathbf{A}_j) \cdot (\mathbf{A}_i - \mathbf{A}_j)$$

$$=2(n+1)^2R^2-2\sum_i \mathbf{A}_i \cdot \sum_j \mathbf{A}_j$$

$$=2(n+1)^2(R^2-OG^2)$$

$$=\sum_i \sum_j \{(\mathbf{A}_i-\mathbf{G})-(\mathbf{A}_j-\mathbf{G})\}^2$$

$$=2(n+1)\sum GA_i^2.$$

$V'=V$ iff $GA_i=$ constant or equivalently O and G coincide. For $n=2$, this implies the triangle is equilateral. This special case is known and is ascribed to Janic [1]. For $n=3$, the tetrahedron must be isosceles (opposite sides are congruent).

Also solved by L. GERBER (St. John's University), who additionally poses the problem of determining the limit of the volume (for $n\geq 3$) if the process is repeated indefinitely.

REFERENCE

[1] O. BOTTEMA, R. Z. DJORDJEVIĆ, R. R. JANIĆ, D. S. MITRINOVIĆ AND P. M. VASIĆ, *Geometric Inequalities*, Walters-Noordhoff, Groningen, 1969.

Inequality for a Simplex

Problem 85-26, *by* M. S. KLAMKIN (University of Alberta).

If O, I, R and r denote the circumcenter, incenter, circumradius, and inradius, respectively, of an n-dimensional simplex, prove that

$$R^2 \geq n^2r^2 + \overline{OI}^2$$

and with equality iff the simplex is regular.

Solution by the proposer.

First we derive a more general inequality. Let F_i denote the content of the $(n-1)$-dimensional face opposite vertex A_i of a given simplex, $i=0,1,\cdots,n$. If P is an arbitrary point, then by Cauchy's inequality,

(1) $$\sum F_i \sum F_i \overline{PA_i}^2 \geq \left\{\sum F_i \overline{PA_i}\right\}^2$$

and with equality if $PA_0=PA_1=\cdots=PA_n$, i.e., if P coincides with 0. Letting $\mathbf{P}$, $\mathbf{A}_i$, $\mathbf{I}$ denote vectors from 0 to P, R_i, and I, respectively, we get

$$\sum F_i \overline{PA_i}^2 = \sum F_i(\mathbf{P}-\mathbf{A}_i)\cdot(\mathbf{P}-\mathbf{A}_i) = \sum F_i\left(R^2+P^2-2\mathbf{P}\cdot\mathbf{A}_i\right)$$

$$=F\left(R^2+P^2-2\mathbf{P}\cdot\sum F_i\mathbf{A}_i/\sum F_i\right)$$

$$=F\left(R^2+\overline{OP}^2-2\mathbf{P}\cdot\mathbf{I}\right)$$

(here $F=\sum F_i$ and $\mathbf{I}=\sum F_i\mathbf{A}_i/F$). Then since $2\mathbf{P}\cdot\mathbf{I}=P^2+I^2-(\mathbf{P}-\mathbf{I})^2$,

$$\sum F_i \sum F_i\overline{PA_i}^2=F^2\left(R^2+\overline{PI}^2-\overline{OI}^2\right).$$

Consequently,

$$(2) \qquad \sum F_i \overline{PA}_i \le F\left\{ R^2 + \overline{PI}^2 - \overline{OI}^2 \right\}^{1/2}$$

and with equality **iff** P coincides with 0. Now if h_i and r_i denote the distances from A_i and P, respectively, to the face F_i, $PA_i \ge h_i - r_i$. Thus,

$$(3) \qquad \sum F_i \overline{PA}_i \ge \sum F_i(h_i - r_i) = \sum F_i h_i - \sum F_i r_i = (n+1)nV - nV = n^2 V$$

where V is the volume of the simplex. Coupling (2) and (3) and using $nV = rF$, we obtain

$$(4) \qquad R^2 \ge n^2 r^2 + \overline{OI}^2 - \overline{PI}^2.$$

Finally, letting P coincide with I, we obtain the desired inequality

$$(5) \qquad R^2 \ge n^2 r^2 + \overline{OI}^2.$$

For equality, $PA_i = h_i - r_i$ for all i, or equivalently, the simplex is orthocentric with the orthocenter coinciding with the circumcenter. This requires the simplex to be regular since

$$A_i^2 = R^2 \quad \text{and} \quad A_i \cdot (A_j - A_k) = 0 \Rightarrow (A_i - A_j)^2 = (A_i - A_k)^2$$

for all i, j, k with $i \ne j, k$.

An Isoperimetric Inequality

Problem 83-5, *by* M. S. Klamkin (University of Alberta)
   Given that A_1 is an interior point of the regular tetrahedron $ABCD$ and that A_2 is an interior point of tetrahedron $A_1 BCD$, it is conjectured that

$$\text{I.Q.} (A_1 BCD) > \text{I.Q.}(A_2 BCD)$$

where the isoperimetric quotient of a tetrahedron T is defined by

$$\text{I.Q.} (T) = \frac{\text{Vol}(T)}{[\text{Area}(T)]^{3/2}}.$$

Also, one could replace Area$^{3/2}$ by total edge length cubed.
   Prove or disprove the conjecture and also consider the analogous problems in E^n for a simplex for which there are many different isoperimetric quotients.

   Editorial note. The 2-dimensional version of the problem was set by the proposer in the 1982 U.S.A. Mathematical Olympiad.

Solution by NOAM ELKIES (Harvard University).
   We extend the previous solution [April, 1984] for tetrahedra to simplexes. Let S be a regular n-dimensional simplex of unit edge length and vertices $AB_1 B_2 \cdots B_n$. Let S' be the simplex $A'B_1 B_2 \cdots B_n$ where A' is an interior point of S. We will show that

$$[\text{I.Q.}(S')]^{n-1} = \frac{[\text{Vol.}(S')]^{n-1}}{[\text{Area}(S')]^n}$$

is an increasing function of each of the base angles of S'. By a base angle of S', we mean the angle between the $(n-1)$-dimensional face B_0 opposite A' and any other $(n-1)$-dimensional face. By the vol. and area of S', we mean its n-dimensional content and the sum of the $(n-1)$-dimensional contents of all of its faces, respectively (see D.M.Y. Sommerville, *An Introduction to the Geometry of N Dimensions*, Dover, New York, 1958).

Let h_0 be the altitude of a unit $(n-1)$-dimensional regular simplex and V_0 be the $(n-2)$-dimensional content of a unit $(n-2)$ regular simplex. Then the content of each $(n-1)$-dimensional face of $S = h_0 V_0/(n-1)$. Also let θ_0 be the angle between any two $(n-1)$-dimensional faces of S. Since $-\cos\theta_0$ is the inner product of any two of the unit vectors $V_0, V_1, \cdots, V_n$ normal to the faces of S, we have as well known that $\Sigma V_i = 0$ and

$$\cos\theta_0 = -V_0 \cdot V_1 = \frac{-1}{n(n+1)} \sum_{i \neq j} V_i \cdot V_j = \frac{1}{n(n+1)} \left\{ \sum |V_i|^2 - \left| \sum V_i \right|^2 \right\} = \frac{1}{n}.$$

Let B_i' be the face of S' opposite B_i, θ_i' be the angle between B_i' and B_0, and H' be the altitude of S' from A'. Since A' is an interior point of S, $0 < \theta_i' < \theta_0$.

If P' is the projection of A' on B_0, the distance from P' to the $(n-2)$-dimensional face of B_0 opposite B_i is $H' \cot\theta_i'$. Since B_0 is regular, $H' = h_0/\Sigma \cot\theta_i'$. Also, the $(n-1)$-dimensional content of B_i' is $H'V_0(\csc\theta_i')/(n-1)$. Thus,

$$\text{Area}(S') = \text{Area } B_0 + \Sigma \text{ Area } B_i = \left(\frac{V_0}{n-1} \right) \left(h + H' \sum \csc\theta_i' \right)$$

$$= \frac{V_0 h_0}{n-1} \frac{\Sigma \cot(\theta_i'/2)}{\Sigma \cot\theta_i'},$$

$$\text{Vol}(S') = V_0 h_0 H'/(n(n-1)).$$

Consequently,

$$[\text{I.Q.}(S')]^{n-1} = \frac{(n-1)h_0^{n-2}}{n^{n-1}V_0} \left\{ \sum \cot\theta_i' \right\} \left\{ \sum \cot(\theta_i'/2) \right\}^{-n}.$$

We now show that $F(\theta_1, \theta_2, \cdots, \theta_n) \equiv \{\Sigma \cot\theta_i\}\{\Sigma \cot(\theta_i/2)\}^{-n}$ is an increasing function in each of the variables θ_i for $0 < \theta_i < \cos^{-1}(1/n)$.

By symmetry, it suffices to show that $\partial F/\partial\theta_1 > 0$. Here,

$$\left\{ \sum \cot(\theta_i/2) \right\}^{n+1} \frac{\partial F}{\partial\theta_1} = \frac{n}{2} \csc^2(\theta_1/2) \left\{ \sum \cot\theta_i \right\} - \csc^2\theta_1 \left\{ \sum \cot(\theta_i/2) \right\}.$$

Clearly, $\{\Sigma \cot(\theta_i/2)\}^{n+1} > 0$, and furthermore

$$\frac{n}{2} \csc^2(\theta_1/2) - (n+1)\csc^2\theta_1 = \csc^2\theta_1 \{2n\cos^2(\theta_1/2) - n - 1\}$$

$$= \frac{n\cos\theta_1 - 1}{\sin^2\theta_1} > 0.$$

Since also

$$(n+1)\sum \cot\theta_i - \sum \cot(\theta_i/2) = \sum \frac{n\cos\theta_i - 1}{\sin\theta_i} > 0,$$

it follows that $\partial F/\partial \theta_1 > 0$.

Editorial note. There is a corresponding result for simplexes S'' which contain S. Let A'' be a point exterior to S such that the simplex S'': $A''B_1B_2 \cdots B_n$ covers S. It then follows from the previous analysis that I.Q. (S'') is a decreasing function of each of the base angles θ_i'. Note that here $\theta_i' > \cos^{-1}(1/n)$.   [M.S.K.]

Supplementary References
Geometry

[1] H. ABELSON AND A. DISESSA, *Turtle Geometry: The Computer as a Medium for Exploring Mathematics*, M.I.T. Press, Cambridge, 1981.

[2] H. AHRENS, *Geodesic Mathematics and How to Use it*, University of California Press, Berkeley, 1976.

[3] F. J. ALMGREN, JR., AND J. E. TAYLOR, *"The geometry of soap films and soap bubbles,"* Sci. Amer. (1976) pp. 82–93.

[4] F. L. ALT, *"Digital pattern recognition by moments,"* J.A.C.M. (1962) pp. 240–258.

[5] D. L. ANDERSON AND A. M. DZIEWONSKI, *"Seismic tomography,"* Sci. Amer. (1984) pp. 60–68.

[6] H. G. APSIMON, *"Wrapping a parcel,"* BIMA (1980) pp. 160–164.

[7] I. I. ARTOBOLEVSKII, *Mechanisms for the Generation of Plane Curves*, Pergamon, Oxford, 1964.

[8] E. ASHER AND A. JANNER, *"Algebraic aspects of crystallography: Space groups as extensions,"* Helv. Phys. Acta (1965) pp. 551–572.

[9] J. G. ASPINHALL, *"Implementation of a hidden line remover,"* UMAP J. (1984) pp. 431–461.

[10] F. ATTNEAVE AND M. D. ARNOULT, *"The quantitative study of shape and pattern,"* Psych. Bull. (1956) pp. 452–471.

[11] T. E. AVERY AND G. D. BERLIN, *Interpretation of Aerial Photographs*, Burgess, Minneapolis, 1985.

[12] M. S. BARTLETT, *"Stochastic geometry: An introduction and reading list,"* Int. Statist. Rev. (1982) pp. 179–193.

[13] M. BERMAN, *The Statistical Analysis of Spatial Pattern*, Chapman & Hall, London, 1975.

[14] G. BINNIG AND H. ROHRER, *"The scanning tunneling microscope,"* Sci. Amer. (1985) pp. 50–56.

[15] A. A. BLANK, *"The Luneberg theory of binocular visual space,"* J. Optical Soc. Amer. (1953) pp. 717–727.

[16] F. J. BLOORE AND H. R. MORTON, *"Advice on hanging pictures,"* AMM (1985) pp. 309–321.

[17] E. D. BOLKER AND H. CRAPO, *"Bracing rectangular frameworks I,"* SIAM J. Appl. Math. (1979) pp. 473–490.

[18] ———, *"How to brace a one storey building?"* Environment and Planning (1977) pp. 125–152.

[19] E. D. BOLKER, *"Bracing rectangular frameworks II,"* SIAM J. Appl. Math. (1979) pp. 491–508.

[20] M. BORN AND E. WOLFE, *Principles of Optics*, Pergamon, Oxford, 1980.

[21] O. BOTTEMA AND B. ROTH, *Theoretical Kinematics*, North-Holland, Amsterdam, 1979.

[22] J. B. BOYLING, *"Munro's problem: How to count mountains,"* Math. Scientist (1984) pp. 117–124.

[23] J. W. BOYSE, *"Interference detection among solids and surfaces,"* Comm. Assoc. Comp. Mach. (1979) pp. 3–9.

[24] E. BROOKNER, *"Phased-array radars,"* Sci. Amer. (1985) pp. 94–102.

[25] A. R. BROWN, *Optimum Packing and Depletion*, Elsevier, N.Y., 1971.

[26] C. R. CALLADINE, *"Buckminster Fuller's 'tensegrity' structures and Clerk Maxwell's rules for the construction of stiff frames,"* Int. J. Solids and Structures (1978) pp. 161–172.

[27] T. M. CANNON AND B. R. HUNT, *"Image processing by computer,"* Sci. Amer. (1981) pp. 214–225.

[28] I. CARLBOM AND J. PACIOREK, *"Planar geometric projections and viewing transformations,"* ACM Comp. Surveys (1978) pp. 465–503.

[29] J. L. CHERMANT, ed., *Quantitative Analysis of Microstructures in Biology, Material Science, and Medicine*, Riederer, Stuttgart, 1978.

[30] E. J. COCKAYNE, *"On the Steiner problem,"* Canad. Math. Bull. (1967) pp. 431–450.

[31] D. E. COLE, *"The Wankel engine,"* Sci. Amer. (1972) pp 14–23.

[32] J. J. COLEMAN, *"Kinematics of tape recording,"* Amer. J. Phys. (1982) pp. 184–185.

[33] R. COLEMAN, *An Introduction to Mathematical Stereology*, Memoir Series, Aarhus, 1979.

[34] R. CONNELLY, *"The rigidity of certain cabled frameworks and the second order rigidity of arbitrarily triangulated convex surfaces,"* Adv. in Math. (1980) pp. 272–298.

[35] ———, *"The rigidity of polyhedral surfaces,"* MM (1979) pp. 275–283.

[36] L. A. COOPER AND R. N. SHEPARD, *"Turning something over in the mind,"* Sci. Amer. (1984) pp. 106–114.

[37] H. CRAPO, *"Structural rigidity,"* Struct. Topology (1979) pp. 26–45.

[38] H. CRAPO AND W. WHITELEY, *"Statics of frameworks and motions of panel structures, a projective geometric introduction,"* Struct. Topology (1982) pp. 43–82.

[39] L. CREMONA, *Graphical Statics*, Clarendon Press, Oxford, 1890.

[40] L. M. CRUZ-ORIVE, *"Particle size-shape distributions: The general spheroid problem I, II,"* J. Microscopy (1976) pp. 235–253, (1978) pp. 153–167.

[41] H. M. CUNDY, *"Getting it taped,"* Math. Gaz. (1971) pp. 43–47.

[42] H. M. CUNDY AND A. P. ROLLETT, *Mathematical Models,* Oxford University Press, Oxford, 1961.

[43] R. T. DE HOFF AND F. N. RHINES, eds. *Quantitative Microscopy,* McGraw-Hill, N.Y., 1968.

[44] G. H. DE VISME, *"The length of unexposed film left in a cassette,"* Math. Gaz. (1969) pp. 139–141.

[45] J. C. DAVIS AND M. J. McCULLAGH, *Display and Analysis of Spatial Data,* Wiley, Chichester, 1975.

[46] R. F. DeMAR, *"The problem of the shortest network joining n points,"* MM (1968) pp. 225–231.

[47] A. K. DEWDNEY, *"Computer Recreations" (on fractals),* Sci. Amer. (1985) pp. 16–24.

[48] J. B. DeVELIS AND G. O. REYNOLDS, *Theory and Applications of Holography,* Addison-Wesley, Reading, 1967.

[49] W. W. DOLAN, *"Early sundials and the discovery of the conic sections,"* MM (1972) pp. 8–12.

[50] ———, *"The ellipse in eighteenth century sundial design,"* MM (1972) pp. 205–209.

[51] R. O. DUDA AND P. E. HART, *Pattern Classification and Scene Analysis,* Wiley, N.Y., 1973.

[52] A. M. DZIEWONSKI AND D. L. ANDERSON, *Seismic tomography of the earth's interior,"* Amer. Sci. (1984) pp. 483–494.

[53] P. J. ELL AND B. L. HOLMAN, eds., *Computed Emission Tomography,* Oxford University Press, Oxford, 1982.

[54] P. FAZIO, G. HAIDER AND A. BIRIN, *Proceedings of the IASS World Conference on Space Enclosures,* Concordia University, Montreal, 1976.

[55] D. Z. FREEDMAN AND P. VAN NIEUWENHUIZEN, *"Hidden dimensions of spacetime,"* Sci. Amer. (1985) pp. 74–81.

[56] J. G. FREEMAN, *"A method of determining the north and latitude,"* Math. Gaz. (1975) pp. 41–44.

[57] ———, *"How to make a portable sun-dial,"* Math. Gaz. (1975) pp. 261–264.

[58] B. FULLER, *Explorations in the Geometry of Thinking,* Macmillan, N.Y., 1975.

[59] H. FUCHS, Z. M. KEDEM AND S. P. USELTON, *"Optimal surface reconstruction from planar contours,"* Comm. Assoc. Comp. Mach. (1977) pp. 693–702.

[60] P. L. GAILBRAITH AND D. A. PRAEGAR, *"Mathematical sailmanship,"* Math. Gaz. (1975) pp. 94–97.

[61] R. S. GARFINKLE, *"Minimizing wallpaper waste, Part I: A class of travelling salesman problems,"* Oper. Res. (1977) pp. 741–751.

[62] A. GHEORGHIU AND V. DRAGOMIR, *Geometry of Structural Forms,* Applied Science Publ., London, 1978.

[63] E. N. GILBERT AND H. O. POLLAK, *"Steiner minimal trees,"* SIAM Rev. (1968) pp. 1–29.

[64] E. N. GILBERT, *"Distortion in maps,"* SIAM Rev. (1974) pp. 47–62.

[65] B. GILLAM, *"Geometrical illusions,"* Sci. Amer. (1980) pp. 102–11, 162.

[66] H. E. GOHEEN, *"On space frames and plane trusses,"* AMM (1950) pp. 481–516.

[67] M. GOLDBERG, *"A six-plate linkage in three dimensions,"* Math. Gaz. (1974) pp. 287–289.

[68] ———, *"A three-dimensional analogy of a plane Kempe linkage,"* J. Math. Phys. (1946) pp. 96–110.

[69] ———, *"New five-bar and six-bar linkages in three dimensions,"* A.S.M.E. (1943) pp. 649–661.

[70] ———, *"Polyhedral linkages,"* MM (1942) pp. 323–332.

[71] ———, *"Tubular linkages,"* J. Math. Phys. (1947) pp. 10–21.

[72] ———, *"Unstable polyhedral structures,"* MM (1978) pp. 165–170.

[73] R. GREENLER, *Rainbows, Halos, and Glories,* Cambridge University Press, N.Y., 1980.

[74] R. L. GREGORY, *The Intelligent Eye,* McGraw-Hill, N.Y., 1970.

[75] ———, *"Visual illusions,"* Sci. Amer. (1968) pp. 66–76.

[76] U. GRENANDER, *Pattern Synthesis.* Lectures in Pattern Theory I, Springer-Verlag, N.Y., 1976.

[77] W. E. L. GRIMSON, *From Images to Surfaces: A Computational Study of the Human Early Visual System,* M.I.T. Press, Cambridge, 1981.

[78] B. GRUNBAUM AND G. C. SHEPHARD, *"Satin and twills: An introduction to the geometry of fabrics,"* MM (1980) pp. 139–161.

[79] D. HARRIS, *Computer Graphics and Applications,* Chapman and Hall, N.Y., 1984.

[80] J. G. HAYES, *"Numerical methods for curve and surface fitting,"* BIMA (1974) pp. 144–152.

[81] R. M. HAZEN AND L. W. FINGER, *"Crystals at high pressure,"* Sci. Amer. (1985) pp. 110–117.

[82] R. S. HEATH, *Treatise on Geometrical Optics,* Cambridge University Press, Cambridge, 1895.

[83] G. T. HERMAN, ed., *Image Reconstruction from Projections: Implementation and Applications,* Springer-Verlag, N.Y., 1979.

[84] J. HIGGINS, *"Getting it taped,"* Math. Gaz. (1971) pp. 47–48.

[85] T. P. HILL, *"Determining a fair border,"* AMM (1983) pp. 438–442.

[86] J. E. HOCHBERG, *Perception,* Prentice-Hall, N.J., 1978.

[87] D. D. HOFFMAN, *"The interpretation of visual illusions,"* Sci. Amer. (1984) pp. 154–162.

[88] W. C. HOFFMAN, *"Visual illusions of angle as an application of Lie transformation groups,"* SIAM Rev. (1971) pp. 169–184.

[89] A. HOLDEN, *Shapes, Space and Symmetry,* Columbia University Press, N.Y., 1971.

[90] H. HONDA, P. B. TOMLINSON AND J. B. FISHER, *Two geometrical models of branching of botanical trees,*" Ann. Bot. (1982) pp. 1–11.

[91] B. K. P. HORN AND K. IKEUCHI, *The mechanical manipulation of randomly oriented parts,*" Sci. Amer. (1984) pp. 100–111.

[92] B. K. P. HORN, R. J. WOODHAM AND W. M. SILVER, *Determining shape and reflectance using multiple images,*" M.I.T. A.I. Laboratory, Memo 490; August, 1978.

[93] H. S. HORN, *The Adaptive Geometry of Trees,* Princeton University Press, Princeton, 1971.

[94] M. HOTINE, *Mathematical Geodesy,* U.S. Dept. of Commerce, Washington, D.C., 1969.

[95] M. K. HU, *Visual pattern recognition of moment invariants,*" I. R. E. Trans. Inform. Theory (1962) pp. 179–187.

[96] K. H. HUNT, *Kinematic Geometry of Mechanisms,* Clarendon Press, Oxford, 1978.

[97] P. HUYBERS, *Polyhedral housing units,*" Int. J. Housing Sci. Appl. (1979) pp. 215–225.

[98] C. ISENBERG, *The Science of Soap Films and Soap Bubbles,* Tieto Ltd., England, 1978.

[99] A. J. JAKEMAN AND R. S. ANDERSSEN, *Abel type integral equations in stereology. I. General discussion,*" J. Microscopy (1975) pp. 121–134.

[100] M. A. JASWON, *Introduction to Mathematical Crystallography,* Elsevier, N.Y., 1965.

[101] R. V. JEAN, *Mathematical Approach to Pattern and Form in Plant Growth,* Wiley, N.Y., 1984.

[102] E. B. JENSEN, H. J. G. GUNDERSEN AND R. OSTERBY, *Determination of membrane thickness distribution from orthogonal intercepts,*" J. Microscopy (1979) pp. 19–33.

[103] B. K. JOHNSON, *Optics and Optical Instruments,* Dover, N.Y., 1960.

[104] D. KALMAN, *A model for playing time,*" MM (1981) pp. 247–250.

[105] W. M. KAULA, *Theory of Satellite Geodesy,* Blaisdell, Waltham, 1967.

[106] J. M. KAPPRAFF, *A course in the mathematics of design,*" Struct. Topology (1983) pp. 37–52.

[107] J. B. KELLER, *Parallel reflection of light by plane mirrors,*" Quart. Appl. Math. (1953) pp. 216–219.

[108] E. KHULAR, K. THYAGARAJAN AND A. K. GHATAK, *A note on mirage formation,*" Amer. J. Phys. (1977) pp. 90–92.

[109] E. KRETSCHMER, ed., *Studies in Theoretical Cartography,* Deuticke, Vienna, 1977.

[110] T. L. KUNII, ed., *Computer Graphics–Theory and Applications,* Springer-Verlag, Tokyo, 1983.

[111] G. LAMAN, *On graphs and rigidity of plane skeletal structures,*" J. Eng. Math. (1970) pp. 331–340.

[112] C. LANCZOS, *Space Through the Ages,* Academic Press, N.Y., 1970.

[113] A. L. LOEB, *Color and Symmetry,* Wiley, N.Y., 1971.

[114] ———, *Space Structures, Their Harmony and Counterpoint,* Addison-Wesley, Reading, 1976.

[115] M. LUCKIESH, *Visual Illusions: Their Causes, Characteristics, and Applications,* Dover, N.Y., 1965.

[116] R. K. LUNEBERG, *The metric of binocular visual space,*" J. Optical Soc. Amer. (1950) pp. 627–642.

[117] H. E. MALDE, *Panoramic photographs,*" Amer. Sci. (1983) pp. 132–140.

[118] B. B. MANDELBROT, *The Fractal Geometry of Nature,* Freeman, San Francisco, 1982.

[119] ———, *Physical objects with fractional dimension: seacoasts, galaxy clusters, turbulence, and soap,*" BIMA (1977) pp. 189–196.

[120] L. MARCH AND P. STEADMAN, *The Geometry of Environment,* M.I.T. Press, Cambridge, 1971.

[121] J. C. MAXWELL, *On reciprocal figures, frames, and diagrams of forces,*" Trans. Royal Soc. Edinburgh (1869–1872) pp. 1–40.

[122] D. MARR, *Vision: A Computational Investigation into the Human Representation and Processing of Visual Information,* Freeman, San Francisco, 1982.

[123] J. P. McKELVEY, *Dynamics and kinematics of textile machinery,*" Amer. J. Phys. (1982) pp. 1160–1162.

[124] ———, *Kinematics of tape recording,*" Amer. J. Phys. (1981) pp. 81–83.

[125] R. J. Y. McLEOD AND E. L. WACHSPRESS, eds., *Frontiers of Applied Geometry,* Math. Modelling (1980) pp. 271–403.

[126] R. K. MELLUISH, *An Introduction to the Mathematics of Map Projection,* Cambridge University Press, Cambridge, 1931.

[127] T. W. MELNYK, O. KNOP AND W. P. SMITH, *Extremal arrangements of points and unit charges on a sphere: Equilibrium configurations revisited,*" Canad. J. Chem. (1977) pp. 1745–1761.

[128] W. MERZKIRCH, ed., *Flow Visualization* 2, Hemisphere, N.Y., 1982.

[129] H. R. MILLS, *Positional Astronomy and Astro-Navigation Made Easy,* S. Thornes, Halsted, N.Y., 1978.

[130] J. MILNOR, *A problem in cartography,*" AMM (1969) pp. 1101–1112.

[131] M. MINNAERT, *Light and Colour in the Open Air,* Dover, N.Y., 1954.

[132] P. MOON, *The Scientific Basis of Illuminating Engineering,* Dover, N.Y., 1936.

[133] I. I. MUELLER, *Introduction to Satellite Geodesy,* Ungar, N.Y., 1964.

[134] D. E. NASH, *Rotary engine geometry,*" MM (1977) pp. 87–89.

[135] L. NEUWIRTH, *The theory of knots,*" Sci. Amer. (1979) pp. 110–124.

[136] H. M. NUSSENZVIEG, *The theory of the rainbow,*" Sci. Amer. (1977) pp. 116–127.

[137] T. O'NEILL, *A mathematical model of a universal joint,*" UMAP J. (1982) pp. 201–219.

[138] F. Otto, ed., *Tensile Structures: Vol. I—Pneumatic Structures, Vol. II—Cables, Nets, and Membranes*, M.I.T. Press, Cambridge, 1967.

[139] E. W. Parkes, *Braced Frameworks*, Pergamon, Oxford, 1965.

[140] D. Pedoe, "A geometrical proof of the equivalence of Fermat's principle and Snell's law," AMM (1964) pp. 543–544.

[141] ———, *Geometry and the Visual Arts*, Penguin, England, 1976.

[142] P. Pearce, *Structure in Nature is a Strategy for Design*, M.I.T. Press, Cambridge, 1978.

[143] C. E. Pearson, "Optimal mapping from a sphere onto a plane," SIAM Rev. (1982) pp. 469–475.

[144] T. Poggio, "Vision by man and machine," Sci. Amer. (1984) pp. 106–116.

[145] W. K. Pratt, *Digital Image Processing*, Wiley, N.Y., 1978.

[146] A. Pugh, *An Introduction to Tensegrity*, University of California Press, Berkeley, 1976.

[147] D. Rawlins, "Doubling your sunsets or how anyone can measure the earth's size with a wristwatch and meterstick," Amer. J. Phys. (1979) pp. 126–128.

[148] A. Recski, "A network theory approach to the rigidity of skeletal structures. Part III. An electric model of planar networks," Struct. Topology (1984) pp. 59–71.

[149] W. P. Reid, "Line of flight from shock recordings," MM (1968) pp. 59–63.

[150] A. P. Roberts and D. Sprevak, "Research on a shoestring," BIMA (1980) pp. 215–218.

[151] I. Rosenholz, "Calculating surface area from a blueprint," MM (1979) pp. 252–256.

[152] I. Rosenberg, "Structural rigidity: Foundations and rigidity," Ann. Discrete Math. (1980) pp. 143–161.

[153] A. Rosenfeld, ed., *Image Modeling*, Academic Press, N.Y., 1981.

[154] B. Roth, "Rigid and flexible frameworks," AMM (1981) pp. 6–21.

[155] B. Roth and W. Whiteley, "Rigidity of tensegrity frameworks," Trans. AMS (1981) pp. 419–445.

[156] W. H. Ruckle, *Geometric Games and their Applications*, Pitman, London, 1983.

[157] M. Salvadori, *Structure in Architecture*, Prentice-Hall, N.J., 1975.

[158] ———, *Why Buildings Stand Up*, Norton, N.Y., 1980.

[159] D. Schattschneider and W. Walker, *M. C. Escher Kaleidocycles*, Ballantine, N.Y., 1977.

[160] B. F. Schutz, *Geometrical Methods of Mathematical Physics*, Cambridge University Press, Cambridge, 1980.

[161] L. A. Shepp, *Computed Tomography*, AMS, 1983.

[162] L. A. Shepp and J. B. Kruskal, "Computerized tomography: The new medical x-ray technology, AMM (1978) pp. 420–439.

[163] A. V. Shubnikov and N. V. Belov, *Colored Symmetry*, Pergamon, Oxford, 1964.

[164] C. C. Slama, C. Theurer and S. W. Henriksen, eds., *Manual of Photogrammetry*, Amer. Soc. Photogrammetry, Falls Church, 1980.

[165] N. J. A. Sloane, "Multiplexing methods in spectroscopy," MM (1979) pp. 71–80.

[166] ———, "The packing of spheres," Sci. Amer. (1984) pp. 116–125.

[167] N. J. A. Sloane and M. Harwit, "Masks for Hadamard transform optics and weighing designs," Appl. Optics (1976) pp. 107–114.

[168] W. M. Smart, *Text-Book on Spherical Astronomy*, Cambridge University Press, Cambridge, 1965.

[169] D. A. Smith, "A descriptive model for perception of optical illusions," J. Math. Psych. (1978) pp. 64–85.

[170] ———, "Descriptive models for perception of optical illusions: 1," UMAP J. (1985) pp. 59–95.

[171] J. V. Smith, *Geometrical and Structural Crystallography*, John Wiley, N.Y., 1982.

[172] J. P. C. Southall, *Introduction to Physiological Optics*, Dover, N.Y., 1961.

[173] J. A. Steers, *Introduction to the Study of Map Projections*, University of London Press, London, 1965.

[174] H. Steinhaus, *Mathematical Snapshots*, Oxford University Press, N.Y., 1969.

[175] P. Stevens, *Patterns in Nature*, Little Brown, Boston, 1974.

[176] G. C. Steward, *The Symmetrical Optical System*, Cambridge University Press, Cambridge, 1958.

[177] R. S. Stichartz, "What's up moonface?" UMAP J. (1985) pp. 3–34.

[178] G. Strang, "The width of a chair," AMM (1982) pp. 529–534.

[179] W. J. Supple, ed., *Proceedings of the Second International Conference on Space Structures*, University of Surrey, Guilford, 1975.

[180] W. Swindell and H. H. Barrett, "Computerized tomography: Taking sectional x-rays," Phys. Today (1977) pp. 32–41.

[181] J. L. Synge, "Reflection in a corner formed by three plane mirrors," Quart. Appl. Math. (1946) pp. 166–176.

[182] P. M. L. Tammes, "On the origin of number and arrangement of places of exit on the surface of pollen-grains," Recueil des Travaux Botanique Neerlandais (1930) pp. 1–84.

[183] T. Tarnei, "Simultaneous static and kinematic indeterminacy of space trusses with cyclic symmetry," Int. J. Solids and Structures (1980) pp. 347–359.

[184] ———, "Spherical circle-packing in nature, practice and theory," Struct. Topology (1984) pp. 39–58.

[185] D. E. Thomas, "Mirror images," Sci. Amer. (1980) pp. 206–228.

[186] W. P. Thurston and J. R. Weeks, "The mathematics of three-dimensional manifolds," Sci. Amer. (1984) pp. 108–120.

[187] S. TIMOSHENKO AND D. H. YOUNG, *Theory of Structures*, McGraw-Hill, N.Y., 1965.

[188] S. TOLANSKY, *Optical Illusions*, Pergamon, Oxford, 1964.

[189] P. B. TOMLINSON, *"Tree architecture,"* Amer. Sci. (1983) pp. 141–149.

[190] A. TORMEY AND J. F. TORMEY, *"Renaissance intarsia: The art of geometry,"* Sci. Amer. (1982) pp. 136–143.

[191] L. TORNHEIM, *"Determination of large parallelepipeds,"* SIAM Rev. (1962) pp. 223–226.

[192] G. TOUSSAINT, *"Generalizations of pi: Some applications,"* Math. Gaz. (1974) pp. 289–291.

[193] R. A. R. TRICKER, *Introduction to Meteorological Optics*, Mills and Boon, London, 1970.

[194] E. E. UNDERWOOD, *Quantitative Stereology*, Addison-Wesley, Reading, 1970.

[195] W. R. UTTAL, *Visual Form Detection in 3-Dimensional Space*, Erlbaum Associates, 1983.

[196] H. WALLACH, *"Perceiving a stable environment,"* Sci. Amer. (1985) pp. 118–124.

[197] H. WALLACH AND J. BACON, *"Two kinds of adaptation in the constancy of visual direction and their different effects on the perception of shape and visual direction,"* Perception & Psychophysics (1977) pp. 227–242.

[198] WATTS BROS. TOOL WORKS, *How to Drill Square, Hexagon, Octagon, Pentagon Holes*, Wilmerding, PA, 1966.

[199] A. E. WAUGH, *Sundials, Their Theory and Construction*, Dover, N.Y., 1973.

[200] E. R. WEIBEL, *Stereological Methods I, II*, Academic Press, N.Y., 1980.

[201] A. F. WELLS, *Models in Structural Inorganic Chemistry*, Oxford University Press, Oxford, 1970.

[202] S. J. WILLIAMSON AND H. Z. CUMMINS, *Light and Color*, John Wiley, N.Y., 1983.

[203] D. H. WILSON, *"The geometry of the Wankel engine,"* BIMA (1972) pp. 323–325.

[204] A. T. WINFREE, *The Geometry of Biological Time*, Springer-Verlag, Berlin, 1980.

[205] J. M. WOLFE, *"Hidden visual processes,"* Sci. Amer. (1983) pp. 94–103, 154.

[206] R. W. WOOD, *Physical Optics*, Macmillan, N.Y., 1944.

[207] J. H. WOODHOUSE AND A. M. DZIEWONSKI, *"Mapping the upper mantle: Three dimensional modelling by inversion of seismic waveforms,"* J. Geophys. Res. (1984) pp. 5953–5986.

[208] E. W. WOODLAND AND G. M. CLEMENCE, *Spherical Astronomy*, Academic Press, N.Y., 1966.

[209] L. A. WHYTE, *"Unique arrangements of points on a sphere,"* AMM (1952) pp. 606–612.

[210] T. T. WU, *"Some mathematical problems in fiber x-ray crystallography,"* SIAM Rev. (1972) pp. 420–432.

[211] I. M. YAGLOM, *A Very Simple Non-Euclidean Geometry and its Physical Basis*, Springer-Verlag, Heidelberg, 1979.

[212] J. I. YELLOT, JR., *"Binocular depth inversion,"* Sci.Amer. (1981) pp. 148–159.

[213] W. M. ZAGE, *"The geometry of binocular visual space,"* MM (1980) pp. 289–294.

[214] H. ZIMMER, *Geometrical Optics*, Springer-Verlag, Heidelberg, 1970.

[215] L. ZUSNE, *"Moments of area and of the perimeter of visual forms as predictors of discrimination performance,"* J. Exp. Psych. (1965) pp. 213–220.

17. POLYNOMIALS

Positivity of the Coefficients of a Polynomial

Problem 81-13*, *by* Ko-Wei Lih *and* E. T. H. Wang (Institute of Mathematics, Academia Sinica, Taipei, Taiwan).

Prove or disprove that all the coefficients of the polynomial $P_n(x)$ are positive, where

$$P_n(x) = \sum_{k=0}^{n} \binom{n}{k}^2 (1 + x)^{2n-2k}(1 - x)^{2k}.$$

The problem arose as we attempted to prove the monotonicity of the permanent function on a certain line segment in the convex polytope of all $n \times n$ doubly stochastic matrices.

Solution by R. E. Shafer (Palo Alto, CA).

$$\sum_{k=0}^{n} \binom{n}{k}^2 (1 + x)^{2n-2k}(1 - x)^{2k} = (1 + x)^{2n}\, {}_2F_1\left[-n, -n; 1; \left(\frac{1 - x}{1 + x}\right)^2\right]$$

$$= 2^{2n}\, {}_2F_1[-n, \tfrac{1}{2}; 1; 1 - x^2] \qquad \text{H.T.F., 2.11 (36)}$$

$$= \binom{2n}{n}\, {}_2F_1[-n, \tfrac{1}{2}; -n + \tfrac{1}{2}; x^2] \qquad \text{H.T.F., 2.10 (1)}$$

$$= \sum_{k=0}^{n} \binom{2n - 2k}{n - k}\binom{2k}{k} x^{2k}.$$

H.T.F. refers to A. Erdélyi et al. (1953), *Higher Transcendental Functions*, vol. I, McGraw-Hill, New York, Chapter 2.

Editorial note. W. W. Meyer and O. G. Ruehr both note the generalization to the polynomial

$$\sum_{k=0}^{n} \binom{n + \alpha}{k}\binom{n + \alpha}{n - k}(1 + x)^{2n-2k}(1 - x)^{2k}$$

$$= \binom{n + \alpha}{n}(1 + x)^{2n}\, {}_2F_1\left[-n, -n - \alpha; \alpha + 1; \left(\frac{1 - x}{1 + x}\right)^2\right],$$

$\alpha > -\tfrac{1}{2}$, which has positive coefficients by virtue of the same quadratic and linear transformations as used by Shafer (above).   [C.C.R.]

521

A Factorial Polynomial Inequality

Problem 83-1, *by* A. FRANSÉN (National Defense Research Institute, Stockholm, Sweden).

If

$$(x + 1)(x + 2) \cdots (x + k) = a_1 + a_2 x + \cdots + a_{k+1} x^k,$$

prove that

$$a_{i+1} \leq \frac{k!}{(i + 1)!} \binom{k + 1}{i}$$

with equality iff $i = 0$, $k - 1$ and k, and strict inequality for $1 \leq i \leq k - 2$ (for $k \geq 3$). Also, is there a better simple general upper bound?

Solution by W. A. J. LUXEMBURG (California Institute of Technology).

In the usual notation for the Stirling numbers

$$(x + 1)(x + 2) \cdots (x + k) = \sum_{i=0}^{k} s(k + 1, i + 1) x^i$$

we need to show that

$$s(k + 1, i + 1) \leq \frac{k!}{(i + 1)!} \binom{k + 1}{i}, \qquad 0 \leq i \leq k - 1.$$

For $k = 0, 1, 2$, this is trivial. Using the relation

$$s(k + 1, i + 1) = s(k, i) + ks(k, i + 1),$$

we obtain by induction that

$$s(k + 1, i + 1) \leq \frac{(k - 1)!}{i!} \binom{k}{i - 1} + k \frac{(k - 1)!}{(i + 1)!} \binom{k}{i}$$

$$= \frac{k!}{(i + 1)!} \left(\binom{k}{i} + \frac{i + 1}{k} \binom{k}{i - 1} \right)$$

$$\leq \frac{k!}{(i + 1)!} \left(\binom{k}{i} + \binom{k}{i - 1} \right) = \frac{k!}{(i + 1)!} \binom{k + 1}{i}.$$

The equality cases $i = 0$, $k - 1$ and k are trivial.

Remark. For fixed i and k large the upper bound is not very precise. See the interesting article of L. Moser and M. Wyman: *Asymptotic development of the Stirling numbers of the first kind,* J. London Math. Soc., 33 (1958), pp. 133–146.

Inequality for the Roots of a Hypergeometric Polynomial

Problem 84-6*, *by* JERRY L. FIELDS (University of Alberta) *and* YUDELL L. LUKE (University of Missouri at Kansas City).

For $p, q \in \mathbb{Z}$ such that $p \geq q \geq 0$, consider the hypergeometric polynomials

$$P_n(z) = {}_{p+2}F_q \left(\begin{matrix} -n, n + \lambda, a_1, \cdots, a_p \\ b_1, \cdots, b_q \end{matrix} \middle| z \right)$$

$$= \sum_{j=0}^{n} A_j z^j = A_n \prod_{j=1}^{n} (z - z_{j,n}),$$

$$A_n = \frac{(-1)^n (n+\lambda)_n \, \Pi_{k=1}^p \, (a_k)_n}{\Pi_{k=1}^q \, (b_k)_n}, \qquad (w)_k = \frac{\Gamma(w+k)}{\Gamma(w)},$$

$$1 - a_k, 1 - b_k \notin \mathbb{Z}^+, \quad n + 1 \in \mathbb{Z}^+,$$

where the complex parameters a_k, b_k, λ are independent of n.

Prove or disprove the following conjecture which arose in various discussions between the authors over the last few years.

Conjecture. There exists a constant M, independent of n, such that

(*) $$\qquad\qquad |z_{j,n}| \le M n^{q-p-1}, \qquad j = 1, 2, \cdots, n.$$

If (*) were to hold, the symmetric sums $\sigma_k = \sum_{j=1}^n (z_{j,n})^k$ would satisfy the inequality

(**) $$\qquad\qquad |\sigma_k| \le M^k n^{1+k(q-p-1)}.$$

Although the first few σ_k can be computed explicitly, e.g.,

$$\sigma_1 = \frac{-A_{n-1}}{A_n} = \frac{n \, \Pi_{k=1}^q \, (n-1+b_k)}{(2n+\lambda-1) \, \Pi_{k=1}^p \, (n-1+a_k)},$$

it is hard to establish the general estimate (**). With these estimates, it would be easy to derive an asymptotic expansion for $P_n(z)$; i.e., for $|z| \ge m > 0$, m independent of n, we would have for n sufficiently large,

$$P_n(z) = A_n z^n \exp\left\{ \sum_{j=1}^n \log(1 - z_{j,n}/z) \right\}$$

$$= A_n z^n \exp\left\{ -\sum_{k=1}^\infty \frac{\sigma_k}{kz^k} \right\}$$

$$= A_n z^n \exp\left\{ \frac{-\sigma_1}{z} + O\left(\frac{1}{nz^2}\right) \right\}, \qquad n \to +\infty.$$

For $p = q = 0$, the $P_n(z)$ reduce essentially to the Bessel polynomials, and the above argument goes back to Obrechkoff [2], and Dočev [1]. For the Bessel polynomials, the zeros can actually be shown to be in the left half-plane, and more precisely still, that their reciprocals lie in a parabolic region centered on the negative axis; see Saff and Varga [3]. In this special case, (*) can be improved to

$$|z_{j,n}| \le \frac{1}{n-1+\mathrm{Re}(\lambda-1)}.$$

REFERENCES

[1] K. Dočev, *On the generalized Bessel polynomials*, Bulgar. Akad. Nauk. Izv. Mat. Inst., 6 (1962), pp. 89–94. (In Bulgarian.)

[2] N. Obrechkoff, *Two theorems on the approximation of linear forms*, Bulgar. Akad. Nauk. Izv. Mat. Inst., 2 (1956), pp. 45–68. (In Bulgarian.)

[3] E. B. Saff and R. S. Varga, *Zero-free parabolic regions for sequences of polynomials*, SIAM J. Math. Anal., 7 (1976), pp. 344–357.

A Polynomial Problem

Problem 80-16*, *by* P. B. BORWEIN (The Mathematical Institute, Oxford, England).

Does there exist a sequence of real algebraic polynomials $\{P_n(x)\}$, where $P_n(x)$ is of degree n, so that

$$\lim_{n\to\infty} \text{measure } \{x \,|\, P'_n(x)/P_n(x) \geq 2n\} = 1?$$

One can show that

$$\text{measure } \{x \,|\, P'_n(x)/P_n(x) \geq 2n\} \leq 1,$$

and L. Loomis (Bull. Amer. Math. Soc, 50 (1946), pp. 1082–1086) has shown that if $P_n(x)$ has all real roots, then

$$\text{measure } \{x \,|\, P'_n(x)/P_n(x) \geq n\} = 1.$$

Solution by G. K. KRISTIANSEN (Ibsgården 48, DK4000 Roskilde, Denmark).

It is possible to define a sequence of polynomials $(P_m \,|\, \text{degree}(P_m) = m \text{ odd})$, so that

$$\lim_{m\to\infty} \text{measure } \{x \,|\, P'_m(x)/P_m(x) \geq 2m\} = 1.$$

We shall use the following notation:

We set $m = 2n - 1$. The nth degree Laguerre polynomial (with arbitrary normalization) is written as

$$L_n(x) = Q_{n-1}(x)(x - x_n),$$

where x_n is the largest root.

Then P_m is defined as the polynomial solution to the differential equation

(1) $$P'_m(x) - 2mP_m(x) = (Q_{n-1}(2mx))^2(x_n - 2mx).$$

In integral form,

(2) $$P_m(x) = \frac{1}{2m} e^{2mx} \int_{2mx}^{\infty} dy \, e^{-y} L_n(y) Q_{n-1}(y).$$

From the orthogonality properties of the Laguerre polynomials, we deduce that $P_m(0) = 0$.

From (2), it follows that

(3) $$\frac{d}{dx} (e^{-2mx} P_m(x)) = e^{-2mx} (Q_{n-1}(2mx))^2 (x_n - 2mx).$$

The function $f(x) = e^{-2mx} P_m(x)$ vanishes for $x = 0$ and for $x \to \infty$. If f assumed negative values in $(0, \infty)$, it would have a minimum there, which is incompatible with (3).

From (1) it follows, then, that $P'_m(x)/P_m m(x) \geq 2m$ for $0 < x < x_n/2m$. From the asymptotic formula in [1, p. 200], it easily follows that $x_n \sim 4n \sim 2m$ for $n \to \infty$, which completes the proof of the conjecture.

REFERENCE

[1] MAGNUS ERDÉLYI AND TRICOMI OBERHETTINGER, *Higher Transcendental Functions*, vol. 2, McGraw-Hill, New York, 1953.

Problem 73-8. *A Polynomial Diophantine Equation*, by M. S. KLAMKIN (Ford Motor Company).

Determine all real solutions of the polynomial Diophantine equation

$$(1) \qquad\qquad P(x)^2 - P(x^2) = x\{Q(x)^2 - Q(x^2)\}.$$

Solution by O. P. LOSSERS (Technological University, Eindhoven, the Netherlands).

From the given equation, it follows that

$$P(x^4) - x^2 Q(x^4) = P^2(x^2) - x^2 Q^2(x^2)$$
$$= \{P(x^2) - xQ(x^2)\}\{P(x^2) + xQ(x^2)\}.$$

Letting $F(x) = P(x^2) - xQ(x^2)$, we have

$$(2) \qquad\qquad F(x^2) = F(x)F(-x).$$

Conversely, any solution of (1) may be obtained from a solution of (2) by taking

$$P(x) = \frac{1}{2}\{F(\sqrt{x}) + F(-\sqrt{x})\},$$

$$Q(x) = \frac{1}{2x}\{-F(\sqrt{x}) + F(-\sqrt{x})\}.$$

Polynomial solutions of (2) may be written in the form

$$F(x) = C(x - \alpha_1)(x - \alpha_2)\cdots(x - \alpha_n) \qquad (C \text{ is a constant}).$$

Then

$$F(-x) = (-1)^n C(x + \alpha_1)(x + \alpha_2)\cdots(x + \alpha_n),$$

so that

$$F(x)F(-x) = (-1)^n C^2 (x - \alpha_1)(x + \alpha_1)(x - \alpha_2)(x + \alpha_2)\cdots(x - \alpha_n)(x + \alpha_n).$$

On the other hand, taking β_i such that $\beta_i^2 = \alpha_i (i = 1, \cdots, n)$, we find

$$F(x^2) = C(x - \beta_1)(x + \beta_1)(x - \beta_2)(x + \beta_2)\cdots(x - \beta_n)(x + \beta_n).$$

Therefore, in view of (2), excluding the trivial case $C = 0$, we obtain $C = (-1)^n$ and $(\alpha_i)_{i=1}^n$ is a permutation of $(\beta_i)_{i=1}^n$.

Finite, squaring-invariant subsets of the complex plane can only contain 0 and roots of unity of odd order. The irreducible polynomials corresponding to these roots are

$$\lambda_0(x) = x, \qquad \lambda_k(x) = \prod_{(2k-1,l)=1} [x - \exp[2\pi i l/(2k-1)], \qquad k = 1, 2, 3, \cdots,$$

(the cyclotomic polynomials). Since for all $k = 1, 2, 3, \cdots$, the set $\{\exp[2\pi i l/(2k-1)]\}_{(l, 2k-1)=1}$ is squaring-invariant and the set of solutions of (2) is closed under multiplication, the general polynomial solution of (2) is

$$F(x) = (-1)^{\deg F} \prod_{k=0}^{\infty} (\lambda_k(x))^{n_k},$$

the n_k being nonnegative integers, $n_k \neq 0$, for a finite number of indices k. These polynomials all have integral coefficients.

Also solved by the proposer, who notes that one can give extensions by considering higher order roots of unity. For example, letting $\omega^3 = 1$, consider $F(x^3) = F(x)F(\omega x)F(\omega^2 x)$, where $F(x) = P(x^3) + \omega x Q(x^3) + \omega^2 x^2 R(x^3)$.

A Set of Linearly Independent Polynomials

Problem 80-6*, *by* P. W. SMITH[1] (University of Alberta), (corrected).
    Let $k \geq 2$ be a given positive integer and suppose that $t_1 < t_2 < \cdots$ is an infinite sequence of real numbers. Define the polynomials $\{P_i(x)\}_{i=0}^{\infty}$ by

$$P_i(x) = \prod_{j=1}^{k-1} (x - t_{i+j}).$$

Show that for any selection of k distinct indices $0 \leq i_1 < i_2 < \cdots < i_k$, the set of k polynomials $\{P_{i_1}, P_{i_2}, \cdots, P_{i_k}\}$ is linearly independent.
    Remark. For $k = 2, 3$, this is easy to see.

Zeros of a Polynomial

Problem 80-12*, *by* S. RIEMENSCHNEIDER, A. SHARMA *and* P. W. SMITH[1] (University of Alberta), (corrected).
    Prove or disprove that the polynomial

$$P(z) = \begin{vmatrix} 1 & 1 & \cdots & 1 \\ z^{m_1} & (z+1)^{m_1} & \cdots & (z+n)^{m_1} \\ \vdots & \vdots & & \vdots \\ z^{m_n} & (z+1)^{m_n} & \cdots & (z+n)^{m_n} \end{vmatrix}$$

has all its zeros in the left-half plane Re $(z) < 0$, where the m_i are integers such that

$$0 < m_1 < m_2 \cdots < m_n.$$

The result has been proved for the case $n = 2$ corresponding to $m_1 = 2$, $m_2 = 3$ and also other small values of m_1, m_2.
    Note that the cases when $P(z)$ reduce to a constant are to be excluded. For example, this occurs when $m_i = i$, $(i = 1, 2, \cdots, n)$, in which case $P(z)$ is a Vandermonde determinant reducing to a constant.

Location of the Zeros of a Polynomial

Problem 81-2, *by* E. DEUTSCH (Polytechnic Institute of New York).
    Let $f_1(x)$, $f_2(x)$, $\cdots$ be a sequence of real, monic, orthogonal polynomials associated with a given distribution. It is known that $f_n(x)$ is of degree n $(n = 1, 2, \cdots)$, $f_n(x)$ has only real, simple zeros and the zeros of $f_{n-1}(x)$ interlace strictly with those of $f_n(x)$ $(n = 2, 3, \cdots)$. Show that if $x_1 < x_2 < \cdots < x_n$ denote the zeros of $f_n(x)$, then the zeros of $f_{n-1}(x)$ $(n \geq 3)$ are located in the open interval

---

[1] Currently at Old Dominion University.

$$\left(x_1+\frac{(x_2-x_1)f_{n-1}(x_1)}{f'_n(x_1)},\ x_n-\frac{(x_n-x_{n-1})f_{n-1}(x_n)}{f'_n(x_n)}\right).$$

Solution by P. G. NEVAI (Ohio State University).

The solution is contained in the following statement.

PROPOSITION. *Let* $P(x)=\prod_{k=1}^{n}(x-a_k)$ *and* $R(x)=\prod_{k=1}^{n-1}(x-b_k)$, *where*

(1) $a_1<b_1<a_2<b_2<\cdots<a_{n-1}<b_{n-1}<a_n.$

Then

(2) $b_{n-1}<a_n-\dfrac{(a_n-a_{n-1})R(a_n)}{P'(a_n)}$

and

(3) $b_1>a_1+\dfrac{(a_2-a_1)R(a_1)}{P'(a_1)}.$

Proof. By (1)

$$\prod_{k=1}^{n-2}(a_n-b_k)<\prod_{k=1}^{n-2}(a_n-a_k),$$

so that

$$\frac{R(a_n)}{a_n-b_{n-1}}<\frac{P'(a_n)}{a_n-a_{n-1}},$$

which is equivalent to (2). The proof of (3) is similar.

Remark. Even though orthogonality of P and R is not required when proving (2) and (3), it is easy to see that P and R are orthogonal polynomials whenever (1) is satisfied.

Problem 75-14*, Simultaneous Iteration towards All Roots of a Complex Polynomial, by M. W. GREEN, A. J. KORSAK and M. C. PEASE (Stanford Research Institute).

It has been found in practice that the following very simple (but very effective) procedure always converges for any n starting trial roots:

$$x'_i=x_i-P(x_i)\Big/\prod_{j\neq i}(x_i-x_j),\qquad\qquad i=1,2,\cdots,n,$$

where $P(x)$ is an arbitrary (complex coefficient) monic polynomial in x of degree n. In fact, even when $P(x)$ has multiple roots, the above procedure still converges, but only linearly (as opposed to quadratically in the distinct root case). Show that this procedure is globally convergent outside of a set of measure zero in the starting space and describe this set, for $n=2$. Show the same result, if possible for arbitrary n. It might be worth mentioning that this iteration was used by Weierstrass in 1903 as part of a more involved globally convergent procedure to prove constructively the fundamental theorem of algebra. The method has also recently appeared in various technical papers and even has been stated as an algorithm in *Comm. ACM*, but with no reference to possible global convergence.

Partial solution by R. D. SMALL (University of New Brunswick, New Brunswick, Canada).

We examine the case $n = 2$ and let the complex roots be A and B. Then the algorithm is

(1)
$$x_1' = x_1 - \frac{(x_1 - A)(x_1 - B)}{x_1 - x_2},$$

$$x_2' = x_2 - \frac{(x_2 - A)(x_2 - B)}{x_2 - x_1}.$$

Summing the equations gives

(2)
$$x_1' + x_2' = A + B.$$

Thus all the computed approximations satisfy this condition. If we let x_1 and x_2 be the first computed approximations, derived from some original ones, then they satisfy (2) as well, and we eliminate x_2 from (1), giving

(3)
$$x_1' = x_1 - \frac{(x_1 - A)(x_1 - B)}{2x_1 - A - B}.$$

For the case $A = B$, (3) becomes

$$x_1' - A = \frac{x_1 - A}{2},$$

with geometrical convergence of x_1 to A.

Now we assume that A and B are distinct. The convergence proof is simplified by sending A and B to ± 1 by the linear transformation

(4)
$$x_1 = \tfrac{1}{2}(A - B)Z + \tfrac{1}{2}(A + B).$$

The sequence of approximations satisfies

(5)
$$Z_{n+1} = \tfrac{1}{2}(Z_n + Z_n^{-1}), \qquad n = 1, 2, \cdots,$$

and thus the limit points satisfy a quadratic equation with roots ± 1. [*Editorial note*: Subsequently, the author shows that $\operatorname{Re} Z_1 > 0 \Rightarrow \text{limit} = +1$, $\operatorname{Re} Z_1 < 0 \Rightarrow \text{limit} = -1$ and $\operatorname{Re} Z_1 = 0 \Rightarrow$ no limit. A simpler proof follows easily from the explicit solution of (5), i.e.,

$$Z_{n+1} = \frac{(Z_1 + 1)^{2^n} + (Z_1 - 1)^{2^n}}{(Z_1 + 1)^{2^n} - (Z_1 - 1)^{2^n}.}]$$

Thus the imaginary axis represents the set of Z that does not converge and (4) maps this onto the right bisector of A and B in the original coordinates.

The original trial roots did not necessarily satisfy (2). The set of such roots, that give x_1' and x_2' on the right bisector of A and B, is found by taking

$$x_1' = iC(A - B) + \tfrac{1}{2}(A + B)$$

for C real. Then using the definition of x_1' and eliminating A and B by means of

$$P(x_i) = (x_i - A)(x_i - B), \qquad i = 1, 2,$$

we obtain

$$\frac{1}{4C^2} = \frac{4P(x_1)P(x_2)}{[P(x_1)+P(x_2)-(x_1-x_2)^2]^2} - 1,$$

contingent upon $x_1 \neq x_2$. Thus trial roots making the expression on the right positive real form a set of measure zero that does not give convergence. Of course, $x_1 = x_2$ must join this set as (1) is not defined.

Problem 69-5, *A Quartic*, by H. HOLLOWAY and M. S. KLAMKIN (Ford Scientific Laboratory).

Solve the quartic equation

$$x^4 + (1 - b)x^3 + (1 - 3a - b + 3a^2 + 3ab + b^2)x^2 + b(2 - 3a - 2b)x + b^2 = 0.$$

Also, show that if

$$0 < a + b < 1, \qquad a, b > 0,$$

then all the roots have absolute value less than unity.

The quartic arises from an analysis of x-ray diffraction by faulted cubic close-packed crystals. The crystal lattice may be regarded as an array of close-packed $\{1, 1, 1\}$ planes which can occupy three sets of stacking positions. In a faulted crystal the stacking has randomness and the diffraction problem requires specification of the probable stacking relationships between pairs of layers. This can be done by using geneological tables of stacking arrangements to generate difference equations which give the probability for a given stacking relationship as a function of the fault probability and of the separation between the layers. Growth faulting and intrinsic faulting both generate second order difference equations [1], [2]. A more complex model for growth faulting [3] gives a fourth order difference equation, and the quartic which results was solved numerically. Analysis of diffraction by crystals with both intrinsic and extrinsic faulting also yields a fourth order difference equation whose solution involves solution of the above quartic, where *a* and *b* are the intrinsic and extrinsic fault probability.

REFERENCES

[1] A. J. C. WILSON, *Imperfections in the structure of cobalt. II: Mathematical treatment of proposed structure*, Proc. Roy. Soc. London Sect. A, 180 (1942), pp. 277–285.

[2] M. S. PATERSON, *X-ray diffraction by face-centered cubic crystals with deformation faults*, J. Appl. Phys., 23 (1952), pp. 805–811.

[3] H. JAGODZINSKI, *Eindimensionale Fehlordnung in Kristallen und ihr Einfluss auf die Röntgeninterferenzen. I*, Acta Cryst., 2 (1949), pp. 201–207.

Solution by the Proposers.

The given quartic can be factored into the following complex form:

$$\{x^2 + [\omega(1 - b) - (\omega + \omega^2)a]x + \omega b\}\{x^2 + [-\omega^2(1 - b)$$

$$+ (\omega + \omega^2)a]x - \omega^2 b\} = 0,$$

where $\omega = (1 + i\sqrt{3})/2$ (a cube root of -1). Thus the four roots are

$$\tfrac{1}{2}\{\omega^2(1 - b) - (\omega + \omega^2)a \pm [-\omega(1 - b)^2 - 3a^2 + 2(1 + \omega)a(1 - b) + 4\omega^2 b]^{1/2}\},$$

$\frac{1}{2}\{-\omega(1 - b) + (\omega + \omega^2)a \pm [\omega^2(1 - b)^2 - 3a^2 + 2(1 - \omega^2)a(1 - b) - 4\omega b]^{1/2}\}.$

To show that the absolute value of the roots are less than one, we use the following theorem of Cauchy [M. MARDEN, *Geometry of Polynomials*, American Mathematical Society, Providence, R.I., 1966, p. 122]:
All the roots of

$$a_0 + a_1 z + \cdots + a_n z^n = 0, \qquad\qquad a_n \neq 0,$$

lie in the circle $|z| \leq r$, where r is the positive root of the equation

$$|a_0| + |a_1|z + \cdots + |a_{n-1}|z^{n-1} - |a_n|z^n = 0.$$

Taking one of the quadratic factors above, r is the positive root of

$$b + |\omega(1 - b) - (\omega + \omega^2)a|z - z^2 = 0$$

or

$$z^2 = z\left\{\left(\frac{1 - b}{2}\right)^2 + 3\left(\frac{1 - b}{2} - a\right)\right\}^{1/2} + b.$$

r will be less than one if

$$(1 - b)^2 > \left(\frac{1 - b}{2}\right)^2 + 3\left(\frac{1 - b}{2} - a\right)^2$$

or

$$3a(1 - b - a) > 0$$

(and similarly for the other quadratic term).
For applications of the solution, see the authors' paper, *Diffraction by fcc Crystals with Intrinsic and Extrinsic Faults*, J. Appl. Phys., **40** (1969), pp. 1681–1689.

Factorization of a Homogeneous Polynomial

Problem 81-8, (revised problem statement) *by* A. M. GLEASON (Harvard University).
Let P be a nonzero homogeneous polynomial in $\mathbb{R}[X_1, X_2, \cdots, X_n]$. Suppose that for any two real sequences $\langle a_1, a_2, \cdots, a_n \rangle$ and $\langle b_1, b_2, \cdots, b_n \rangle$,

$$P(a_1 s + b_1 t, a_2 s + b_2 t, \cdots, a_n s + b_n t)$$

is either identically zero or the product of linear factors in $\mathbb{R}[s, t]$. Prove that P is the product of linear factors in $\mathbb{R}[X_1, X_2, \cdots, X_n]$.

Solution by the proposer.
If P factors at all, then each of its factors must satisfy the hypothesis. Hence it is sufficient to prove that if P is irreducible and satisfies the hypothesis, then P has degree one. Suppose, on the contrary, that P is irreducible and $k = $ degree $P > 1$.
Consider the surface S in real projective $(n - 1)$-space given by $P = 0$. The hypothesis means that each projective line either lies wholly in S or meets S in k points counting multiplicity. It follows from the irreducibility of P that S has a regular point, that is, a point q at which not all the partial derivatives of P vanish, so there is a hyperplane π properly tangent to S at q. Now π does not lie wholly in S, for then P would

be divisible by the linear polynomial for π; so there is a line L in π passing through q but not lying wholly in S. Let r be a point of L not on S, and let M be an affine plane containing $L - \{r\}$ but not lying wholly in π. Then $S \cap M$ is a curve and L is tangent to it at q. A small parallel displacement of L produces a line L' meeting S in two imaginary points (at least), contrary to the hypothesis. This contradiction establishes the result.

Subsequent editorial note. W. Dahmen (Universität Bielefeld, Belfeld, West Germany) comments that this problem is solved in greater generality in his joint paper with C. A. Micchelli, *On the entire functions of affine lineage*, Proc. Amer. Math. Soc., 84 (1982), pp. 344-346. The paper arose from questions relating to multivariale splines. It also related to the work of Motzkin abd Schoenberg (which is referred to).

Problem 65-3, On the Zeros of a Set of Polynomials, by M. N. S. SWAMY (University of Saskatchewan, Regina).

The polynomials $B_n(x)$ and $b_n(x)$ defined by the following relations appear in electrical network theory:

$$b_n(x) = xB_{n-1}(x) + b_{n-1}(x), \qquad\qquad n > 0,$$

$$B_n(x) = (x + 1)B_{n-1}(x) + b_{n-1}(x), \qquad\qquad n > 0,$$

$$b_0(x) = B_0(x) = 1.$$

Determine the polynomials $B_n(x)$ and $b_n(x)$ and show that all of their zeros are real, negative, and distinct.

Solution by A. J. STRECOK (Argonne National Laboratory). It follows that

$$b_0 = 1, \qquad\qquad\qquad B_0 = 1,$$

$$b_1 = y - 1 + y^{-1}, \qquad\qquad B_1 = y + y^{-1},$$

$$b_{n+1} = (y + y^{-1})b_n - b_{n-1}, \qquad B_{n+1} = (y + y^{-1})B_n - B_{n-1},$$

where $y + y^{-1} = x + 2$.

Then using mathematical induction, it can be shown that

$$b_n = y^n - y^{n-1} + \cdots - y^{-(n-1)} + y^{-n} = y^{-n}(y^{2n+1} + 1)/(y + 1)$$

and

$$B_n = y^n + y^{n-2} + \cdots + y^{-(n-2)} + y^{-n} = y^{-n}(y^{2n+2} - 1)/(y^2 - 1).$$

The n distinct roots of b_n can be determined by taking $y^{2n+1} = e^{(2j-1)\pi i}$ for $j = 1, 2, \cdots, n$. Consequently the roots in terms of x are determined from

$$x + 2 = \exp\left(\frac{2j - 1}{2n + 1} \pi i\right) + \exp\left(\frac{-2j + 1}{2n + 1} \pi i\right) = 2 \cos\frac{2j - 1}{2n + 1} \pi < 2,$$

which shows that all roots are real, negative, and distinct.

The n distinct roots of B_n can be determined by taking $y^{2n+2} = e^{2j\pi i}$ for $j = 1, 2, \cdots, n$. Consequently the roots in terms of x are determined from

$$x + 2 = \exp\left(\frac{j\pi i}{n + 1}\right) + \exp\left(\frac{-j\pi i}{n + 1}\right) = 2 \cos\frac{j\pi}{n + 1} < 2,$$

again showing that all roots are real, negative, and distinct.

Finally, it is possible to write

$$b_n(x) = \prod_{j=1}^{n} \left(x + 2 - 2 \cos \frac{2j-1}{2n+1} \pi \right)$$

and

$$B_n(x) = \prod_{j=1}^{n} \left(x + 2 - 2 \cos \frac{j\pi}{n+1} \right).$$

Solution by H. E. FETTIS (Wright-Patterson AFB, Ohio).
Eliminating $b_n(x)$ between the two relations gives

$$B_{n+1}(x) - 2 \left(1 + \frac{x}{2} \right) B_n(x) + B_{n-1}(x) = 0,$$

$$B_0(x) = 1,$$

$$B_1(x) = x + 2.$$

The recurrence relation and the initial values are identical to those of the
Chebyshev polynomials $U_{n+1}(z)$, where

$$\sqrt{1 - z^2}\, U_n(z) = \sin(n \cos^{-1} z)$$

and $z = 1 + x/2$. Hence,

$$B_n(x) = U_{n+1} \left(1 + \frac{x}{2} \right).$$

The zeros of the Chebyshev polynomials are real, distinct, and lie in the interval
$-1 < z < 1$. It follows from this that if x_i is a zero of $B_n(x)$, then

$$-1 < 1 + \frac{x_i}{2} < 1,$$

or that

$$-4 < x_i < 0.$$

This proves the first assertion concerning the polynomials $B_n(x)$. For the polynomials $b_n(x)$, we have

$$b_{n-1}(x) = B_n(x) - (x + 1)B_{n-1}(x),$$

and also

$$b_{n-1}(x) = b_n(x) - xB_{n-1}(x).$$

Hence,

$$b_n(x) = B_n(x) - B_{n-1}(x) = U_{n+1} \left(1 + \frac{x}{2} \right) - U_n \left(1 + \frac{x}{2} \right);$$

or after some simplification,

$$(\sqrt{x+4})b_n(x) = 2 \cos \left[\left(n + \frac{1}{2} \right) \cos^{-1} \left(1 + \frac{x}{2} \right) \right].$$

Thus the zeros of $b_n(x)$ are given by

$$\left(n + \frac{1}{2} \right) \cos^{-1} \left(1 + \frac{x}{2} \right) = \pi/2,\ 3\pi/2,\ \cdots,\ (2n-1)\pi/2,$$

or by

$$x_i = 2 \left(\cos \frac{2i-1}{2n+1} \pi - 1 \right), \qquad i = 1, 2, \cdots, n,$$

which are clearly distinct and negative.

W. W. HOOKER (Philco Corporation) and W. G. TUEL, JR. (IBM, San Jose) in their solutions use Sturm's theorem to establish the indicated properties of the zeros. It would be of interest to obtain a general class of second order difference equations whose solutions are polynomials whose roots cannot be obtained explicitly but whose roots can all be shown to be negative by means of Sturm's theorem.

L. MAGAGNA (University of Toronto) notes in his solution that both $\{B_n\}$ and $\{b_n\}$ are special cases of a sequence of polynomials studied by W. LEDERMANN AND G. E. H. REUTER, *Spectral theory of the differential equations of birth-death processes*, Philos. Trans. Royal Soc. London, Ser. A 246 (1953–54). He also notes that recursion relations of the above type are satisfied by the characteristic determinant of certain ladder systems, be they ladder networks, simple queues, or distillation towers. The roots found above then become eigenvalues of the physical system. Spectral properties of a class of such systems, in particular the behavior of the eigenvalues under variation in matrix entries, are studied in his dissertation, *On a system described by a tridiagonal matrix and its control by parameter variation* (soon to be available at the library of the University of Toronto).

E. P. MERKES (University of Cincinnati) in his solution notes that b_n and B_n are also the $2n$th and $(2n+1)$st denominators, respectively, of the periodic continued fraction

$$\frac{1}{1+} \frac{1}{x+} \frac{1}{1+} \frac{1}{x+} \cdots .$$

Also, since the latter is easily converted into an S-fraction, the results on the zeros of $b_n(x)$ and $B_n(x)$ follow from Stieltjes' general theory of such continued fractions (H. S. Wall, *Analytic Theory of Continued Fractions*, Van Nostrand, New York, 1948, pp. 119–120).

Problem 64-16, Factorization of a Set of Polynomials, by N. MULLINEUX (College of Advanced Technology, Birmingham, England).

A system of polynomial equations occurring in control system theory can be reduced to the form

$$
\begin{aligned}
L_1: \quad & y && = 0, \\
L_2: \quad & y^2 - 1 && = 0, \\
L_3: \quad & y^3 - 2y && = 0, \\
L_4: \quad & y^4 - 3y^2 + 1 && = 0, \\
L_5: \quad & y^5 - 4y^3 + 3y && = 0,
\end{aligned}
$$

where the rows L_n satisfy the recurrence relation

(1) $$L_{n+1} = yL_n - L_{n-1}, \qquad n > 2.$$

By induction, one can show that

$$(2) \qquad L_n = \sum_{r=0}^{[n/2]} (-1)^r \binom{n-r}{r} y^{n-2r}.$$

It is of interest to the electrical engineer to be able to factorize L_n and consideration of a few cases for small n has led to the conjecture* that

$$(3) \qquad L_n = \prod_{k=1}^{n} \left\{ y + 2 \cos \frac{k\pi}{n+1} \right\}.$$

Prove the conjecture.

Solution by EVELYN FRANK (University of Illinois).

The polynomial equation $L_n = 0$ can also be written as the tridiagonal determinantal equation

$$L_n \equiv \begin{vmatrix} y & -1 & 0 & \cdots & & 0 \\ -1 & y & -1 & \cdots & & 0 \\ 0 & -1 & y & \cdots & & 0 \\ \vdots & \vdots & \vdots & & & \vdots \\ 0 & 0 & 0 & \cdots & -1 & y \end{vmatrix} = 0,$$

since it is easily shown that the determinant satisfies the same recurrence relation. Also, the given recurrence relation is the known one for Chebyshev polynomials (C. LANCZOS, *Applied Analysis*, Prentice Hall, Englewood Cliffs, New Jersey, 1956). It is well known that the solutions of such Chebyshev polynomial equations are $y = 2 \cos [k\pi/(n+1)]$, $k = 1, 2, \cdots, n$. Hence,

$$L_n = \prod_{k=1}^{n} \left(y - 2 \cos \frac{k\pi}{n+1} \right) = \prod_{k=1}^{n} \left(y + 2 \cos \frac{k\pi}{n+1} \right).$$

Editorial Note. Brenner refers to D. E. RUTHERFORD, *Some continuant determinants arising in physics and chemistry*, Proc. Roy. Soc. Edinburgh Sect. A, 62, 229–236. Carlitz refers to J. RIORDAN, *An Introduction to Combinatorial Analysis*, Wiley, New York, 1958, p. 222. Lind refers to SIR THOMAS MUIR, *Theory of Determinants*, Dover, New York, 1960, vol. 4, p. 401, which gives the following evaluation for the nth order continuant:

$$\begin{vmatrix} a & b & 0 & \cdot & \cdot & \cdot \\ c & a & b & \cdot & \cdot & \cdot \\ 0 & c & a & \cdot & \cdot & \cdot \\ \cdot & \cdot & \cdot & & & \\ \cdot & \cdot & \cdot & & a & b \\ \cdot & \cdot & \cdot & & c & a \end{vmatrix} = \prod_{k=1}^{n} \left\{ a - 2\sqrt{bc} \cos \frac{k\pi}{n+1} \right\}.$$

Navot refers to B. J. DASHER AND M. F. MOAD, *Analysis of four-terminal cascade networks*, IEEE Trans. Circuit Theory, CT-11 (1964), pp. 260–267, and V. O. MOWERY, *Fibonacci numbers and Tchebycheff polynomials in ladder networks*, Ibid., CT-8 (1961), pp. 167–168.

An Identity

Problem 85-10, *by* M. S. KLAMKIN (University of Alberta) *and* O. G. RUEHR (Michi-

---

* By J. F. Young, College of Advanced Technology, Birmingham, England.

gan Technological University).
Let

$$S(x,\ y,\ z,\ m,\ n,\ r) \equiv \frac{x^{m+1}}{m!} \sum_{j=0}^{n} \sum_{k=0}^{r} \frac{y^j z^k (j+k+m)!}{j! k!}.$$

Show that if $x+y+z=1$, then

$$S(x,\ y,\ z,\ m,\ n,\ r) + S(y,\ z,\ x,\ n,\ r,\ m) + S(z,\ x,\ y,\ r,\ m,\ n) = 1.$$

Solution by A. J. BOSCH *and* F. W. STEUTEL (Eindhoven University of Technology).
First let $x>0$, $y>0$, $z>0$ (and $x+y+z=1$) and consider the following probabilistic model.

Three urns labelled I, II and III contain m, n and r balls respectively. We perform independent drawings by choosing I, II and III with probabilities x, y and z and taking one ball from the urn chosen (without replacement). Let P_{I} be the probability defined by
$P_{\mathrm{I}} = P$ (I is the first urn to be chosen when empty). Then by elementary combinatorics we have

$$P_{\mathrm{I}} = x \sum_{j=0}^{n} \sum_{k=0}^{r} \frac{(m+j+k)!}{m! j! k!} x^m y^j z^k.$$

The probabilities P_{II} and P_{III} are defined similarly. Since eventually an empty urn will be chosen, we have $P_{\mathrm{I}} + P_{\mathrm{II}} + P_{\mathrm{III}} = 1$. Finally, a polynomial in x and y that is identically one for $x>0$, $y>0$, $x+y<1$ is also identically one without restriction.

Editorial note: Both Damjanovic and the proposers found the generalization

$$(*) \quad \sum_{i=1}^{n} x_i \sum_{m_1=0}^{p_1} \frac{x_1^{m_1}}{m_1!} \sum_{m_2=0}^{p_2} \frac{x_2^{m_2}}{m_2!} \cdots \sum_{m_n=0}^{p_n} \frac{x_n^{m_n}}{m_n!} \delta_{m_i,p_i} (m_1+m_2+\cdots+m_n)! = 1,$$

using the following generating function:

$$(**) \quad \sum_{i=1}^{n} x_i \left\{ \frac{1}{\prod_{j=1, j \neq 1}^{n}(1-u_j)} \right\} \left\{ \frac{1}{1-\sum_{s=1}^{n} u_s x_s} \right\} = \frac{1}{\prod_{i=1}^{n}(1-u_i)}.$$

The assertion is proved by repeated use of the following elementary identities:

$$\sum_{p=0}^{\infty} u^p \sum_{m=0}^{p} \frac{x^m}{m!} (m+k)! = \frac{1}{(1-u)} \frac{1}{(1-xu)^{k+1}},$$

$$\frac{1}{1-u} \sum_{q=0}^{\infty} v^q \sum_{s=0}^{q} \frac{y^s (s+k)!}{s!(1-xu)^{s+k+1}} = \frac{k!}{(1-u)(1-v)(1-xu-yv)^{k+1}}.$$

Finally to show that $(**)$ is an elementary algebraic identity, let

$$P_N = \prod_{i=1}^{N}(1-u_i), \quad L_N = \sum_{i=1}^{N} x_i, \quad R_N = \sum_{i=1}^{N} u_i x_i.$$

Then, if $L_N = 1$, we have

$$\frac{1}{P_N} = \frac{L_N - 1 + 1 - R_N}{P_N(1 - R_N)} = \sum_{i=1}^{N} \frac{x_i(1 - u_i)}{P_N(1 - R_N)}$$

$$= \sum_{i=1}^{N} x_i \left\{ \frac{1}{\prod_{j=1, j \neq 1}^{N}(1 - u_j)} \right\} \left\{ \frac{1}{1 - \sum_{s=1}^{N} u_s x_s} \right\},$$

which is (∗∗).

A simpler derivation of (∗) can be easily obtained by extending the probabilistic argument in the featured solution to n urns.

Supplementary References
Polynomials

[1] R.Askey, *Orthogonal Polynomials and Special Functions*, SIAM, Philadelphia, 1975.

[2] J.W.Archbold, *Algebra*, Pitman, London,

[3] E.J.Barbeau, *Polynomials*, Springer-Verlag, N.Y., 1989.

[4] S. Barnett, *Polynomials and Linear Control Systems*, Dekker, N.Y., 1983.

[5] M.Bocher, *An Introduction to Higher Algebra*, MacMillan, N.Y., 1935.

[6] W.S.Burnside and A.W.Panton, *The Theory of Equations, I,II*, Dover, N.Y., 1960.

[7] T.S.Chihara, *An Introduction to Orthogonal Polynomials*, Gordon and Breach, N.Y., 1978.

[8] D.K.Fadeev and I.S. Sominskii, *Problems in Higher Algebra*, Freeman, San Francisco, 1965.

[9] R.P.Feinerman and D.J.Newman, *Polynomial Approximation*, Williams & Wilkins, Baltimore, 1974.

[10] E.Grosswald, *Bessel Plynomials*, Springer-Verlag, Berlin, 1978.

[11] H.Lausch and W.Nobauer, *Algebra of Polynomials*, North-Holland, Amsterdam, 1973.

[12] I.G.MacDonald, *Symmetric Functions and Hall Polynomials*, Clarendon Press, Oxford, 1979.

[13] M.Marden, *Geometry of Polynomials*, AMS, Providence, 1966.

[14] A.P.Mishina and I.V.Proskurykov, *Higher Algebra*, Pergamon, London, 1965.

[15] A.Mostowski and M.Stark, *Introduction to Higher Algebra*, Pergamon, London, 1964.

[16] A.Schinzel, *Selected Topics on Polynomials*, University of Michigan Press, Ann Arbor, 1982.

[17] G.Szego, *Orthogonal Polynomials*, AMS, Providence, 1975.

[18] J.V.Uspensky, *Theory of Equations*, McGraw-Hill, N.Y., 1948.

[19] L.Weisner, *Introduction to the Theory of Equations*, Macmillan, N.Y.,

18. SIMULTANEOUS EQUATIONS

Problem 60–9*, A Set of Convolution Equations, by WALTER WEISSBLUM (AVCO Research and Advanced Development Division).

In simulating the operation of a certain radar detection system on a computer, it was necessary to have a method of producing a stationary sequence of normal random variables with mean zero, and with a prescribed autocorrelation sequence

$$\cdots, 0, 0, 0, C_n, C_{n-1}, \cdots, C_1, C_0, C_1, \cdots, C_{n-1}, C_n, 0, 0, 0, \cdots$$

That is, the sequence X_i of random variables produced must satisfy $E(X_i) = 0$ and

$$E(X_i \cdot X_j) = C_{|i-j|} \quad \text{for} \quad |i - j| \leqq n,$$

$$E(X_i \cdot X_j) = 0 \qquad \text{for} \quad |i - j| > n.$$

The following method was proposed: First produce a sequence Y_i of independent normal random variables with mean zero and standard deviation 1 and then define

$$X_i = \sum_{j=0}^{n} a_j Y_{i-n+j}$$

with suitable constants $a_0, a_1, \cdots, a_n$. This construction clearly provides stationarity, normality, mean zero, and zero correlation for $|i - j| > n$. The only question is the existence of suitable real $\{a_i\}$ to produce $E(X_i \cdot X_j) = C_{|i-j|}$ for $|i - j| \leqq n$. This reduces to determining the existence of a real solution to the following set of convolution equations:

(1)
$$\begin{aligned}
a_0^2 + a_1^2 + \cdots + a_n^2 &= C_0, \\
a_0 a_1 + a_1 a_2 + \cdots + a_{n-1} a_n &= C_1, \\
a_0 a_2 + a_1 a_3 + \cdots + a_{n-2} a_n &= C_2, \\
\vdots \qquad \vdots \qquad\qquad \vdots \quad \vdots \\
a_0 a_n &= C_n.
\end{aligned}$$

Determine a necessary and sufficient condition on the $\{C_i\}$ such that real $\{a_i\}$ exist, and give a method for finding them.

I. Solution by A. W. MCKINNEY (Sandia Corporation).

In order that there exist a sequence of real normally distributed random variables $\{X_j\}$ with means equal to zero and with covariance function C_j, where $C_n \neq 0$ and $C_j = 0$ for $j > n$, it is necessary and sufficient that the sequence

537

$\{C_j\}$ be positive definite [1]. The sequence $\{C_j\}$ is positive definite if and only if the function

$$G(\lambda) = 2C_0 + \frac{2}{\pi} \sum_{h=1}^{n} C_h \frac{\sin 2\pi h\lambda}{h}$$

is monotone nondecreasing for $0 \leq \lambda \leq \frac{1}{2}$ [1], thus if and only if the function

$$F(\lambda) = \frac{1}{2}G'(\lambda) = C_0 + 2\sum_{h=1}^{n} C_h \cos 2\pi h\lambda$$

is nonnegative for $0 \leq \lambda \leq \frac{1}{2}$. Since $\cos 2\pi h \left(\frac{1}{2} + \lambda\right) = \cos 2\pi h \left(\frac{1}{2} - \lambda\right)$ for integers h and all real λ, it follows that the sequence $\{C_j\}$ is positive definite if and only if the trigonometric polynomial

$$(2) \qquad\qquad F(\lambda) = C_0 + 2\sum_{h=1}^{n} C_h \cos 2\pi h\lambda$$

is nonnegative for all real λ. By a theorem of L. Fejér and F. Riesz [2], a trigonometric polynomial $T(\lambda)$ is nonnegative for all real λ if and only if it can be written in the form

$$T(\lambda) = \left| \sum_{h=0}^{n} a_h e^{2\pi i h\lambda} \right|^2,$$

where the coefficients $a_0, \cdots, a_n$ are real or complex numbers. An examination of the proof given by Szego for this theorem [2] yields as an immediate consequence the fact that the coefficients $a_0, \cdots, a_n$ can be taken to be all real if and only if in the trigonometric polynomial

$$T(\lambda) = C_0 + 2\sum_{h=1}^{n} (C_h \cos 2\pi h\lambda + D_h \sin 2\pi h\lambda),$$

the coefficients D_h are equal to zero for $h = 1, \cdots, n$. Since this is the case for $F(\lambda)$ in (2), it follows that the sequence $\{C_j\}$ is positive definite if and only if there exist real numbers $a_0, \cdots, a_n$ such that

$$(3) \qquad\qquad C_0 + 2\sum_{j=1}^{n} C_j \cos 2\pi j\lambda = \left| \sum_{h=0}^{n} a_h e^{2\pi i\lambda} \right|^2,$$

or

$$(4) \qquad C_0 + 2\sum_{j=1}^{n} C_j \cos 2\pi j\lambda = \sum_{h=0}^{n} a_h^2 + 2\sum_{j=1}^{n} \sum_{h=0}^{n-j} (a_h a_k + j \cos 2\pi j\lambda).$$

Multiplying each member of (4) by $\cos 2\pi k\lambda$ and integrating over the range $\lambda = 0$ to $\lambda = \frac{1}{2}$, it follows that

$$C_k = \sum_{h=0}^{n-k} a_h a_{h+k} \text{ for } k = 0, \cdots, n.$$

Therefore, *there exist real numbers* $a_0, \cdots, a_n$ *such that* $\sum_{h=0}^{n-k} a_h a_{h+k} = C_k$ *if and only if the sequence* $\{C_j\}$ *is positive definite.* If a sequence $\{C_j\}$ is to be tested for positive definiteness, the most convenient approach on a digital computer probably is to compute the polynomial $F(\lambda)$ defined by (2) for many values of λ between 0 and $\frac{1}{2}$, and see if $F(\lambda)$ remains positive. If a sequence $\{C_j\}$ is pre-

sumed to be positive definite, and if the coefficients a_0, $\cdots$, a_n are to be found, the only feasible methods are iterative. It should be noted that there are, in general, several real sets of coefficients a_0, $\cdots$, a_n which satisfy the required equations—in fact, if a_0, $\cdots$, a_n is one such sequence, then so is $b_0 = a_n$, $b_1 = a_{n-1}$, $\cdots$, $b_n = a_0$—and therefore, the choice of a starting value may well affect the rate of convergence of Newton's method. (Of course, Newton's method always converges if the starting values are close enough to the required solution.)†
One possible choice of starting values is to set $a_h = C_h$ for $h = 0, 1, \cdots, n$. It may prove desirable in an iterative solution to make use of the fact (easily deduced from (3) by putting $\lambda = 0$ or $\frac{1}{2}$, respectively) that

$$a_0 + a_1 + a_2 + \cdots + a_n = (C_0 + 2C_1 + 2C_2 + \cdots + 2C_n)^{1/2},$$

$$a_0 - a_1 + a_2 - \cdots \pm a_n = (C_0 - 2C_1 + 2C_2 - \cdots \pm 2C_n)^{1/2}.$$

REFERENCES

1. J. L. Doob, *Stochastic Processes*, John Wiley and Sons, Inc., New York, 1953, pp. 72, 474.
2. U. Grenander and G. Szego, *Toeplitz Forms and their Applications*, University of California Press, Berkeley, 1958, pp. 20, 21.

IV. Solution by Henry D. Friedman (Sylvania Electronics Systems).

Let

$$S(z) = \sum_{i=0}^{n} a_{n-i} z^i,$$

$$P(z) = \sum_{k=-n}^{n} C_k z^{n+k}.$$

The expression $S(z)\,S(1/z)z^n = P(z)$ generates the given convolution equations. Thus to solve for real $\{a_i\}$ we seek a polynomial $S(z)$, with real coefficients, satisfying $S(z)\,S(1/z)z^n = P(z)$. This expression hears a useful similarity to the algebraic equation $x^2 = a$. In analogy with the ordinary Newton iterative method for finding a root of $x^2 = a$, we make an initial guess $S_0(z)$ of $S(z)$, divide into $P(z)$, and obtain a quotient polynomial $S_0^*(z)$ (and a remainder, which we will ignore). If $S_0(z)$ is sufficiently close to $S(z)$ then $S_0^*(z)$ will be close to $S(1/z)z^n$, which is simply $S(z)$ with coefficients in reverse order. In the simplest analogy with Newton's method, we would average corresponding coefficients to obtain the next approximation $S_1(z)$ to $S(z)$, and repeat the procedure.

A more efficient method may be obtained by analyzing the propagation of error. We first express the coefficients of $P(z)$ in terms of the $\{a_i\}$, then divide $P(z)$ by $S_0(z) = \sum (a_{n-i} + \epsilon_{n-i})z^i$, where the $\{\epsilon_{n-i}\}$ are error terms. In dividing by $a_0 + \epsilon_0$ to determine the coefficients of the quotient polynomial, we use $(a_0 + \epsilon_0)^{-1} \approx a_0^{-1}(1 - \epsilon_0/a_0)$, and we ignore all terms of second or higher degree in the ϵ's. The resulting approximation to the quotient polynomial takes the form $S_0^*(z) \approx \sum (a_j + \delta_j)z^i$, where

$$\delta_j = \sum_{r=0}^{n-j} A_{n-j-r} \epsilon_r,$$

and where the $\{A_{n-j-r}\}$ are functions of the $\{a_i\}$ only.

---

† *Editorial Note:* There are pathological cases where this is not true.

Let $b_i = a_i + \epsilon_i$, $d_j = a_j + \delta_j$. Since, in practice, we would guess (initially) at the $\{b_i\}$, and determine by division the $\{d_j\}$, these values are known, while the $\{a_i\}$ and $\{\epsilon_i\}$ are unknown. Using $a_i = b_i - \epsilon_i = d_i - \delta_i$, we obtain $n + 1$ equations in the $\{\epsilon_i\}$ given by

$$b_i - \epsilon_i = d_i - \sum_{r=0}^{n-i} A_{n-i-r}\epsilon_r .$$

If, finally (again in analogy with Newton's method), we substitute the known $\{b_i\}$ for the unknown $\{a_i\}$ in the functions $\{A_{n-i-r}\}$, we may solve these equations for the $\{\epsilon_i\}$. The new approximation $S_1(z)$ to $S(z)$ will have as coefficients $\{b_i - \epsilon_i\}$. The iterative algorithm is thus complete, and can be applied repeatedly.

A third method, a compromise between the two methods given above, is developed by ignoring all error terms which appear below the quotient in the division of $P(z)$ by $S_0(z)$. For example, by synthetic division we have

$$
\begin{array}{llll}
a_0 + \epsilon_0, & a_1 + \epsilon_1, & a_2 + \epsilon_2 & \dfrac{a_2 - a_2\,\epsilon_0/a_0 \quad a_1 - a_1\,\epsilon_0/a_0 \quad a_0 - \epsilon_0}{|a_0\,a_2 \qquad\qquad a_0\,a_1 + a_1\,a_2 \quad a_0^2 + a_1^2 + a_2^2 \cdots}
\end{array}
$$

$a_0 + \epsilon_0,$	$a_1 + \epsilon_1,$	$a_2 + \epsilon_2$	$a_2 - a_2\,\epsilon_0/a_0$	$a_1 - a_1\,\epsilon_0/a_0$	$a_0 - \epsilon_0$
			$\underline{|a_0\,a_2}$	$a_0\,a_1 + a_1\,a_2$	$a_0^2 + a_1^2 + a_2^2 \cdots$
			$a_0\,a_2$	$a_1\,a_2$	$a_0^2 + a_1^2 \quad \cdots$
			$\underline{0}$	$a_0\,a_1$	$a_0^2 + a_1^2 \quad \cdots$
				$a_0\,a_1$	$a_1^2 \qquad \cdots$
				$\underline{0}$	$a_0^2 \qquad \cdots$
					$a_0^2 \qquad \cdots$
					$\underline{0}$

For general n, the quotient polynomial is given by

$$S_0^*(z) \approx \sum_{j=0}^{n} a_j \left(\frac{1 - \epsilon_0}{a_0}\right) z^j.$$

Again, using $b_i = a_i + \epsilon_i$, $d_j = a_j(1 - \epsilon_0/a_0)$, we have the $n + 1$ relations $a_i = b_i - \epsilon_i = d_i(1 - \epsilon_0/a_0)^{-1}$. Multiplying both sides by $(1 - \epsilon_0/a_0)$, ignoring the $\epsilon_0\epsilon_1$ term, and using b_0 in place of a_0, we obtain $\epsilon_i \approx b_i(1 - \epsilon_0/b_0) - d_i$. At $i = 0$, $\epsilon_0 = (b_0 - d_0)/2$, and all the $\{\epsilon_i\}$ can be found. The corrected coefficients of the new trial divisor $S_1(z)$ are given by

$$b_i' = b_i - \epsilon_i = d_i + b_i\epsilon_0/b_0 ,$$

whichs completes the iterative algorithm.

The last method has been programmed and used successfully on a number of problems. In methods of this type, which are designed to converge when the $\{\epsilon_i\}$ are small, one seldom expects a general guarantee of convergence. (The best known iterative methods for extracting a quadratic factor from a polynomial have the same deficiency.) Nevertheless, the method has been applied, thus far without failure, to a number of different $\{C_i\}$ sequences satisfying the positive definite criterion, for values of n as high as 16, and for widely different starting values of the $\{b_i\}$. An interesting feature is that where more than one answer $S(z)$ exists, the method has always converged to the answer with highest leading coefficient a_0, or to its negative. For example, if $P(z)$ is chosen to have the roots $-1, -2, -3, -4$ and their reciprocals, we obtain $C_3 = 24$, $C_2 = 242$, $C_1 = 867$, $C_0 = 1334$. Using the known roots, it is a simple matter to find the 16 possible answer sequences a_0, a_1, a_2, a_3, which are

$$
\begin{array}{rrrr}
6, & 29, & 21, & 4 \\
8, & 30, & 19, & 3 \\
12, & 31, & 15, & 2 \\
24, & 26, & 9, & 1
\end{array}
$$

together with their reverses and negatives. On the computer, the method always converges to the answer sequence 24, 26, 9, 1 (or its negative), regardless of how close to one of the other answers the initial guess is taken (except when the initial trial falls exactly on another answer).

All three of the above methods can be improved, both at the initial guess and at each iteration, by performing the iteration only on (roughly) the first half of the answer sequence. The rest of the answer sequence can be filled in during the division. For instance, having assumed an initial value b_0 (or having calculated a later value b_0') for a_0, a relatively good substitute for a_n is the first coefficient $d_n = C_n/b_0 = a_0 a_n/b_0$ of the quotient polynomial. It is always possible to perform the division so that the first $(n + 1)/2$ or $(n + 2)/2$ b_i's (n odd or even, respectively) determine the other b_i's. This can also be done using the original convolution equations, but the polynomial division algirithm is a more natural vehicle for the problem.

Comment by M. J. LEVIN (RCA Missile Electronics and Controls Division).

A similar problem arises in the design of phased array antennas. A solution for complex $\{C_i\}$ is given by S. SILVER, *Microwave Antenna theory and Design*, Vol. 12, Radiation Laboratory Series, McGraw-Hill, 1949, pp. 280–1.

This problem also arose in connection with a method of filter synthesis suggested by R. THORTON of M. I. T. who has a solution similar to the one above.

Problem 73-6. Determination of a Sequence, by MICHAEL A. NARCOWICH (Polytechnic Institute of Brooklyn).

Determine the sequence A_n, where

$$
-A_{2k} = \sum_{m=1}^{2k-1} (-1)^m A_m A_{2k-m}, \qquad k = 1, 2, 3, \cdots,
$$

$$
A_{2k+1} = 2A_{2k},
$$

and

$$
A_1 = 1.
$$

The problem arose in determining a solution of Schrödinger's equation for a slowly moving two-state atom in the presence of a certain plane wave.

Solution by D. R. BREACH (University of Canterbury, Christchurch, New Zealand).

Let $G(x)$ denote the generating function

$$
G(x) = \sum_{n=0}^{\infty} A_n x^n.
$$

Then the last two conditions are satisfied if $2A_0 = 1$, $A_1 = 1$ and $G(x) = (2x + 1)F(x)$, where $F(x)$ is even. The first condition requires $\frac{1}{2} - F(x)$

$= [G(x) - \frac{1}{2}][G(-x) - \frac{1}{2}]$ or $4(1 - 4x^2)F^2(x) = 1.$
Thus, $2G(x) = (1 + 2x)(1 - 4x^2)^{-1/2}$ and

$$A_{2n} = \binom{2n-1}{n}.$$

Problem 61-8, *On a Set of Linear Equations,* by B. W. Rosen (General Electric Company).

Solve the following set of n linear equations:

$$-\lambda X_1 + X_2 = A_1,$$

$$X_{r-1} - 2X_r + X_{r+1} = A_r, \qquad r = 2, 3, \cdots, n-1,$$

$$X_{n-1} - \lambda X_n = A_n.$$

This problem has arisen in the analysis of manned re-entry vehicles subject to moderately high heating rates for long-time periods. The use of a series of parallel plates in radiation equilibrium is considered as a heat shielding technique. It is assumed that the top plate is subject to a constant heat flux while the bottom plate radiates to a zero temperature medium. It is further assumed that the thermal properties are constant and are the same for each plate. X_r denotes the fourth power of the temperature of the rth plate. The A_r terms arise from the fact that each plate is also assumed to be a heat sink due to a heat absorbing material which may be used on some of the layers.

Solution by N. A. Lindberger (IBM Nordiska Laboratorier, Sweden).

The difference equation

$$\Delta^2 X_{r-1} = X_{r-1} - 2X_r + X_{r+1} = A_r \qquad (r = 2, 3, \cdots, n)$$

has a particular solution

$$X_1{}^p = 0, \qquad X_r{}^p = \sum_{k=1}^{r-1} \sum_{i=1}^{k} A_i,$$

and a complementary solution
$$X_r{}^c = \alpha r + \beta \qquad\qquad (r = 1, 2, \cdots, n).$$

The general solution will then be

$$X_r = \sum_{i=1}^{r} (r - i)A_i + \alpha r + \beta, \qquad (r = 1, 2, \cdots, n),$$

after having rewritten the double sum. The constants α and β are determined by the boundary conditions represented by the first and last equations of the problem. It then follows that

$$\alpha = [\lambda(n - 1) - (n - 3)]^{-1} \sum_{i=1}^{n} \{(n - i)(1 - \lambda) - 1\}A_i,$$

$$\beta = \frac{2 - \lambda}{1 - \lambda}\,\alpha,$$

unless $\lambda = 1$ or $\lambda = \dfrac{n-3}{n-1}$. These are the two values of λ that make the system determinant

$$\Delta = (-1)^n (\lambda - 1) [\lambda(n-1) - (n-3)]$$

equal to zero, in which case the problem has generally no solution.

Editorial note: It is to be noted that the first solution is incomplete since no mention is made of the critical values of λ. Also, while the second solution notes the critical values of λ, it doesn't consider the special cases when there is a solution for these critical values.

The solution for these critical values (as well as in general) is given by ALAN G. KONHEIM (IBM), and E. G. KOGBETLIANTZ (N. Y. C.), i.e.,

(1) $$\lambda = 1.$$

A solution exists if and only if $\sum_1^n A_j = 0$, and in which case

$$X_r = \sum_{j=1}^{r-1} (r-j)A_j + X_1 \qquad (r = 2, 3, \cdots, n),$$

with X_1, arbitrary.

(2) $$\lambda = \frac{n-3}{n-1}.$$

A solution exists if and only if

$$\sum_{j=1}^n \left\{ 1 - 2\frac{n-j}{n-1} \right\} A_j = 0,$$

in which case

$$X_r = \sum_{j=1}^{r-1} (r-j)A_j + \left\{ r - (r-1)\frac{n+1}{n-1} \right\} X_1 \qquad (r = 2, 3, \cdots, n),$$

with X_1 arbitrary.

KURT EISEMANN (Remington Rand Univac) indicated the solution to the more general system.

$$C_r X_{r-1} + D_r X_r + E_r X_{r+1} = A_r.$$

One can extend the previous results to matrices whose rows are the alternating binomial coefficients. Let us just consider the case of 4th order coefficients, i.e.,

(1) $$\sum_1^n a_i X_i = A_1,$$

(2) $$\sum_1^n b_i X_i = A_2,$$

(3) $$X_{r-2} - 4X_{r-1} + 6X_r - 4X_{r+1} + X_{r+2} = A_r, \qquad (r = 3, 4, \cdots, n-2),$$

(4) $$\sum_1^n c_i X_i = A_{n-1},$$

(5) $$\sum_1^n d_i X_i = A_n.$$

The complementary solution of Eqs. (3) $(\Delta^4 X_{r-2} = A_r)$ is given by

$$X_r^c = \alpha_1 + \alpha_2 r + \alpha_3 r^2 + \alpha_4 r^3, \qquad (r = 1, 2, \cdots, n).$$

A particular solution is gotten by expanding out the inverse operator of Δ^4 i.e.,

$$X_{r-2}^p = \frac{1}{\Delta^4} A_r = \frac{1}{E^4} \left\{ 1 - \frac{1}{E} \right\}^{-4} A_r,$$

or

$$X_{r-2}^p = \frac{1}{3!} \{ 3 \cdot 2 \cdot 1 A_{r-4} + 4 \cdot 3 \cdot 2 A_{r-5} + \cdots + (r-2)(r-3)(r-4)A_1 \}.$$

The general solution will then be

$$X_r = X_r^c + X_r^p, \qquad (r = 1, 2, \cdots, n),$$

and the four arbitrary constants α_1, α_2, α_3, α_4, are determined from Eqs. (1), (2), (4), and (5). Again there will be special cases when the coefficient determinant of the system vanishes.

Problem 64–4, *A Set of Linear Equations*, by M. J. Pascual and J. E. Zweig (Watervliet Arsenal).

If

$$\sum_{n=1}^{\infty} \frac{A_n}{2m - 2n + 1} = \frac{1}{2m}, \qquad m = 1, 2, 3, \cdots,$$

and

$$\sum_{n=1}^{\infty} \frac{2nB_n}{2n - 2m + 1} = \frac{1}{2m - 1}, \qquad m = 1, 2, 3, \cdots.$$

determine A_n and B_n and show that

$$B_n = \sum_{n=1}^{\infty} \frac{2(2n - 1)A_n}{(2m - 2n + 1)(2m + 2n - 1)}.$$

This problem has arisen in determining a series solution of a heat conduction problem in an infinite slab contained between two parallel planes in which a semi-infinite plane filament (which is a perfect insulator) is imbedded parallel to the bounding planes of the slab.

Solution by R. A. Hurd (National Research Council, Canada).

We consider the integral

$$(1) \qquad I = \int_C \frac{f(z)}{2m - z} \frac{\Gamma(\tfrac{1}{2}z)}{\Gamma(\tfrac{1}{2}z + \tfrac{1}{2})} \tan\left(\tfrac{1}{2}\pi z\right) \cdot \frac{dz}{z}, \qquad m = 1, 2, \cdots,$$

where C is a circle of infinite radius which encloses but does not cut the poles of $\tan \tfrac{1}{2}\pi z$. Let $f(z)$ be regular within C. The integrand has simple poles at $z = 0$ and at $z = 2n - 1$, $n = 1, 2, \cdots$. If the integrand is $O(z^{-1-\delta})$ at infinity, with $\delta > 0$, then $I = 0$. A residue calculation then gives, for $m = 1, 2, \cdots$,

(2) $\qquad -\dfrac{2}{\pi} \displaystyle\sum_{n=1}^{\infty} \dfrac{f(2n-1)}{2m-2n+1} \dfrac{\Gamma(n-\tfrac{1}{2})}{\Gamma(n)} \dfrac{1}{2n-1} + \dfrac{\pi f(0)}{2m\Gamma(\tfrac{1}{2})} = 0.$

Upon comparison with

$$\sum_{n=1}^{\infty} \frac{A_n}{2m-2n+1} = \frac{1}{2m},$$

we obtain

(3) $\qquad A_n = 2\pi^{-\tfrac{1}{2}} \cdot \dfrac{\Gamma(n-\tfrac{1}{2})}{\Gamma(n)} \cdot \dfrac{f(2n-1)}{f(0)} \cdot \dfrac{1}{2n-1}, \qquad n = 1, 2, \cdots.$

To determine $f(z)$, we examine the integrand for large $|z|$. By Stirling's formula,

$$\frac{\Gamma(\tfrac{1}{2}z)}{\Gamma(\tfrac{1}{2}z+\tfrac{1}{2})} = O(z^{-1/2}),$$

for large $|z|$, z not negative real. For $z = -|z|$, the relation $\Gamma(x)\Gamma(-x) = \pi \csc \pi x / x$ yields

$$\frac{\Gamma(-\tfrac{1}{2}|z|)}{\Gamma(-\tfrac{1}{2}|z|+\tfrac{1}{2})} = -\frac{2}{|z|} \cot\,(\tfrac{1}{2}\pi\,|z|)\,\frac{\Gamma(\tfrac{1}{2}+\tfrac{1}{2}|z|)}{\Gamma(\tfrac{1}{2}|z|)},$$

which is also $O(|z|^{-1/2})$ for $|z|$ large. Thus, for the integrand to be $O(z^{-1-\delta})$, we must have

$$f(z) = O(z^{\tfrac{3}{2}-\delta}), \qquad \delta > 0.$$

Then by the extension to Liouville's theorem, $f(z)$ is a first degree polynomial:

(4) $\qquad\qquad\qquad f(z) = f(0)[1+az],$

so that

(5) $\qquad A_n = 2\pi^{-\tfrac{1}{2}} \cdot \dfrac{[1+(2n-1)a]}{2n-1} \cdot \dfrac{\Gamma(n-\tfrac{1}{2})}{\Gamma(n)},$

where a is arbitrary. (Presumably there are some physical conditions which will serve to determine a.)

The second equation

(6) $\qquad \displaystyle\sum_{n=1}^{\infty} \dfrac{2nB_n}{2n-2m+1} = \dfrac{1}{2m-1}, \qquad m = 1, 2, \cdots,$

is handled similarly. The integral used is

(7) $\qquad \displaystyle\int_{c} \dfrac{g(z)}{z-2m+1} \dfrac{\Gamma(\tfrac{1}{2}z+\tfrac{1}{2})}{\Gamma(\tfrac{1}{2}z)} \cot\,(\tfrac{1}{2}\pi z)\, \dfrac{dz}{z},$

where C is a large contour enclosing, but not cutting, the poles of $\cot(\tfrac{1}{2}\pi z)$, and $g(z)$ is regular. We obtain

(8) $\qquad \dfrac{2}{\pi} \displaystyle\sum_{n=1}^{\infty} \dfrac{g(2n)}{2n-2m+1} \dfrac{\Gamma(n+\tfrac{1}{2})}{\Gamma(n)} \dfrac{1}{2n} + \dfrac{\Gamma(\tfrac{1}{2})g(0)}{\pi(1-2m)} = 0.$

Thus

$$(9) \qquad B_n = \frac{1}{2\sqrt{\pi}\, n^2} \frac{\Gamma(n + \frac{1}{2})}{\Gamma(n)} \frac{g(2n)}{g(0)}.$$

For large $|z|$ the integrand is $O[g(z) \cdot z^{-\frac{3}{2}}]$, so that in that case $g(z) = g(0)$, and we have

$$(10) \qquad B_n = \frac{1}{2\sqrt{\pi} \cdot n^2} \frac{\Gamma(n + \frac{1}{2})}{\Gamma(n)}.$$

To show that

$$(11) \qquad B_m = \sum_{n=1}^{\infty} \frac{2(2n - 1)A_n}{(2m - 2n + 1)(2m + 2n - 1)},$$

we modify (1) to

$$(12) \qquad \int_c \frac{f(z)}{4m^2 - z^2} \cdot \frac{\Gamma(\frac{1}{2}z)}{\Gamma(\frac{1}{2}z + \frac{1}{2})} \cdot \tan\left(\frac{1}{2}\pi z\right) dz,$$

which has poles at $z = -2m$ and at $z = 2n - 1$, $m = 1, 2, \cdots$. A residue calculation yields

$$(13) \qquad -\frac{2}{\pi} \sum_{n=1}^{\infty} \frac{f(2n - 1)}{4m^2 - (2n - 1)^2} \cdot \frac{\Gamma(n - \frac{1}{2})}{\Gamma(n)} + \frac{f(-2m)}{4m^2} \frac{\Gamma(m + \frac{1}{2})}{\Gamma(m)} = 0,$$

which with (3) and (10) gives

$$\sum_{n=1}^{\infty} \frac{2(2n - 1)A_n}{4m^2 - (2n - 1)^2} = \frac{f(-2m)}{f(0)} B_m,$$

which reduces to the required relation only if $a = 0$.

Editorial Note: Hansen and Trench based their solutions on the known identity

$$\sum_{n=1}^{\infty} \frac{\Gamma(n - a)}{\Gamma(n)\Gamma(1 - a)} \cdot \frac{1}{z + n - 1} = \frac{\Gamma(z)\Gamma(a)}{\Gamma(z + a)}.$$

It also follows from this identity that the homogeneous set of equations

$$(i) \qquad \sum_{n=1}^{\infty} \frac{A_n}{2m - 2n + 1} = 0, \qquad m = 1, 2, 3, \cdots,$$

has a solution

$$A_n = \frac{k\Gamma(n - \frac{1}{2})}{\Gamma(n)}.$$

Similar results hold for B_n.

The proposers also raise some conjectures relating to the solution of a finite set of equations of the form (i). These can be answered by inverting the coefficient matrix which is a special case of

$$M \equiv \left\| \frac{1}{a_r + b_s} \right\|, \qquad r, s, = 1, 2, \cdots, n,$$

where $a_r \neq a_s$ and $b_r \neq b_s$ for $r \neq s$. The inversion of matrix M is given in J. Edwards, *A Treatise on the Integral Calculus*, Vol. II, Chelsea, New York,

1954, p. 900, and by S. Schechter, *On the inversion of certain matrices*, MTAC, 13 (1959), pp. 73–77.

A compact way of obtaining the inverse is as follows. Consider the set of equations

$$\sum_{r=1}^{n} \frac{x_r}{a_r + b_s} = \delta_{sj} , \qquad\qquad s = 1, 2, \cdots, n.$$

The relation in y of

$$\sum_{i=1}^{n} \frac{x_i}{a_i + y} = \prod_{r=1}^{n} \left\{ \frac{a_r + b_j}{a_r + y} \right\} \cdot \prod_{\substack{r=1 \\ r \neq j}}^{n} \left\{ \frac{y - b_r}{b_j - b_r} \right\}$$

is satisfied by $y = b_s$, $s = 1, 2, \cdots, n$, and since this relation can be expressed as a polynomial of degree $(n - 1)$ in y it must be an identity.

Thus, by letting $y \to -a_i$,

$$x_i = \frac{1}{a_i + b_j} \prod_{\substack{r=1 \\ r \neq i}}^{n} \left\{ \frac{a_r + b_j}{a_r - a_i} \right\} \cdot \prod_{\substack{r=1 \\ r \neq j}}^{n} \left\{ \frac{b_r + a_i}{b_r - b_j} \right\} ,$$

which is the (i, j) term of the inverse matrix.

Problem 68-7, An Infinite Set of Simultaneous Equations, by L. CARLITZ (Duke University).

Find the general solution of

(1)
$$U_m U_n = \sum_{r=0}^{\min(m,n)} \binom{m}{r}\binom{n}{r} r! 2^r V_{m+n-2r},$$

$$U_0 = V_0 = 1, \quad m, n = 0, 1, 2, \cdots.$$

Solution by the proposer.
If we take $n = 0$ in (1), we get

$$U_m = V_m, \qquad m = 0, 1, 2, \cdots.$$

Put

$$F(x) = \sum_{n=0}^{\infty} U_n \frac{x^n}{n!}.$$

It follows from (1) that

$$\sum_{m=0}^{\infty} U_m \frac{x^m}{m!} \sum_{n=0}^{\infty} U_n \frac{y^n}{n!} = \sum_{r=0}^{\infty} \frac{2^r x^r y^r}{r!} \sum_{m,n=0}^{\infty} U_{m+n} \frac{x^m y^n}{m!n!}$$

$$= e^{2xy} \sum_{k=0}^{\infty} U_k \frac{(x + y)^k}{k!},$$

so that

$$F(x)F(y) = e^{2xy}F(x + y).$$

If we differentiate with respect to y and then put $y = 0$, we get

(2)
$$F'(x) = F(x)(2u - 2x),$$

where $U_1 = 2u$. Solving (2), we get

$$F(x) = \exp(2ux - x^2).$$

Since

$$\exp(2ux - x^2) = \sum_{n=0}^{\infty} H_n(u)\frac{x^n}{n!},$$

where $H_n(u)$ is the Hermite polynomial of degree n, it follows that the general solution of (1) is furnished by $U_n = V_n = H_n(u)$.

Remark. The formula

$$H_m(x)H_n(x) = \sum_{r=0}^{\min(m,n)} \binom{m}{r}\binom{n}{r} r! 2^r H_{m+n-2r}(x)$$

was first proved by Nielsen (*Det kgl. Danske Videnskabernes Selskab*, Math.-fysiske Meddelelser, I, 6 (1918), pp. 1–78, in particular pp. 31–33) are rediscovered by Feldheim (J. London Math. Soc., 13 (1938), pp. 22–29).

Problem 72-4, A Set of Nonlinear Equations, by L. CARLITZ (Duke University).

Find the general solution of

$$u_m u_n = \sum_{r=0}^{\min(m,n)} \begin{bmatrix} m \\ r \end{bmatrix}\begin{bmatrix} n \\ r \end{bmatrix} (q)_r a^r v_{m+n-2r}, \qquad u_0 = v_0 = 1, \qquad m, n = 0, 1, 2, \cdots,$$

where

$$(q)_r = (1 - q)(1 - q^2)\cdots(1 - q^r), \qquad (q)_0 = 1, \qquad \begin{bmatrix} n \\ r \end{bmatrix} = \frac{(q)_n}{(q)_r (q)_{n-r}}.$$

Solution by GEORGE E. ANDREWS (Pennsylvania State University).

By setting $m = 0$, we observe that $u_n = v_n$ for all n. Let $u_1 = v_1 = b$. Then setting $m = 1$, we observe that

$$u_n u_1 = u_{n+1} + (1 - q^n)a u_{n-1},$$

or

(1) $$u_{n+1} = b u_n - (1 - q^n)a u_{n-1}.$$

Thus, (1) together with $u_0 = 1$, $u_2 = b$ completely defines the u_n sequence. We note now that the u_n are closely related to the q-Hermite polynomials. Namely, if

$$H_n(t) = \sum_{m=0}^{n} \begin{bmatrix} n \\ m \end{bmatrix} t^m$$

(see [1, p. 359]), and

$$h_n(x, y) = \sum_{m=0}^{n} \begin{bmatrix} n \\ m \end{bmatrix} x^m y^{n-m}$$

(see [2, (3), p. 337]), then

$$h_n(x, y) = y^n H_n(x/y).$$

Furthermore, as Carlitz has noted [1, (1.6), p. 360], $H_0(x) = 1$, $H_1(x) = 1 + x$ and

$$(2) \qquad H_{n+1}(x) = (1 + x)H_n(x) - (1 - q^n)xH_{n-1}(x).$$

Therefore, the $h_n(x, y)$ are uniquely defined by $h_0(x, y) = 1$, $h_1(x, y) = x + y$ and

$$(3) \qquad h_{n+1}(x, y) = (x + y)h_n(x, y) - (1 - q^n)xyh_{n-1}(x, y).$$

Choose $x = (b + \sqrt{b^2 - 4a})/2 = \rho_1$, $y = (b - \sqrt{b^2 - 4a})/2 = \rho_2$. Then, $h_0(\rho_1, \rho_2) = 1$, $h_1(\rho_1, \rho_2) = b$ and

$$(4) \qquad h_{n+1}(\rho_1, \rho_2) = bh_n(\rho_1, \rho_2) - (1 - q^n)ah_{n-1}(\rho_1, \rho_2).$$

Thus, we see that the only possible solution of the given equation is

$$(5) \qquad \begin{aligned} u_n = v_n &= h_n(\rho_1, \rho_2) \\ &= h_n(\tfrac{1}{2}(b + \sqrt{b^2 - 4a}), \tfrac{1}{2}(b - \sqrt{b^2 - 4a})). \end{aligned}$$

All that remains is to prove that (5) does in fact yield a solution.

Now Carlitz [1 (1.7), p. 360] has shown that

$$H_m(x)H_n(x) = \sum_{r=0}^{\min(m,n)} \begin{bmatrix} m \\ r \end{bmatrix} \begin{bmatrix} n \\ r \end{bmatrix} (q)_r x^r H_{m+n-2r}(x).$$

Hence,

$$h_m(x, y)h_n(x, y) = \sum_{r=0}^{\min(m,n)} \begin{bmatrix} m \\ r \end{bmatrix} \begin{bmatrix} n \\ r \end{bmatrix} (q)_r x^r y^r h_{m+n-2r}(x, y).$$

Therefore, since $\rho_1\rho_2 = a$, we see that

$$h_m(\rho_1, \rho_2)h_n(\rho_1, \rho_2) = \sum_{r=0}^{\min(m,n)} \begin{bmatrix} m \\ r \end{bmatrix} \begin{bmatrix} n \\ r \end{bmatrix} (q)_r a^r h_{m+n-2r}(\rho_1, \rho_2).$$

Thus (5) does yield a solution; in fact the solution (5) is unique once $u_1 = b$ is specified.

REFERENCES

[1] L. CARLITZ, *Some polynomials related to theta functions*, Ann. Math. Pura Appl. (4), 41 (1956), pp. 359–373.

[2] L. J. ROGERS, *On the expansion of some infinite products*, Proc. London Math. Soc. (1), 24 (1893), pp. 337–352.

Problem 70-22, A Unique Solution, by D. J. NEWMAN (Yeshiva University).

Show that the only solution to the set of equations

$$\sum_{r=0}^{n} \binom{n}{r} x_r y_{n-r} = 2^n, \qquad n = 0, 1, 2, \cdots,$$

where

$$x_0 = x_1 = y_0 = y_1 = 1, \qquad x_n, y_n \geq 0,$$

is given by

$$x_n = y_n = 1.$$

Solution by L. CARLITZ (Duke University).

We shall prove the following more general result.

THEOREM. *Let $a > 0$. Then the only factorizations*

(1) $$e^{az} = f(z)g(z),$$

where $f(z)$, $g(z)$ are formal power series,

$$f(z) = \sum_{n=0}^{\infty} b_n \frac{z^n}{n!}, \qquad g(z) = \sum_{n=0}^{\infty} c_n \frac{z^n}{n!},$$

with

$$b_0 = c_0 = 1, \quad b_n \geq 0, \quad c_n \geq 0, \qquad n = 1, 2, 3, \cdots,$$

are given by

$$b_n = b^n, \quad c_n = c^n, \quad b + c = a, \quad b \geq 0, \quad c \geq 0.$$

Proof. Clearly (1) is equivalent to

$$\sum_{k=0}^{n} \binom{n}{k} b_k c_{n-k} = a^n, \qquad n = 0, 1, 2, \cdots.$$

This evidently implies that $b_n \leq a^n, c_n \leq a^n$. It follows that $f(z)$, $g(z)$ are entire and

$$|f(z)| \leq e^{a|z|}, \qquad |g(z)| \leq e^{a|z|}.$$

Thus $f(z)$, $g(z)$ are entire functions of order 1. Moreover, by (1), they have no zeros. Therefore by Hadamard's factorization theorem (see, for example, Titch-marsh's *Theory of Functions*, 2nd ed., Oxford University Press, London, 1939, p. 250) we have

$$f(z) = e^{bz}, \qquad g(z) = e^{cz}.$$

This evidently proves the stated theorem.

In particular when $a = 2$, $b_1 = c_1 = 1$, it is clear that the theorem reduces to the statement of Problem 70–22.

Editorial note. Jagöst and Shepp, in their solutions, used a theorem of Raikov that the Poisson law has only Poisson factors.

Problem 70-1, *A Set of Linear Equations*, by PETER H. ASTOR (North Carolina State University at Raleigh).

If x_i satisfy the set of n linear equations

$$\sum_{i=1}^{n} \frac{x_i}{b_i + t(j - i)} = A, \quad j = 1, 2, \cdots, n, \quad A, t \neq 0,$$

show that

$$\sum_{i=1}^{n} x_i = A \sum_{i=1}^{n} b_i.$$

This problem is a generalization of one that arose in a study of ship stability.

Solution by J. W. RAINEY (University of Tulsa).

Let

$$P(z) = \prod_{k=1}^{n} (z + b_k - kt).$$

From the given system one easily deduces the relation

$$\sum_{i=1}^{n} \frac{x_i P(z)}{z + b_i - it} - AP(z) = -A \prod_{j=1}^{n} (z - jt).$$

Equating coefficients of z^{n-1}, we have

$$\sum_{i=1}^{n} x_i - A\left[\sum_{i=1}^{n} (b_i - it)\right] = -A\left[-\sum_{i=1}^{n} it\right],$$

and therefore,

$$\sum_{i=1}^{n} x_i = A\left\{\sum_{i=1}^{n} (b_i - it) + \sum_{i=1}^{n} it\right\} = A\sum_{i=1}^{n} b_i.$$

R. C. CAVANAGH (Bolt, Beranek and Newman) considered the following generalization: Suppose the x_i satisfy the equations

(5) $G_1(k)x_1 + \cdots + G_n(k)x_n = CG(k), \qquad k = 0, 1, \cdots, n - 1,$

where the $G_i(k)$ are monic polynomials of degree $n - 1$ and $G(k)$ is a monic polynomial of degree n with the coefficient of k^{n-1} equal to B. Then

$$\sum_{i=1}^{n} x_i = C\left(B + \binom{n}{2}\right).$$

The proof is established easily with the aid of the well-known identities

$$\Delta^{n-1}k^j = \begin{cases} 0 & \text{for } j = 0, 1, \cdots, n - 2, \\ (n - 1)! & \text{for } j = n - 1, \\ n!k + (n - 1)!\binom{n}{2} & \text{for } j = n, \end{cases}$$

where Δ is the forward difference operator $(\Delta G(k) = G(k + 1) - G(k))$ and Δ^{n-1} its $(n - 1)$st iterate. For then $\Delta^{n-1}G$ and $\Delta^{n-1}G_i$ are defined at $k = 0$, and

$$(\Delta^{n-1}G_i)(0) = (n - 1)! \quad \text{for } i = 1, 2, \cdots, n,$$

$$(\Delta^{n-1}G)(0) = (n - 1)!\left(B + \binom{n}{2}\right).$$

Finally, from (5), we have

$$\Delta^{n-1}\left[\sum_{i=1}^{n} G_i(k)x_i\right] = C\Delta^{n-1}G(k) \quad \text{at } k = 0;$$

that is,

$$\sum_{i=1}^{n} x_i = C\left(B + \binom{n}{2}\right).$$

For Problem 70-1, let

$$G(k) = \left(\frac{1}{t^n}\right)(b_1 + kt)(b_2 + (k-1)t) \cdots (b_n + (k-n+1)t),$$

$$G_i(k) = \left(\frac{1}{t^{n-1}}\right)\left(\frac{G(k)}{b_i + (k-i+1)t}\right) \quad \text{for } i = 1, 2, \cdots, n, \qquad t \neq 0.$$

Then $B = (1/t)\sum_{i=1}^{n}b_i - \binom{n}{2}$, $C = At$, and we obtain

$$\sum_{i=1}^{n} x_i = (At)\left(\left(\frac{1}{t}\right)\sum_{i=1}^{n}b_i - \binom{n}{2} + \binom{n}{2}\right) = A\sum_{i=1}^{n}b_i, \qquad \text{[H.E.F.]}$$

A Nonlinear Difference Equation

Problem 86-18*, *by* G. C. WAKE (Massey University, Palmerston North, New Zealand).
(1) Find a closed form expression for C_n defined by

$$C_n = -\sum_{k=0}^{n-1} C_{n-1-k}C_k/2n(2k+3), \qquad n \geq 1, \quad C_0 = 1.$$

(2) Find the sum of

$$\sum_{n=0}^{\infty} C_n z^n.$$

These series arise in thermal ignition theory as follows:

The steady-state heat conduction equation with heat production caused by an exothermic chemical or nuclear reaction which is exponentially dependent on the local temperature rise (see Boddington, Gray and Wake [1]) can be written as

$$\nabla^2 y + \lambda e^y = 0, \quad \text{in the region,}$$
$$\text{with } y = 0 \qquad \text{on the boundary.}$$

For spherically-symmetric regions, this equation becomes

(*)
$$\frac{d^2y}{dr^2} + \frac{2}{r}\frac{dy}{dr} + \lambda e^y = 0, \qquad 0 < r < 1,$$
$$\text{with } y'(0) = 0, \qquad y(1) = 0.$$

The problem is to examine the multiplicity of solutions for various values of λ. We write $m = y(0)$ and so we wish to determine the structure of the bifurcation diagram $\lambda = \lambda(m)$ and, in particular, its turning points which represent ignition or extinction points in parameter-space for the spherically-shaped thermal regime.

Letting

$$p = \lambda r^2 e^y, \qquad q = \frac{r}{y}\frac{dy}{dr},$$

the problem ($*$) becomes

$$r\frac{dp}{dr} = pq, \qquad p(0) = 0,$$

$$r\frac{dq}{dr} = 2 - p - q, \qquad q(0) = 2,$$

which have solutions

$$p(r) = \sum_{n=0}^{\infty} C_n \lambda^{n+1} e^{mn} r^{2(n+1)},$$

$$q(r) = 2 - \sum_{n=0}^{\infty} \frac{C_n \lambda^{n+1} e^{mn}}{2n+3} r^{2(n+1)}$$

where (C_n) is as above. The elusive $\lambda(m)$ is thus given by the condition $p(1) = \lambda$ or the implicit relationship

$$\sum_{n=0}^{\infty} C_n \lambda^n e^{mn} = e^{-m}.$$

REFERENCE

[1] T. BODDINGTON, P. GRAY AND G. C. WAKE, *Criteria for thermal explosions with and without reactant consumption*, Proc. Roy. Soc. London Ser. A, 357 (1977), pp. 403–422.

Problem 66-1, *A Recursion Relation*, by E. L. PUGH (System Development Corporation).

Let $C_{i,j}$ be defined for $i = 2, 3, 4, \cdots$ and $j = 1, 2, 3, \cdots, (i-1)$ with $C_{i,1} = 1$ for all i, and

$$C_{i,j} = \sum_{\nu=1}^{j-1} \frac{(i-j)^{j-\nu} C_{j,\nu}}{(j-\nu)!}, \qquad\qquad j \geq 2.$$

Find an explicit expression for $C_{i,j}$.

This problem arose in the study of the distribution of the Kolmogorov statistic:

$$D_n = \sup_x |F(x) - F_n(x)|,$$

where $F(x)$ is a continuous distribution function and $F_n(x)$ is the empirical distribution function of a sample of size n taken from $F(x)$. D_n has the distribution (unknown in explicit form) of the random variable:

$$\max_{i=1,\cdots,n} \left[\max\left(X_{(i)} - \frac{i-1}{n}, \frac{i}{n} - X_{(i)} \right) \right],$$

where $X_1, \cdots, X_n$ are independent and uniformly distributed on $[0, 1]$, and $X_{(1)} < X_{(2)} < \cdots < X_{(n)}$ are the $X_1, \cdots, X_n$ arranged in order.

Solution by L. CARLITZ (Duke University).

It follows that

$$C_{n,2} = n - 2, \qquad\qquad n > 2,$$

$$C_{n,3} = \tfrac{1}{2}(n - 1)(n - 3), \qquad\qquad n > 3,$$

$$C_{n,4} = \tfrac{1}{6}(n - 1)^2(n - 4), \qquad\qquad n > 4.$$

This suggests the formula

$$(1) \qquad\qquad C_{n,j} = \frac{(n - j)(n - 1)^{j-2}}{(j - 1)!}, \qquad\qquad i \leqq j < n.$$

Assuming the truth of (1) for fixed n and $1 \leqq j < k$, we have

$$\sum_{r=1}^{k-1} \frac{(n - k)^{k-r} C_{k,r}}{(k - r)!} = \sum_{r=1}^{k-1} \frac{(n - k)^{k-r}(k - r)(k - 1)^{r-2}}{(k - r)!\,(r - 1)!}$$

$$= \sum_{r=0}^{k-2} \frac{(n - k)^{k-r-1}(k - 1)^{r-1}}{(k - r - 2)!\,r!}$$

$$= \frac{1}{(k - 1)!} \sum_{r=0}^{k-2} \binom{k - 2}{r} (n - k)^{k-r-1}(k - 1)^r$$

$$= \frac{(n - k)(n - 1)^{k-2}}{(k - 1)!}.$$

Hence (1) holds for the value k, thus completing the induction.

Supplementary References
Simultaneous Equations

[1] J.M.Ortega, *Iterative Solution of Nonlinear Equations in Several Variables*, Academic Press, N.Y., 1970.

[2] A.Ostrowski, *Solution of Equations and Systems of Equations*, Academic Press, N.Y., 1966.

[3] J.F.Traub, *Iterative Methods for the Solution of Equations*, Prentice-Hall, N.J., 1964.

19. IDENTITIES

Problem 70-4*, Some Algebraic Identities, by STAN KAPLAN (University of Southern California).

The following questions arise in connection with the interpretation of certain isotope distribution experiments in physiology.

Suppose we are given a set of distinct real numbers $\{\lambda_1, \cdots, \lambda_n\}$ and a second set of positive numbers $\{N_1, \cdots, N_n\}$ satisfying

$$(1) \qquad \sum_{j=1}^{n} 1/N_j = 1.$$

These numbers, λ_i, N_j, are to be regarded as experimental measurements. We now define the theoretical quantities $\{\alpha_2, \cdots, \alpha_n\}$ as the roots of the equation

$$(2) \qquad \sum_{j=1}^{n} 1/N_j(\alpha_i + \lambda_j) = 0, \qquad\qquad i = 2, \cdots, n.$$

For each of these α_i we define

$$(3) \qquad \beta_i = \left[\sum_{j=1}^{n} 1/N_j(\alpha_i + \lambda_j)^2 \right]^{-1}, \qquad\qquad i = 2, \cdots, n,$$

and finally define

$$(4) \qquad \alpha_1 = - \sum_{j=1}^{n} \lambda_j/N_j.$$

For some rather subtle conditions on the λ_i, N_j, it is known that the α's and β's defined by (2)–(4) necessarily satisfy the following relations:

$$(5) \qquad \sum_{j=1}^{n} \frac{1}{N_j(\alpha_i + \lambda_j)(\alpha_k + \lambda_j)} = 0, \qquad i, k = 2, 3, \cdots, n, \text{ and } i \neq k,$$

$$(6) \qquad 1 + \sum_{i=2}^{n} \left[\frac{\beta_i}{(\alpha_i + \lambda_j)(\alpha_i + \lambda_k)} \right] = N_j \delta_{jk}$$

(where δ_{jk} is the Kronecker delta) and

$$(7) \qquad \sum_{i=2}^{n} \frac{\beta_i}{(\alpha_i + \lambda_j)} = \alpha_1 + \lambda_j \qquad\qquad \text{for all } j.$$

The question is whether these relations hold for *all* such sets of λ's and N's. That is, are the relations (5)–(7) simply algebraic identities following from the definitions (2)–(4) or do they depend on the fact that the λ's and N's are associated with a physical system of a certain structure?

Solution by ALSTON S. HOUSEHOLDER (University of Tennessee).

It is to be assumed that when the proposer writes $1/N_j(\alpha_i + \lambda_j)$, he means $1/[N_j(\alpha_i + \lambda_j)]$. The stated relations are all algebraic identities that are even mildly more general than is implied, and neither requiring nor imposing additional constraints on the variables. They are some, and not altogether the simplest, of a large class of identities that follow fairly directly from appropriate application of the Lagrangean interpolation formula.

Some mild notational changes will make the formulations a bit neater. First, n will be changed to $n + 1$, and the ranges of the indices will be

$$h, j = 0, 1, 2, \cdots, n; \qquad i, k = 1, 2, \cdots, n.$$

This convention will be followed throughout, and not further indicated. Let

(8) $$v_j = N_i^{-1}, \qquad r_j = -\lambda_j$$

(equations (1)–(7) are those in the statement of the problem). Let

(9) $$f(x) = (x - r_0)(x - r_1) \cdots (x - r_n).$$

Presumably the λ_j, hence the r_j, are distinct. If every $N_j > 0$, then the α_i are distinct and strictly separate the r_j. The requirement $N_j > 0$ is not necessary, but it will be assumed that the α_i are distinct. Then

(10) $$\phi(x) = f(x) \sum \phi(r_j)/[(x - r_j)f'(r_j)],$$

where $\phi(x)$ is a polynomial of degree n or less. If the $\phi(r_j)$ are chosen arbitrarily, then (10) defines the corresponding interpolating polynomial; if $\phi(x)$ is given, then (10) is an identity. In particular, if

(11) $$\phi(r_j) = v_j f'(r_j),$$

where the v_j are assigned, and if

$$\sum v_j = 1,$$

which is condition (1), then $\phi(x)$ is a monic polynomial of degree n exactly. Then if

(12) $$\psi(x) = \phi(x)/f(x),$$

the α_i are the roots of $\psi(x) = 0$, hence of $\phi(x) = 0$, and

(13) $$\phi(x) = (x - \alpha_1)(x - \alpha_2) \cdots (x - \alpha_n).$$

Evidently

$$\psi'(x) = -\sum v_j/(x - r_j)^2 ;$$

hence,

(14) $$\beta_i = -f(\alpha_i)/\phi'(\alpha_i).$$

It is, of course, equally true that

(15) $$\omega(x) = \phi(x) \sum \omega(\alpha_i)/[(x - \alpha_i)\phi'(\alpha_i)]$$

is a polynomial of degree $n - 1$ at most, and if the values of $\omega(\alpha_i)$ are assigned, this defines the interpolating polynomial, whereas if $\omega(x)$ is given, (15) is an identity. All stated identities follow from those of the form (10) where, possibly, $\phi(x)$ is replaced by other polynomials of degree n at most, and from (15) when $\omega(x)$ is suitably defined.

An identity that is not among those stated can be obtained by taking $x = 0$

in (10); a set of n identities can be obtained by setting $x = \alpha_i$:

(16)
$$0 = \sum_j v_j/(\alpha_i - r_j),$$

in view of (11). This can be regarded as a companion set to (7).

Let $\phi_i(x) = \phi(x)/(x - \alpha_i)$. This is of degree $n - 1$, hence

(17)
$$\phi(x)/(x - \alpha_i) = f(x) \sum_j [\phi(r_j)/(r_j - \alpha_i)]/[(x - r_j)f'(r_j)].$$

On setting $x = \alpha_i$, the left member becomes $\phi'(\alpha_i)$. Hence (17) implies that

(18)
$$\sum_j v_j(\alpha_i - r_j)^{-2} = \beta_i^{-1},$$

which is not among the stated identities. But for $x = \alpha_k$, $k \neq i$, the result is

(19)
$$\sum_j v_j/[(\alpha_i - r_j)(\alpha_k - r_j)] = 0,$$

which is (5).

Now let

$$\eta(x) = x\phi(x) - f(x).$$

This is of degree n, and

$$\eta(r_j) = r_j\phi(r_j).$$

Hence

$$x\phi(x) - f(x) = f(x) \sum_j v_j r_j/(x - r_j).$$

Examination of the leading coefficient shows that

(20)
$$\sum r_j - \sum \alpha_i = \sum v_j r_j.$$

This is not one of the stated identities, but will be required below. In the proposer's notation this is α_1. It might be remarked in passing that for $x = \alpha_i$ the result is

(21)
$$1 = \sum_j v_j r_j/(r_j - \alpha_i),$$

which may be put beside (16).

Now turn to (15), and let

$$\omega_h(x) = f(x)/(x - r_h) - \phi(x).$$

This is of degree $n - 1$, and

$$\omega_h(\alpha_i) = f(\alpha_i)/(\alpha_i - r_h).$$

Hence in view of (14),

$$\omega_h(x) = \phi(x) \sum_i \beta_i/[(\alpha_i - r_h)(\alpha_i - x)].$$

First observe the leading terms. In view of (20), the comparison gives

(22)
$$-1 = \sum_i \beta_i/(\alpha_i - r_h) = \sum v_j r_j - r_h,$$

which is (7). Next, set $x = r_j$, $j \neq h$, and obtain

(23) $$-1 = \sum_i \beta_i/[(\alpha_i - r_j)(\alpha_i - r_h)], \qquad j \neq h,$$

which is part of (6). For $x = r_h$, the result is

(24) $$N_h - 1 = \sum_i \beta_i/(\alpha_i - r_h)^2,$$

which is the other part of (6). This completely establishes the stated relations (and some others) as identities.

Inequality or Identity

Problem 81-19*, *by* E. A. VAN DOORN (Netherlands Postal and Telecommunications Services, Leidschendam, the Netherlands).

Let $p_k \equiv p_k(\alpha)$ and $q_k \equiv q_k(\alpha)$ denote the roots of the equation $k(x - 1)^2 = \alpha x$, and let

$$F(\alpha) = \sum_{k=1}^{\infty} \frac{k^{k-1}}{k!} \left(\frac{p_k^k \exp(-kp_k)}{1 + p_k} + \frac{q_k^k \exp(-kq_k)}{1 + q_k} \right).$$

It can be shown that $\alpha F(\alpha) \leq 1 - \exp(-\alpha)$ for $\alpha > 0$. Prove or disprove that equality holds for all positive α.

Solution by OTTO G. RUEHR (Michigan Technological University).

The equality indeed holds, as we will show. Let $I(s)$ denote the Laplace transform of $F(\alpha)/\alpha^{1/2}$ and integrate term-by-term using the Weierstrass test, with the scale change $\alpha \to \alpha k$.

$$I(s) = \int_0^{\infty} e^{-s\alpha} \frac{F(\alpha)}{\sqrt{\alpha}} d\alpha = \sum_{k=1}^{\infty} \frac{k^{k-1/2}}{k!} \int_0^{\infty} \frac{e^{-sk} d\alpha}{\sqrt{\alpha}} \left[\frac{(pe^{-p})^k}{p + 1} + \frac{(qe^{-q})^k}{q + 1} \right].$$

Now p and q are independent of k and satisfy $(x - 1)^2 = \alpha x$, say

$$p = 1 + \frac{\alpha}{2} + \sqrt{\frac{\alpha^2}{4} + \alpha}, \qquad q = 1 + \frac{\alpha}{2} - \sqrt{\frac{\alpha^2}{4} + \alpha}.$$

Let p and q be the variables of integration, noting that as α runs from zero to infinity, p goes from one to infinity and q runs from one to zero. We have

$$I(s) = \sum_{k=1}^{\infty} \frac{k^{k-1/2}}{k!} \int_0^{\infty} x^{k-3/2} \exp\left\{ -kx - \frac{sk(x - 1)^2}{x} \right\} dx$$

$$= 2 \sum_{k=1}^{\infty} \frac{k^{k-1/2}}{k!} e^{2sk} \left(\frac{s}{1 + s} \right)^{(2k-1)/4} K_{k-1/2}(2k\sqrt{s(s + 1)})$$

where $K_{k-1/2}(z)$ is a modified spherical Bessel function of the third kind. It is well known [1, p. 85] that

$$K_{k-1/2}(z) = (-1)^{k-1} \sqrt{\frac{\pi}{2}} z^{k-1/2} \left(\frac{1}{z} \frac{d}{dz} \right)^{k-1} \frac{e^{-z}}{z}.$$

We take $z = 2k\sqrt{w}$, where $w = s(s + 1)$ in this representation, to write

$$I(s) = -\frac{\sqrt{\pi}}{\sqrt{s}} \sum_{k=1}^{\infty} \frac{(-s)^k}{k!} \frac{d^{k-1}}{dw^{k-1}} \frac{e^{2k[s - \sqrt{w}]}}{\sqrt{w}} \qquad \{w = s(s + 1)\}.$$

Finally, we are prepared to apply Lagrange's theorem [2, p. 133] directly to sum the series. Briefly, the general expansion of Lagrange is

$$f(z) = f(w) + \sum_{k=1}^{\infty} \frac{t^k}{k!} \frac{d^{k-1}}{dw^{k-1}} [f'(w)\phi\{(w)\}^k]$$

where z is a suitable solution of the equation $z - w = t\phi(z)$. We choose $f(w) = 2\sqrt{w}$, $\phi(w) = \exp[2(s - \sqrt{w})]$, and $t = -s$ to find that $z = s^2$ and consequently

$$I(s) = -(\pi/s)^{1/2}[f(z) - f(w)] = 2\sqrt{\pi}(\sqrt{s+1} - \sqrt{s}).$$

Now we take the inverse Laplace transform of $I(s)$ to obtain

$$F(\alpha)/\sqrt{\alpha} = (1 - e^{-\alpha})/\alpha^{3/2},$$

as required.

Similar identities have been discovered by Chihara and Ismail [3, p. 17, l. 3.16; p. 23, l. 4.12].

REFERENCES

[1] W. Magnus, F. Oberhettinger, and R. P. Soni, *Special Functions of Mathematical Physics*, Springer-Verlag, New York, 1966.
[2] E. T. Whittaker and G. N. Watson, *Modern Analysis*, Cambridge, Univ. Press, London, 1969.
[3] T. S. Chihara and M. E. H. Ismail, *Orthogonal polynomials suggested by a queueing model*, Adv. Appl. Math., to appear.

Conjectured Trigonometrical Identities

Problem 86-5*, by M. Henkel (University of Bonn, W. Germany) *and* J. Lacki (University of Geneva, Switzerland).

In computing low-temperature series for a Z_n-symmetric Hamiltonian [1] (in statistical mechanics), we apparently discovered some new trigonometrical identities. These were verified up to $n = 100$ on a computer. Prove or disprove the following conjectured identities for general n.

In what follows, we use the notation,

$$\omega = \exp\left(\frac{2\pi i}{n}\right), \qquad \theta(x) = \begin{cases} 1 & \text{if } x > 0, \\ 0 & \text{if } x \leq 0, \end{cases}$$

and

$$\sum_r \text{ stands for } \sum_{k_r=1}^{n-1}, \qquad \sum_{r \neq 1} \text{ stands for } \sum_{k_r=1, k_r \neq k_1}^{n-1}.$$

(1) $$\sum_1 \sum_{2 \neq 1} [(1 - \omega^{k_1})(1 - \omega^{k_2 - k_1})(1 - \omega^{-k_2})]^{-1} = 0.$$

(2) $$\sum_1 \sum_{2 \neq 1} \sum_{3 \neq 2} [(1 - \omega^{k_1})(1 - \omega^{k_2 - k_1})(1 - \omega^{k_3 - k_2})(1 - \omega^{-k_3})]^{-1} = \frac{(n^2 - 1)(n^2 - 4)}{180}.$$

(3) $$\sum_1 \sum_2 \frac{1}{1 + \theta(k_2 - k_1)} \left[\sin^2 \frac{\pi k_1}{n} \sin^2 \frac{\pi k_2}{n}\right]^{-1} = \frac{(n^2 - 1)(4n^2 - 1)}{45}.$$

(4)
$$\sum_1 \sum_2 \sum_3 \frac{1}{1+\theta(k_2-k_1)+\theta(k_3-k_2)}\left[\sin^2\frac{\pi k_1}{n}\sin^2\frac{\pi k_2}{n}\sin^2\frac{\pi k_3}{n}\right]^{-1}$$

$$=\frac{(n^2-1)(22n^4-13n^2+15)}{945}.$$

(5)
$$\sum_{k=1}^{n-1} k\sin^{-2}\frac{\pi k}{n}=\frac{n(n^2-1)}{6}.$$

(6)
$$\sum_1 \sum_{2\neq1} (k_1+k_2)[(1-\omega^{k_1})(1-\omega^{k_2-k_1})(1-\omega^{-k_2})]^{-1}=0.$$

(7)
$$\sum_1 \sum_2 \frac{k_1+k_2}{1+\theta(k_1-k_2)}\left[\sin^2\frac{\pi k_1}{n}\sin^2\frac{\pi k_2}{n}\right]^{-1}=\frac{n(n^2-1)(4n^2-1)}{45}.$$

REFERENCE

[1] M. HENKEL AND J. LACKI, Bonn preprint, submitted to J. Phys. A.

Editorial note. The proposers also had eleven more similar conjectured identities with triple, quadruple and quintuple summations.

Composite solution by S. W. GRAHAM *and* O. RUEHR (Michigan Technological University), O. P. LOSSERS (Eindhoven University of Technology, Eindhoven, the Netherlands), *and* M. RENARDY (University of Wisconsin).

All of the conjectured identities are true. Identities 1 and 6 follow from symmetry arguments alone. The remaining identities reduce to evaluation of the sums

$$S_p=\sum_{k=1}^{n-1}\csc^{2p}\frac{\pi k}{n}, \qquad p=1,2,3.$$

The required sums,

$$S_1=\frac{n^2-1}{3}, \qquad S_2=\frac{(n^2-1)(n^2+11)}{45},$$

$$S_3=\frac{(n^2-1)(2n^4+23n^2+191)}{945},$$

may be found in [1] or may be computed using any one of several methods. For example, S_p may be derived by considering the contour integral

$$\int_C \cot z\,\csc^{2p}\frac{z}{n}\,dz$$

where C consists of the vertical line $\mathrm{Re}\,z=-\delta$ and $\mathrm{Re}\,z=n\pi-\delta$, $0<\delta<\pi$, traversed in opposite directions. Clearly, the integral vanishes, so by residue calculus one has

$$S_p=-\operatorname*{Re}_{z=0} s\,\cot z\,\csc^{2p}\frac{z}{n}.$$

We now turn to the conjectured identities. Let $T_1, T_2, \cdots, T_7$ denote the given sums.

To prove the first identity, note that the summand changes sign if we reverse the roles of k_1 and k_2. Thus the terms cancel in pairs and the sum vanishes. The same argument proves the sixth identity as well.

To prove the second identity, use the trigonometric formula $\csc x \csc y = \csc(x+y)[\cot x + \cot y]$ to write the summand as

$$\frac{1}{16}\csc\frac{\pi k_1}{n}\csc\frac{\pi k_3}{n}\csc\frac{\pi(k_2-k_1)}{n}\csc\frac{\pi(k_2-k_3)}{n}$$

$$=\frac{1}{16}\csc^2\frac{\pi k_2}{n}\left(\cot\frac{\pi k_1}{n}+\cot\frac{\pi(k_2-k_1)}{n}\right)\left(\cot\frac{\pi k_3}{n}+\cot\frac{\pi(k_2-k_3)}{n}\right).$$

Since $\sum_{k=1}^{n-1}\cot\pi k/n=0$, we have

$$\sum_{1\ne2}\cot\frac{\pi k_1}{n}=\sum_{1\ne2}\cot\frac{\pi(k_2-k_1)}{n}=-\cot\frac{\pi k_2}{n}$$

and

$$\sum_{3\ne2}\cot\frac{\pi k_3}{n}=\sum_{3\ne2}\cot\frac{\pi(k_2-k_3)}{n}=-\cot\frac{\pi k_2}{n}.$$

Thus

$$T_2=\frac{1}{4}\sum_{k=1}^{n-1}\csc^2\frac{\pi k}{n}\cot^2\frac{\pi k}{n}$$

$$=\frac{1}{4}(S_2-S_1)=\frac{(n^2-1)(n^2-4)}{180}.$$

To prove the third identity, write

$$T_3=\sum_1\sum_2 W(k_1,k_2)\csc^2\frac{\pi k_1}{n}\csc^2\frac{\pi k_2}{n}$$

where

$$W(k_1,k_2)=\frac{1}{2}\left\{\frac{1}{1+\theta(k_2-k_1)}+\frac{1}{1+\theta(k_1-k_2)}\right\}=\begin{cases}\frac{3}{4} & \text{if } k_1\ne k_2,\\ 1 & \text{if } k_1=k_2.\end{cases}$$

It follows that

$$T_3=\frac{3}{4}S_1^2+\frac{1}{4}S_2=\frac{(n^2-1)(4n^2-1)}{45}.$$

Identity 4 may be established similarly. Write

$$T_4=\sum_1\sum_2\sum_3 W(k_1,k_2,k_3)\csc^2\frac{\pi k_1}{n}\csc^2\frac{\pi k_2}{n}\csc^2\frac{\pi k_3}{n}$$

where

$$W(k_1,k_2,k_3)=\frac{1}{6}\sum\frac{1}{1+\theta(k_2-k_1)+\theta(k_3-k_2)},$$

the sum extending over all permutations of k_1, k_2, k_3. It is readily found that

$$W(k_1, k_2, k_3) = \begin{cases} \frac{5}{9} & \text{if } k_1, k_2, k_3 \text{ are distinct,} \\ \frac{2}{3} & \text{if exactly two of } k_1, k_2, k_3 \text{ are equal,} \\ 1 & \text{if } k_1 = k_2 = k_3. \end{cases}$$

We thus find

$$T_4 = \frac{5}{9}S_1^3 + \frac{1}{3}S_1 S_2 + \frac{1}{9}S_3 = \frac{(n^2 - 1)(22n^4 - 13n^2 + 15)}{945}.$$

Finally, we make use of the fact that $\sin \pi k/n = \sin \pi(n - k)/n$ to reduce identities 5 and 7 to known sums. In particular, to evaluate T_5, write the sum a second time with k replaced by $n - k$ and add the two sums. We thus find $2T_5 = nS_1$, so

$$T_5 = \frac{n(n^2 - 1)}{6}.$$

Apply the same technique to T_7, replacing k_1 by $n - k_1$ and k_2 by $n - k_2$. Thus $2T_7 = 2nT_3$, so

$$T_7 = \frac{n(n^2 - 1)(4n^2 - 1)}{45}.$$

REFERENCE

[1] M. E. Fisher, *Solution to Problem 69-14**, *Sums of Inverse Powers of Cosines*, by L. A. Gardiner, Jr., this Review, 13 (1971), pp. 116–119.

Editorial note. A variety of methods was proposed for evaluating S_p. Different starting points included the Mittag–Leffler expansion of $\csc^2 z$ (A. A. Jagers), Chebyshev polynomials (W. van Assche), eigenvalues of the matrix

$$D = \begin{bmatrix} 2 & -1 & & & 0 \\ -1 & 2 & & & \\ & & \ddots & \ddots & \\ & & & & -1 \\ 0 & & & -1 & 2 \end{bmatrix}$$

(D. Foulser) and the formula $\pi^2 \csc^2 \pi x = \psi'(1 - x) + \psi'(x)$, $0 < x < 1$, where ψ denotes the digamma function (R. Richberg).

An Identity for Complex Numbers

Problem 80-15, *by* M. S. Klamkin *and* A. Meir (University of Alberta).

Given that z_1, z_2, z_3 are complex numbers such that $|z_1| = |z_2| = |z_3| = 1$ and

$$0 \leqq \arg z_1 \leqq \arg z_2 \leqq \arg z_3 \leqq \pi,$$

prove that

$$(-z_3 z_1 + z_1 z_2 + z_2 z_3)\{|z_3^2 - z_1^2| + |z_1^2 - z_2^2| + |z_2^2 - z_3^2|\}$$

$$= z_2^2|z_3^2 - z_1^2| + z_3^2|z_1^2 - z_2^2| + z_1^2|z_2^2 - z_3^2|.$$

Solution by O. G. Ruehr (Michigan Technological University).

The identity can be rewritten in the form

$$I \equiv (z_1 - z_2)(z_2 - z_3)|z_3^2 - z_1^2| + (z_2 - z_3)(z_1 + z_3)|z_1^2 - z_2^2| + (z_1 + z_3)(z_2 - z_1)|z_2^2 - z_3^2| = 0.$$

LEMMA. *If* $|z| = |w| = 1$ *and* $0 \leq \arg z \leq \arg w \leq \pi$, *then*

$$izw|z^2 - w^2| = w^2 - z^2.$$

Then, using the lemma, we have

$$I = \frac{(z_3 + z_1)(z_1 - z_2)(z_2 - z_3)}{iz_1 z_2 z_3}\{z_2(z_3 - z_1) - z_3(z_1 + z_2) + z_1(z_2 + z_3)\} = 0.$$

Editorial note. It follows immediately by barycentric coordinates that the complex number representation of the incenter I of a triangle whose vertices are z_1^2, z_2^2, z_3^2 is given by

$$I = \frac{z_1^2|z_2^2 - z_3^2| + z_2^2|z_3^2 - z_1^2| + z_3^2|z_1^2 - z_2^2|}{|z_2^2 - z_3^2| + |z_3^2 - z_1^2| + |z_1^2 - z_2^2|}.$$

The proposed problem is then equivalent to also showing that

$$I = z_1 z_2 + z_2 z_3 - z_3 z_1.$$

This latter representation was given by Frank Morley and F. V. Morley (*Inversive Geometry*, Chelsea, New York, 1954, pp. 193–194) immediately after noting that I was the intersection point of the three angle bisectors which are given by the joins of the three pairs of points $(z_1^2, z_2 z_3)$, $(z_2^2, -z_3 z_1)$, $(z_3^2, z_1 z_2)$. Although it is not immediately apparent to me that the intersection point is $z_1 z_2 + z_2 z_3 - z_3 z_1$, this can be verified easily by using the criterion

$$\begin{vmatrix} 1 & 1 & 1 \\ w_1 & w_2 & w_3 \\ \overline{w_1} & \overline{w_2} & \overline{w_3} \end{vmatrix} = 0$$

for the collinearity of the points w_1, w_2, w_3 and also noting that $\overline{z_1} = 1/z_1$, etc.

[M.S.K.]

20. ZEROS

A Conjecture on the Number of Real Zeros

Problem 82-14*, *by* T. C. Y. LAU (University of Hong Kong).
   It is conjectured that the function

$$P_m(x) - e^{-x} \prod_{i=1}^{n} (1 + b_i x)$$

cannot have more than $m + 2$ real zeros on the nonnegative real axis. Here, $P_m(x)$ is any polynomial of degree at most m and b_i, $i = 1, 2, \cdots, n$ are n positive real numbers. Prove or disprove.
   The problem arose in trying to approximate exponential functions by rational polynomials with real poles. The special case $m = 0$ has been solved.

Solution by G. WANNER (Université de Genève).
   This conjecture is true. It is an extension of [1, Thm. 8] and was first formulated in the present form in [2, Thm. 3]. In order to outline the proof, we replace, in accordance with these two papers, x by $-x$ and write

(1) $$R(x) = \frac{P_m(-x)}{\prod_{i=1}^{n} (1 - b_i x)} , \qquad S(x) = \frac{R(x)}{e^x}$$

so that the given equation becomes $R(x) = e^x$, $S(x) = 1$; i.e., its zeros are the *interpolation points* of $R(x)$. The main idea of the proof is then the study of the "order star"

(2) $$A = \{z \in \mathbb{C}; |S(z)| > 1\}$$

in the complex plane.
   As an example, Fig. 1 shows the set A for $R(z) = (1 + z(1 - s_1) + z^2 (\frac{1}{2} - s_1 + s_2))/(1 - s_1 z + s_2 z^2)$, $s_1 = b_1 + b_2$, $s_2 = b_1 b_2$, $b_1 = 0.175$, $b_2 = 0.27$. $R(z)$ possesses two poles at $1/b_1 = 5.714$, $1/b_2 = 3.704$, and a pair of complex zeros at $z = -2.714 \pm 1.554i$. There are five interpolation points at $z = -1.370$, $z = 5.917$ and a triple point at $z = 0$.
   At each real interpolation point the boundary of A crosses the real axis vertically; at q-fold points this degenerates to a star-like configuration with regularly distributed sectors of angle π/q ([1], [2]; see Fig. 2).
   Due to the behaviour of $\exp(z)$ for $z \to \infty$, only *one* boundary curve of A reaches infinity [1], all others form loops and must enclose (since arg(S) rotates monotonically on ∂A [1]) the same number of *zeros* or *poles* of R as there are interpolation points on this loop [2].
   Since, by hypothesis, all poles $1/b_i$ are positive, there cannot be more than m (the number of zeros of R) plus one (for the boundary curve to infinity) plus one (for the first loop surrounding a pole), i.e. $m + 2$, interpolation points on the nonpositive real axis.

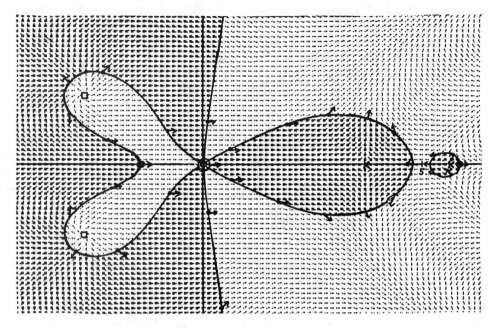

FIG. 1. *Example of an order star.* ○—*interpolation points;* x—*poles;* □—*zeros of R(z);* ↗ —*indication of* arg(S).

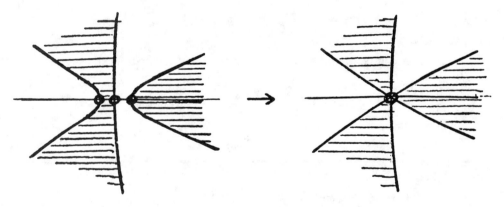

FIG. 2. *Degeneration of A for a triple interpolation point.*

REFERENCES

[1] G. WANNER, E. HAIRER AND S. P. NØRSETT, *Order stars and stability theorems,* BIT, 18 (1978), pp. 475–489.

[2] A. ISERLES, *Generalised order star theory,* in Padé Approximation and Its Applications Amsterdam 1980, M. G. de Bruin and H. van Rossum, eds., Lecture Notes in Mathematics 888, Springer-Verlag, New York, 1981, pp. 228–238.

Problem 76-22, *A Zero of Maximum Multiplicity*, by N. LIRON (Weizmann Institute of Science) and L. A. RUBENFELD (Rensselaer Polytechnic Institute).

Define $F(x) = B_m(x^2) \sin x - x A_n(x^2) \cos x$, where $B_m(z)$ and $A_n(z)$ are polynomials of orders m and n respectively, $B_m(0) = 1$ and where $m - n = 0$ or 1. Prove that the coefficients in the polynomials B_m and A_n can be uniquely chosen so that $F(x)$ vanishes to maximum order at $x = 0$, and the order of the zero is $2(m + n) + 3$. This result provides a criterion for characterizing the Bessel functions of positive half-order, as has been shown in [1].

REFERENCE

[1] N. LIRON, Math. Comp., 25 (1971), p. 769.

Solution by W. B. JORDAN (Scotia, NY).

We change the notation for subscripts. The problem is equivalent to finding A_j and B_j such that $A_j(x^2)/B_j(x^2)$ osculates with $x^{-1} \tan x$ to as high an order as possible at $x = 0$. This is solved by the various convergents to the continued fraction for the tangent:†

$$x^{-1} \tan x = \frac{1}{1-} \frac{x^2}{3-} \frac{x^2}{5-} \frac{x^2}{7-} \cdots .$$

We use p_j/q_j to denote the jth convergent and find

$$p_1 = 1, \qquad\qquad q_1 = 1,$$
$$p_{j+1} = (2j + 1)p_j - x^2 p_{j-1}, \qquad q_{j+1} = (2j + 1)q_j - x^2 q_{j-1},$$

with $p_0 = 0$ and $q_0 = 1$ the starters for the recursion.

p and q can be identified with A and B if they are first normalized to give $B(0) = 1$:

$$p_j = 1 \cdot 3 \cdot 5 \cdots (2j - 1)A_j, \qquad q_j = 1 \cdot 3 \cdot 5 \cdots (2j - 1)B_j,$$

so that

$$A_1 = 1, \qquad\qquad B_1 = 1,$$
$$A_{j+1} = A_j - x^2 A_{j-1}/(4j^2 - 1), \qquad B_{j+1} = B_j - x^2 B_{j-1}/(4j^2 - 1),$$

with the starters $A_0 = 0$ and $B_0 = 1$. We compute

$$A_2 = 1, \qquad\qquad B_2 = 1 - x^2/3,$$
$$A_3 = 1 - x^2/15, \qquad B_3 = 1 - 2x^2/5,$$
$$A_4 = 1 - 2x^2/21, \qquad B_4 = 1 - 3x^2/7 + x^4/105, \qquad \text{etc.}$$

The pair A_j, B_j gives a zero of order $2j + 1$ in F; this is equivalent to the formula given by the proposer. The requirement $m - n = 0$ or 1 is also met.

Distinct Zeros of a Function

Problem 81–11*, *by* K. SCHRADER (University of Missouri-Columbia).

Let M, N be positive integers and $y(x)$ a twice continuously differentiable real-valued function on $[0, 1]$ satisfying $|y(x)| \leq M$ and $|y'(x)| \leq M$. Prove or disprove the conjecture that if

$$\int_0^1 |y''(x)| \, dx$$

is sufficiently large, depending on M, N, then there exists constants a, b so that the equation

$$y(x) = a + bx$$

has at least N distinct solutions for $0 \leq x \leq 1$.

Zeros of a Definite Integral

Problem 84-2*, *by* TIAN JINGHUANG (Chengdu Branch, Academia Sinica, Chengdu, China).

Prove that the function

$$F(x, k) = \int_0^{\arccos(x-1)} \sin \{k \arccos (x - \cos \lambda)\} \, d\lambda$$

has exactly $k - 1$ simple zeros in the interval $(0, 2)$ for $k = 1, 2, 3, \cdots$.

Zero of Least Modulus

Problem 88-3*, *by* M. L. GLASSER (Clarkson University).

The zeros of $F(z) \equiv J_0^2(z) + J_1^2(z)$ are of interest with respect to the properties of water waves on a sloping beach [1]. One zero of $F(z)$ is approximately given by $z = 2.977 + 1.266i$. Does this correspond to the zero of least modulus?

REFERENCE

[1] G. CARRIER AND H. GREENSPAN, *Water waves of finite amplitude on a sloping beach*, J. Fluid Mech., 4 (1958), pp. 97–110.

Solution by. MICHAEL RENARDY (Virginia Polytechnic Institute).
The answer is yes. To show this, we simply need to evaluate

$$\frac{1}{2\pi i} \int_C \frac{F'(z)}{F(z)} dz,$$

where C is a circle about the origin of radius r. This expression gives the number of zeros of F inside the circle. Approximate values for the integral can easily be found numerically. I used Taylor expansion with 30 terms for the Bessel functions, and the IMSL routine DQDAG for the integral. I found no zeros for $r = 3.2$ and four for $r = 3.3$. The number four results from symmetry: If z is a root, so are $-z$, $\bar{z}$, and $-\bar{z}$.

Also solved by W. B. JORDAN (Scotia, New York), who applied the Graeffe root-squaring procedure to the Maclaurin series for F.

Editorial note. Another effective numerical procedure is to examine the radius of convergence of the series for $1/F$.

Denote the zeros of F in the first quadrant by $z_n = x_n + iy_n$, ordered in increasing magnitude. The asymptotic behavior for n large is as follows:

$$(*) \quad x_n \cong n\pi - \frac{(L_n - 1/4)}{(2n\pi)}, \quad y_n \cong L_n + \frac{(8L_n^2 - 12L_n + 7)}{(32n^2\pi^2)} \quad \text{where } L_n = \frac{1}{2}\log(4n\pi).$$

This gives four correct digits for $n = 1$ and 10 correct digits for $n = 10$; thus $(*)$ coupled with Newton's method is numerically efficient, e.g.,

$x_1 = 2.9803\ 82414\ 79048\ 78703,$

$y_1 = 1.2796\ 02540\ 29914\ 66533,$

$x_2 = 6.1751\ 53070,$

$y_2 = 1.6187\ 17384,$

$x_3 = 9.3419\ 60983,$

$y_3 = 1.8188\ 72787.$ [O. G. R.]

Supplementary References
Zeros

[1] M.Marden, *Geometry of Polynomials*, AMS, Providence, 1966.
[2] A.Ostrowski, *Solution of Equations and Systems of Equations*, Academic Press, 1966.
[3] G.Polya, *Location of Zeros*, MIT Press, Cambridge, 1974.

21. FUNCTIONAL EQUATIONS

Problem 71–1, A Functional Equation, by L. A. SHEPP (Bell Telephone Laboratories).
Prove that there is only one nonnegative function F for which

$$F\left\{\int_0^x F(u)\,du\right\} = x, \qquad x \geq 0,$$

namely $F(x) = Ax^n$ for appropriate values of A and n.

Solution by A. C. HINDMARSH (Lawrence Radiation Laboratory).

This problem can be reduced to Problem E2105 in the Amer. Math. Monthly, 76 (1969), p. 696.

I must assume that F is at least measurable and locally bounded. Then $f(x) \equiv \int_0^x F(u)\,du$ is continuous, and cannot be bounded since $F(f(x)) = x$ is not. It follows that for any $x \geq 0$, $f(y) = x$ has a solution $y \geq 0$. We take $y = 0$ for $x = 0$.

If $x_1 < x_2$ and $f(y_i) = x_i$, then

$$x_2 - x_1 = \int_{y_1}^{y_2} F(u)\,du.$$

Thus $y_1 < y_2$, and $F(x_1) = F(f(y_1)) = y_1 < y_2 = F(x_2)$, or F is strictly increasing. Also, if $x_2 \to x_1$ (or $x_1 \to x_2$), we must have (using $F(u) > 0$ for $u > 0$) $y_2 \to y_1$ (or $y_1 \to y_2$). Hence F is continuous, and $f'(x) = F(x)$.

Since F is clearly unbounded, we see that it is a strictly increasing continuous map of $[0, \infty)$ onto itself, and so has a unique inverse, which must be f since $F(f(x)) = x$. The map f has the same properties as F, and in addition is C^1 with $f' = f^{-1}$. This completes the required reduction. The unique solution $f(x) = ax^c$, $c = (1 + \sqrt{5})/2$, $a = c^{1-c}$, yields the unique F of the same form, from the solution of E2105.

Erratum. In line 5 of the solution cited above, replace "$f'(f(x)) = x$" by "$f(f'(x)) = x$."

Problem 79–6*, A Functional Equation by L. B. KLEBANOV (Civil Engineering Institute, Leningrad, U.S.S.R.).

Let $f(x)$, $g(x)$ be two probability densities on R^1 with $g(x) > 0$. Suppose that the condition

$$\int_{-\infty}^{\infty} (u - c) \prod_{j=1}^{n} f(x_j - u) g(u)\,du = 0$$

holds for all $x_1, x_2, \cdots, x_n$ such that $\sum_{j=1}^n x_j = 0$ where $n \geq 3$ and c is some constant.

Prove that

$$f(x) = \frac{1}{\sqrt{2\pi}\sigma} \exp\{-(x-a)^2/2\sigma^2\}.$$

Editorial note. This problem is related to a known theorem characterizing the normal distribution, i.e., if

$$\int_{-\infty}^{\infty} u \prod_{j=1}^{n} f(x_j - u) g(u) \, du \div \int_{-\infty}^{\infty} \prod_{j=1}^{n} f(x_j - u) g(u) \, du = c_1 + c_2 \sum_{j=1}^{n} x_j$$

for all values of x_j, then the density f must be a normal density (A. M. Kagan, Y. V. Linnik, C. R. Rao, *Characterization Problems in Mathematical Statistics*, Wiley-Interscience, New York, 1973, p. 480, Thm. B.2.1.) The proposer notes that he can establish the result of his problem under further assumptions of regularity conditions. However, the assertion of the problem as is may not be correct. Nevertheless, even a proof of this would be of interest.

A Functional Equation

Problem 81-1, *by* K. R. PARTHASARATHY (Indian Statistical Institute, New Delhi, India).

Let $f: \mathcal{R}^k \to \mathcal{R}$ be a continuous map satisfying the relation

$$f(\mathbf{u} + \mathbf{v}) = f(\mathbf{u}) + f(\mathbf{v})$$

whenever $\mathbf{u} \perp \mathbf{v}$. Show that there exist a constant c and a vector $\mathbf{v}_0 \in \mathcal{R}^k$ such that

$$f(\mathbf{u}) = c\|\mathbf{u}\|^2 + \mathbf{u} \cdot \mathbf{v}_0$$

for all $\mathbf{u}$.

Solution by M. ST. VINCENT (Melrose, Massachusetts).

If $f: \mathcal{R}^k \to \mathcal{R}$ is continuous and such that $f(\mathbf{u} + \mathbf{v}) = f(\mathbf{u}) + f(\mathbf{v})$ whenever $\mathbf{u} \cdot \mathbf{v} = 0$, then its even and odd parts have these properties also. Consequently, it is sufficient to prove that $f(\mathbf{u}) = c\|\mathbf{u}\|^2$ if f is even, and $f(\mathbf{u}) = \mathbf{u} \cdot \mathbf{v}_0$ if f is odd. Before proceeding, it should be pointed out that the problem can be generalized slightly. It will be clear from the following that everything remains true if the domain of f is any real Hilbert space of dimension > 1.

For f even, consider any $\mathbf{u}$, $\mathbf{v}$ with $\|\mathbf{u}\| = \|\mathbf{v}\|$ and let $\mathbf{w}_1 = (\mathbf{u} + \mathbf{v})/2$, $\mathbf{w}_2 = (\mathbf{u} - \mathbf{v})/2$. Then $\mathbf{w}_1 \cdot \mathbf{w}_2 = 0$, so

$$f(\mathbf{u}) = f(\mathbf{w}_1) + f(\mathbf{w}_2)$$

and

$$f(\mathbf{v}) = f(\mathbf{w}_1) + f(-\mathbf{w}_2),$$

which implies that $f(\mathbf{u}) = f(\mathbf{v})$ since f is even. As a result, it follows that $f(\mathbf{u}) = g(\|\mathbf{u}\|^2)$ where g is continuous. By considering appropriate $\mathbf{u}$, $\mathbf{v}$ for which $\mathbf{u} \cdot \mathbf{v} = 0$, it is easily verified that $g(x + y) = g(x) + g(y)$ for all x, $y \geq 0$. This implies that

$$g(nx) = ng(x),$$

and so also

$$g\left(\frac{x}{n}\right) = \frac{g(x)}{n},$$

for all $x \geq 0$ and all integers $n > 0$. Consequently, $g(r) = rg(1)$ for all rational $r > 0$ and so, since g is continuous,

(1) $$g(x) = xg(1)$$

for all real $x \geq 0$. This establishes that f has the desired form when f is even.

For f odd, it is sufficient to show that f must be linear. To this end, it will first be shown that

(2) $$f(n\mathbf{u}) = nf(\mathbf{u})$$

for all $\mathbf{u}$ and all integers $n > 0$. It's obviously true for $n = 1$, so assume it's true for $n = k$. If $\mathbf{u} = 0$, then there's nothing to do, since $f(\mathbf{0}) = 0$ by virtue of f being odd and continuous. For any $\mathbf{u} \neq 0$, let $\mathbf{v}$ be such that $\mathbf{v} \cdot \mathbf{u} = 0$ and $\|\mathbf{v}\| = k^{1/2}\|\mathbf{u}\|$. Then $(k\mathbf{u} - \mathbf{v}) \cdot (\mathbf{u} + \mathbf{v}) = 0$, so

$$f[(k+1)\mathbf{u}] = f(k\mathbf{u} - \mathbf{v}) + f(\mathbf{u} + \mathbf{v})$$
$$= f(k\mathbf{u}) + f(-\mathbf{v}) + f(\mathbf{u}) + f(\mathbf{v})$$
$$= (k+1)f(\mathbf{u}),$$

since f is odd and $f(k\mathbf{u}) = kf(\mathbf{u})$ by the induction hypothesis. So (2) is established and it now follows, from reasoning similar to that used to prove (1) and the fact that f is odd, that

$$f(t\mathbf{u}) = tf(\mathbf{u})$$

for all $\mathbf{u}$ and all real t. To prove that f is linear, it only remains to show that $f(\mathbf{u} + \mathbf{v}) = f(\mathbf{u}) + f(\mathbf{v})$ for all $\mathbf{u}, \mathbf{v}$. For any $\mathbf{u}, \mathbf{v}$ with $\mathbf{u} \neq 0$, let $a = (\mathbf{u} \cdot \mathbf{v})/\|\mathbf{u}\|^2$. (If $\mathbf{u} = \mathbf{v} = \mathbf{0}$, then it is immediate, since $f(\mathbf{0}) = 0$.) Then $\mathbf{u} \cdot (\mathbf{v} - a\mathbf{u}) = 0$, so

$$f(\mathbf{u} + \mathbf{v}) = f[(1 + a)\mathbf{u} + (\mathbf{v} - a\mathbf{u})]$$
$$= (1 + a)f(\mathbf{u}) + f(\mathbf{v} - a\mathbf{u})$$
$$= f(\mathbf{u}) + af(\mathbf{u}) + f(\mathbf{v} - a\mathbf{u})$$
$$= f(\mathbf{u}) + f(\mathbf{v}).$$

Thus f is linear, and so must have the desired form.

Subsequent comment by B. R. EBANKS (Texas Tech University).
The solution of Problem 81-1 (this Review, 23 (1982), p. 78) is contained in K. Sundaresan, *Orthogonality and nonlinear functionals on Banach spaces*, Proc. Amer. Math. Soc., 34 (1972), pp. 187–190. A related result is contained in S. Gudder and D. Strawther, *A converse of Pythagoras' theorem*, Amer. Math. Monthly, 84 (1977), pp. 551–553. Recently, Jürg Rätz (Mathematische Institut, Universität Bern) has found the general form of such (not necessarily continuous) "orthogonally additive" mappings in a very general setting (announcement in Proc. 18th International Symposium on Functional Equations, Univ. of Waterloo, 1980, pp. 22–23).

Supplementary References
Functional Equations

[1] J.Aczel, *Lectures on Functional Equations and their Applications,* Academic Press, N.Y., 1966.

[2] J.Aczel and J.Dhombres, *Functional Equations in Several Variables,* Cambridge University Press, Cambridge, 1989.

[3] R.Bellman, *Dynamic Programming,* Princeton University Press, Princeton, 1957.

[4] G.Boole, *A Treatise on the Calculus of Finite Differences,* Stechert, N.Y., 1946.

[5] W.Eichhorn, *Functional Equations in Economics,* Addison-Wesley, Reading, 1978.

[6] T.Fort, *Finite Differences and Difference Equations in the Real Domain,* Oxford University Press, Oxford, 1948.

[7] M.Kuczma, *Functional Equations in a Single Variable,* Polish Scientific Publishers, Warsaw, 1968.

[8] T.L.Saaty, *Modern Nonlinear Equations,* McGraw-Hill, N.Y., 1967.

22. MISCELLANEOUS

Problem 74-2*, A Normalization Constant, by M. L. Mehta (Centre d'Etudes
Nucleaires de Saclay, Gif-sur-Yvette, France).
  For all positive integers n and k, calculate the constant

$$A(n, k) = \left(\sum_{j=1}^{n} \frac{\partial^2}{\partial x_j^2} \right)^p \prod_{n \geq j > l \geq 1} (x_j - x_l)^{2k},$$

where $p \equiv n(n - 1)k/2$. It is conjectured that the correct value is

$$A(n, k) = 2^p p!(k!)^{-n} \prod_{j=1}^{n} (kj)!.$$

  The conjectured value has been verified for $k = 1$ and $k = 2$ (all n) and for $n = 2$
and $n = 3$ (all k). Thus, the smallest values for which the conjecture may be false
are $k = 3$, $n = 4$.
  Editorial note. The constant $A(n, k)$ is related to the normalization constant
$C_{n\beta}$ of $P(x_1, x_2, \cdots, x_n)$ for the case where $\beta = 2k$. Also, it is related to the partition
function of a certain Coulomb gas. The reduction of the problem of computing
the normalization integral to the combinatorial problem given above was given
by Mehta and Dyson in [1], and it is discussed in [2, Chap. 4]. [C.C.R.]

REFERENCES

[1] M. L. Mehta and F. J. Dyson, *Statistical theory of energy levels in complex systems. V*, J. Math.
    Phys., 4 (1963), pp. 713–719.
[2] M. L. Mehta, *Random Matrices and the Statistical Theory of Energy Levels*, Academic Press,
    New York, 1967.

Problem 69-1, Sequences of Absolute Differences, by John M. Hammersley
(University of Oxford, England).

  In Table 1 the entries in the second column are obtained as differences
(ignoring sign) of successive entries in the first column, the first element of the
column being regarded as the successor of the last element, i.e., $21 = 37 - 16$,
$29 = 37 - 8$, $82 = 90 - 8$ and $74 = 90 - 16$. The third column is obtained from
the second column in the same fashion, and so on. Eventually, we get columns
whose entries are all zero. Is this a particular property of the four numbers in the
first column or will zero columns eventually arise from an arbitrary first column?

TABLE 1

16	21	8	45	0	0	$\cdots$
37	29	53	45	0	0	$\cdots$
8	82	8	45	0	0	$\cdots$
90	74	53	45	0	0	$\cdots$

If the table has only three rows, we can get a case where no column consists entirely of zeros, as in Table 2. In this example the columns eventually follow a cyclic pattern with three columns to the cycle. Is this true for all three-rowed tables? What happens eventually in a table with m rows?

TABLE 2

16	21	8	13	5	2	3	1	0	1	1	0	$\cdots$
37	29	21	8	3	5	2	1	1	0	1	1	$\cdots$
8	8	13	5	8	3	1	2	1	1	0	1	$\cdots$

Editorial note. This problem also appears in the highly recommended provocative paper by the proposer, *On the enfeeblement of mathematical skills by "modern mathematics" and by similar soft intellectual trash in schools and universities,* Bulletin of the Institute of Mathematics and its Applications, (1968), pp. 66–85. The paper is an expanded version of a lecture given at the annual meeting of the I.M.A. in London, 1967 as well as the colloquium on *How to Teach Mathematics so as to be Useful,* sponsored by the International Commission on Mathematical Instruction in Utrecht, Netherlands, 1967.

Solution by L. CARLITZ and R. SCOVILLE (Duke University).

Let k denote the number of rows. We may evidently assume $k \geq 3$.

We suppose first that the k numbers in the first column are rational and so may be assumed to be integers. We shall show that after a finite number of steps we get either all zeros or some zeros and some a's, where a is a positive integer; if $k = 2^m$ then we certainly get all zeros ultimately. The number of steps required may be arbitrarily large.

Let $a_1, a_2, \cdots, a_k$ be the elements in the first column. Then we assert that applying a certain number of the permitted operations the maximum element decreases (until we reach one of the final stages described in the previous paragraph). Let a denote a maximal element. The only favorable cases are those of the following types:

$$ba00 \cdots 0c, \quad b0 \cdots 0ac, \quad b0 \cdots \cdot 0a0 \cdots 0c,$$

where b, c are positive. Consider for example $a000c$; we get successively $a00c$, $a0c, ac, a - c$. Thus unless all the nonzero elements are equal there will be a decrease in the maximum. This proves that ultimately we shall get all zeros or some zeros and some a's.

Next let $k = 2^m$. The allowed operation can be described in matric terms as follows. Put ($k = 4$)

$$\sigma = \begin{bmatrix} 0 & 1 & 0 & 0 \\ 0 & 0 & 1 & 0 \\ 0 & 0 & 0 & 1 \\ 1 & 0 & 0 & 0 \end{bmatrix}, \quad a = \begin{bmatrix} a_1 \\ a_2 \\ a_3 \\ a_4 \end{bmatrix}, \quad b = \begin{bmatrix} b_1 \\ b_2 \\ b_3 \\ b_4 \end{bmatrix}.$$

The vector a denotes the first column, the vector b the second column. Then clearly

$$(I + \sigma)a \equiv b \quad (\text{mod } 2).$$

Since $\sigma^k = I$, it follows that

$$(I + \sigma)^k = (I + \sigma)^{2^m} \equiv I + \sigma^{2^m} \equiv 0 \quad (\text{mod } 2).$$

Thus after k steps each element of the $(k + 1)$st column is divisible by 2. Hence we shall eventually get all zeros.

We show next that the number of steps necessary can be arbitrarily large. Take first $k = 3$. Let

$$F_0 = 0, \quad F_1 = 1, \quad F_n = F_{n-1} + F_{n-2}, \qquad n \geq 2,$$

so that the F_n are the Fibonacci numbers. Consider the following array.

$$\begin{array}{cccc} F_n & F_{n-2} & F_{n-4} & F_{n-3} \cdots \\ F_{n-1} & F_{n-3} & F_{n-2} & F_{n-4} \cdots \\ F_{n-2} & F_{n-1} & F_{n-3} & F_{n-5} \cdots. \end{array}$$

Clearly it will take $n - 3$ steps to reach $(F_2, F_1, F_0) = (1, 1, 0)$ or a cyclic permutation.

For $k = 4$ we define

$$u_0 = u_1 = 0, \quad u_2 = 1, \quad u_n = u_{n-1} + u_{n-2} + u_{n-3}, \qquad n \geq 3.$$

Consider

$$\begin{array}{cccc} u_n & u_{n-2} + u_{n-3} & u_{n-3} + u_{n-5} & 2u_{n-5} \\ u_{n-1} & u_{n-3} + u_{n-4} & u_{n-4} + u_{n-6} & 2u_{n-2} \\ u_{n-2} & u_{n-4} + u_{n-5} & u_{n-1} + u_{n-3} & 2u_{n-3} \\ u_{n-3} & u_{n-1} + u_{n-2} & u_{n-2} + u_{n-4} & 2u_{n-4}. \end{array}$$

Since the fourth column is a cyclic permutation of $u_{n-2}, u_{n-3}, u_{n-4}, u_{n-5}$, it is clear that we require approximately $3n/2$ steps.

For arbitrary $k \geq 3$, define

(1)
$$u_0 = \cdots = u_{k-3} = 0, \qquad u_{k-2} = 1,$$

$$u_n = u_{n-1} + \cdots + u_{n-k+1}, \qquad n \geq k - 1.$$

For the first column take

$$u_n, u_{n-1}, \cdots, u_{n-k+1}.$$

Then the entries in the second column are

$$u_n - u_{n-1}, u_{n-1} - u_{n-2}, \cdots, u_{n-k+2} - u_{n-k+1}, u_n - u_{n-k+1}.$$

By (1) we have

$$u_n - u_{n-k+1} = u_{n-1} + \cdots + u_{n-k+2} = u_{n+1} - u_n.$$

Thus, after a cyclic permutation, the second column becomes

$$\Delta u_n, \Delta u_{n-1}, \cdots, \Delta u_{n-k+1},$$

where $\Delta u_n = u_{n+1} - u_n$; every element is positive for $n \geq 2k - 3$. Since, by (1)

$$\Delta u_n = u_{n-1} + u_{n-2} + \cdots + u_{n-k+2},$$

$$\Delta^2 u_n = \Delta \cdot \Delta u_n = \Delta u_{n-1} + \cdots + \Delta u_{n-k+2} > 0$$

for n sufficiently large. The entries in the third column are

$$\Delta^2 u_{n-1}, \cdots, \Delta^2 u_{n-k+1}, \Delta u_n - \Delta u_{n-k+1}.$$

Since

$$\Delta u_n - \Delta u_{n-k+1} = \Delta(u_n - u_{n-k+1})$$

$$= \Delta \cdot \Delta u_n = \Delta^2 u_n,$$

the entries, after a cyclic permutation, are

$$\Delta^2 u_n, \Delta^2 u_{n-1}, \cdots, \Delta^2 u_{n-k+1}.$$

Proceeding in this way we see that the entries in the rth column are

$$\Delta^{r-1} u_n, \Delta^{r-1} u_{n-1}, \cdots, \Delta^{r-1} u_{n-k+1},$$

provided these numbers are all positive. This will evidently be the case provided $n \geq r(k - 1)$.

We conclude that the number of steps required for the first column u_n, $u_{n-1}, \cdots, u_{n-k+1}$ is approximately $n/(k - 1)$. This proves the assertion that the number of steps may be arbitrarily large.

Finally we show that when the entries in the first column are irrational the process need not end in a finite number of steps. Define the real number α by means of

$$\alpha^{k-1} = \alpha^{k-2} + \cdots + \alpha + 1, \qquad \alpha > 1.$$

For first column take

$$\alpha^{k-1}, \alpha^{k-2}, \cdots, \alpha, 1.$$

Then the entries in the second column are

$$(\alpha^{k-3} + \cdots + 1), \alpha^{-1}(\alpha^{k-3} + \cdots + 1), \alpha^{-k+2}(\alpha^{k-3} + \cdots + 1), \alpha(\alpha^{k-3} + \cdots + 1).$$

Multiplying each element by $\alpha^{k-2}/(\alpha^{k-3} + \cdots + 1)$ and permuting cyclically, we get again

$$\alpha^{k-1}, \alpha^{k-2}, \cdots, \alpha, 1.$$

Hence the process continues indefinitely.

Editorial note. R. E. GREENWOOD (University of Texas) refers to R. Sprague, *Recreation in Mathematics*, Blackie, London, 1963, pp. 5, 26–28 and B. Freeman, *The four numbers game*, Scripta Math., 14 (1948), pp. 35–47. Freeman shows that if the number of members (k) of the original set is a power of 2, then the process of differencing leads ultimately to a set where every member is zero. He also shows that if $k \neq 2^m$, then only in special cases will the differencing lead to an all zero set.

Another reference found after the problem was published is M. Lotan, *A problem in difference sets*, Amer. Math. Monthly, 56 (1949), pp. 535–541. Lotan shows that the differencing operation when applied to any four real numbers, excepting those of the form $1, q, q^2, q^3$ where q is the positive solution of $q^3 = q^2 + q + 1$, will lead to a set of zeros. [M.S.K.]

Problem 76-5*, An Arithmetic Conjecture, by D. J. NEWMAN (Yeshiva University).

To determine positive integers $a_1, a_2, \cdots, a_n$ such that $S_n \equiv \sum_{i=1}^{n} 1/a_i < 1$ and S_n is a maximum, it is conjectured that at each choice one picks the smallest integer still satisfying the inequality constraint, for example, for $n = 4$, one would choose

$$\frac{1}{2} + \frac{1}{3} + \frac{1}{7} + \frac{1}{43}.$$

Editorial note. P. Erdös notes that this problem was raised by Kellog in 1921 and solved by Curtiss (*On Kellog's Diophantine equation*, Amer. Math. Monthly, 29 (1922), pp. 380–387). Curtiss shows that if $\{u_n\}$ is defined by $u_1 = 1$, $u_{k+1} = u_k(u_k + 1)$ (giving rise to the sequence 1, 2, 6, 42, 1806, $\cdots$) and if $1/F_n = 1 - S_n$, then the maximum finite value of F_{n-1}, for all positive values of $a_1, a_2, \cdots, a_{n-1}$ is u_n and also there is but one set of the a's which give this maximum value, namely, $a_k = u_{k+1}$, $k = 1, 2, \cdots, n-1$.

It would be of interest to solve the following extension of the problem: we wish the stated conjecture to still be valid if the a_i's are further restricted to be members of a given infinite sequence $\{b_k\}$ with $\sum 1/b_k = \infty$. Characterize all such sequences $\{b_k\}$. In particular, is the conjecture valid for $b_k = 2k$; for $b_k = 2k + 1$, $k = 1, 2, 3, \cdots$? [M.S.K.]

Problem 70-8*, Palindromic Biquadrates, by G. J. SIMMONS and D. RAWLINSON (Sandia Laboratories).

A number is a palindrome if it is unchanged by reversal; i.e., 121, 14541, etc.

The fourth power of an $i + 2$ digit decimal number of the form $1\overset{i}{\overbrace{0\cdots0}}1$ where $i \geq 0$ is the $4i + 5$ digit palindrome $1\overset{i}{\overbrace{0\cdots0}}4\overset{i}{\overbrace{0\cdots0}}6\ \overset{i}{\overbrace{0\cdots0}}4\overset{i}{\overbrace{0\cdots0}}1$. An exhaustive computer search has shown that the only palindromic biquadrates less than 2.8×10^{14} have roots of this type. Does there exist any number greater than 1 and not of the form $1\,0\cdots0\,1$ whose fourth power is a palindrome?

Problem 62–10, A Binary Multiplication, by W. C. McGEE (Ramo-Wooldridge).

The designer of a certain binary digital computer proposes to mechanize the multiply instruction by the familiar technique of repeated addition. Three registers are required for this purpose, each having a capacity of n bits; the E register, which holds the multiplicand throughout the process; the A register, in which the most-significant part of the product is developed; and the P register, from which the multiplier is shifted and in which the least-significant part of the product is developed.

The process is carried out in a sequence of distinct steps. On each step:

(a) If the low-order digit of P is zero, the contents of A and P are shifted one position to the right, in such a way that the low-order digit of A enters the high order position of P. A zero enters the high-order position of A and the low-order digit of P is lost.

(b) If the low-order digit of P is one, the contents of E are added into A (carry out the left end not being lost) and the contents of A and P are then shifted one position to the right exactly as above, with any high-order carry from the addition "shifting" into the high-order position of A.

If we let

$$e = \text{contents of } E,$$

$$\left. \begin{array}{l} a_k = \text{contents of } A \text{ after the } k\text{th step} \\ p_k = \text{contents of } P \text{ after the } k\text{th step} \end{array} \right\} \quad k = 0, 1, 2, \cdots, n,$$

and if e, a_k, and p_k are regarded as positive integers, it can be shown that the procedure described above produces on the $(k + 1)$st step the positive integers

$$a_{k+1} = [s_k/2],$$

$$p_{k+1} = (s_k \bmod 2) \cdot 2^{n-1} + [p_k/2],$$

where

$$s_k = \begin{cases} a_k & \text{if } p_k \bmod 2 = 0, \\ a_k + e & \text{if } p_k \bmod 2 = 1, \end{cases}$$

and $[x]$ denotes the integral part of x.

It is the designer's contention that, after the nth step of the process described above, the A register and the P register contain the n most-significant and n least-significant bits, respectively, of the positive integer

$$a_0 + p_0 e.$$

Prove the designer's contention.

Solution by ALVIN L. SCHREIBER (Datarol Corporation).

Let N_k denote the $2n$-bit integer whose n most significant bits are the contents of the A register, whose next k bits are the k most significant bits of the P register, and whose remaining $n - k$ bits are zeros.

Let b_k denote the k-bit integer whose most significant bit is the kth least significant bit of p_0, and whose remaining $k - 1$ bits are zeros.

Each step of the process described consists of calculating N_k from N_{k-1} under the following rule,

$$N_k = N_{k-1} + b_k e,$$

and putting N_k in the A register and the k high order positions of the P register considered as one $(n + k)$-bit register. it is clear that the unused portion of p_0 is not destroyed in this process.

Then

$$N_n = N_{n-1} + b_n e$$

$$= N_{n-2} + b_{n-1}e + b_n e$$

$$\vdots$$

$$= N_0 + \sum_{k=1}^{n} b_k e$$

$$= a_0 + p_0 e.$$

Problem 63-13*, *An Infinite Permutation*, by M. S. KLAMKIN (Ford Scientific Laboratory).

Consider the infinite permutation

$$P \equiv \begin{pmatrix} 1 & 2 & 3 & 4 & 5 & 6 & \cdots & n & \cdots \\ 1 & 3 & 2 & 5 & 7 & 4 & \cdots & f(n) & \cdots \end{pmatrix},$$

where

$$f(3n - 2) = 4n - 3,$$

(1) $$\qquad f(3n - 1) = 4n - 1,$$

$$f(3n) = 2n.$$

We now write P as a product of cycles:

$$P \equiv (1)(2, 3)(4, 5, 7, 9, 6)(8, 11, 15, \cdots) \cdots .$$

It is conjectured that the cycle $(8, 11, 15, \cdots)$ is infinite. Other problems concerning P are:

(a) Can P be expressed as a product of a finite number of cycles?

(b) Are there any finite cycles other than those indicated?

Editorial note. For a similar problem where there are cycles of every length, see [Problem 5109, Amer. Math. Monthly, 70(1963), pp. 572–573]. A major difference between the two permutations apparently is that in the latter case, the ratio of odd to even numbers of $f(n)$ approaches 1, whereas in the former case, the ratio approaches 2.

Comment by A. O. L. ATKIN (Atlas Computer Laboratory, Chilton, England).

There are finite cycles of lengths 1, 2, 5 and 12, respectively, given by

(2) $$\qquad (1), (2, 3), (4, 5, 7, 9, 6) \quad \text{and} \quad (44, 59, \cdots, 66),$$

the last of these being announced in the previous comment on this problem by D. Shanks [this Journal, 7 (1965), pp. 284–286]. We give below a method applicable in principle to the problem of finding all cycles (if any) of *given* period p, although the computation required becomes formidable if q/p is a good approximation to $\log_2 3$. The method utilizes a quantitative refinement of the "approximate theory" given in the previous comment. We show that there are no cycles of period less than 200 other than (2); in particular, there are none of periods 41 and 53 which are denominators of convergents to $\log_2 3$.

Suppose that there is a cycle (a_r) of period p, and that m is its least term. If there are $p - k$ transformations of the form $f(3n) = 2n$, and k transformations of the other two kinds, then

$$1 = \left(\frac{2}{3}\right)^{p-k} \left(\frac{4}{3}\right)^k \prod_{r=1}^{k} \left(\frac{3f(a_r)}{4a_r}\right),$$

reordering the a_r if necessary.

Also for all r, $1 \leq r \leq k$, we have

$$|1 - 3f(a_r)/4a_r| \leq 1/4m.$$

Hence

$$(1 - 1/4m)^k \leq 3^p/2^{p+k} \leq (1 + 1/4m)^k.$$

Now for $0 < x < 1$, we have

$$\log(1 + x) < x,$$

$$\log(1 - x) > -x - x^2/2(1 - x),$$

so that (taking $m \geq 8$, $x \leq \frac{1}{32}$ in view of (2))

$$-63k/248m \leq p \log(3/2) - k \log 2 \leq k/4m.$$

Thus for a given p, we must have

$$m \leq \frac{63}{248} \Big/ \min_k \left|\frac{p}{k} \log(1.5) - \log 2\right| = g(p), \quad \text{say.}$$

A program was run on the I.C.T. Atlas 1 computer of the Science Research Council at Chilton, to show that (other than (2)) all cycles with least terms less than 5000 have at least 342 terms in their periods. Next, for $p \leq 341$, a tabulation of $g(p)$ showed that $g(p) < 5000$ except when $p = 200, 253, 306$. Hence the only possible periods of new cycles are $p = 200, 253, 306$ and $p > 341$.

A similar run was performed for the permutations obtained by permuting the right-hand sides of (1). In some of these cases we obtain periods which are not denominators of convergents to $\log_2 3$. For instance, with $f(3n) = 4n + 3$, $f(3n + 1) = 2n$, $f(3n + 2) = 4n + 1$, there is a cycle of period 94, least term 140. While 149/94 is a good approximation to $\log_2 3$ it is not a convergent. As the referee points out, however, there is more chance of such a (nondenominator) period here since, for example, $(4n + 3)/3n$ is further from $\frac{4}{3}$ than $(4n + 3)/(3n + 2)$.

My general conjecture, on a probability basis, is that for any "congruence" permutation of this kind, the number of finite cycles is finite, since (here) the "expected" value of $f(t)/t$ is about $(\frac{2}{3} \cdot \frac{4}{3} \cdot \frac{4}{3})^{1/3}$, and that of $f^{-1}(t)/t$ is $(\frac{3}{2} \cdot \frac{3}{2} \cdot \frac{3}{4} \cdot \frac{3}{4})^{1/4}$, so that most cycles tend to infinity in both directions. Dr. D. A. Burgess of Nottingham University has given an elegant proof that these expected ratios cannot be unity for any congruence permutation.

Editorial note. R. Eddy (David Taylor Model Basin) notes that there are "near" closures at periods 41, 53 and at "counterexample" periods 17 and 29 for the original permutation. Here

$$36 \to 37 : 17 \text{ transforms,}$$
$$46 \to 47 : 17 \text{ transforms,}$$
$$78 \to 77 : 29 \text{ transforms,}$$
$$50 \to 49 : 41 \text{ transforms,}$$
$$554 \to 553 : 53 \text{ transforms.}$$

Postscript. For an interesting up to date treatment of this problem, see J.C. Lagarias, The $3x + 1$ Problem and its Generalizations, Amer. Math. Monthly 92 (1985) 3–23.

Problem 76-1, *The Game of Slash*, by D. N. BERMAN (University of Waterloo).

The board used here consists of a single row of positions $(1, 2, \cdots, n)$, ordered from left to right, in which a given number of pieces are placed in some fashion among the positions. Only one piece may ever occupy a given position. Alternating play between two players is made by moving any one of the pieces as far to the left as desired but still remaining to the right of the piece immediately on its left. The winner is the player who leaves his opponent no possible move.

Another variation of the game allows the players to move as far to the left as desired to an unoccupied position.

Determine a winning strategy for the game.

Solution by D. J. WILSON (University of Melbourne).

Considering each unoccupied position in turn, count the number of *pieces* lying to its *right*. If this number is odd, place a match in the position as a marker; if it is even, leave the position empty. Thus, between any pair of successive pieces either every position will be occupied by a match or they will all be empty. Regard every collection of matches lying between a pair of successive pieces as a single pile, and consider the position which these piles form in the game of *Nim* [1]. The player whose turn it is to move can make sure of winning if and only if this Nim position is *unsafe*.

A winning move is to take matches away from one of the piles so as to leave a *safe* position in Nim and then make the number of spaces between the corresponding pair of pieces equal to the number of matches in the reduced pile. This can always be done by moving the right-hand piece of the pair.

After the next move the number of matches in exactly one pile will have to be changed. Whether this number is increased or decreased, it is easy to show the position left by the move must again correspond to an unsafe position of Nim. Since the final position of the game corresponds to a safe position of Nim it follows that a player can be sure of winning if he always leaves his opponent with such a position.

REFERENCE

[1] J. C. HOLLADAY, *Cartesian products of termination games*, Contributions to the Theory of Games, vol. III, Annals of Mathematics Studies, no. 39, Princeton University Press, Princeton, N.J., 1957, pp. 189–200.

Partial solution of the second variation of Slash by the proposer.

A winning strategy for the second variation of Slash in the case of four pieces can be given in terms of Nim. This includes the strategy for two (three) pieces simply by placing two (one) pieces initially in positions 1, 2(1).

Suppose the pieces are in positions $P_1, P_2, P_3, P_4, P_i \neq P_j, i \neq j$. Call this a Slash position. Construct four Nim piles of sizes $P_1 - 1, P_2 - 1, P_3 - 1, P_4 - 1$ and call the Slash position safe if the corresponding Nim position is balanced. The player faced with an unsafe position will win by moving to a safe position at each turn. The proof is a consequence of the following three assertions.

(a) Any move from a safe position always results in an unsafe position.

(b) For any unsafe position, there is a move that makes it safe.

(c) The final Slash position is safe.

Assertion (a) follows directly from Nim. Each move in Slash corresponds to reducing a Nim pile thereby making an unbalanced Nim position. If we start with an unbalanced Nim position we can always balance the piles by reducing one pile. But Nim allows two piles to be the same size and this is not allowed in Slash. However the four piles corresponding to any unsafe position must be of different sizes, and it is easily verified that the resulting balanced position cannot have two equal piles. This proves (b). The Nim piles 0, 1, 2, 3 corresponding to the final Slash position 1, 2, 3, 4 is balanced as asserted in (c).

The above strategy is not a winning strategy for more than four pieces since (b) is not always true. forming two equal Nim piles may be required in moving from an unsafe position to a safe position, even though the original Nim piles were of different sizes.

Editorial note. W. C. Davidon indicated that a program for playing Nim or Slash with an HP-65 pocket calculator is being submitted to the Hewlett-Packard User's Library. It analyzes Nim with up to five heaps or Slash positions with up to ten pieces; the number of each Nim heap, or in alternate Slash intervals, can be from zero through 99.

Subsequent comment by R. Silber (North Carolina State University).

This problem also appears in J. Conway's new book, *On Numbers and Games.* It also was discussed by M. Gardner in the September, 1976 issue of Scientific American, where it is called the "Silver Dollar Game without the Dollar." The generalized game is deep and is also solved (see C. P. Welter, *The theory of a class of games on a sequence of squares in terms of the advancing operation on a special group*, Proc. Roy. Acad. Amsterdam Ser. A, 57 (1954), pp. 304–314). Welter's game is discussed briefly in T. H. O'Beirne, *Puzzles and Paradoxes*, Chap. 9, as well as in the aforementioned book of Conway where Welter's results are presented and simplified.

Problem 70-5, Conformal Mapping of a Cross Slit Strip, by B. J. CERIMELE (North Carolina State University).

Determine a transformation which conformally maps the horizontal strip $|\text{Im}(z)| < \pi$ with the symmetric cross slit

$$|\text{Re}(z)| \leq \lambda, \qquad \text{Im}(z) = 0,$$

$$|\text{Im}(z)| \leq \lambda, \qquad \text{Re}(z) = 0,$$

where $\lambda < \pi$, upon a ring, i.e., the interior of an annulus.

Solution by the proposer.

Because of the symmetry of the region, the Schwarz symmetry principle yields that the mapping of the quarter-region, $0 < \text{Im}(z) < \pi$, $0 < \text{Re}(z) < \infty$, establishes the mapping of the entire region.

The function

(1) $q = \cosh z$

maps the quarter-strip onto the upper half-plane. This half-plane is converted by the homographic transformation

(2) $$r = \frac{(\cosh \lambda - \cos \lambda)(1 - q)}{(2 - \cos \lambda - \cosh \lambda)q - (\cosh \lambda - 2\cos \lambda \cosh \lambda + \cos \lambda)}$$

into a form tractable by an elliptic integral. In particular, the inverse Jacobian elliptic function

(3) $s = \mathrm{sn}^{-1}(r; k),$

where $k = (1 - \cos \lambda \cosh \lambda)/(\cosh \lambda - \cos \lambda)$, maps the upper half-plane onto the interior of the rectangle with vertices $\pm K$, $\pm K + iK'$. This rectangle is translated and rotated by

(4) $t = e^{-\pi i/2}(s - k) = -i(s - K)$

into a rectangle with vertices 0, K', $K' + 2iK$, $2iK$, whose interior in turn is finally transformed under

(5) $w = e^{\pi t/(4K)}$

into a quarter-ring with inner radius unity and outer radius $e^{\pi K'/(4K)}$. Thus, the sequence of mappings (1) through (5) effects the conformal mapping of the entire cross slit strip upon the ring $1 < |w| < e^{\pi K'/(4K)}$. See Fig. 1.

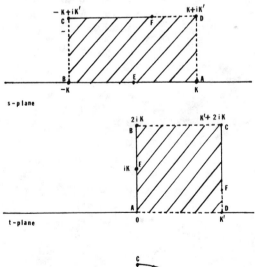

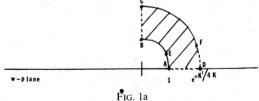

FIG. 1a

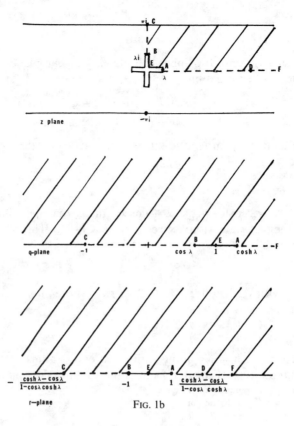

FIG. 1b

23. APPENDIX

References

[1] E.T. Bell, *The Development of Mathematics*, 2nd ed., McGraw-Hill, N.J.

[2] C.B. Boyer, *A History of Mathematics*, Wiley, N.Y., 1968.

[3] J. Fauvel and J. Gray, ed., *The History of Mathematics*, Macmillan, Hong Kong, 1988.

[4] M. Kline, *Mathematical Thought from Ancient to Modern Times*, Oxford University Press, N.Y., 1972.

[5] E.E. Kramer, *The Nature and Growth of Modern Mathematics*, Hawthorn, N.Y., 1970.

[6] O. Neugebauer, *The Exact Sciences of Antiquity*, Princeton University Press, Princeton, 1952.

[7] B.L. van der Waerden, *Science Awakening*, Noordhoff, Groningen, 1954.

[8] L.E. Dickson, *History of the Theory of Numbers, II*, Stechert, N.Y., 1934.

[9] Bull. *Amer. Math. Soc.*, 8(1901-02)437-479.

[10] D.M.Campbell and D.M. Higgins, *Mathematics—People-Problems-Results*, Wadsworth, Belmont, 1984, Vol. I, pp.273-278, 300-304.

[11] F.E. Browder, ed., *Mathematical Developments Arising from Hilbert Problems*, Proceedings of symposia in pure mathematics; v.28, I and II, Amer. Math. Soc., 1976.

[12] G. Polya, *How To Sove It*, Princeton University Press, Princeton, 1973.

[13] G. Polya, *Mathematics and Plausible Reasoning I,II*, Princeton University Press, Princeton, 1954.

[14] G. Polya, *Mathematical Discovery*, Princeton University Press, Princeton, 1962, 1965.

[15] B.A. Fusaro, *Mathematical Competition in Modeling*, Math. Modelling 6(1985)473-484.

[16] A.M. Gleason, R.E. Greenwood, and L.M. Kelly, *The William Lowell Putnam Mathematical Competition*, Problems and Solutions 1938-1964, MAA, USA, 1980.

[17] G.L. Alexanderson, L.F. Klosinski, and L.G. Larson, *The William Lowell Putnam Mathematical Competition*, Problems and Solutions, 1965-1984, MAA, USA, 1985.

Problem Collection Books

[18] B. Andrasfai, *Introductory Graph Theory*, Pergamon, N.Y., 1977.

[19] M. Berger, P. Pansu, J.P. Berry, and X. Saint-Raymond, *Problems in Geometry*, Springer-Verlag, Harrisonburg, 1982.

[20] P. Biler and A. Witowski, *Problems in Mathematical Analysis*, Dekker, N.Y., 1990.

[21] J.L. Brenner, *Problems in Differential Equations*, Freeman, San Francisco, 1963.

[22] J.C. Burkill and H.M. Cundy, *Mathematical Scholarship Problems*, Cambridge University Press, Cambridge, 1961.

[23] T. Cacoullos, *Exercisesin Probability*, Springer-Verlag, N.Y., 1989.

[24] M. Capobianco and J. C. Molluzo, *Examples and Counterexamples in Graph Theory*, Elsevier North-Holland, N.Y., 1978.

[25] J.D. Dixon, *Problems in Group Theory*, Dover, N.Y., 1973.

[26] H. Dorrie, *100 Great Problems in Elementary Mathematics*, Dover, N.Y., 1965.

[27] H.G. Eggleston, *Problems in Euclidean Space*, Pergamon, N.Y., 1957.

[28] P. Erdos and R.L. Graham, *Old and New Problems and Results in Comb- inatorial Number Theory*, L'Enseignment Mathematique, Geneva, 1980.

[29] H. Eves, and E.P. Starke, ed., *The Otto Dunkel Memorial Problem Book*, MAA, USA, 1957.

[30] B.R. Gelbaum, *Problems in Analysis*, Springer-Verlag, N.Y., 1982.

[31] B.R. Gelbaum, and J.M.H. Olmsted, *Counterexamples in Analysis*, Holden- Day, San Francisco, 1964.

[32] I.M. Glazman and J.I. Ljubic, *Finite-Dimensional Linear Analysis: A Systematic Presentation in Problem Form*, M.I.T. Press, Cambridge,1974

[33] R.C. Gunning, ed., *Problems in Analysis*, Princeton University Press, Princeton, 1976.

[34] R.K. Guy, *Unsolved Problems in Number Theory*, Springer-Verlag, N.Y., 1981.

[35] H. Hadwiger, H. Debrunner, and V. Klee, *Combinatorial Geometry in the Plane*, Holt, Rinehart and Winston, N.Y., 1964.

[36] P.R. Halmos, *A Hilbert Space Problem Book*, Van Nostrand, Princeton, 1967.

[37] J. Hammer, *Unsolved Problems Concerning Lattice Points*, Pitman, London, 1977.

[38] K. Hardy and K.S. Williams, *The Green Book*, Integer Press, Ottawa, 1985.

[39] K. Hardy and K.S. Williams, *The Red Book*, Integer Press, Ottawa, 1988.

[40] W.K. Hayman, *Research Problems in Function Theory*, Athlone Press, London, 1967.

[41] R. Honsberger, *Mathematical Gems, Vol. I,II,III*, MAA, USA, 1973, 1976, 1985.

[42] R. Honsberger, *Mathematical Morsels*, MAA, USA, 1978.

[43] R. Honsberger, ed., *Mathematical Plums*, MAA USA, 1979.

[44] M.G. Kendall, *Exercises in Statistics*, Hafner, N.Y., 1956.

[45] A.A. Kirillov and A.D. Gvishiani, *Theory and Problems in Functional Analysis*, Springer-Verlag, N.Y., 1982.

[46] G. Klambauer, *Problems and Propositions in Analysis*, Dekker, N.Y., 1979.

[47] K. Knopp, *Problem Book in the Theory of Functions, I,II*, Dover,N.Y.,1948.

[48] M.L. Krasnov, A.I. Kiselev, and G.I. Makarenko, *A Book of Problems in Ordinary Differential Equations*, Mir, Moscow, 1983.

[49] M.L. Krasnov, A.I. Kiselev, and G.I. Makarenko, *Problems and Exercises in the Calculus of Variations*, Mir, Moscow, 1984.

[50] J.G. Krzyz, *Problems in Complex Number Theory*, American Elsevier, N.Y., 1971.

[51] L.C. Larson, *Problem-Solving Through Problems*, Springer-Verlag, N.Y., 1983.

[52] N.N. Lebedev, I.P. Skalakaya, and I.P. Uflyand, *Worked Problems in Applied Mathematics*, Dover, N.Y., 1965.

[53] G. Lefort, *Algebra and Analysis, Problems and Solutions*, Saunders, Philadelphia, 1966.

[54] L. Lovasz, *Combinatorial Problems and Exercises*, North-Holland, Amsterdam, 1979.

[55] R.D. Mauldin, ed., *The Scottish Book, Mathematics from the Scottish Cafe*, Birkhauser, Boston, 1981.

[56] Z.A. Melzak, *Companion to Concrete Mathematics, I,II*, Wiley, N.Y., 1973, 1976.

[57] Z.A. Melzak, *Bypasses*, Wiley, N.Y., 1983.

[58] V.P. Minorsky, *Problems in Higher Mathematics*, Mir Moscow, 1975.

[59] D.S. Mitrinovic, *Elementary Inequalities*, Noordhoff, Groningen, 1964.

[60] D.S. Mitrinovic, E.S. Barnes, and J.R.M. Radok, *Functions of a Complex Variable*, Noordhoff, Groningen, 1965.

[61] D.S. Mitrinovic and J.H. Michael, *Calculus of Residues*, Noordhoff, Groningen, 1966.

[62] D.S. Mitrinovic and R.B. Potts, *Elementary Matrices*, Noordhoff, Groningen, 1965.

[63] W.O. Moser and J. Pach, *100 Research Problems in Discrete Geometry*, McGill University, Montreal, 1986.

[64] D.J. Newman, *A Problem Seminar*, Springer-Verlag, N.Y., 1982.

[65] I. Niven, *Maxima and Minima without Calculus*, MAA, Washington,D,C., 1981.

[66] G. Polya and G. Szego, *Problems and Theorems in Analysis, I,II*, Springer-Verlag, N.Y., 1972, 1976.

[67] H. Rademacher and O. Toeplitz, *The Enjoyment of Mathematics*, Princeton University Press, Princeton, 1957.

[68] I.J. Schoenberg, *Mathematical Time Exposures*, MAA, USA, 1982.

[69] B.A. Sevastyanov, V.P. Chistykov, and A.M. Zubkov, *Problems in the Theory of Probability*, Mir, Moscow, 1985.

[70] W. Sierpinski, *A Selection of Problems in the Theory of Numbers*, Pergamon, Oxford, 1964.

[71] W. Sierpinski, *250 Problems in Elementary Number Theory*, American Elsevier, N.Y., 1970.

[72] D.O. Shklyarsky, N.N. Chentsov, and I.M. Yaglom, *Selected Problems and Theorems in Elementary Mathematics*, Mir, Moscow, 1979.

[73] R.A. Silverman, ed., *Worked Problems in Applied Mathematics*, Dover, N.Y., 1965.

[74] A. Soifer, *Mathematics as Problem Solving*, Center for Excellence in Math. Educ., Colorado Springs, 1987.

[75] J.E. Spencer and R.C. Geary, *Exercises in Mathematical Economics and Econometrics with Outlines of Theory*, Hafner, N.Y., 1961.

[76] A.T. Starr, *Scholarship Mathematics, I-Analysis, II-Geometry*, Pitman, London, 1961.

[77] L.A. Steen, and J.A. Seebach, *Counterexamples in Topology*, Springer- Verlag, N.Y., 1978.

[78] H. Steinhaus, *One Hundred Problems in Elementary Mathematics*, Basic Books, N.Y., 1964.

[79] H. Steinhaus, *Mathematical Snapshots*, Oxford University Press, N.Y., 1969.

[80] J.M. Stoyanov, *Counterexamples in Probability*, Wiley, N.Y., 1988.

[81] A.A. Sveshnikov, *Problems in Probability Theory, Mathematical Statistics and Theory of Random Functions*, Dover, N.Y., 1968.

[82] G. Szasz, et al (eds.), *Contests in Higher Mathematics, Hungary 1949-1961*, Akademiai Kiado, Budapest, 1968.

[83] H. Tietze, *Famous Problems of Mathematics*, Graylock Press, N.Y., 1965.

[84] I. Tomescu, *Problems in Combinatorics and Graph Theory*, Wiley, N.Y., 1985.

[85] I. Tomescu, *Introduction to Combinatorics*, Collet's, Romania, 1972.

[86] N.Y. Vilenkin, *Combinatorics*, Academic Press, N.Y., 1971.

[87] L.I. Volkovyskii, G.L. Lunts, and I.G. Aramanovich, *A Collection of Problems on Complex Analysis*, Addison-Wesley, Reading, 1965.

[88] E. Wentzel and L. Ovcharov, *Applied Problems in Probability Theory*, Mir, Moscow, 1986.

[89] J.C. Weston, and H.J. Godwin, *Some Exercises in Pure Mathematics with Expository Comments*, Cambridge University Press, Cambridge, 1968.

[90] W.A. Whitworth, *Choice and Chance*, Hafner, N.Y., 1948.

[91] J. Wolstenholme, *Mathematical Problems*, Macmillan, London, 1891.

[92] I.M. Yaglom and V.G. Boltyanskii, *Convex Figures*, Holt, Rinehart, and Winston, N.Y., 1961.

[93] A.M. Yaglom and I.M.Yaglom, *Challenging Mathematical Problems with Elementary Solutions*, *I,II*, Holden-Day, San Francisco, 1964, 1967.

Journals With Problem Sections

[1]   *Aequationes Mathematicae*

[2]   *AMATYC Review*

[3]   *American Mathematics Monthly*

[4]   *Archimede* (Italian)

[5]   *College Mathematics Journal*

[6]   *Crux Mathematicorum*

[7]   *Discrete Mathematics*

[8]   *Elemente der Mathematik* (German)

[9]   *Fibonacci Quarterly*

[10] *Gaceta Matematica* (Spanish)

[11] *Gazeta Matematica* (Romanian)

[12] *Graphs and Combinatorics*

[13] *Journal of Recreational Mathematics*

[14] *K ozepiskolai Matematikai Lapok* (Hungarian)

[15] *Kvantovaya Èlektronika* (Russian)

[16] *Matematika v Shkole* (Russian)

[17] *Matematikai Lapok* (Hungarian)

[18] *Matematische Semesterberichte* (German)

[19] *Mathematical Digest*

[20] *Mathematical Intelligencer*

[21] *Mathematical Questions and Solutions from the Educational Times*

[22] *Mathematical Spectrum*

[23] *Mathematics and Computer Education*

[24] *Mathematics Magazine*

[25] *Mathesis* (French)

[26] *Nieuw Archief voor Wiskunde* (Dutch)

[27] *Nordisk Matematisk Tidskrift* (Norwegian)

[28] *Pentagon*

[29] *Pi Mu Epsilon Journal*

[30] Quantum
[31] *School Science and Mathematics*
[32] *SIAM Review*
[33] *Wiadomosc Matematyczne* (Polish)

Journals 11, 14, 15, 16, 19, and 30 are secondary school journals; 30 is an English version of 15. Nevertheless, the problems are challenging Olympiad type problems.